**Laura Kay** thanks her wife, M.P.M. She dedicates this book to her late uncle, Lee Jacobi, for an early introduction to physics, and to her late colleagues at Barnard College, Tally Kampen and Sally Chapman.

**Stacy Palen** thanks her husband, John Armstrong, for his patient support during this project.

**George Blumenthal** gratefully thanks his wife, Kelly Weisberg, and his children, Aaron and Sarah Blumenthal, for their support during this project. He also wants to thank Professor Robert Greenler for stimulating his interest in all things related to physics.

# Brief Contents

# Contents

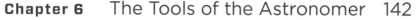

## PART II   The Solar System

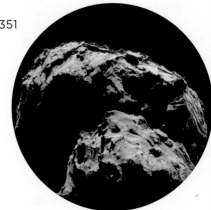

## PART III    Stars and Stellar Evolution

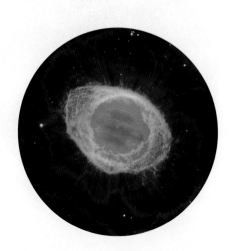

# Working It Out

CHAPTERS 15–23 ARE NOT INCLUDED IN THIS EDITION.

# AstroTours

CHAPTERS 15–23 ARE NOT INCLUDED IN THIS EDITION.

AstroTour animations are available from the free Student Site at the Digital Landing Page, and they are also integrated into assignable Smartwork5 exercises. Offline versions of the animations for classroom presentation are available from the Instructor's Resource USB Drive. digital.wwnorton.com/astro5.

# Astronomy in Action Videos

CHAPTERS 15–23 ARE NOT INCLUDED IN THIS EDITION.

digital.wwnorton.com/astro5

# Nebraska Simulations

**CHAPTERS 15–23 ARE NOT INCLUDED IN THIS EDITION.**

digital.wwnorton.com/astro5

# Preface

## Dear Student

Why is it a good idea to take a science course, and in particular, why is astronomy a course worth taking? Many people choose to learn about astronomy because they are curious about the universe. Your instructor likely has two basic goals in mind for you as you take this course. The first is to understand some basic physical concepts and how they apply to the universe around us. The second is to think like a scientist and learn to use the scientific method not only to answer questions in this course but also to make decisions in your life. We have written the fifth edition of *21st Century Astronomy* with these two goals in mind.

Throughout this book, we emphasize not only the content of astronomy (for example, the differences among the planets, the formation of chemical elements) but also *how* we know what we know. The scientific method is a valuable tool that you can carry with you and use for the rest of your life. One way we highlight the process of science is the **Process of Science Figures**. In each chapter, we have chosen one discovery and provided a visual representation illustrating the discovery or a principle of the process of science. In these figures, we try to illustrate that science is not a tidy process, and that discoveries are sometimes made by different groups, sometimes by accident, but always because people are trying to answer a question and show why or how we think something is the way it is.

The most effective way to learn something is to "do" it. Whether playing an instrument or a sport or becoming a good cook, reading "how" can only take you so far. The same is true of learning astronomy. We have written this book to help you "do" as you learn. We have created several tools in every chapter to make reading a more active process. At the beginning of each chapter, we have provided a set of Learning Goals to guide you as you read. There is a lot of information in every chapter, and the Learning Goals should help you focus on the most important points. We present a big-picture question in association with the chapter-opening figure at the beginning of each chapter. For each of these, we have tried to pose a question that is not only relevant to its chapter but also something you may have wondered about. We hope that these questions, plus the photographs that accompany them, capture your attention as well as your imagination.

In addition, there are **Check Your Understanding** questions at the end of each chapter section. These questions are designed to be answered quickly if you have understood the previous section. The answers are provided in the back of the book so you can check your answer and decide if further review is necessary.

As a citizen of the world, you make judgments about science, distinguishing between good science and pseudoscience. You use these judgments to make decisions in the grocery store, pharmacy, car dealership, and voting booth. You may base these decisions on the presentation of information you receive through the media, which is very different from the presentation in class. One important skill is the ability to recognize what is credible and to question what is not. To help you

Process of Science   CONVERGING LINES OF INQUIRY
Astronomers asked: Why is the Solar System a disk, with all planets orbiting in the same direction?

Stellar astronomers test the nebular hypothesis, seeking evidence for or against.

Stellar astronomers find dust and gas around young stars.

Stellar astronomers observe this gas and dust to be in the shape of disks.

Mathematicians suggest the nebular hypothesis: a collapsing rotating cloud formed the Solar System.

Planetary scientists test the nebular hypothesis, seeking evidence for or against.

Planetary scientists study meteorites that show the Solar System bodies formed from many smaller bodies.

Beginning from the same fundamental observations about the shape of the Solar System, theorists, planetary scientists, and stellar astronomers converge on the nebular theory that stars and planets form together from a collapsing cloud of gas and dust.

## CHECK YOUR UNDERSTANDING 7.4

Suppose that astronomers found a rocky, terrestrial planet beyond the orbit of Neptune. What is the most likely explanation for its origin? (a) It formed close to the Sun and migrated outward. (b) It formed in that location and was not disturbed by migration. (c) It formed later in the Sun's history than other planets. (d) It is a captured planet that formed around another star.

hone this skill, we have provided **Reading Astronomy News** sections at the end of every chapter. These features include a news article with questions to help you make sense of how science is presented to you. It is important that you learn to be critical of the information you receive, and these features will help you do that.

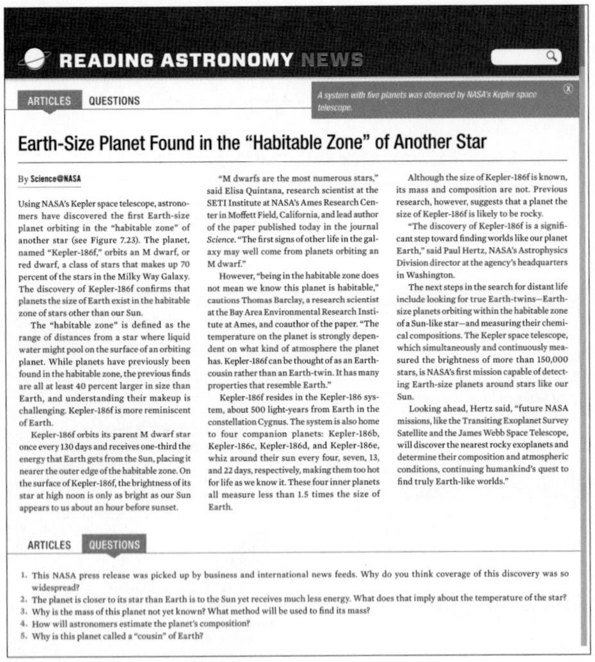
READING ASTRONOMY NEWS

ARTICLES QUESTIONS

A system with five planets was observed by NASA's Kepler space telescope.

Earth-Size Planet Found in the "Habitable Zone" of Another Star

By Science@NASA

Using NASA's Kepler space telescope, astronomers have discovered the first Earth-size planet orbiting in the "habitable zone" of another star (see Figure 7.23). The planet, named "Kepler-186f," orbits an M dwarf, or red dwarf, a class of stars that makes up 70 percent of the stars in the Milky Way Galaxy. The discovery of Kepler-186f confirms that planets the size of Earth exist in the habitable zone of stars other than our Sun.

The "habitable zone" is defined as the range of distances from a star where liquid water might pool on the surface of an orbiting planet. While planets have previously been found in the habitable zone, the previous finds are all at least 40 percent larger in size than Earth, and understanding their makeup is challenging. Kepler-186f is more reminiscent of Earth.

Kepler-186f orbits its parent M dwarf star once every 130 days and receives one-third the energy that Earth gets from the Sun, placing it nearer the outer edge of the habitable zone. On the surface of Kepler-186f, the brightness of its star at high noon is only as bright as our Sun appears to us about an hour before sunset.

"M dwarfs are the most numerous stars," said Elisa Quintana, research scientist at the SETI Institute at NASA's Ames Research Center in Moffett Field, California, and lead author of the paper published today in the journal *Science*. "The first signs of other life in the galaxy may well come from planets orbiting an M dwarf."

However, "being in the habitable zone does not mean we know this planet is habitable," cautions Thomas Barclay, a research scientist at the Bay Area Environmental Research Institute at Ames, and coauthor of the paper. "The temperature on the planet is strongly dependent on what kind of atmosphere the planet has. Kepler-186f can be thought of as an Earth-cousin rather than an Earth-twin. It has many properties that resemble Earth."

Kepler-186f resides in the Kepler-186 system, about 500 light-years from Earth in the constellation Cygnus. The system is also home to four companion planets: Kepler-186b, Kepler-186c, Kepler-186d, and Kepler-186e, whiz around their sun every four, seven, 13, and 22 days, respectively, making them too hot for life as we know it. These four inner planets all measure less than 1.5 times the size of Earth.

Although the size of Kepler-186f is known, its mass and composition are not. Previous research, however, suggests that a planet the size of Kepler-186f is likely to be rocky.

"The discovery of Kepler-186f is a significant step toward finding worlds like our planet Earth," said Paul Hertz, NASA's Astrophysics Division director at the agency's headquarters in Washington.

The next steps in the search for distant life include looking for true Earth-twins—Earth-size planets orbiting within the habitable zone of a Sun-like star—and measuring their chemical compositions. The Kepler space telescope, which simultaneously and continuously measured the brightness of more than 150,000 stars, is NASA's first mission capable of detecting Earth-size planets around stars like our Sun.

Looking ahead, Hertz said, "future NASA missions, like the Transiting Exoplanet Survey Satellite and the James Webb Space Telescope, will discover the nearest rocky exoplanets and determine their composition and atmospheric conditions, continuing humankind's quest to find truly Earth-like worlds."

ARTICLES QUESTIONS

1. This NASA press release was picked up by business and international news feeds. Why do you think coverage of this discovery was so widespread?
2. The planet is closer to its star than Earth is to the Sun yet receives much less energy. What does that imply about the temperature of the star?
3. Why is the mass of this planet not yet known? What method will be used to find its mass?
4. How will astronomers estimate the planet's composition?
5. Why is this planet called a "cousin" of Earth?

While we know a lot about the universe, science is an ongoing process, and we continue to search for new answers. To give you a glimpse of what we don't know, we provide an **Unanswered Questions** feature near the end of each chapter. Most of these questions represent topics that scientists are currently studying.

### ? UNANSWERED QUESTIONS

- How typical is the Solar System? Only within the past few years have astronomers found other systems containing four or more planets, and so far the observed distributions of large and small planets in these multiplanet systems have looked different from those of the Solar System. Computer simulations of planetary system formation suggest that a system with an orbital stability and a planetary distribution like those of the Solar System may develop only rarely. Improved supercomputers can run more complex simulations, which can be compared with the observations to understand better how solar systems are configured.
- How Earth-like must a planet be before scientists declare it to be "another Earth"? An editorial in the science journal *Nature* cautioned that scientists should define "Earth-like" in advance—before multiple discoveries of planets "similar" to Earth are announced and a media frenzy ensues. Must a planet be of similar size and mass, be located in the habitable zone, and have spectroscopic evidence of liquid water before we call it "Earth 2.0"?

The language of science is mathematics, and it can be as challenging to learn as any other language. The choice to use mathematics as the language of science is not arbitrary; nature "speaks" math. To learn about nature, you will need to speak its language. We don't want the language of math to obscure the concepts, so we have placed this book's mathematics in **Working It Out** boxes to make it clear when we are beginning and ending a mathematical argument, so that you can spend time with the concepts in the chapter text and then revisit the mathematics of the concept to study the formal language of the argument. You will learn to work with data and identify when data aren't quite right. We want you to be comfortable reading, hearing, and speaking the language of science, and we will provide you with tools to make it easier.

### 7.3 Working It Out Estimating the Radius of an Extrasolar Planet

The masses of extrasolar planets can often be estimated using Kepler's laws and the conservation of angular momentum. When planets are detected by the transit method, astronomers can estimate the radius of an extrasolar planet. In this method, astronomers look for planets that eclipse their stars and observe how much the star's light decreases during this eclipse (see Figure 7.19). In the Solar System when Venus or Mercury transits the Sun, a black circular disk is visible on the face of the circular Sun. During the transit, the amount of light from the transited star is reduced by the area of the circular disk of the planet divided by the area of the circular disk of the star:

$$\text{Percentage reduction in light} = \frac{\text{Area of disk of planet}}{\text{Area of disk of star}}$$

$$= \frac{\pi R^2_{\text{planet}}}{\pi R^2_{\text{star}}} = \frac{R^2_{\text{planet}}}{R^2_{\text{star}}}$$

Then, to solve for the radius of the planet, astronomers need an estimate of the radius of the star and a measurement of the percentage reduction in light during the transit. The radius of a star is estimated from the surface temperature and the luminosity of the star.

Let's consider an example. Kepler-11 is a system of at least six planets that transit a star. The radius of the star, $R_{\text{star}}$, is estimated to be 1.1 times the radius of the Sun, or $1.1 \times (7.0 \times 10^5\ \text{km}) = 7.7 \times 10^5\ \text{km}$. The light from planet Kepler-11c is observed to decrease by 0.077 percent, or 0.00077 (see Figure 7.19). What is Kepler-11c's size?

$$0.00077 = \frac{R^2_{\text{Kepler-11c}}}{R^2_{\text{star}}} = \frac{R^2_{\text{Kepler-11c}}}{(7.7 \times 10^5\ \text{km})^2}$$

$$R^2_{\text{Kepler-11c}} = 4.5 \times 10^8\ \text{km}^2$$

$$R_{\text{Kepler-11c}} = 2.1 \times 10^4\ \text{km}$$

Dividing Kepler-11c's radius by the radius of Earth (6,400 km) shows that the planet Kepler-11c has a radius of 3.3 $R_{\text{Earth}}$.

Each chapter concludes with an **Origins** section, which relates material or subjects found in the chapter to questions about the origin of the universe and the origin of life in the universe and on Earth. Astrobiologists have made much progress in recent years on understanding how conditions in the universe may have helped or hindered the origin of life, and in each Origins we explore an example from its chapter that relates to how the universe and life formed and evolved.

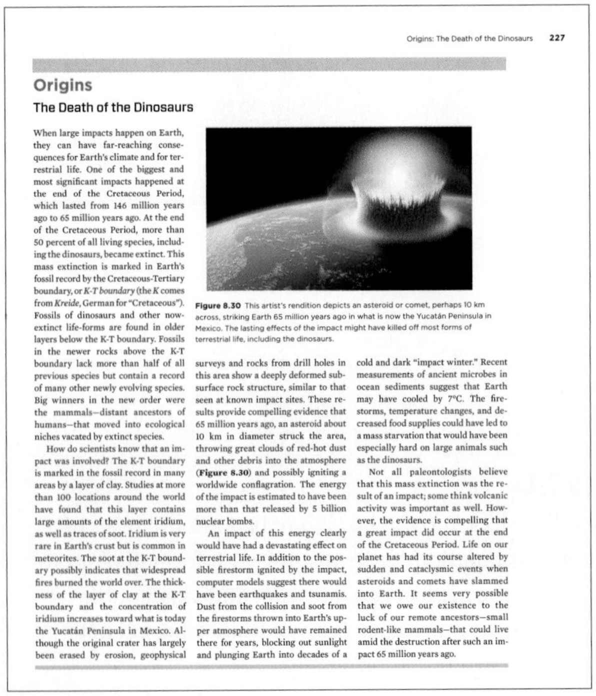
Origins: The Death of the Dinosaurs 227

Origins

The Death of the Dinosaurs

When large impacts happen on Earth, they can have far-reaching consequences for Earth's climate and for terrestrial life. One of the biggest and most significant impacts happened at the end of the Cretaceous Period, which lasted from 146 million years ago to 65 million years ago. At the end of the Cretaceous Period, more than 50 percent of all living species, including the dinosaurs, became extinct. This mass extinction is marked in Earth's fossil record by the Cretaceous-Tertiary boundary, or *K-T boundary* (the *K* comes from *Kreide*, German for "Cretaceous"). Fossils of dinosaurs and other now-extinct life-forms are found in older layers below the K-T boundary. Fossils in the newer rocks above the K-T boundary lack more than half of all previous species but contain a record of many other newly evolving species. Big winners in the new order were the mammals—distant ancestors of humans—that moved into ecological niches vacated by extinct species.

How do scientists know that an impact was involved? The K-T boundary is marked in the fossil record in many areas by a layer of clay. Studies at more than 100 locations around the world have found that this layer contains large amounts of the element iridium, as well as traces of soot. Iridium is very rare in Earth's crust but is common in meteorites. The soot at the K-T boundary possibly indicates that widespread fires burned the world over. The thickness of the layer of clay at the K-T boundary and the concentration of iridium increases toward what is today the Yucatán Peninsula in Mexico. Although the original crater has largely been erased by erosion, geophysical surveys and rocks from drill holes in this area show a deeply deformed subsurface rock structure, similar to that seen at known impact sites. These results provide compelling evidence that 65 million years ago, an asteroid about 10 km in diameter struck the area, throwing great clouds of red-hot dust and other debris into the atmosphere (**Figure 8.30**) and possibly igniting a worldwide conflagration. The energy of the impact is estimated to have been more than that released by 5 billion nuclear bombs.

An impact of this energy clearly would have had a devastating effect on terrestrial life. In addition to the possible firestorm ignited by the impact, computer models suggest there would have been earthquakes and tsunamis. Dust from the collision and soot from the firestorms thrown into Earth's upper atmosphere would have remained there for years, blocking out sunlight and plunging Earth into decades of a cold and dark "impact winter." Recent measurements of ancient microbes in ocean sediments suggest that Earth may have cooled by 7°C. The firestorms, temperature changes, and decreased food supplies could have led to a mass starvation that would have been especially hard on large animals such as the dinosaurs.

Not all paleontologists believe that this mass extinction was the result of an impact; some think volcanic activity was important as well. However, the evidence is compelling that a great impact did occur at the end of the Cretaceous Period. Life on our planet has had its course altered by sudden and cataclysmic events when asteroids and comets have slammed into Earth. It seems very possible that we owe our existence to the luck of our remote ancestors—small rodent-like mammals—that could live amid the destruction after such an impact 65 million years ago.

**Figure 8.30** This artist's rendition depicts an asteroid or comet, perhaps 10 km across, striking Earth 65 million years ago in what is now the Yucatán Peninsula in Mexico. The lasting effects of the impact might have killed off most forms of terrestrial life, including the dinosaurs.

At the end of each chapter, we have provided several types of questions, problems, and activities for you to practice your skills. The Test Your Understanding questions focus on more detailed facts and concepts from the chapter. Thinking about the Concepts questions ask you to synthesize information and explain the "how" or "why" of a situation. Applying the Concepts problems give you a chance to practice the quantitative skills you learned in the chapter and to work through a situation mathematically. The **Using the Web** questions and **Explorations** represent other opportunities to "learn by doing." Using the Web sends you to websites of space missions, observatories, experiments, or archives to access recent observations, results, or press releases. Other sites are for "citizen science" projects in which you can contribute to the analysis of new data.

Explorations show you how to use the concepts and skills you learned in an interactive way. Most of the book's Explorations ask you to use animations and simulations on the Student Site, while the others are hands-on, paper-and-pencil activities that use everyday objects such as ice cubes or balloons.

The resources outside of the book (at the Student Site) can help you understand and visualize many of the physical concepts described in the book. **AstroTours** and **Nebraska Simulations** are represented by icons in the margins of the book. There is also a series of short **Astronomy in Action** videos that are represented by icons in the margins and available at the Student Site. These videos feature one of the authors (and several students) demonstrating physical concepts at work. Your instructor might assign these videos to you or you might choose to watch them on your own to create a better picture of each concept in your mind.

Astronomy gives you a sense of perspective that no other field of study offers. The universe is vast, fascinating, and beautiful, filled with a wealth of objects that, surprisingly, can be understood using only a handful of principles. By the end of this book, you will have gained a sense of your place in the universe.

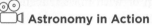 **Astronomy in Action**

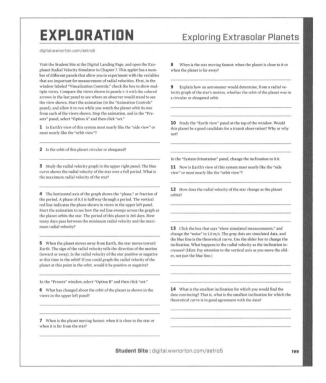

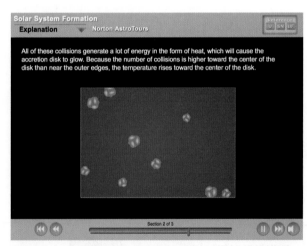

Solar System Formation
Explanation          Norton AstroTours

All of these collisions generate a lot of energy in the form of heat, which will cause the accretion disk to glow. Because the number of collisions is higher toward the center of the disk than near the outer edges, the temperature rises toward the center of the disk.

Section 2 of 3

 **AstroTour**

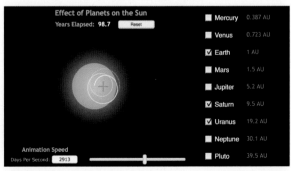

Effect of Planets on the Sun
Years Elapsed: 98.7    Reset

| | |
|---|---|
| ☐ Mercury | 0.387 AU |
| ☐ Venus | 0.723 AU |
| ☑ Earth | 1 AU |
| ☐ Mars | 1.5 AU |
| ☐ Jupiter | 5.2 AU |
| ☑ Saturn | 9.5 AU |
| ☑ Uranus | 19.2 AU |
| ☐ Neptune | 30.1 AU |
| ☐ Pluto | 39.5 AU |

Animation Speed
Days Per Second:  2913

 **Nebraska Simulation**

# Dear Instructor

We wrote this book with a few overarching goals: to inspire students, to make the material interactive, and to create a useful and flexible tool that can support multiple learning styles.

As scientists and as teachers, we are passionate about the work we do. We hope to share that passion with students and inspire them to engage in science on their own. Through our own experience, familiarity with education research, and surveys of instructors, we have come to know a great deal about how students learn and what goals teachers have for their students. We have explicitly addressed many of these goals and learning styles in this book, sometimes in large, immediately visible ways such as the inclusion of features but also through less obvious efforts such as questions and problems that relate astronomical concepts to everyday situations or a fresh approach to organizing material.

For example, many teachers state that they would like their students to become "educated scientific consumers" and "critical thinkers" or that their students should "be able to read a news story about science and understand its significance." We have specifically addressed these goals in our Reading Astronomy News feature, which presents a news article and a series of questions that guide a student's critical thinking about the article, the data presented, and the sources.

In nearly every chapter, we have Visual Analogy figures that compare astronomy concepts to everyday events or objects. Through these analogies, we strive to make the material more interesting, relevant, and memorable.

Education research shows that the most effective way to learn is by doing. Exploration activities at the end of each chapter are hands-on, asking students to take the concepts they've learned in the chapter and apply them as they interact with animations and simulations on the Student Site or work through pencil-and-paper activities. Many of these Explorations incorporate everyday objects and can be used either in your classroom or as activities at home. The Using the Web problems direct students to "citizen science" projects, where they can contribute to the analysis of new astronomical data. Other problems send students to websites of space missions, observatories, collaborative projects, and catalogs to access the most current observations, results, and news releases. These Web problems can be used for homework, lab exercises, recitations, or "writing across the curriculum" projects.

We also believe students should be exposed to the more formal language of science—mathematics. We have placed the math in Working It Out boxes, so it does not interrupt the flow of the text or get in the way of students' understanding of conceptual material. But we've gone further by beginning with fundamental ideas in early Working It Out boxes and slowly building in complexity through the book. We've also worked to remove some of the stumbling blocks that affect student confidence by providing calculator hints, references to earlier Working It Out boxes, and detailed, fully worked examples. Many chapters include problems on reading and interpreting graphs. Appendix 1, "Mathematical Tools," has also been reorganized and expanded.

Discussion of basic physics is contained in Part I to accommodate courses that use the *Solar System* or *Stars and Galaxies* volumes. A "just-in-time" approach to introducing the physics is still possible by bringing in material from Chapters 2–6 as needed. For example, the sections on tidal forces in Chapter 4 can be taught along with the moons of the Solar System in Part II, or with mass transfer in binary stars in Part III, or with galaxy interactions in Part IV. Spectral lines in Chapter 5 can be taught with planetary atmospheres in Part II or with stellar spectral types in Part III, and so on.

In our overall organization, we have made several efforts to encourage students to engage with the material and build confidence in their scientific skills as they proceed through the book. For planets, stars and galaxies, we have organized the material to cover the general case first and then delve into more details with specific examples. Thus, you will find "planetary systems" before our own Solar System, "stars" before the Sun, and "galaxies" before the Milky Way. This allows us to avoid frustrating students by making assumptions about what they know about stars or galaxies or forward-referencing to basic definitions and overarching concepts. This organization also implicitly helps students understand their place in the universe: our galaxy and our star are each one of many. They are specific examples of a physical universe in which the same laws apply everywhere. Planets have been organized comparatively to emphasize that science is a process of studying individual examples that lead to collective conclusions. All of these organizational choices were made with the student perspective in mind and a clear sense of the logical hierarchy of the material.

Even our layout has been designed to maximize student engagement—one wide text column is interrupted as seldom as possible. Material from the earlier edition's Connections boxes has been streamlined and incorporated into the text.

We have continued to respond to commentary from you, our colleagues. We have reorganized the material in the first half of Part IV to reflect user feedback. We begin in Chapter 19 by introducing galaxies as a whole and our measurements of them, including recession velocities. Then we address the Milky Way in Chapter 20—a specific example of a galaxy that we can discuss in detail. This follows the repeating motif of moving from the general to the specific that exists throughout the text and gives students a basic grounding in the concepts of spiral galaxies, supermassive black holes, and dark matter before they need to apply those concepts to the specific example of our own galaxy. Chapter 21, "The Expanding Universe," covers the cosmological principle, the Hubble expansion, and the observational evidence for the Big Bang.

We revised each chapter, streamlining some topics, and updating the science to reflect the progress in the field. When appropriate, we have updated the Origins sections, which often illustrate how astrobiologists and other scientists approach the study of a scientific question from the chapter related to the origin of the universe and of life. We have enhanced the material on exoplanets and incorporated material about exoplanets into other chapters when appropriate. We include new images of Mars, Ceres, Comet 67P/Churyumov-Gerasimenko, and Pluto. We note the discovery of our new home supercluster, Laniakea. We've updated the cosmology sections on the highest-redshift objects and the first stars and galaxies.

Many professors find themselves under pressure from accrediting bodies or internal assessment offices to assess their courses in terms of learning goals. To help you with this, we've revised each chapter's Learning Goals and organized the end-of-chapter Summary to correspond to the chapter's Learning Goals. In Smartwork5, questions and problems are tagged and can be sorted by Learning Goal. Smartwork5 contains more than 2,000 questions and problems that are tied directly to this text, including the Check Your Understanding questions and versions of the Reading Astronomy News and Exploration questions. Any of these could be used as a reading quiz to be completed before class or as homework. Every question in Smartwork5 has hints and answer-specific feedback so that students are coached to work toward the correct answer. An instructor can easily modify any of the provided questions, answers, and feedback or can create his or her own questions.

We've also created a series of 23 videos explaining and demonstrating concepts from the text, accompanied by questions integrated into Smartwork5. You might assign these videos prior to lecture—either as part of a flipped modality or as a "reading quiz." In either case, you can use the diagnostic feedback from the questions in Smartwork5 to tailor your in-class discussions. Or you might show them in class, to stimulate discussion. Or you might simply use them as a jumping-off point—to get ideas for activities to do with your own students.

We continue to look for better ways to engage students, so please let us know how these features work for your students.

# Ancillaries for Students
# digital.wwnorton.com/astro5

## Smartwork5

Steven Desch, Guilford Technical Community College
Violet Mager, Penn State Wilkes-Barre
Todd Young, Wayne State College

More than 2,000 questions support *21st Century Astronomy, Fifth Edition*—all with answer-specific feedback, hints, and ebook links. Questions include Check Your Understanding, Test Your Understanding, Reading Astronomy News, and versions of the Explorations (based on AstroTours and the University of Nebraska simulations). New ranking, sorting, and labeling tasks are designed to get students thinking visually. Also new to this edition, Astronomy in Action video questions focus on getting students to come to class prepared and on overcoming common misconceptions. Rounding out the Smartwork5 course, Process of Science Guided Inquiry Assignments help students apply the scientific method to important questions in astronomy, challenging them to think like scientists.

## Student Site

W. W. Norton's free and open student website features the following:

- Thirty AstroTour animations. These animations, some of which are interactive, use art from the text to help students visualize important physical and astronomical concepts. All are now tablet-compatible.

- Nebraska Simulations (sometimes called applets or NAAPs, for Nebraska Astronomy Applet Programs). These simulations allow students to manipulate variables and see how physical systems work.

- Twenty-three Astronomy in Action videos that feature author Stacy Palen demonstrating the most important concepts in a visual, easy to understand, and memorable way.

## Learning Astronomy by Doing Astronomy: Collaborative Lecture Activities

Stacy Palen, Weber State University
Ana Larson, University of Washington

Students learn best by doing. Devising, writing, testing, and revising suitable in-class activities that use real astronomical data, illuminate astronomical concepts, and pose probing questions that ask students to confront misconceptions can be challenging and time consuming. In this workbook, the authors draw on their experience teaching thousands of students in many different types of courses (large in-class, small in-class, hybrid, online, flipped, and so forth) to bring 30 field-tested activities that can be used in any classroom today. The activities have been designed to require no special software, materials, or equipment and to take no more than 50 minutes to do.

## Starry Night Planetarium Software (College Version) and Workbook

Steven Desch, Guilford Technical Community College
Michael Marks, Bristol Community College

Starry Night is a realistic, user-friendly planetarium simulation program designed to allow students in urban areas to perform observational activities on a computer screen. Norton's unique accompanying workbook offers observation assignments that guide students' virtual explorations and help them apply what they've learned from the text reading assignments.

# For Instructors

## Instructor's Manual

Ben Sugerman, Goucher College

This resource includes brief chapter overviews; suggested discussion points; notes on the AstroTour animations, Nebraska Simulations, and Astronomy in Action videos contained on the Instructor Resource USB Drive (described later); and worked solutions to all end-of-chapter questions and problems, including answers to all Reading Astronomy News and Check Your Understanding questions found in the textbook.

## PowerPoint Lecture Slides

Jack Hughes, Rutgers University
Jack Brockway, Radford University

These ready-made lecture slides integrate selected textbook art, all Check Your Understanding and Working It Out questions from the text, and links to the AstroTour animations. Designed with accompanying lecture outlines, these lecture slides are fully editable and are available in Microsoft PowerPoint format.

## Test Bank

Joshua Thomas, Clarkson University
Parviz Ghavamian, Towson University
Adriana Durbala, University of Wisconsin–Stevens Point

The Test Bank has been revised using Bloom's Taxonomy and provides a quality bank of more than 2,400 multiple-choice and short-answer questions. Each chapter of the Test Bank consists of six question levels classified according to Bloom's Taxonomy:

Remembering

Understanding

Applying

Analyzing

Evaluating

Creating

Questions are further classified by section and difficulty, making it easy to construct tests and quizzes that are meaningful and diagnostic. The Test Bank assesses a common set of Learning Objectives consistent with the textbook and Smartwork5 online homework.

## Norton Instructor's Resource Site

This Web resource contains the following resources to download:

- Test Bank, available in ExamView, Word RTF, and PDF formats

- Instructor's Manual in PDF format

- Lecture PowerPoint slides with lecture notes

- All art and tables in JPEG and PPT formats

- Starry Night College, W. W. Norton Edition, Instructor's Manual

- AstroTour animations

- Selected Nebraska Simulations

- Coursepacks, available in BlackBoard, Angel, Desire2Learn, and Moodle formats

## Coursepacks

Norton's Coursepacks, available for use in various Learning Management Systems (LMSs), feature all Test Bank questions, links to the AstroTours and Nebraska Simulations, worksheets based on the Explorations and Astronomy in Action videos, and automatically graded versions of the end-of-chapter Test Your Understanding multiple-choice questions. Coursepacks are available in BlackBoard, Canvas, Desire2Learn, and Moodle formats.

## Instructor Resource USB Drive

This USB drive contains all instructor resources found on the Instructor's Resource Site, including offline versions of the Astronomy in Action videos, AstroTour animations, and Nebraska Simulations.

# Acknowledgments

The authors would like to acknowledge the extraordinary efforts of the staff at W. W. Norton: Arielle Holstein, who kept things flowing smoothly; Diane Cipollone, who shepherded manuscript through the redesigned layout; Jane Miller and Trish Marx for managing the numerous photos; and the copy editor, Christopher Curioli, who made sure that all the grammar and punctuation survived the multiple rounds of the editing process. We would especially like to thank Becky Kohn for the developmental editing process, and our editor Erik Fahlgren for his degree of commitment to the project.

Andy Ensor managed the production. Hope Miller Goodell was the design director. Rob Bellinger and Julia Sammaritano worked on the media and supplements, and Stacy Loyal will help get this book into the hands of people who can use it.

We gratefully acknowledge the contributions of the authors who worked on previous editions of *21st Century Astronomy*: Dave Burstein, Ron Greeley, Jeff Hester, Brad Smith, Howard Voss, and Gary Wegner, with special thanks to Dave for starting the project, to Jeff for leading the original authors through the first edition, and to Brad for leading the second and third editions.

Laura Kay
Stacy Palen
George Blumenthal

And we would like to thank the reviewers, whose input at every stage improved the book:

## Reviewers for the Fifth Edition

James Applegate, Columbia University
Matthew Bailey, University of Nevada–Reno
Fabien Baron, Georgia State University
Bob Becker, University of California–Davis
Miles Blanton, Bowling Green State University
Jean Brodie, University of California–Santa Cruz
Gerald Cecil, University of North Carolina at Chapel Hill
Damian Christian, California State University–Northridge
Micol Christopher, Mt. San Antonio College
Bethany Cobb, George Washington University
Kate Dellenbusch, Bowling Green State University
Karna Desai, Indiana University–Bloomington
Matthias Dietrich, The Ohio State University
Yuri Efremenko, University of Tennessee
Robert Egler, North Carolina State University
Jason Ferguson, Wichita State University
Jay Gallagher, University of Wisconsin–Madison
Richard Gelderman, Western Kentucky University
Douglas Gobeille, University of Rhode Island
Greg Gowens, University of West Georgia
Aaron Grocholski, Louisiana State University
Peter Hahn, George Washington University

Javier Hasbun, University of West Georgia
Lynnette Hoerner, Red Rocks Community College
Michael Hood, Mt. San Antonio College
Douglas Ingram, Texas Christian University
Bill Keel, University of Alabama
Charles Kerton, Iowa State University
David Kirkby, University of California–Irvine
Mark Kruse, Duke University
Silas Laycock, University of Massachusetts–Lowell
Alexandre Lazarian, University of Wisconsin–Madison
Lauren Likkel, University of Wisconsin–Eau Claire
Catherine Lovekin, Mount Allison University
Lori Lubin, University of California–Davis
Loris Magnani, University of Georgia
Joseph McMullin, Pima Community College
David Menke, Pima Community College
Bahram Mobasher, University of California–Riverside
Edward M. Murphy, University of Virginia
Robert Mutel, University of Iowa
Harold Nations, College of Southern Nevada–Charleston
Richard Nolthenius, Cabrillo College
Chris Packham, University of Texas at San Antonio

Jay Pehl, Indiana University–Purdue University Indianapolis
Michael Reid, University of Toronto
Edward Rhoads, Indiana University–Purdue University Indianapolis
Russell Robb, University of Victoria
James Roberts, University of North Texas
Lindsay Rocks, Front Range Community College
Eric Schlegel, University of Texas at San Antonio
Ohad Shemmer, University of North Texas
Allyn Smith, Austin Peay State University
Inseok Song, University of Georgia
James Sowell, Georgia Institute of Technology
Adriane Steinacker, University of California–Santa Cruz
Gregg Stiesberg, Ithaca College
Michael Strauss, Princeton University
Edmund Sutton, University of Illinois at Urbana-Champaign
Glenn Tiede, Bowling Green State University
Christy Tremonti, University of Wisconsin–Madison
Chandra Vanajakshi, College of San Mateo
Frederick Walter, Stony Brook University
Kevin Williams, Buffalo State
Kurtis Williams, Texas A&M University–Commerce
David Wittman, University of California–Davis
Brian Woodahl, Indiana University–Purdue University Indianapolis
Garett Yoder, Eastern Kentucky University

## Reviewers of Previous Editions

Scott Atkins, University of South Dakota
Timothy Barker, Wheaton College
Peter A. Becker, George Mason University
Timothy C. Beers, Michigan State University
David Bennum, University of Nevada–Reno
Edwin Bergin, University of Pittsburgh
William Blass, University of Tennessee
Steve Bloom, Hampden Sydney College
Daniel Boice, University of Texas at San Antonio
Bram Boroson, Clayton State University
David Branning, Trinity College
Julie Bray-Ali, Mt. San Antonio College
Jack Brockway, Radford University
Suzanne Bushnell, McNeese State University
Paul Butterworth, George Washington University
Juan E. Cabanela, Minnesota State University–Moorhead
Amy Campbell, Louisiana State University
Michael Carini, West Kentucky University
Gerald Cecil, University of North Carolina at Chapel Hill
Supriya Chakrabarti, Boston University
Damian Christian, California State University–Northridge

Micol Christopher, Mt. San Antonio College
Robert Cicerone, Bridgewater State College
David Cinabro, Wayne State University
Judith Cohen, California Institute of Technology
Eric M. Collins, California State University–Northridge
John Cowan, University of Oklahoma–Norman
Debashis Dasgupta, University of Wisconsin–Milwaukee
Kate Dellenbusch, Bowling Green State University
Robert Dick, Carleton University
Matthias Dietrich, The Ohio State University
Gregory Dolise, Harrisburg Area Community College
Yuri Efremenko, University of Tennessee
Tom English, Guilford Technical Community College
David Ennis, The Ohio State University
Jason Ferguson, Wichita State University
John Finley, Purdue University
Matthew Francis, Lambuth University
Kevin Gannon, College of Saint Rose
Todd Gary, O'More College of Design
Christopher Gay, Santa Fe College
Parviz Ghavamian, Towson University
Martha Gilmore, Wesleyan University
Greg Gowens, University of West Georgia
Bill Gutsch, St. Peter's College
Karl Haisch, Utah Valley University
Javier Hasbun, University of West Georgia
Charles Hawkins, Northern Kentucky University
Sebastian Heinz, University of Wisconsin–Madison
Barry Hillard, Baldwin Wallace College
Paul Hintzen, California State University–Long Beach
Paul Hodge, University of Washington
William A. Hollerman, University of Louisiana at Lafayette
Hal Hollingsworth, Florida International University
Olencka Hubickyj-Cabot, San Jose State University
Kevin M. Huffenberger, University of Miami
James Imamura, University of Oregon
Douglas Ingram, Texas Christian University
Adam Johnston, Weber State University
Steven Kawaler, Iowa State University
Bill Keel, University of Alabama
Charles Kerton, Iowa State University
Monika Kress, San Jose State University
Jessica Lair, Eastern Kentucky University
Alex Lazarian, University of Wisconsin–Madison
Kevin Lee, University of Nebraska–Lincoln
Matthew Lister, Purdue University
M. A. K. Lodhi, Texas Tech University
Leslie Looney, University of Illinois at Urbana–Champaign
Jack MacConnell, Case Western Reserve University
Kevin Mackay, University of South Florida
Dale Mais, Indiana University–South Bend

Michael Marks, Bristol Community College
Norm Markworth, Stephen F. Austin State University
Kevin Marshall, Bucknell University
Stephan Martin, Bristol Community College
Justin Mason, Ivy Tech Community College
Amanda Maxham, University of Nevada–Las Vegas
Chris McCarthy, San Francisco State University
Ben McGimsey, Georgia State University
Charles McGruder, West Kentucky University
Janet E. McLarty-Schroeder, Cerritos College
Stanimir Metchev, Stony Brook University
Chris Mihos, Case Western University
Milan Mijic, California State University–Los Angeles
J. Scott Miller, University of Louisville
Scott Miller, Sam Houston State University
Kent Montgomery, Texas A&M University–Commerce
Andrew Morrison, Illinois Wesleyan University
Edward M. Murphy, University of Virginia
Kentaro Nagamine, University of Nevada–Las Vegas
Ylva Pihlström, University of New Mexico
Jascha Polet, California State Polytechnic University
Dora Preminger, California State University–Northridge
Daniel Proga, University of Nevada–Las Vegas
Laurie Reed, Saginaw Valley State University
Judit Györgyey Ries, University of Texas
Allen Rogel, Bowling Green State University
Kenneth Rumstay, Valdosta State University
Masao Sako, University of Pennsylvania
Samir Salim, Indiana University–Bloomington
Ata Sarajedini, University of Florida

Eric Schlegel, University of Texas at San Antonio
Paul Schmidtke, Arizona State University
Ann Schmiedekamp, Pennsylvania State University
Jonathan Secaur, Kent State University
Ohad Shemmer, University of North Texas
Caroline Simpson, Florida International University
Paul P. Sipiera, William Rainey Harper College
Ian Skilling, University of Pittsburgh
Tammy Smecker-Hane, University of California–Irvine
Allyn Smith, Austin Peay State University
Jason Smolinski, State University of New York at Oneonta
Roger Stanley, San Antonio College
Ben Sugerman, Goucher College
Neal Sumerlin, Lynchburg College
Angelle Tanner, Mississippi State University
Christopher Taylor, California State University–Sacramento
Donald Terndrup, The Ohio State University
Todd Thompson, The Ohio State University
Glenn Tiede, Bowling Green State University
Frances Timmes, Arizona State University
Trina Van Ausdal, Salt Lake Community College
Walter Van Hamme, Florida International University
Karen Vanlandingham, West Chester University
Nilakshi Veerabathina, University of Texas at Arlington
Paul Voytas, Wittenberg University
Ezekiel Walker, University of North Texas
James Webb, Florida International University
Paul Wiita, Georgia State University
Richard Williamon, Emory University
David Wittman, University of California–Davis

# About the Authors

**Laura Kay** is a professor of Physics and Astronomy at Barnard College, where she has taught since 1991. She received a BS degree in physics and an AB degree in feminist studies from Stanford University, and MS and PhD degrees in astronomy and astrophysics from the University of California–Santa Cruz. As a graduate student she spent 13 months at the Amundsen Scott station at the South Pole in Antarctica. She studies active galactic nuclei, using ground-based and space telescopes. She teaches courses in astronomy, astrobiology, women and science, and polar exploration. At Barnard she has served as chair of the Physics & Astronomy Department, chair of the Women's Studies Department, chair of Faculty Governance, and interim associate dean for Curriculum and Governance.

**Stacy Palen** is an award-winning professor in the physics department and the director of the Ott Planetarium at Weber State University. She received her BS in physics from Rutgers University and her PhD in physics from the University of Iowa. As a lecturer and postdoc at the University of Washington, she taught Introductory Astronomy more than 20 times over 4 years. Since joining Weber State, she has been very active in science outreach activities ranging from star parties to running the state Science Olympiad. Stacy does research in formal and informal astronomy education and the death of Sun-like stars. She spends much of her time thinking, teaching, and writing about the applications of science in everyday life. She then puts that science to use on her small farm in Ogden, Utah.

**George Blumenthal** is chancellor at the University of California–Santa Cruz, where he has been a professor of astronomy and astrophysics since 1972. He received his BS degree from the University of Wisconsin–Milwaukee and his PhD in physics from the University of California–San Diego. As a theoretical astrophysicist, George's research encompasses several broad areas, including the nature of the dark matter that constitutes most of the mass in the universe, the origin of galaxies and other large structures in the universe, the earliest moments in the universe, astrophysical radiation processes, and the structure of active galactic nuclei such as quasars. Besides teaching and conducting research, he has served as Chair of the UC–Santa Cruz Astronomy and Astrophysics Department, has chaired the Academic Senate for both the UC–Santa Cruz campus and the entire University of California system, and has served as the faculty representative to the UC Board of Regents.

THE SOLAR SYSTEM

FIFTH EDITION

21ST CENTURY
# ASTRONOMY

# 1 Thinking Like an Astronomer

**T**his is a fascinating time to be studying this most ancient of the sciences. Loosely translated, the word **astronomy** means "patterns among the stars." But modern astronomy—the astronomy we will talk about in this book—is about far more than looking at the sky and cataloging the visible stars. The contents of the universe, the origin and fate of the universe, and the nature of space and time have become the subjects of rigorous scientific investigation. Humans have long speculated about our *origins*. How and when did the Sun, Earth, and Moon form? Are other galaxies, stars, planets, and moons similar to our own? The answers that scientists are finding to these questions are changing not only our view of the cosmos but also our view of ourselves.

## LEARNING GOALS

In this chapter, we will begin the study of astronomy by exploring our place in the universe and the methods of science. By the conclusion of this chapter, you should be able to:

**LG 1**  Describe the size and age of the universe and Earth's place in it.

**LG 2**  Use the scientific method to study the universe.

**LG 3**  Demonstrate how scientists use mathematics, including graphs, to find patterns in nature.

The first view of Earth seen from deep space. In December 1968, *Apollo 8* astronauts photographed Earth above the Moon's limb. ▶ ▶ ▶

What is your
cosmic
address?

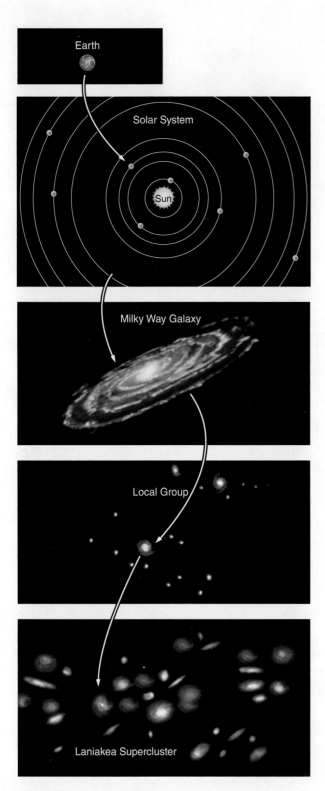

**Figure 1.1** Our cosmic address is Earth, Solar System, Milky Way Galaxy, Local Group, Laniakea Supercluster. We live on Earth, a planet orbiting the Sun in our Solar System, which is a star in the Milky Way Galaxy. The Milky Way is a large galaxy within the Local Group of galaxies, which in turn is located in the Laniakea Supercluster.

# 1.1 Earth Occupies a Small Place in the Universe

Astronomers contemplate our place in the universe by studying Earth's position in space and time. Locating Earth in the larger universe is the first step in learning the science of astronomy. In this section, you will get a feel for the neighborhood in which Earth is located. You will also begin to explore the scale of the universe in space and time.

## Our Place in the Universe

Most people receive their postal mail at an address—building number, street, city, state, and country. We can expand our view to include the enormously vast universe we live in. What is our "cosmic address"? We reside on a planet called Earth, which is orbiting under the influence of gravity around a star called the Sun. The **Sun** is a typical, middle-aged star and seems extraordinary only because of its importance to us within our own **Solar System**. Our Solar System consists of eight planets: Mercury, Venus, Earth, Mars, Jupiter, Saturn, Uranus, and Neptune. It also contains many smaller bodies, such as dwarf planets, asteroids, and comets. All of these objects are bound to the Sun by gravity.

The Sun is located about halfway out from the center of the **Milky Way Galaxy**, a flattened collection of stars, gas, and dust. Our Sun is just one among several hundred billion stars scattered throughout our galaxy, and many of these stars are themselves surrounded by planets.

The Milky Way is a member of a collection of a few dozen galaxies called the **Local Group**. Most galaxies in this group are much smaller than the Milky Way. Looking farther outward, the Local Group is part of a vastly larger collection of thousands of galaxies—a **supercluster**—called the Laniakea Supercluster. There are millions of superclusters in the observable universe.

We can now define our cosmic address—Earth, Solar System, Milky Way Galaxy, Local Group, Laniakea Supercluster—as illustrated in **Figure 1.1**. Yet even this address is not complete, as the Laniakea Supercluster encompasses only the *local universe*. The part of the universe that we can see—the *observable universe*—extends to 50 times the size of Laniakea in every direction. Within this volume, there are about as many galaxies as there are stars in the Milky Way—several hundred billion. The universe is not only much larger than the local universe but also contains much more than the observed planets, stars, and galaxies. Up to 95 percent of the mass of the universe is made up of matter that does not interact with light, known as *dark matter*, and a form of energy that permeates all of space, known as *dark energy*. Neither of these is well understood, and they are among the many exciting areas of research in astronomy.

## The Scale of the Universe

As you saw in Figure 1.1, the size of the universe completely dwarfs our human experience. We can start by comparing astronomical sizes and distances to something more familiar. For example, the diameter of our Moon is about equal to the distance between the offices of the first two authors of this book, in New York, New York, and Ogden, Utah (**Figure 1.2a**). The distance from Earth to the Moon is about 100 times the Moon's diameter, and the planet Saturn with its majestic rings would fill much of that distance (Figure 1.2b). The distance from Earth to the Sun is about 400 times the Earth–Moon distance, and the distance to the planet Neptune is about 30 times the Earth–Sun distance.

But as we move out from the Solar System to the stars, the distances become so enormous that they are difficult to comprehend. The nearest star is about 9,000 times farther away from the Sun than the Sun's distance to the planet Neptune. The diameter of our Milky Way Galaxy is 30,000 times the distance to that nearest star. The Andromeda Galaxy, the nearest similar large galaxy to the Milky Way, is about 600,000 times farther away than that nearest star. The diameter of the Local Group of galaxies is about 4 times the distance to Andromeda, and the diameter of the recently identified Laniakea Supercluster, which includes the Local Group and many other galaxy groups, is 50 times larger than the Local Group. As noted earlier, this is just one of millions of superclusters in the observable universe.

To get a better sense of these distances, imagine a model in which the objects and distances in the universe are 1 billion times smaller than they really are. In this model, Earth is about the size of a marble or a peanut M&M (about 1.3 centimeters, or half an inch), the Moon is 38 centimeters (cm) away, and the Sun is 150 meters away. Neptune is 4.5 kilometers (km) from the Sun, and the nearest star to the Sun is about 40,000 km away (or about the length of the circumference of the real Earth). The model Milky Way Galaxy would fill the Solar System nearly to the orbit of Saturn. The distance between the model Milky Way and Andromeda galaxies would fill the Solar System 20 times farther, out beyond humanity's most distant space probe. The model Laniakea Supercluster would fill the Solar System and go about one-eighth of the way to the nearest star.

When thinking about the distances in the universe, it can be helpful to discuss the time it takes to travel to various places. If someone asks you how far it is to the nearest city, you might say 100 km or you might say 1 hour. In either case, you will have given that person an idea of how far the city is. In astronomy, the speed of a car on the highway is far too slow to be useful. Instead, we use the fastest speed in the universe—the speed of light. Light travels at 300,000 kilometers per second (km/s). Light can circle Earth, a distance of 40,000 km, in just under $\frac{1}{7}$ of a second. So we say that the circumference of Earth is $\frac{1}{7}$ of a light-second. Even relatively small distances in astronomy are so vast that they are measured in units of **light-years (ly)**: the distance light travels in 1 year, about 9.5 trillion km, or 6 trillion miles.

Because light takes time to reach us, we see astronomical objects as they were in the past: the extent back in time depends on the object's distance from us. Because light takes $1\frac{1}{4}$ seconds to reach us from the Moon, we see the Moon as it was $1\frac{1}{4}$ seconds ago. Because light takes $8\frac{1}{3}$ minutes to reach us from the Sun, we see the Sun as it was $8\frac{1}{3}$ minutes ago. We see the nearest star as it was more than 4 years ago and objects across the Milky Way as they were tens of thousands of years ago. The light from the Virgo Cluster of galaxies has been traveling 50 million years to reach us. The light from the most distant observable objects has been traveling for almost the age of the universe—nearly 13.8 billion

(a)

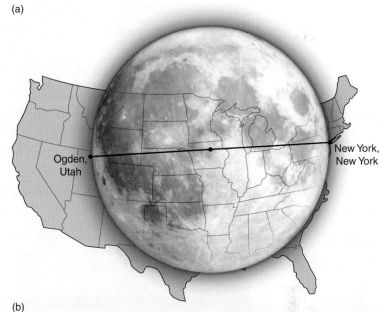

(b)

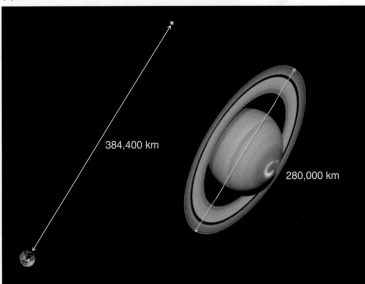

**Figure 1.2** (a) The diameter of the Moon is about the same as the distance between New York, New York, and Ogden, Utah. (b) The size of Saturn, including the rings, is about 70 percent of the distance between Earth and the Moon.

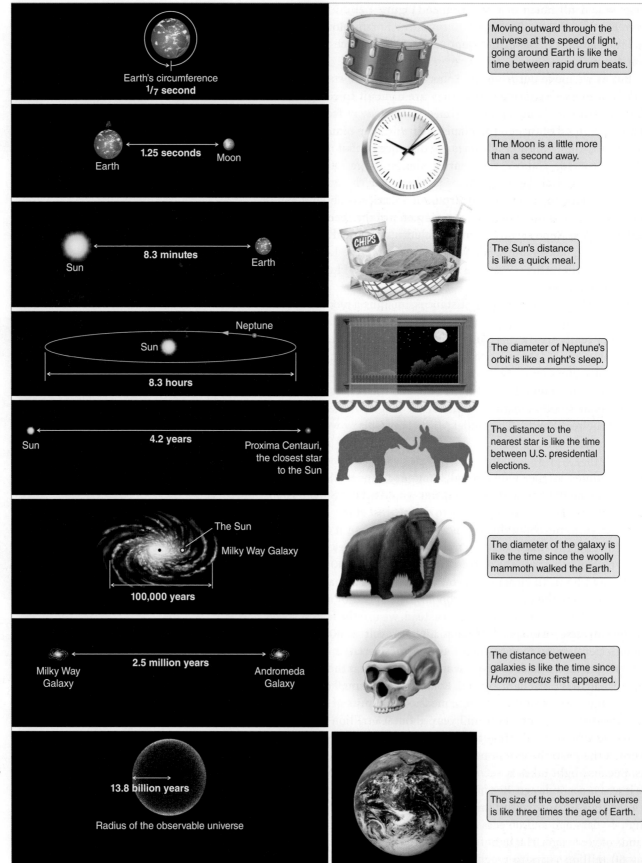

**Figure 1.3** Thinking about the time it takes for light to travel between objects helps us comprehend the vast distances in the universe. (Figures such as this one, with "Visual Analogy" tags, are images that make analogies between astronomical phenomena and everyday objects more concrete.)

Earth's circumference
**1/7 second**

Moving outward through the universe at the speed of light, going around Earth is like the time between rapid drum beats.

Earth **1.25 seconds** Moon

The Moon is a little more than a second away.

Sun **8.3 minutes** Earth

The Sun's distance is like a quick meal.

Neptune
Sun
**8.3 hours**

The diameter of Neptune's orbit is like a night's sleep.

Sun **4.2 years** Proxima Centauri, the closest star to the Sun

The distance to the nearest star is like the time between U.S. presidential elections.

The Sun
Milky Way Galaxy
**100,000 years**

The diameter of the galaxy is like the time since the woolly mammoth walked the Earth.

Milky Way Galaxy **2.5 million years** Andromeda Galaxy

The distance between galaxies is like the time since *Homo erectus* first appeared.

**13.8 billion years**
Radius of the observable universe

The size of the observable universe is like three times the age of Earth.

**VISUAL ANALOGY**

years. **Figure 1.3** begins with Earth and progresses outward to the observable universe.

The vast distances from Earth to other objects in the universe tell us that we occupy a very small part of the space in the universe and a very small part of time. Earth and the Solar System are only about one-third the age of the universe. Animals have existed on Earth for even less time. Imagine the age of the universe and the important events in it as if they took place within a single day, as illustrated in **Figure 1.4**. In this timeline, the Big Bang begins the cosmic day at midnight, and the original light chemical elements are created within the first 2 seconds. The first stars and galaxies appear within the first 10 minutes. Our Solar System formed from recycled gas and dust left over from previous generations of stars, at about 4 P.M. on this cosmic clock. The first bacterial life appears on Earth at 5:20 P.M., the first animals at 11:20 P.M., and modern humans at 11:59:59.8 P.M.—with only a fifth of a second to go in this cosmic day. We humans appeared quite recently in the history of the universe.

### CHECK YOUR UNDERSTANDING 1.1

Rank the following in order of size: (a) a light-minute, (b) a light-year, (c) a light-hour, (d) the radius of Earth, (e) the distance from Earth to the Sun, (f) the radius of the Solar System.

## 1.2 Science Is a Way of Viewing the Universe

Humans have long paid attention to the sky and the stars and developed the dynamic science of astronomy. New discoveries happen frequently, and ideas about the universe are evolving rapidly. To view the universe through the eyes of an astronomer, you will need to understand how science itself works. Throughout this book, we will emphasize not only scientific discoveries but also the process of science. In this section, we will examine the scientific method.

### The Scientific Method

The **scientific method** is a systematic way of testing new ideas or explanations. Often, scientists begin with a fact—an observation or a measurement. For example, you might observe that the weather changes in a predictable way each year and wonder why that happens. You then create a **hypothesis**, a testable explanation of the observation: "I think that it is cold in the winter and warm in the summer because Earth is closer to the Sun in the summer." You and your colleagues come up with a test: if it is cold in the winter and warm in the summer because Earth is closer to the Sun in the summer, then it will be cold in the winter everywhere on the planet—Australia should have winter at the same time of year as the United States. This test can be used to check your hypothesis. You travel from the United States to Australia in January and find that it is summer in Australia. Your hypothesis has just been proved incorrect, so we say that it has been **falsified**. This is different than the meaning in common usage, where one might think of "falsified" evidence as having been manipulated to misrepresent the truth. There are two important elements of your test that all scientific tests share. Your observation is reproducible: anyone who goes to Australia will find the same result.

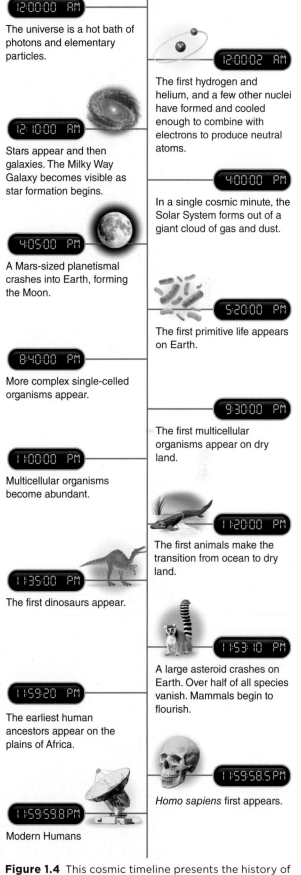

**12:00:00 AM** The universe is a hot bath of photons and elementary particles.

**12:00:02 AM** The first hydrogen and helium, and a few other nuclei have formed and cooled enough to combine with electrons to produce neutral atoms.

**12:10:00 AM** Stars appear and then galaxies. The Milky Way Galaxy becomes visible as star formation begins.

**4:00:00 PM** In a single cosmic minute, the Solar System forms out of a giant cloud of gas and dust.

**4:05:00 PM** A Mars-sized planetismal crashes into Earth, forming the Moon.

**5:20:00 PM** The first primitive life appears on Earth.

**8:40:00 PM** More complex single-celled organisms appear.

**9:30:00 PM** The first multicellular organisms appear on dry land.

**11:00:00 PM** Multicellular organisms become abundant.

**11:20:00 PM** The first animals make the transition from ocean to dry land.

**11:35:00 PM** The first dinosaurs appear.

**11:53:10 PM** A large asteroid crashes on Earth. Over half of all species vanish. Mammals begin to flourish.

**11:59:20 PM** The earliest human ancestors appear on the plains of Africa.

**11:59:58.5 PM** *Homo sapiens* first appears.

**11:59:59.8 PM** Modern Humans

**Figure 1.4** This cosmic timeline presents the history of the universe as a 24-hour day.

**Nebraska Simulation:** Lookback Time Simulator

Your result is also repeatable: if you conducted a similar test next year or the year after, you would get the same result. Because you have falsified your hypothesis, you must revise or completely change it to be consistent with the new data.

Any idea that is not testable—that is not falsifiable—must be accepted or rejected based on intuition alone, so it is not a scientific idea. A falsifiable hypothesis or idea does not have to be testable using current technology, but we must be able to imagine an experiment or observation that could prove the idea wrong if we could carry it out. As continuing tests support a hypothesis by failing to disprove it, scientists come to accept the hypothesis as a theory. A **theory** is a well-developed idea or group of ideas that is tied to known physical laws and makes testable predictions. As in the previous paragraph, the scientific meaning is different than the meaning in common usage. In everyday language, theory may mean a guess: "Do you have a theory about who did it?" In everyday language, a theory can be something we don't take too seriously. "After all," people say, "it's only a theory."

In stark contrast, scientists use the word *theory* to mean a carefully constructed proposition that takes into account every piece of relevant data as well as our entire understanding of how the world works. A theory has been used to make testable predictions, and all of those predictions have come true. Every attempt to prove it false has failed. A classic example is Einstein's theory of relativity, which we cover in some depth in Chapter 18. For more than a century, scientists have tested the predictions of the theory of relativity and have not been able to falsify it. Even after 100 years of verification, if a prediction of the theory of relativity failed tomorrow, the theory would require revision or replacement. As Einstein himself noted, a theory that fails only one test is proved false. In this sense, all scientific knowledge is subject to challenge.

In the loosely defined hierarchy of scientific knowledge, an *idea* is a notion about how something might be. Moving up the hierarchy we come to a *fact*, which is an observation or measurement. For example, the measured value of Earth's radius is a fact. A *hypothesis* is an idea that leads to testable predictions. A hypothesis may be the forerunner of a scientific theory, or it may be based on an existing theory, or both. At the top of the hierarchy is a *theory*: an idea that has been examined carefully, is consistent with all existing theoretical and observational knowledge, and makes testable predictions. Ultimately, the success of the predictions is the deciding factor between competing theories. A scientific *law* is a series of observations that leads to an ability to make predictions but has no underlying explanation of why the phenomenon occurs. So we might have a "law of daytime" that says the Sun rises and sets once each day. We could have a "theory of daytime" that says the Sun rises and sets once each day because Earth spins on its axis. Scientists themselves can be sloppy about the way they use these words, and you will sometimes see them used differently than we have defined them here.

As shown in the **Process of Science Figure**, the scientific method follows a specific sequence. Scientists begin with an observation or idea, followed by careful analysis, followed by a hypothesis, followed by prediction, followed by further observations or experiments to test the prediction. A hypothesis may lead to a scientific theory, or it may be based on an existing theory, or both. Ultimately, the basis for deciding among competing theories is the success of their predictions. Scientists can use theories to take their knowledge a step further by building theoretical models. A **theoretical model** is a detailed description of the properties of a particular object or system in terms of known physical laws or theories, which are used to connect the theoretical model to the behavior of a complex system.

The construction of new theories is often guided by scientific **principles**, which are general ideas or a sense about the universe that will guide the

# Process of Science

## THE SCIENTIFIC METHOD

The scientific method is a formal procedure used to test the validity of scientific hypotheses and theories.

**Start with an observation or idea.**

**Suggest a hypothesis.**

**Make a prediction.**

**If the test or observation does not support the hypothesis, make more observations, revise the hypothesis, or choose a new one.**

**Perform a test, experiment, or additional observation and analyze the data.**

**If the test provides evidence to support the hypothesis, make additional predictions and test them.**

An idea or observation leads to a falsifiable hypothesis that is either accepted as a tested theory or rejected on the basis of observational or experimental tests of its predictions. The blue loop goes on indefinitely as scientists continue to test the theory.

construction of new theories. Two principles used in the study of astronomy are the *cosmological principle* and *Occam's razor*.

The **cosmological principle** is the testable assumption that the physical laws that apply here and now also apply everywhere and at all times. This principle also encompasses the assumption that there are no special locations or directions in the universe. By extension, the cosmological principle asserts that matter and energy obey the same physical laws throughout space and time as they do today on Earth. This principle means that the same physical laws that we observe and apply in laboratories on Earth can be used to understand what goes on in the centers of stars or in the hearts of distant galaxies. Each new theory that comes from applying the cosmological principle to observations of the universe around us adds to our confidence in the validity of this principle, which we will discuss in more detail in Chapter 21.

**Occam's razor** states that when we are faced with two hypotheses that explain all the observations equally well, we should use the one that requires the fewest assumptions until we have evidence to the contrary. For example, a hypothesis that atoms are constructed differently in the Andromeda Galaxy and in the Milky Way Galaxy would violate the cosmological principle. This hypothesis would also require a large number of assumptions to explain how atoms in the Andromeda Galaxy are constructed differently and yet still appear to behave in the same way as atoms in the Milky Way. For example, suppose you assume that the center of an atom in Andromeda is negatively charged, opposite to that in the Milky Way, where the center of an atom is positively charged. However, that assumption would require you to make an assumption about the location of the boundary between Andromeda-like matter and Milky Way–like matter and about why atoms on the boundary between the two regions did not destroy each other. You would also need an assumption about why atoms in the two regions are constructed so differently. If reasonable experimental evidence ever challenges the validity of the cosmological principle, scientists will construct a new description of the universe that takes this new data into account. Until then, the cosmological principle is the hypothesis that has the fewest assumptions, satisfying Occam's razor. To date, the cosmological principle has repeatedly been tested and remains valid.

In many sciences, researchers can conduct controlled experiments to test different hypotheses. This experimental method is often not available to astronomers: we cannot change the tilt of Earth or the temperature of a star to see what happens. Instead, astronomers work from observations or existing models. They make multiple observations using various methods and create mathematical and physical models based on established science to explain the observations.

For example, when astronomers first discovered planets orbiting other stars, these new *extrasolar planets* were most often giant gaseous planets (similar to Jupiter) in short-period orbits very close to their star. These planets are the easiest to discover because they have a strong pull on their star, as we will discuss in Chapter 7. However, their proximity to their star is completely unlike the situation in our own Solar System, where the giant planets are far from the Sun. These discoveries challenged existing ideas about how our Solar System formed. As different observers using multiple telescopes found more and more of these planets, astronomers realized they needed new explanations of planet formation to explain how such large planets could wind up so close to their star. Astronomers could not build different Solar Systems—but they could create computer simulations of planetary systems using the known laws of physics. When they did, they found that planets can migrate, moving to orbits closer to or farther from their star. Planetary scientists search for evidence to test this idea in our own Solar

System, where this may have occurred early in the life of the Sun. In this example, the observations occurred before the theory was constructed.

Alternatively, an astronomer might make predictions from an existing successful mathematical or physical model and then conduct observations and analyze data to test the predictions. One example is the discovery of black holes. In the late 18th century, two scientists hypothesized the existence of "dark stars": massive objects having such strong gravity that light could not escape. At that time, the scientists did not have a way to test this hypothesis. More than 100 years later, in the early 20th century, Karl Schwarzschild studied Einstein's relativity equations and calculated that these collapsed dark stars would be very small, with a radius of only a few kilometers. Fifty years later, these objects were named *black holes*. There was still no evidence of their existence until the 1970s and 1980s, when the new technology of space-based X-ray telescopes made possible the observations needed to test the hypothesis. Einstein's existing theory made the discovery of black holes possible.

The scientific method provides the rules for testing whether an idea is false, but it offers no insight into where the idea came from in the first place or how an experiment was designed. Scientists discussing their work use words such as *insight, intuition,* and *creativity*. Scientists speak of a beautiful theory in the same way that an artist speaks of a beautiful painting or a musician speaks of a beautiful performance. Science has an aesthetic that is as human and as profound as any found in the arts.

Yet science is not the same as art or music in one important respect. Whereas art and music are judged by a human jury alone, the validity of a scientific hypothesis or theory is subject to the natural world. Nature alone provides the final decisions about which theories can be kept and which theories must be discarded. It does not matter what we want to be true. In the history of science, many a beautiful and beloved theory has been abandoned.

## Scientific Revolutions

Scientific inquiry is necessarily dynamic. Scientists do not have all the answers and must constantly refine their ideas in response to new data and new insights. The vulnerability of knowledge may seem like a weakness. "Gee, you really don't know anything," the cynical person might say. But this vulnerability is actually the greatest strength of the scientific process: it means that science self-corrects. Incorrect ideas are eventually overturned by new information. In science, even our most cherished ideas about the nature of the physical world remain fair game, subject to challenge by new evidence. Many of history's best scientists earned their status by falsifying a universally accepted idea. This is a powerful motivation for scientists to challenge old ideas constantly—to formulate and test new explanations for their observations.

For example, Sir Isaac Newton developed classical physics in the 17th century to explain motion, forces, and gravity. Newtonian physics (discussed in detail in Chapters 3 and 4) withstood the scrutiny of scientists for more than 200 years. Yet during the late 19th and early 20th centuries, a series of scientific revolutions completely changed our understanding of the nature of reality. The work of Albert Einstein (**Figure 1.5**) is representative of these scientific revolutions. Einstein's special and general theories of relativity replaced Newton's mechanics. Einstein did not prove Newton wrong but rather showed that Newton's theories were a special case of a far more general and powerful set of physical laws. Einstein's new ideas unified the concepts of mass and energy and destroyed the conventional notion of space and time as separate concepts.

**Figure 1.5** Albert Einstein is perhaps the most famous scientist of the 20th century, and he was *Time* magazine's selection for Person of the Century. Einstein helped to usher in two different scientific revolutions.

Throughout this text, you will encounter many other discoveries that forced scientists to abandon accepted theories. Einstein himself never embraced the view of the world offered by *quantum mechanics*—a second revolution he helped start. Yet quantum mechanics, a statistical description of the behavior of particles smaller than atoms, has held up for more than 100 years. In science, all authorities are subject to challenge, even Einstein.

Science is a way of thinking about the world. It is a search for the relationships that make our world what it is. Scientific inquiry assumes that nature operates by consistent, explicable, inviolate rules. Scientific knowledge is an accumulated collection of ideas about how the universe works, yet scientists are always aware that what is known today may be superseded tomorrow. A scientist assumes that there is an order in the universe and that the human mind is capable of grasping the essence of the rules underlying that order. Scientists build on these assumptions to make predictions and then test those predictions, finding the underlying rules that allow humanity to solve problems, invent new technologies, or find a new appreciation for the natural world. In the final analysis, science has found such a central place in our civilization because *science works*.

## CHECK YOUR UNDERSTANDING 1.2

The scientific method is a process by which scientists: (a) prove theories to be known facts; (b) gain confidence in theories by failing to prove them wrong; (c) show all theories to be wrong; (d) survey what the majority of people think about a theory.

## 1.3 Astronomers Use Mathematics to Find Patterns

Scientific thinking allows scientists to make predictions. Once a pattern has been observed, for example the daily rising and setting of the Sun, scientists can predict what will happen next.

Imagine that the patterns in your life became disrupted, so that the world became entirely unpredictable. For example, what would life be like if you could not predict whether an object you dropped would fall up or down? Or what if one morning the Sun rises in the east and sets in the west, and the next day it rises in the west and sets in the east? In fact, objects do fall toward the ground. The Sun rises, sets, and then rises again at predictable times and in predictable locations. Spring turns into summer, summer turns into autumn, autumn turns into

**Figure 1.6** Since ancient times, people recognized that patterns in the sky change with the seasons. These and other patterns shape our lives. These star maps show the sky in the Northern Hemisphere during each season.

winter, and winter turns into spring. The rhythms of nature produce patterns in our lives, and these patterns give us clues about the nature of the physical world. Astronomers identify and characterize these patterns and use them to understand the world around us. Some of the most easily identified patterns in nature are those we see in the sky. What in the sky will look different or the same a week from now? A month from now? A year from now? As you can see in **Figure 1.6**, patterns in the sky mark the changing of the seasons. Because planting and harvesting times are determined by the seasons, it is no surprise that astronomy—which studies these patterns that are so important to agriculture—is the oldest of all sciences. We will see many other examples of patterns in the sky in the next chapter.

Astronomers use mathematics to analyze patterns and to communicate complex material compactly and accurately. Because the study of patterns in nature is so important to science, it should come as no surprise that mathematics is the language of science. Many people find mathematics to be a major obstacle that prevents them from appreciating the beauty and elegance of the world as seen through the eyes of a scientist. It is our goal in this book to explain any necessary math in everyday language. We will describe what equations mean and help you use them in a way that allows you to connect scientific concepts to the world. Your responsibility is to accept the challenge and make an honest effort to understand the material. **Working It Out 1.1** and **Working It Out 1.2** in this chapter review some basics of mathematical tools and graphs. At the back of the book, Appendix 1 explains some essential mathematical tools, and Appendix 2 contains physical constants of nature. Other appendixes contain data tables with key information about planets, moons, and stars.

## CHECK YOUR UNDERSTANDING 1.3

When you see a pattern in nature, it is usually evidence of: (a) a theory being displayed; (b) a breakdown of random clustering; (c) an underlying physical law.

## 1.1 Working It Out Mathematical Tools

Mathematics provides scientists many of the tools that they need to understand the patterns they observe and to communicate that understanding to others. Following are a few tools that will be useful in our study of astronomy:

**Scientific notation.** Scientific notation is how we handle numbers of vastly different sizes. Writing out 7,540,000,000,000,000,000,000 in standard notation is very inefficient. Scientific notation uses the first few digits (the *significant* ones) and counts the number of decimal places to create the condensed form $7.54 \times 10^{21}$. Similarly, rather than writing out 0.000000000005, we write $5 \times 10^{-12}$. The exponent on the 10 is positive or negative depending on the direction that the decimal point was moved. For example, the average distance to the Sun is 149,600,000 km, but astronomers usually express it as $1.496 \times 10^8$ km.

**Ratios.** Ratios are a useful way to compare things. A star may be "10 times as massive as the Sun" or "10,000 times as luminous as the Sun." These expressions are ratios.

**Proportionality.** Often, understanding a concept amounts to understanding the *sense* of the relationships that it predicts or describes. "If you have twice as far to go, it will take you twice as long to get here." "If you have half as much money, you will be able to buy only half as much gas." These are examples of proportionality.

Appendix 1 has a more detailed explanation of mathematical tools used in this book.

# 1.2  Working It Out  Reading a Graph

Scientists often convey complex information and mathematical patterns in graphical form. Reading graphs is a skill that is important not only in astronomy but also in life. Economists, social and political scientists, mortgage brokers, financial analysts, retirement planners, doctors, and scientists all use graphs to evaluate and communicate important information.

Graphs typically have two axes: a horizontal axis (the *x*-axis) and a vertical axis (the *y*-axis). Typically, the *x*-axis shows an independent variable, which is the one you might have control over in an experiment. The *y*-axis shows the dependent variable, which—in many experiments—is the variable a researcher is studying.

Graphs can take different shapes. Suppose we plot the distance a car travels over a period of time, as shown in **Figure 1.7a**. In a linear graph, each interval on an axis represents the same-sized step. Each step on the horizontal axis of the graph in Figure 1.7a represents 5 minutes. Each step on the vertical axis represents a distance of 5 km traveled by the car. Data are plotted on the graph, with one dot for each observation; for example, the distance the car has traveled after 20 minutes is 20 km.

Drawing a line through these data indicates the trend of the data. To understand what the trend means, scientists often find the slope of the line, which is the relationship of the line's rise along the *y*-axis to its movement along the *x*-axis. To find the slope, we look at the change between two points on the vertical axis divided by the change between two points on the horizontal axis; for example, finding the slope of the line gives

$$\text{Slope} = \frac{\text{Change in vertical axis}}{\text{Change in horizontal axis}}$$

$$= \frac{(15 - 10)\,\text{km}}{(15 - 10)\,\text{min}}$$

$$= 1\,\text{km/min}$$

In this case, the trend tells us that the car is traveling at 1 kilometer per minute (km/min), or 60 kilometers per hour (km/h). The slope of a line often contains extra information that is useful.

Many observations of natural processes do not result in a straight line on a graph. An example of this is an exponential process. Think about what happens when you catch a cold. When you get up in the morning at 7:00 A.M. you feel fine. At 9:00 A.M. you feel a little tired. By 11:00 A.M. you have a bit of a sore throat or a sniffle and think, "I wonder if I'm getting sick," and by 1:00 P.M. you have a runny nose and congestion and fever and chills. This is an exponential process, because the virus that has infected you reproduces exponentially.

For the sake of this discussion, suppose the virus produces one copy of itself each time it invades a cell. (In fact, viruses produce between 1,000 and 10,000 copies each time they invade a cell, so the exponential curve is actually much steeper.) One virus infects a cell and multiplies, so now there are two viruses—the original and a copy. These viruses invade two new cells, and each one produces a copy. Now there are four viruses. After the next cell invasion, there are eight. Then 16, 32, 64, 128, 256, 512, 1,024, 2,048, and so on. This behavior is plotted in Figure 1.7b.

It can be difficult to see what's happening in the early stages of an exponential curve, because the later numbers are so much larger than the earlier ones. For this reason, we sometimes plot this type of data *logarithmically*, by putting the logarithm (the power of 10) of the data on the vertical axis, as shown in Figure 1.7c. Now each step on the axis represents 10 times as many viruses as the previous step. Even though we draw all the steps the same size on the page, they represent different-sized steps in the data—the number of viruses, for example. We often use this technique in astronomy because it has a second, related advantage: very large variations in the data can easily fit on the same graph.

Each time you see a graph, you should first understand the axes—what data are plotted on this graph? Then you should check whether the axes are linear or logarithmic. Finally, you can look at the actual data or lines in the graph to understand how the system behaves.

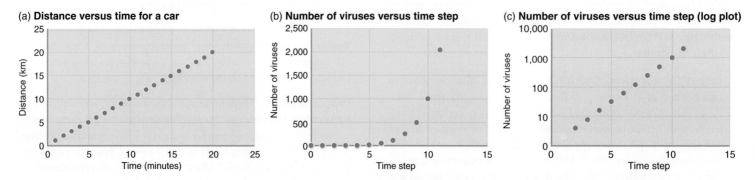

(a) **Distance versus time for a car**

(b) **Number of viruses versus time step**

(c) **Number of viruses versus time step (log plot)**

**Figure 1.7** Graphs like these show relationships between quantities. (a) The time and distance traveled. (b, c) These graphs show the relationship between time and the number of viruses.

# Origins

## An Introduction

How and when did the universe begin? What combination of events led to the existence of humans as sentient beings living on a small rocky planet orbiting a typical middle-aged star? Are there others like us scattered throughout the galaxy?

Earlier in the chapter we mentioned the theme of origins. Throughout this book, you will see that this theme involves much more than how humans came to be on Earth. In these Origins sections, which conclude each chapter, we'll look into the origin of the universe and the origin of life on Earth. We will also examine the possibilities of life elsewhere in the Solar System and beyond—a subject called **astrobiology**. Our origins theme will include the discovery of planets around other stars and how they compare with the planets of our own Solar System.

Later in the book, we will present observational evidence that supports the **Big Bang** theory, which states that the universe started expanding from an infinitesimal size about 13.8 billion years ago. Only the lightest chemical elements were found in substantial amounts in the early universe: hydrogen and helium, and tiny amounts of lithium, and beryllium. However, we live on a planet with a central core consisting mostly of very heavy elements such as iron and nickel, surrounded by outer layers made up of rocks containing large amounts of silicon and various other elements, all heavier than the original elements. The human body contains carbon, nitrogen, oxygen, calcium, phosphorus, and a host of other chemical elements—except for hydrogen itself, all are heavier than hydrogen and helium. If these heavier elements that make up Earth and our bodies were not present in the early universe, where did they come from?

The answer to this question lies within the stars (**Figure 1.8**). In the core of a star, light elements, such as hydrogen, combine to form more massive atoms, which eventually leads to atoms such as carbon. When a star nears the end of its life, it often loses much of its material—including some of the new atoms formed in its interior—by blasting it back into interstellar space. This material combines with material lost from other stars—some of which produced even more massive atoms as they exploded—to form large clouds of dust and gas. Those clouds go on to make new stars and planets, similar to our Sun and Solar System. Prior "generations" of stars supplied the building blocks for the chemical processes that we see in the universe, including life. The atoms that make up much of what we see were formed in the cores of stars. The phrase "We are stardust" is not just poetry. We are actually made of recycled stardust.

The study of origins also provides examples of the process of science. Many of the physical processes in chemistry, geology, planetary science, physics, and astronomy that are seen on Earth or in the Solar System are observed across the galaxy and throughout the universe. But as of this writing, the only biology we know about is that existing on Earth. Thus, at this point in human history, much of what scientists can say about the origin of life on Earth and the possibility of life elsewhere is reasoned extrapolation and educated speculation. In these Origins sections, we'll address some of these hypotheses and try to be clear about which are speculative and which have been tested.

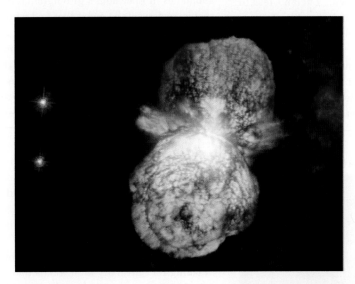

**Figure 1.8** You and everything around you are composed of atoms forged in the interiors of stars that lived and died before the Sun and Earth were formed. The supermassive star Eta Carinae, shown here, is currently ejecting a cloud of chemically enriched material just as earlier generations of stars once did to enrich the gas that would become our Solar System.

*This article illustrates how direct and indirect observations both contribute to new discoveries.*

# Probe Detects Southern Sea Under Ice on Saturnian Moon Enceladus

By **ALAN BOYLE, NBC News**

NASA's *Cassini* orbiter has detected the faint signature of a hidden southern ocean beneath the ice of the Saturnian moon Enceladus, confirming past suspicions and sparking fresh speculation about extraterrestrial marine life.

"It makes the interior of Enceladus a very attractive potential place to look for life," said Jonathan Lunine, a planetary scientist at Cornell University and a member of the science team reporting the discovery in this week's issue of the journal *Science*.

Astrobiologists have had Enceladus on their list since 2006, when *Cassini* detected geysers of water spewing up from fissures in the southern hemisphere (**Figure 1.9**). However, it took much more subtle observations to confirm the source.

"It was not a surprise to find a water reservoir . . . but the mass and geometry of this reservoir were unknown," Luciano Iess of Sapienza University of Rome, the *Science* study's lead author, told reporters during a teleconference.

Iess and his colleagues say the reservoir is a sea of liquid water, buried under 19 to 25 miles (30 to 40 kilometers) of ice. The sea is at least 6 miles (10 kilometers) deep and extends at least halfway up from the south pole toward the equator in every direction.

"This means that it is as large, or larger, than Lake Superior," said Caltech's David Stevenson, another coauthor of the *Science* study.

### How scientists know

It took masterful feats of observation and calculation to figure all that out.

Astronomers began by measuring slight variations in *Cassini*'s velocity as it sped past the 310-mile-wide (500-kilometer-wide) moon on three occasions between 2010 and 2012. Those changes amounted to mere millimeters per second, and could be detected only by analyzing the Doppler shifts in the radio transmissions from the spacecraft. (A classic example of Doppler shift is the rise and fall in the pitch of a train's whistle as it zooms past you.)

The velocity variations were caused by anomalies in Enceladus's gravitational field; that is, regions of the moon that had more or less mass than average. Astronomers had already known about a huge depression in Enceladus's southern hemisphere, so they expected the mass concentration to be less in the south. But after taking everything they knew into account, researchers determined that the concentration was more massive than it should have been.

The best way to explain the extra mass was to assume the existence of a sea in the south, lying between Enceladus's rocky core and icy shell (**Figure 1.10**). Liquid water is denser than water ice, as illustrated by the ice cubes floating in a glass of water.

Planetary physicist William McKinnon of Washington University in St. Louis told *Science* that the interpretation made sense. "You could create a model without water, but people wouldn't find it satisfying," he said.

### Looking for life

The gravity measurements mesh nicely with the presence of those water geysers spewing out from Enceladus's southern fissures, which

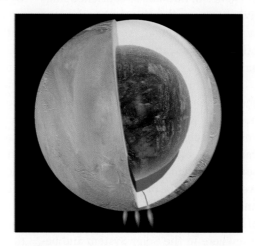

**Figure 1.10** An illustration of one model of the interior of Enceladus, showing the rocky core and a southern sea, with water making its way up through cracks in the moon's icy shell and spewing out as jets of water vapor and ice.

are also known as "tiger stripes." Enceladus's core undergoes tidal flexing as it circles Saturn, and that flexing is thought to generate heat that's concentrated at the poles.

Astronomers suggest that there's enough heat at the south pole to melt the ice and push seawater up to the surface through the fissures.

That scenario is exciting for astrobiologists, because it means the sea could be in contact with organic-rich silicate material at the bottom, at the right temperature for sustaining life.

Earlier observations from *Cassini* have shown that the water in Enceladus's geysers contains salts as well as organic molecules such as methane and ethane. However, the spacecraft's instruments aren't designed to detect the heavier organic molecules that would constitute evidence for life, Lunine said.

The easiest way to check for life would be to send a probe with the right kind of instruments through Enceladus's geysers to look for the right chemicals.

Enceladus isn't the only game in town when it comes to the search for life, however.

**Figure 1.9** A backlit view of Enceladus from the *Cassini* orbiter shows illuminated jets of water spewing out from surface fissures.

Scientists say that Europa, one of Jupiter's moons, also appears to have an ice-covered sea. Last December, researchers reported evidence that Europa is also spewing geysers of water into space. Such findings have led NASA to seek $15 million to start work on a mission to Europa.

The Europa mission alone would require years to plan and at least $1 billion in funding. Right now, Enceladus is a lower priority for future exploration, and it's not clear when the moon's southern ocean will get a closer look. But the latest findings suggest that places like Enceladus and Europa (and perhaps Ganymede and Callisto, two more ice-covered moons of Jupiter) could represent astrobiological frontiers at least as promising as Mars.

"I look at this as a cornucopia of habitable environments in the outer solar system," Lunine said.

## ARTICLES   QUESTIONS

1. How large is Enceladus? How does this compare with the size of our Moon or with the distance between Chicago and St. Louis or between Santa Cruz and Los Angeles? How large is the ocean?
2. Was the original discovery of geysers on Enceladus in 2006 from an observation, a hypothesis, or a model?
3. How did scientists make this new discovery? Did they directly observe the water? If not, how did they conclude that it exists?
4. Why is it important to have multiple observations leading to a conclusion?
5. Why are scientists excited about the discovery of water on another world?

"Probe detects Southern Sea under ice on Saturnian moon Enceladus," by Alan Boyle. NBC News, April 3, 2014. © NBCUniversal Archives. Reprinted by permission.

## Summary

Astronomy seeks answers to many compelling questions about the universe. It uses all available tools to follow the scientific method. The process of science is based on objective reality, physical evidence, and testable hypotheses. Scientists continually strive to improve their understanding of the natural world and must be willing to challenge accepted truths as new information becomes available. We are a product of the universe: the very atoms we're made of were formed in stars that died long before the Sun and Earth were formed.

**LG 1** **Describe the size and age of the universe and Earth's place in it.** We reside on a planet orbiting a star at the center of a solar system in a vast galaxy that is one of many galaxies in the universe. We occupy a very tiny part of the universe in space and time.

**LG 2** **Use the scientific method to study the universe.** The scientific method is an approach to learning about the physical universe. It includes observations, forming hypotheses, making predictions to enable the testing and refining of those hypotheses, and repeated testing of theories. All scientific knowledge is provisional. Like art, literature, and music, science is a creative human activity; it is also a remarkably powerful, successful, and aesthetically beautiful way of viewing the world.

**LG 3** **Demonstrate how scientists use mathematics, including graphs, to find patterns in nature.** Mathematics provides many of the tools that astronomers need to understand the patterns we see and to communicate that understanding to others.

## ? UNANSWERED QUESTIONS

- What makes up the universe? We have listed planets, stars, and galaxies as making up the cosmos, but astronomers now have evidence that 95 percent of the universe is composed of dark matter and dark energy, which we do not understand. Scientists are using the largest telescopes and particle colliders on Earth, as well as telescopes and experiments in space, to explore what makes up dark matter and what constitutes dark energy.

- Does life as we know it exist elsewhere in the universe? At the time of this writing, there is no scientific evidence that life exists on any other planet. Our universe is enormously large and has existed for a great length of time. What if life is too far away or existed too long ago for us ever to "meet"?

# Questions and Problems

## Test Your Understanding

1. Rank the following in order of increasing size.
   - a. Local Group
   - b. Milky Way
   - c. Solar System
   - d. universe
   - e. Sun
   - f. Earth
   - g. Laniakea Supercluster
   - h. Virgo Supercluster

2. If an event were to take place on the Sun, how long would it take for the light it generates to reach us?
   - a. 8 minutes
   - b. 11 hours
   - c. 1 second
   - d. 1 day
   - e. It would reach us instantaneously.

3. *Understanding* in science means that
   - a. we have accumulated lots of facts.
   - b. we are able to connect facts through an underlying idea.
   - c. we are able to predict events on the basis of accumulated facts.
   - d. we are able to predict events on the basis of an underlying idea.

4. The cosmological principle states that
   - a. on a large scale, the universe is the same everywhere at a given time.
   - b. the universe is the same at all times.
   - c. our location is special.
   - d. all of the above

5. The Sun is part of
   - a. the Solar System.
   - b. the Milky Way Galaxy.
   - c. the universe.
   - d. all of the above

6. A light-year is a measure of
   - a. distance.
   - b. time.
   - c. speed.
   - d. mass.

7. Occam's razor states that
   - a. the universe is expanding in all directions.
   - b. the laws of nature are the same everywhere in the universe.
   - c. if two hypotheses fit the facts equally well, the simpler one is the more likely to apply.
   - d. patterns in nature are really manifestations of random occurrences.

8. The circumference of Earth is $\frac{1}{7}$ of a light-second. Therefore,
   - a. if you were traveling at the speed of light, you would travel around Earth 7 times in 1 second.
   - b. light travels a distance equal to Earth's circumference in $\frac{1}{7}$ of a second.
   - c. neither a nor b
   - d. both a and b

9. According to the graphs in Figures 1.7b and c, by how much did the number of viruses increase in four time steps?
   - a. It doubled.
   - b. It tripled.
   - c. It quadrupled.
   - d. It went up more than 10 times.

10. Any explanation of a phenomenon that includes a supernatural influence is not scientific because
    - a. it does not have a hypothesis.
    - b. it is wrong.
    - c. people who believe in the supernatural are not credible.
    - d. science is the study of the natural world.

11. "All scientific knowledge is provisional." In this context, *provisional* means
    - a. "wrong."
    - b. "relative."
    - c. "temporary."
    - d. "incomplete."

12. When we observe a star that is 10 light-years away, we are seeing that star
    - a. as it is today.
    - b. as it was 10 days ago.
    - c. as it was 10 years ago.
    - d. as it was 20 years ago.

13. Which of the following was *not* made in the Big Bang?
    - a. hydrogen
    - b. lithium
    - c. beryllium
    - d. carbon

14. "We are stardust" means that
    - a. Earth exists because of the collision of two stars.
    - b. the atoms in our bodies have passed through (and in many cases formed in) stars.
    - c. Earth is primarily formed of material that used to be in the Sun.
    - d. Earth and the other planets will eventually form a star.

15. The following astronomical events led to the formation of you. Place them in order of their occurrence over astronomical time.
    - a. Stars die and distribute heavy elements into the space between the stars.
    - b. Hydrogen and helium are made in the Big Bang.
    - c. Enriched dust and gas gather into clouds in interstellar space.
    - d. Stars are born and process light elements into heavier ones.
    - e. The Sun and planets form from a cloud of interstellar dust and gas.

## Thinking about the Concepts

16. Suppose you lived on the planet named "Tau Ceti e" that orbits Tau Ceti, a nearby star in our galaxy. How would you write your cosmic address?

17. Imagine yourself living on a planet orbiting a star in a very distant galaxy. What does the cosmological principle tell you about the physical laws at this distant location?

18. If the Sun suddenly exploded, how soon after the explosion would we know about it?

19. If a star exploded in the Andromeda Galaxy, how long would it take that information to reach Earth?

20. Give an example of a scientific theory that has been superseded by a newer theory. As scientists developed this new theory, where on the Process of Science Figure did a change occur so that the old theory became invalid and the new theory was accepted?

21. Some people have proposed the theory that Earth was visited by extraterrestrials (aliens) in the remote past. Can you think of any tests that could support or refute that theory? Is it falsifiable? Would you regard this proposal as science or pseudoscience?

22. What does the word *falsifiable* mean? Give an example of an idea that is not falsifiable. Give an example of an idea that is falsifiable.

23. Explain how the word *theory* is used differently in the context of science than in common everyday language.

24. What is the difference between a *hypothesis* and a *theory* in science?

25. Suppose the tabloid newspaper at your local supermarket claimed that children born under a full Moon become better students than children born at other times.
    a. Is this theory falsifiable?
    b. If so, how could it be tested?

26. A textbook published in 1945 stated that light takes 800,000 years to reach Earth from the Andromeda Galaxy. In this book, we assert that it takes 2,500,000 years. What does this difference tell you about a scientific "fact" and how our knowledge evolves with time?

27. Astrology makes testable predictions. For example, it predicts that the horoscope for your star sign on any day should fit you better than horoscopes for other star signs. Read the daily horoscopes for all of the astrological signs in a newspaper or online. How many of them might fit the day you had yesterday? Repeat the experiment every day for a week and keep a record of which horoscopes fit your day each day. Was your horoscope sign consistently the best description of your experiences?

28. A scientist on television states that it is a known fact that life does not exist beyond Earth. Would you consider this scientist reputable? Explain your answer.

29. Some astrologers use elaborate mathematical formulas and procedures to predict the future. Does this show that astrology is a science? Why or why not?

30. Why can it be said that we are made of stardust? Explain why this statement is true.

## Applying the Concepts

31. Review Working It Out 1.1. Convert the following numbers to scientific notation:
    a. 7,000,000,000
    b. 0.00346
    c. 1,238

32. Review Working It Out 1.1. Convert the following numbers to standard notation:
    a. $5.34 \times 10^8$
    b. $4.1 \times 10^3$
    c. $6.24 \times 10^{-5}$

33. If a car is traveling at 35 km/h, how far does it travel in
    a. 1 hour?
    b. half an hour?
    c. 1 minute?

34. Review Appendix 1.7. The surface area of a sphere is proportional to the square of its radius. How many times larger is the surface area if the radius is
    a. doubled?
    b. tripled?
    c. halved (divided by 2)?
    d. divided by 3?

35. The average distance from Earth to the Moon is 384,400 km. How many days would it take you, traveling at 800 km/h—the typical speed of jet aircraft—to reach the Moon?

36. The average distance from Earth to the Moon is 384,400 km. In the late 1960s, astronauts reached the Moon in about 3 days. How fast (on average) must they have been traveling (in kilometers per hour) to cover this distance in this time? Compare this speed to the speed of a jet aircraft (800 km/h).

37. (a) If it takes about 8 minutes for light to travel from the Sun to Earth, and Neptune is 30 times farther from Earth than the Sun is, how long does it take light to reach Earth from Neptune? (b) Radio waves travel at the speed of light. What does this fact imply about the problems you would have if you tried to conduct a two-way conversation between Earth and a spacecraft orbiting Neptune?

38. The distance from Earth to Mars varies from 56 million km to 400 million km. How long does it take a radio signal traveling at the speed of light to reach a spacecraft on Mars when Mars is closest and when Mars is farthest away?

39. The surface area of a sphere is proportional to the square of its radius. The radius of the Moon is only about one-quarter that of Earth. How does the surface area of the Moon compare with that of Earth?

40. A remote Web page may sometimes reach your computer by going through a satellite orbiting approximately $3.6 \times 10^4$ km above Earth's surface. What is the minimum delay, in seconds, that the Web page takes to show up on your computer?

41. Imagine that you have become a biologist, studying rats in Indonesia. Most of the time, Indonesian rats maintain a constant population. Every half century, however, these rats suddenly begin to multiply exponentially! Then the population crashes back to the constant level. Sketch a graph that shows the rat population over two of these episodes.

42. New York is 2,444 miles from Los Angeles. What is that distance in car-hours? In car-days? (Assume a travel speed of 70 mph.)

43. Some theorize that a tray of hot water will freeze more quickly than a tray of cold water when both are placed in a freezer.
    a. Does this theory make sense to you?
    b. Is the theory falsifiable?
    c. Do the experiment yourself. Note the results. Was your intuition borne out?

44. A pizzeria offers a 9-inch-diameter pizza for $12 and an 18-inch-diameter pizza for $24. Are both offerings equally economical? If not, which is the better deal? Explain your reasoning.

45. The circumference of a circle is given by $C = 2\pi r$, where $r$ is the radius of the circle.
    a. Calculate the approximate circumference of Earth's orbit around the Sun, assuming that the orbit is a circle with a radius of $1.5 \times 10^8$ km.
    b. Noting that there are 8,766 hours in a year, how fast, in kilometers per hour, does Earth move in its orbit?
    c. How far along in its orbit does Earth move in 1 day?

## USING THE WEB

46. Go to the interactive "Scale of the Universe" Web page at the Astronomy Picture of the Day website (http://apod.nasa.gov/apod/ap140112.html). Start at $10^0$ (human size) and scale upward; clicking on an object gives you its exact size. What astronomical bodies are about the size of the United States? What objects are about the size of Earth? What stars are larger than the distance from Earth to the Sun? How many light-days is the distance of *Voyager 1* to the Earth? What objects are about the size of the distance from the Sun to the nearest star? How much larger is the Milky Way than the size of the Solar System? How much larger is the Local Group

than the Milky Way? How much larger is the observable universe than the Local Group?

47. a. For a video representation of the scale of the universe, view the short video *The Known Universe* at the Hayden Planetarium website (http://www.haydenplanetarium.org/universe), which takes the viewer on a journey from the Himalayan mountains to the most distant galaxies. How far have broadcast radio programs from Earth traveled? Is the Sun a particularly luminous star compared to others? Do you think the video is effective for showing the size and scale of the universe?
    b. A similar film produced in 1996 in IMAX format, *Cosmic Voyage*, can be found online at http://topdocumentaryfilms.com/cosmic-voyage. Watch the "powers of ten" zoom out to the cosmos, starting at the 7-minute mark, for about 5 minutes. Do the "powers of ten" circles add to your understanding of the size and scale of the universe? (The original film *Powers of Ten* can be viewed online at http://apod.nasa.gov/apod/ap150324.html, but notably it extends a few powers of ten less than the newer film.)

48. Go to the Astronomy Picture of the Day (APOD) app or website (http://apod.nasa.gov/apod) and click on "Archive" to look at the recent pictures and videos. Submissions to this website come from all around the world. Pick one and read the explanation. Was the image or video taken from Earth or from space? Is it a combination of several images? Does it show Earth, our Solar System, objects in our Milky Way Galaxy, more distant galaxies, or something else? Is the explanation understandable to someone who has not studied astronomy? Do you think this website promotes a general interest in astronomy?

49. Throughout this book, we will examine how discoveries in astronomy and space are covered in the media. Go to your favorite news website (or to one assigned by your instructor) and find a recent article about astronomy or space. Does this website have a separate section for science? Is the article you selected based on a press release, on interviews with scientists, or on an article in a scientific journal? Use Google News or the equivalent to see how widespread the coverage of this story is. Have many newspapers carried it? Has it been picked up internationally? Has it been discussed in blogs? Do you think this story was interesting enough to be covered?

50. Go to a blog about astronomy or space. Is the blogger a scientist, a science writer, a student, or an enthusiastic amateur astronomer? What is the current topic of interest? Is it controversial? Are readers making many comments? Is this blog something you would want to read again?

## smartwork5

If your instructor assigns homework in Smartwork5, access your assignments at digital.wwnorton.com/astro5.

digital.wwnorton.com/astro5

Logic is fundamental to the study of science and to scientific thinking. A logical fallacy is an error in reasoning, which good scientific thinking avoids. For example, "because Einstein said so" is not an adequate argument. No matter how famous the scientist is (even if he is Einstein), he or she must still supply a logical argument and evidence to support a claim. Anyone who claims that something must be true because Einstein said it has committed the logical fallacy known as an *appeal to authority*. There are many types of logical fallacies, but a few of them crop up often enough in discussions about science that you should be aware of them.

**Ad hominem.** In an ad hominem fallacy, you attack the person who is making the argument, instead of the argument itself. Here is an extreme example of an ad hominem argument: "A famous politician says Earth is warming. But I think this politician is an idiot. So Earth can't be warming."

**Appeal to belief.** This fallacy has the general pattern "Most people believe X is true; therefore X is true." For example, "Most people believe Earth orbits the Sun. Therefore, Earth orbits the Sun." Note that even if the conclusion is correct, you may still have committed a logical fallacy in your argument.

**Begging the question.** In this fallacy, also known as circular reasoning, you assume the claim is true and then use this assumption as evidence to prove the claim is true. For example, "I am trustworthy; therefore I must be telling the truth." No real evidence is presented for the conclusion.

**Biased sample.** If a sample drawn from a smaller pool has a bias, then conclusions about the sample cannot be applied to a larger pool. For example, imagine you poll students at your university and find that 30 percent of them visit the library one or more times per week. Then you conclude that 30 percent of Americans visit the library one or more times per week. You have committed the biased sample fallacy, because university students are not a representative sample of the American public.

***Post hoc ergo propter hoc.*** *Post hoc ergo propter hoc* is Latin for "after this, therefore because of this." Just because one thing follows another doesn't mean that one caused the other. For example, "There was an eclipse and then the king died. Therefore, the eclipse killed the king." This fallacy is often connected to related inverse reasoning: "If we can prevent an eclipse, the king won't die."

**Slippery slope.** In this fallacy, you claim that a chain reaction of events will take place, inevitably leading to a conclusion that no one could want. For example, "If I don't get A's in all of my classes, I will not ever be able to get into graduate school, and then I won't ever be able to get a good job, and then I will be living in a van down by the river until I'm old and starve to death." None of these steps actually follows inevitably from the one before.

Following are some examples of logical fallacies. Identify the type of fallacy represented. Each of the fallacies we just discussed is represented once.

**1** You get a chain email threatening terrible consequences if you break the chain. You move it to your spam box. Later that day you get in a car accident. The following morning, you retrieve the chain email and send it along.

_____

_____

**2** If I get question 1 on the assignment wrong, then I'll get question 2 wrong as well, and before you know it, I will never catch up in the class.

_____

_____

**3** All my friends love the band Degenerate Electrons. Therefore, all people my age love this band.

_____

_____

**4** Eighty percent of Americans believe in the tooth fairy. Therefore, the tooth fairy exists.

_____

_____

**5** My professor says that the universe is expanding. But my professor is a geek, and I don't like geeks. So the universe can't be expanding.

_____

_____

**6** When applying for a job, you use a friend as a reference. Your prospective employer asks you how she can be sure your friend is trustworthy, and you say, "I can vouch for him."

_____

_____

# 2 Patterns in the Sky—Motions of Earth and the Moon

**A**ncient peoples learned that they could use the patterns they observed in the sky to predict the changing length of day, the change of seasons, and the changes in the appearance of the Moon. Some people understood these patterns well enough to create complicated calendars and predict rare eclipses. But now we can see these patterns with the perspective of centuries of modern science, and we can explain these changes as a consequence of the motions of Earth and the Moon. Discovering what causes these patterns has shown us the way outward into the universe.

## LEARNING GOALS

In this chapter, we will examine the patterns in the sky and on Earth and the underlying motions that cause these patterns. By the conclusion of this chapter, you should be able to:

**LG 1** Describe how Earth's rotation about its axis and revolution around the Sun affect our perception of celestial motions as seen from different places on Earth.

**LG 2** Explain why there are different seasons throughout the year.

**LG 3** Describe the factors that create the phases of the Moon.

**LG 4** Sketch the alignment of Earth, the Moon, and the Sun during eclipses of the Sun and the Moon

The changing seasons bring noticeable variations to the landscape. ▶ ▶ ▶

What causes
the seasons?

23

(a)

(b)

(c)

**Figure 2.1** (a) One of the suspected uses of Stonehenge 4,000 years ago was to keep track of celestial events. (b) The Mayan El Caracol at Chichén Itzá in Mexico (906 CE) is believed to have been designed to align with the planet Venus. (c) The Beijing Ancient Observatory in China (1442 CE) includes ancient astronomical instruments as well as some brought by European Jesuits in the 17th and 18th centuries.

▶❚❚ **AstroTour:** The Celestial Sphere and the Ecliptic

# 2.1 Earth Spins on Its Axis

Ancient humans may not have known that they were "stardust," but they did sense that there was a connection between their lives on Earth and the sky above. Before our modern technological civilization, people's lives were more attuned to the ebb and flow of nature, which includes the patterns in the sky. By watching the repeating patterns of the Sun, Moon, and stars in the sky, people found that they could predict when the seasons would change and when the rains would come and the crops would grow. Ancient astronomers with knowledge of the sky—priests and priestesses, natural philosophers and explorers—all had knowledge of the world that others did not, and knowledge of the world was power. Some of these early observations and ideas about these patterns live on today in the names of stars and the apparent grouping of stars we call **constellations**, in calendars based on the Moon and Sun that are still in use by many cultures, and in the astronomical names of the days of the week.

From Mesopotamia to Africa, from Europe to Asia, from the Americas to the British Isles, the archaeological record holds evidence that ancient cultures built structures that were sometimes used to study astronomical positions and events. **Figure 2.1** shows some examples of these. Pre-telescopic astronomical observatories from the 8th through 17th centuries were used to study the sky for timekeeping and navigation. Many of these structures and observatories are now national historical or UNESCO World Heritage sites.

## The Celestial Sphere

Long before Christopher Columbus sailed, Aristotle and other Greek philosophers knew that Earth is a sphere. However, because Earth seems stationary, they did not realize that the changes they observed in the sky from day to day and year to year are caused by Earth's motions. As we will see in this subsection, Earth's rotation on its axis determines the passage of day and night, which dictates the rhythm of life on Earth.

One reason the ancients did not suspect that Earth rotates was that they could not perceive Earth's spinning motion. As Earth rotates about its axis, its surface is moving quite fast—about 1,674 kilometers per hour (km/h) at the equator. We do not "feel" that motion any more than we would feel the motion of a car with a perfectly smooth ride cruising down a straight highway. Nor do we feel the direction of Earth's spin, although the hourly motion of the Sun, Moon, and stars across the sky reveals it. Earth's **North Pole** is at the north end of Earth's rotation axis. Imagine you are in space far above Earth's North Pole. From there you would see Earth complete a counterclockwise rotation, once each 24-hour period, as shown in **Figure 2.2**. As the rotating Earth carries an observer on the surface from west to east, objects in the sky appear to move in the other direction, from east to west. As seen from Earth's surface, the path each object takes across the sky is called its **apparent daily motion**.

To help visualize the apparent daily motions of the Sun and stars, it is useful to think of the sky as a huge sphere with the stars painted on its surface and Earth at its center. From ancient Greek times to the Renaissance, most people believed this to be a true representation of the heavens. Astronomers call this imaginary sphere, shown in **Figure 2.3**, the **celestial sphere**. The celestial sphere is a useful concept because it is easy to visualize, but never forget that it is, in fact, imaginary.

Each point on the celestial sphere indicates a direction in space. Directly above Earth's North Pole is the **north celestial pole (NCP)**. Directly above Earth's **South Pole**, which is at the south end of Earth's rotation axis, is the **south celestial pole (SCP)**. Directly above Earth's **equator** is the **celestial equator**, an imaginary circle that divides the sky into a northern half and a southern half. Just as the north celestial pole is the projection of the direction of Earth's North Pole into the sky, the celestial equator is the projection of the plane of Earth's equator into the sky. Just as Earth's North Pole is 90° away from Earth's equator, the north celestial pole is 90° away from the celestial equator. If you are in the Northern Hemisphere and you point one arm toward the celestial equator and one arm toward the north celestial pole, your arms will always form a right angle, so the north celestial pole is 90° away from the celestial equator. If you are in the Southern Hemisphere, the same holds true there: the angle between the celestial equator and the south celestial pole is always 90° as well.

Between the celestial poles and the equator, objects have positions on the celestial sphere with coordinates analogous to latitude and longitude on Earth. **Latitude** is an indication of distance north or south from Earth's equator. On the celestial sphere, **declination** similarly indicates the distance of an object north or south of the celestial equator (from 0° to ±90°). On Earth, **longitude** measures how far east or west you are from the Royal Observatory in Greenwich, England. **Right ascension** on the celestial sphere is similar to longitude on Earth and measures the angular distance of a celestial body eastward along the celestial equator from the point where the Sun's path crosses the celestial equator from south to north. These coordinates are used to locate objects in the sky quickly. The **ecliptic** is the path of the Sun in the sky throughout the year. Detailed descriptions and illustrations of latitude and longitude, and of celestial coordinates used with the celestial sphere, can be found in Appendix 7.

The **zenith** is the point in the sky directly above you wherever you are, as shown in **Figure 2.3a**. You can find the **horizon** by standing up and pointing your right hand at the zenith and your left hand straight out from your side. Turn in a complete circle. Your left hand has traced out the entire horizon. You can divide the sky into an east half and a west half with a line that runs from the horizon at due north through the zenith to the horizon at due south. This imaginary north–south line is called the **meridian**, shown as a dashed line in Figure 2.3a. Figure 2.3b shows these locations on the celestial sphere. The meridian line continues around the far side of the celestial sphere, through the **nadir** (the point directly below you), and back to the starting point due north.

Take a moment to visualize all these locations in space. To see how to use the celestial sphere, consider the Sun at noon and at midnight. Local noon occurs when the Sun crosses the meridian at your location. This is the highest point above the horizon that the Sun will reach on any given day. The highest point is almost never the zenith. You have to be in a specific place on a specific day for the Sun to be directly over your head at noon, for example, at a latitude 23.5° north of the

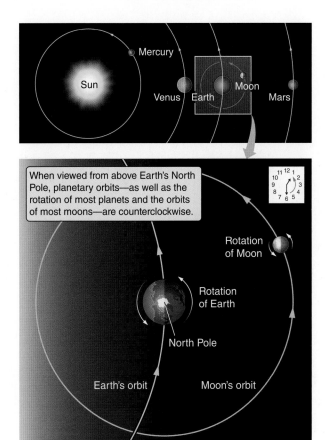

**Figure 2.2** Motions in the Solar System, as viewed from above Earth's North Pole. (Not drawn to scale.)

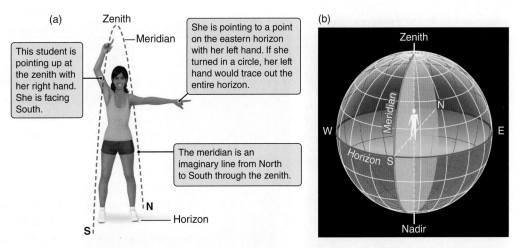

**Figure 2.3** (a) The meridian is a line on the celestial sphere that runs from north to south, dividing the sky into an east half and a west half. (b) At any location on Earth, the sky is divided into an east half and a west half by an imaginary meridian projected onto the celestial sphere.

**Astronomy in Action:** Vocabulary of the Celestial Sphere

**Nebraska Simulations:** Celestial and Horizon Systems Comparison; Rotating Sky Explorer

▶❚❚ **AstroTour:** The View from the Poles

**Figure 2.4** As viewed from (a) Earth's North Pole, (b) stars move throughout the night in counterclockwise, circular paths about the zenith. (c) The same half of the sky is always visible from the North Pole.

**(a)**

North celestial pole (NCP)

North Pole

This disk represents the horizon, the boundary between the part of the sky you can see and the part that is blocked from view by Earth.

Equator

South Pole

From the North Pole looking directly overhead, the **north celestial pole (NCP)** is at the zenith.

**(b)**

North celestial pole at the zenith

As Earth rotates, the stars appear to move in a counterclockwise direction around the **NCP**.

**(c)**

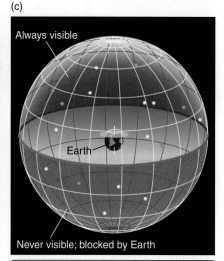

Always visible

Earth

Never visible; blocked by Earth

From the North Pole, you always see the same half of the sky.

equator on June 20. Local midnight occurs when the Sun is precisely opposite from its position at local noon. From our perspective on Earth, the celestial sphere appears to rotate, carrying the Sun across the sky to its highest point at noon, over toward the west to set in the evening. In *reality*, the Sun remains in the same place in space through the entire 24-hour period, and Earth rotates so that any given location on Earth faces a different direction at every moment. When it is noon where you live, Earth has rotated so that you face most directly toward the Sun. Half a day later, at midnight, your location on Earth has rotated to face most directly away from the Sun.

## The View from the Poles

The apparent daily motions of the stars and the Sun depend on where you live. For example, the apparent daily motions of celestial objects in a northern place such as Alaska are quite different from the apparent daily motions seen from a tropical island such as Hawaii. To understand why your location matters, let's examine the view of the stars from the poles—and then use these to guide our thinking about the view of the stars from other latitudes.

Imagine that you are standing on the North Pole watching the sky as in **Figure 2.4a**. At the North Pole, the north celestial pole is directly overhead at the zenith. Ignore the Sun for the moment and pretend that you can always see stars in the sky. You are standing where Earth's axis of rotation intersects its surface, which is like standing at the center of a rotating carousel. As Earth rotates, the spot directly above you remains fixed over your head while everything else in the sky appears to revolve in a counterclockwise direction around this spot. Figure 2.4b depicts this overhead view.

No matter where you are on Earth, you can see only half of the sky at any one time. The horizon is the boundary between the part of the sky you can see and the other half of the sky that is blocked by Earth. Except at the poles, the visible half of the sky changes constantly as Earth rotates, because the zenith points to different locations in the sky as Earth carries you around. In contrast, if you are standing at the North Pole, the zenith is always in the same location in space, so the objects visible from the North Pole follow circular paths that always have the same **altitude**, or angle above the horizon. Objects close to the zenith appear to follow small circles, while objects near the horizon follow the largest circles (Figure 2.4b). The view from the North Pole is special because from there, nothing rises or sets each day as Earth turns: from there you will always see the *same* half of the celestial sphere (Figure 2.4c).

The view from Earth's South Pole is much the same—with two major differences. First, the South Pole is on the opposite side of Earth from the North Pole, so the visible half of the sky at the South Pole is precisely the half that is hidden from view at the North Pole. The second difference is that stars appear to move clockwise around the south celestial pole rather than counterclockwise as they do at the north celestial pole. To visualize why these motions are different, stand up and spin around from right to left. As you look at the ceiling, things appear to move in a counterclockwise direction, but as you look at the floor, they appear to be moving clockwise.

No matter where you are on Earth, stars appear to rotate about a point called the:
(a) zenith; (b) celestial pole; (c) nadir; (d) meridian.

## The View Away from the Poles

Suppose that you leave the North Pole to travel south to lower latitudes. Imagine a line from the center of Earth to your location on the surface of the planet, as in **Figure 2.5**. Now imagine a second line from the center of Earth to the point on the equator closest to you. The angle between these two lines is your latitude. At the North Pole, for example, these two imaginary lines form a 90° angle. At the equator, they form a 0° angle. So the latitude of the North Pole is 90° north, and the latitude of the equator is 0°. The South Pole is at latitude 90° south.

Your latitude determines the part of the sky that you can see throughout the year. As you move south from the North Pole, your zenith moves away from the north celestial pole, and so the horizon moves as well. At the North Pole, the horizon makes a 90° angle with the north celestial pole, which is at the zenith. At a latitude of 60° north, as in Figure 2.5, your horizon is tilted 60° from the north celestial pole. The angle between your horizon and the north celestial pole is equal to your latitude no matter where you are on Earth. The situation is the same in the Southern Hemisphere—your latitude is the altitude of the south celestial pole. At the equator, at a latitude of 0°, the north and south celestial poles would be at the northern and southern horizons, respectively.

One way to solidify your understanding of the view of the sky at different latitudes is to draw pictures like the one in Figure 2.5. If you can draw a picture like this for any latitude—filling in the values for each angle in the drawing and imagining what the sky looks like from that location—then you will be well on your way to developing a working knowledge of the appearance of the sky. That knowledge will prove useful later, when we discuss a variety of phenomena, such as the changing of the seasons.

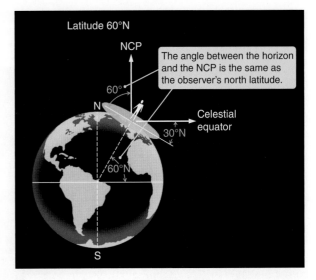

**Figure 2.5** Your perspective on the sky depends on your location on Earth. The locations of the celestial poles and the celestial equator in an observer's sky depend on the observer's latitude. In this case, an observer at latitude 60° north sees the north celestial pole at an altitude of 60° above the northern horizon and the celestial equator 30° above the southern horizon.

### Motions of the Stars and the Celestial Poles

**Figure 2.6** shows two time-lapse views of the sky from different latitudes. The apparent motions of the stars about the celestial poles also differs from latitude to latitude. The visible part of the sky constantly changes, as stars rise and set with Earth's rotation. From this perspective the horizon appears fixed, while the stars appear to move. From these different latitudes, if we focus our attention on the north celestial pole, we see much the same thing we saw from Earth's North Pole. The north celestial pole remains fixed in the sky, and all of the stars appear to move throughout the night in counterclockwise, circular paths around that point. But because the north celestial pole is no longer directly overhead as it was at the North Pole, the apparent circular paths of the stars are now tipped relative to the horizon. (More correctly, your horizon is now tipped relative to the apparent circular paths of the stars.)

**Figure 2.6** Time exposures of the sky showing the apparent motions of stars through the night. Note the difference in the circumpolar portion of the sky as seen from the two different northern latitudes.

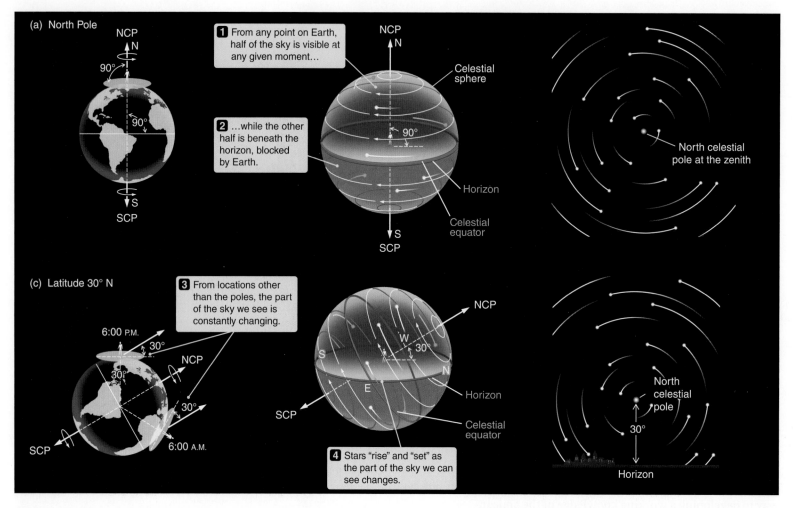

**(a) North Pole**

**1** From any point on Earth, half of the sky is visible at any given moment…

**2** …while the other half is beneath the horizon, blocked by Earth.

Celestial sphere

Horizon

Celestial equator

North celestial pole at the zenith

**(c) Latitude 30° N**

**3** From locations other than the poles, the part of the sky we see is constantly changing.

6:00 P.M.

6:00 A.M.

Horizon

Celestial equator

**4** Stars "rise" and "set" as the part of the sky we can see changes.

North celestial pole

Horizon

**Figure 2.7** The celestial sphere is shown here as viewed by observers at four different latitudes. At all locations other than the poles, stars rise and set as the part of the celestial sphere that we see changes during the day.

From the vantage point of an observer in the Northern Hemisphere, stars near the north celestial pole never dip below the horizon and thus stay visible all night. Recall from Figure 2.5 that the latitude is equal to the altitude of the north celestial pole. Stars closer to the north celestial pole than this angle never dip below the horizon as they complete their apparent paths around the pole. These stars are called **circumpolar** ("around the pole") stars. Another group of stars, near the south celestial pole, never rise above the horizon in the Northern Hemisphere. Stars between those that never rise and those in the circumpolar region can be seen for *only part* of each 24-hour day. These stars appear to rise and set as Earth turns. The only place on Earth where you can see the entire sky over the course of 24 hours is the equator. From the equator, the north and south celestial poles sit on the northern and southern horizons, respectively, and all of the stars move through the sky each 24-hour day. (Even though the Sun lights the sky for roughly half of this time, the stars are still there.)

**Figure 2.7** shows the orientation of the sky as seen by observers at four different latitudes. For an observer at the North Pole (Figure 2.7a), the celestial equator lies exactly along the horizon. The north celestial pole is at the zenith, and the

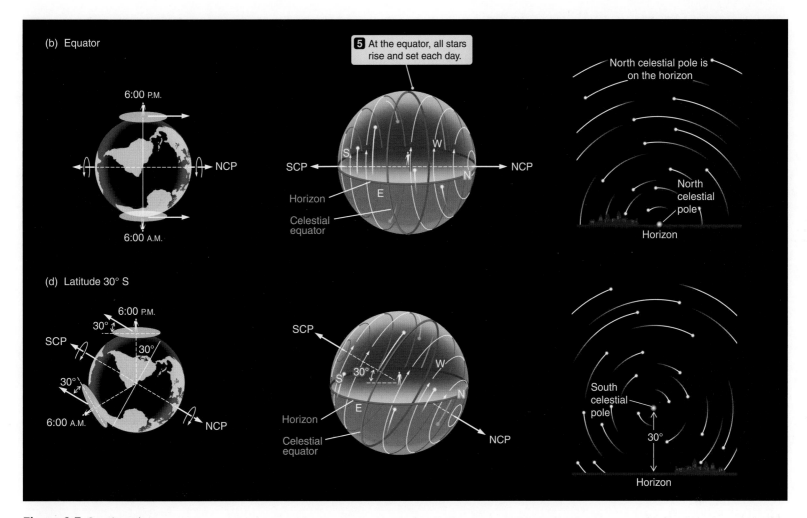

**Figure 2.7** Continued.

southern half of the sky is never visible. Stars neither rise nor set; their paths form circles parallel to the horizon. For an observer at the equator (Figure 2.7b), the celestial poles are both at the horizon, and all the stars are visible in a 24-hour period, rising straight up and setting straight down each day.

At other latitudes, the celestial equator intersects the horizon due east and due west. Therefore, a star on the celestial equator rises due east and sets due west. Stars located north of the celestial equator rise north of east and set north of west. Stars located south of the celestial equator rise south of east and set south of west.

From everywhere else on Earth (except at the poles), half of the celestial equator is always visible above the horizon. Therefore, any object located on the celestial equator is visible half of the time—above the horizon for 12 hours each day. Objects that are not on the celestial equator are above the horizon for differing amounts of time. Figures 2.7c and d show that stars in the observer's hemisphere are visible for more than half the day because more than half of each star's path in the sky is above the horizon. In contrast, stars in the opposite hemisphere are visible for less than half the day because less than half of each star's path in the sky is above the horizon.

**Nebraska Simulations:** Meridional Altitude Simulator

**Nebraska Simulation:** Declination Ranges Simulator

**Nebraska Simulations:** Big Dipper Clock

For example, as seen from the Northern Hemisphere, stars north of the celestial equator remain above the horizon for more than 12 hours each day. The farther north the star is, the longer it stays up. Circumpolar stars are the extreme example of this phenomenon; they are always above the horizon. In contrast, stars south of the celestial equator are above the horizon for less than 12 hours each day. The farther south a star is, the less time it is visible. For an observer in the Northern Hemisphere, stars located close to the south celestial pole never rise above the horizon.

**Using the Stars for Navigation** Since ancient times, travelers have used the stars for navigation. They would find the north or south celestial poles by recognizing the stars that surround them. In the Northern Hemisphere, a moderately bright star happens to be located close to the north celestial pole (**Figure 2.8a**). This star is called Polaris, the "North Star." The altitude of Polaris in the sky is nearly equal to your latitude. If you are in Phoenix, Arizona, for example (latitude 33.5° north), the north celestial pole has an altitude of 33.5°. In Fairbanks, Alaska (latitude 64.6° north), the north celestial pole sits much higher, with an altitude of 64.6°. Similarly in the Southern Hemisphere, the constellation Crux (commonly called the Southern Cross) points to a star near the south celestial pole (Figure 2.8b). A navigator who has located a pole star can identify north and south, and therefore east, west, and her latitude. This enables the navigator to determine which direction to travel. The location of the north celestial pole in the sky was used to measure the size of Earth, as described in **Working It Out 2.1**. Determining your longitude by astronomical methods is much more complicated because of Earth's rotation. Longitude cannot be determined from astronomical observation alone.

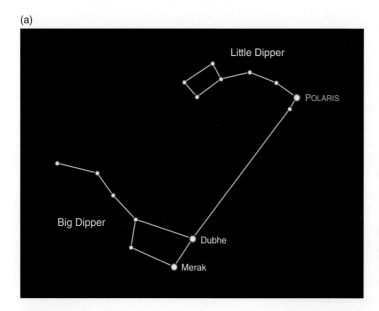

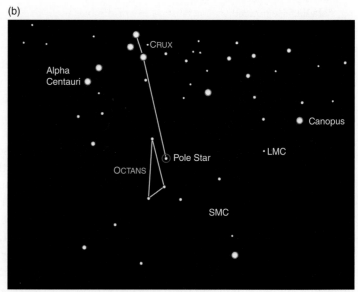

**Figure 2.8** Groups of stars near the pole stars in the sky can be used to locate a pole star. (a) Two bright "pointer stars" stars in the cup of the Big Dipper point toward Polaris, the "North Star." (b) In the Southern Hemisphere, the constellation Crux and two of its bright "pointer stars" can be used to locate the relatively faint southern pole star.

## 2.1 Working It Out How to Estimate the Size of Earth

We can use the location of the north celestial pole in the sky to estimate the size of Earth. Suppose we start out in Phoenix, Arizona, and we observe the north celestial pole to be 33.5° above the horizon. If we head north, by the time we reach the Grand Canyon, about 290 kilometers (km) from Phoenix, we notice that the north celestial pole has risen to about 36° above the horizon. This difference between 33.5° and 36° (2.5°) is 1/144 of the way around a circle. (A circle is 360°, and 2.5°/360° = 1/144.)

This means that we must have traveled 1/144 of the way around the circumference, $C$, of Earth by traveling the 290 km between Phoenix and the Grand Canyon. In other words,

$$\frac{1}{144} \times C = 290 \text{ km}$$

Rearranging the expression, the circumference of Earth is given by

$$C = 144 \times 290 \text{ km} \approx 42,000 \text{ km}$$

The actual circumference of Earth is just over 40,000 km, so our simple calculation was close. The circumference of a circle is equal to $2\pi$ multiplied by its radius. So, the radius of Earth is given by

$$\text{Radius} = \frac{C}{2\pi} = \frac{40,000 \text{ km}}{2\pi} = 6,400 \text{ km}$$

It was in much this same way that the Greek astronomer Eratosthenes (276–194 BCE) made the first accurate measurements of the size of Earth in about 230 BCE. As illustrated in **Figure 2.9**, Eratosthenes used the distance between his home city of Alexandria and the city of Syene (currently Aswân, in Egypt), which was 5,000 "stadia." He noticed that on the first day of summer in Syene, the sunlight reflected directly off the water in a deep well, so the Sun must have been nearly at the zenith. By measuring the shadow of the Sun from an upright stick in Alexandria, he saw that the Sun was about 7.2° south of the zenith on the same date. Assuming Earth was spherical and Syene was directly south of Alexandria, he determined the

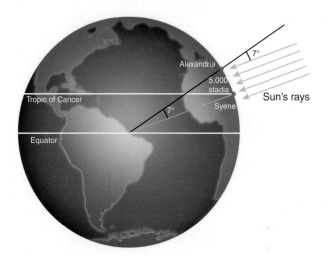

**Figure 2.9** Eratosthenes estimated the size of Earth using observations and basic calculations.

distance between the two cities to be 7.2° divided by 360, or 1/50 of the circumference of Earth.

It was difficult to estimate distances accurately in those days, and although historians know Eratosthenes concluded that the cities were 5,000 stadia apart, they are still not at all sure of the value of his stadion unit. If the stadion was 185 meters, then Eratosthenes would have worked the math in a similar way:

$$\frac{1}{50} \times C = 5,000 \text{ stadia} \times 185 \text{ meters/stadion}$$

$$= 925,000 \text{ meters} = 925 \text{ km}$$

Eratosthenes would have found the circumference of Earth to be

$$C = 50 \times 925 \text{ km} = 46,250 \text{ km}$$

only about 16 percent higher than the modern value.

## Relative Motions and Frame of Reference

Why don't we feel the motion of Earth as it spins on its axis and moves through space in its orbit around the Sun? Astronomers use the concept of a **frame of reference**, which is a coordinate system within which an observer measures positions and motions. The difference in motion between two individual frames of reference is called the **relative motion**. For example, imagine that you are riding in a car traveling down a straight section of highway at a constant speed. If you are not looking out the window or feeling road vibrations, there is no experiment you can easily do to tell the difference between riding in a car down a straight section of highway at constant speed and sitting in the car while it is parked in

(a) Frame of reference: Viewer on the street

**1** A ball is thrown directly at the slower car from the faster car.

Slower car

**3** ...resulting in a total motion relative to the ground.

Faster car

**2** The ball shares the forward motion of the car from which it is thrown...

(b) Frame of reference: Viewer in faster car

In the frame of reference of the faster car, the ball misses because the slower car is moving backward.

Slower car

Faster car

**Figure 2.10** The motion of an object depends on the frame of reference of the observer.

(c) Frame of reference: Viewer in slower car

In the frame of reference of the slower car, the ball misses because the ball and the faster car are moving forward.

Slower car

Faster car

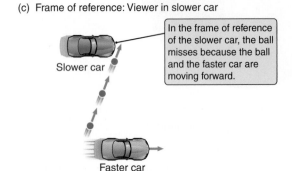

The ground near the equator is like the faster car in Figure 2.10. The ground at higher latitudes is like the slower car.

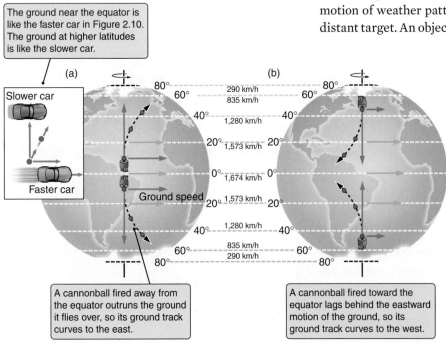

Slower car

Faster car

Ground speed

(a)

80°
60°
40°
20°
0°
20°
40°
60°
80°

290 km/h
835 km/h
1,280 km/h
1,573 km/h
1,674 km/h
1,573 km/h
1,280 km/h
835 km/h
290 km/h

(b)

80°
60°
40°
20°
0°
20°
40°
60°
80°

A cannonball fired away from the equator outruns the ground it flies over, so its ground track curves to the east.

A cannonball fired toward the equator lags behind the eastward motion of the ground, so its ground track curves to the west.

**Figure 2.11** The Coriolis effect causes objects to be deflected as they move across the surface of Earth. The green dashed line shows the path of the cannonball as measured by a local observer.

your driveway. Because everything in the car is moving together, the relative motions between objects in the car are all that can be measured, and they are all zero, no motion is observed. Similarly, the resulting relative motions between objects that are near each other on Earth are zero. This is why we do not notice Earth's motion.

Now imagine that two cars are driving down the road at different speeds, as shown in **Figure 2.10a**. Ignoring for the moment any real-world complications, like wind resistance, if you were to throw a ball from the faster-moving car directly out the side window at the slower-moving car as the two cars passed, you would miss. The ball shares the forward motion of the faster car, so the ball outruns the forward motion of the slower car. As shown in Figure 2.10b, from your perspective in the faster car, the slower car lagged behind the ball. From the slower car's perspective, represented in Figure 2.10c, your car and the ball sped on ahead.

Although we cannot feel Earth's rotation, it influences things as diverse as the motion of weather patterns on Earth and how an artillery gunner must aim at a distant target. An object on the surface of Earth moves in a circle each day around Earth's rotation axis. This circle is larger for objects near Earth's equator and smaller for objects closer to one of Earth's poles. But all objects must complete their circular motion in exactly 1 day. As you can see in **Figure 2.11**, the surface of Earth moves faster at the equator than at higher latitudes. An object closer to the equator has farther to go each day than does an object nearer a pole. Therefore, the object nearer the equator must be moving faster than the object at higher latitude. If an object starts out at one latitude and then moves to another, its apparent motion over the surface of Earth is influenced by this difference in speed.

Now consider two locations at different latitudes. Imagine that a cannonball is launched directly north from a point in the Northern Hemisphere, as shown in Figure 2.11a. Because the cannon is located nearer to the equator than its target is, the cannonball is moving toward the east faster than its target. Even though the cannonball is fired toward the north, it shares the eastward velocity of the cannon itself. This means that the cannonball is *also* moving toward the east faster than its target. Recall how the ball thrown from the faster car outpaced the slower-moving car in Figure 2.10. Similarly, as the cannonball flies north, it moves toward the east faster than the ground underneath it does. To an observer on the ground, the cannonball appears to curve toward the east as it outruns the eastward motion of the ground it is crossing. The farther north the cannonball flies, the greater the difference between its eastward velocity and the eastward velocity of the ground. Thus, the cannonball follows a path that appears to curve more and more to the east the farther north it goes. If you are located in the Northern Hemisphere and fire a cannonball *south* toward the equator, as shown in Figure 2.11b, the opposite effect will occur. Now the cannon is moving toward the east more slowly than its target. As the cannonball flies toward the south, its eastward motion lags behind that of the ground underneath it, and the cannonball appears to curve toward the west.

This curving motion of objects from the difference in Earth's rotation speeds at different latitudes is called the **Coriolis effect**. In the Northern Hemisphere, the Coriolis effect causes a cannonball fired north to drift to the east as seen from the surface of Earth. In other words, the cannonball appears to curve to the right. A cannonball fired south appears to curve to the west, which also gives it the appearance of curving to the right. In the Northern Hemisphere, the Coriolis effect seems to deflect things to the *right*. If you think through this example for the Southern Hemisphere, you will see that south of the equator, the Coriolis effect seems to deflect things to the *left*. In between, at the equator itself, the Coriolis effect vanishes.

On Earth the effect is enough to deflect a fly ball hit north or south into deep left field in a stadium in the northern United States by about a half a centimeter. At some time or other, the Coriolis effect from the rotation of Earth has probably determined the outcome of a ball game.

## CHECK YOUR UNDERSTANDING 2.2

If the star Polaris has an altitude of 35°, then we know that: (a) our longitude is 55° east; (b) our latitude is 55° north; (c) our longitude is 35° west; (d) our latitude is 35° north.

## 2.2 Revolution about the Sun Leads to Changes during the Year

Earth orbits (or **revolves**) around the Sun in the same direction that Earth spins about its axis—counterclockwise as viewed from above Earth's North Pole (see **Figure 2.12**). A **year** is the time it takes for Earth to complete one revolution around the Sun. The motion of Earth around the Sun is responsible for many of the patterns of change we see in the sky and on Earth, including changes in the stars we see overhead. Because of this motion, the stars in the night sky change throughout the year, and Earth experiences seasons.

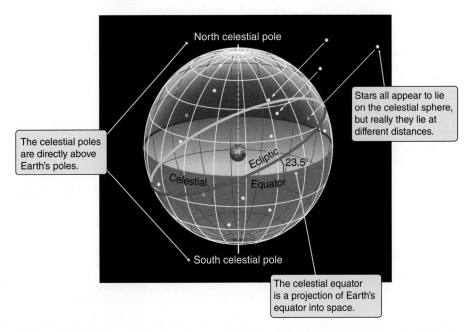

**Figure 2.12** The celestial sphere is a useful fiction for thinking about the appearance and apparent motion of the stars in the sky.

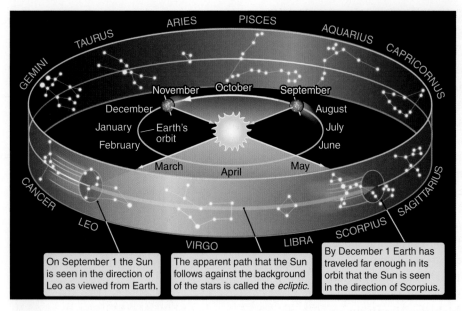

On September 1 the Sun is seen in the direction of Leo as viewed from Earth.

The apparent path that the Sun follows against the background of the stars is called the *ecliptic*.

By December 1 Earth has traveled far enough in its orbit that the Sun is seen in the direction of Scorpius.

**Figure 2.13** As Earth orbits around the Sun, the Sun's apparent position against the background of stars changes. The imaginary circle traced by the annual path of the Sun is called the ecliptic. The blue band shows 12 of the zodiacal constellations along the ecliptic.

**Nebraska Simulations:** Ecliptic (Zodiac) Simulator

## Constellations and the Zodiac

As shown in **Figure 2.13**, as Earth orbits the Sun, our view of the night sky changes. Six months from now, Earth will be on the other side of the Sun. The stars that are overhead at midnight 6 months from now are those that are near overhead at noon today. In order to follow the patterns of the Sun and the stars, early humans grouped together stars that formed recognizable patterns, called **constellations**. But people from different cultures saw different patterns and projected ideas from their own cultures onto what they saw in the sky. Constellations named for winged horses, dragons, and other imaginary images, and the stories that go with them, are creations of the human imagination. If you look at the sky, no obvious pictures of these images emerge. Instead, there is only the random pattern of stars—about 5,000 of them visible to the naked eye—spread out across the sky.

Modern constellations visible from the Northern Hemisphere draw heavily from the list of constellations compiled 2,000 years ago by the Alexandrian astronomer Ptolemy. Modern constellation names in the southern sky come from European explorers visiting the Southern Hemisphere during the 17th and 18th centuries. Today, astronomers use an officially sanctioned set of 88 constellations as a kind of road map of the sky. Every star in the sky lies within the borders of a single constellation, and the names of constellations are used in naming the stars that lie within their boundaries. For example, Sirius, the brightest star in the sky, lies within the boundaries of the constellation Canis Major (meaning "big dog"). Following the Greek alphabet, the brightest star in a constellation is called *alpha*, the second brightest is called *beta*, and so forth. The official name of Sirius is Alpha Canis Majoris, indicating that it is the brightest star in Canis Major. Appendix 6 provides sky maps showing the constellations.

If you could note the position of the Sun relative to the stars each day for a year, you would find that the Sun traces out a great circle against the background of the stars. On September 1, the Sun appears to be in the direction of the constellation Leo. Six months later, on March 1, Earth is on the other side of the Sun, and the Sun appears from our perspective on Earth to be in the direction of the constellation Aquarius. Recall that the apparent path that the Sun follows against the background of the stars is called the ecliptic and is illustrated as the yellow band in Figure 2.13. The 13 constellations that lie along the ecliptic through which the Sun appears to move are called the constellations of the **zodiac**.

## The Tilt of Earth's Axis and the Seasons

We have discussed the rotation of Earth on its axis and the revolution of Earth around the Sun. To understand why the seasons change, we need to consider the combined effects of these two motions. Many people believe that Earth is closer to the Sun in the summer and farther away in the winter, and this change in distance causes the seasons. Can this hypothesis be falsified? We can make a prediction that if the distance from Earth to the Sun caused the seasons, then all of Earth should experience summer at the same time of year. But the United States experiences

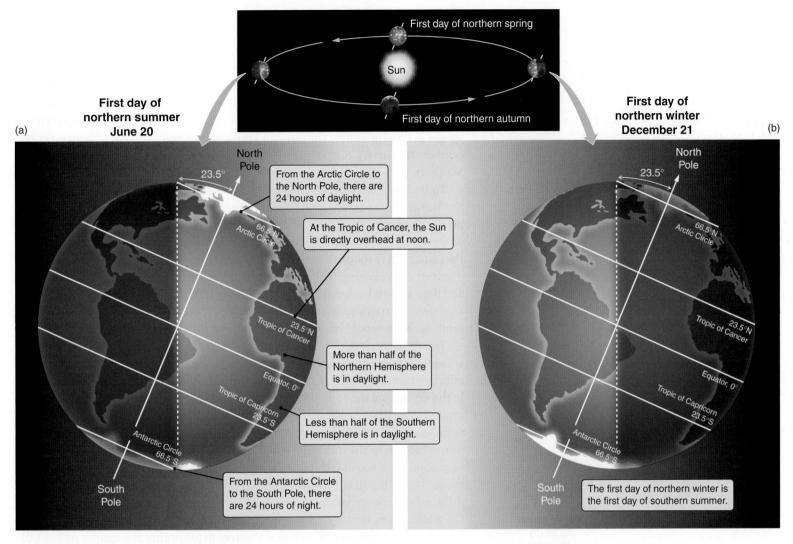

**Figure 2.14** (a) On the first day of the northern summer (around June 20, the summer solstice), the northern end of Earth's axis is tilted most nearly toward the Sun, while the Southern Hemisphere is tipped away. (b) Six months later, on the first day of the northern winter (around December 21, the winter solstice), the situation is reversed. Seasons are opposite in the two hemispheres.

summer in June, while Australia experiences summer in December. In modern times, we can directly measure the distance, and we find that Earth is actually closest to the Sun at the beginning of January. We have just falsified this hypothesis, and we need to look for another one that explains all of the available facts.

We observe that as Earth orbits the Sun, the Sun appears to move along the ecliptic, which is tilted 23.5° with respect to the celestial equator. This occurs because Earth's axis of rotation is tilted by 23.5° from the perpendicular to Earth's orbital plane. We notice that during the summer, the days are longer than in winter, and the Sun is higher in the sky as it crosses the meridian in summer than in winter.

**Figure 2.14** shows that as Earth moves around the Sun, its axis always points towards Polaris, in the same direction in space. During its orbit, sometimes Earth is on one side of the Sun, and sometimes on the other side. Therefore, sometimes Earth's North Pole is tilted more toward the Sun, and other times the South Pole is tilted more towards the Sun. When Earth's North Pole is tilted toward the Sun, an observer on Earth views the Sun north of the celestial equator; for observers in the Northern Hemisphere, the Sun is above the horizon more than 12 hours each day, thus the days are longer than 12 hours. Six months later, when Earth's North

▶‖ **AstroTour:** The Earth Spins and Revolves

**Nebraska Simulations:** Seasons and Ecliptic Simulator

**Astronomy in Action:** The Cause of Earth's Seasons

**Nebraska Simulations:** Daylight Hours Explorer

(a)

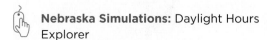

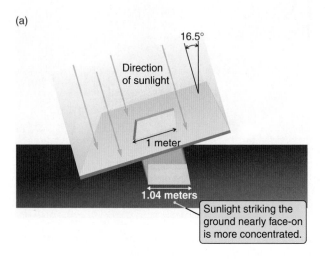

16.5°

Direction of sunlight

1 meter

1.04 meters

Sunlight striking the ground nearly face-on is more concentrated.

(b)

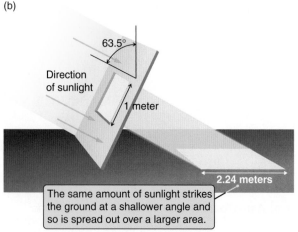

63.5°

Direction of sunlight

1 meter

2.24 meters

The same amount of sunlight strikes the ground at a shallower angle and so is spread out over a larger area.

**Figure 2.15** Local noon at latitude 40° north. (a) On the first day of northern summer, sunlight strikes the ground almost face-on. (b) On the first day of northern winter, sunlight strikes the ground more obliquely, and less than half as much sunlight falls on each square meter of ground each second.

Pole is tilted away from the Sun, an observer in the same place views the Sun south of the celestial equator.

In the preceding paragraph, we were careful to specify the *Northern* Hemisphere because seasons are opposite in the Southern Hemisphere. Look again at Figure 2.14. Around June 20, while the Northern Hemisphere is enjoying the long days and short nights of summer, Earth's South Pole is tilted away from the Sun. It is winter in the Southern Hemisphere; less than half of the Southern Hemisphere is illuminated by the Sun, and the days are shorter than 12 hours. On December 21, Earth's South Pole is tilted toward the Sun. It is summer in the Southern Hemisphere; the days are long and the nights are short there.

To understand how the combination of Earth's axial tilt and its path around the Sun creates seasons, consider a limiting case. If Earth's spin axis were exactly perpendicular to the plane of Earth's orbit (the **ecliptic plane**), then the Sun would always be on the celestial equator. At every latitude, the Sun would follow the same path through the sky every day, rising due east each morning and setting due west each evening. The Sun would be above the horizon exactly half the time, and days and nights would always be exactly 12 hours long everywhere on Earth. In short, if Earth's spin axis were exactly perpendicular to the plane of Earth's orbit, each day would be just like the last, and there would be no seasons.

The differing length of days through the year is part of the explanation for seasonal temperature changes, but it is not the whole story. Another important effect relates to the angle at which the Sun's rays strike Earth. The Sun is higher in the sky during the summer than it is during the winter, and sunlight strikes the ground *more directly* during the summer than during the winter. To see why this is important, study **Figure 2.15**. During the summer, Earth's surface is more nearly face-on to the incoming sunlight. More energy falls on each square meter of ground each second; the light is concentrated and bright. During the winter, the surface of Earth is more tilted with respect to the sunlight, so the light is more diffuse. Less energy falls on each square meter of the ground each second. This is the main reason why it is hotter in the summer and colder in the winter. As you can see in the **Process of Science Figure**, determining the causes of seasonal change requires accounting for all the known facts.

We can compare the average temperatures found at different latitudes on Earth to see the effect of the height of the Sun in the sky. Near the equator, the Sun passes high overhead every day, regardless of the season. As a result, the average temperatures are warm throughout the year. At high latitudes, however, the Sun is *never* high in the sky, and the average temperatures can be cold and harsh even during the summer. In between, at latitude 40° north, which stretches across the United States from northern California to New Jersey, more than twice as much solar energy falls on each square meter of ground per second at noon on June 20 as falls there at noon on December 21. These two effects—the directness of sunlight and the differing length of the night—mean that the Sun heats a hemisphere more during summer than winter.

## The Solstices and the Equinoxes

Four days during Earth's orbit mark unique moments in the year. The day when the Sun is highest in the sky as it crosses the meridian—the line from due north to due south that passes overhead—is called the **summer solstice**. On this day, the Sun rises farthest north of east and sets farthest north of west. This occurs each year near June 20, the first day of summer in the Northern Hemisphere. This orientation of Earth and Sun is shown in Figure 2.14a.

**Many people misunderstand the phenomenon of changing seasons because they do not account for all the relevant facts.**

### Take 1

### The Hypothesis

We have seasons because Earth is closer to the Sun in summer and farther away in winter.

### The Test

If this is true, both the Northern and Southern hemispheres would have summer in July.
The Northern and Southern hemispheres experience opposite seasons.

### The Conclusion

The hypothesis is falsified.

### Take 2

### The Hypothesis

We have seasons because the tilt of Earth's axis causes one hemisphere to be significantly closer to the Sun than the other.

### The Test

If this is true, the distances must be very different to cause such a large effect. Earth is tiny compared to its distance from the Sun: the difference in distance between hemispheres is less than 0.004 percent of the distance from the Sun.

### The Conclusion

The hypothesis is falsified.

### Take 3

### The Hypothesis

We have seasons because Earth's tilt changes the distribution of energy—one hemisphere receives more light than the other.

### The Test

If this is true, the amount of sunlight striking the ground in the summer should be more than in the winter, and the days should be longer in summer.

### The Conclusion

Seasons are caused primarily by a change in illumination due to Earth's tilt. During winter, less energy falls on each square meter of ground per second.

New information often challenges misconceptions.

(a)    **Motion of Earth around the Sun**

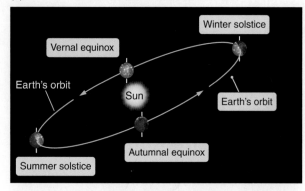

(b)    **Apparent motion of the Sun seen from Earth**

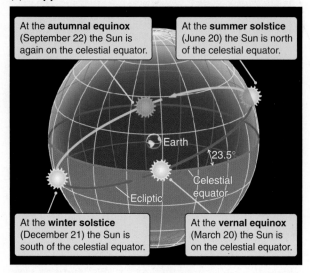

At the **autumnal equinox** (September 22) the Sun is again on the celestial equator.

At the **summer solstice** (June 20) the Sun is north of the celestial equator.

At the **winter solstice** (December 21) the Sun is south of the celestial equator.

At the **vernal equinox** (March 20) the Sun is on the celestial equator.

Earth

23.5°

Celestial equator

Ecliptic

**Figure 2.16** The motion of Earth around the Sun is shown from the frame of reference of (a) the Sun and (b) Earth.

**Astronomy in Action:** The Earth-Moon-Sun System

Six months after the summer solstice, the North Pole is tilted away from the Sun. This day is the **winter solstice** in the Northern Hemisphere, shown in Figure 2.14b. The winter solstice occurs each year about December 21, the shortest day of the year and the first day of winter in the Northern Hemisphere. Almost all cultural traditions in the Northern Hemisphere include a major celebration of some sort in late December. These winter festivals celebrate the return of the source of Earth's light and warmth. The days have stopped growing shorter and are beginning to get longer, and spring will come again.

Between the two solstices, there are two days when the ecliptic crosses the celestial equator. On these days, the Sun lies directly above Earth's equator. We call these days *equinoxes*, which means "equal night," because the entire Earth experiences 12 hours of daylight and 12 hours of darkness. Halfway between summer solstice and winter solstice, the **autumnal equinox** marks the beginning of fall in the Northern Hemisphere; it occurs around September 22. Halfway between winter solstice and summer solstice, the **vernal equinox** marks the beginning of spring in the Northern Hemisphere; it occurs around March 20.

**Figure 2.16** shows the solstices and equinoxes from two perspectives. Figure 2.16a shows Earth in orbit around a stationary Sun, and Figure 2.16b shows the Sun's apparent motion along the celestial sphere, which is how it appears to observers on Earth. In both cases, we are looking at the plane of Earth's orbit from the side, so that it is shown in perspective and looks quite flattened. We have also tilted the images so that the North Pole of Earth points straight up. The equinoxes correspond to the points in the sky where the celestial equator meets the ecliptic. Practice shifting between these two perspectives. You will know that you understand these differing perspectives when you are able to look at a position in either panel and predict the corresponding positions of the Sun and Earth in the other panel.

Just as it takes time for a pot of water on a stove to heat up when the burner is turned up and time for the pot to cool off when the burner is turned down, it takes time for Earth to respond to changes in heating from the Sun. The hottest months of northern summer are usually July and August, which come after the summer solstice, when the days are growing shorter. Similarly, the coldest months of northern winter are usually January and February, which occur after the winter solstice, when the days are growing longer. Temperature changes on Earth lag behind changes in the amount of heating we receive from the Sun.

This picture of the seasons must be modified somewhat near Earth's poles. At latitudes north of 66.5° north and south of 66.5° south, the Sun is circumpolar for

**Figure 2.17** This composite photo shows the midnight Sun, which can be seen in latitudes above 66.5° north (or south). In the 360-degree panoramic view, the Sun moves 15° each hour.

a part of the year surrounding the first day of summer. These lines of latitude are the **Arctic Circle** and the **Antarctic Circle** (see Figure 2.14). When the Sun is circumpolar, it is above the horizon 24 hours a day, earning the polar regions the nickname "land of the midnight Sun" (**Figure 2.17**). There is an equally long period surrounding the first day of winter when the Sun never rises and the nights are 24 hours long. The Sun never rises high in the Arctic or Antarctic sky, so the sunlight is never very direct. Even with the long days at the height of summer, the Arctic and Antarctic regions remain relatively cool.

In contrast, on the equator, *all* stars are above the horizon 12 hours a day, and the Sun is no exception. On the equator, days and nights are 12 hours long throughout the year. The Sun passes directly overhead on the first day of spring and the first day of autumn because these are the days when the Sun is on the celestial equator. Sunlight is most direct, perpendicular to the ground, at the equator on these days. On the summer solstice, the Sun is at its northernmost point along the ecliptic. On this day, and on the winter solstice, the Sun is farthest from the zenith at noon, and therefore sunlight is least direct at the equator.

As shown in Figure 2.14, latitude 23.5° north is called the Tropic of Cancer, and latitude 23.5° south is called the Tropic of Capricorn. The band between these two latitudes is called the **Tropics**. If you live in the tropics—in Rio de Janeiro or Honolulu, for example—the Sun will be directly overhead at noon twice during the year.

## Precession of the Equinoxes

When the Alexandrian astronomer Ptolemy and his associates were formalizing their knowledge of the positions and motions of objects in the sky 2,000 years ago, the Sun appeared in the constellation Cancer on the first day of northern summer and in the constellation Capricornus on the first day of northern winter. Today, the Sun is in Taurus on the first day of northern summer and in Sagittarius on the first day of northern winter. Why have the constellations in which solstices appear changed? There are actually two motions associated with Earth and its axis: Earth spins on its axis, but its axis also wobbles like the axis of a spinning top (**Figure 2.18**). The wobble is very slow: it takes about 26,000 years for the north celestial pole to complete one trip around a large circle centered on the north ecliptic pole. Currently, Polaris is the star we see near the north celestial pole. However, if you could travel several thousand years into the past or future, you would find that the point about which the northern sky appears to rotate is no longer near Polaris, but instead the stars rotate about another point on the path shown in Figure 2.18b. This figure shows the path of the north celestial pole through the sky during one cycle of this wobble.

The celestial equator is perpendicular to Earth's axis. Therefore, as Earth's axis wobbles, the celestial equator must also wobble. As the celestial equator wobbles, the locations where it crosses the ecliptic—the equinoxes—change as well. During each 26,000-year wobble of Earth's axis, the locations of the equinoxes make one complete circuit around the celestial equator. This change of the position of the equinox, due to the wobble of Earth's axis, is called the **precession of the equinoxes**.

## CHECK YOUR UNDERSTANDING 2.3

If Earth's axis were tilted by 45°, instead of its actual tilt, how would the seasons be different than they are currently? (a) The seasons would remain the same. (b) Summers would be colder. (c) Winters would be shorter. (d) Winters would be colder.

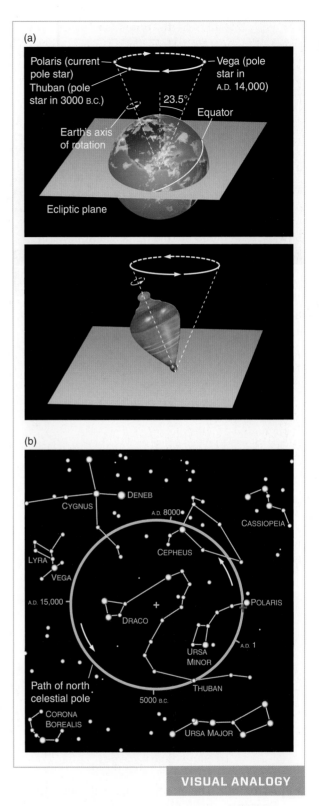

**VISUAL ANALOGY**

**Figure 2.18** (a) Earth's axis of rotation changes orientation in the same way that the axis of a spinning top changes orientation. (b) This precession causes the projection of Earth's rotation axis to move in a circle, centered on the north ecliptic pole (orange cross in the center). The red cross marks the projection of Earth's axis on the sky in the early 21st century.

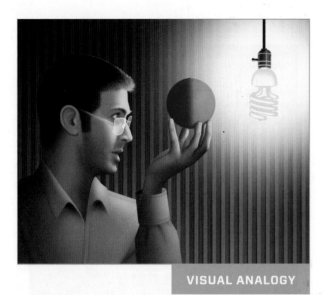

**VISUAL ANALOGY**

**Figure 2.19** An orange and a lamp can help you visualize the changing phases of the Moon.

Astronomy in Action: Phases of the Moon

▶❚❚ AstroTour: The Moon's Orbit: Eclipses and Phases

# 2.3 The Moon's Appearance Changes as It Orbits Earth

The most prominent object in our sky after the Sun is the Moon. Just as Earth orbits around the Sun, the Moon orbits around Earth once every 27.3 days. In this section, we will discuss the phases of the Moon as seen from Earth.

## The Changing Phases of the Moon

The Moon and its changing aspects have long fascinated humans. We speak of the "man in the Moon," the "harvest Moon," and sometimes a "blue Moon." In mythology, the Moon was the Roman goddess Diana, the Greek goddess Artemis, and the Inuit god Igaluk. The Moon has been the frequent subject of mythology, art, literature, and music.

Unlike the Sun, the Moon has no light source of its own; it shines by reflected sunlight. As the Moon orbits Earth, our view of the illuminated portion of the Moon is constantly changing. These different appearances of the Moon are called **phases** of the Moon. As the Moon orbits Earth, our view of the illuminated portion of the Moon is constantly changing. During a new Moon, when the Moon is between Earth and the Sun, the side facing away from us is illuminated, and during a full Moon, when the Earth is between the Sun and the Moon, the side facing toward us is illuminated. The rest of the time, only part of the illuminated portion can be seen from Earth. Sometimes the Moon appears as a circular disk in the sky. Other times it is nothing more than a thin sliver or its face appears dark.

To help you visualize the changing phases of the Moon, use an orange, a lamp, and your head. Your head is Earth, the orange is the Moon, and the lamp is the Sun (**Figure 2.19**). Turn off all the other lights in the room, and step back as far from the lamp as you can. Hold up the orange slightly above your head so that it is illuminated from one side by the lamp. Move the orange clockwise around your head and watch how the appearance of the orange changes. When you are between the orange and the lamp, the face of the orange that is toward you is fully illuminated. The orange appears to be a bright, circular disk. As the orange moves around its circle, you will see a progression of lighted shapes, depending on how much of the bright side and how much of the dark side of the orange you can see. This progression of shapes exactly mimics the changing phases of the Moon.

**Figure 2.20** shows the changing phases of the Moon. The **new Moon** occurs when the Moon is between Earth and the Sun. The far side is illuminated, but the near side is in darkness and we cannot see it. The new Moon appears close to the Sun in the sky, so it is up in the daytime with the Sun: it rises in the east at sunrise, crosses the meridian near noon, and sets in the west near sunset. A new Moon is never above the horizon in the nighttime sky.

A few days after a new Moon, as the Moon orbits Earth, a sliver of its illuminated half, called a **waxing crescent Moon**, becomes visible. *Waxing* here means "growing in size and brilliance"; the name refers to the fact that the Moon appears to be "filling out" from night to night at this time. From our perspective, the Moon has also moved away from the Sun in the sky. Because the Moon travels around Earth in the same direction in which Earth rotates, we now see the Moon trailing the Sun, so it is east of the Sun in the sky. A waxing crescent Moon is

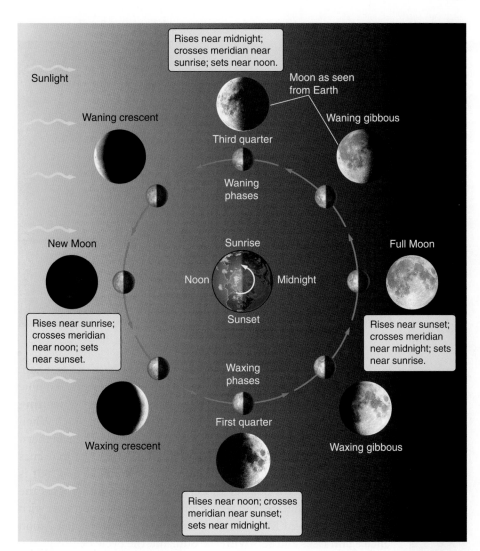

Sunlight

Waning crescent

Rises near midnight; crosses meridian near sunrise; sets near noon.

Moon as seen from Earth

Waning gibbous

Third quarter

Waning phases

New Moon

Sunrise

Full Moon

Noon     Midnight

Sunset

Rises near sunrise; crosses meridian near noon; sets near sunset.

Rises near sunset; crosses meridian near midnight; sets near sunrise.

Waxing phases

First quarter

Waxing crescent

Waxing gibbous

Rises near noon; crosses meridian near sunset; sets near midnight.

**Figure 2.20** The inner circle of images (connected by blue arrows) shows the Moon as it orbits Earth, as seen by an observer far above Earth's North Pole. The Sun is on the left. The outer ring of images shows the corresponding phases of the Moon as seen from Earth.

visible in the western sky in the evening, near the setting Sun but remaining above the horizon after the Sun sets. The "horns" of the crescent always point directly away from the Sun.

As the Moon moves farther along in its orbit, and the angle between the Sun and Moon grows, more and more of its near side becomes illuminated. About a week after the new Moon, half of the near side of the Moon is illuminated and half is in darkness. This phase is called a **first quarter Moon** because the Moon has moved a quarter of the way around Earth and has completed the first quarter of its cycle from new Moon to new Moon. A look at Figure 2.20 shows that the first quarter Moon rises at noon, crosses the meridian at sunset, and sets at midnight.

As the Moon moves beyond first quarter, more than half of its near side is illuminated. This phase is called a **waxing gibbous Moon**, from the Latin *gibbus*, meaning, "hump." The waxing gibbous Moon continues nightly to "grow" until finally we see the entire near side of the Moon illuminated—a **full Moon**. Earth is now between the Sun and the Moon, which appear opposite each other in the sky when viewed from Earth. The full Moon rises as the Sun sets, crosses the meridian at midnight, and sets in the morning as the Sun rises.

(a)　　　(b)

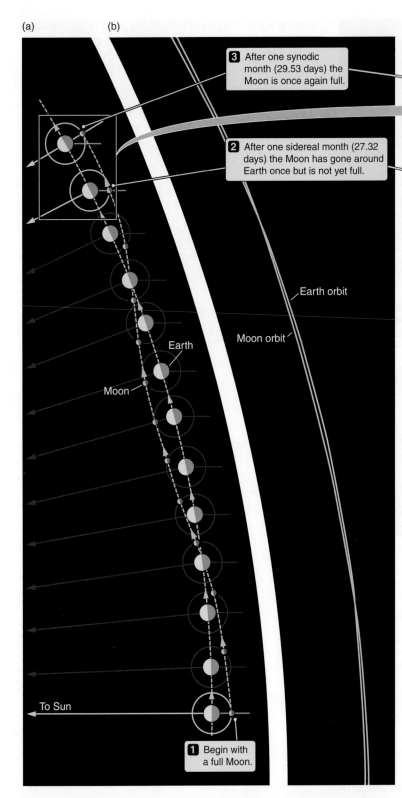

3 After one synodic month (29.53 days) the Moon is once again full.

2 After one sidereal month (27.32 days) the Moon has gone around Earth once but is not yet full.

Earth orbit

Moon orbit

Earth

Moon

To Sun

1 Begin with a full Moon.

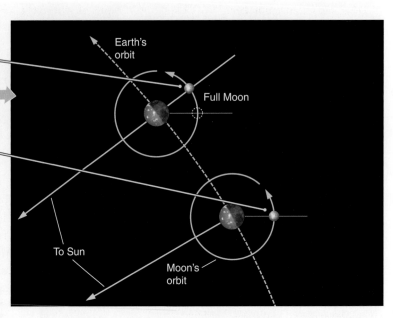

Earth's orbit

Full Moon

To Sun

Moon's orbit

**Figure 2.21** (a) The Moon completes one sidereal orbit in 27.32 days, but the synodic period (the period between phases seen from Earth) from one full Moon to the next is 29.53 days. The horizontal orange line to the right of the Moon indicates a fixed direction in space. (b) The orbits of Earth and the Moon are shown here to scale although the sizes of Earth and the Moon are not.

The second half of the Moon's orbit is the reverse of the first half. The Moon continues in its orbit, again appearing gibbous but now becoming smaller each night. This phase is called a **waning gibbous Moon** because *waning* means "becoming smaller." When the Moon is waning, the left side—as viewed from the Northern Hemisphere—appears illuminated. A **third quarter Moon** occurs when half of the near side is illuminated by sunlight and half is in darkness. A third quarter Moon rises at midnight, crosses the meridian near sunrise, and sets at noon. The cycle continues with a **waning crescent Moon** in the morning sky, until the new Moon once again rises and sets with the Sun, and the cycle begins again. Notice that when the Moon is farther from the Sun than Earth is, it is in gibbous (or full) phases. When the Moon is closer to the Sun than Earth is, it is in crescent (or new) phases.

You can always tell a waxing Moon from a waning Moon because the side that is illuminated is always the side facing the Sun. When the Moon is waxing, it appears in the evening sky, so its western side is illuminated. This is the right side as viewed from the Northern Hemisphere. Conversely, when the Moon is waning, the eastern side appears bright. This is the left side as viewed from the Northern Hemisphere.

**Figure 2.21** illustrates two types of lunar periods, the first one based on the Moon's orbit in space, and the second based on the alignment of the moving Moon, Earth, and Sun. The Moon completes one orbit around Earth in 27.32 days: this **sidereal period** is how long it takes to return to the same location in its orbit. However, because of the changing relationships among Earth, the Moon, and the Sun due to Earth's orbital motion, it takes 29.53 days to go from one full Moon to the next. This is known as its **synodic period** and is the basis for our "month" because it is what we can easily observe from Earth.

Do not try to memorize all possible combinations of where the Moon is in the sky at what phase and at what time of day. Instead, work on understanding the motion and phases of the Moon, and then use your understanding to figure out the specifics of any given case. To study the phases of the Moon, draw a picture like Figure 2.20, and use it to follow the Moon around its orbit. From your drawing, figure out what phase you would see and where it would appear in the sky at a given time of day. You might also try the simulations described in "Exploration: The Phases of the Moon" at the end of the chapter.

## The Visible Face of the Moon

Although the Moon's illumination varies at different parts of its orbit, one aspect of the Moon's appearance that does not change is the face of the Moon that we see. If we were to go outside next week or next month, or 20 years from now, or 20,000 centuries from now, we would still see the same side of the Moon that we see tonight. This happens because the Moon rotates on its axis exactly once for each revolution that it makes around Earth.

Imagine walking around the Washington Monument while keeping your face toward the monument at all times. By the time you complete one circle around the monument, your head has turned completely around once. When you were south of the monument, you were facing north; when you were east of the monument, you were facing west; and so on. But someone looking at you from the monument would always see your face. The Moon does exactly the same thing, rotating on its axis once per revolution around Earth, always keeping the same face toward Earth (**Figure 2.22**). When an object's revolution and rotation are synchronized (or in sync) with each other, its called **synchronous rotation**. We will see other examples of this in our Solar System.

The Moon's *far side*, facing away from Earth, is often called the "dark side of the Moon." In fact, there is no side of the Moon that is always dark. At any given time, half of the Moon is in sunlight and half is in darkness—just as at any given time, half of Earth is in sunlight and half is in darkness. The side of the Moon that faces away from Earth, the "far side," spends just as much time in sunlight as the side of the Moon that faces toward Earth does.

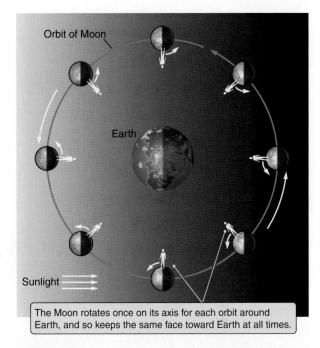

The Moon rotates once on its axis for each orbit around Earth, and so keeps the same face toward Earth at all times.

**Figure 2.22** The Moon rotates once on its axis for each orbit around Earth—an effect called synchronous rotation. In this illustration, the Sun is far to the left of the Earth-Moon system.

---

**CHECK YOUR UNDERSTANDING 2.4**

You see the Moon rising just as the Sun is setting. What phase is the Moon in?

 **Nebraska Simulations:** Lunar Phase Simulator

# 2.4 Calendars Are Based on the Day, Month, and Year

Archeologists tell us that the development of agriculture was crucial for the rise of human civilization, and keeping track of the seasons and best times of the year to plant and harvest was critical to successful farming. Records going back to the dawn of humanity suggest that people kept track of time by following the patterns in the sky, especially those of the Sun, the Moon and the stars. Some anthropologists have speculated that notches on fragments of bone found in southern France represent a 33,000-year-old lunar calendar. In this section, we will examine some different calendars.

**Figure 2.23** The ancient Egyptian calendar used a system of 12 months, plus festival days.

## Lunar and Lunisolar Calendars

As civilizations developed around the globe, different cultures tried to solve the "problem" of the calendar. The rates of rotation of Earth and revolutions of the Moon around Earth and Earth around the Sun are not even multiples of each other. A lunar cycle (full Moon to full Moon) is 29.5 days, and a solar cycle—the time it takes for the Sun to appear to move from and return to its highest possible point in the sky at noon on the summer solstice—is 365.24 days. One solar cycle has 12.38 lunar cycles. These fractions of days and months are what make calendars complicated.

Some of the oldest known calendars come from the Egyptians, the Babylonians, and the Chinese. The ancient Egyptians used a system of 12 months of 30 days each—which added up to 360 days—and then added five "festival days" to the end of the year (**Figure 2.23**). Without leap years, this 365-day year led to a drift of the seasons, so an extra month was added when necessary. When we consider how we celebrate the days between the modern December holidays and the New Year, an end-of-year calendar break for festivals seems like a good solution!

The Babylonians started the 24-hour day and 7-day week—7 for the Sun, the Moon, and the 5 planets visible with the naked eye. They created the first *lunisolar* calendar, in which a month began with the first sighting of the lunar crescent, and a 13th month was added when needed to catch up to the solar year. As the Babylonians developed mathematics, they discovered that 235 lunar months equals 19 solar years (and 6,940 days). Then they created a calendar cycle that consisted of 19 years, in which 12 of the years have 12 months and 7 of the years have 13 months, and then the cycle repeats. The ancient Hebrew calendar adopted this cycle, and the Jewish calendar still uses it today. This type of calendar keeps holidays in the same season from year to year, even though the dates are different.

The ancient Chinese calendar dating back several thousand years occasionally added a 13th month. By about 500 BCE, the Chinese were using a year of 365.25 days and a system similar to the Babylonians of adding a 13th month into some years. A few cultures used stellar calendars, for example following the position of a bright star like Sirius or certain prominent groups of stars in their sky, such as the Pleiades or the Big Dipper, to mark out a year.

The Islamic calendar is a purely lunar calendar, with no 13th lunar month added in. Their 12 months of 29 or 30 days each add up to 354 days—11.24 days short of the solar year. For this reason, the Islamic New Year and all other holidays drift earlier in each successive solar year. In the Islamic calendar, a holiday may fall in the winter in some years, and then a few years later it will have moved back to autumn.

## The Modern Civil Calendar

The international civil calendar used today is a solar calendar known as the **Gregorian calendar**. It is based on the **tropical year**, which measures the 365.242 solar days from one vernal equinox to the next. A **solar day** is the 24-hour period of Earth's rotation that brings the Sun back to the same local meridian. This is in contrast to the **sidereal day**, which is the time it takes for Earth to make one rotation and face the exact same star on the meridian. The sidereal day is about 23 hours 56 minutes and differs from the solar day because of Earth's motion around the Sun.

The Gregorian calendar includes a system of **leap years**—years in which a 29th day is added to the month of February—decreed by Julius Caesar in 45 BCE to make up for the extra fraction of a day. Leap years prevent the seasons from slowly sliding through the year to become increasingly out of sync with the months, so we don't end up experiencing winter in December one year and in August other years.

The Gregorian calendar is named for Pope Gregory XIII. He was concerned that the Easter holiday, which falls on the first Sunday after the first full Moon following March 21, was drifting away from the vernal (spring) Equinox. Julius Caesar's rule of one leap year every 4 years resulted in an average year of 365.25 days, but the actual year is 365.242 days. This difference of 0.008 day is about 11.5 minutes per year, or 3 days every 400 years, and by Gregory's time it had caused the date of the vernal equinox to drift in the Julian calendar by about 10 days. So in 1582, Pope Gregory decreed that 10 days would be deleted from the calendar to move the vernal equinox back to March 21. To make this work out better in the future, he declared that only century years divisible by 400 are leap years, thereby deleting 3 leap years (and the 3 days) every 400 years. Catholic countries followed this system immediately, but Protestant countries did not adopt it until the 1700s. Eastern Orthodox countries, including Russia, did not switch from the Julian to the Gregorian calendar until the 1900s. One slight further revision—making years divisible by 4,000 into common 365-day years—has been proposed so the modern Gregorian calendar will slip by about only 1 day in 20,000 years.

Despite international adoption of the Gregorian calendar, billions of people still celebrate holidays and festivals according to a lunisolar or lunar calendar. Chinese New Year, Passover, Easter, Ramadan, Rosh Hashanah, and Diwali, among others, have dates that change from one year to the next because they are based on lunar months from lunisolar or lunar calendars. The astronomy of people from long ago is still in use today.

 **Nebraska Simulations:** Synodic Lag

---

### CHECK YOUR UNDERSTANDING 2.5

Suppose that the astronomical cycles were even multiples of each other, so that a month was precisely 30 days, and a year was precisely 12 months. How would this change the dates of "wandering" holidays such as Chinese New Year or Ramadan?

........................................................................................

## 2.5 Eclipses Result from the Alignment of Earth, Moon, and the Sun

For ancient peoples attuned to the patterns of the sky, it must have been terrifying to look up to see the Sun being eaten away as if by a giant dragon or the full Moon turning the color of blood. An **eclipse** is the total or partial obscuration of one celestial body, or the light from that body, by another celestial body. Archaeological evidence suggests that ancient peoples put great effort into trying to figure out the pattern of eclipses and thereby bring them into the orderly scheme of the heavens. Stonehenge, pictured in Figure 2.1a, may have enabled its builders to predict when eclipses might occur. Ancient Chinese, Babylonian, and Greek astronomers had figured out that eclipses occur in cycles, and they were able to use their knowledge to make predictions about when and where eclipses would occur. In this section, we will describe the different types of eclipses and their frequency.

**Figure 2.24** Different parts of the Sun are blocked at different places within the Moon's shadow. An observer on Earth in the umbra (point A) sees a total solar eclipse, observers in the penumbra (points B and C) see a partially eclipsed Sun, and observers at point D see an annular solar eclipse.

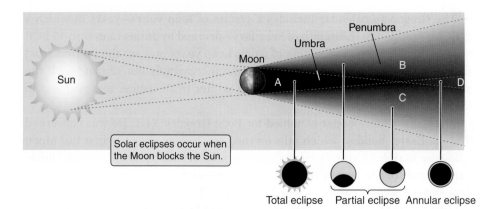

Solar eclipses occur when the Moon blocks the Sun.

Total eclipse    Partial eclipse    Annular eclipse

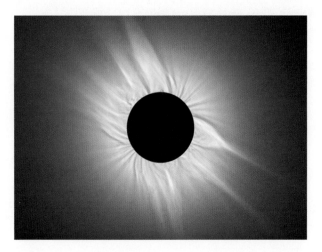

**Figure 2.25** The full spectacle of a total eclipse of the Sun.

## Solar Eclipses

A **solar eclipse** occurs when the Moon passes between Earth and the Sun; observers on Earth in the shadow of the Moon will see the eclipse. There are three different types of solar eclipses: *total, partial,* and *annular.* Consider the structure of the shadow of the Sun cast by a round object such as the Moon, as shown in **Figure 2.24**. An observer at point A would be unable to see any part of the surface of the Sun. This darkest, inner part of the shadow is called the **umbra**. If a location on Earth passes through the Moon's umbra, the Sun's light is totally blocked by the Moon, and a **total solar eclipse** will be observed (**Figures 2.25 and 2.26a**). At points B and C in Figure 2.24, an observer can see one side of the disk of the Sun but not the other. This outer region, which is only partially in shadow, is the **penumbra**. If a location on the surface of Earth passes through the Moon's penumbra, viewers at that location will observe a **partial solar eclipse**, in which the disk of the Moon blocks the light from a portion of the Sun's disk.

**Figure 2.26** Time sequences of images of the Sun taken (a) during a total solar eclipse and (b) during the annular solar eclipse of May 20, 2012. The Sun set during the ending phases.

(a) Solar eclipse geometry (not to scale)

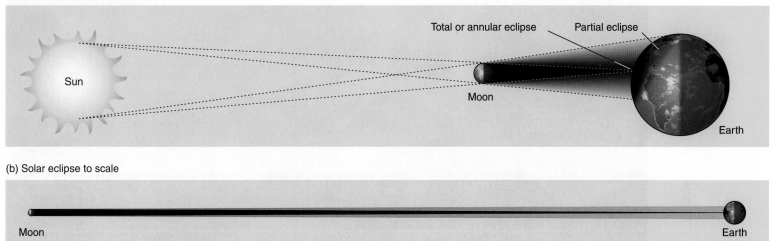

(b) Solar eclipse to scale

**Figure 2.27** (a, b) A solar eclipse occurs when the shadow of the Moon falls on the surface of Earth. Note that (b) is drawn to proper scale.

In the third type of solar eclipse, called an **annular solar eclipse**, the Sun appears as a bright ring surrounding the dark disk of the Moon (Figure 2.26b). An observer at point D in Figure 2.24 is far enough from the Moon that the Moon's apparent size in the sky is smaller than the Sun's. The apparent size of an object in the sky depends on the object's actual size and its distance from us. The Sun is about 400 times the diameter of the Moon, and the distance between the Sun and Earth is about 400 times more than the distance between the Moon and Earth. As a result, the Moon and Sun have almost exactly the same apparent size in the sky. Another factor is that the Moon's orbit is not a perfect circle. When the Moon and Earth are a bit closer together than average, the Moon appears slightly larger in the sky than the Sun. An eclipse occurring at that time will be total for some observers. When the Moon and Earth are farther apart than average, the Moon appears smaller than the Sun, so eclipses occurring during this time will be annular for some observers. Among all solar eclipses, one-third are total at some location on the surface of Earth, one-third are annular, and one-third are seen only as a partial eclipse.

**Figure 2.27** shows the geometry of a solar eclipse when the Moon's shadow falls on the surface of Earth. Figures like this usually show Earth and the Moon much closer together than they really are. The page is too small to draw them correctly and still see the critical details. The relative sizes and distances between Earth and the Moon are roughly equivalent to the difference between a basketball and a tennis ball placed 7 meters apart. Figure 2.27b shows the geometry of a solar eclipse with Earth, the Moon, and the separation between them drawn to the correct scale. Compare this drawing to Figure 2.27a and you will understand why drawings of Earth and the Moon are rarely drawn to the correct scale. If the Sun were drawn to scale in Figure 2.27a, it would be bigger than your head and located almost 64 meters off the left side of the page.

From any particular location, you are more likely to observe a partial solar eclipse than a total solar eclipse. Where the Moon's penumbra touches Earth, it has a diameter of almost 7,000 km—large enough to cover a substantial fraction of Earth. Thus, a partial solar eclipse is often visible from many locations on

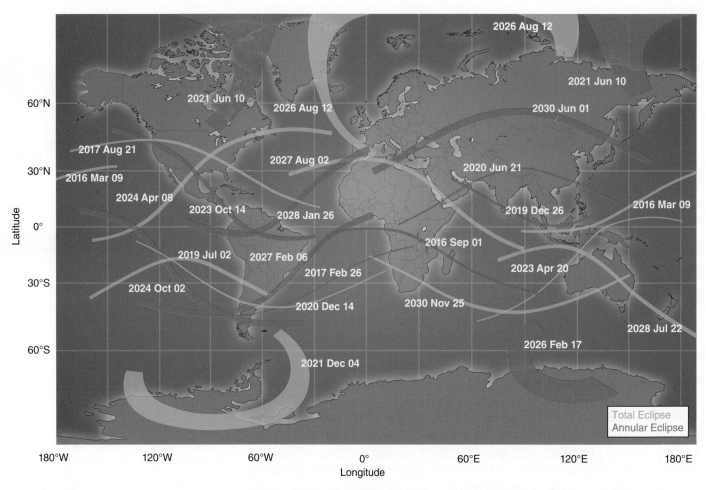

**Figure 2.28** The paths of total solar eclipses through 2020. Solar eclipses occurring in Earth's polar regions cover more territory because the Moon's shadow hits the ground obliquely.

Earth. In contrast, the path along which a total solar eclipse can be seen, shown in **Figure 2.28**, covers only a tiny fraction of Earth's surface. Even when the distance between Earth and the Moon is at a minimum, the umbra is only 270 km wide at the surface of Earth. As the Moon moves along in its orbit, this tiny shadow sweeps across the face of Earth at speeds of a few thousand kilometers per hour. Additionally, the Moon's shadow falls on the curved surface of Earth, causing the region shaded by the Moon during a solar eclipse to be elongated by differing amounts. The curvature can even cause an eclipse that started out as annular to become total.

The result is that a total solar eclipse can never last longer than 7½ minutes and is usually significantly shorter. Even so, it is one of the most amazing sights in nature. People all over the world flock to the most remote corners of Earth to witness the fleeting spectacle of the bright disk of the Sun blotted out of the daytime sky. Perhaps you saw some of the annular eclipse that was visible from much of the United States in May 2012. The first total solar eclipse since 1979 that will be visible in the continental United States will take place in August 2017 (followed by another in 2024). Annular eclipses will be visible from parts of the United States in 2021 and 2023. Viewing a solar eclipse should be on your lifetime to-do list!

(a) Lunar eclipse geometry (not to scale)

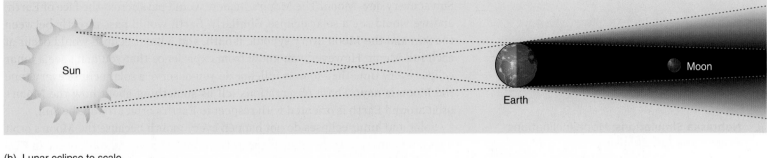

(b) Lunar eclipse to scale

**Figure 2.29** (a, b) A lunar eclipse occurs when the Moon passes through Earth's shadow. Note that (b) is drawn to proper scale.

## Lunar Eclipses

**Lunar eclipses** occur when the Moon moves through the shadow of Earth. The geometry of a lunar eclipse is shown in **Figure 2.29a** and is drawn to scale in Figure 2.29b. Here Earth is between the Sun and the Moon. Because Earth is much larger than the Moon, the dark umbra of Earth's shadow at the distance of the Moon is more than 2½ times the diameter of the Moon. A **total lunar eclipse**, when the Moon is entirely within Earth's shadow, lasts as long as 1 hour 40 minutes. In a total lunar eclipse, the Moon often appears red (**Figure 2.30a**). This "blood-red Moon," as it has been called in literature and poetry, occurs because the Moon is being illuminated by red light from the Sun that is bent as it travels through Earth's atmosphere and hits the Moon. Other colors of light are absorbed or scattered away from the Moon by Earth's atmosphere and therefore do not illuminate it.

A **penumbral lunar eclipse** occurs when the Moon passes through the penumbra of Earth's shadow; these are noticeable only from a very dark location or when the Moon passes within about 1,000 km of the umbra. If Earth's shadow incompletely covers the Moon, some of the disk of the Moon remains bright and some of it is in shadow. This is called a **partial lunar eclipse**. Figure 2.30b shows a composite of images taken at different times during a partial lunar eclipse. In the center image, the Moon is nearly completely eclipsed by Earth's shadow.

Many more people have observed a total lunar eclipse than have observed a total solar eclipse. To see a total solar eclipse, you must be located within that very narrow band of the Moon's shadow as it moves across Earth's surface. In contrast, when the Moon is immersed in Earth's shadow, anyone located in the hemisphere of Earth that is facing the Moon can see it. As a result, from any location, total eclipses of the Moon are relatively common, and you may have seen at least one.

## Frequency of Eclipse Seasons

How did some people in ancient cultures successfully predict eclipses? From their understanding of lunar and solar cycles for making calendars, they were able to compute cycles of eclipses. Imagine Earth, the Moon, and the Sun all sitting on the same flat tabletop. If the Moon's orbit were in exactly the same plane

(a)

(b)

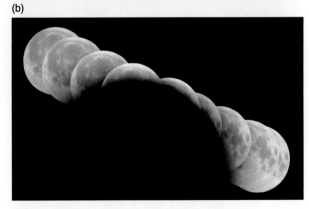

**Figure 2.30** (a) During a total lunar eclipse, the Moon often appears blood red. (b) A time-lapse series of photographs of a partial lunar eclipse clearly shows Earth's shadow. Note the size of Earth's shadow compared to the size of the Moon.

**Nebraska Simulations:** Moon Inclinations; Eclipse Shadow Simulator

as the orbit of Earth, then the Moon would pass directly between Earth and the Sun at every new Moon. The Moon's shadow would pass across the face of Earth, and we would see a solar eclipse. Similarly, Earth would pass directly between the Sun and the Moon every synodic month, and a lunar eclipse would occur at each full Moon. However, you know from experience that you don't see a lunar eclipse every time the Moon is full, nor do you observe a solar eclipse every time the Moon is new. These observations tell us something about how the Moon's orbit around Earth is oriented with respect to Earth's orbit around the Sun.

Solar and lunar eclipses do not happen every month because the Moon's orbit does not lie in exactly the same plane as the orbit of Earth. As you can see in **Figure 2.31**, the plane of the Moon's orbit around Earth is inclined by about 5.2° with respect to the plane of Earth's orbit around the Sun. The line along which the orbital plane of the Sun and the orbital plane of the Moon intersect is called the **line of nodes**. For part of the year, the line of nodes points in the general direction of the Sun. During these times, called **eclipse seasons**, a new Moon passes directly between the Sun and Earth, casting its shadow on Earth's surface and causing a solar eclipse. Similarly, a full Moon occurring during an eclipse season passes through Earth's shadow, and a lunar eclipse results. An eclipse season lasts only 38 days. That's how long the Sun is close enough to the line of nodes for eclipses to occur. Most of the time the line of nodes points farther away from the Sun, and Earth, Moon, and Sun cannot line up closely enough for an eclipse to occur. At these times, a solar eclipse cannot take place because the shadow of a new Moon passes "above" or "below" Earth. Similarly, no lunar eclipse can occur because a full Moon passes "above" or "below" the shadow of Earth.

If the plane of the Moon's orbit always had the same orientation, then eclipse seasons would occur twice a year, as suggested by Figure 2.31. In actuality, eclipse

**Figure 2.31** Eclipses are possible only when the Sun, Moon, and Earth lie along (or very close to) an imaginary line known as the line of nodes. When the Sun does not lie along the line of nodes, Earth passes under or over the shadow of a new Moon, and a full Moon passes under or over the shadow of Earth.

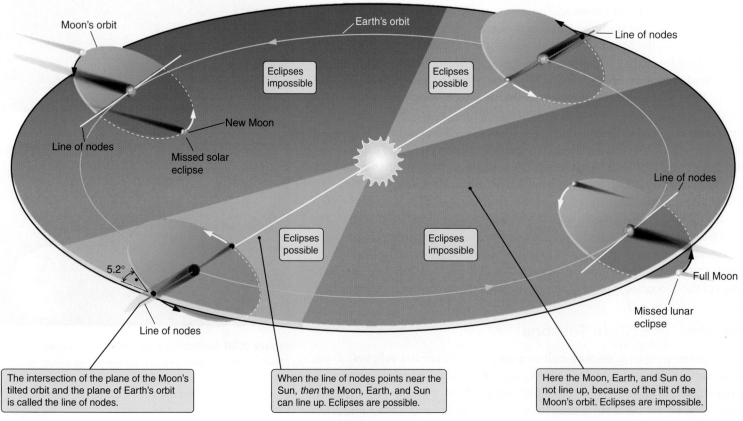

The intersection of the plane of the Moon's tilted orbit and the plane of Earth's orbit is called the line of nodes.

When the line of nodes points near the Sun, *then* the Moon, Earth, and Sun can line up. Eclipses are possible.

Here the Moon, Earth, and Sun do not line up, because of the tilt of the Moon's orbit. Eclipses are impossible.

seasons occur about every 5 months 20 days. The roughly 10-day difference is due to the fact that the plane of the Moon's orbit slowly wobbles, much like the wobble of a spinning plate balanced on the end of a circus performer's stick. As it does so, the line of nodes changes direction. This wobble rotates in the direction opposite the direction of the Moon's motion in its orbit. That is, the line of nodes moves clockwise as viewed from above Earth's orbital plane. One wobble of the Moon's orbit takes 18.6 years, so we say that the line of nodes regresses at a rate of 360° every 18.6 years, or 19.4° per year. This amounts to about a 20-day regression each year. If January 1 marks the middle of an eclipse season, the next eclipse season will be centered around June 20, and the one after that around December 10.

 **Nebraska Simulations:** Eclipse Table

### CHECK YOUR UNDERSTANDING 2.6

If the Moon were in its same orbital plane but twice as far from Earth, which of the following would happen? (a) The Moon would not go through phases. (b) Total eclipses of the Sun would not be possible. (c) Annular eclipses of the Sun would not be possible. (d) Total eclipses of the Moon would not be possible.

# Origins
## The Obliquity of Earth

The various motions of Earth give rise to the most basic of patterns faced by life on Earth. Earth's rotation is responsible for the cycle of night and day. Earth's axial tilt and its passage around the Sun bring the change of the seasons. As life evolved on Earth, it had to adapt to these patterns.

The range of climate on Earth based on distance from the equator likely has contributed to the broad diversity of life of our planet. Earth's biodiversity includes life that adapted to the long, cold polar nights and life in equatorial latitudes that adapted to much higher temperatures. Earth's life adapted to seasonal patterns in rain and drought, leading to acquired seasonal patterns of migration and reproduction. If Earth had no axial tilt, the poles would continually be in winter and probably too cold for humans. Midlatitudes would not have the cool winters that are needed for many food crops. Latitudes near the equator would be even consistently warmer than they are now.

Life might have been affected by periodic changes in Earth's axial tilt. If Earth's tilt were larger than 23.5°, the seasonal variation would be even stronger. If the tilt were smaller, the seasonal variation would be weaker. Chinese, Indian, Greek, and Arabic records going back 3,000 years indicate that the tilt was estimated in ancient times by measuring the length of the shadow from a vertical pole on the day of the solstice. We now know that for the past few million years, Earth's axial tilt actually varied from 22.1° to 24.5° over a 41,000-year cycle. The Moon's gravity is responsible for maintaining the tilt within this small range over the past half-billion years—about the time since animal life greatly diversified on Earth. Currently, the tilt is about midway between the two extremes and getting smaller. It will reach its minimum value of 22.1° in about 10,000 years. Scientists are studying whether this variation in tilt correlates with periods of temperature change on Earth, especially the times of ice ages.

 **Nebraska Simulation:** Obliquity Simulator

*The first total solar eclipse in the continental United States in decades will take place in August 2017. In this story, one town claims that it will be the best place to see the eclipse.*

# Thousands Expected in Hopkinsville for 2017 Solar Eclipse

By **ADAM GHASSEMI**

HOPKINSVILLE, Kentucky—Hopkinsville has a population of a little more than 30,000 but in 2017 it's expected to more than double for just a few days.

"It's going to be a really big deal," said Cheryl Cook with the Hopkinsville-Christian County Convention and Visitors Bureau who started planning seven years ago for the biggest event they've ever seen. "It could cause traffic jams from here to Nashville," she said.

Visitors are coming from places as far away as Germany, Australia, and Japan to see a total eclipse for 2:40 of darkness.

It's so rare, a map of the projected path has been hanging in the office of Austin Peay Astronomy Professor Spencer Buckner for 15 years.

"The best place on Earth that will have the longest period of totality is actually just north and west of Hopkinsville," Buckner said Friday. "I don't know of another one that will happen in the next few hundred years."

City leaders want to make sure they don't miss any details, like visitors have protective eyewear to see the eclipse, or cutting electricity off in certain parts of the city. That way when things do go dark, lights won't automatically come on and disrupt the view.

"We're hoping for a bright, sunny, warm day. No clouds in the sky," Cook said.

Scientists are planning how they'll capture and study the eclipse along its cross-country path from Oregon to South Carolina, while "eclipse chasers" have their trips to Hopkinsville already confirmed.

"Over 400 rooms booked," said Chairman of the Hopkinsville-Christian County Convention and Visitors Bureau Board Jeff Smith. "I've never seen anything like this."

When the Moon perfectly aligns with the Sun there won't be any better place in the world to see it on August 21, 2017.

"The world's coming to Hopkinsville," Cook went on to say.

Spectators will be able to see the eclipse from a number of places across Southern Kentucky and Middle Tennessee, but people just 70 miles from Hopkinsville in Nashville will get 44 fewer seconds to witness it.

1. Why are people excited about this solar eclipse?
2. How long is this eclipse as seen from Hopkinsville? Explain why people in Nashville observe a much shorter eclipse.
3. Would you predict that there will be a lunar eclipse in summer 2017? If so, what are the possible dates?
4. Look at the map on the NASA eclipse page: http://eclipse.gsfc.nasa.gov/SEgoogle/SEgoogle2001/SE2017Aug21Tgoogle.html. Will you be able to see the eclipse from your school?
5. The claim in this story was challenged by people who argued that Hopkinsville may have the shadow of the Moon passing closest to Earth's center, but a town in Illinois will have 0.1 seconds more of totality. Will most people care about this 0.1-second difference in totality? What other factors will likely affect where people go to see this eclipse?

# Summary

The motions of Earth and the Moon are responsible for many of the repeating patterns that can be observed in the sky. Calendars keep track of time using these patterns. Earth's rotation about its axis causes daily patterns of rising and setting. Earth's revolution around the Sun causes yearly patterns of the stars in the sky and the passage of the seasons. The tilt of Earth on its axis changes both the length of daytime and the intensity of sunlight, causing the seasons. The Moon's revolution around Earth causes the month-long pattern of the phases of the Moon. Occasionally, alignments of Earth, the Moon, and the Sun cause eclipses. The tilt of Earth's axis causes the variation in climate. This tilt varies slightly over tens of thousands of years. Life on Earth adapted to these seasonal variations.

**LG 1** **Describe how Earth's rotation about its axis and revolution around the Sun affect our perception of celestial motions as seen from different places on Earth.** The daily rotation of Earth on its axis causes the apparent daily motion of the Sun, Moon, and stars. Our location on Earth and Earth's location in its orbit around the Sun determine which stars we see at night. You can determine your latitude from the altitude of the pole star. When observing objects in our sky, we need to consider the relative motion of Earth and other objects. The ecliptic is the path that the Sun appears to take through the stars.

**LG 2** **Explain why there are different seasons throughout the year.** A year is the time it takes for Earth to complete one revolution around the Sun. Constellations are patterns of stars that reappear in the same place in the sky at the same time of night each year. The tilt of Earth's axis determines the seasons by changing the angle at which sunlight strikes the surface of Earth in different locations. The changing angle of sunlight and the differing length of the day cause the seasonal variations on Earth. The changing seasons are marked by equinoxes and solstices.

**LG 3** **Describe the factors that create the phases of the Moon.** The relative locations of the Sun, Earth, and Moon determine the phases of the Moon. The Moon takes one sidereal month to complete one revolution around Earth and one synodic month to go through a cycle of phases. The Moon's motion around Earth causes it to be illuminated differently at different times. When the Moon is farther from the Sun than Earth is, it is in gibbous phases. When the Moon is closer to the Sun than Earth is, it is in crescent phases.

**LG 4** **Sketch the alignment of Earth, the Moon, and the Sun during eclipses of the Sun and the Moon.** A solar eclipse occurs when the new Moon is in the plane of Earth and the Sun and the shadow of the Moon falls on Earth. A lunar eclipse occurs when the full Moon is in the plane of Earth and the Sun and the shadow of Earth falls on the Moon. Twice a year, at new or at full Moon, the Moon is exactly in line between Earth and the Sun. At these times, eclipses occur.

## ? UNANSWERED QUESTION

- How long will Earth continue to have total solar eclipses? These occur because the Moon and the Sun are coincidentally the same size in our sky, but will that always be the case? The observed size of an object in the sky depends on its actual diameter and its distance from us. One or both of these can change. The Moon is slowly moving away from Earth by about 4 meters per century. Over time, the Moon will appear smaller in the sky, and it won't be able to cover the full disk of the Sun. While we can measure the current rate of the Moon's movement away from Earth, we are less certain of how this rate may change with time. A lesser and more uncertain effect comes from the Sun—which will continue to brighten slowly, as it has throughout its history. With this brightening, the actual diameter of the Sun may slightly increase, and it will appear larger in our sky. A more distant Moon and a larger Sun will eventually result in an end to total eclipses on Earth.

# Questions and Problems

## Test Your Understanding

1. Constellations are groups of stars that
   a. are close to each other in space.
   b. are bound to each other by gravity.
   c. are close to each other in Earth's sky.
   d. all have the same composition.

2. Where on Earth can you stand and, over the course of a year, see the entire sky?
   a. only at the North Pole
   b. at either pole
   c. at the equator
   d. anywhere

3. Day and night are caused by
   a. the tilt of Earth on its axis.
   b. the rotation of Earth on its axis.
   c. the revolution of Earth around the Sun.
   d. the revolution of the Sun around Earth.

4. Polaris, the North Star, is unique because
   a. it is the brightest star in the night sky.
   b. it is the only star in the sky that doesn't move throughout the night.
   c. it is always located at the zenith, for any observer.
   d. it has a longer path above the horizon than any other star.

5. There is an angle between the ecliptic and the celestial equator because
   a. Earth's axis is tilted with respect to its orbit.
   b. Earth's orbit is tilted with respect to the orbits of other planets.
   c. the Sun follows a rising and falling path through space.
   d. the Sun's orbit is tilted with respect to Earth's.

6. The tilt of Earth's axis causes the seasons because
   a. one hemisphere of Earth is closer to the Sun in summer.
   b. the days are longer in summer.
   c. the rays of light strike the ground more directly in summer.
   d. both a and b
   e. both b and c

7. Which is *not* true on the vernal and autumnal equinoxes?
   a. Every place on Earth has 12 hours of daylight and 12 hours of darkness.
   b. The Sun rises due east and sets due west.
   c. The Sun is located on the celestial equator.
   d. The motion of the stars in the sky is different than on other days.

8. We always see the same side of the Moon because
   a. the Moon does not rotate on its axis.
   b. the Moon rotates on its axis once for each revolution around Earth.
   c. when the other side of the Moon is facing Earth, it is unlit.
   d. when the other side of the Moon is facing Earth, it is on the opposite side of Earth.

9. You see the Moon on the meridian at sunrise. The phase of the Moon is
   a. waxing gibbous.
   b. full.
   c. first quarter.
   d. third quarter.

10. A lunar eclipse occurs when _____ shadow falls on _____.
    a. Earth's; the Moon
    b. the Moon's; Earth
    c. the Sun's; the Moon
    d. the Sun's; Earth

11. Different cultures created different calendars because
    a. they had measured different lengths of the day, month, and year.
    b. they used different definitions of the day, month, and year.
    c. the number of days in a month and the number of days and months in a year are not integers.
    d. calendars are completely arbitrary.

12. Which stars we see at night depends on
    a. our location on Earth.
    b. Earth's location in its orbit.
    c. the time of the observation.
    d. all of the above

13. On the summer solstice in June, the Sun will be directly above _____ and all locations north of _____ will experience daylight all day.
    a. the Tropic of Cancer; the Antarctic Circle
    b. the Tropic of Capricorn; the Arctic Circle
    c. the Tropic of Cancer; the Arctic Circle
    d. the Tropic of Capricorn; the Antarctic Circle

14. The Sun, Moon, and stars
    a. appear to move each day because the celestial sphere rotates about Earth.
    b. change their relative positions over time.
    c. rise north or south of west and set north or south of east, depending on their location on the celestial sphere.
    d. always remain in the same position relative to each other.

15. You see the first quarter Moon on the meridian. Where is the Sun?
    a. on the western horizon
    b. on the eastern horizon
    c. below the horizon
    d. on the meridian.

## Thinking about the Concepts

16. Polaris was used for navigation by seafarers such as Columbus as they sailed from Europe to North America. When Magellan sailed the South Seas, he could not use Polaris for navigation. Explain why.

17. If you were standing at Earth's North Pole, where would you see the north celestial pole relative to your zenith?

18. Observers in the Northern Hemisphere see the zodiacal constellation Gemini in the winter. Why do they not see it in the summer?

19. Imagine that you are flying along in a jetliner.
    a. Describe ways to tell that you are moving.
    b. If you look down at a building, which way is it moving relative to you?

20. Astronomers are sometimes asked to serve as expert witnesses in court cases. Suppose you are called in as an expert witness, and the defendant states that he could not see the pedestrian because the full Moon was casting long shadows across the street at midnight. Is this claim credible? Why or why not?

21. Imagine that one person was developing a theory of seasons as described in the three "takes" in the Process of Science Figure in this chapter. Compare this process with the flowchart of the Process of Science Figure in Chapter 1. Describe what the development of this theory would look like on that flowchart.

22. Why is the winter solstice *not* the coldest time of year?

23. Earth spins on its axis and wobbles like a top.
    a. How long does it take to complete one spin?
    b. How long does it take to complete one wobble?

24. What is the approximate time of day when you see the full Moon near the meridian? At what time is the first quarter (waxing) Moon on the eastern horizon? Use a sketch to help explain your answers.

25. Assume that the Moon's orbit is circular. Imagine that you are standing on the side of the Moon that faces Earth.
    a. How would Earth appear to move in the sky as the Moon made one revolution around Earth?
    b. How would the "phases of Earth" appear to you compared to the phases of the Moon as seen from Earth?

26. If people on Earth were observing a lunar eclipse, what would you see from the Moon?

27. From any given location, why are you more likely to witness a partial eclipse of the Sun than a total eclipse?

28. Why do we not see a lunar eclipse each time the Moon is full or witness a solar eclipse each time the Moon is new?

29. How would Earth's temperature variation be different if it was tilted 90° on its axis (like the planet Uranus)?

30. Explain how a cyclic change in Earth's tilt could affect its seasonal temperatures.

## Applying the Concepts

31. Earth is spinning along at 1,674 km/h at the equator. Use this fact, along with the length of the day, to calculate Earth's equatorial diameter.

32. Determine the latitude where you live. Draw and label a diagram showing that your latitude is the same as (a) the altitude of the north celestial pole and (b) the angle (along the meridian) between the celestial equator and your local zenith. What is the altitude of the Sun at noon as seen from your home at the times of the winter solstice and the summer solstice?

33. Using a protractor, you estimate an angle of 40° between your zenith and Polaris. What is the altitude of Polaris? What is your latitude? Are you in the continental United States or Canada?

34. The southernmost star in a group of stars known as the Southern Cross lies approximately 65° south of the celestial equator. What is the farthest-north latitude for which the entire Southern Cross is visible? Can it be seen in any U.S. states? If so, which ones?

35. Imagine that you are standing on the South Pole at the time of the southern summer solstice.
    a. How far above the horizon will the Sun be at noon?
    b. How far above (or below) the horizon will the Sun be at midnight?

36. Suppose the tilt of Earth's equator relative to its orbit were 10° instead of 23.5°. At what latitudes would the Arctic and Antarctic circles and the Tropics of Cancer and Capricorn be located?

37. The Moon's orbit is tilted by about 5° relative to Earth's orbit around the Sun. What is the highest altitude in the sky that the Moon can reach, as seen in Philadelphia (latitude 40° north)?

38. Suppose you would like to witness the midnight Sun (when the Sun appears just above the northern horizon at midnight), but you don't want to travel any farther north than necessary.
    a. How far north (that is, to which latitude) would you have to go?
    b. At what time of year would you make this trip?

39. If, as some historians believe, the Egyptian stadion was about 157.5 meters, then what would Eratosthenes have computed for the size of Earth? How close is this to the modern value?

40. a. The vernal equinox is now in the zodiacal constellation Pisces. The precession of Earth's axis will eventually cause the vernal equinox to move into Aquarius. How long, on average, does the vernal equinox spend in each of the 12 zodiacal constellations?
    b. Stonehenge was erected roughly 4,000 years ago. Referring to the zodiacal constellations shown in Figure 2.14, identify the constellation in which these ancient builders saw the vernal equinox.

41. Referring to Figure 2.19, estimate when Vega, the fifth-brightest star in our sky (excluding the Sun), will once again be the northern pole star.

42. The apparent diameter of the Moon in the sky is approximately ½°. How long does it take the Moon to move 360°? About how long does it take the Moon to move a distance equal to its own diameter across the sky?

43. The apparent size of an object in the sky is proportional to its actual diameter divided by its distance. The Moon has a radius of 1,737 km, with an average distance of $3.780 \times 10^5$ km from Earth. The Sun has a radius of 696,000 km, with an average distance of $1.496 \times 10^8$ km from Earth. Show that the apparent sizes of the Moon and Sun in our sky are approximately the same.

44. Earth has an average radius of 6,371 km. If you were standing on the Moon, how much larger would Earth appear in the lunar sky than the Moon appears in our sky?

45. How would the length of the eclipse season change if the plane of the Moon's orbit were inclined less than its current 5.2° to the plane of Earth's orbit? Explain your answer.

to the next? Bring up the "Duration of Days/Darkness Table for One Year" page for your location. Are the days getting longer or shorter? When do the shortest and the longest days occur? Look up a location in the opposite hemisphere (Northern or Southern). When are the days shortest and longest?

47. Go to the "Earth and Moon Viewer" website (http://fourmilab.ch/earthview). Under "Viewing the Earth," click on "latitude, longitude and altitude" and enter your approximate latitude and longitude, and 40,000 for altitude; then select "View Earth." Are you in daytime or nighttime? Now play with the locations; keep the same latitude but change to the opposite hemisphere (Northern or Southern). Is it still night or day? Go back to your latitude, and this time enter 180° minus your longitude, and change from west to east, or from east to west, so that you are looking at the opposite side of Earth. Is it night or day there? What do you see at the North Pole (latitude 90° north) and the South Pole (latitude 90° south)? At the bottom of your screen you can play with the time. Move back 12 hours. What do you observe at your location and at the poles?

48. Go to the U.S. Naval Observatory website (USNO "Data Services," at http://aa.usno.navy.mil/data). Look up the Moon data for the current day. When will it rise and set? What is the phase? How will it change over the next 4 weeks. Enter one day at a time or look at the yearlong tables for moonrise and moonset and for the dates of primary phases. What time of day does a third quarter Moon rise? When (and in what phases) can you see the Moon in the daytime?

49. Using the times of moonrise and moonset that you located in question 48, make a plan to observe the Moon directly at least once a day for a week. Take a picture of the Moon (or make a sketch) every day. How is the brightness of the Moon changing? If it's daytime, how far is the Moon from the Sun in the sky? If it's nighttime, are the stars that are near the Moon in the sky the same every night?

50. Go to the "NASA Eclipse" website (http://eclipse.gsfc.nasa.gov/eclipse.html). When is the next lunar eclipse? Will it be visible at your location if the skies are clear? Is it a total or partial eclipse? When is the next solar eclipse? Will it be visible at your location? Compare the fraction of Earth that the solar eclipse will affect with the fraction for the lunar eclipse. Why are lunar eclipses visible in so many more locations?

## USING THE WEB

46. Go to the U.S. Naval Observatory website (USNO "Data Services," at http://aa.usno.navy.mil/data). Look up the times for sunrise and sunset for your location for the current week. (You can change the dates one at a time or bring up a table for the entire month.) How are the times changing from one day

# smartwork5

If your instructor assigns homework in Smartwork5, access your assignments at digital.wwnorton.com/astro5.

digital.wwnorton.com/astro5

Visit the Student Site at the Digital Landing Page, and open the Lunar Phase Simulator applet in Chapter 2.

Study the diagrams shown in the simulator. The largest window shows a view of the Earth-Moon system as seen from above Earth's North Pole. The Sun is far off the screen to the left. An observer stands on Earth. The small window at upper right shows the appearance of the Moon as seen from the Northern Hemisphere. The small window at lower right shows the observer's location, with the Sun and Moon pictured as flat disks in the sky.

**1** Given the relative positions of the observer and the Sun, approximately what time is it for this observer: 6:00 A.M., noon, 6:00 P.M., or midnight?

_____

**2** Where is the Moon in the observer's sky: on the eastern horizon, on the western horizon, below the horizon, or crossing the meridian?

_____

**3** What is the phase of the Moon?

_____

**4** Imagine yourself on the Moon in the image shown in the larger window. If you looked toward Earth, what phase of Earth would you see?

_____

Now select "start animation." Allow the animation to run until the Moon is 20 percent illuminated (as shown in the upper small window on the right).

**5** In which direction does the Moon orbit Earth: clockwise or counterclockwise?

_____

**6** For observers in the Northern Hemisphere, which side of the Moon is illuminated first after a new Moon: right or left?

_____

**7** If you observe a crescent Moon with the horns of the crescent pointing right, is the Moon waxing or waning?

_____

Grab the Moon with your mouse, and drag it to first quarter. Drag the observer so that her local time is approximately midnight.

**8** Where is the first quarter Moon in the observer's sky: on the eastern horizon, on the western horizon, below the horizon, or crossing the meridian?

_____

These three things are related: the time, the phase of the Moon, and the Moon's location in the sky.

**9** Arrange the observer and the Moon so that the Moon is full and crossing the meridian. What time is it for the observer: 6:00 A.M., noon, 6:00 P.M., or midnight?

_____

**10** Arrange the observer and the Moon so that the Moon is in third quarter and the time for the observer is approximately noon (the Sun is on the meridian). Where is the Moon in the observer's sky: on the eastern horizon, on the western horizon, below the horizon, or crossing the meridian?

_____

**11** Arrange the observer and the Moon so that it is approximately 6:00 P.M. (sunset) for the observer, and the Moon is just rising on the eastern horizon. What is the phase of the Moon?

_____

There are many other combinations of the time, phase of the Moon, and Moon's location to play with. Challenge yourself to be able to set up any two of the three and find the third. When you can do this without the simulator, just by making the picture in your head, you will really understand the phases of the Moon.

# 3 Motion of Astronomical Bodies

The birth of modern astronomy dates back to the time when astronomers and mathematicians discovered regular patterns in the motions of the planets. A successful theory of how Earth moved and how it fit in with its neighbors in the Solar System was the first step toward understanding Earth's place in the universe.

## LEARNING GOALS

In this chapter, we will examine how astronomers came to understand that Earth and other planets orbit the Sun. By the conclusion of this chapter, you should be able to:

**LG 1** Describe and contrast the geocentric and heliocentric models of the Solar System.

**LG 2** Use Kepler's laws to describe the motion of objects in the Solar System.

**LG 3** Explain how Galileo's astronomical discoveries provided empirical evidence for the heliocentric model.

**LG 4** Describe the work of Galileo and Newton, which led them to discover the physical laws that govern the motion of all objects.

The shadows of a few of the Galilean moons of Jupiter fall on the planet. ▶ ▶ ▶

**Why doesn't
Jupiter orbit
Earth?**

# 3.1 The Motions of Planets in the Sky

**Nebraska Simulation:** Ptolemaic Orbit of Mars

**Figure 3.1** In the Ptolemaic view of the heavens, Earth is at the center, orbited by the Moon, Mercury, Venus, the Sun, Mars, Jupiter, and Saturn.

When people in ancient times looked up at the sky, they saw that the Sun, Moon, and stars rose in the east and set in the west and appeared to be moving around Earth. The ancient peoples were aware of five planets (*planet* means "wandering star") because they moved in a generally eastward direction from one night to the next among the stars, whose positions were fixed on the celestial sphere. But they did not know that Earth was similar to these planets. A successful theory of how Earth and the planets move and how Earth fits in with its neighbors in the Solar System was the first step to understanding Earth's place in the universe. The history of the progression of ideas—from Earth at the center of all things to Earth as a tiny, insignificant rock—is a prime example of the self-correcting nature of science.

## The Geocentric Model

As you saw in the previous chapter, hard evidence of the motions of Earth was remarkably difficult to come by. Largely because the motion of Earth through space cannot be felt, early astronomers developed a **geocentric** (Earth-centered) model of the Solar System to explain what they observed in the sky. When people looked up at the sky, the Sun, Moon, planets, and stars appeared to be moving around Earth. Before the 17th century, most educated people believed that the Sun, the Moon, and the known planets all moved in circles around a stationary Earth. **Figure 3.1** illustrates Ptolemy's geocentric model.

However, the geocentric model did not account for all observations. Ancient astronomers knew that the planets would occasionally do something unusual. Most of the time, the planets have an eastward **prograde motion**, in which each night they move a little eastward compared to the background stars. But sometimes the ancients observed apparent **retrograde motion**, in which the planets appear to move westward for a period of time before resuming their normal eastward travel. This retrograde motion is shown for Mars in **Figure 3.2**.

This odd behavior of the five "naked eye" planets—Mercury, Venus, Mars, Jupiter, and Saturn—created a puzzling problem for the geocentric model as it was summarized in 150 CE by the Alexandrian astronomer Ptolemy (Claudius Ptolemaeus, 90–168 CE). Some astronomers and philosophers in ancient times, for example Aristarchus of Samos (310–230 BCE), hypothesized that the Sun might be the center of the Solar System, but they did not have the tools to test the hypothesis or the mathematical insight to formulate a more complete and testable model. But most other astronomers of the time were skeptical, because they thought that if Earth moved around the Sun, they should feel Earth's motion. Therefore they preferred the geocentric model, in which the Sun, Moon, and planets all moved in perfect circles around a stationary Earth, with the "fixed stars" being located somewhere way out beyond the planets.

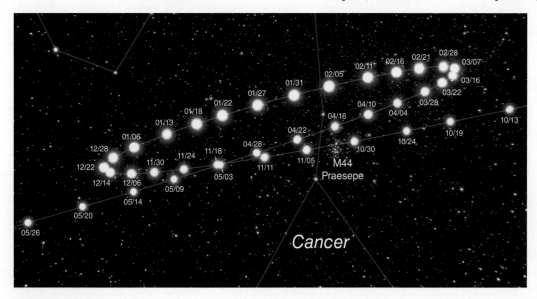

**Figure 3.2** This time-lapse photographic series shows Mars as it moves in apparent retrograde motion.

Because the geocentric model in its simplest form failed to explain retrograde motion of the planets, Ptolemy added an embellishment called an *epicycle*, a small circle superimposed on each planet's larger circle, as illustrated in **Figure 3.3**. In this model, as the planet travels along its larger circle around Earth, it is also moving along a smaller circle. When its motion along the smaller circle was in a direction opposite to that of the forward motion of the larger circle, its forward motion would be reversed. Ptolemy's model had many of these epicycles, but they made the model *work*, in the sense that this model was reasonably successful at predicting the positions of planets in the sky. For nearly 1,500 years, Ptolemy's model of the heavens was the accepted paradigm in the Western world.

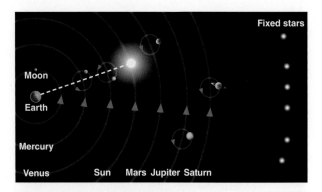

**Figure 3.3** To reconcile retrograde motion with the geocentric model of the Solar System, additional loops called epicycles were added to each planet's circular orbit around Earth.

## CHECK YOUR UNDERSTANDING 3.1

How did the ancients know the planets were different from the stars?

## Copernicus Proposes a Heliocentric Model

Nicolaus Copernicus (1473–1543—**Figure 3.4**) is famous for placing the Sun rather than Earth at the center of the Solar System. He was not the first person to consider the idea that Earth orbited the Sun, and probably he had read the ideas of ancient Greek and medieval Arab astronomers who considered putting the Sun at the center of the Solar System. However, he was the first to develop a comprehensive mathematical model that could be tested by later astronomers. This work was the beginning of the Copernican Revolution. Through the work of 16th- and 17th-century scientists such as Tycho Brahe, Galileo Galilei, Johannes Kepler, and Isaac Newton, the **heliocentric** (Sun-centered) theory of the Solar System became one of the best-corroborated theories in all of science.

**Figure 3.4** Nicolaus Copernicus rejected the ancient Greek model of an Earth-centered universe and replaced it with a model that centered on the Sun.

Copernicus was multilingual and highly educated: he studied philosophy, canon (Catholic) law, medicine, economics, mathematics, and astronomy in his native Poland and in Italy. Copernicus conducted astronomical observations from a small tower, and sometime around 1514 he started writing about heliocentricity. Eighteen years later, he completed his manuscript. He did not publish the book because he knew his ideas would be controversial: philosophical and religious views of the time held that humanity and thus Earth must be the center of the universe. Late in his life, Copernicus was finally persuaded to publish his ideas, and his great work *De revolutionibus orbium coelestium* ("On the Revolutions of the Heavenly Spheres") appeared in 1543, the year of his death. This work pointed the way toward the modern cosmological principle introduced in Chapter 1—the idea that our location in the universe is not special.

 **Nebraska Simulation:** Retrograde Motion

**Figure 3.5** shows Copernicus's model with the planets orbiting around the Sun. This model explained the observed motions of Earth, the Moon, and the planets, including retrograde motion, much more simply than the geocentric model did. Think about when you were in a car or train and you passed a slower-moving car or train, and it seemed as if the other vehicle was moving backward. It can be hard to tell which vehicle is moving and in which direction without an external frame of reference. Copernicus provided that frame of reference for the Sun and its planets.

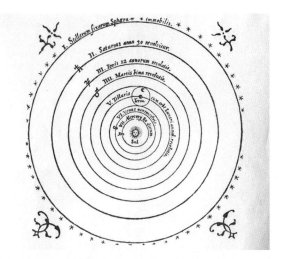

**Figure 3.5** This illustration shows the Copernican heliocentric view of the Solar System (II–VII) and the fixed stars (I). The Sun is at the center and is orbited by Mercury, Venus, Earth, Mars, Jupiter, and Saturn. The Moon orbits Earth.

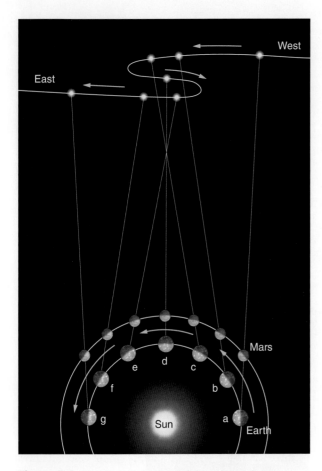

**Figure 3.6** The Copernican model explains the apparent retrograde motion of Mars (see Figure 3.1) as seen in Earth's sky when Earth passes Mars in its orbit. The figure is not to scale.

In the Copernican model, the planets farther from the Sun undergo apparent retrograde motion when Earth overtakes them in their orbits. **Figure 3.6** illustrates this for Mars. Conversely, the planets closer to the Sun than Earth is—Mercury and Venus—move in apparent retrograde motion when overtaking Earth. Except for the Sun, all Solar System objects exhibit apparent retrograde motion. The magnitude of the effect diminishes with increasing distance from Earth. Retrograde motion is an illusion caused by the relative motion between Earth and the other planets.

Copernicus still conceived of the planets as moving in perfectly circular orbits, and as a result he needed to use some epicycles to match the observations. His model made testable predictions of the location of each planet on a given night, which were at least as accurate, but not more accurate, than those of the geocentric model. But this heliocentric model was, overall, simpler than the geocentric model and became the basis for further refinements in understanding how Earth moved. As copies of *De Revolutionibus* and Copernicus's ideas slowly spread across Europe, other scientists were excited by the heliocentric model, and a scientific revolution began.

## Scaling the Solar System

Copernicus's model assumed that the planets traveled around the Sun in circular orbits with constant speeds. From his observations he deduced the correct order of the planets and concluded that planets closer to the Sun traveled faster than planets farther from the Sun. He also realized he needed to consider two categories of planets: **inferior** planets are closer to the Sun than Earth is; and **superior** planets are farther from the Sun than Earth is.

In Copernicus's model, periodically Earth, another planet, and the Sun line up in space to form either a straight line or a right triangle. As shown in **Figure 3.7a**, when a superior planet is in line with the Sun and Earth but on the other side of the Sun from Earth, we call the configuration a *conjunction*. A superior planet in conjunction will rise and set in the sky with the Sun. Note that when a superior planet is in conjunction, it is the farthest away from Earth that it gets and

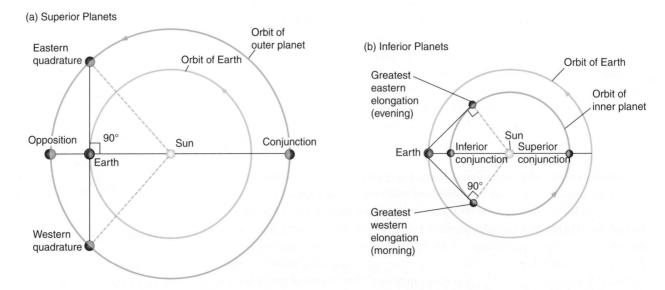

**Figure 3.7** The diagrams show planetary configurations for (a) superior (outer) planets and (b) inferior (inner) planets.

therefore is at its faintest. You won't see the planet at all exactly at conjunction, because it's on the far side of the Sun.

In contrast, when a superior planet is in line with the Sun and Earth on the same side of the Sun as Earth, we call the configuration an *opposition*. At opposition, the superior planet is "opposite" the Sun in the sky: like a full Moon, it rises when the Sun sets and sets when the Sun rises. When the superior planet is in opposition, it is the closest it gets to Earth and thus is at its brightest; therefore, opposition is the best time to observe the planet in the sky. Opposition is also the time when the planet exhibits retrograde motion, because that is exactly when Earth is overtaking the planet in its orbit. *Quadrature* is when Earth, the Sun, and a superior planet form a right triangle in space.

For an inferior planet, the configurations are slightly different (Figure 3.7b). If the inferior planet is between Earth and the Sun, we call the configuration an *inferior conjunction*: this is when the planet is closest to Earth. If the inferior planet is on the other side of the Sun from Earth, we call the configuration a *superior conjunction*: this is when the planet is farthest from Earth. When the inferior planet forms a right triangle with Earth and the Sun—and thus is the farthest it gets from the Sun in the sky—we call the configuration the *greatest elongation*. When you are standing on Earth and looking toward the inner planets, you always see them close to the Sun in the sky, so you see Mercury and Venus only within a few hours of sunrise or sunset. The best time to observe these planets is at greatest elongation because they will have the greatest separation from the Sun in the sky.

Copernicus used the geometry of these alignments along with his observations of the positions of the planets in the sky, including their altitudes and the times they rose and set, to estimate the planet–Sun distances in terms of the Earth–Sun distance. He realized there were two types of orbital periods. Similar to the terms used for lunar orbits, a planet's *sidereal period* is how long it takes the planet to make one orbit around the Sun with respect to the stars and return to the same point in space. A planet's *synodic period* is how long it takes the planet to return to the same configuration with the Sun and Earth, such as inferior conjunction to inferior conjunction or opposition to opposition. The synodic period is what can be observed from Earth; for example, from opposition to opposition.

As shown in **Figure 3.8a**, Earth and the superior planet are in opposition at point A. Superior planets move around the Sun more slowly than Earth does, so Earth will complete one orbit around the Sun and then catch up to the superior planet to form the next opposition at point B. In Figure 3.8b, Earth and the inferior planet are in inferior conjunction at point A. An inferior planet moves around the Sun faster than Earth does, so it completes one sidereal period and then must continue in its orbit to catch up to Earth for the next inferior conjunction at point B. The numerical details are shown in **Working It Out 3.1**. **Table 3.1** shows that these relative distances calculated by Copernicus are remarkably close to distances obtained by modern methods. Copernicus's model not only predicted planetary positions in the sky but also could be used to compare the distances between the planets and the Sun accurately and thus to set the scale of the Solar System.

## CHECK YOUR UNDERSTANDING 3.2

The planet Uranus will be observed in retrograde motion when: (a) Uranus is closest to the Sun; (b) Uranus is farthest from the Sun; (c) Earth overtakes Uranus in its orbit; (d) Uranus overtakes Earth in its orbit.

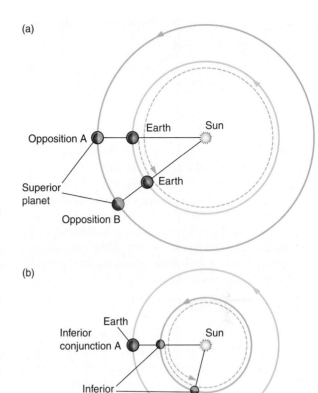

**Figure 3.8** The synodic periods of planets indicate how long it takes them to return to the same configuration with Earth and the Sun. (a) Earth completes one orbit around the Sun first and then catches up to the superior planet. (b) Inferior planets complete a full orbit around the Sun first and then catch up to Earth.

 **Nebraska Simulation:** Planetary Configurations Simulator

| TABLE 3.1 | Copernicus's Scale of the Solar System | |
|---|---|---|
| Planet | Copernicus's Value (AU) | Modern Value (AU) |
| Mercury | 0.38 | 0.39 |
| Venus | 0.72 | 0.72 |
| Earth | 1.00 | 1.00 |
| Mars | 1.52 | 1.52 |
| Jupiter | 5.22 | 5.20 |
| Saturn | 9.17 | 9.58 |

AU = astronomical unit.

## Orbital Periods

Copernicus was able to calculate the sidereal period from observations of the synodic period. He didn't know the actual Earth–Sun distance in miles or kilometers, so he let the Earth–Sun distance equal 1. We call this distance 1 astronomical unit (AU). Let $P$ be the sidereal period and $S$ the synodic period of a planet. $E$ is the sidereal period of Earth, which equals 1 year, or 365.25 days. By thinking about the distance that Earth and the planet move in one synodic period, and noting that an inferior planet orbits the Sun in less time than Earth does, it can be shown that

$$\frac{1}{P} = \frac{1}{E} + \frac{1}{S}$$

for an inferior planet, with $P$, $E$, and $S$ all in the same units of days or years. Similarly, Earth orbits the Sun in less time than a superior planet does, so the planet has traveled only part of its orbit around the Sun after 1 Earth year. The equation for a superior planet is

$$\frac{1}{P} = \frac{1}{E} - \frac{1}{S}$$

Let's consider Saturn as an example. The time that passes between one opposition—the date of maximum brightness—and the next shows that Saturn's synodic period is 378 days, or $378 \div 365.25 = 1.035$ years. Then to compute its sidereal period, $P$ of Saturn in years, we use $S = 1.035$ and $E = 1$:

$$\frac{1}{P} = \frac{1}{1\,\text{yr}} - \frac{1}{1.035\,\text{yr}}$$

$$= 1 - 0.966 = 0.034\,\text{yr}^{-1}$$

and thus

$$P = \frac{1}{0.034\,\text{yr}^{-1}} = 29.4\,\text{yr}$$

The sidereal period of Saturn is 29.4 years, meaning that it takes Saturn 29.4 years to travel around the Sun and return to where it started in space.

 **Nebraska Simulation:** Synodic Period Calculator

## Scaling the Solar System

Copernicus used the configurations of the planets shown in Figure 3.7 along with the sidereal periods of the planets to compute the relative distances of the planets. For the superior planets, he measured the fraction of the circular orbit that the planet completed in the time between opposition and quadrature, and then used trigonometry to solve for the planet–Sun distance in astronomical units (see Figure 3.7a). For the inferior planets, he had a right triangle at the point of greatest elongation, and then he used right-triangle trigonometry to solve for the planet–Sun distance in astronomical units (see Figure 3.7b). Copernicus's values are impressively similar to modern values (see Table 3.1). Copernicus still did not know the actual value of the astronomical unit in miles, but he was able to compute accurately the relative distances of the planets from the Sun for the first time.

# 3.2 Kepler's Laws Describe Planetary Motion

Copernicus did not understand *why* the planets move about the Sun, but he realized that his heliocentric picture provided a way to compute the relative distances of the planets. His theory is an example of **empirical science**, which seeks to describe patterns in nature with as much accuracy as possible. Copernicus's work was revolutionary because he was able to challenge the accepted geocentric model and propose that Earth is one planet among many. His conclusions paved the way for other great empiricists, including Tycho Brahe and Johannes Kepler.

## Tycho Brahe's Observations

Tycho Brahe (1546–1601—**Figure 3.9**) was a Danish astronomer of noble birth who entered university at age 13 to study philosophy and law. After seeing a partial solar eclipse in 1560, Tycho (conventionally referred to by his first name) became interested in astronomy. A few years later, he observed Jupiter and Saturn near each other in the sky, but not in the exact positions predicted by the astro-

**Figure 3.9** Tycho Brahe, known commonly as Tycho, was one of the greatest astronomical observers before the invention of the telescope.

nomical tables based on Ptolemy's model. Tycho gave up studying law and devoted himself to making better tables of the positions of the planets in the sky.

The king of Denmark granted Tycho the island of Hven, located between Sweden and Denmark, to build an observatory. Tycho designed and built new instruments, operated a printing press, and taught students and others how to conduct observations. With the assistance of his sister Sophie, Tycho carefully measured the precise positions of planets in the sky over several decades, developing the most comprehensive set of planetary data available at that time. He created his own geo-heliocentric model, shown in **Figure 3.10**. In Tycho's model, the planets orbit the Sun, and the Sun and planets orbit Earth. This model gained limited acceptance among people who preferred to keep Earth at the center for philosophical or religious reasons. Tycho lost his financial support when the king died, and in 1600 he relocated to Prague.

## Kepler's Laws

In 1600, Tycho hired a more mathematically inclined astronomer, Johannes Kepler (1571–1630—**Figure 3.11**), as his assistant. Kepler, who had studied the ideas of Copernicus, was responsible for the next major step toward understanding the motions of the planets. Upon Tycho's death, Kepler inherited the records of his observations. Working first with Tycho's observations of Mars, Kepler deduced three empirical rules, now generally referred to as **Kepler's laws**, which accurately describe the motions of the planets. These laws are empirical: they use prior data to make predictions about future behavior but do not include an underlying theory of why the objects behave as they do.

**Kepler's First Law** When Kepler compared Tycho's extensive planetary observations with predictions from Copernicus's heliocentric model, he expected the data to confirm circular orbits for planets orbiting the Sun. Instead he found disagreements between his predictions and Tycho's observations. He was not the first to notice such discrepancies. Rather than discarding Copernicus's model, Kepler made some revisions.

Kepler discovered that if he replaced Copernicus's circular orbits with *elliptical* orbits, he could predict the positions of planets for any day, and found that his predictions fit Tycho's observations almost perfectly. An **ellipse** is a shape that looks like an elongated circle. It is symmetric from right to left and from top to bottom. As shown in **Figure 3.12a**, you can draw an ellipse by attaching the two

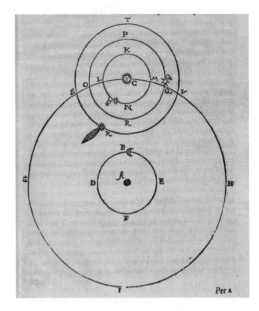

**Figure 3.10** This reproduction shows Tycho's geo-heliocentric model with the Moon and Sun orbiting Earth, and the other planets orbiting the Sun.

**Figure 3.11** Johannes Kepler explained the motions of the planets with three empirically based laws.

▶❚❚ **AstroTour:** Kepler's Laws

🖐 **Nebraska Simulation:** Eccentricity Demonstrator

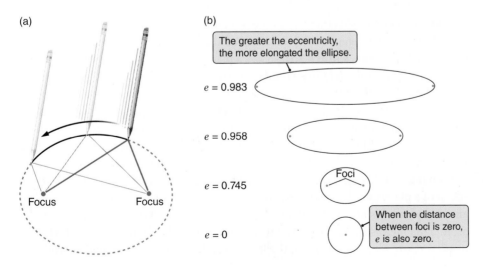

(a)

Focus    Focus

(b)

The greater the eccentricity, the more elongated the ellipse.

e = 0.983

e = 0.958

e = 0.745

Foci

e = 0

When the distance between foci is zero, e is also zero.

**Figure 3.12** (a) We can draw an ellipse by attaching a length of string to a piece of paper at two points (called foci) and then pulling the string around as shown. (b) Ellipses range from circles to elongated eccentric shapes. e = eccentricity.

## Kepler's First Law

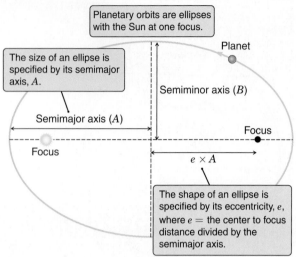

Planetary orbits are ellipses with the Sun at one focus.

The size of an ellipse is specified by its semimajor axis, $A$.

Planet

Semiminor axis ($B$)

Semimajor axis ($A$)

Focus

Focus

$e \times A$

The shape of an ellipse is specified by its eccentricity, $e$, where $e$ = the center to focus distance divided by the semimajor axis.

**Figure 3.13** Planets move on elliptical orbits with the Sun at one focus. The eccentricity is given by the center-to-focus distance divided by the semimajor axis.

(a)

1 AU

Sun

$e = 0.017$

Earth's orbit is only slightly eccentric.

Circular orbit with same semimajor axis, centered on Sun

(b)

39.2 AU

Sun

$e = 0.249$

Pluto's eccentric orbit is both noticeably elongated and noticeably offset as compared with a circle centered on the Sun.

Circular orbit with same semimajor axis, centered on Sun

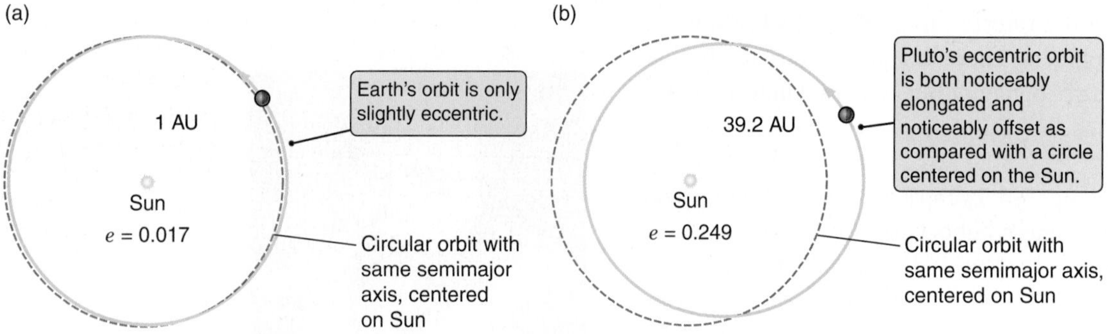

**Figure 3.14** The orbits of (a) Earth and (b) Pluto compared with circles around the Sun. $e$ = eccentricity.

## Kepler's Second Law

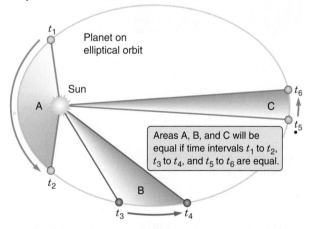

$t_1$

Planet on elliptical orbit

Sun

A

C

$t_6$

$t_5$

Areas A, B, and C will be equal if time intervals $t_1$ to $t_2$, $t_3$ to $t_4$, and $t_5$ to $t_6$ are equal.

$t_2$

B

$t_3$

$t_4$

**Figure 3.15** An imaginary line between a planet and the Sun sweeps out an area as the planet orbits. If the three intervals of time shown are equal, then the three areas A, B, and C will be the same.

ends of a piece of string to a piece of paper, stretching the string tight with the tip of a pencil, and then drawing around those two points while keeping the string tight. Each of the points at which the string is attached is a **focus** (plural: *foci*) of the ellipse. In Figure 3.12b, you can see that as the two foci become closer together, the ellipse becomes more circular. As the two foci move farther apart, the ellipse becomes more and more elongated. The **eccentricity (*e*)** of an ellipse measures this elongation; it is determined by the separation between the two foci divided by the length of the long axis. A circle has an eccentricity of 0 because the two foci coincide at the center of the circle. The more elongated the ellipse becomes, the closer its eccentricity gets to 1.

**Kepler's first law** of planetary motion states that the orbit of each planet is an ellipse with the Sun located at one focus (see the **Process of Science Figure**). There is nothing but empty space at the other focus. **Figure 3.13** illustrates Kepler's first law and shows how the features of an ellipse match observed planetary motions. The dashed lines in Figure 3.13 represent the two main axes of the ellipse. Half of the length of the long axis (the major axis) of the ellipse is called the **semimajor axis**, often denoted by the letter $A$. The semimajor axis of an orbit is the average distance between one focus and the ellipse itself. The average distance between the Sun and a planet equals the semimajor axis of the planet's orbit.

The eccentricities of planetary orbits vary widely, but most planetary objects in our Solar System have nearly circular orbits. As shown in **Figure 3.14a**, Earth's orbit is very nearly a circle centered on the Sun; with an eccentricity of 0.017, the distance variation is small. In contrast, dwarf planet Pluto's orbit, as shown in Figure 3.14b, has an eccentricity of 0.249. The orbit is noticeably elongated, with the Sun offset from center.

**Kepler's Second Law**  From a close analysis of Tycho's observational data of changes in the positions of the planets, Kepler found that a planet moves fastest when it is closest to the Sun and slowest when it is farthest from the Sun. For example, we now measure Earth's average speed in its orbit around the Sun at 29.8 kilometers per second (km/s). When Earth is closest to the Sun, it travels at 30.3 km/s. When it is farthest from the Sun, it travels at 29.3 km/s.

Kepler found an elegant way to describe the changing speed of a planet in its orbit around the Sun. **Figure 3.15** shows a planet at six different points in its orbit ($t_1$ to $t_6$). Imagine a straight line connecting the Sun with this planet. We can think of this line as "sweeping out" an area as it moves with the planet from one point to another. Area A (in orange) is swept out between times $t_1$ and $t_2$, area B (in blue) is swept out between times $t_3$ and $t_4$, and area C (in green) is swept out between times $t_5$ and $t_6$. When the planet is closest to the Sun (area A), it is moving the fastest, but the distance between the planet and the Sun is small. Kepler realized that changes in the distance between the Sun and a planet and changes in the speed of a planet work together so that the area swept out by a planet in the same amount of time is always the same, regardless of the location of the planet in its orbit. This means that if the three time intervals in the figure are equal (that is, $t_1 \rightarrow t_2 = t_3 \rightarrow t_4 = t_5 \rightarrow t_6$), then the three areas A, B, and C will be equal as well.

Early astronomers studied the motions of the planets but did not understand why they behave as they do.

**The Big Idea:**
Copernicus proposes that planets move in circular heliocentric orbits, with epicycles.

**The Observation:**
Tycho observes and collects lots of data about planet positions.

**The Prediction:**
Kepler uses Copernicus's model to predict where planets should be.

**The Test:**
Kepler compares predictions with data— they disagree. Copernicus's idea is falsified!

**The New Big Idea:**
Planet orbits are not circular. They are elliptical.

**Model gains acceptance with a physical understanding of Newton's laws.**

Scientific theories must be, in principle, falsifiable. Disproving an old theory always leads to deeper understanding.

**Kepler's second law**, also called Kepler's **law of equal areas**, states that the imaginary line connecting a planet to the Sun sweeps out equal areas in equal times, regardless of where the planet is in its orbit. This law applies to only one planet at a time. The area swept out by Earth in a given time is always the same. Likewise, the area swept out by Mars in a given time is always the same. But the area swept out by Earth and the area swept out by Mars in a given time are *not* the same. This law can be used to find the speed of a planet anywhere in its orbit.

## CHECK YOUR UNDERSTANDING 3.3

Kepler's second law says that if a planet is in an elliptical orbit around a star, then the planet moves fastest when the planet is: (a) farthest from the star; (b) closest to the star; (c) located at one of the foci; (d) closest to another planet.

**Kepler's Third Law** Kepler looked for patterns in the orbital periods of the planets. He found that compared to planets closer to the Sun, planets farther from the Sun travel on longer orbits and they move more slowly in their orbits around the Sun. Kepler discovered a mathematical relationship between the sidereal period of a planet's orbit—how many years it takes to go around the Sun and return to the same position in space—and its average distance from the Sun in astronomical units. **Kepler's third law** states that in these units, the square of the sidereal period ($P$) of a planet's orbit is equal to the cube of the semimajor axis ($A$) of the planet's orbit.

**Table 3.2** shows the periods and semimajor axes of the orbits of the eight classical planets and three of the dwarf planets, along with the values of the ratio $P^2$ divided by $A^3$. These data are also plotted in **Figure 3.16**. Kepler referred to this relationship as his **harmonic law** or, more poetically, as the "Harmony of the Worlds." Kepler's third law is explored further in **Working It Out 3.2**. Kepler's laws enhanced the heliocentric mathematical model of Copernicus and led to its greater acceptance.

 **Nebraska Simulation:** Planetary Orbit Simulator

---

## 3.2 | Working It Out | Kepler's Third Law

Kepler's third law states that the square of the period of a planet's orbit, measured in years $P_{\text{years}}$, is equal to the cube of the semimajor axis of the planet's orbit, measured in astronomical units $A_{\text{AU}}$. Translated into math, the law says

$$(P_{\text{years}})^2 = (A_{\text{AU}})^3$$

Kepler used units based on Earth—astronomical units and years—as a matter of convenience. If other units were used, then $P^2$ would still be proportional to $A^3$, but the constant of proportionality would not be 1.

As an example of using this law, suppose that you want to know the average radius of Neptune's orbit in astronomical units. First you need to find out how long Neptune's period is in Earth years, which

you can determine by observing the synodic period and then computing its sidereal period from that (using Working It Out 3.1). Neptune's sidereal period is 165 years. Plugging this number into Kepler's third law gives this result:

$$(P_{\text{years}})^2 = (165)^2 = 27{,}225 = (A_{\text{AU}})^3$$

To solve this equation, you must first square 165 to get 27,225 and then take its cube root (see Appendix 1 for calculator hints). Then

$$A_{\text{AU}} = \sqrt[3]{27{,}225} = 30.1$$

The semimajor axis of Neptune's orbit, that is, the average distance between Neptune and the Sun, is 30.1 AU.

**TABLE 3.2** Kepler's Third Law: $P^2 = A^3$

| Planet | Period $P$ (years) | Semimajor Axis $A$ (AU) | $\dfrac{P^2}{A^3}$ |
|---|---|---|---|
| Mercury | 0.241 | 0.387 | $\dfrac{0.241^2}{0.387^3} = 1.00$ |
| Venus | 0.615 | 0.723 | $\dfrac{0.615^2}{0.723^3} = 1.00$ |
| Earth | 1.000 | 1.000 | $\dfrac{1.000^2}{1.000^3} = 1.00$ |
| Mars | 1.881 | 1.524 | $\dfrac{1.881^2}{1.524^3} = 1.00$ |
| Ceres | 4.599 | 2.765 | $\dfrac{4.559^2}{2.765^3} = 1.00$ |
| Jupiter | 11.86 | 5.204 | $\dfrac{11.86^2}{5.204^3} = 1.00$ |
| Saturn | 29.46 | 9.582 | $\dfrac{29.46^2}{9.582^3} = 0.99^*$ |
| Uranus | 84.01 | 19.201 | $\dfrac{84.01^2}{19.201^3} = 1.00$ |
| Neptune | 164.79 | 30.047 | $\dfrac{164.79^2}{30.047^3} = 1.00$ |
| Pluto | 247.68 | 39.236 | $\dfrac{247.68^2}{39.236^3} = 1.02^*$ |
| Eris | 557.00 | 67.696 | $\dfrac{557.00^2}{67.696^3} = 1.00$ |

*Slight perturbations from the gravity of other planets are the reason that these ratios are not exactly 1.00.

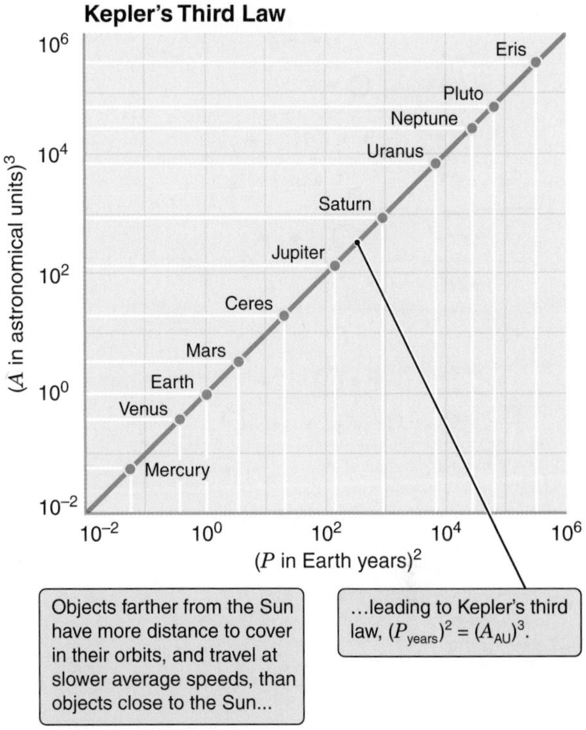

**Kepler's Third Law**

Objects farther from the Sun have more distance to cover in their orbits, and travel at slower average speeds, than objects close to the Sun...

...leading to Kepler's third law, $(P_{years})^2 = (A_{AU})^3$.

**Figure 3.16** A plot of $A^3$ versus $P^2$ for objects in our Solar System shows that they obey Kepler's third law. (Note that by plotting powers of 10 on each axis, we are able to fit both large and small values on the same plot. We will do this frequently.)

## CHECK YOUR UNDERSTANDING 3.4

Place the following in order from largest to smallest semimajor axis: (a) a planet with a period of 84 Earth days; (b) a planet with a period of 1 Earth year; (c) a planet with a period of 2 Earth years; (d) a planet with a period of 0.5 Earth year.

# 3.3 Galileo's Observations Supported the Heliocentric Model

Galileo Galilei (1564–1642—**Figure 3.17**)—one of the heroes of astronomy—was the first to use a telescope to conduct and report on significant discoveries about astronomical objects. Galileo's telescopes were relatively small, yet sufficient for him to observe spots on the Sun, the uneven surface and craters of the Moon, and the large number of stars in the band of light in the sky called the Milky Way.

**Figure 3.17** Galileo Galilei laid the physical framework for Newton's laws.

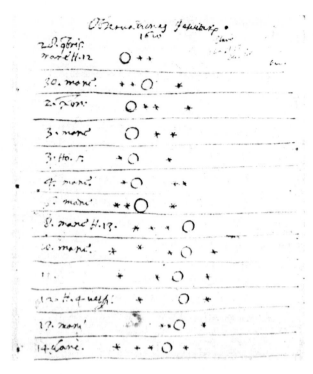

**Figure 3.18** This page from Galileo's notebook shows his observations of the four largest moons of Jupiter.

 **Nebraska Simulations:** Phases of Venus; Ptolemaic Phases of Venus

 **AstroTour:** Velocity, Acceleration, Inertia

**Figure 3.19** Modern photographs of the phases of Venus show that when we see Venus more illuminated, it also appears smaller, implying that Venus is farther away at that time.

## Galileo's Observations

Galileo provided the first observational evidence that some objects in the sky do not orbit Earth. When Galileo turned his telescope to the planet Jupiter, he observed four "stars" in a line near the planet. Over time he observed that the objects changed position from night to night (**Figure 3.18**). Galileo hypothesized that these objects were moons in orbit around Jupiter. These four moons are the largest of Jupiter's many moons and are still called the Galilean moons. Galileo also estimated the relative distance of each moon from Jupiter and the periods of their orbits, and he was able to show that they followed Kepler's third law.

Galileo also observed that the planet Venus went through phases like the Moon. He noticed that the phases of Venus were correlated with the size of the image of Venus in his telescope. In a geocentric model in which Venus orbits Earth like the Moon does, the apparent size of Venus would be constant, like we see for the Moon. In the heliocentric model, the Earth–Venus distance varies, and the size of Venus changes accordingly. When Venus is in its gibbous to full phases, it is farther away, on the other side of the Sun than the side Earth is on, and it is smaller in the sky. When Venus is in its crescent to new phases, it is closer, on the same side of the Sun as the side Earth is on, and it is larger in the sky (**Figure 3.19**). The observations of Jupiter's moons and the phases of Venus in particular convinced Galileo that Copernicus was correct in placing the Sun at the center of the Solar System.

In addition to his astronomical observations, Galileo did important work on the motion of objects. Unlike natural philosophers, who thought about objects in motion but did not actually experiment with them, Galileo conducted experiments with falling and rolling objects. As with his telescopes, Galileo improved or developed new technology to enable him to conduct these experiments. For example, by carefully rolling balls down an inclined plane and by dropping various objects from a height, he found that the distance traveled by a falling object is proportional to the square of the time it has been falling. If he simultaneously dropped two objects of different masses, they reached the ground at the same time, demonstrating that all objects falling to Earth accelerate at the same rate, independent of their mass.

Galileo's observations and experiments with many types of moving objects, such as carts and balls, led him to disagree with the Greek philosophers about when and why objects continue to move or come to rest. Prior to Galileo, it was thought that the natural state of an object was to be at rest. But Galileo found that the natural state of an object is to keep doing what it was doing until a force acts on it. That is, an object in motion continues moving along a straight line with a constant speed until a force acts on it to change its state of motion. This idea of *inertia*, which was later adopted by Newton as his first law of motion, has implications for not only the motion of carts and balls but also the orbits of planets.

## Dialogue Concerning the Two Chief World Systems

Much has been written about the considerable danger Galileo faced because of his work. His later life was consumed by conflict with the Catholic Church over his support of the Copernican system. In 1632, Galileo published his best-selling book, *Dialogo sopra i due massimi sistemi del mondo* ("Dialogue Concerning the

Two Chief World Systems"). The *Dialogo* presents a brilliant philosopher named Salviati as the champion of the Copernican heliocentric view of the universe. The defender of an Earth-centered universe, Simplicio—who uses arguments made by the classical Greek philosophers and the pope—sounds silly and ignorant.

Galileo, a religious man who had two daughters in a convent, thought he had the tacit approval of the Catholic Church for his book. But when he placed a number of the pope's geocentric arguments in the unflattering mouth of Simplicio, the perceived attack on the pope attracted the attention of the church. Galileo was put on trial for heresy, sentenced to prison, and eventually placed under house arrest. To escape a harsher sentence, Galileo was forced publicly to recant his belief in the Copernican theory that Earth moves around the Sun. According to one story, as he left the courtroom, Galileo stamped his foot on the ground and muttered, "And yet it moves!"

The *Dialogo* was placed on the pope's Index of Prohibited Works, along with Copernicus's *De Revolutionibus*, but it traveled across Europe, was translated into other languages, and was read by other scientists. (Two centuries later, in 1835, the church finally removed the uncensored version of the *Dialogo* from its prohibited list.) Galileo spent his final years compiling his research on inertia and other ideas into the book *Discourses and Mathematical Demonstrations Relating to Two New Sciences*, which was published in 1638 in Holland, outside the jurisdiction of the Catholic Church.

---

**CHECK YOUR UNDERSTANDING 3.5**

Which of Galileo's astronomical observations were best explained by a heliocentric model? (a) sunspots; (b) craters on the Moon; (c) moons of Jupiter; (d) phases of Venus

---

## 3.4 Newton's Three Laws Help to Explain the Motion of Celestial Bodies

Empirical laws, like Kepler's laws, describe what happens, but they do not explain why. Kepler described the orbits of planets as ellipses, but he did not explain why they should be so. To take that next step in the scientific process, scientists use basic physical principles and the tools of mathematics to derive the empirically determined laws. Alternatively, a scientist might start with physical laws and predict relationships, which are then verified or falsified by experiment and observation. If these predictions are verified by experiment and observation, the scientist may have determined something fundamental about how the universe works.

Sir Isaac Newton (1642–1727—**Figure 3.20**) took this next step in explaining the nature of motion. Newton was a student of mathematics at Cambridge University when it closed down because of the Great Plague and students were sent home to the safer countryside. Over the next 2 years, he continued to study on his own, and at the age of 23 he invented calculus, which would become crucial to his development of the physics of motion. The German mathematician Gottfried Leibniz independently developed calculus around the same time.

Building on the work of Kepler, Galileo, and others, Newton proposed three physical laws that govern the motions of all objects in the sky and on Earth. In this section, we will examine these three laws, which are essential to an understanding of the motions of the planets and all other celestial bodies.

**Figure 3.20** Sir Isaac Newton formulated three laws of motion.

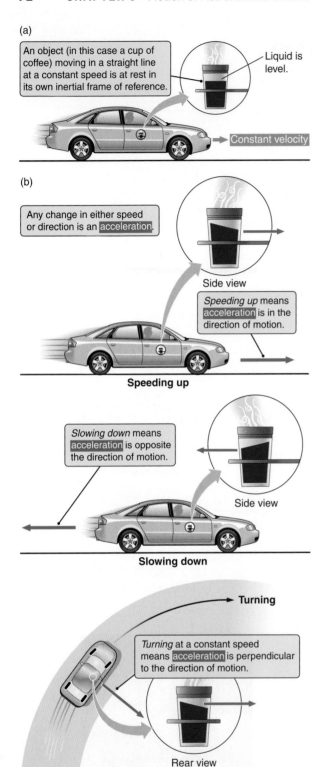

(a)

An object (in this case a cup of coffee) moving in a straight line at a constant speed is at rest in its own inertial frame of reference.

Liquid is level.

Constant velocity

(b)

Any change in either speed or direction is an acceleration.

Side view

*Speeding up* means acceleration is in the direction of motion.

**Speeding up**

*Slowing down* means acceleration is opposite the direction of motion.

Side view

**Slowing down**

Turning

*Turning* at a constant speed means acceleration is perpendicular to the direction of motion.

Rear view

**Figure 3.21** (a) An object moving in a straight line at a constant speed is at rest in its own inertial frame of reference. (b) Any change in the velocity of an object is an acceleration. When you are driving, for example, any time your speed changes or you follow a curve in the road, you are experiencing an acceleration. (Throughout the text, velocity arrows will be shown as red, and acceleration arrows will be shown as green.)

## Newton's First Law: Objects at Rest Stay at Rest; Objects in Motion Stay in Motion

A **force** is a push or a pull on an object. It is possible for two or more forces to oppose one another in such a way that they are perfectly balanced and cancel out. For example, gravity pulls down on you as you sit in your chair. But the chair pushes up on you with an exactly equal and opposite force. So you remain motionless. Forces that cancel out have no effect on an object's motion. When forces add together to produce an effect, we often use the term *net force*, or sometimes just *force*.

Imagine that you are driving a car, and your phone is on the seat next to you. A rabbit runs across the road in front of you, and you press the brakes hard. You feel the seat belt tighten to restrain you. At the same time, your phone flies off the seat and hits the dashboard. You have just experienced what Newton describes in his first law of motion. **Inertia** is the tendency of an object to maintain its state—either of motion or of rest—until it is pushed or pulled by a net force. In the case of the stopping car, you did not hit the dashboard because the force of the seat belt slowed you down. The phone did hit the dashboard because no such force acted upon it.

**Newton's first law of motion** describes inertia and states that an object in motion tends to stay in motion, in the same direction, until a net force acts upon it; and an object at rest tends to stay at rest until a net force acts upon it. Galileo's law of inertia became the cornerstone of physics as Newton's first law.

Recall from Section 2.1 of Chapter 2 the concept of a frame of reference. Within a frame of reference, only the relative motions between objects have any meaning. Without external clues, you cannot tell the difference between sitting still and traveling at constant speed in a straight line. For example, if you close your eyes while riding in the passenger seat of a quiet car on a smooth road, you would feel as though you were sitting still. Returning to the earlier example, your phone was "at rest" beside you on the front seat of your car, but a person standing by the side of the road would see the phone moving past at the same speed as the car. People in a car approaching you would see the phone moving quite fast—at the speed they are traveling plus the speed you are traveling! All of these perspectives are equally valid, and all of these speeds of the phone are correct when measured in the appropriate reference frame.

A reference frame moving in a straight line at a constant speed is an **inertial frame of reference**. Any inertial frame of reference is as good as another. As illustrated in **Figure 3.21a**, in the inertial frame of reference of a cup of coffee, the cup is at rest in its own frame even if the car is moving quickly down the road.

## Newton's Second Law: Motion Is Changed by Forces

What if a net force does act? In the earlier example, you were traveling in the car, and your motion was slowed when the force of the seat belt acted upon you. Forces change an object's motion—by changing either the speed or the direction. This reflects **Newton's second law of motion**: if a net force acts on an object, then the object's motion changes.

As an example of changes in an object's motion, think about a car. When you are in the driver's seat of a car, you have a number of controls, including a gas pedal and a brake pedal. You use these to make the car speed up or slow down. A *change in speed* is one way the car's motion can change. But you also have the steering wheel in your hands. When you are moving down the road and you turn the wheel, your speed does not necessarily change, but the direction of your motion does. A *change in direction* is also a kind of change in motion.

Together, the combined speed and direction of an object's motion are called the object's **velocity**. "Traveling at 50 kilometers per hour (km/h)" indicates speed; "traveling north at 50 km/h" indicates velocity. The rate at which the velocity of an object changes is called **acceleration**. Acceleration tells you how rapidly a change in velocity happens. For example, if you go from 0 to 100 km/h in 4 seconds, you feel a strong push from the seat back as it shoves your body forward, causing you to accelerate along with the car. However, if you take 2 minutes to go from 0 to 100 km/h, the acceleration is so slight that you hardly notice it.

Partly because the gas pedal on a car is often called the accelerator, some people think *acceleration* always means that an object is speeding up. But we need to stress that, as used in physics, any change in speed or direction is an acceleration. Figure 3.21b illustrates the point by showing what happens to the coffee in a coffee cup as the car speeds up, slows down, or turns. Slamming on your brakes and going from 100 to 0 km/h in 4 seconds is just as much acceleration as going from 0 to 100 km/h in 4 seconds. Similarly, the acceleration you experience as you go through a fast, tight turn at a constant speed is every bit as real as the acceleration you feel when you slam your foot on the gas pedal or the brake pedal. Speeding up, slowing down, turning left, turning right—if you are not moving in a straight line at a constant speed, you are experiencing an acceleration.

Newton's second law of motion says that a net force causes acceleration. The acceleration an object experiences depends on two things. First, as shown in **Figure 3.22**, the acceleration depends on the strength of the net force acting on the object to change its motion. If the forces acting on the object do *not* add up to zero, then there is a net force and the object accelerates (Figure 3.22a). The stronger the net force, the greater the acceleration. Push on something twice as hard and it experiences twice as much acceleration (Figure 3.22b). Push on something 3 times as hard and its acceleration is 3 times as great. The resulting change in motion occurs in the direction the net force points. Push an object away from you, and it will accelerate away from you.

The acceleration that an object experiences also depends on its inertia. You can push some objects easily, for example, an empty box from a refrigerator delivery. However, the actual refrigerator, even though it is about the same size, is not easily shoved around. The greater the mass, the greater the inertia, and the *less* acceleration will occur in response to the same net force, as shown in Figure 3.22c. This relationship among acceleration, force, and mass is expressed mathematically in **Working It Out 3.3**.

## Newton's Third Law: Whatever Is Pushed, Pushes Back

Imagine that you are standing on a skateboard and pushing yourself along with your foot. Each shove of your foot against the ground sends you faster along your way. But why does this happen? Your muscles flex, and your foot exerts a force on the ground. (Earth itself does not noticeably accelerate, because its great mass gives it great inertia.) Yet this does not explain why *you* experience an acceleration. The fact that you accelerate means that as you push on the ground, the ground must be pushing back on you.

Part of Newton's genius was his ability to see patterns in such everyday events. Newton realized that *every* time one object exerts a force on another, the second

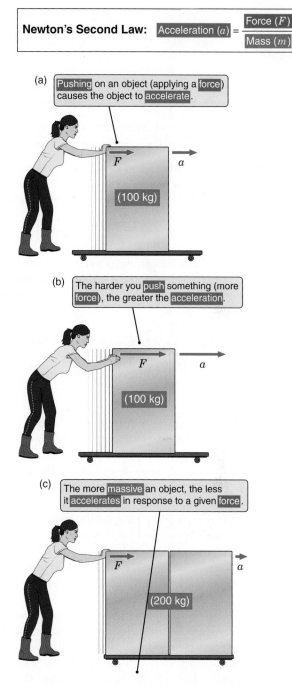

**Newton's Second Law:** Acceleration ($a$) = $\dfrac{\text{Force } (F)}{\text{Mass } (m)}$

(a) Pushing on an object (applying a force) causes the object to accelerate.

$F$    $a$    (100 kg)

(b) The harder you push something (more force), the greater the acceleration.

$F$    $a$    (100 kg)

(c) The more massive an object, the less it accelerates in response to a given force.

$F$    $a$    (200 kg)

**Figure 3.22** Newton's second law of motion says that the acceleration experienced by an object is determined by the force acting on the object divided by the object's mass. (Throughout the text, force arrows will be shown as blue.)

**Astronomy in Action:** Velocity, Force, and Acceleration

▶❙ **AstroTour:** Velocity, Acceleration, Inertia

## 3.3 Working It Out Using Newton's Laws

Your acceleration is determined by how much your velocity changes, divided by how long it takes for that change to happen:

$$\text{Acceleration} = \frac{\text{How much velocity changes}}{\text{How long the change takes to happen}}$$

For example, if an object's speed goes from 5 to 15 meters per second (m/s), then the change in velocity is 10 m/s. If that change happens over the course of 2 seconds, then the acceleration is given by

$$a = \frac{15\,\text{m/s} - 5\,\text{m/s}}{2\,\text{s}} = 5\,\text{m/s}^2$$

If we want to know how an object's motion is changing, we need to know two things: what net force is acting on the object, and what is the resistance of the object to that force? We can put the idea into equation form as follows:

$$\begin{pmatrix} \text{The} \\ \text{acceleration} \\ \text{experienced} \\ \text{by an object} \end{pmatrix} = \frac{\text{The force acting to change the object's motion}}{\text{The object's resistance to that change}} = \frac{\text{Force}}{\text{Mass}}$$

Newton's second law above is often written as Force = mass × acceleration, or $F = ma$. The units of force are called **newtons (N)**, so that $1\,\text{N} = 1\,\text{kg m/s}^2$.

As a simple example, suppose you are holding two blocks of the same size. The block in your right hand has twice the mass of the block in your left hand. When you drop the blocks, they both fall with the same acceleration, as shown by Galileo, and they hit your two feet at the same time. Which will hit with more force: the block falling onto your right foot or the one falling onto your left foot? The block in your right hand, with twice the mass, will hit your right foot with twice the force that the other block hits your left foot.

To see how Newton's three laws of motion work together, study **Figure 3.23**. An astronaut is adrift in space, motionless with respect to the nearby space shuttle. With no tether to pull on, how can the astronaut get back to the ship? Suppose the 100-kg astronaut throws a 1-kg wrench directly away from the shuttle at a speed of 10 m/s. Newton's second law says that in order to cause the motion of the wrench to change, the astronaut has to apply a force to it in the direction away from the shuttle. Newton's third law says that the wrench must therefore push back on the astronaut with as much force but in the opposite direction. The force of the wrench on the astronaut causes the astronaut to begin drifting toward the shuttle. How fast will the astronaut move? Turn to Newton's second law again. Because the astronaut has more mass, she will accelerate less than the wrench will. A force that causes the 1-kg wrench to accelerate to 10 m/s will

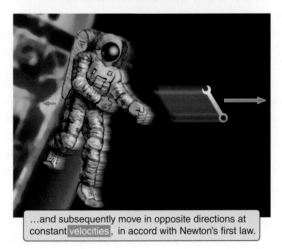

An astronaut adrift in space pushes on a wrench, which, according to Newton's third law, pushes back on the astronaut.

Space shuttle

While in contact with each other, the wrench and the astronaut experience accelerations proportional to the inverse of their masses…

…and subsequently move in opposite directions at constant velocities, in accord with Newton's first law.

**Figure 3.23** According to Newton's laws, if an astronaut adrift in space throws a wrench, the two will move in opposite directions. Their speeds will depend on their masses: the same force will produce a smaller acceleration of a more massive object than of a less massive object. (Acceleration and velocity arrows are not drawn to scale.)

have much less effect on the 100-kg astronaut. Because acceleration equals force divided by mass, the 100-kg astronaut will experience only 1/100 as much acceleration as the 1-kg wrench. The astronaut will drift toward the shuttle, but only at the leisurely rate of 1/100 × 10 m/s, or 0.1 m/s.

object exerts a matching force on the first. That second force is exactly as strong as the first force but is in exactly the opposite direction. When you are accelerating yourself on the skateboard, you push backward on Earth, and Earth pushes you forward. As shown in **Figure 3.24**, a woman pulling a load on a cart pulls on the rope, and the rope pulls back. A car tire pushes back on the road, and the road pushes forward on the tire. Earth pulls on the Moon, and the Moon pulls on Earth. A rocket engine pushes hot gases out of its nozzle, and those hot gases push back on the rocket, propelling it into space.

All of these force pairs are examples of **Newton's third law of motion**, which says that forces always come in pairs, and the forces of a pair are always equal in strength but opposite in direction. The forces in these pairs always act on two different objects. Your weight pushes down on the floor, and the floor pushes back up on your feet with the same amount of force. For every force there is *always* an equal force in the opposite direction.

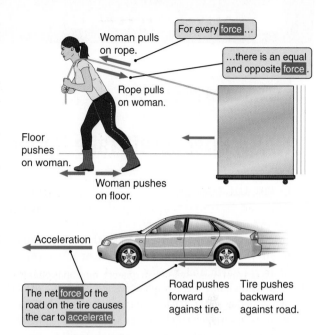

**Figure 3.24** Newton's third law states that for every force there is always an equal and opposite force. These opposing forces always act on the two different objects in the same pair.

## CHECK YOUR UNDERSTANDING 3.6

Imagine a planet moving in a perfectly circular orbit around the Sun. Is this planet experiencing acceleration? (a) Yes, because it is changing its speed all the time. (b) Yes, because it is changing its direction of motion all the time. (c) No, because its speed is not changing all the time. (d) No, because planets do not experience accelerations.

# Origins

## Planets and Orbits

In addition to the planets in our own Solar System, several thousand planets have been detected orbiting stars other than our own. Their orbits can be calculated and understood by applying the same three Kepler's laws that we have discussed here. Astrobiologists think that the orbit of a planet around its star affects its chances of developing life.

Consider the average distance of the planet from its star. You might intuitively guess that a planet close to its star will receive more energy from its star than that received by a planet far from its star. If Earth were closer to the Sun, it would be hotter throughout the year—perhaps so hot that water would evaporate and not exist as a liquid. If Earth were farther from the Sun, it would be colder and perhaps all water would freeze. We know that liquid water was a crucial element for the formation of life

on Earth. So some astronomers look for planets at a distance from their star such that liquid water can exist (this distance will vary depending on the temperature and size of the star).

We might also think about the eccentricity of a planet's orbit: recall from Figure 3.14a that Earth's orbit differs from a circle by less than 2 percent. Thus, Earth's distance from the Sun does not vary much throughout the year; and as we saw in Chapter 2, seasonal variation on Earth is caused by the tilt of Earth's axis, not by the slight changes in its distance from the Sun. However, if we look at our neighboring planet Mars, which has about the same axial tilt as Earth, we see a greater seasonal variation because of its more eccentric orbit. The distance between Mars and the Sun varies from 1.38 AU (207 million km) to 1.67 AU (249 million km)—an

eccentricity of 9 percent. As a result, the seasons on Mars are not equal. They are shorter when Mars is closer to the Sun and moving faster and longer when Mars is farther from the Sun and moving slower. The inequality of the seasons on Mars has an effect on the overall stability of its temperature and climate. When we look at planets orbiting other stars, we see that many have orbital eccentricities even higher than that of Mars and therefore large variations in temperature.

Earth is at the right distance from the Sun to have temperatures that permit water to be liquid, and its orbital eccentricity is low enough that the average planetary temperature does not change much over the course of its orbit. These orbital characteristics have contributed to making the conditions on Earth suitable for the development of life.

# NASA Spacecraft Take Spring Break at Mars

By **MIKE WALL**, Space.com

NASA's robotic Mars explorers are taking a cosmic break for the next few weeks, thanks to an unfavorable planetary alignment of Mars, the Earth, and the Sun.

Mission controllers won't send any commands to the agency's *Opportunity* rover, *Mars Reconnaissance Orbiter* (*MRO*), or *Mars Odyssey* orbiter from today (April 9) through April 26. The blackout is even longer for NASA's car-size *Curiosity* rover, which is slated to go solo from April 4 through May 1.

The cause of the communications moratorium is a phenomenon called a Mars solar conjunction, during which the Sun comes between Earth and the Red Planet (**Figure 3.25**). Our star can disrupt and degrade interplanetary signals in this formation, so mission teams won't be taking any chances.

"Receiving a partial command could confuse the spacecraft, putting them in grave danger," NASA officials explain in a video posted last month by the agency's Jet Propulsion Laboratory (JPL) in Pasadena, California.

*Opportunity* and *Curiosity* will continue performing stationary science work, using commands already beamed to the rovers.

*Curiosity* will focus on gathering weather data, assessing the martian radiation environment, and searching for signs of subsurface water and hydrated minerals, officials said Monday (April 8).

*MRO* and *Odyssey* will also keep studying the Red Planet from above, and they'll continue to serve as communications links between the rovers and Earth. The conjunction will also affect the European Space Agency's *Mars Express* orbiter, officials have said.

*Odyssey* will send rover data home as usual during conjunction, though the orbiter may have to relay information multiple times due to dropouts. *MRO*, on the other hand, entered record-only mode on April 4. The spacecraft will probably have about 52 gigabits of data to relay when it's ready to start transmitting again on May 1, *MRO* officials have said.

Mars solar conjunctions occur every 26 months, so NASA's Red Planet veterans have dealt with them before. This is the fifth conjunction for *Opportunity*, in fact, and the sixth for *Odyssey*, which began orbiting Mars in 2001.

But it'll be the first for *Curiosity*, which touched down on August 5, kicking off a two-year surface mission to determine if the Red

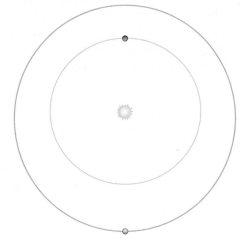

**Figure 3.25** Every 26 months, Mars and Earth are on opposite sides of the Sun, making communication between the two planets impossible.

Planet has ever been capable of supporting microbial life.

"The biggest difference for this 2013 conjunction is having *Curiosity* on Mars," *Odyssey* mission manager Chris Potts, of NASA's Jet Propulsion Laboratory in Pasadena, California, said in a statement last month.

1. How often do these Mars solar conjunctions occur? Does this interval correspond to the sidereal period or the synodic period of Mars?
2. Give two reasons that conjunction is a bad time to view Mars in the sky.
3. Make a sketch of the Mars solar conjunction, showing Mars, Earth, and the Sun. Using the orbital periods in Table 3.2, add in the positions of Mars and Earth at the next Mars conjunction. Will it take place at the same location in space?
4. View this short video from NASA: http://www.jpl.nasa.gov/video/?id=1204. Do you think it explains the concepts well to someone who is not reading an astronomy textbook?
5. If people were on a mission on Mars, a loss of contact would be very troubling. How might NASA plan to avoid losing contact with astronauts on Mars during conjunctions?

# Summary

Early astronomers hypothesized that Earth was stationary at the center of the Solar System. Later astronomers realized that a Sun-centered Solar System was much simpler and could explain the observations. Planets, like Jupiter, orbit the Sun, not Earth. Kepler's laws describe the elliptical orbits of planets around the Sun, including details about how fast a planet travels at various points in its orbit. These laws helped Newton to advance science by developing his laws of motion, which govern the motion of all objects (not just orbiting ones). Orbital semimajor axis, eccentricity, and stability may affect a planet's suitability to foster life.

**LG 1**  **Describe and contrast the geocentric and heliocentric models of the Solar System.** Earth's motion is hard to detect, so prior to the Copernican Revolution, most people accepted a geocentric model of the Solar System, in which all objects orbit around Earth. In particular, apparent retrograde motion of the planets was difficult to understand in this model. Copernicus created the first comprehensive mathematical model of the Solar System with the Sun at the center, called a heliocentric model. His model explained apparent retrograde motion as a visual illusion seen when an inner planet passes an outer planet in their orbits.

**LG 2**  **Use Kepler's laws to describe the motion of objects in the Solar System.** Using Tycho's observational data, Kepler developed empirical rules to describe the motions of the planets. Kepler's three laws state that (1) planets move in elliptical orbits around the Sun, (2) planets move fastest when closest to the Sun and slowest when farthest from the Sun, so that the planets sweep out equal areas in equal times, and (3) the orbital period of a planet squared equals the semimajor axis of its orbit cubed, or $P^2 = A^3$.

**LG 3**  **Explain how Galileo's astronomical discoveries provided empirical evidence for the heliocentric model.** Galileo invented astronomical telescopes and used them to observe moons in orbit around Jupiter. He also saw Venus going through phases like the Moon, but changing its apparent size in each phase. These astronomical observations were difficult to explain with a geocentric model.

**LG 4**  **Describe the work of Galileo and Newton, which led them to discover the physical laws that govern the motion of all objects.** Galileo studied the physics of falling objects and discovered the principle of inertia. Newton's laws state that (1) objects do not change their motion unless they experience a net force, (2) Force = mass × acceleration, and (3) "every force has an equal and opposite force." Net forces cause accelerations; that is, changes in motion. Inertia resists changes in motion.

## ? UNANSWERED QUESTIONS

- Would the history of scientific discoveries in physics and astronomy have been different if the Catholic Church had not prosecuted Galileo? Galileo wrote the *Dialogo* after being ordered by the Catholic Church in 1616 not to "hold or defend" the idea that Earth moves and the Sun is still. And he wrote his equally famous *Discorsi e Dimostrazioni Matematiche* (often shortened in English to "Two New Sciences") while under house arrest after his trial. However undeterred Galileo appeared to be, the effects of the decrees, prohibitions, and prosecutions might have dissuaded other scientists in Catholic countries from pursuing this type of work. Indeed, after Galileo's experiences, the center of the scientific revolution moved north to Protestant Europe.

- What percentage of planets are in unstable orbits? In younger planetary systems, planets might migrate in their orbits because of the presence of other massive planets nearby. We will see in Chapter 7 that Uranus and Neptune might have migrated in this way. Some planets have been discovered moving through the galaxy without any obvious orbit around a star—and therefore are not in the stable orbits we see in our own Solar System.

# Questions and Problems

## Test Your Understanding

1. An *empirical science* is one that is based on
   a. hypothesis.
   b. calculus.
   c. computer models.
   d. observed data.

2. When Earth catches up to a slower-moving outer planet and passes it in its orbit, the planet
   a. exhibits retrograde motion.
   b. slows down because it feels Earth's gravitational pull.
   c. decreases in brightness as it passes through Earth's shadow.
   d. moves into a more elliptical orbit.

3. Copernicus's model of the Solar System was superior to Ptolemy's because
   a. it had a mathematical basis that could be used to predict the positions of planets.
   b. it was much more accurate.
   c. it did not require any epicycles.
   d. it fit the telescopic data better.

4. An inferior planet is one that is
   a. smaller than Earth.
   b. larger than Earth.
   c. closer to the Sun than Earth is.
   d. farther from the Sun than Earth is.

5. The time it takes for a planet to come back to the same position relative to the Sun is called its _____ period.
   a. synodic
   b. sidereal
   c. heliocentric
   d. geocentric

6. Suppose a planet is discovered orbiting a star in a highly elliptical orbit. While the planet is close to the star it moves _____, but while it is far away it moves _____.
   a. faster; slower
   b. slower; faster
   c. retrograde; prograde
   d. prograde; retrograde

7. If a superior planet is observed from Earth to have a synodic period of 1.2 years, what is its sidereal period?
   a. 0.54 years
   b. 1.8 years
   c. 4.0 years
   d. 6.0 years

8. A net force must be acting when an object
   a. accelerates.
   b. changes direction but not speed.
   c. changes speed but not direction.
   d. all of the above

9. For Earth, $P^2/A^3 = 1.0$ (in appropriate units). Suppose a new dwarf planet is discovered that is 14 times as far from the Sun as Earth is. For this planet,
   a. $P^2/A^3 = 1.0$.
   b. $P^2/A^3 > 1.0$.
   c. $P^2/A^3 < 1.0$.
   d. one can't know the value of $P^2/A^3$ without more information.

10. Galileo observed that Venus had phases that correlated with its size in his telescope. From this information, you may conclude that Venus
    a. is the center of the Solar System.
    b. orbits the Sun.
    c. orbits Earth.
    d. orbits the Moon.

11. Kepler's second law says that
    a. planetary orbits are ellipses with the Sun at one focus.
    b. the square of a planet's orbital period equals the cube of its semimajor axis.
    c. net forces cause changes in motion.
    d. planets move fastest when they are closest to the Sun.

12. Suppose you read in the newspaper that a new planet has been found. Its average speed in orbit is 33 km/s. When it is closest to its star it moves at 31 km/s, and when it is farthest from its star it moves at 35 km/s. This story is in error because
    a. the average speed is far too fast.
    b. Kepler's third law says the planet has to sweep out equal areas in equal times, so the speed of the planet cannot change.
    c. Kepler's second law says the planet must move fastest when it is closest, not when it is farthest away.
    d. using these numbers, the square of the orbital period will not be equal to the cube of the semimajor axis.

13. Galileo observed that Jupiter has moons. From this information, you may conclude that
    a. Jupiter is the center of the Solar System.
    b. Jupiter orbits the Sun.
    c. Jupiter orbits Earth.
    d. some things do not orbit Earth.

14. If you start from rest and accelerate at 10 mph/s and end up traveling at 60 mph, how long did it take?
    a. 1 second
    b. 6 seconds
    c. 60 seconds
    d. 0.6 seconds

15. Planets with high eccentricity may be unlikely candidates for life because
    a. the speed varies too much.
    b. the period varies too much.
    c. the temperature varies too much.
    d. the orbit varies too much.

## Thinking about the Concepts

16. Study Figure 3.1. During normal motion, does Mars move toward the east or west? Which direction does it travel when moving retrogradely? For how many days did Mars move retrogradely? If one of the martian missions were photographing *Earth* in the sky during these days, what would it have observed?

17. Copernicus and Kepler engaged in what is called empirical science. What do we mean by *empirical*?

18. Explain why the synodic period of Saturn is very close to 1 Earth year (a sketch may help).

19. Make a sketch of Earth, Venus, and the Sun in a geocentric model and in a heliocentric model. Label the Sun and Earth, and then show the changing phases of Venus over the course of one orbit of Venus. What would we observe in each model? Why was the invention of the telescope necessary to distinguish between these models?

20. Experiment with falling objects as Galileo did. Drop pairs of objects of different masses—do they reach the ground at the same time? Do they hit the ground with the same force? Does this work with a sheet of paper or a tissue—why or why not?

21. The speed of a planet in its orbit varies in its journey around the Sun. At what point in its orbit is the planet moving the fastest? At what point is it moving the slowest?

22. The orbit of the Moon around Earth also is elliptical, with an eccentricity of 0.05. How does this compare with the eccentricity of Earth's orbit? How do these elliptical orbits explain the types of solar eclipses discussed in Chapter 2?

23. Galileo came up with the concept of inertia. What do we mean by *inertia*?

24. If Kepler had lived on Mars, would he have deduced the same empirical laws for the motion of the planets? Explain.

25. What is the difference between speed and velocity? between velocity and acceleration?

26. When involved in an automobile collision, a person not wearing a seat belt will move through the car and often strike the windshield directly. Which of Newton's laws explains why the person continues forward, even though the car stopped?

27. When riding in a car, we can sense changes in speed or direction through the forces that the car applies on us. Do we wear seat belts in cars and airplanes to protect us from speed or from acceleration? Explain your answer.

28. An astronaut standing on Earth can easily lift a wrench having a mass of 1 kg, but not a scientific instrument with a mass of 100 kg. In the International Space Station, she is quite capable of manipulating both, although the scientific instrument responds much more slowly than the wrench. Explain why.

29. The Process of Science Figure illustrates that scientific ideas are always open to challenge. Construct an argument that this constant process of challenging and falsifying ideas is a strength of science, rather than a weakness.

30. How might you expect conditions on Earth to be different if the eccentricity of its orbit was 0.17 instead of 0.017?

## Applying the Concepts

31. Study the graph in Figure 3.16. Is this graph linear or logarithmic? From the data on the graph, find the approximate semimajor axis and period of Saturn. Show your work.

32. Study Figure 3.19, which shows that the apparent size of Venus changes as it goes through phases. Approximately how many times larger is Venus in the sky at the tiniest crescent than at the gibbous phase shown? Therefore, approximately how many times closer is Venus to us at the phase of that tiniest crescent than at the gibbous phase?

33. Suppose a new dwarf planet is discovered orbiting the Sun with a semimajor axis of 50 AU. What would be the orbital period of this new dwarf planet?

34. Planet Neptune's orbital period is 164.8 years. What is the semimajor axis of its orbit? How much time passes between oppositions of Neptune?

35. Dwarf planet Ceres is located at 2.77 AU from the Sun. Its synodic period is 1.278 years.
    a. Use Working It Out 3.1 to find the sidereal period in years.
    b. Use Kepler's law to find the sidereal period in years.
    c. Compare your results for (a) and (b).

36. Suppose you read online that "experts have discovered a new planet with a distance from the Sun of 2 AU and a period of 3 years." Use Kepler's third law to argue that this is impossible.

37. Show, as Galileo did, that Kepler's third law applies to the four moons of Jupiter that he discovered by calculating $P^2$ divided by $A^3$ for each moon. (Data on the moons can be found in Appendix 4.)

38. In a period of 3 months, a planet travels 30,000 km with an average speed of 3.8 m/s. Some time later, the same planet travels 65,000 km in 3 months. How fast is the planet traveling at this later time? During which period is the planet closer to the Sun?

39. If you were on Mars, how often would you see retrograde motion of *Earth* in the martian night sky? (You can view a simulation at http://mars.jpl.nasa.gov/allaboutmars/nightsky/retrograde/.)

40. The elliptical orbit of a comet recently visited by a spacecraft is 1.24 AU from the Sun at its closest approach and 5.68 AU from the Sun at its farthest.
    a. Sketch the orbit of the comet. When is it moving fastest? When is it moving slowest?
    b. What is the semimajor axis of its orbit? How long does it take to go around the Sun?
    c. What is the distance from the Sun to the "center" of the ellipse? What is the eccentricity of the comet's orbit?

41. You are driving down a straight road at a speed of 90 km/h, and you see another car approaching you at a speed of 110 km/h along the road.
    a. Relative to your own frame of reference, how fast is the other car approaching you?
    b. Relative to the other driver's frame of reference, how fast are you approaching the other driver's car?

42. During the latter half of the 19th century, a few astronomers thought there might be a planet circling the Sun inside Mercury's orbit. They even gave it a name: Vulcan. We now know that Vulcan does not exist. If a planet with an orbit one-fourth the size of Mercury's actually existed, what would be its orbital period relative to that of Mercury?

43. Suppose you are pushing a small refrigerator of mass 50 kg on wheels. You push with a force of 100 N.
    a. What is the refrigerator's acceleration?
    b. Assume the refrigerator starts at rest. How long will the refrigerator accelerate at this rate before it gets away from you (that is, before it is moving faster than you can run—of the order 10 m/s)?

44. If a 100-kg astronaut pushes on a 5,000-kg satellite and the satellite experiences an acceleration of 0.1 m/s², what is the acceleration experienced by the astronaut in the opposite direction?

45. Sketch the orbit of Mars using the information provided in the "Origins: Planets and Orbits" section of the chapter for the closest and farthest distances of Mars from the Sun.
    a. What is its major axis? What is its semimajor axis?
    b. What is the distance from the "center" of the orbit to the Sun? Compute the eccentricity of the orbit. Compare this with the eccentricity of Earth's orbit.

## USING THE WEB

46. Go to the Web page "This Week's Sky at a Glance" (http://skyandtelescope.com/observing/ataglance) at the *Sky & Telescope* magazine website. Which planets are visible in your sky this week? Why are Mercury and Venus visible only in the morning before sunrise or in the evening just after sunset? Before telescopes, how did people know the planets were different from the stars?

47. Look up the dates for the next opposition of Mars, Jupiter, or Saturn. One source is the NASA "Sky Events Calendar" at http://eclipse.gsfc.nasa.gov/SKYCAL/SKYCAL.html. Check only the "Planet Events" box in "Section 2: Sky Events"; and in Section 3, generate a Sky Events Calendar for the year. If you are coming up on an opposition, take pictures of the planet over the next few weeks. Can you see its position move in retrograde motion with respect to the background stars?

48. Refer to the Web page from question 47 to find the current observational positions of all the planets.
    a. Which ones are in or near to conjunction, opposition, or greatest elongation?
    b. Which are visible in the morning sky? in the evening sky?
    c. To connect when we see the planets on Earth with the physical alignments of the planets with Earth and the Sun in space, sketch the Solar System with Earth, Sun, and planets as it looks today from "above." Check your result using NASA's "Solar System Simulator" (http://space.jpl.nasa.gov): set it for "Show Me Solar System" as seen from above, and look at the field of view of 2°, 20°, and 45° to see the inner and then the outer planets. Does the simulator agree with your sketch?

49. Go to the Museo Galileo website and view the exhibit on Galileo's telescope (http://www.museogalileo.it/en/explore/exhibitions/pastexhibitions/galileostelescope.html). What did his telescope look like? What other instruments did he use? From the museum page you can link to short videos (in English) on his science and his trial (http://catalogue.museogalileo.it/index/VideoIndexByThematicArea.html#s7). For example, click on "Galileo's micrometer": How did he measure the separation of the moons from Jupiter? How did this measurement allow him to show that the moons obeyed Kepler's law? Why is Galileo often considered the first modern scientist? Why is his middle finger on display in the museum?

50. Go to the online "Extrasolar Planets Encyclopedia" (http://exoplanet.eu/catalog).
    a. Find a planet with an orbital period similar to that of Earth. What is the semimajor axis of its orbit? If it is very different from 1 AU, then the mass of the star is different from that of the Sun. Click on the star name in the first column to see the star's mass. What is the orbital eccentricity?
    b. Click on "Planet" to sort by name, and select a star with multiple planets. Verify that Kepler's third law applies by showing that the value of $P^2/A^3$ is about the same for each of the planets of this star. How eccentric are the orbits of the multiple planets?

# smartwork5

If your instructor assigns homework in Smartwork5, access your assignments at digital.wwnorton.com/astro5.

digital.wwnorton.com/astro5

In this Exploration, we will examine how Kepler's laws apply to the orbit of Mercury. Visit the Student Site at the Digital Landing page, and open the Planetary Orbit Simulator applet. This simulator animates the orbits of the planets, enabling you to control the simulation speed, as well as a number of other parameters. Here we focus on exploring the orbit of Mercury, but you may wish to spend some time examining the orbits of other planets as well.

## Kepler's First Law

To begin exploring the simulation, in the "Orbit Settings" panel, use the drop-down menu next to "set parameters for" to select "Mercury" and then click "OK." Click the "Kepler's 1st Law" tab at the bottom of the control panel. Use the radio buttons to select "show empty focus" and "show center."

**1** How would you describe the shape of Mercury's orbit?

_____

_____

_____

Deselect "show empty focus" and "show center," and select "show semiminor axis" and "show semimajor axis." Under "Visualization Options," select "show grid."

**2** Use the grid markings to estimate the ratio of the semiminor axis to the semimajor axis.

_____

_____

_____

**3** Calculate the eccentricity of Mercury's orbit from this ratio using $e = [1 - (\text{Ratio})^2]^{1/2}$.

_____

_____

_____

## Kepler's Second Law

Click on "reset" near the top of the control panel, set parameters for Mercury, and click "OK." Then click on the "Kepler's 2nd Law" tab at the bottom of the control panel. Slide the "adjust size" slider to the right, until the fractional sweep size is $\frac{1}{8}$.

Click on "start sweeping." The planet moves around its orbit, and the simulation fills in area until one-eighth of the ellipse is filled. Click on "start sweeping" again as the planet arrives at the rightmost point in its orbit (that is, at the point in its orbit farthest from the Sun). You

may need to slow the animation rate using the slider under "Animation Controls." Click on "show grid" under the visualization options. (If the moving planet annoys you, you can pause the animation.) One easy way to estimate an area is to count the number of squares.

**4** Count the number of squares in the yellow area and in the red area. You will need to decide what to do with fractional squares. Are the areas the same? Should they be?

_____

_____

## Kepler's Third Law

Click on "reset" near the top of the control panel, set parameters for Mercury, and then click on the "Kepler's 3rd Law" tab at the bottom of the control panel. Select "show solar system orbits" in the "Visualization Options" panel. Study the graph. Use the eccentricity slider to change the eccentricity of the simulated planet. Make the eccentricity first smaller and then larger.

**5** Did anything in the graph change?

_____

_____

**6** What do your observations of the graph tell you about the dependence of the period on the eccentricity?

_____

_____

Set parameters back to those for Mercury. Now use the semimajor axis slider to change the semimajor axis of the simulated planet.

**7** What happens to the period when you make the semimajor axis smaller?

_____

_____

**8** What happens when you make it larger?

_____

_____

**9** What do these results tell you about the dependence of the period on the semimajor axis?

_____

_____

# 4 Gravity and Orbits

In this chapter, we explore the physical laws that explain the regular patterns in the motions of the planets. Because the Sun is far more massive than all the other parts of the Solar System combined, its gravity shapes the motions of every object in its vicinity, from the almost circular orbits of some planets to the extremely elongated orbits of comets.

## LEARNING GOALS

By the conclusion of this chapter, you should be able to:

**LG 1** Explain the elements of Newton's universal law of gravitation.

**LG 2** Use the laws of motion and gravitation to explain planetary orbits.

**LG 3** Explain how tidal forces from the Sun and Moon create Earth's tides.

**LG 4** Describe the effects of tidal forces on solid bodies.

The International Space Station in orbit around Earth. ▶ ▶ ▶

What keeps a
space station
in orbit?

# 4.1 Gravity Is a Force between Any Two Objects Due to Their Masses

In Chapter 3, we explored Kepler's work on the movement of the planets around the Sun and Newton's laws of motion. In this chapter, we build on those concepts to look at Newton's universal law of gravitation. Although some of the properties of gravity were observed before Newton, his work connected the everyday phenomenon of falling objects to the motion of the planets around the Sun. Newton's theory of gravity combined Kepler's empirical laws and Newton's own laws of motion.

## Gravity, Mass, and Weight

Many forces that we see in everyday life involve direct contact between objects. The cue ball on a pool table slams into the eight ball, knocking it into the pocket. The shoe of the child pushing a scooter presses directly against the surface of the pavement. When there is physical contact between two objects, the source of the forces between them is easy to see.

If you drop a ball, it falls toward the ground. The ball picks up speed as it falls, accelerating downward toward Earth. Newton's second law says that where there is acceleration, there is force. But where is the force that causes the ball to accelerate? The ball falling toward Earth is an example of a different kind of force, one that acts at a distance across the intervening void of space. The ball falling toward Earth is accelerating in response to the force of gravity. **Gravity**, one of the fundamental forces of nature, is the mutually attractive force between objects with mass.

Recall from Chapter 3 that Galileo discovered that all freely falling objects accelerate toward Earth at the same rate, regardless of their mass. Drop a marble and a book at the same time and from the same height, and they will hit the ground together. Note that air resistance becomes a factor at higher speeds, but it is negligible for dense, slow objects. The acceleration of falling objects due to gravity near the surface of Earth, also measured experimentally by Galileo, is usually written as $g$ (lowercase) and has an average value across the surface of Earth of 9.8 meters per second squared ($m/s^2$). The value of $g$ varies slightly across Earth's surface, ranging from 9.78 $m/s^2$ at the equator to 9.83 $m/s^2$ at the poles. This variation exists because Earth is not a perfect sphere: its rotation makes it flatter at the poles, so the radius of Earth is smaller at the poles.

After working out the laws governing the motion of objects, Newton realized that if all objects, no matter what their mass, fall with the same acceleration, then the gravitational *force* on an object must be determined by the object's *mass*.

Recall Newton's second law from Chapter 3: acceleration equals force divided by mass, or $a = F/m$. The acceleration due to gravity can be the same for all objects only if the value of the force divided by the mass is the same for all objects. In other words, an object twice as massive has double the gravitational force acting on it; an object 3 times as massive has triple the gravitational force acting on it, and so on.

The gravitational force acting on an object attracted by a planet is called the object's **weight**. On the surface of Earth, weight equals mass multiplied by the acceleration of gravity at Earth's surface, $g$. In common language, we often use weight and mass interchangeably. To be more scientifically precise, astronomers use *mass* to refer to the amount of stuff in an object and *weight* to refer to the force exerted on that object by the planet's gravitational pull. Your mass is the same no matter what planet or moon you are on, but your weight will be different:

$$F_{weight} = m \times g$$

where $F_{weight}$ is an object's weight in newtons (N), the metric unit of force; $m$ is the object's mass in kilograms (kg); and $g$ is Earth's constant for acceleration due to gravity, 9.8 m/s². On Earth, an object with a *mass* of 1 kg has a *weight* of 9.8 N. As illustrated in **Figure 4.1**, on the Moon the acceleration due to gravity is about 6 times lower at 1.6 m/s², so the 1-kg mass would have a weight of 1.6 N. On the Moon, your weight would be about one-sixth of your weight on Earth.

## Newton's Law of Gravity

As Newton told the story, he saw an apple fall from a tree to the ground, and he reasoned that if gravity is a force that depends on mass, then there should be a gravitational force between *any* two masses, including between a falling apple and Earth. This great insight came from applying his third law of motion to gravity. Recall that Newton's third law states that for every force there is an equal and opposite force. Therefore, if Earth exerts a force of 9.8 N on a 1-kg mass sitting on its surface, then that 1-kg mass must also exert a force of 9.8 N on Earth. Drop a 7-kg bowling ball and it falls toward Earth, but at the same time Earth falls toward the 7-kg bowling ball. The reason we do not notice the motion of Earth is that Earth is very massive, so it has a lot of resistance to changes in its motion. In the time it takes a 7-kg bowling ball to fall to the ground from a height of 1 kilometer (km), Earth has "fallen" toward the bowling ball by only a tiny fraction of the size of an atom.

Newton reasoned that if doubling the mass of any object doubles the gravitational force between the object and Earth, then doubling the mass of Earth ought to do the same thing. In short, the gravitational force between Earth and an object must be equal to the product of the two masses multiplied by *something*:

Gravitational force = Something × Mass of Earth × Mass of object

If the mass of the object is 2 times greater, then the force of gravity will be 2 times greater. Likewise, if the mass of Earth happened to be 3 times what it is, the force of gravity would also have to be 3 times greater. If both the mass of Earth and the mass of the object were greater by these amounts, the gravitational force would increase by a factor of 2 × 3, or 6 times. Because objects fall toward the center of Earth, we know that gravity is an attractive force acting along a line between the two masses.

If gravity is a force that depends on mass, then there should be a gravitational force between *any* two masses. Suppose we have two masses—call them mass 1 and mass 2, or $m_1$ and $m_2$ for short. The gravitational force between them is *something* multiplied by the product of the masses:

Gravitational force between two objects = Something × $m_1$ × $m_2$

We have gotten this far just by combining Galileo's observations of falling objects with (1) Newton's laws of motion and (2) Newton's belief that Earth is a

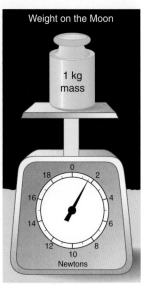

**Figure 4.1** On the Moon, a mass of 1 kg has $\frac{1}{6}$ the weight (displayed in newtons) that it has on Earth.

mass just like any other mass. But what about that "something" in the previous expression?

Kepler had already thought about this question. He reasoned that because the Sun is the focal point for planetary orbits, the Sun must exert an influence over the motions of the planets. Kepler speculated that this influence must grow weaker with distance from the Sun, because the planets closer to the Sun moved much faster than the farther ones. Kepler did not know about forces or inertia or gravity as the cause of celestial motion, but he thought that geometry alone suggested how this solar "influence" might change for planets progressively farther from the Sun.

To see why the influence must become weaker, imagine you have a certain amount of paint to spread over the surface of a sphere. If the sphere is small, you will get a thick coat of paint. But if the sphere is larger, the paint has to spread farther, and you will get a thinner coat. The surface area of a sphere depends on the square of the sphere's radius. Double the radius of a sphere, and the sphere's surface area becomes 4 times what it was. If you paint this new, larger sphere, the paint must cover 4 times as much area, and the thickness of the paint will be only a fourth of what it was on the smaller sphere. Triple the radius of the sphere: the sphere's surface will be 9 times larger and the coat of paint will be only one-ninth as thick.

Kepler reasoned that as the influence of the Sun extended farther and farther into space, it would have to spread out to cover the surface of a larger and larger imaginary sphere centered on the Sun. The influence of the Sun should diminish with the square of the distance from the Sun—a relationship known as an **inverse square law**.

Kepler had an interesting idea, but not a scientific hypothesis with testable predictions. He lacked an explanation for how the Sun influences the planets and the mathematical tools to calculate how an object would move under such an influence. Newton had both. If gravity is a force between *any* two objects, then there should be a gravitational force between the Sun and each of the planets. If this gravitational force were the same as Kepler's "influence," then gravity might behave according to an inverse square law.

Newton's expression for gravity came to look like this:

$$\text{Gravitational force between two objects} = \text{Something} \times \frac{m_1 \times m_2}{(\text{Distance between objects})^2}$$

There is still a "something" left in this expression, and that something is a constant of proportionality. This constant determines the strength of gravity between objects, and it is the same for all pairs of objects. Newton named it the **universal gravitational constant**, written as $G$ (uppercase). Newton estimated this gravitational constant $G$ by using Galileo's measurement of $g$, estimates of Earth's radius, and a guess at the mass of Earth by assuming it had about the same density as typical rocks. It was not until many years later that the actual value of $G$ was first measured. Today the value of $G$ is accepted as $6.67 \times 10^{-11}$ N m$^2$/kg$^2$ or its equivalents: $6.67 \times 10^{-11}$ m$^3$/(kg s$^2$) or $6.67 \times 10^{-20}$ km$^3$/(kg s$^2$).

## A Universal Law of Gravitation

Newton's **universal law of gravitation**, illustrated in **Figure 4.2**, states that gravity is a force between any two objects having mass and has the following properties:

1. It is an attractive force acting along a straight line between the two objects.

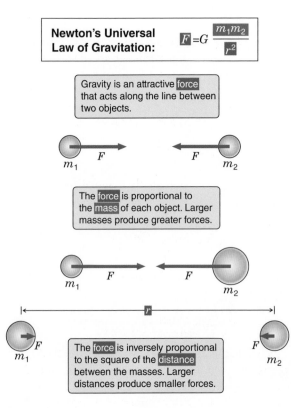

Newton's Universal Law of Gravitation: $F = G \dfrac{m_1 m_2}{r^2}$

Gravity is an attractive force that acts along the line between two objects.

$m_1$  $F$       $F$  $m_2$

The force is proportional to the mass of each object. Larger masses produce greater forces.

$m_1$  $F$       $F$  $m_2$

$r$

$F$ $m_1$    The force is inversely proportional to the square of the distance between the masses. Larger distances produce smaller forces.    $F$ $m_2$

**Figure 4.2** Gravity is an attractive force between two objects. The force of gravity depends on the masses of the objects, $m_1$ and $m_2$, and the distance, $r$, between them.

2. It is proportional to the mass of one object ($m_1$) multiplied by the mass of the other object ($m_2$). If we double $m_1$, then the force ($F$) increases by a factor of 2. Likewise, if we double $m_2$, $F$ increases by a factor of 2.

3. It is inversely proportional to the square of the distance $r$ between the centers of the two objects: As seen in **Figure 4.3**, if we double $r$, $F$ decreases by a factor of 4. If we triple $r$, $F$ falls by a factor of 9.

Written as a mathematical formula, the universal law of gravitation states

$$F_{grav} = G \times \frac{m_1 \times m_2}{r^2}$$

where $F_{grav}$ is the force of gravity between two objects, $m_1$ and $m_2$ are the masses of objects 1 and 2, $r$ is the distance between the centers of mass of the two objects, and $G$ is the universal gravitational constant. The relationship between the force of gravity and the masses and separation distance between two objects is further explored in **Working It Out 4.1**.

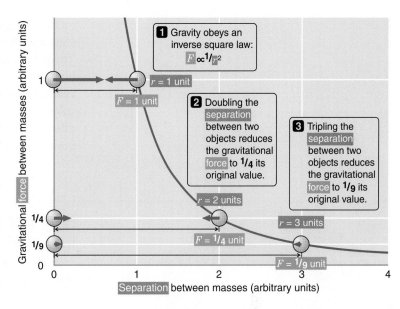

**1** Gravity obeys an inverse square law: $F \propto 1/r^2$

**2** Doubling the separation between two objects reduces the gravitational force to $1/4$ its original value.

**3** Tripling the separation between two objects reduces the gravitational force to $1/9$ its original value.

$r$ = 1 unit
$F$ = 1 unit
$r$ = 2 units
$F$ = $1/4$ unit
$r$ = 3 units
$F$ = $1/9$ unit

Gravitational force between masses (arbitrary units)

Separation between masses (arbitrary units)

**Figure 4.3** As two objects move apart, the gravitational force between them decreases by the inverse square of the distance between them.

## 4.1 Working It Out  Playing with Newton's Laws of Motion and Gravitation

For any two objects, the force of gravity is directly proportional to the masses and inversely proportional to the *square* of the distance between them. Let's look at a few examples of how to use this equation:

### Changing the Distance

How would the gravitational force between Earth and the Moon change if the distance between them were doubled? In this example, the masses of the Sun and Moon stay the same, and $r$ becomes $2r$. We can calculate how the force changes by writing the equation for distance $r$ and again for distance $2r$, and then taking a ratio to compare them:

$$\frac{F_{grav\ at\ distance\ 2r}}{F_{grav\ at\ distance\ r}} = \frac{G \times \frac{M_{Earth}M_{Moon}}{(2r)^2}}{G \times \frac{M_{Earth}M_{Moon}}{r^2}}$$

We can cancel out the constant $G$ and the masses of Earth and the Moon, which do not change. Then you need to multiply both the numerator and denominator of the fraction by $r^2$, and remember that both the 2 and the $r$ get squared in $(2r)^2 = 4r^2$. The equation becomes

$$\frac{F_{grav\ at\ distance\ 2r}}{F_{grav\ at\ distance\ r}} = \frac{\cancel{G} \times \cancel{M_{Earth}} \times \cancel{M_{Moon}}}{\cancel{G} \times \cancel{M_{Earth}} \times \cancel{M_{Moon}}} \times \frac{r^2}{(2r)^2}$$

$$= \frac{\cancel{r^2}}{4\cancel{r^2}} = \frac{1}{4}$$

Doubling the distance reduced the force by a factor of 4; that is, the force is $\frac{1}{4}$ as strong.

 **Nebraska Simulation:** Gravity Algebra

### Gravitational Acceleration

There are two ways to think about the gravitational force that Earth exerts on an object with mass $m$ located on the surface of Earth. Recall Newton's second law of motion: $F = m \times a$. Here we are considering the gravitational force and the acceleration due to gravity, or $F_{grav} = m \times g$. The other way to think about the force is from the perspective of the universal law of gravitation, which says

$$F_{grav} = G \times \frac{M_{Earth} \times m}{R_{Earth}^2}$$

Here, $M_{Earth}$ is the mass of Earth, and $R_{Earth}$ is the radius of Earth. The two expressions describing this force must be equal to each other. Therefore,

$$\cancel{m} \times g = G \times \frac{M_{Earth} \times \cancel{m}}{R_{Earth}^2}$$

The mass $m$ is on both sides of the equation, so we can cancel it out. The equation then becomes

$$g = G \times \frac{M_{Earth}}{R_{Earth}^2}$$

The expression shows that the gravitational acceleration ($g$) experienced by an object of mass $m$ on the surface of Earth is determined by the mass of Earth and by the radius of Earth. The mass of the object itself ($m$) appears nowhere in this expression, so changing $m$ has no effect on the gravitational acceleration of an object on Earth.

Gravity pulls you toward the center of Earth. Gravity holds the planets and stars together and keeps the thin blanket of air we breathe close to Earth's surface. The planets, including Earth, orbit around the Sun, and gravity holds them in orbit. Gravity caused a vast interstellar cloud of gas and dust to collapse 4.5 billion years ago to form the Sun, Earth, and the rest of the Solar System. Gravity binds colossal groups of stars into galaxies. Gravity shapes space and time, and it can affect the ultimate fate of the universe. We will return often to the concept of gravity and find it central to an understanding of the universe.

(a)

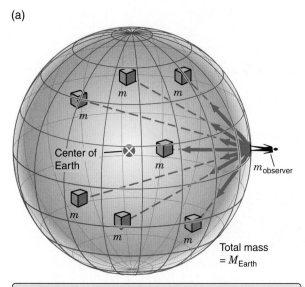

Center of Earth

$m$

$m_{observer}$

Total mass = $M_{Earth}$

An object on the surface of a spherical mass (such as Earth) feels a gravitational attraction toward each small part of the sphere.

(b)

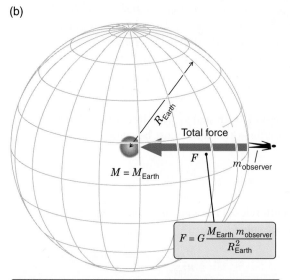

$R_{Earth}$

Total force

$F$

$M = M_{Earth}$

$m_{observer}$

$$F = G \frac{M_{Earth}\, m_{observer}}{R_{Earth}^2}$$

The net force is the same as if we scooped up the mass of the entire sphere and concentrated it at a point at the center.

**Figure 4.4** Outside a sphere, the net gravitational force due to a spherical mass is the same as the gravitational force from the same mass concentrated at a point at the center of the sphere.

---

## CHECK YOUR UNDERSTANDING 4.1

If the distance between Earth and the Sun were cut in half, the gravitational force between these two objects would: (a) decrease by a factor of 4; (b) decrease by a factor of 2; (c) increase by a factor of 2; (d) increase by a factor of 4.

---

## Gravity Differs from Place to Place within an Object

You can think of Earth as a collection of small masses, each of which feels a gravitational attraction toward every other small part of Earth. The mutual gravitational attraction that occurs among all parts of the same object is called self-gravity.

As you sit reading this book, you are exerting a gravitational attraction on every other fragment of Earth, and every other fragment of Earth is exerting a gravitational attraction on you. Your gravitational interaction is strongest with the parts of Earth closest to you. The parts of Earth that are on the other side of our planet are much farther from you, so their pull on you is correspondingly less.

The net effect of all these forces is to pull you (or any other object) toward the center of Earth. If you drop a hammer, it falls directly toward the ground. Because Earth is nearly spherical, for every piece of Earth pulling you toward your right, a corresponding piece of Earth is pulling you toward your left with just as much force. For every piece of Earth pulling you forward, a corresponding piece of Earth is pulling you backward. As you can see in **Figure 4.4a**, because Earth is almost spherically symmetric, all of these "sideways" forces cancel out, leaving behind an overall force that points toward Earth's center.

Some parts of Earth are closer to you and others are farther away, but there is an average distance between you and all of the small fragments of Earth that are pulling on you. This average distance is the distance between you and the center of Earth. As illustrated in Figure 4.4b, the overall pull that you feel is the same as it would be if all the mass of Earth was concentrated at a single point located at the very center of the planet.

This relationship is true for any spherically symmetric object. Outside the object, the gravity from such an object behaves as if all the mass of that object were concentrated at a point at its center. This relationship will be important in many applications. For example, when you estimate your weight on another planet, you are calculating the force of gravity between you and the planet. The "distance" in the gravitational equation will be the distance between you and the center of the planet, which is just the radius of the planet.

## CHECK YOUR UNDERSTANDING 4.2

If Earth shrank to a smaller radius but kept the same mass, would the gravitational force between Earth and the Moon become: (a) smaller; (b) larger; or (c) stay the same? Would everyone's weight on Earth: (a) increase; (b) decrease; or (c) stay the same?

# 4.2 An Orbit Is One Body "Falling around" Another

Kepler's laws on the motions of planets enabled astronomers to predict the positions of the planets accurately, but these laws did not explain why the planets behave as they do. Newton's work explained why planets orbit the Sun.

Newton used his laws of motion and his proposed law of gravity to predict the paths of planetary orbits. His calculations showed that these orbits should be ellipses with the Sun at one focus, that planets should move faster when closer to the Sun, and that the square of the period of a planet's orbit should vary as the cube of the semimajor axis of that elliptical orbit. Newton's universal law of gravitation *predicted* that planets should orbit the Sun in just the way that Kepler's empirical laws described. By explaining Kepler's laws, Newton found important corroboration for his law of gravitation.

## Gravity and Orbits

Newton's laws tell us how forces change an object's motion and how objects interact with each other through gravity. To know where an object will be at any given time, we carefully have to "add up" the object's motion over time. Newton invented calculus to do this, but we will aim just for a conceptual understanding.

In Newton's time, the closest thing to making a heavy object fly was shooting cannonballs out of a cannon, so he used cannonballs in "thought experiments" about planetary motions. If one drops a cannonball, it falls directly to the ground, like any other mass does. However, as you can see in **Figure 4.5a**, a cannonball fired out of a cannon that is level with the ground behaves differently. The cannonball still falls to the ground in the same amount of time as it does when it is dropped, but while falling it also travels over the ground, following a curved path that carries it a horizontal distance before it finally lands. As shown in Figure 4.5b, the faster the cannonball moves when it is fired from the cannon, the farther it will go before it hits the ground.

In the real world this experiment reaches a natural limit. To travel through air, the cannonball must push the air out of its way—an effect normally referred to as *air resistance*—which slows it down. But because this is only a thought experiment, we can ignore such real-world complications. Instead imagine that, having inertia, the cannonball continues along its course until it runs into something. The faster the cannonball moves when it is fired, the farther it goes before hitting the ground. If the cannonball flies far enough, Earth's surface curves out from under it, as shown in Figure 4.5c. As illustrated in Figure 4.5d, eventually a point is reached where the cannonball is flying so fast that the surface of Earth curves away from the cannonball at exactly the same rate at which the cannonball is falling toward Earth. When this occurs, the cannonball, which always falls

(a) A cannonball travels over the ground as it falls toward Earth.

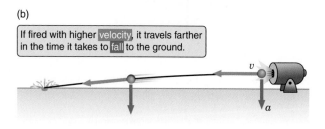

(b) If fired with higher velocity, it travels farther in the time it takes to fall to the ground.

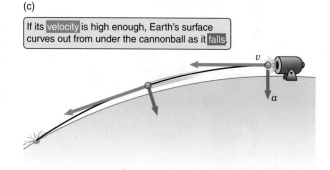

(c) If its velocity is high enough, Earth's surface curves out from under the cannonball as it falls.

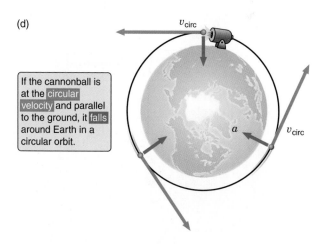

(d) If the cannonball is at the circular velocity and parallel to the ground, it falls around Earth in a circular orbit.

**Figure 4.5** Newton realized that a cannonball fired at the right speed would fall around Earth in a circle. Velocity (*v*) is indicated by a red arrow and acceleration (*a*) by a green arrow.

▶❙❙ **AstroTour:** Newton's Laws and Universal Gravitation

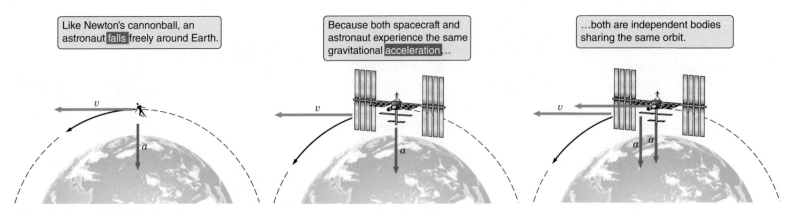

Like Newton's cannonball, an astronaut falls freely around Earth.

Because both spacecraft and astronaut experience the same gravitational acceleration...

...both are independent bodies sharing the same orbit.

**Figure 4.6** A "weightless" astronaut has not escaped Earth's gravity. Rather, an astronaut and a spacecraft share the same orbit as they fall around Earth together.

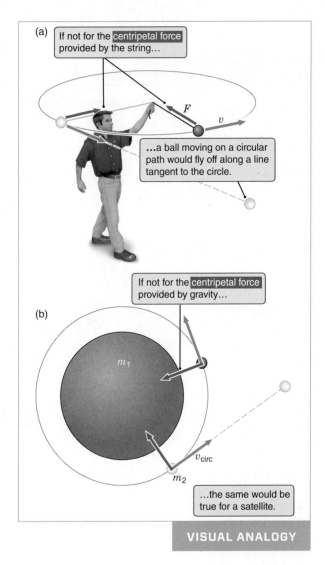

(a)

If not for the centripetal force provided by the string...

...a ball moving on a circular path would fly off along a line tangent to the circle.

If not for the centripetal force provided by gravity...

(b)

$m_1$

$v_{circ}$

$m_2$

...the same would be true for a satellite.

**VISUAL ANALOGY**

**Figure 4.7** (a) A string provides the centripetal force that keeps a ball moving in a circle. (We are ignoring the smaller force of gravity that also acts on the ball.) (b) Similarly, gravity provides the centripetal force that holds a satellite in a circular orbit.

toward the center of Earth, is, in a sense, falling around the world. An **orbit** is the path of one object that freely falls around another.

Why do astronauts appear to float freely about the cabin of a spacecraft? It is not because they have escaped Earth's gravity; it is Earth's gravity that holds them in their orbit. Instead the answer lies in Galileo's early observation that all objects fall with the same acceleration, regardless of their mass. The astronauts and the spacecraft are both in orbit around Earth, moving in the same direction, at the same speed, and experiencing the same gravitational acceleration, so they fall around Earth together. **Figure 4.6** demonstrates this point. The astronaut is orbiting Earth just as the spacecraft is orbiting Earth. On the surface of Earth, your body tries to fall toward the center of Earth, but the ground gets in the way. You experience your weight when you are standing on Earth because the ground pushes on you to oppose the force of gravity, which pulls you downward. In the spacecraft, however, nothing interrupts the astronaut's fall, because the spacecraft is falling around Earth in just the same orbit. The astronaut is in **free fall**, falling freely in Earth's gravity. The **Process of Science Figure** illustrates the universality of Newton's law of gravitation.

## What Velocity Is Needed to Reach Orbit?

How fast must Newton's cannonball be fired for it to fall around the world? The cannonball would be in **uniform circular motion**, which means it moves along a circular path at constant speed. This type of motion is discussed in more depth in Appendix 8. Another example of uniform circular motion is a ball whirling around your head on a string, illustrated in **Figure 4.7a**. If you let go of the string, the ball will fly off in a straight line in whatever direction it is traveling at the time, just as Newton's first law predicts for an object in motion. The string prevents the ball from flying off by constantly changing the direction the ball is traveling. The central force of the string on the ball is called a **centripetal force**: a force toward the center of the circle. Using a more massive ball, speeding up its motion, or making the string shorter so that the turn is tighter all increase the force needed to keep a ball moving in a circle.

In the case of Newton's cannonball (or a satellite), there is no string to hold the ball in its circular motion. Instead, the force is provided by gravity, as illustrated in Figure 4.7b. The force of gravity must be just enough to keep the satellite moving on its circular path. Because this force has a specific strength, it follows that

## UNIVERSALITY

The laws of physics are the same everywhere and at all times.
The principle underlies our understanding of the natural world.

Galileo determined that
all objects have the same
gravitational acceleration.

Newton's law of universal
gravitation extended this
observation to the Solar System.

*Apollo 15* commander
David Scott tested the law
with a feather and a hammer
on the Moon. With no air
resistance, the feather and
the hammer fell at the
same rate.

The same physical laws apply to falling objects, to planets orbiting the Sun, to stars orbiting within
the galaxy, and to galaxies orbiting each other.

the satellite must be moving at a particular speed around the circle, which we call its **circular velocity** ($v_{circ}$). If the satellite were moving at any other velocity, it would not be moving in a circular orbit. Remember the cannonball: if it is moving too slowly, it will drop below the circular path and hit the ground. Similarly, if the cannonball is moving too fast, its motion will carry it above the circular orbit. Only a cannonball moving at just the right velocity—the circular velocity—will fall around Earth on a circular path (see Figure 4.5d).

Newton's thought experiment became a reality in 1957, when the Soviet Union launched the first human-made object to orbit Earth. They used a rocket to lift Sputnik 1, an object about the size of a basketball, high enough above Earth's upper atmosphere that air resistance wasn't an issue. Sputnik 1 was given a high enough speed that it fell around Earth, just as Newton's imaginary cannonball.

When one object is falling around a much more massive object, we say that the less massive object is a **satellite** of the more massive object. Planets are satellites of the Sun, and moons are natural satellites of planets. Newton's imaginary cannonball and Sputnik 1 were satellites (*sputnik* means "satellite" in Russian). A spacecraft orbiting Earth and the astronauts inside of it are independent satellites of Earth that conveniently happen to share the same orbit.

## The Shape of Orbits

Some Earth satellites travel a circular path at constant speed. Just like the ball on the string, satellites traveling at the circular velocity remain the same distance from Earth at all times, neither speeding up nor slowing down in orbit. But what if the satellite then fired its rockets and started traveling *faster* than the circular velocity? The pull of Earth is as strong as ever, but because the satellite has greater speed, its path is not bent by Earth's gravity sharply enough to hold it in a circle. So the satellite begins to climb above a circular orbit.

As the distance between the satellite and Earth increases, the satellite slows down. Think about a ball thrown upwards into the air, illustrated in **Figure 4.8a**. As the ball climbs higher, the pull of Earth's gravity opposes its motion, slowing the ball down. The ball climbs more and more slowly until its vertical motion

**Nebraska Simulation:** Earth Orbit Plot

▶❙❙ **AstroTour:** Elliptical Orbit

**Figure 4.8** (a) A ball thrown into the air slows as it climbs away from Earth and then speeds up as it heads back toward Earth. (b) A planet on an elliptical orbit around the Sun does the same thing. (Although no planet has an orbit as eccentric as the one shown here, the orbits of comets can be far more eccentric.)

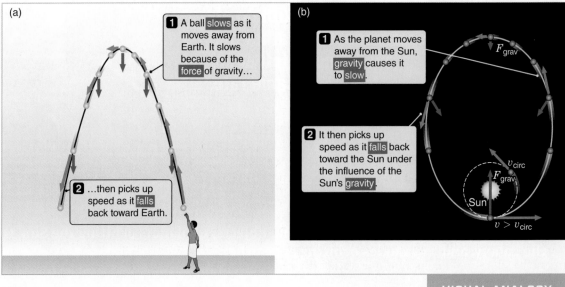

VISUAL ANALOGY

stops for an instant and then is reversed; the ball then begins to fall back toward Earth, speeding up along the way. A satellite does exactly the same thing. As the satellite climbs above a circular orbit and begins to move away from Earth, Earth's gravity opposes the satellite's outward motion, slowing the satellite down. The farther the satellite is from Earth, the more slowly the satellite moves—just like the ball thrown into the air. Also just like the ball, the satellite reaches a maximum height on its curving path and then begins falling back toward Earth, while Earth's gravity speeds it up as it gets closer and closer to Earth. The satellite's orbit has changed from circular to elliptical.

Any object in an elliptical orbit, including a planet orbiting the Sun, will therefore move faster when it is closer to the object it is orbiting due to gravity. Recall from Chapter 3 that Kepler's second law says that a planet moves fastest when it is closest to the Sun and slowest when it is farthest from the Sun. As shown in Figure 4.8b, planets lose speed as they pull away from the Sun and then gain that speed back as they fall inward toward the Sun.

Newton's laws do more than explain Kepler's laws: they predict different types of orbits beyond Kepler's empirical experience. **Figure 4.9a** shows a series of satellite orbits, each with the same point of closest approach to Earth but with different velocities at that point, as indicated in Figure 4.9b. The greater the speed a satellite has at its closest approach to Earth, the farther the satellite is able to pull away from Earth, and the more eccentric its orbit becomes. As long as it remains elliptical, no matter how eccentric, the orbit is called a **bound orbit** because the satellite is gravitationally bound to the object it is orbiting.

In this sequence of faster and faster satellites there comes a point at which the satellite is moving so fast that gravity is unable to reverse its outward motion, so the object travels away from Earth, never to return. The lowest speed at which an object can permanently leave the gravitational grasp of another mass is called the **escape velocity**, $v_{esc}$. Once a satellite's velocity at closest approach equals or exceeds $v_{esc}$, and it is no longer gravitationally bound to the object it was orbiting, we say it is in an **unbound orbit**.

As Figure 4.9 shows, an object with a velocity *less* than the escape velocity ($v_{esc}$) will be on an elliptically shaped orbit and will follow the same path over and over again. Unbound orbits do not close like an ellipse (see Figure 4.9a). An object such as a comet on an unbound orbit makes only a single pass around the Sun and then continues away from the Sun into deep space, never to return. Circular velocity and escape velocity are further explored in **Working It Out 4.2**.

## Measuring Mass Using Newton's Version of Kepler's Law

Newton's calculations opened up an entirely new way of investigating the universe. He showed that the same physical laws that describe the flight of a cannonball on Earth—or the legendary apple falling from a tree—also describe the motions of the planets through the heavens. His laws of motion and gravitation predict all three of Kepler's empirical laws of planetary motion. Newton's version of Kepler's laws can be used to measure the mass of the Sun from the orbit of Earth, as seen in **Working It Out 4.3**.

When a much lower mass object such as Earth is orbiting a much more massive object such as the Sun, the Sun's gravity has a strong influence on Earth, but Earth's gravity has little effect on the Sun. Therefore, it is a good approximation to say that the Sun remains motionless as Earth orbits around it.

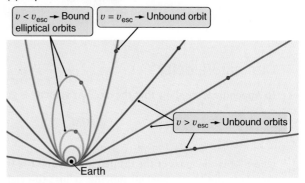

(a) Representative orbits

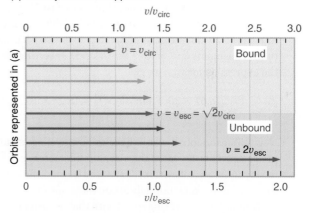

(b) Velocity at closest approach

**Figure 4.9** (a) A range of different orbits that share the same point of closest approach but differ in velocity at that point. (b) Closest-approach velocities for the orbits in (a). An object's velocity at closest approach determines the orbit shape and whether the orbit is bound or unbound. $v_{circ}$ = circular velocity; $v_{esc}$ = escape velocity.

## 4.2 Working It Out  Circular Velocity and Escape Velocity

### Circular Velocity

In Appendix 7, we show that the circular velocity is given by

$$v_{circ} = \sqrt{\frac{GM}{r}}$$

where $M$ is the mass of the orbited object, and $r$ is the radius of the circular orbit. A cannonball moving at just the right velocity—the circular velocity—will fall around Earth on a circular path.

We can use this equation to show how fast Newton's cannonball would have to travel to stay in its circular orbit. The average radius of Earth is 6,370 km, the mass of Earth is $5.97 \times 10^{24}$ kg, and the gravitational constant is $6.67 \times 10^{-20}$ km$^3$/(kg s$^2$). Inserting these values into the expression for $v_{circ}$, we get

$$v_{circ} = \sqrt{\frac{[6.67 \times 10^{-20}\,\text{km}^3/(\text{kg}\,\text{s}^2)] \times (5.97 \times 10^{24}\,\text{kg})}{6,370\,\text{km}}} = 7.9\,\text{km/s}$$

Newton's cannonball would have to be traveling about 8 kilometers per second (km/s)—more than 28,000 kilometers per hour (km/h)—to stay in its circular orbit. That's well beyond the reach of a typical cannon, but rockets routinely attain these speeds.

Now let's compare this speed with that needed to launch a satellite from the Moon into orbit just above the lunar surface. The radius of the Moon is 1,740 km, and its mass is $7.35 \times 10^{22}$ kg. These values give the following circular velocity:

$$v_{circ} = \sqrt{\frac{[6.67 \times 10^{-20}\,\text{km}^3/(\text{kg}\,\text{s}^2)] \times (7.35 \times 10^{22}\,\text{kg})}{1,740\,\text{km}}} = 1.7\,\text{km/s}$$

The velocity needed to launch a satellite into a low circular orbit is considerably lower on the Moon than on Earth.

### Escape Velocity

Sending a spacecraft to another planet requires launching it with a velocity greater than Earth's escape velocity. The escape velocity is a factor of $\sqrt{2}$, or approximately 1.41, times the circular velocity. This relation can be expressed as

$$v_{esc} = \sqrt{2} \times v_{circ} = \sqrt{\frac{2GM}{R}}$$

Using the numbers in the above example, we can calculate the escape velocity from the surface of Earth:

$$v_{esc} = \sqrt{2} \times v_{circ} = 1.41 \times 7.9\,\text{km/s} = 11.2\,\text{km/s}$$

To leave Earth, a rocket must have a speed of 11.2 km/s, or 40,300 km/h.

As with weight, the escape velocity from other astronomical objects will be different than the escape velocity from Earth. Ida is a small asteroid orbiting the Sun between the orbits of Mars and Jupiter. Ida has an average radius of 15.7 km and a mass of $4.2 \times 10^{16}$ kg. Therefore,

$$v_{esc} = \sqrt{\frac{2 \times [6.67 \times 10^{-20}\,\text{km}^3/(\text{kg}\,\text{s}^2)] \times (4.2 \times 10^{16}\,\text{kg})}{15.7\,\text{km}}}$$

$$v_{esc} = 0.019\,\text{km/s} = 68\,\text{km/h}$$

A baseball thrown at about 130 km/h would easily escape from Ida's surface and fly off into interplanetary space.

**Astronomy in Action:** Center of Mass

However, later we will see that sometimes the two objects are closer to having the same mass; for example, dwarf planet Pluto and its moon Charon, or a large planet and a star, or two stars. In these examples, both objects experience significant accelerations in response to their mutual gravitational attraction. The two objects are both orbiting about a common point located between them, called the **center of mass**, so we now must think of them as falling around each other. Each mass is moving on its own elliptical orbit around the two objects' mutual center of mass. From measuring the size and period of any orbit, we can calculate the *sum* of the masses of the orbiting objects. Almost all knowledge about the masses of astronomical objects comes directly from the application of Newton's version of Kepler's third law.

### CHECK YOUR UNDERSTANDING 4.3

If we wanted to increase the Hubble Space Telescope's altitude above Earth and keep it in a stable orbit, we also would need to: (a) increase its orbital speed; (b) increase its weight; (c) decrease its weight; (d) decrease its orbital speed.

## 4.3 Working It Out Calculating Mass from Orbital Periods

### Newton's Version of Kepler's Third Law

The time it takes for a planet to complete one orbit around the Sun equals the distance traveled divided by the planet's speed. For simplicity, let's assume the orbit is circular. Thus, the time it takes an object to make one trip around the Sun is the circumference of the circle ($2\pi r$) divided by the object's speed ($v$). The speed of the planet must be equal to the circular velocity discussed in Working It Out 4.2. Putting these relationships together, we have

$$\text{Orbital period } (P) = \frac{\text{Circumference of orbit}}{\text{Circular velocity}} = \frac{2\pi r}{\sqrt{\dfrac{GM_{\text{Sun}}}{r}}}$$

Squaring both sides of the equation gives

$$P^2 = \frac{4\pi^2 r^2}{\dfrac{GM_{\text{Sun}}}{r}} = \frac{4\pi^2}{GM_{\text{Sun}}} \times r^3$$

The square of the period of an orbit is equal to a constant ($4\pi^2/GM_{\text{Sun}}$) multiplied by the cube of the radius of the orbit. This is Kepler's third law ($P^2 = constants \times A^3$) applied to circular orbits. When Kepler used Earth units of years and astronomical units for the planets, he was taking a ratio of their periods and orbital radii with those for Earth, so the constants cancelled out. Using calculus, Newton was similarly able to derive Kepler's third law for elliptical orbits with semimajor axis $A$ instead of radius $r$.

### Mass of the Sun

If we can measure the size and period of any orbit, then we can use Newton's universal law of gravitation to calculate the mass of the object being orbited. To do so, we rearrange Newton's form of Kepler's third law above to read

$$M = \frac{4\pi^2}{G} \times \frac{A^3}{P^2}$$

Everything on the right side of this equation is either a constant (4, $\pi$, and $G$) or a quantity that we can measure (the semimajor axis $A$ and period $P$ of an orbit). The left side of the equation is the mass of the object at the focus of the ellipse. For example, we can find the mass of the Sun by noting the period and semimajor axis of the orbit of a planet around the Sun. Let's use the numbers for Earth. Whenever we have an equation with $G$, it is best to put everything else into the same units as $G$ (km, kg, s). So first we must compute the number of seconds in 1 year: $P = 1$ yr $= 365.24$ days/yr $\times$ 24 h/day $\times$ 60 min/h $\times$ 60 s/min $= 3.16 \times 10^7$ s. The semimajor axis $A = 1$ AU $= 1.5 \times 10^8$ km. Then the mass of the Sun can be computed:

$$M_{\text{Sun}} = \frac{4\pi^2}{G} \times \frac{A^3}{P^2} = \frac{4\pi^2}{6.67 \times 10^{-20}\,\text{km}^3/(\text{kg}\,\text{s}^2)} \times \frac{(1.5 \times 10^8\,\text{km})^3}{(3.16 \times 10^7\,\text{s})^2}$$

$$M_{\text{Sun}} = 2.00 \times 10^{30}\,\text{kg}$$

We could have used the period and semimajor axis of any Solar System planet to get the same result.

## 4.3 Tidal Forces Are Caused by Gravity

The rise and the fall of the oceans are called Earth's **tides**. Coastal dwellers long ago noted that the strength of the tides varies with the phase of the Moon. Tides are strongest during a new or a full Moon and are weakest during first quarter or third quarter Moon. In this section, we see how tides result from differences between the strength of the gravitational pull of the Moon and Sun on one part of Earth in comparison to their pull on other parts of Earth.

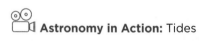 **Astronomy in Action:** Tides

### Tides and the Moon

Figure 4.4 demonstrated that each small part of an object feels a gravitational attraction toward every other small part of the object, and this self-gravity differs from place to place. In addition, each small part of an object feels a gravitational attraction toward every other mass in the universe, and these external forces differ from place to place within the object as well.

▶❚❚ **AstroTour:** Tides and the Moon

The Moon's gravity pulls on Earth as if the mass of the Moon is concentrated at the Moon's center. The side of Earth that faces the Moon is closer to the Moon than is the rest of Earth, so it feels a stronger-than-average gravitational

 **Nebraska Simulation:** Tidal Bulge Simulation

(a)      (b)          (c)          (d)

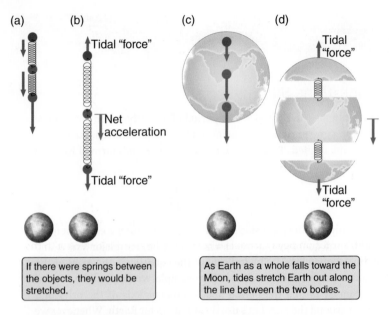

If there were springs between the objects, they would be stretched.

As Earth as a whole falls toward the Moon, tides stretch Earth out along the line between the two bodies.

**Figure 4.10** (a) Imagine three objects connected by springs. (b) The springs are stretched as if there were forces pulling outward on each end of the chain. (c) Similarly, three locations on Earth experience different gravitational attractions toward the Moon. (d) The difference in the Moon's gravitational attraction across Earth causes of Earth's tides.

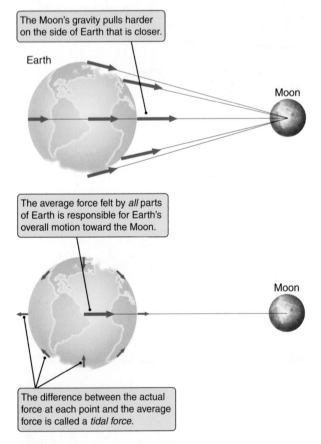

The Moon's gravity pulls harder on the side of Earth that is closer.

Earth

Moon

The average force felt by *all* parts of Earth is responsible for Earth's overall motion toward the Moon.

Moon

The difference between the actual force at each point and the average force is called a *tidal force*.

**Figure 4.11** Tidal forces stretch Earth along the line between Earth and the Moon but compress Earth perpendicular to this line.

attraction toward the Moon. In contrast, the side of Earth facing away from the Moon is farther than average from the Moon, so it feels a weaker-than-average attraction toward the Moon. The pull of the Moon on the near side of Earth is about 7 percent greater than its pull on the far side of Earth.

To understand the consequence of this variation in the pull of the Moon, imagine three rocks being pulled by gravity toward the Moon. A rock closer to the Moon feels a stronger force than a rock farther from the Moon. Now suppose the three rocks are connected by springs (**Figure 4.10a**). As the rocks are pulled toward the Moon, the purple rock pulls away from the red rock, and the red rock pulls away from the blue rock. Therefore, the differences in the gravitational forces they feel will stretch *both* of the springs (Figure 4.10b). Now instead of springs, imagine that the rocks are at different places on Earth (Figure 4.10c). On the side of Earth away from the Moon, the force is smaller (as indicated by the shorter arrow), so that part gets left behind (Figure 4.10d). These differences in the Moon's gravitational attraction on different parts of Earth are called **tidal forces**.

**Figure 4.11** shows how Earth is stretched, causing a tidal bulge. The Moon is not pushing the far side of Earth away; rather, it simply is not pulling on the far side of Earth as hard as it is pulling on the planet as a whole. The far side of Earth is "left behind" as the rest of the planet is pulled more strongly toward the Moon. Figure 4.11 shows that there is also a net force squeezing inward on Earth in the direction perpendicular to the line between Earth and the Moon. Together, the stretching by tidal forces along the line between Earth and the Moon and the squeezing by tidal forces perpendicular to this line distort the shape of Earth like a rubber ball caught in the middle of a tug-of-war.

If the surface of Earth was perfectly smooth and covered with a uniform ocean and Earth did not rotate, then the Moon would pull our oceans into an elongated **tidal bulge** like that shown in **Figure 4.12a**. The water would be at its deepest on the side toward the Moon and on the side away from the Moon and at its shallowest midway between. However, Earth is *not* covered with perfectly uniform oceans, and Earth does rotate. As any point on Earth rotates through the ocean's tidal bulges, that point experiences the ebb and flow of the tides. In addition, friction between the spinning Earth and its tidal bulge drags the oceanic tidal bulge around in the direction of Earth's rotation, as illustrated in Figure 4.12b.

Follow along in Figure 4.12c as you imagine riding on Earth throughout the course of a day. You begin as the rotating Earth carries you through the tidal bulge on the Moonward side of the planet. Because Earth's rotation drags the tidal bulge, the Moon is not exactly overhead but is instead high in the western sky. When you are at the high point in the tidal bulge, the ocean around you has risen higher than average—called a *high tide*. About $6\frac{1}{4}$ hours later, somewhat after the Moon has settled beneath the western horizon, the rotation of Earth carries you through a point where the ocean is lower than average—called a *low tide*. If you wait another $6\frac{1}{4}$ hours, it is again high tide. You are now passing through the region where ocean water is "left behind" (relative to Earth as a whole) in the tidal bulge on the side of Earth that is away from the Moon. The Moon, which is responsible for the tides you see, is itself at that time hidden from view on the far side of Earth. About $6\frac{1}{4}$ hours later, sometime after the Moon has risen above the

(a)

The Moon's tidal forces stretch Earth and its oceans into an elongated shape. The departure from spherical is called Earth's *tidal bulge*.

(b)

Because of friction, Earth's rotation drags its tidal bulge around, out of perfect alignment with the Moon.

(c)

Ocean tides rise and fall as the rotation of Earth carries us through the ocean's tidal bulges.

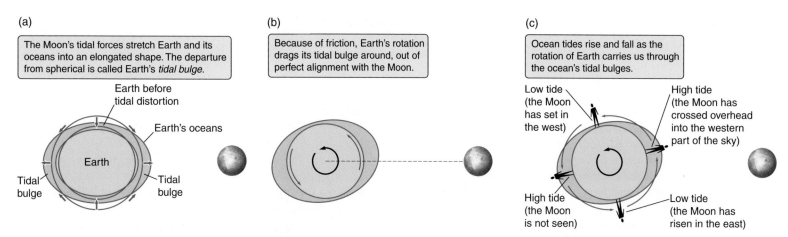

**Figure 4.12** (a) Tidal forces pull Earth and its oceans into a tidal bulge. (b) Earth's rotation pulls its tidal bulge slightly out of alignment with the Moon. (c) As Earth's rotation carries us through these bulges, we experience the ocean tides. The magnitude of the tides has been exaggerated in these diagrams for clarity. In these figures, the observer is looking down from above Earth's North Pole. Sizes and distances are not to scale.

eastern horizon, it is low tide. About 25 hours after you started this journey—the amount of time the Moon takes to return to the same point in the sky from which it started—you again pass through the tidal bulge on the near side of the planet. This is the age-old pattern by which mariners have lived their lives for millennia: the twice-daily coming and going of high tide, shifting through the day in lock-step with the passing of the Moon.

Local geography, such as the shapes of Earth's shorelines and ocean basins, complicate the simple picture of tides. In addition, there are oceanwide oscillations similar to water sloshing around in a basin. As they respond to the tidal forces from the Moon, Earth's oceans flow around the various landmasses that break up the water covering our planet. Some places, like the Mediterranean Sea and the Baltic Sea, are protected from tides by their relatively small sizes and the narrow passages connecting these bodies of water with the larger ocean. In other places, the shape of the land funnels the tidal surge from a large region of ocean into a relatively small area, concentrating its effect, as at the Bay of Fundy (**Figure 4.13**). Tidal effects from the Sun and Moon are very slight even in the Great Lakes, where the water is more affected by local weather and geography.

## Solar Tides

The tides resulting from the pull of the Moon are called **lunar tides**. The Sun also influences Earth's tides. The gravitational pull of the Sun causes Earth to stretch along a line pointing approximately in the direction of the Sun. The side of Earth closer to the Sun is pulled toward the Sun more strongly than is the side of Earth away from the Sun, just as the side of Earth closest to the Moon is pulled more strongly toward the Moon. Tides on Earth due to differences in the gravitational pull of the Sun are called **solar tides**. Although the absolute strength of the Sun's pull on Earth is nearly 200 times greater than the strength of the Moon's pull on Earth, the Sun's gravitational attraction does not change by much from one side of Earth to the other, because the Sun is much farther away than the Moon. As a result, solar tides are only about half as strong as lunar tides (**Working It Out 4.4**).

(a)

(b)

**Figure 4.13** The world's most extreme tides are found in the Bay of Fundy in eastern Canada. Water rocks back and forth in this bay with a period of about 13 hours, close to the 12.5-hour period of the tides. The shape of the basin amplifies the tides so that difference in water depth between low tide (a) and high tide (b) is extreme; typically about 14.5 meters and as much as 16.6 meters.

## 4.4  Working It Out  Tidal Forces

Earlier you learned that the strength of the gravitational force between two bodies is proportional to their masses and inversely proportional to the square of the distance between them. The strength of tidal forces caused by one body acting on another is also proportional to the mass of the body that is raising the tides, but it is inversely proportional to the *cube* of the distance between them.

The equation for the tidal force comes from the difference of the gravitational force on one side of a body compared with the force on the other side. The tidal force acting on Earth by the Moon is given by

$$F_{\text{tidal}}(\text{Moon}) = \frac{2GM_{\text{Earth}}M_{\text{Moon}}R_{\text{Earth}}}{d_{\text{Earth-Moon}}^3}$$

where $R_{\text{Earth}}$ is Earth's radius and $d_{\text{Earth-Moon}}$ is the distance between Earth and the Moon.

As an example, let's compare the tidal force acting on Earth by the Moon with the tidal force acting on Earth by the Sun. $F_{\text{tidal}}$ from the Moon is given in the preceding equation; $F_{\text{tidal}}$ from the Sun is given by

$$F_{\text{tidal}}(\text{Sun}) = \frac{2GM_{\text{Earth}}M_{\text{Sun}}R_{\text{Earth}}}{d_{\text{Earth-Sun}}^3}$$

We know the Moon is much closer to Earth than the Sun is, but the Sun is much more massive than the Moon. To compare the lunar and solar tides, we can take a ratio of the tidal forces and proceed in a similar way to our comparison of gravitational forces in Working It Out 4.1:

$$\frac{F_{\text{tidal}}(\text{Moon})}{F_{\text{tidal}}(\text{Sun})} = \frac{\dfrac{2GM_{\text{Earth}}M_{\text{Moon}}R_{\text{Earth}}}{d_{\text{Earth-Moon}}^3}}{\dfrac{2GM_{\text{Earth}}M_{\text{Sun}}R_{\text{Earth}}}{d_{\text{Earth-Sun}}^3}}$$

Canceling out the constant $G$ and the terms common in both equations ($M_{\text{Earth}}$ and $R_{\text{Earth}}$) gives

$$\frac{F_{\text{tidal}}(\text{Moon})}{F_{\text{tidal}}(\text{Sun})} = \frac{M_{\text{Moon}}}{M_{\text{Sun}}} \times \frac{d_{\text{Earth-Sun}}^3}{d_{\text{Earth-Moon}}^3}$$

$$\frac{F_{\text{tidal}}(\text{Moon})}{F_{\text{tidal}}(\text{Sun})} = \frac{M_{\text{Moon}}}{M_{\text{Sun}}} \times \left(\frac{d_{\text{Earth-Sun}}}{d_{\text{Earth-Moon}}}\right)^3$$

Using the values from Appendix 4, $M_{\text{Moon}} = 7.35 \times 10^{22}$ kg, $M_{\text{Sun}} = 2 \times 10^{30}$ kg, $d_{\text{Earth-Moon}} = 384,400$ km, and $d_{\text{Earth-Sun}} = 1.5 \times 10^8$ km gives

$$\frac{F_{\text{tidal}}(\text{Moon})}{F_{\text{tidal}}(\text{Sun})} = \frac{7.35 \times 10^{22}\,\text{kg}}{2 \times 10^{30}\,\text{kg}} \times \left(\frac{1.5 \times 10^8\,\text{km}}{384,400\,\text{km}}\right)^3 = 2.2$$

So the tidal force from the Moon is 2.2 times stronger than the tidal force from the Sun, which is why we often hear that tides are caused by the Moon. But the Sun is an important factor, too, and that's why the tides change depending on the alignment of the Moon and the Sun with Earth.

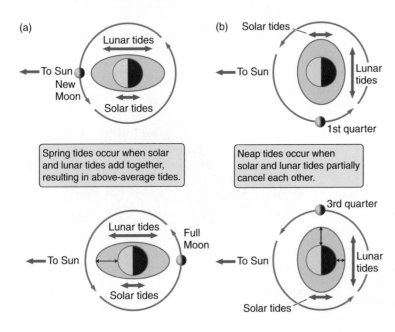

**Figure 4.14** Solar tides are about half as strong as lunar tides. The interactions of solar and lunar tides result in either (a) spring tides when they are added together or (b) neap tides when they partially cancel each other.

Solar and lunar tides interact. As shown in **Figure 4.14a**, when the Moon and the Sun are lined up with Earth, at either new or full Moon, the lunar and solar tides on Earth overlap. This creates more extreme tides ranging from extra-high high tides to extra-low low tides. The extreme tides near the new or full Moon are called **spring tides**—not because of the season, but because the water appears to spring out of the sea. Conversely—illustrated in Figure 4.14b—when the Moon, Earth, and Sun make a right angle, at the Moon's first and third quarters, the lunar and solar tidal forces stretch Earth in different directions, creating less extreme tides known as **neap tides**. The word *neap* is derived from the Saxon word *neafte*, which means "scarcity": at these times of the month, shellfish and other food gathered in the tidal region are less accessible because the low tide is higher than at other times. Neap tides are only about half as strong as average tides and only a third as strong as spring tides.

### CHECK YOUR UNDERSTANDING 4.4

Rank in order of the strongest tides: (a) new Moon in July; (b) first quarter Moon in July; (c) full Moon in January; (d) third quarter Moon in January.

# 4.4 Tidal Forces Affect Solid Bodies

In the previous section, we focused on the movements of the liquid of Earth's oceans in response to the tidal forces from the Moon and Sun. But these tidal forces also affect the solid body of Earth. As Earth rotates through its tidal bulge, the solid body of the planet is constantly being deformed by tidal forces. Earth is somewhat elastic (like a rubber ball), and tidal stresses cause a vertical displacement of about 30 centimeters (cm) between high tide and low tide, or roughly a third of the displacement of the oceans. It takes energy to deform the shape of a solid object. (If you want a practical demonstration of this fact, hold a rubber ball in your hand and squeeze and release it a few dozen times.) This energy from the deformation is converted into thermal energy by friction in Earth's interior. This friction opposes and takes energy from the rotation of Earth, causing Earth to gradually slow. Earth's internal friction adds to the slowing caused by friction between Earth and its oceans as the planet rotates through the tidal bulge of the oceans. As a result, Earth's days are currently lengthening by about 1.7 milliseconds (ms) every century. This sounds small but it adds up: when dinosaurs ruled, the day was closer to 23 hours long, and 200 million years into the future, the day will be close to 25 hours.

## Tidal Locking

Other solid bodies besides Earth experience tidal forces. For example, the Moon has no bodies of liquid to make tidal forces obvious, but its shape is distorted in the same manner as Earth. Because of Earth's much greater mass and the Moon's smaller radius, the tidal effects of Earth on the Moon are about 20 times as great as the tidal effects of the Moon on Earth. Given that the average tidal deformation of Earth is about 30 cm, the average tidal deformation of the Moon should be about 6 meters. However, what we actually observe on the Moon is a tidal bulge of about 20 meters. This unexpectedly large displacement exists because the Moon's tidal bulge was "frozen" into its relatively rigid crust at an earlier time, when the Moon was closer to Earth and tidal forces were much stronger than they are today. Planetary scientists sometimes call this deformation the Moon's *fossil tidal bulge*.

Recall from Chapter 2 that the Moon's rotation period exactly equals its orbital period. This synchronous rotation of the Moon is a result of **tidal locking**. Early in its history, the period of the Moon's rotation was almost certainly different from its orbital period. As the Moon rotated through its extreme tidal bulge, however, friction within the Moon's crust was tremendous, rapidly slowing the Moon's rotation. After a fairly short time, the period of the Moon's rotation equaled the period of its orbit. When its orbital and rotation periods became equalized, the Moon no longer rotated with respect to its tidal bulge. Instead, the Moon and its tidal bulge rotated *together*, in lockstep with the Moon's orbit around Earth. As illustrated in **Figure 4.15**, this scenario continues today as the tidally distorted Moon orbits Earth, always keeping the same face and the long axis of its tidal bulge toward Earth.

Tidal forces affect not only the rotations of the Moon and Earth, but also their orbits. Because of its tidal bulge, Earth is not a perfectly spherical body. Therefore, the material in Earth's tidal bulge on the side nearer the Moon pulls on the Moon more strongly than does material in the tidal bulge on the back side of

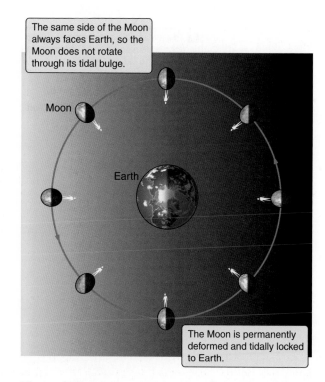

The same side of the Moon always faces Earth, so the Moon does not rotate through its tidal bulge.

Moon

Earth

The Moon is permanently deformed and tidally locked to Earth.

**Figure 4.15** Tidal forces due to Earth's gravity lock the Moon's rotation to its orbital period.

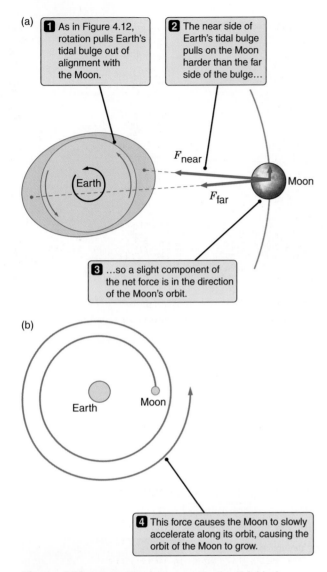

(a)

1 As in Figure 4.12, rotation pulls Earth's tidal bulge out of alignment with the Moon.

2 The near side of Earth's tidal bulge pulls on the Moon harder than the far side of the bulge...

Earth

$F_{near}$

$F_{far}$

Moon

3 ...so a slight component of the net force is in the direction of the Moon's orbit.

(b)

Earth

Moon

4 This force causes the Moon to slowly accelerate along its orbit, causing the orbit of the Moon to grow.

**Figure 4.16** Interaction between Earth's tidal bulge and the Moon causes the Moon to accelerate in its orbit and the Moon's orbit to grow.

Earth. Because the tidal bulge on the Moonward side of Earth "leads" the Moon somewhat, as shown in **Figure 4.16**, the gravitational attraction of the bulge causes the Moon to accelerate slightly along the direction of its orbit around Earth. The rotation of Earth is dragging the Moon along with it. The acceleration of the Moon in the direction of its orbital motion causes the orbit of the Moon to grow larger. At present, the Moon is drifting away from Earth at a rate of 3.83 cm per year.

As the Moon grows more distant, the length of the lunar month increases by about 0.014 second each century. If this increase in the radius of the Moon's orbit were to continue long enough (about 50 billion years), Earth would become tidally locked to the Moon, just as the Moon is now tidally locked to Earth. At that point, the period of rotation of Earth, the period of rotation of the Moon, and the orbital period of the Moon would all be exactly the same—about 47 of our current days—and the Moon would be about 43 percent farther from Earth than it is today. However, this situation will never actually occur—at least not before the Sun itself has burned out.

The effects of tidal forces can be seen throughout the Solar System. Most of the moons in the Solar System are tidally locked to their parent planets, and in the case of dwarf planet Pluto and its largest moon, Charon, each is tidally locked to the other.

Tidal locking is only one way that orbits and rotations can be coupled together. Tidal forces have coupled the planet Mercury's rotation to its very elliptical orbit around the Sun. Yet unlike the Moon with its synchronous rotation, Mercury spins on its axis three times for every two trips around the Sun. The period of Mercury's orbit—87.97 Earth days—is exactly $1\frac{1}{2}$ times the 58.64 days that it takes Mercury to spin once on its axis. When Mercury comes to the point in its orbit that is closest to the Sun, one hemisphere faces the Sun, and then in the next orbit the other hemisphere faces the Sun.

## Tidal Forces on Many Scales

We normally think of the effects of tidal forces as small compared to the force of gravity holding an object together, yet tidal effects can be extremely destructive. Consider for a moment the fate of a small moon, asteroid, or comet that wanders too close to a massive planet such as Jupiter or Saturn. All objects in the Solar System larger than about a kilometer in diameter are held together by their self-gravity. However, the self-gravity of a small object such as an asteroid, a comet, or a small moon is feeble. In contrast, the tidal forces close to a massive object such as Jupiter can be very strong. If the tidal forces trying to tear an object apart become greater than the self-gravity trying to hold the object together, the object will break into pieces.

The **Roche limit** is the distance at which a planet's tidal forces are greater than the self-gravity of a smaller object—such as a moon, asteroid, or comet—causing the object to break apart. For a smaller object having the same density as the planet, the Roche limit is about 2.45 times the planet's radius. Such an object bound together solely by its own gravity can remain intact when it is outside a planet's Roche limit, but not when it is inside the limit. Objects such as the International Space Station and other Earth satellites are not torn apart, even though they orbit well within Earth's Roche limit, because chemical bonds hold them together, not just self-gravity.

We have concentrated on the role that tidal forces play on Earth and the Moon. We find tidal forces throughout the Solar System and the universe. Any time two objects of significant size or two collections of objects interact gravitationally, the gravitational forces will differ from one place to another within the objects, giving rise to tidal effects. Tidal disruption of small bodies is the source of the particles that make up the rings of the giant planets. Tidal interactions can cause material from one star in a binary pair to be pulled onto the other star. Tidal effects can strip stars from clusters consisting of thousands of stars. Galaxies can pass close enough together to strongly interact gravitationally. When this happens, as in **Figure 4.17**, tidal effects can grossly distort both galaxies taking part in the interaction. Tidal forces even play a role in shaping huge collections of galaxies—the largest known structures in the universe.

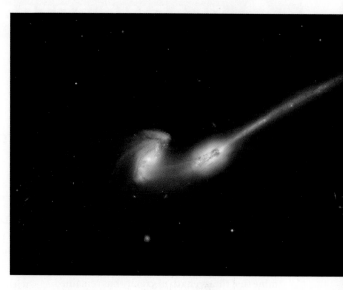

**Figure 4.17** The tidal "tails" seen here are characteristic of tidal interactions between galaxies.

## CHECK YOUR UNDERSTANDING 4.5

The Moon always keeps the same face toward Earth because of: (a) tidal locking; (b) tidal forces from the Sun; (c) tidal forces from Earth; (d) tidal forces from the Earth and Sun.

---

# Origins

## Tidal Forces and Life

In this chapter, we noted that Earth's rotation is slowing down as the Moon slowly moves away into a larger orbit. In the distant past the Moon was closer and Earth rotated faster, so tides would have been stronger and the interval between high tides would have been shorter. The tides are also affected by the configuration of the continents and oceans on Earth, so it is not known precisely how much faster Earth rotated billions of years ago. But the stronger and more frequent tides would have provided additional energy to the oceans of the young Earth.

Scientists debate whether life on Earth originated deep in the ocean, on the surface of the ocean, or on land (see Chapter 24). The tides shaped the regions in the margins between land and ocean, such as tide pools and coastal flats. Some think that these border

regions, which alternate between wet and dry with the tides, could have been places where concentrations of biochemicals periodically became high enough for more complex reactions to take place. These complex reactions were important to the earliest life. Later, these border regions may have been important as advanced life moved from the sea to the land.

Elsewhere in the Solar System, the giant planets Jupiter and Saturn are far from the Sun, and thus very cold. Jupiter and Saturn each have many moons, and the closest moons would experience strong tidal forces from their respective planet. As you saw in Reading Astronomy News in Chapter 1, several of these moons are thought to have a liquid ocean underneath an icy surface. Tidal forces from Jupiter or Saturn provide the heat to keep the water in a

liquid state. Astrobiologists think that these subsurface liquid oceans are perhaps the most probable location for life elsewhere in the Solar System.

We have seen that on Earth, the tidal forces from the Sun are about half as strong as those from the Moon. A planet with a closer orbit would experience much stronger tidal forces from its star. In Chapter 7, you will see that many of the planets detected outside of the Solar System have orbits very close to their stars. These planets experience strong tidal forces, and they might be tidally locked so that they have synchronous rotation as the Moon does around Earth, with one side of the planet always facing the star and one side facing away. How might life on Earth have evolved differently if half of the planet was in perpetual night and half in perpetual day?

# Exploding Stars Prove Newton's Law of Gravity Unchanged over Cosmic Time

By **Swinburne Media Centre**

Australian astronomers have combined all observations of supernovae ever made to determine that the strength of gravity has remained unchanged over the last 9 billion years.

Newton's gravitational constant, known as $G$, describes the attractive force between two objects, together with the separation between them and their masses. It had previously been suggested that $G$ could have been slowly changing over the 13.8 billion years since the Big Bang.

If $G$ had been decreasing over time, for example, this would mean the Earth's distance to the Sun was slightly smaller in the past, meaning that we would experience longer seasons now compared to much earlier points in the Earth's history.

But researchers at Swinburne University of Technology in Melbourne have now analysed the light given off by 580 supernova explosions in the nearby and far universe and have shown that the strength of gravity has not changed.

"Looking back in cosmic time to find out how the laws of physics may have changed is not new," said Professor Jeremy Mould. "But supernova cosmology now allows us to do this with gravity."

A Type Ia supernova marks the violent death of a star called a white dwarf, which is as massive as our Sun but packed into a ball the size of our Earth.

Telescopes can detect the light from this explosion and use its brightness as a "standard candle" to measure distances in the universe, a tool that helped Australian astronomer Professor Brian Schmidt in his 2011 Nobel Prize–winning work discovering the mysterious force Dark Energy.

Professor Mould and his PhD student Syed Uddin at the Swinburne Centre for Astrophysics and Supercomputing and the ARC Centre of Excellence for All-sky Astrophysics (CAASTRO) assumed that these supernova explosions happen when a white dwarf reaches a critical mass or after colliding with other stars to "tip it over the edge."

"This critical mass depends on Newton's gravitational constant $G$ and allows us to monitor it over billions of years of cosmic time—instead of only decades as was the case in previous studies," Professor Mould said.

Despite these vastly different time spans, their results agree with findings from the Lunar Laser Ranging Experiment that has been measuring the distance between the Earth and the Moon since NASA's *Apollo* missions in the 1960s and has been able to monitor possible variations in $G$ at very high precision.

"Our cosmological analysis complements experimental efforts to describe and constrain the laws of physics in a new way and over cosmic time," Mr Uddin said.

In their current publication, the Swinburne researchers were able to set an upper limit on the change in Newton's gravitational constant of 1 part in 10 billion per year over the past 9 billion years.

This research was published in *Publications of the Astronomical Society of Australia*.

1. Explain how observing very distant supernova means observing back in time.
2. Why are physicists still measuring $G$, centuries after Newton described this constant?
3. What quantities that you studied in the chapter would change if the value of $G$ changed over time?
4. Why is it important to the process of science that this result agrees with a result from a totally different experiment?
5. If the distance to the Sun were slightly smaller in the past, would solar tides have been weaker or stronger? Would you expect this to have been as important to tides on Earth as changes in the distance of the Moon?

# Summary

Objects stay in orbit because of gravity. Newton's laws of motion and his proposed law of gravity predict the paths of planetary orbits and explain Kepler's laws. Newton's calculations showed that these orbits should be ellipses with the Sun at one focus, that planets should move faster when closer to the Sun, and that the square of the period of a planet's orbit should vary as the cube of the semi-major axis of that elliptical orbit. Newton also showed mathematically Galileo's conclusion that falling objects have an accelerated motion independent of their mass. Tidal forces provide energy to Earth's oceans. Tide pools on Earth may have been a site of early biochemical reactions. Some moons in the outer Solar System might have liquid water because of tidal heating from their respective planet.

LG 1 **Explain the elements of Newton's universal law of gravitation.** Gravity is a force between any two objects due to their masses. As one of the fundamental forces of nature, gravity binds the universe together. The force of gravity is proportional to the mass of each object and inversely proportional to the square of the distance between them.

LG 2 **Use the concepts of motion and gravitation to explain planetary orbits.** An orbit is one body "falling around" another. Planets orbit the Sun in elliptical orbits. All objects affected by gravity move either on bound elliptical orbits or unbound paths. Orbits are ultimately given their shape by the gravitational attraction of the objects involved, which in turn is a reflection of the masses of these objects.

LG 3 **Explain how tidal forces from the Sun and Moon create Earth's tides.** Tides on Earth are the result of differences between how hard the Moon and Sun pull on one part of Earth in comparison with their pull on other parts of Earth. The primary cause of tides is the Moon, which stretches out Earth. The tides are the strongest when the Sun, Moon, and Earth are aligned. As Earth rotates, tides rise and fall twice each day.

LG 4 **Describe the effects of tidal forces on solid bodies.** Tidal forces lock the Moon's rotation to its orbit around Earth. Tidal forces can break up an object if its gets too close to a more massive object. Tidal forces are observed throughout the universe, in planets and moons, pairs of stars, and interacting galaxies.

## ? UNANSWERED QUESTION

- What range of gravities will support human life? Humans have evolved to live on Earth's surface, but what happens when humans go elsewhere? What are the limits for our hearts, lungs, eyes, and bones? At the higher end of human tolerance, fighter pilots have been trained to experience about 10 times the normal surface gravity on Earth for very short periods of time (too long and they black out). Astronauts who spend several months in near-weightless conditions experience medical problems such as bone loss. On the Moon or Mars, humans will weigh much less than on Earth. Numerous science fiction tales have been written about what happens to children born on a space station or on another planet or moon with low surface gravity: would their hearts and bodies ever be able to adjust to the higher surface gravity of Earth or must they stay in space forever?

# Questions and Problems

## Test Your Understanding

1. In Newton's universal law of gravitation, the force is
   a. proportional to both masses.
   b. proportional to the radius.
   c. proportional to the radius squared.
   d. inversely proportional to the orbiting mass.

2. Rank the following objects in order of their circular velocities, from smallest to largest.
   a. a 5-kg object orbiting Earth halfway to the Moon
   b. a 10-kg object orbiting Earth just above Earth's surface
   c. a 15-kg object orbiting Earth at the same distance as the Moon
   d. a 20-kg object orbiting Earth one-quarter of the way to the Moon

3. An object in a(n) _____ orbit in the Solar System will remain in its orbit forever. An object in a(n) _____ orbit will escape from the Solar System.
   a. unbound; bound
   b. circular; elliptical
   c. bound; unbound
   d. elliptical; circular

4. Compared to your mass on Earth, on the Moon your mass would be
   a. lower because the Moon is smaller than Earth.
   b. lower because the Moon has less mass than Earth.
   c. higher because of the combination of the Moon's mass and size.
   d. the same, mass doesn't change.

5. If you went to Mars, your weight would be
   a. higher because you are closer to the center of the planet.
   b. lower because Mars has two small moons instead of one big moon, so there's less tidal force.
   c. lower because Mars has lower mass and a smaller radius that together produce a lower gravitational force.
   d. the same as on Earth.

6. Venus has about 80 percent of Earth's mass and about 95 percent of Earth's radius. Your weight on Venus will be
   a. 20 percent more than on Earth.
   b. 20 percent less than on Earth.
   c. 10 percent more than on Earth.
   d. 10 percent less than on Earth.

7. The connection between gravity and orbits enables astronomers to measure the _____ of stars and planets.
   a. distances
   b. sizes
   c. masses
   d. compositions

8. If the Moon had twice the mass that it does, how would the strength of lunar tides change?
   a. The highs would be higher, and the lows would be lower.
   b. Both the highs and the lows would be higher.
   c. The highs would be lower, and the lows would be higher.
   d. Nothing would change.

9. If Earth had half of its current radius, how would the strength of lunar tides change?
   a. The highs would be higher, and the lows would be lower.
   b. Both the highs and the lows would be higher.
   c. The highs would be lower, and the lows would be higher.
   d. Both the highs and the lows would be lower.

10. If the Moon were 2 times closer to Earth than it is now, the gravitational force between Earth and the Moon would be
    a. 2 times stronger.
    b. 4 times stronger.
    c. 8 times stronger.
    d. 16 times stronger.

11. If the Moon were 2 times closer to Earth than it is now, the tides would be
    a. 2 times stronger.
    b. 4 times stronger.
    c. 8 times stronger.
    d. 16 times stronger.

12. If two objects are tidally locked to each other,
    a. the tides always stay on the same place on each object.
    b. the objects always remain in the same place in each other's sky.
    c. the objects are falling together.
    d. both a and b

13. Spring tides occur only when
    a. the Sun is near the vernal equinox.
    b. the Moon's phase is new or full.
    c. the Moon's phase is first quarter or third quarter.
    d. it is either spring or fall.

14. If an object crosses from farther to closer than the Roche limit, it
    a. can no longer be seen.
    b. begins to accelerate very quickly.
    c. slows down.
    d. may be torn apart.

15. Self-gravity is
    a. the gravitational pull of a person.
    b. the force that holds objects like people and lamps together.
    c. the gravitational interaction of all the parts of a body.
    d. the force that holds objects on Earth.

## Thinking about the Concepts

16. Both Kepler's laws and Newton's laws tell us something about the motion of the planets, but there is a fundamental difference between them. What is that difference?

17. Explain the difference between circular velocity and escape velocity. Which of these must be larger? Why?

18. Explain the difference between weight and mass.

19. Weight on Earth is proportional to mass. On the Moon, too, weight is proportional to mass, but the constant of proportionality is different on the Moon than it is on Earth. Why? Explain why this difference does not violate the universality of physical law, described in the Process of Science Figure.

20. Two comets are leaving the vicinity of the Sun, one traveling in an elliptical orbit and the other in a unbound orbit. What can you say about the future of these two comets? Would you expect either of them to return eventually?

21. What is the advantage of launching satellites from spaceports located near the equator? Would you expect satellites to be launched to the east or to the west? Why?

22. Explain how to use celestial orbits to estimate an object's mass. What are the observational quantities you need to make this mass estimation?

23. Suppose astronomers discovered an object approaching the Sun in an unbound orbit. What would that say about the origin of the object?

24. What determines the strength of gravity at various radii between Earth's center and its surface?

25. The best time to dig for clams along the seashore is when the ocean tide is at its lowest. What phases of the Moon would be best for clam digging? What would be the best times of day during those phases?

26. The Moon is on the meridian at your seaside home, but your tide calendar does not show that it is high tide. What might explain this apparent discrepancy?

27. We may have an intuitive feeling for why lunar tides raise sea level on the side of Earth facing the Moon, but why is sea level also raised on the side facing away from the Moon?

28. Tides raise and lower the level of Earth's oceans. Can they do the same for Earth's landmasses? Explain your answer.

29. Lunar tides raise the ocean surface less than 1 meter. How can tides as large as 5–10 meters occur?

30. Most commercial satellites are well inside the Roche limit as they orbit Earth. Why are they not torn apart?

## Applying the Concepts

31. Mars has about one-tenth the mass of Earth and about half of Earth's radius. What is the value of gravitational acceleration on the surface of Mars compared to that on Earth? Estimate your mass and weight on Mars compared with your mass and weight on Earth. Do Hollywood movies showing people on Mars accurately portray this difference in weight?

32. Earth speeds along at 29.8 km/s in its orbit. Neptune's nearly circular orbit has a radius of $4.5 \times 10^9$ km, and the planet takes 164.8 years to make one trip around the Sun. Calculate the speed at which Neptune moves along in its orbit.

33. Venus's circular velocity is 35.03 km/s, and its orbital radius is $1.082 \times 10^8$ km. Use this information to calculate the mass of the Sun.

34. At the surface of Earth, the escape velocity is 11.2 km/s. What would be the escape velocity at the surface of a very small asteroid having a radius $10^{-4}$ that of Earth and a mass $10^{-12}$ that of Earth?

35. How long does it take Newton's cannonball, moving at 7.9 km/s just above Earth's surface, to complete one orbit around Earth?

36. When a spacecraft is sent to Mars, it is first launched into an Earth orbit with circular velocity.
    a. Describe the shape of this orbit.
    b. What minimum velocity must we give the spacecraft to send it on its way to Mars?

37. Earth's average radius is 6,370 km and its mass is $5.97 \times 10^{24}$ kg. Show that the acceleration of gravity at the surface of Earth is $9.81$ m/s².

38. Using 6,370 km for Earth's radius, compare the gravitational force acting on a NASA rocket when it is sitting on its launchpad with the gravitational force acting on it when it is orbiting 350 km above Earth's surface.

39. The International Space Station travels on a nearly circular orbit 350 km above Earth's surface. What is its orbital speed?

40. Rearrange the terms in the last equation in Working It Out 4.1 to calculate the mass of Earth, using the measured values of $g$, $G$, and $R_{\text{Earth}}$.

41. As described in Working It Out 4.4, tidal force is proportional to the masses of the two objects and is inversely proportional to the cube of the distance between them. Some astrologers claim that your destiny is determined by the "influence" of the planets that are rising above the horizon at the moment of your birth. Compare the tidal force of Jupiter (mass $1.9 \times 10^{27}$ kg; distance $7.8 \times 10^8$ km) with that of the doctor in attendance at your birth (mass 80 kg, distance 1 meter).

42. The asteroid Ida (mass $4.2 \times 10^{16}$ kg) is attended by a tiny asteroidal moon, Dactyl, which orbits Ida at an average distance of 90 km. Neglecting the mass of the tiny moon, what is Dactyl's orbital period in hours?

43. Suppose you go skydiving.
    a. Just as you fall out of the airplane, what is your gravitational acceleration?
    b. Would this acceleration be bigger, smaller, or the same if you were strapped to a flight instructor, and so had twice the mass?
    c. Just as you fall out of the airplane, what is the gravitational force on you? (Assume your mass is 70 kg.)
    d. Would the gravitational force be bigger, smaller, or the same if you were strapped to a flight instructor, and so had twice the mass?

44. Assume that a planet just like Earth is orbiting the bright star Vega at a distance of 1 astronomical unit (AU). The mass of Vega is twice that of the Sun.
    a. How long in Earth years will it take to complete one orbit around Vega?
    b. How fast is the Earth-like planet traveling in its orbit around Vega?

45. Suppose in the past the Moon was 80 percent of the distance from Earth that it is now. Calculate how much stronger the lunar tides would have been. How would the neap and spring tides be different from now?

## USING THE WEB

46. Go to NASA's "Apollo 15 Hammer-Feather Drop" Web page (http://nssdc.gsfc.nasa.gov/planetary/lunar/apollo_15_feather_drop.html) and watch the video from *Apollo 15* of astronaut David Scott dropping the hammer and falcon feather on the Moon. (You might find a better version on YouTube.) What did this experiment show? What would happen if you tried this on Earth with a feather and a hammer? Would it work? Suppose instead you dropped the hammer and a big nail. How would they fall? How does the acceleration of falling objects on the Moon compare to the acceleration of falling objects on Earth?

47. Go to the Exploratorium's "Your Weight on Other Worlds" Web page (http://exploratorium.edu/ronh/weight), which will calculate your weight on other planets and moons in our Solar System. On which objects would your weight be higher than it is on Earth? What difficulties would human bodies have in a higher-gravity environment? For example, would it be easy to get up out of bed and walk? What are the possible short-term and long-term effects of lower gravity on the human body? Can you think of some types of life on Earth that might adapt well to a different gravity?

48. a. Watch the first 12 minutes of the episode "Gravity" in the *Universe* series (http://www.history.com/shows/the-universe/videos/the-universe-gravity) to see several illustrated examples of gravity. Why does the tennis ball appear to float when dropped at the top of the roller coaster? How was Newton able to imagine a satellite orbiting Earth centuries before it was possible? Why is it the speed of the cannonball that determines whether it goes into orbit? What was the technical difficulty in launching a satellite?

    b. In the same video as in part (a), watch the trip on the zero-G plane, at 23–29 minutes. How does the plane simulate zero-G? Why does it last for only 20–30 seconds? How is this similar to the roller coaster in part (a)?

49. Go to a website that will show you the times for high and low tides; for example, http://saltwatertides.com. Pick a location and bring up the tide table for today and the next 14 days. Why are there two high tides and two low tides every day? What is the difference in the height of the water between high and low tides? In the last few columns of the table, the times of moonrise and moonset are indicated, as well as the percent of lunar illumination. Does the time of the high tide lead or follow the highest position of the Moon in the sky? Compare with Figure 4.12c: what phases of the Moon have the greatest differences in the height of high and low tides?

50. Figure 4.17 shows two galaxies pulling at each other, most likely after they have already collided. Go to http://www.cita.utoronto.ca/~dubinski/nbody/ and scroll down to "Movie 2" to see a simulation of this interaction. Galaxies are not solid objects: they contain stars and gas and dust and a lot of empty space. Explain how these tidal tails can result from this type of interaction.

# smartw⊛rk**5**

If your instructor assigns homework in Smartwork5, access your assignments at digital.wwnorton.com/astro5.

digital.wwnorton.com/astro5

In the Exploration of Chapter 3, we used the Planetary Orbit Simulator to explore Kepler's laws for Mercury. Now that we know how Newton's laws explain why Kepler's laws describe orbits, we will revisit the simulator to explore the Newtonian features of Mercury's orbit. Visit the Student Site at the Digital Landing Page, and open the Planetary Orbit Simulator applet.

## Acceleration

To begin exploring the simulation, set parameters for "Mercury" in the "Orbit Settings" panel and then click "OK." Click the "Newtonian Features" tab at the bottom of the control panel. Select "show solar system orbits" and "show grid" under "Visualization Options." Change the animation rate to 0.01, and select the "start animation" button.
   Examine the graph at the bottom of the panel.

**1** Where is Mercury in its orbit when the acceleration is smallest?

_____

_____

**2** Where is Mercury in its orbit when the acceleration is largest?

_____

_____

**3** What are the values of the largest and smallest accelerations?

_____

_____

_____

In the "Newtonian Features" graph, mark the boxes for vector and line that correspond to the acceleration. Specifying these parameters will insert an arrow that shows the direction of the acceleration and a line that extends the arrow.

**4** To what Solar System object does the arrow point?

_____

_____

**5** In what direction is the force on the planet?

_____

_____

## Velocity

Examine the graph at the bottom of the panel again.

**6** Where is Mercury in its orbit when the velocity is smallest?

_____

_____

**7** Where is Mercury in its orbit when the velocity is largest?

_____

**8** What are the values of the largest and smallest velocities?

_____

Add the velocity vector and line to the simulation by clicking on the boxes in the graph window. Study the resulting arrows carefully.

**9** Are the velocity and the acceleration always perpendicular (is the angle between them always 90°)?

_____

_____

**10** If the orbit were a perfect circle, what would be the angle between the velocity and the acceleration?

_____

_____

## Hypothetical Planet

In the "Orbit Settings" panel, change the semimajor axis to 0.8 AU.

**11** How does this imaginary planet's orbital period now compare to Mercury's?

_____

_____

_____

Now change the semimajor axis to 0.1 AU.

**12** How does this planet's orbital period now compare to Mercury's?

_____

_____

_____

**13** Summarize your observations of the relationship between the speed of an orbiting object and the semimajor axis.

_____

_____

_____

# 5 Light

Our knowledge of the universe beyond Earth comes from light emitted, absorbed, or reflected by astronomical objects. Light carries information about the temperature, composition, and speed of the objects. Light also tells us about the nature of the material that the light passed through on its way to Earth. Yet light plays a far larger role in astronomy than that of being a messenger. Light is one of the primary means by which energy is transported throughout the universe. Stars, planets, and vast clouds of gas and dust filling the space between the stars heat up as they absorb light and cool off as they emit light. Light carries energy generated in the heart of a star outward through the star and off into space. Light transports energy from the Sun outward through the Solar System, heating the planets; and light carries energy away from each planet, allowing each one to cool. The balance between these two processes establishes each planet's temperature and therefore a planet's possible suitability for life.

## LEARNING GOALS

An astronomer must try to understand the universe by the light and other particles that reach Earth from distant objects. By the conclusion of this chapter, you should be able to:

**LG 1** Describe the wave and particle properties of light, and describe the electromagnetic spectrum.

**LG 2** Describe how to measure the chemical composition of distant objects using the unique spectral lines of different types of atoms.

**LG 3** Describe the Doppler effect and how it can be used to measure the motion of distant objects.

**LG 4** Explain how the spectrum of light that an object emits depends on its temperature.

**LG 5** Differentiate luminosity from brightness, and illustrate how distance affects each.

The visible part of the electromagnetic spectrum is laid out in the colors of this rainbow. ▶ ▶ ▶

What is light?

# 5.1 Light Brings Us the News of the Universe

Since the earliest investigations of light, there has been disagreement over the question of whether light is composed of particles or if it is a disturbance that travels from one point to another, called a **wave**. Scientists have since come to understand that light sometimes acts like a wave and at other times acts like a particle. We will begin this section with a discussion of how fast light travels. We will then discuss its wavelike properties. After that we will look at how light behaves as a particle.

## The Speed of Light

In the 1670s, Danish astronomer Ole Rømer (1644–1710) studied the movement of the moons of Jupiter, measuring the times when each moon disappeared behind the planet. To his amazement, the observed times did not follow the regular schedule that he predicted from Kepler's laws. Sometimes the moons disappeared behind Jupiter sooner than expected, and at other times they disappeared behind Jupiter later than expected. Rømer realized that the difference depended on where Earth was in its orbit. If he began tracking the moons when Earth was closest to Jupiter, by the time Earth was farthest from Jupiter the moons were almost 17 minutes "late." When Earth was once again closest to Jupiter, the moons again passed behind Jupiter at the predicted times.

Rømer correctly concluded that his observations were not a failure of Kepler's laws. Instead, he was seeing the first clear evidence that light travels at a finite speed. As shown in **Figure 5.1**, the moons appeared "late" when Earth was farther from Jupiter because of the time needed for light to travel the extra distance between the two planets. Over the course of Earth's yearly trip around the Sun, the distance between Earth and Jupiter changes by 2 astronomical units (AU). The speed of light equals this distance divided by Rømer's 16.7-minute delay, or about $3 \times 10^5$ kilometers per second (km/s). The value that Rømer actually announced in 1676 was a bit on the low side—$2.25 \times 10^5$ km/s—because the size of

**Figure 5.1** Danish astronomer Ole Rømer realized that apparent differences between the predicted and observed orbital motions of Jupiter's moons depend on the distance between Earth and Jupiter. He used these observations to measure the speed of light. (The superscript letters in the expression "$16^m40^s$" stand for minutes and seconds of time, respectively.)

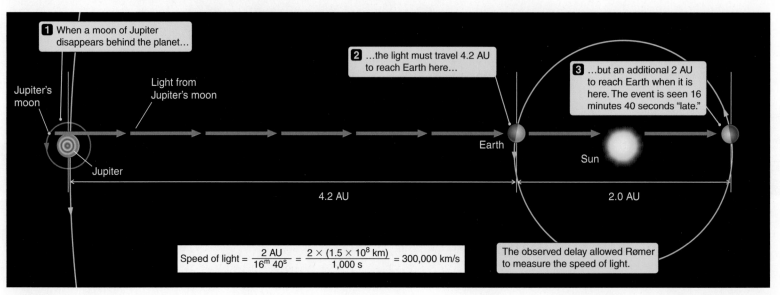

1 When a moon of Jupiter disappears behind the planet...

2 ...the light must travel 4.2 AU to reach Earth here...

3 ...but an additional 2 AU to reach Earth when it is here. The event is seen 16 minutes 40 seconds "late."

Jupiter's moon

Light from Jupiter's moon

Jupiter

Earth

Sun

4.2 AU

2.0 AU

$$\text{Speed of light} = \frac{2 \text{ AU}}{16^m \, 40^s} = \frac{2 \times (1.5 \times 10^8 \text{ km})}{1{,}000 \text{ s}} = 300{,}000 \text{ km/s}$$

The observed delay allowed Rømer to measure the speed of light.

Earth's orbit was not well known. Modern measurements of the speed of light give a value of $2.99792458 \times 10^5$ km/s in a **vacuum** (a region of space devoid of matter). The speed of light in a vacuum is one of nature's fundamental constants, usually written as $c$ (lowercase). The speed of light through any medium, such as air or glass, is always less than $c$.

The International Space Station moves around Earth at a speed of about 28,000 kilometers per hour (km/h), taking 91 minutes to complete one orbit. Light travels almost 40,000 times faster than this and can circle Earth in only $\frac{1}{7}$ of a second. Because light is so fast, the travel time of light is a convenient way of expressing cosmic distances. It takes light $1\frac{1}{4}$ seconds to travel between Earth and the Moon, so we say that the Moon is $1\frac{1}{4}$ light-*seconds* from Earth. The Sun is $8\frac{1}{3}$ light-*minutes* away, and the next-nearest star is $4\frac{1}{3}$ light-*years* distant. Thus, a *light-year* is defined as the distance traveled by light in 1 year, or about 9.5 trillion km. Although it is sometimes misused as a measure of time, the light-year is a measure of distance.

As light travels at this high speed, it carries energy from place to place. **Energy** is the ability to do work, and it comes in many forms. **Kinetic energy** is the energy of moving objects. **Thermal energy** is closely related to kinetic energy and is the sum of all the random motion of atoms, molecules, and particles, by which we measure their temperature. For example, when light from the Sun strikes the pavement, the pavement heats up. That energy was carried from the Sun to the pavement by light. Rømer knew how long it took for light to travel a given distance, but it would take more than 200 years for physicists to figure out what light actually is.

## Light as an Electromagnetic Wave

In the late 19th century, the Scottish physicist James Clerk Maxwell (1831–1879) introduced the concept that electricity and magnetism are two components of the same physical phenomenon. An **electric force** is the push or pull between electrically charged particles that make up atoms, such as protons and electrons, arising from their electric charges. Particles with opposite charges attract, and those with like charges repel. A **magnetic force** is a force between electrically charged particles arising from their motion.

To describe these electric and magnetic forces, Maxwell considered what happens when charged particles move in *electric fields* and *magnetic fields*. An **electric field** is a measure of the electric force on a charge at any point in space. Similarly, a **magnetic field** is a measure of the magnetic force acting on a small magnet at any point in space.

Maxwell summarized the behavior of electric fields and magnetic fields in four elegant equations. Among other things, these equations say that a changing electric field causes a magnetic field, and that a changing magnetic field causes an electric field. A change in the motion of a charged particle causes a changing electric field, which causes a changing magnetic field, which causes a changing electric field, and so on. You can see this interaction in **Figure 5.2**. Once the process starts, a self-sustaining procession of oscillating electric and magnetic fields moves out in all directions through space. In other words, an accelerating charged particle gives rise to an **electromagnetic wave**. These electromagnetic waves, and the accelerating charges that generate them, are the sources of electromagnetic radiation. Maxwell's equations also predict the speed at which an electromagnetic wave should travel, which agrees with the measured speed of light ($c$).

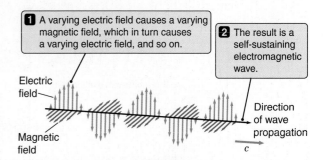

**1** A varying electric field causes a varying magnetic field, which in turn causes a varying electric field, and so on.

**2** The result is a self-sustaining electromagnetic wave.

Electric field

Magnetic field

Direction of wave propagation

$c$

**Figure 5.2** An electromagnetic wave consists of oscillating electric and magnetic fields that are perpendicular both to each other and to the direction in which the wave travels.

Maxwell's wave description of light also gives us an idea of how light originates and how it interacts with matter. When a drop of water falls from the faucet into a sink full of water, it causes a disturbance, or wave, like the one shown in **Figure 5.3a**. The wave moves outward as a ripple on the surface of the water. As shown in Figure 5.3b, electromagnetic waves resulting from periodic changes in the strength of the electric and magnetic fields move out through space away from their source in much the same way. However, the ripples in the sink are distortions of the water's surface, and they require a **medium**: a substance to travel through. Light waves move through empty space—what we call a vacuum— in the absence of a medium.

Now imagine that a soap bubble is floating in the sink, illustrated in **Figure 5.4a**. The bubble remains stationary until the ripple from the dripping faucet reaches it. As the ripple passes by, the rising and falling water causes the bubble to rise and fall. This can only happen if the wave is carrying energy—a conserved quantity that gives objects and particles the ability to do work. Light waves similarly carry energy through space and cause electrically charged particles to vibrate, as in Figure 5.4b.

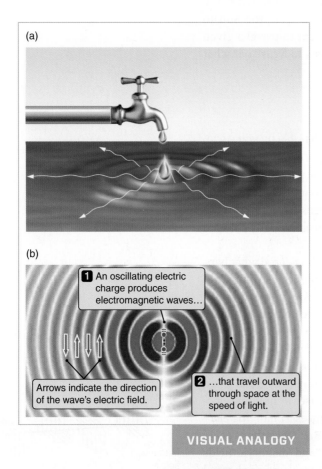

**Figure 5.3** (a) A drop falling into water generates waves that move outward across the water's surface. (b) In similar fashion, an oscillating (accelerated) electric charge generates electromagnetic waves that move away at the speed of light.

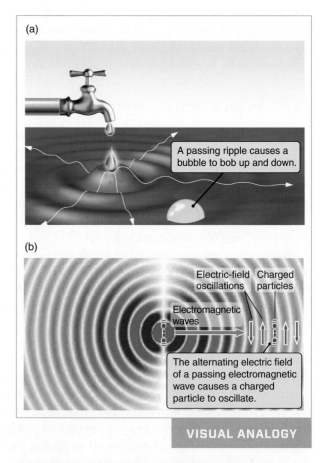

**Figure 5.4** (a) When waves moving across the surface of water reach a bubble, they cause the bubble to bob up and down. (b) Similarly, a passing electromagnetic wave causes an electric charge to oscillate in response to the wave.

## Characterizing Waves

In this book you will learn about several kinds of waves, including electromagnetic waves crossing the vast expanse of the universe and earthquakes traveling through Earth. Waves are generally characterized by four quantities: *amplitude*, *speed*, *frequency*, and *wavelength*. Each of these quantities is illustrated in **Figure 5.5**. The **amplitude** of a wave is the height of the wave above the undisturbed position (Figure 5.5a). For water waves, the amplitude is how far the water is lifted up by the wave. In the case of light, the amplitude of a light wave is related to the brightness of the light. A water wave travels at a particular speed, *v* (Figure 5.5b), through the water. The water itself doesn't travel; it just moves up and down at the same location. For waves like those in water, this speed is variable and depends on the density of the substance the wave moves through, among other things. Light, in contrast, always moves through a vacuum at the same speed, $c \approx 300{,}000$ km/s.

The distance from one crest of a wave to the next is the **wavelength**, usually denoted by the Greek letter lambda, $\lambda$ (Figure 5.5c). The number of wave crests passing a point in space each second is called the wave's **frequency**, *f*. The unit of frequency is cycles per second, which is called **hertz** (abbreviated **Hz**) after the 19th century physicist Heinrich Hertz (1857–1894), who was the first to experimentally confirm Maxwell's predictions about electromagnetic radiation. Panels (c) and (d) of Figure 5.5 show that waves with longer wavelengths have lower frequencies, and waves with shorter wavelengths have higher frequencies. Higher-frequency waves carry more energy. Think about standing on an ocean beach: if the ocean waves are more frequent, they will be more energetic.

Waves travel a distance of one wavelength each cycle, so the speed of a wave can be found by multiplying the frequency and the wavelength. Translating this idea into math, we have $v = \lambda f$. The speed of light in a vacuum is always *c*, so once the wavelength of a wave of light is known, its frequency is known, and vice versa. Because light travels at constant speed, its wavelength and frequency are inversely proportional to each other: if the wavelength increases, the frequency decreases. A tremendous amount of information can be carried by waves; for example, complex and beautiful music, which travels by sound waves. As you continue your study of the universe, time and time again you will find that the information you receive, whether about the interior of Earth or about a distant star or galaxy, rides in on a wave.

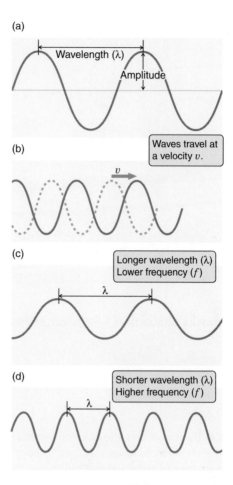

**Figure 5.5** A wave is characterized by the distance from one peak to the next (wavelength, $\lambda$), the frequency of the peaks (*f*), the maximum deviations from the medium's undisturbed state (amplitude), and the speed (*v*) at which the wave pattern travels from one place to another. In an electromagnetic wave, the amplitude is the maximum strength of the electric field, and the speed of light is written as *c*.

## The Electromagnetic Spectrum

Most light signals are made up of many wavelengths. You have almost certainly seen a rainbow, spread out across the sky, as in the chapter opening figure. A rainbow is created when white light interacts with water droplets and is spread out into its component colors. Light spread out by wavelength is called a **spectrum**. At the long-wavelength (and therefore low-frequency) end of the visible spectrum is red light. At the other end is violet light. A commonly used unit for the wavelength of visible light is the **nanometer**, abbreviated **nm**. A nanometer is one-billionth ($10^{-9}$) of a meter. Human eyes can see light between violet (about 380 nm) and red (750 nm). Stretched out between the two, in a rainbow, is the rest of the visible spectrum.

The light-sensitive cells in our eyes respond to visible light. But this is only a small sample of the range of possible wavelengths for electromagnetic radiation.

Light can have wavelengths that are much shorter or much longer than your eyes can perceive. The whole range of different wavelengths of light is collectively called the **electromagnetic spectrum**, illustrated in **Figure 5.6**. Most of the electromagnetic spectrum, and therefore most of the information in the universe, is invisible to the human eye. To detect light outside the visible, we must use specialized detectors of various kinds, as we will discuss in Chapter 6.

Refer to Figure 5.6 as we take a tour of the electromagnetic spectrum, beginning with the shortest wavelengths and working our way to the longest ones. The very shortest wavelengths of light are called **gamma rays**, or sometimes gamma radiation. Because this light has the shortest wavelengths, it has the highest frequency and the highest energy, so it penetrates matter easily. Wavelengths between 0.1 nm and 40 nm are called **X-rays**. You have probably encountered X-rays at the dentist or in an emergency room—X-ray light has enough energy to penetrate through skin and muscle but is stopped by denser bone. **Ultraviolet (UV) radiation** has wavelengths between 40 and about 380 nm—longer than X-rays but shorter than visible light. You are familiar with this type of light from sunburns: UV light has enough energy to penetrate into your skin, but not much farther.

**Infrared (IR) radiation** has longer wavelengths than the reddest wavelengths in the visible range. You are familiar with a small wavelength range of this kind of radiation because you often feel it as heat. When you hold your hand next to a hot stove, some of the heat you feel is carried to your hand by infrared radiation emitted from the stove. In this sense, you could think of your skin as being a giant infrared eyeball—it is sensitive to infrared wavelengths. Infrared radiation is also used in television remote controls, and night vision goggles detect infrared radiation from warm objects such as animals. A useful unit for infrared light is the **micron** (abbreviated **$\mu$m**, where $\mu$ is the Greek letter mu). One micron is 1,000 nm, or one-millionth ($10^{-6}$) of a meter. Infrared wavelengths are longer than red light and shorter than 500 microns.

**Microwave radiation** has even longer wavelengths than infrared radiation. The microwave in your kitchen heats the water in food using light of these

**Nebraska Simulation:** EM Spectrum Module

**Figure 5.6** By convention, the electromagnetic spectrum is divided into loosely defined regions ranging from gamma rays to radio waves. Throughout the book, we use the following labels to indicate the form of radiation used to produce astronomical images, with an icon to remind you: G = gamma rays; X = X-rays; U = ultraviolet; V = visible; I = infrared; R = radio. If more than one region is represented, multiple labels are highlighted.

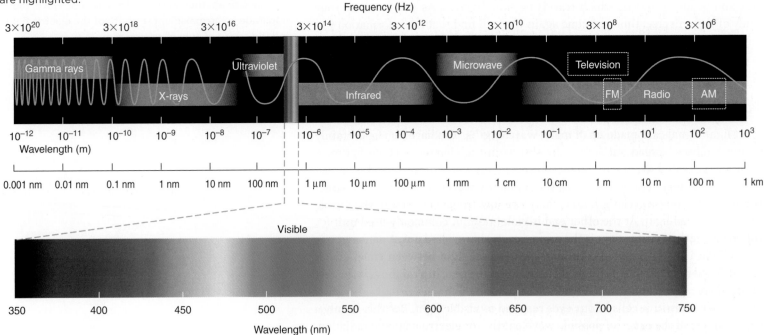

wavelengths. The longest-wavelength light, which has wavelengths longer than a few centimeters, is called **radio waves**. Light of these wavelengths in the form of FM, AM, television, and cell phone signals is used to transmit information from place to place.

## CHECK YOUR UNDERSTANDING 5.1

Rank the following in order of decreasing wavelength: (a) gamma rays; (b) visible light; (c) infrared light; (d) ultraviolet light; (e) radio waves.

## Light as a Particle

Although the wave theory of light describes many observations, it does not provide a complete picture of the properties of light. Many of the difficulties with the wave model of light have to do with the way in which light interacts with small particles such as atoms and molecules.

The work of Albert Einstein and other scientists modified our understanding of light to show that light sometimes acts like a wave and sometimes acts like a particle. In 1905, Einstein explained the *photoelectric effect*, in which electrons are emitted when surfaces are illuminated by electromagnetic radiation greater than a certain frequency. He showed that the rate at which electrons are emitted depends only on the amount of incoming light, and that the speed of the electrons depends only on the frequency of the incoming radiation. This work earned Einstein a Nobel Prize in 1921. In the particle model, light is made up of massless particles called **photons** (*phot-* means "light," as in *photograph*; and *-on* signifies a particle). Photons always travel at the speed of light (**Process of Science Figure**), and they carry energy. A photon is a *quantum* of light; *quantized* means that something is subdivided into individual units. **Quantum mechanics** is a branch of physics that deals with the quantization of energy and of other properties of matter.

The energy of a photon and the frequency of the electromagnetic wave are directly proportional to each other: the higher the frequency of the light wave, the greater the energy each photon carries. This relationship connects the particle and the wave concepts of light. For example, photons with higher frequencies carry more energy than that carried by photons with lower frequencies. The constant of proportionality between the energy, $E$, and the frequency, $f$, is called Planck's constant, $h$, which is equal to $6.63 \times 10^{-34}$ joule-seconds (a joule is a unit of energy). Specifically, $E = hf$, where $E$ = the energy of the photon, $h$ = Planck's constant, and $f$ = frequency. Because the wavelength and frequency of electromagnetic waves are inversely proportional, this also means that the photon energy is inversely proportional to the wavelength. In visible light, high-energy light is blue, and low-energy light is red. **Working It Out 5.1** explores the relationships among wavelength, frequency, and energy of light.

In the particle description of light, the electromagnetic spectrum is a spectrum of photon energies. The higher the frequency of the electromagnetic wave, the greater the energy carried by each photon. Photons of shorter wavelength (higher frequency) carry more energy than that carried by photons of longer wavelength (lower frequency). For example, photons of blue light carry more energy than that carried by photons of longer-wavelength red light. Ultraviolet photons carry more energy than that carried by photons of visible light, and X-ray photons carry more energy than that carried by ultraviolet photons. The lowest-energy photons are radio wave photons.

▶❙❙ **AstroTour:** Light as a Wave, Light as a Photon

# Process of Science

Scientists working on very different problems in different fields all find the same result: light has a speed and can be measured.

**Ole Rømer studies eclipses of Jupiter's moons.**

**Rømer calculates the speed of light from eclipse delays of Jupiter's moons.**

**James Bradley studies the apparent motion of stars, which appear to make small circles because of the relative motion of Earth.**

**A half century after Rømer's measurement, Bradley's motion studies lead to a more accurate measurement of the speed of light.**

**James Clerk Maxwell studies electricity and magnetism.**

**Einstein builds on Maxwell's theory in 1905 to claim that the speed of light is the same for all observers.**

Astronomers and physicists converge on an understanding that photons travel at the speed of light—a fundamental constant of the universe that is the same for all observers.

## 5.1 Working It Out  Working with Electromagnetic Radiation

### Wavelength and Frequency

When you tune to a radio station at, say, 770 AM, you are receiving an electromagnetic signal that travels at the speed of light and is broadcast at a frequency of 770 kilohertz (kHz), or $7.7 \times 10^5$ Hz. We can use the relationship between wavelength and frequency, $c = \lambda f$ to calculate the wavelength of the AM signal:

$$\lambda = \frac{c}{f} = \left( \frac{3 \times 10^8 \, \text{m/s}}{7.7 \times 10^5 \text{/s}} \right) = 390 \, \text{m}$$

This AM wavelength is about 4 times the length of a football field. FM wavelengths are much shorter than AM wavelengths.

The human eye is most sensitive to light in green and yellow wavelengths, about 500–590 nm. If we examine green light with a wavelength of 530 nm, we can compute its frequency:

$$f = \frac{c}{\lambda} = \left( \frac{3 \times 10^8 \, \text{m/s}}{530 \times 10^{-9} \, \text{m}} \right) = 5.66 \times 10^{14} \text{/s} = 5.66 \times 10^{14} \, \text{Hz}$$

This frequency corresponds to 566 *trillion* wave crests passing by each second.

### Photon Energy

Let's compare the energy of an X-ray photon with a wavelength of 1 nm and the energy of a visible light photon with a wavelength of 530 nm as used in the previous calculation. The equation for the energy of a photon is $E = hf$. Because $f = c/\lambda$, substituting $c/\lambda$ for $f$ yields the inverse relationship, $E = hc/\lambda$, with $c$ = the speed of light = $3 \times 10^8$ m/s, and $h$ = Planck's constant. Because we are making a *comparison*, we can take a ratio, and then the constants $h$ and $c$ cancel out:

$$\frac{E_{\text{X-ray photon}}}{E_{\text{visible photon}}} = \frac{hc/\lambda_{1\,\text{nm}}}{hc/\lambda_{500\,\text{nm}}} = \frac{\cancel{hc}}{\cancel{hc}} \times \frac{530\,\text{nm}}{1\,\text{nm}} = 530$$

The X-ray photon has 530 times the energy of the visible light photon.

---

The *total* amount of energy that a beam of the light carries is called its **intensity**. A beam of red light can be just as intense as a beam of blue light—that is, it can carry just as much energy—but because the energy of a red photon is less than the energy of a blue photon, maintaining that same intensity requires more red photons than blue photons. This relationship, illustrated in **Figure 5.7**, is a lot like money: $10 is $10, but it takes a lot more pennies (low-energy photons) than quarters (high-energy photons) to make up $10.

### CHECK YOUR UNDERSTANDING 5.2

As wavelength increases, the energy of a photon _____ and its frequency _____ . (a) increases; decreases (b) increases; increases (c) decreases; decreases (d) decreases; increases

## 5.2 The Quantum View of Matter Explains Spectral Lines

Light and matter interact, and this interaction allows us to detect matter even at great distances in space. To understand this interaction, we must understand the building blocks of matter. In this section, we will review atomic structure and the process by which astronomers identify the chemical elements in astronomical objects.

### Atomic Structure

**Matter** is anything that occupies space and has mass. Atoms are composed of a central massive **nucleus**, which contains **protons** with a positive charge and **neutrons**, which have no charge. A cloud of negatively charged **electrons**

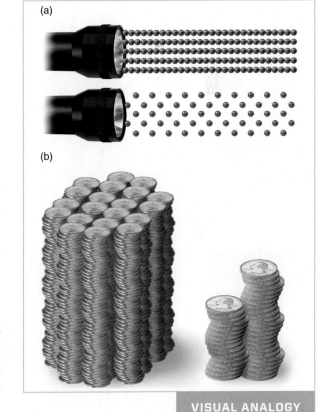

(a)

(b)

**VISUAL ANALOGY**

**Figure 5.7** (a) Red light carries less energy than that carried by blue light, so it takes more red photons than blue photons to make a beam of a particular intensity. (b) Similarly, pennies are worth less than quarters, so it takes more pennies than quarters to add up to $10.

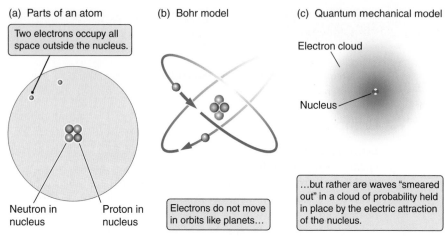

(a) Parts of an atom

Two electrons occupy all space outside the nucleus.

Neutron in nucleus

Proton in nucleus

(b) Bohr model

Electrons do not move in orbits like planets...

(c) Quantum mechanical model

Electron cloud

Nucleus

...but rather are waves "smeared out" in a cloud of probability held in place by the electric attraction of the nucleus.

**Figure 5.8** (a) An atom (in this case helium) is made up of a nucleus consisting of positively charged protons and electrically neutral neutrons and surrounded by less massive negatively charged electrons. (b) Atoms are often drawn as miniature "solar systems," but this model is incorrect. (c) Electrons are actually smeared out around the nucleus in quantum mechanical clouds of probability.

▶❙❙ **AstroTour:** Atomic Energy Levels and the Bohr Model

surrounds the nucleus. Atoms with the same number of protons are all of the same type, known as an **element**. For example, an atom with two protons, shown in **Figure 5.8a**, is the element helium. An atom with six protons is the element carbon; one with eight protons is the element oxygen; and so forth. An element may have many **isotopes**: atoms with the same number of protons but different numbers of neutrons. **Molecules** are groups of atoms bound together by shared electrons. A single teaspoon of water contains about $10^{23}$ atoms—about as many atoms as there are stars in the observable universe.

For an atom to be electrically neutral, it must have the same number of electrons as protons. Electrons have much less mass than protons or neutrons have, so almost all the mass of an atom is found in its nucleus. This description led to a model of an atom with the massive nucleus sitting in the center and the smaller electrons orbiting around it, much as planets orbit around the Sun (Figure 5.8b). It is called the **Bohr model** after the Danish physicist Niels Bohr (1885–1962), who proposed it in 1913.

However, Bohr's model is not a complete description of an atom. Just as waves of light have particle-like properties, particles of matter also have wavelike properties. With this realization, the Bohr model of the atom was modified so that the positively charged nucleus is surrounded by electron "clouds" or "waves," as shown in Figure 5.8c. In this model, it is not possible to know precisely where the electron is in its orbit. The wave characteristics of particles make it impossible to pin down simultaneously their exact location and their exact velocity; there will always be some uncertainty. This is why a featureless cloud is used to represent electrons in orbit around an atomic nucleus.

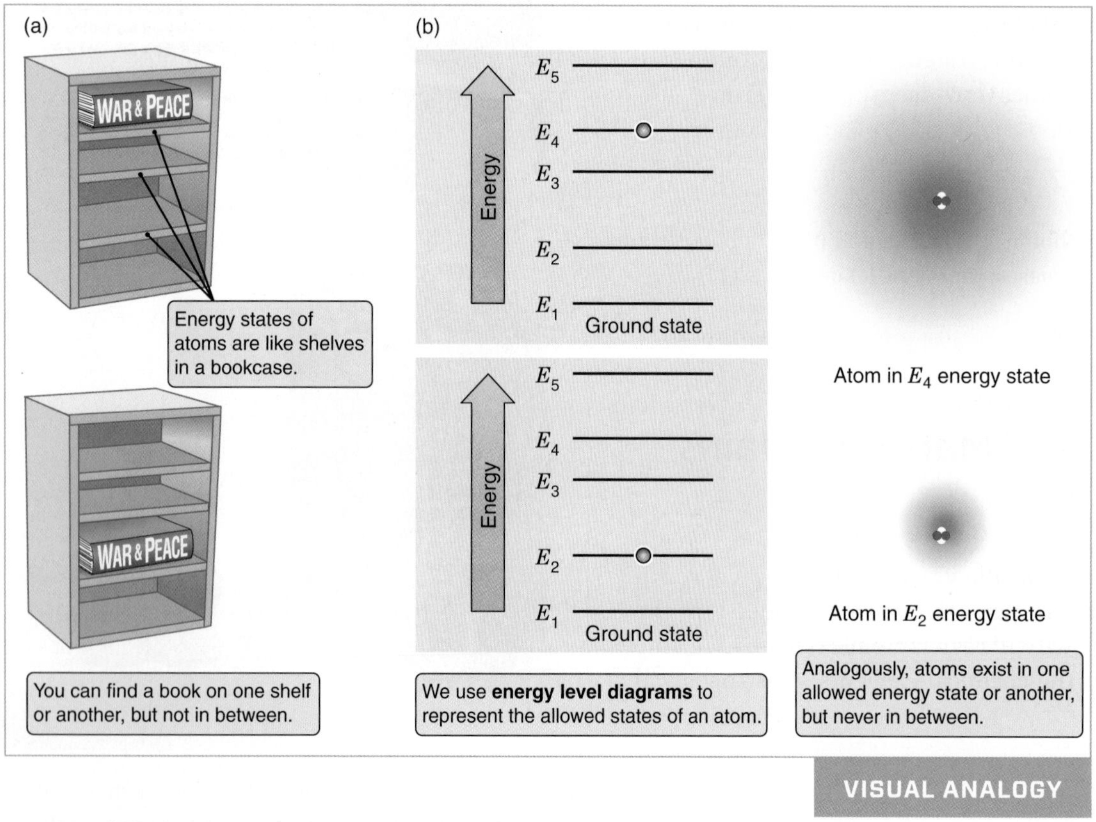

## Atomic Energy Levels

Because of their wavelike properties, electrons in an atom can take on only certain specific energies that depend on the energy states of the atom. The form that the electron waves take depends on the possible energy states of atoms. We can imagine the energy states of atoms as being like a bookcase with a set of shelves, as depicted in **Figure 5.9a**. The energy of an atom might correspond to the energy of one state or to

(a)

WAR & PEACE

Energy states of atoms are like shelves in a bookcase.

WAR & PEACE

You can find a book on one shelf or another, but not in between.

(b)

$E_5$

$E_4$

$E_3$

$E_2$

$E_1$   Ground state

Energy

$E_5$

$E_4$

$E_3$

$E_2$

$E_1$   Ground state

Energy

We use **energy level diagrams** to represent the allowed states of an atom.

Atom in $E_4$ energy state

Atom in $E_2$ energy state

Analogously, atoms exist in one allowed energy state or another, but never in between.

**VISUAL ANALOGY**

**Figure 5.9** (a) Energy states of an atom are analogous to shelves in a bookcase. You can move a book from one shelf to another, but books can never be placed between shelves. (b) Atoms exist in one allowed energy state or another but never in between. There is no level below the ground state.

the energy of the next state, but the energy of the atom is never found between the two states, just as a book can be on only one shelf at a time and cannot be partly on one shelf and partly on another. A given atom may have many different energy states available to it, but these states are *discrete*.

Astronomers keep track of the allowed states of an atom using energy level diagrams, as shown in Figure 5.9b, where each energy level is like a shelf on the bookcase. Both the bookcase and the energy level diagram are simplifications of the possible energies of a three-dimensional system. The lowest possible energy state for a system (or part of a system) such as an atom is called the **ground state**. When the atom is in the ground state, the electron has its minimum energy. It can't give up any more energy to move to a lower state, because there isn't a lower state. An atom will remain in its ground state forever unless it gets energy from outside. In the bookcase analogy, a book sitting on the bottom shelf at the floor is in its ground state. It has nowhere left to fall, and it cannot jump to one of the higher shelves of its own accord.

Energy levels above the ground state are called **excited states**. Just as a book on an upper shelf might fall to a lower shelf, an atom in an excited state might **decay** to a lower state by getting rid of some of its extra energy. An important difference between the atom and the book on the shelf, however, is that whereas a snapshot might catch the book falling between the two shelves, the atom will never be caught between two energy states. When the transition from one higher state to a lower one occurs, the difference in energy between the two states is carried off all at once. A common way for an atom to do this is to emit a photon. The photon emitted by the atom carries away exactly the amount of energy lost by that atom as it goes from the higher energy state to the lower energy state. In a similar fashion, atoms moving from a lower energy state to a higher energy state can absorb only certain specific energies.

To make an analogy with money, suppose you have a penny (1 cent), a nickel (5 cents), and a dime (10 cents), totaling 16 cents. Now imagine that you give away the nickel and are left with 11 cents. You never had exactly 13 cents or 13.6 cents. You had 16 cents, and then 11 cents. Atoms don't accept and give away money to change energy states, but they do accept and give away photons with well-defined energy.

Emission Spectra  Imagine a hypothetical atom that has only two available energy states. The energy of the lower energy state (the ground state) is $E_1$, and the energy of the higher energy state (the excited state) is $E_2$. The energy levels of this atom can be represented in an energy level diagram like the one in **Figure 5.10a**. An atom in the excited state moves to the ground state by getting rid of the "extra" energy all at once. It does this when the electron emits a photon. The atom goes from one energy state to another, but it never has an amount of energy in between.

In Figure 5.10b, the downward arrow indicates that the atom went from the higher state with energy $E_2$ to the lower state with energy $E_1$. The atom lost an amount of energy equal to the difference between the two states, or $E_2 - E_1$. Because energy is never truly lost or created, the energy lost by the atom has to show up somewhere. In this case, the energy shows up in the form of a photon that is emitted by the atom. The energy of the photon emitted exactly matches the energy lost by the atom; that is, $E_{\text{photon}} = E_2 - E_1$.

An atom can emit photons with energies corresponding *only* to the difference between two of its allowed energy states. Because the energy of a photon is related to the frequency or wavelength of electromagnetic radiation, in the

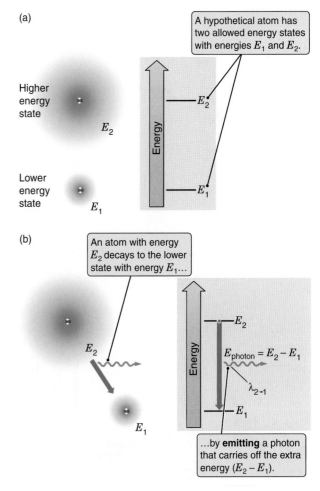

**Astronomy in Action:** Emission and Absorption

(a) A hypothetical atom has two allowed energy states with energies $E_1$ and $E_2$.

(b) An atom with energy $E_2$ decays to the lower state with energy $E_1$...

$E_{\text{photon}} = E_2 - E_1$

$\lambda_{2\rightarrow1}$

...by **emitting** a photon that carries off the extra energy $(E_2 - E_1)$.

**Figure 5.10** (a) The energy levels of a hypothetical two-level atom. (b) A photon with energy $E_{\text{photon}} = E_2 - E_1$ is emitted when an atom in the higher energy state decays to the lower energy state.

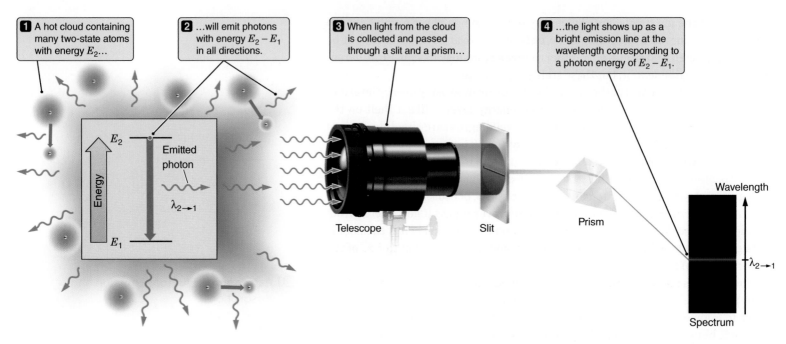

**Figure 5.11** A cloud of gas containing atoms with two energy states, $E_1$ and $E_2$, emits photons with an energy $E_{photon} = E_2 - E_1$, which appear in the spectrum (right) as a single bright *emission line*.

transition from level 2 to level 1, a photon of energy $E_{photon} = E_2 - E_1$ has a specific wavelength $\lambda_{2 \to 1}$ and a specific frequency $f_{2 \to 1}$. Therefore, these emitted photons have a very specific color, and every photon emitted in any transition from $E_2$ to $E_1$ will have this same color defined by a specific wavelength. The energy level structure of an atom determines the wavelengths of the photons it emits—the color of the light that the atom gives off.

**Figure 5.11** illustrates the light coming from a cloud of gas consisting of the hypothetical two-state atoms in Figure 5.10. Any atom in the higher energy state ($E_2$) quickly decays and emits a photon in a random direction, and an enormous number of photons come pouring out of the cloud of gas. Instead of containing photons of all different energies—light of all different colors—this light contains only photons with the specific energy $E_2 - E_1$ and wavelength $\lambda_{2 \to 1}$. In other words, all of the light coming from these atoms in the cloud is the same color. If you spread the light out into its component colors, there would be only one color—a single bright line called an **emission line**.

Why was the atom in the excited state $E_2$ in the first place? An atom sitting in its ground state will remain there unless it absorbs just the right amount of energy to kick it up to an excited state. In general, the atom either absorbs the energy of a photon or it collides with another atom or an unattached electron and absorbs some of the other particle's energy. In a neon sign, an alternating electric field inside the glass tube pushes electrons in the gas back and forth through the neon gas inside the tube. Some of these electrons crash into atoms of the gas, knocking them into excited states. The atoms then drop back down to their ground states by emitting photons, causing the gas inside the tube to glow.

**Absorption Spectra** In the opposite process, an atom in a low energy state can absorb the energy of a passing photon and jump up to a higher energy state as shown in **Figure 5.12**. Once again, the energy required to go from $E_1$ to $E_2$ is the difference in energy between the two states, $E_2 - E_1$. For a photon to cause an

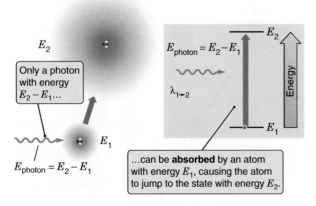

**Figure 5.12** An atom in a lower energy state may absorb a photon of energy $E_{photon} = E_2 - E_1$, leaving the atom in a higher energy state.

atom to jump from $E_1$ to $E_2$, it must provide exactly this much energy. The only photons capable of exciting atoms from $E_1$ to $E_2$ are photons with $E_{photon} = E_2 - E_1$. As with emission, these photons have a corresponding frequency and wavelength $f_{1 \to 2} = E_{photon}/h$ and $\lambda_{1 \to 2} = hc/E_{photon}$. These photons have exactly the same energy—the same color of light—emitted by the atoms when they decay from $E_2$ to $E_1$. This is not a coincidence. The energy difference between the two levels is the same whether the atom is emitting a photon or absorbing one, so the energy of the photon involved will be the same in either case.

In **Figure 5.13a**, white light (with all wavelengths of photons in it) passes directly though a glass prism, which breaks up the light into a rainbow of colors. However, when the white light passes through a cool cloud composed of our hypothetical gas of two-state atoms, as illustrated in Figure 5.13b, some photons will be absorbed. Almost all of the photons will pass through the cloud of gas unaffected, because they do not have the right energy ($E_2 - E_1$) to be absorbed by atoms of the gas. However, photons with just the right amount of energy can be absorbed, and as a result, these photons will be missing in the light passing

**Nebraska Simulation:** Three Views Spectrum Demonstrator

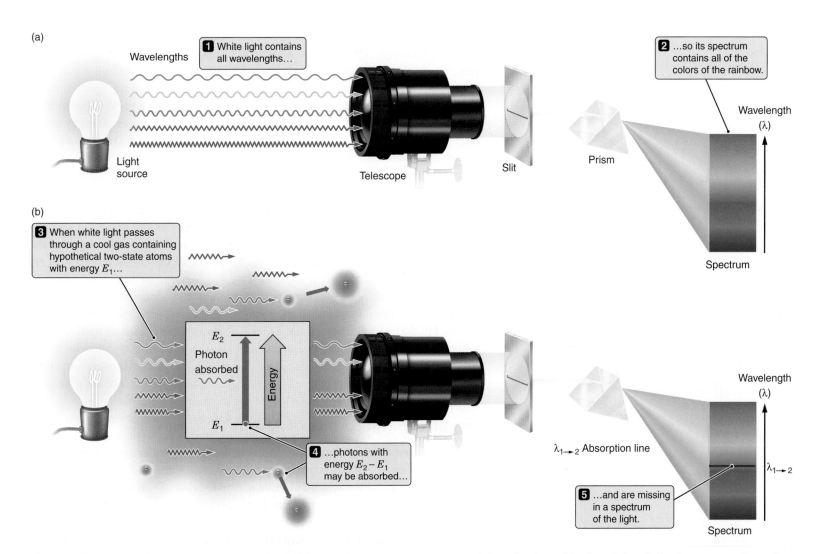

**Figure 5.13** (a) When passed through a prism, white light produces a spectrum containing all colors. (b) When light of all colors passes through a cloud of hypothetical two-state atoms, photons with energy $E_{photon} = E_2 - E_1$ may be absorbed, leading to the dark absorption line in the spectrum.

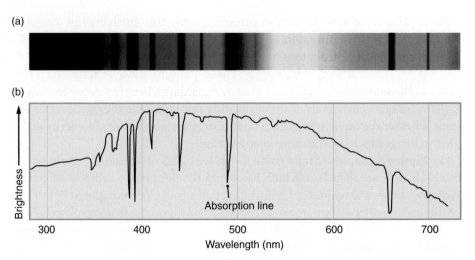

**Figure 5.14** Absorption lines in the spectrum of a star as an image (a) and a graph (b).

**AstroTour:** Atomic Energy Levels and Light Emission and Absorption

**Nebraska Simulation:** Hydrogen Atom Simulator

through the prism. Where the color corresponding to each of these missing photons should be, there is instead a sharp, dark line at the wavelength corresponding to this energy. This process by which atoms capture the energy of passing photons is called **absorption**, and the dark line seen in the spectrum is called an **absorption line**. **Figure 5.14a** shows such absorption lines in the spectrum of a star. The spectrum is shown in two different ways here: as a rainbow with light missing, and then again in Figure 5.14b as a graph of the brightness at every wavelength. Comparing the top and bottom versions of the spectrum, you can see that where there are dark lines, the brightness drops abruptly at a particular wavelength. Places between the dark lines are brighter and therefore higher on the graph than the absorption lines.

When an atom absorbs a photon, it may quickly decay to its previous lower energy state, emitting a photon with the same energy as the photon it just absorbed. If the atom reemits a photon just like the one it absorbed, why does the absorption matter? The photon that was taken out of the passing light was replaced, but all of the absorbed photons were originally traveling in the same direction, whereas the emitted photons are traveling in random directions. In other words, some of the photons with energies equal to $E_2 - E_1$ are diverted from their original paths by their interaction with atoms. If you look at a white light through the cloud in Figure 5.13b, you will observe an absorption line at a wavelength of $\lambda_{1\rightarrow2}$, but if you look at the cloud from another direction, you will observe an emission line at this same wavelength.

## Spectral Fingerprints of Atoms

Spectra of astronomical objects are fundamental to our understanding of the universe. Astronomers who study spectra will say "a spectrum is worth a thousand pictures" because of the wealth of information that can come from it. We will refer to spectra in every chapter in this book. So now let's move beyond the hypothetical atom with two energy levels to see what we can learn from real atoms. Atoms can occupy many more than just two possible energy states; therefore, any given type of atom will be capable of emitting and absorbing photons at many different wavelengths. An atom with three energy states, for example, might jump from state 3 to state 2, or from state 3 to state 1, or from state 2 to state 1. The three

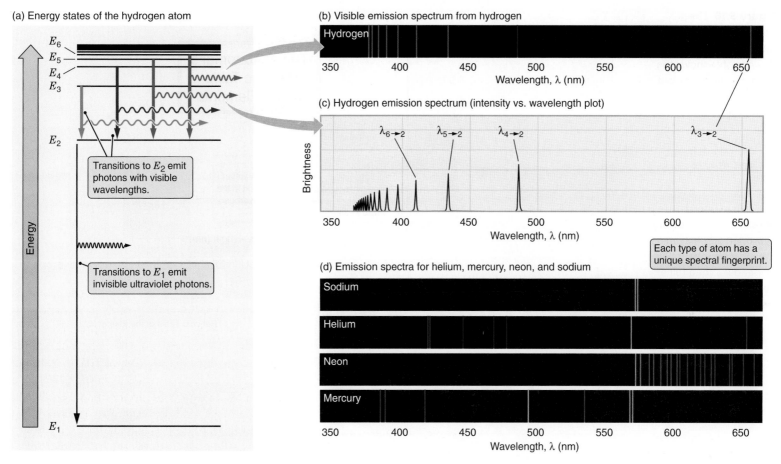

**Figure 5.15** (a) The energy states of the hydrogen atom. Decays to level $E_2$ emit photons in the visible part of the spectrum. (b) This spectrum is what you might see if you looked at the light from a hydrogen lamp projected through a prism onto a screen. (c) This graph of the brightness (intensity) of spectral lines versus their wavelength illustrates how spectra are traditionally plotted. (d) Emission spectra from several other gases: helium, mercury, neon, and sodium.

distinct emission lines in the spectrum from a gas made up of these atoms would have wavelengths of $hc/(E_3 - E_2)$, $hc/(E_3 - E_1)$, and $hc/(E_2 - E_1)$, respectively.

The allowed energy states of an atom are determined by the complex interactions among the electrons and the nucleus. Every neutral hydrogen atom consists of a nucleus containing one proton, plus a single electron in a cloud surrounding the nucleus. Therefore, every hydrogen atom has the same energy states available to it, and all hydrogen atoms have the same emission and absorption lines. **Figure 5.15a** shows the energy level diagram of hydrogen. Figure 5.15b illustrates the visible emission spectrum from hydrogen. Figure 5.15c displays this same information as a graph.

Each different type of atom, that is, each chemical element, has a unique set of available energy states and therefore a unique set of wavelengths at which it can emit or absorb radiation. Figure 5.15d shows the emission spectra of four different kinds of atoms. These unique sets of wavelengths serve as unmistakable spectral "fingerprints" for each chemical element.

Spectral fingerprints are of crucial importance to astronomers. They let astronomers figure out what types of atoms (or molecules) are present in distant objects by simply looking at the spectrum of light from those objects. If the spectral lines of hydrogen, helium, carbon, oxygen, or any other element are visible in

**Figure 5.16** The traditional periodic table of the elements (lower right) shows the chemical elements laid out in ascending order according to the number of protons in the nucleus of each. But the "astronomer's periodic table" displays the abundances of the Sun's elements in boxes of relative size, showing hydrogen and helium as the most abundant. See Appendix 3 for a full periodic table of the elements.

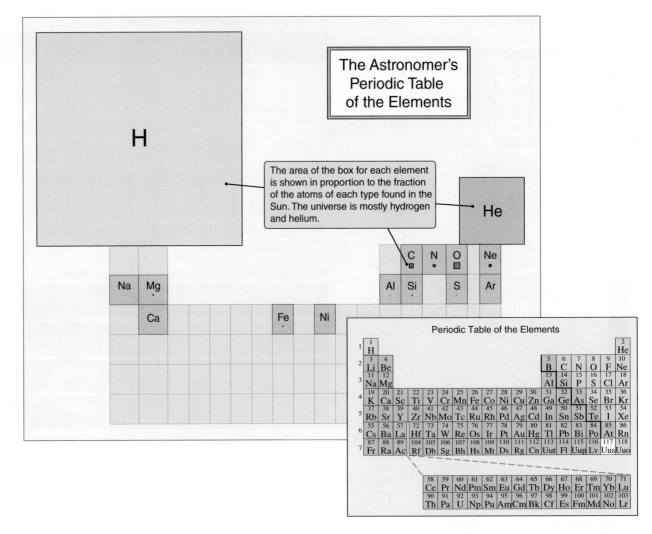

the light from a distant object, then we know that element is present in that object. The strength of a line is determined in part by how many atoms of that type are present in the source. By measuring the strength of the lines from different types of atoms in the spectrum of a distant object, astronomers can often infer the relative amounts of elements that make up the object. Additionally, by looking at the relative strength of different lines from the same element, it is often possible to determine the temperature, density, and pressure of the material as well.

This is how we know what makes up the stars and planets, and that we on Earth are composed of the same elements. Astronomers use the relative abundance of the elements in the Sun as a standard reference, termed **solar abundance**. As illustrated in **Figure 5.16**, hydrogen (H) is the most abundant element in the Sun, followed by helium (He) and 13 others. These 15 elements make up 99.99 percent of the mass of the Sun. The majority of the elements on the regular periodic table (in the lower right) make up less than 0.01 percent of the mass of the Sun.

## Excitement and Decay

Return to the analogy between the emission of a photon and a book falling off a shelf. If a book on a level shelf is not disturbed, it will sit there forever; something must *cause* the book to fall off the shelf. Similarly, physicists have wondered what causes an atom in an excited state to jump down to a lower energy state and emit a photon. Sometimes an atom in a higher energy state can be "stimulated" into emitting a photon—but under most circumstances nothing causes the atom to jump to the lower energy state. Instead, the atom decays *spontaneously*. While scientists can determine on average how long the atom is likely to remain in the excited state, exactly when a given atom will decay cannot be known until after the decay has happened. An atom decays at a random moment that is not influenced by anything in the universe and cannot be known ahead of time.

An example of this phenomenon is in toys that glow in the dark. Photons in sunlight or from a lightbulb are absorbed by certain phosphorescent atoms in the toy, knocking those atoms into excited energy states. The excited states of the atoms in the toy live for many seconds, unlike the excited energy states of many atoms that tend to decay in a small fraction of a second. If on average these atoms tend to remain in their excited state for 1 minute before decaying and emitting a photon, then after 1 minute there is a 50-50 chance that any particular atom in the toy will have decayed and a 50-50 chance that the atom will remain in its excited state. Although it is impossible to say exactly which atoms will decay, about half of the trillions and trillions of atoms in the toy will decay within 1 minute, and the brightness of the glow from the toy will have dropped to half of what it was. After each minute, half of the remaining excited atoms decay, and the glow from the toy drops to half of what it was 1 minute earlier. The glow from the toy slowly fades away.

In deep space, where atoms can remain undisturbed for long periods of time, there are certain excited states of atoms that last, on average, for tens of millions of years or even longer. An atom may have been in such an excited energy state for a few seconds, a few hours, or 50 million years when, in an instant, it decays to the lower energy state without anything causing it to do so. Physicists can only calculate the *probabilities* that certain events would take place.

**CHECK YOUR UNDERSTANDING 5.3**

How can spectra tell us the chemical composition of a distant star?

## 5.3 The Doppler Shift Indicates Motion Toward or Away from Us

You have already seen that light is a tightly packed bundle of information that can reveal a wealth of information about the physical state of material located tremendous distances away. In this section, we shall see how light can be used to measure one of the most straightforward questions about a distant astronomical object: is it moving away from us or toward us, and at what speed?

Have you ever listened to an ambulance speed by with sirens blaring? As the ambulance comes toward you, its siren has a certain high pitch, but as it passes by, the pitch of the siren drops noticeably. If you close your eyes and listen, you have no trouble knowing when the ambulance passed; the change in the pitch of its siren indicates that it has passed you by. You do not even need an ambulance to hear this effect. The sound of normal traffic behaves in the same way. As a car drives past, the pitch of the sound that it makes suddenly drops.

**Astronomy in Action:** Doppler Shift

**AstroTour:** The Doppler Effect

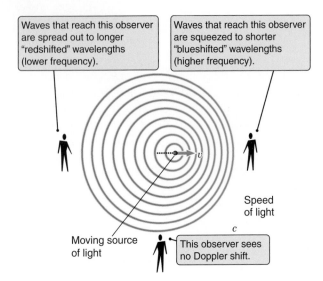

**Figure 5.17** Motion of a light or sound source relative to an observer may cause waves to be spread out (*redshifted*, or lower in pitch) or squeezed together (*blueshifted*, or higher in pitch). A change in the wavelength of light or the frequency of sound is called a *Doppler shift*.

🖐 **Nebraska Simulation:** Doppler Shift Demonstrator

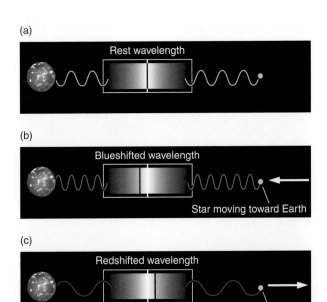

(a)

Rest wavelength

(b)

Blueshifted wavelength

Star moving toward Earth

(c)

Redshifted wavelength

Star moving away from Earth

**Figure 5.18** From their rest wavelength (a), spectral lines of astronomical objects are blueshifted if they are moving toward the observer (b) and redshifted if they are moving away from the observer (c).

The pitch of a sound is like the color of light: it is determined by the wavelength or, equivalently, the frequency of the sound wave. What you perceive as higher pitch corresponds to sound waves with higher frequencies and shorter wavelengths. Sounds that you perceive as lower in pitch are waves with lower frequencies and longer wavelengths. When an object is moving toward you, the waves that it emits, whether light or sound or waves in the water, "crowd together" in front of the object. You can see how this works by looking at **Figure 5.17**, which shows the locations of successive wave crests emitted by a moving object. The waves that reach you have a shorter wavelength and therefore a higher frequency than the waves given off by the object when it is not moving. Conversely, if an object is moving away from you, the waves reaching you from the object are spread out. This change in frequency due to motion is known as the **Doppler effect**, named after physicist Christian Doppler (1803–1853).

The Doppler effect causes a shift in the light emitted from a moving object. If the object were at rest, it would emit light with the **rest wavelength** ($\lambda_{rest}$), as shown in **Figure 5.18a**. If an object such as a star is moving toward you, the light reaching you from the object has a shorter wavelength than its rest wavelength—so we say the light is "bluer" than the rest wavelength, and the light is described as **blueshifted**, as shown by the blue wave in Figure 5.18b. In contrast, light from a source that is moving away from you is shifted to longer wavelengths. The light that you see is "redder" than if the source were not moving away from you, and is described as **redshifted**, as shown by the red waves in Figure 5.18c. The faster the object is moving with respect to you, the larger the shift. The amount by which the wavelength of light is shifted by the Doppler effect is called the **Doppler shift** of the light, and it depends on the speed of the object emitting the light.

The Doppler shift provides information only about the **radial velocity** ($v_r$) of the object, which is the part of the motion that is toward you or away from you. The radial velocity is the rate at which the distance between you and the object is changing: if $v_r$ is positive, the object is getting farther away from you; if $v_r$ is negative, the object is getting closer. At the moment the ambulance is passing you, it is getting neither closer nor farther away, so the pitch you hear is the same as the pitch heard by the crew riding on the truck. Similarly, an object moving across the sky does not move toward or away from you, and so its light will not be Doppler shifted from your point of view.

Doppler shifts become especially useful when you are looking at an object that has emission or absorption lines in its spectrum. These spectral lines enable astronomers to determine how rapidly the object is moving toward or away from Earth. To determine this velocity, astronomers first identify the spectral line as being from a certain chemical element, which has a unique rest wavelength ($\lambda_{rest}$) measured in a lab on Earth. They then measure the observed wavelength ($\lambda_{obs}$) in the spectrum of the distant object. The difference between the rest wavelength and the observed wavelength indicates the object's radial velocity. This is further explored in **Working It Out 5.2**.

## CHECK YOUR UNDERSTANDING 5.4

Which of the following Doppler shifts indicates the fastest approaching object (blueshifted)? (a) 0.04 nm; (b) 0.06 nm; (c) −0.04 nm; (d) −0.06 nm

## 5.2 Working It Out Making Use of the Doppler Effect

The Doppler formula for objects moving at a radial velocity ($v_r$) that is much less than the speed of light is given by

$$v_r = \frac{\lambda_{obs} - \lambda_{rest}}{\lambda_{rest}} \times c$$

A prominent spectral line of hydrogen atoms has a rest wavelength, $\lambda_{rest}$, of 656.3 nm (see Figure 5.15b). Suppose that you measure the wavelength of this line in the spectrum of a distant object and find that instead of seeing the line at 656.3 nm, you see the line at a wavelength, $\lambda_{obs}$, of 659.0 nm. How fast is its radial velocity? Using the above equation,

$$v_r = \frac{659.0\,\text{nm} - 656.3\,\text{nm}}{656.3\,\text{nm}} \times (3 \times 10^5\,\text{km/s})$$

$$v_r = 1{,}200\,\text{km/s}$$

In this way, you determine that the object is moving away from you with a radial velocity of 1,200 km/s.

For another example, suppose you know the velocity and want to compute the wavelength at which you would observe the spectral line? Earth's nearest stellar neighbor, Proxima Centauri, is moving toward us at a radial velocity of −21.6 km/s. What is the observed wavelength, $\lambda_{obs}$, of a magnesium line in Proxima Centauri's spectrum that has a rest wavelength, $\lambda_{rest}$, of 517.27 nm? We can rearrange the above equation to solve for $\lambda_{obs}$:

$$\lambda_{obs} = \left(1 + \frac{v_r}{c}\right) \times \lambda_{rest}$$

$$\lambda_{obs} = \left(1 + \frac{-21.6\,\text{km/s}}{3 \times 10^5\,\text{km/s}}\right) \times 517.27\,\text{nm} = 517.23\,\text{nm}$$

Although the observed Doppler blueshift ($\lambda_{obs} - \lambda_{rest} = 517.23 - 517.27$) is only −0.04 nm, it is easily measured with modern instrumentation.

## 5.4 Temperature Affects the Spectrum of Light That an Object Emits

The temperature of any object results from the balance between heating and cooling in an object. If an object's temperature is constant, then these two must be in balance with each other. In this section, we will examine this balance and see how we can use it to predict the temperatures of planets and stars.

### Equilibrium and Balance

Your body is heated by the release of chemical energy inside it. Sometimes your body is also heated by energy from your surroundings. If you are standing in sunshine on a hot day, the hot air around you and the sunlight falling on you both heat you. In response to this heating, your body cools itself off by perspiration: water seeps from the pores in your skin and evaporates. The energy to evaporate the water comes from your body. As the perspiration evaporates, it cools your body down. For your body temperature to remain stable, the heating must be balanced by the cooling. If there is more heating than cooling, then your body temperature climbs. If there is more cooling than heating, then your body temperature drops.

Imagine two well-matched teams struggling in a tug-of-war contest. Each team pulls steadfastly on the rope, but the force of one team's pull is only enough to match, not overcome, the force exerted by the other team. A picture taken now and another taken 5 minutes from now would not differ in any significant way. In this static equilibrium, opposing forces balance each other exactly.

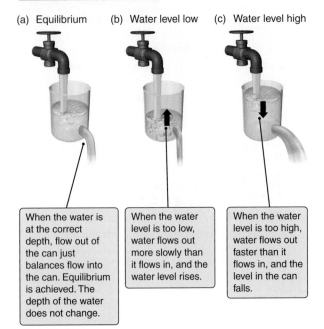

Pressure determines the rate at which water flows out of a hole in a can. The higher the water level, the faster the flow.

(a) Equilibrium (b) Water level low (c) Water level high

When the water is at the correct depth, flow out of the can just balances flow into the can. Equilibrium is achieved. The depth of the water does not change.

When the water level is too low, water flows out more slowly than it flows in, and the water level rises.

When the water level is too high, water flows out faster than it flows in, and the level in the can falls.

**Figure 5.19** Water flowing into and out of a can determines the water level in the can. This is an example of dynamic equilibrium.

**Astronomy in Action:** Changing Equilibrium

Static equilibrium can be stable, unstable, or neutral. A nut in a bowl is in a stable equilibrium: if it moves it will return to its original position at the bottom of the bowl. An example of an unstable equilibrium would be a book standing on its edge, unsupported on either side. If you nudged the book, it would fall over rather than settling back into its original position. When an unstable equilibrium is disturbed, it moves further away from equilibrium rather than back toward it.

Equilibrium can also be dynamic, which means the system is constantly changing so that one source of change is exactly balanced by another source of change, and the configuration of the system remains the same. Examine the demonstration illustrated in **Figure 5.19**. Placing a can with a hole cut in the bottom under an open water faucet provides a simple example of dynamic equilibrium. The depth of the water in the can determines how fast water pours out through the hole in the bottom of the can. When the water reaches just the right depth, as shown in Figure 5.19a, water pours out of the hole in the bottom of the can at exactly the same rate it pours into the top of the can from the faucet. The water leaving the can balances the water entering, and equilibrium is established. If you took a picture now and another picture in a few minutes, little of the water in the can would be the same, but the pictures would be indistinguishable.

If a system is not in equilibrium, its configuration will change. If the level of the water in the can is too low, as shown in Figure 5.19b, water will not flow out of the bottom of the can fast enough to balance the water flowing in. The water level will begin to rise until the amount coming in equals the amount going out. A picture taken now and another taken a short time later would not look the same. Conversely, if the water level in the can is too high (Figure 5.19c), water will flow out of the can faster than it flows into the can. The water level will begin to fall until the amount coming in equals the amount going out. Once again, if the system is not in equilibrium, its configuration will change.

When heating is balanced by cooling, we call it **thermal equilibrium**. Planets have a dynamic but stable thermal equilibrium, and electromagnetic radiation plays a crucial role in maintaining this. Energy from sunlight heats the surface of a planet, driving its temperature up, and the planet emits thermal radiation into space, cooling it down. For a planet to remain at the same average temperature over time, the energy it radiates into space must exactly balance the energy it absorbs from the Sun. **Figure 5.20** illustrates that the equilibrium temperature of a planet is analogous to the water level in Figure 5.19. We will return to planetary equilibrium later in the chapter. There are many kinds of equilibrium besides thermal equilibrium, some of which we will encounter later in the book.

## Temperature

In everyday life, we define hot and cold subjectively: something is hot when it feels hot or cold. When we measure *temperature*, we use degrees on a thermometer, but the way we define a degree is arbitrary. If you grew up in the United States, you probably think of temperatures in degrees Fahrenheit (°F), whereas if you grew up almost anywhere else in the world, you think of temperatures in degrees Celsius (°C).

**Temperature** is a measurement of how energetically the atoms that make up an object are moving about. The air around us is composed of vast numbers of atoms and molecules. Those molecules are moving about every which way. Some move slowly; some move more rapidly. All atoms and molecules are constantly in motion. The average kinetic energy ($E_K$) is given by $E_K = \frac{1}{2}mv^2$, where $m$ is the mass of an atom or molecule and $v$ is its velocity. The more energetically the atoms or molecules are bouncing about, the higher is the object's temperature. In fact, the random motions of atoms and molecules are often called their **thermal motions**, to emphasize the connection between these motions and temperature. **Figure 5.21** illustrates that when the temperature of a gas is increased, the kinetic energy is increased, and therefore the atoms move faster.

The atoms and molecules in a solid body (like you) cannot move about feely but still move back and forth around their normal location, and temperature measures the amount of that movement. If something is hotter than you are, thermal energy flows from that object into you. At the atomic level, that means the object's atoms are bouncing more energetically than are the atoms in your body, so if you touch the object, its atoms collide with your atoms, causing the atoms in your body to move faster. Your body gets hotter as thermal energy flows from the object to you. At the same time, these collisions rob the particles in the object of some of their energy. Their motions slow down, and the hotter object cools. Heating processes increase the average thermal energy of an object's particles, and cooling processes decrease the average thermal energy of those particles.

On the Fahrenheit scale, there are 180 degrees between the freezing point (32°F) and the boiling point (212°F) of water at sea level. On the Celsius scale, water freezes at 0°C and boils at 100°C. Because there is a different number of degrees between freezing and boiling on these two scales, a 1-degree change measured in Fahrenheit is not the same as a 1-degree change measured in Celsius.

There is a lowest possible physical temperature below which no object can fall. As the motions of the particles in an object slow down, the temperature drops lower and lower. The lowest possible temperature, where thermal motions have nearly come to a standstill, is called **absolute zero**. Absolute zero corresponds to −273.15°C, or −459.57°F. Scientists found it useful to define a temperature scale that begins at absolute zero, called the **Kelvin scale**. The size of one unit on the Kelvin scale, called a **kelvin (K)**, is the same as the Celsius degree. Zero kelvin (0 K) is set equal to absolute zero. Other temperatures are equal to Celsius plus 273.15, so water freezes at 273.15 K and water boils at 373.15 K. There are no negative temperatures on the Kelvin scale.

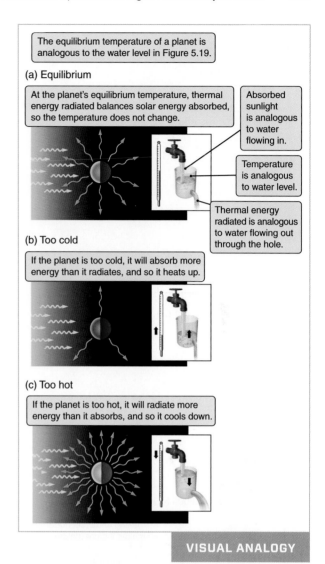

The equilibrium temperature of a planet is analogous to the water level in Figure 5.19.

(a) Equilibrium

At the planet's equilibrium temperature, thermal energy radiated balances solar energy absorbed, so the temperature does not change.

Absorbed sunlight is analogous to water flowing in.

Temperature is analogous to water level.

Thermal energy radiated is analogous to water flowing out through the hole.

(b) Too cold

If the planet is too cold, it will absorb more energy than it radiates, and so it heats up.

(c) Too hot

If the planet is too hot, it will radiate more energy than it absorbs, and so it cools down.

**VISUAL ANALOGY**

**Figure 5.20** Planets are heated by absorbing sunlight (and sometimes by internal heat sources) and cooled by emitting thermal radiation into space. If there are no other sources of heating or means of cooling, then the equilibrium between these two processes determines the temperature of the planet.

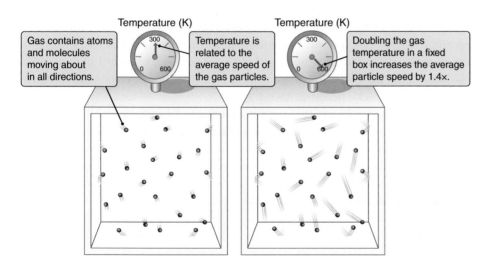

Temperature (K)

Gas contains atoms and molecules moving about in all directions.

Temperature is related to the average speed of the gas particles.

Temperature (K)

Doubling the gas temperature in a fixed box increases the average particle speed by 1.4×.

**Figure 5.21** Hotter gas temperatures correspond to faster motions of atoms.

When temperatures are measured in kelvins, the average thermal energy of particles is proportional to the measured temperature. The average thermal energy of the atoms in an object with a temperature of 200 K is twice the average thermal energy of the atoms in an object with a temperature of 100 K.

## Temperature, Luminosity, and Color

We have seen the way discrete atoms emit and absorb radiation, which leads to a useful understanding of emission lines and absorption lines that tell us about the physical state and motion of distant objects. But not all objects have spectra that are dominated by discrete spectral lines. As you saw in Figure 5.13a, if you pass the light from a lightbulb through a prism, instead of discrete bright and dark bands you will see light spread out smoothly from the blue end of the spectrum to the red. Similarly, if you look closely at the spectrum of the Sun, you will see absorption lines, but mostly you will see light smoothly spread out across all colors of the spectrum—a type of spectrum called a **continuous spectrum**.

We can think of a dense material as being composed of a collection of charged particles that are being jostled as their thermal motions cause them to run into their neighbors. The hotter the material is, the more violently its particles are being jostled. Recall that any time a charged particle is subjected to an acceleration, it radiates. So the jostling of particles due to their thermal motions causes them to give off a continuous spectrum of electromagnetic radiation. This is why any material that is sufficiently dense for its atoms to be jostled by their neighbors emits light simply because of its temperature. Radiation of this sort is called **thermal radiation**.

The radiation from an object changes as the object heats up or cools down. **Luminosity** is the amount of light *leaving* a source; that is, the total amount of light emitted (energy per second, measured in watts, W). The hotter the object, the more energetically the charged particles within it move, and the more energy they emit in the form of electromagnetic radiation. So as an object gets hotter, the light that it emits becomes more intense. Here is our first point about thermal radiation: *an object is more luminous when it is hotter.*

Now let's move to the question of what color light an object emits. As the object gets hotter, the thermal motions of its particles become more energetic, which produce more energetic photons. The average energy of the photons that it emits increases, the average wavelength of the emitted photons gets shorter, and the light from the object gets bluer. Here is our second point about thermal radiation: *hotter objects are bluer.* If you heat a piece of metal, the metal will glow—first a dull red, then orange, then yellow. The hotter the metal becomes, the more the highly energetic blue photons become mixed with the less energetic red photons, and the color of the light shifts from red toward blue. The light becomes more intense and bluer as the metal becomes hotter.

**Blackbodies** are objects that emit electromagnetic radiation only because of their temperature, not their composition. Blackbodies emit just as much thermal radiation as they absorb from their surroundings. Physicist Max Planck (1858–1947) graphed the intensity of the emitted radiation across all wavelengths and obtained the characteristic curves that we now call **Planck spectra** or **blackbody spectra**. **Figure 5.22** shows blackbody spectra for objects at several different temperatures.

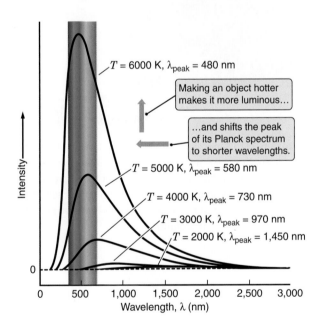

**Figure 5.22** This illustration shows blackbody spectra emitted by sources with temperatures of 2000 K, 3000 K, 4000 K, 5000 K, and 6000 K. At higher temperatures, the peak of the spectrum shifts toward shorter wavelengths, and the amount of energy radiated per second from each square meter of the source increases.

# Blackbody Laws

In the real world, the light from stars such as the Sun and the thermal radiation from a planet often come close to having blackbody spectra. So these objects follow two blackbody laws that relate luminosity with temperature and temperature with color, respectively.

**Stefan-Boltzmann Law**  As the temperature of an object increases, the object gives off more radiation at every wavelength, so the luminosity of the object should increase. Adding up all of the energy in a blackbody spectrum shows that the increase in luminosity is proportional to the fourth power of the temperature: Luminosity $\propto T^4$, known as the **Stefan-Boltzmann law**. This law was discovered in the laboratory by physicist Josef Stefan (1835–1893) and derived mathematically by his student Ludwig Boltzmann (1844–1906).

It is difficult to measure all of the photons emitted by Earth or the Sun in all possible directions, but it is easier to measure the *flux*. The amount of energy radiated by each square meter of the surface of an object each second is called the **flux**, abbreviated $\mathcal{F}$. The flux is proportional to the luminosity. You can find the luminosity by multiplying the flux by the total surface area. The Stefan-Boltzmann law says that the flux is given by the following equation: $\mathcal{F} = \sigma T^4$. The constant $\sigma$ (the Greek letter sigma), which is called the **Stefan-Boltzmann constant**, equals $5.67 \times 10^{-8}$ W/(m$^2$ K$^4$), where 1 watt = 1 joule per second (J/s).

The Stefan-Boltzmann law says that an object rapidly becomes more luminous as its temperature increases. If the temperature of an object doubles, the amount of energy being radiated each second increases by a factor of $2^4$, or 16. If the temperature of an object goes up by a factor of 3, then the energy being radiated by the object each second goes up by a factor of $3^4$, or 81. A lightbulb with a filament temperature of 3000 K radiates 16 times as much light as it would if the filament temperature were 1500 K. Even modest changes in temperature can result in large changes in the amount of luminosity radiated by an object.

**Wien's Law**  Look again at Figure 5.22. The wavelength where the blackbody spectrum is at its peak, $\lambda_{\text{peak}}$, is where the electromagnetic radiation from an object is greatest. As the temperature, $T$, increases, the peak of the spectrum shifts toward shorter wavelengths. For example, compare the peak wavelengths of a 3000 K object and a 6000 K object. Photon energy and wavelength are inversely related; thus, as the peak wavelength becomes shorter, the average photon energy becomes greater. The object becomes bluer. The physicist Wilhelm Wien (1864–1928) found that the peak wavelength in the spectrum is inversely proportional to the temperature of the object. **Wien's law** states that if you double the temperature, the peak wavelength becomes half of what it was. If you increase the temperature by a factor of 3, the peak wavelength becomes a third of what it was. Stefan-Boltzmann's law and Wien's law are further explored in **Working It Out 5.3**. We will return to these laws later in the chapter when we use them to estimate the temperatures of the planets.

 **Astronomy in Action:** Wien's Law

 **Nebraska Simulation:** Blackbody Curves

## CHECK YOUR UNDERSTANDING 5.5

When you look at the sky on a dark night and see stars of different colors, which are the hottest? (a) orange; (b) red-orange; (c) yellow; (d) red; (e) blue

## 5.3  Working It Out  Working with the Stefan-Boltzmann Law and Wien's Law

Stefan-Boltzmann's law can be used to estimate the flux and luminosity of Earth. Earth's average temperature is 288 K, so the flux from its surface is

$$\mathcal{F} = \sigma T^4$$

$$\mathcal{F} = (5.67 \times 10^{-8} \, \text{W/m}^2 \, \text{K}^4) \times (288 \, \text{K})^4$$

$$\mathcal{F} = 390 \, \text{W/m}^2$$

The luminosity is the flux multiplied by the surface area ($A$) of Earth. Surface area is given by $4\pi R^2$, and the radius of Earth is 6,378 km, or $6.378 \times 10^6$ meters. So the luminosity is

$$L = \mathcal{F} \times A = \mathcal{F} \times 4\pi R^2$$

$$L = (390 \, \text{W/m}^2) \times [4\pi (6.378 \times 10^6 \, \text{m})^2]$$

$$L \approx 2 \times 10^{17} \, \text{W}$$

Earth emits the equivalent of the energy used by 2,000,000,000,000,000 (2 million billion) hundred-watt lightbulbs. This is still not anywhere close to the amount emitted by the Sun.

Wien's law also proves useful to astronomers. If they measure the spectrum of an object emitting thermal radiation and find where the peak in the spectrum is, Wien's law can be used to calculate the temperature of the object. Wien's law can be written as

$$T = \frac{2{,}900{,}000 \, \text{nm K}}{\lambda_{\text{peak}}}$$

For example, the spectrum of the light coming from the Sun peaks at a wavelength of $\lambda_{\text{peak}} = 500$ nm, so

$$T = \frac{2{,}900{,}000 \, \text{nm K}}{500 \, \text{nm}} = 5800 \, \text{K}$$

This is how you can know the surface temperature of the Sun.

Suppose you want to calculate the peak wavelength at which Earth radiates. Using Earth's average temperature of 288 K in Wien's law gives

$$\lambda_{\text{peak}} = \frac{2{,}900{,}000 \, \text{nm K}}{288 \, \text{K}} = 10{,}100 \, \text{nm} = 10.1 \, \text{microns}$$

Earth's radiation peaks in the infrared region of the spectrum.

## 5.5  The Brightness of Light Depends on the Luminosity and Distance of the Source

Recall that luminosity refers to the amount of light leaving a source. By contrast, the **brightness** of electromagnetic radiation is the amount of light that is arriving at a particular location. Therefore, brightness depends on the luminosity and the distance of the light source. For example, replacing a 50-W lightbulb with a 100-W bulb makes a room twice as bright because it doubles the light reaching any point in the room. But brightness also depends on the distance from a source of electromagnetic radiation. If you needed more light to read this book, you could replace the bulb in your lamp with a more luminous bulb or you can move the book closer to the light. Conversely, if a light is too bright, you can move away from it. Our everyday experience teaches us that as we move away from a light, its brightness decreases.

The particle description of light provides another way to think about the brightness of radiation and how brightness depends on distance. Suppose you had a piece of cardboard that measured 1 meter by 1 meter. To make the light falling on the cardboard twice as bright, you would need to double the number of photons that hit the cardboard each second. Tripling the brightness of the light would mean increasing the number of photons hitting the cardboard each second by a factor of 3, and so on. Brightness depends on the number of photons falling on each square meter of a surface each second.

Now imagine a lightbulb sitting at the center of a spherical shell, illustrated in **Figure 5.23**. Photons from the bulb travel in all directions and land on the inside of the shell. To find the number of photons landing on each square meter of the shell during each second, that is, to determine the brightness of the light, take the *total* number of photons given off by the lightbulb each second and divide by the number of square meters over which those photons have to be spread. The surface area of a sphere is given by the formula $A = 4\pi r^2$, where $r$ is the distance between the bulb and the surface of the sphere (that is, $r$ = the radius of the sphere). The number of photons striking one square meter each second is equal to the total number of photons emitted each second divided by the surface area $4\pi r^2$.

Now change the size of the spherical shell while keeping the total number of photons given off by the lightbulb each second the same. As the shell becomes larger, the photons from the lightbulb must spread out to cover a larger surface area. Each square meter of the shell receives fewer photons each second, so the brightness of the light decreases. If the shell's surface is moved twice as far from the light, the area over which the light must spread increases by a factor of $2^2 = 2 \times 2 = 4$. The photons from the bulb spread out over 4 times as much area, so the number of photons falling on each square meter each second becomes $\frac{1}{4}$ of what it was. If the surface of the sphere is 3 times as far from the light, the area over which the light must spread increases by a factor of $3^2 = 3 \times 3 = 9$, and the number of photons per second falling on each square meter becomes $\frac{1}{9}$ of what it was originally. This is the same kind of inverse square relationship you saw for gravity in Chapter 4. The brightness of the light from an object is inversely proportional to the square of the distance from the object. Twice as far means one-fourth as bright.

This idea of photons streaming and spreading onto a surface from a light explains why brightness follows an inverse square law. In practice, however, it is usually more convenient to talk about the *energy* coming to a surface each second, rather than the number of photons arriving.

The luminosity of an object is the total number of photons given off by the object multiplied by the energy of each photon. So instead of thinking about how the number of photons must spread out to cover the surface of a sphere, we can think about how the energy carried by the photons must spread out to cover the surface of a sphere. The brightness of the light is the amount of energy falling on a square meter in a second, and it equals the luminosity $L$ divided by the area of the sphere, which depends on the radius squared. This tells us, for example, that the brightness of the Sun will depend on the inverse square of the planet's distance from the Sun. This will factor in as we estimate the equilibrium temperatures of the planets in **Working It Out 5.4**.

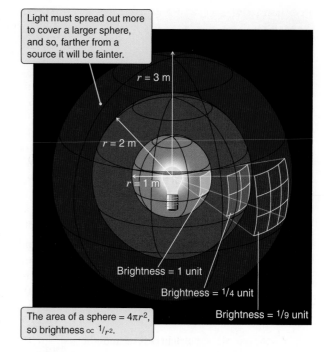

Light must spread out more to cover a larger sphere, and so, farther from a source it will be fainter.

$r = 3$ m

$r = 2$ m

$r = 1$ m

Brightness = 1 unit

Brightness = 1/4 unit

Brightness = 1/9 unit

The area of a sphere = $4\pi r^2$, so brightness $\propto 1/r^2$.

**Figure 5.23** Light obeys an inverse square law as it spreads away from a source. Twice as far means one-fourth as bright.

**Astronomy in Action:** Inverse Square Law

## CHECK YOUR UNDERSTANDING 5.6

The average distance of Mars from the Sun is 1.4 AU. How bright is the Sun on Mars compared with its brightness on Earth? (a) 1.4 times brighter; (b) about 2 times brighter; (c) about 2 times fainter; (d) 1.4 times fainter

## 5.4 Working It Out Using Radiation Laws to Calculate Equilibrium Temperatures of Planets

The temperature of a planet is determined by a balance between the amount of sunlight being absorbed and the amount of energy being radiated back into space. We begin with the amount of sunlight being absorbed. When viewed from the Sun, a planet looks like a circular disk with a radius equal to the radius of the planet, $R_{\text{planet}}$. The area of the planet that is lit by the Sun is

$$\text{Absorbing area of planet} = \pi R_{\text{planet}}^2$$

The amount of energy striking a planet also depends on the brightness of sunlight at the distance at which the planet orbits. The brightness of sunlight at a distance $d$ from the Sun is equal to the luminosity of the Sun ($L_{\text{Sun}}$, in watts) divided by $4\pi d^2$ (we use $d$ here to avoid confusion with the planet's radius, $R_{\text{planet}}$):

$$\text{Brightness of sunlight} = \frac{L_{\text{Sun}}}{4\pi d^2}$$

A planet does not absorb all the sunlight that falls on it. **Albedo**, $a$, is the fraction of the sunlight that reflects from a planet. The corresponding fraction of the sunlight that is absorbed by the planet is 1 minus the albedo. A planet covered entirely in snow would have a high albedo (close to 1), while a planet covered entirely by black rocks would have a low albedo, close to 0:

$$\text{Fraction of sunlight absorbed} = 1 - a$$

We can now calculate the energy absorbed by the planet each second. Writing this relationship as an equation, we say that

$$\begin{pmatrix}\text{Energy absorbed}\\ \text{by the planet}\\ \text{each second}\end{pmatrix} = \begin{pmatrix}\text{Absorbing}\\ \text{area of}\\ \text{the planet}\end{pmatrix} \times \begin{pmatrix}\text{Brightness}\\ \text{of sunlight}\end{pmatrix} \times \begin{pmatrix}\text{Fraction}\\ \text{of sunlight}\\ \text{absorbed}\end{pmatrix}$$

$$= \pi R_{\text{planet}}^2 \times \frac{L_{\text{Sun}}}{4\pi d^2} \times (1 - a)$$

Now let's turn to the other piece of the equilibrium: the amount of energy that the planet radiates away into space each second. We can calculate this amount by multiplying the number of square meters of the planet's total surface area by the energy radiated by each square meter each second. The surface area for the planet is given by $4\pi R_{\text{planet}}^2$. The Stefan-Boltzmann law tells us that the energy radiated by each square meter each second is given by $\sigma T^4$. So we can say that

$$\begin{pmatrix}\text{Energy radiated}\\ \text{by the planet}\\ \text{each second}\end{pmatrix} = \begin{pmatrix}\text{Surface}\\ \text{area of}\\ \text{the planet}\end{pmatrix} \times \begin{pmatrix}\text{Energy radiated}\\ \text{per square meter}\\ \text{per second}\end{pmatrix}$$

$$= 4\pi R_{\text{planet}}^2 \times \sigma T^4$$

If the planet's temperature is to remain stable—not heating up or cooling down—then each second the "Energy radiated" must be equal to "Energy absorbed." When we set these two quantities equal to each other, we arrive at the expression

$$\begin{pmatrix}\text{Energy radiated}\\ \text{by the planet}\\ \text{each second}\end{pmatrix} = \begin{pmatrix}\text{Energy absorbed}\\ \text{by the planet}\\ \text{each second}\end{pmatrix}$$

or

$$4\pi R_{\text{planet}}^2 \sigma T^4 = \pi R_{\text{planet}}^2 \frac{L_{\text{Sun}}}{4\pi d^2}(1 - a)$$

Canceling out $\pi R_{\text{planet}}^2$ on both sides, and rearranging this equation to put $T$ on one side and everything else on the other gives

$$T^4 = \frac{L_{\text{Sun}}(1 - a)}{16\sigma\pi d^2}$$

If we take the fourth root of each side, we get

$$T = \left(\frac{L_{\text{Sun}}(1 - a)}{16\sigma\pi d^2}\right)^{1/4}$$

Putting in the appropriate numbers for the known luminosity of the Sun, $L_{\text{Sun}}$, and the constants $\pi$ and $\sigma$ yields this simpler equation:

$$T = 279\,\text{K} \times \left(\frac{1 - a}{d_{\text{AU}}^2}\right)^{1/4}$$

where $d_{\text{AU}}$ is the distance of the planet from the Sun in astronomical units.

To use this equation, we would need to know a planet's distance from the Sun and its average albedo. For a blackbody ($a = 0$) at 1 AU from the Sun, the temperature is 279 K. For Earth, with an albedo of 0.3 and a distance from the Sun of 1 AU, the temperature is

$$T = 279\,\text{K} \times \left(\frac{1 - 0.3}{1^2}\right)^{1/4} = 255\,\text{K}$$

(Calculator hint: To take a fourth root, you can take the square root twice, or use the $x^y$ button with $y = 0.25$).

Earth is cooler than a blackbody at 1 AU from the Sun because its average albedo is greater than zero. If Earth's albedo changed or the Sun's luminosity changed, that would affect the result. When we examine planets around other stars, we will need to use the luminosity of the particular star in the equation, instead of the Sun's luminosity, so the temperature at 1 AU will be different than what is it for Earth.

# Origins

## Temperatures of Planets

In the previous chapters, we discussed how a planet's axial tilt and its orbital shape can affect its temperature, and thus its prospects for life. Now let's get more specific about the temperatures of planets, using what you learned in this chapter about thermal radiation. For a planet at an equilibrium temperature, the energy radiated by a planet exactly balances the energy absorbed by the planet. If the planet is hotter than this equilibrium temperature, it will radiate energy faster than it absorbs sunlight, and its temperature will fall. If the planet is cooler than this temperature, it will radiate energy slower than it absorbs sunlight, and its temperature will rise.

Planets at different distances from the Sun will have different temperatures, and the temperature should be inversely proportional to the square root of the distance, as you saw in

Working It Out 5.4. **Figure 5.24** plots the actual and predicted temperatures of nine solar system objects. Each vertical orange bar shows the range of temperatures found on the surface of the planet or, in the case of the giant planets, at the top of the planet's clouds. The black dots show the predictions made using the equation in Working It Out 5.4. For most planets, the predictions are not too far off, indicating that our basic understanding of *why* planets have the temperatures they have is probably pretty good. The data for Mercury, Mars, and Pluto agree particularly well.

In some cases, however, the predictions are wrong. For Earth, the actual measured temperature is a bit higher than the predicted temperature, and for Venus the actual surface temperature is much higher than the prediction.

The predicted values assume that the temperature of the planet is the

same everywhere. However, planets are likely to be hotter on the day side than on the night side. The predictions also assume that a planet's only source of energy is sunlight, and that the fraction of sunlight reflected is constant over the surface of each planet. There is also the assumption that the planets absorb and radiate energy into space as blackbodies.

The discrepancies between the calculated and the measured temperatures of some of the planets indicate that for these planets, some or all of these assumptions are incorrect. For example, the planet may have its own source of energy besides sunlight, or it may have an atmosphere. Understanding the temperatures of planets makes it possible to hypothesize why life may have evolved here on Earth, instead of on a different planet in the Solar System.

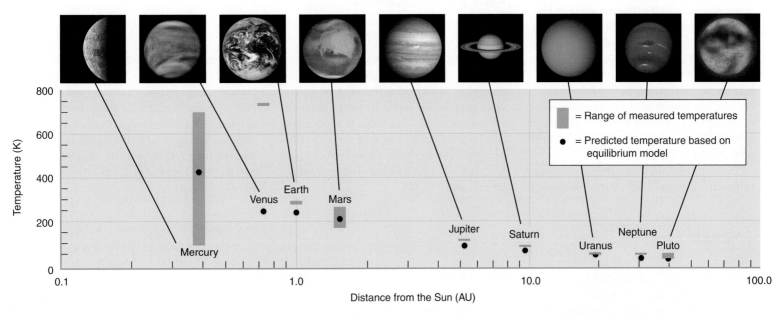

**Figure 5.24** Predicted temperatures for the planets and dwarf planet Pluto are based on the equilibrium between absorbed sunlight and thermal radiation into space. These temperatures are compared with ranges of observed surface temperatures.

# A Study in Scarlet

**ESO**

This new image from ESO's La Silla Observatory in Chile reveals a cloud of hydrogen called Gum 41 (**Figure 5.25**). In the middle of this little-known nebula, brilliant hot young stars are giving off energetic radiation that causes the surrounding hydrogen to glow with a characteristic red hue.

This area of the southern sky, in the constellation of Centaurus (The Centaur), is home to many bright nebulae, each associated with hot newborn stars that formed out of the clouds of hydrogen gas. The intense radiation from the stellar newborns excites the remaining hydrogen around them, making the gas glow in the distinctive shade of red typical of star-forming regions. Another famous example of this phenomenon is the Lagoon Nebula, a vast cloud that glows in similar bright shades of scarlet.

The nebula in this picture is located some 7,300 light-years from Earth. Australian astronomer Colin Gum discovered it on photographs taken at the Mount Stromlo Observatory near Canberra, and included it in his catalog of 84 emission nebulae, published in 1955. Gum 41 is actually one small part of a bigger structure called the Lambda Centauri Nebula, also known by the more exotic name of the Running Chicken Nebula. Gum died at a tragically early age in a skiing accident in Switzerland in 1960.

In this picture of Gum 41, the clouds appear to be quite thick and bright, but this is actually

**Figure 5.25**  The Gum 41 Nebula.

misleading. If a hypothetical human space traveler could pass through this nebula, it is likely that they would not notice it as—even at close quarters—it would be too faint for the human eye to see. This helps to explain why this large object had to wait until the mid-twentieth century to be discovered—its light is spread very thinly and the red glow cannot be well seen visually.

This new portrait of Gum 41—likely one of the best so far of this elusive object—has been created using data from the Wide Field Imager (WFI) on the MPG/ESO 2.2-meter telescope at the La Silla Observatory in Chile. It is a combination of images taken through blue, green, and red filters, along with an image using a special filter designed to pick out the red glow from hydrogen.

**ARTICLES    QUESTIONS**

1. How long has it taken the light from this nebula to reach us?
2. Why are the young stars blue?
3. What type of spectra would you expect to get from the stars and from the gas?
4. Refer to the spectrum of hydrogen in Figure 5.15. Why is the excited hydrogen gas in the image of Gum 41 glowing red?
5. Would you be able to see Gum 41 from your location? Why or why not?

# Summary

Light carries both information and energy throughout the universe. The speed of light in a vacuum is 300,000 km/s; nothing can travel faster. Visible light is only a tiny portion of the entire electromagnetic spectrum. Atoms absorb and emit radiation at unique wavelengths like spectral fingerprints. A planet's temperature depends on its distance from its star, its albedo, and the luminosity of its star.

**LG 1  Describe the wave and particle properties of light, and describe the electromagnetic spectrum.** Light is both a particle and a wave. Light is simultaneously a stream of particles called photons and an electromagnetic wave. Different types of electromagnetic radiation, from gamma rays to visible light to radio waves, are electromagnetic waves that differ in frequency and wavelength.

**LG 2  Describe how to measure the composition of distant objects using the unique spectral lines of different types of atoms.** Nearly all matter is composed of atoms, and light can reveal the identity of the types of atoms that are present in matter. Each type of atom has a different spacing of its electron energy levels, and the emission of photons is related to the electron changing levels. As a result, we can identify different chemical elements and molecules in distant objects.

**LG 3  Describe the Doppler effect and how it can be used to measure the motion of distant objects.** Because of the Doppler effect, light from receding objects is redshifted to longer wavelengths, and light from approaching objects is blueshifted to shorter wavelengths. The wavelength shifts of the spectral lines indicate how fast an astronomical object is moving toward or away from Earth.

**LG 4  Explain how the spectrum of light that an object emits depends on its temperature.** Temperature is a measure of how energetically particles are moving in an object. A light source that emits electromagnetic radiation because of its temperature is called a blackbody. A blackbody emits a continuous spectrum. The total amount of energy emitted is proportional to the temperature to the fourth power, and the peak wavelength, which determines its color, is inversely proportional to the temperature.

**LG 5  Differentiate luminosity from brightness, and illustrate how distance affects each.** The light output, or luminosity, of an object is the amount of light the object emits. The brightness of an object is proportional to its luminosity divided by its distance squared. Thus, the brightness of the Sun is different when measured from each Solar System planet but the luminosity is the same.

## ? UNANSWERED QUESTIONS

- Has the speed of light always been 300,000 km/s? Some theoretical physicists have questioned whether light traveled much faster earlier in the history of our universe. The observational evidence that may test this idea comes from studying the spectra of the most distant objects—whose light has been traveling for billions of years—and determining whether billions of years ago chemical elements absorbed light somewhat differently than they do today. So far, there is no evidence that the speed of light changes.

- Will it ever be possible to travel faster than the speed of light? Our current understanding of the science says no. A staple of science fiction films and stories is spaceships that go into "warp speed" or "hyperdrive"—moving faster than light—to traverse the huge distances of space (and visit a different planetary system every week). If this premise is simply fictional and the speed of light is a true universal limit, then travel between the stars will take many years. Because all electromagnetic radiation travels at the speed of light, even an electromagnetic signal sent to another planetary system would take many years to get there. Interstellar visits (and interstellar conversations) will be quite prolonged.

# Questions and Problems

## Test Your Understanding

1. If the Sun instantaneously stopped giving off light, what would happen on Earth?
   a. Earth would immediately get dark.
   b. Earth would get dark 8 minutes later.
   c. Earth would get dark 27 minutes later.
   d. Earth would get dark 1 hour later.

2. Why is an iron atom a different element from a sodium atom?
   a. A sodium atom has fewer neutrons in its nucleus than an iron atom has.
   b. An iron atom has more protons in its nucleus than a sodium atom has.
   c. A sodium atom is bigger than an iron atom.
   d. A sodium atom has more electrons.

3. Suppose an atom has three energy levels, specified in arbitrary units as 10, 7, and 5. In these units, which of the following energies might an emitted photon have? (Select all that apply.)
   a. 3
   b. 2
   c. 5
   d. 4

4. When a boat moves through the water, the waves in front of the boat bunch up, while the waves behind the boat spread out. This is an example of
   a. the Bohr model.
   b. the wave nature of light.
   c. emission and absorption.
   d. the Doppler effect.

5. As a blackbody becomes hotter, it also becomes _____ and _____ .
   a. more luminous; redder
   b. more luminous; bluer
   c. less luminous; redder
   d. less luminous; bluer

6. Which of the following factors does *not* directly influence the temperature of a planet?
   a. the luminosity of the Sun
   b. the distance from the planet to the Sun
   c. the albedo of the planet
   d. the size of the planet

7. Two stars are of equal luminosity. Star A is 3 times as far from you as star B. Star A appears _____ star B.
   a. 9 times brighter than
   b. 3 times brighter than
   c. the same brightness as
   d. $\frac{1}{3}$ as bright as
   e. $\frac{1}{9}$ as bright as

8. When less energy radiates from a planet, its _____ increases until a new _____ is achieved.
   a. temperature; equilibrium
   b. size; temperature
   c. equilibrium; size
   d. temperature; size

9. How does the speed of light in a medium compare to the speed in a vacuum?
   a. The speed is the same in both a medium and a vacuum, as the speed of light is a constant.
   b. The speed in the medium is always faster than the speed in a vacuum.
   c. The speed in the medium is always slower than the speed in a vacuum.
   d. The speed in the medium may be faster or slower, depending on the medium.

10. When an electron moves from a higher energy level in an atom to a lower energy level,
    a. a continuous spectrum is emitted.
    b. a photon is emitted.
    c. a photon is absorbed.
    d. a redshifted spectrum is emitted.

11. In Figure 5.15, the red photons come from the transition from $E_3$ to $E_2$. These photons will have the _____ wavelengths because they have the _____ energy compared to the other photons.
    a. shortest; least
    b. shortest; most
    c. longest; least
    d. longest; most

12. Star A and star B appear equally bright in the sky. Star A is twice as far away from Earth as star B. How do the luminosities of stars A and B compare?
    a. Star A is 4 times as luminous as star B.
    b. Star A is 2 times as luminous as star B.
    c. Star B is 2 times as luminous as star A.
    d. Star B is 4 times as luminous as star A.

13. What is the surface temperature of a star that has a peak wavelength of 290 nm?
    a. 1000 K
    b. 2000 K
    c. 5000 K
    d. 10,000 K
    e. 100,000 K

14. If a planet is in thermal equilibrium,
    a. no energy is leaving the planet.
    b. no energy is arriving on the planet.
    c. the amount of energy leaving equals the amount of energy arriving.
    d. the temperature is very low.

15. The temperature of an object has a very specific meaning as it relates to the object's atoms. A high temperature means that the atoms
    a. are very large.
    b. are moving very fast.
    c. are all moving together.
    d. have a lot of energy.

## Thinking about the Concepts

16. We know that the speed of light in a vacuum is $3 \times 10^5$ km/s. Is it possible for light to travel at a lower speed? Explain your answer.

17. Is light a wave or a particle or both? Explain your answer.

18. Referring to the Process of Science Figure, if any of these experiments had *not* agreed with the others, what would that mean for the conclusion that light has a finite, constant speed?

19. If photons of blue light have more energy than photons of red light, how can a beam of red light carry as much energy as a beam of blue light?

20. Patterns of emission or absorption lines in spectra can uniquely identify individual atomic elements. Explain how positive identification of atomic elements can be used as one way of testing the validity of the cosmological principle discussed in Chapter 1.

21. An atom in an excited state can drop to a lower energy state by emitting a photon. Is it possible to predict exactly how long the atom will remain in the higher energy state? Explain your answer.

22. Spectra of astronomical objects show both bright and dark lines. Describe what these lines indicate about the atoms responsible for the spectral lines.

23. Astronomers describe certain celestial objects as being *redshifted* or *blueshifted*. What do these terms indicate about the objects?

24. An object somewhere near you is emitting a pure tone at middle C on the octave scale (262 Hz). You, having perfect pitch, hear the tone as A above middle C (440 Hz). Describe the motion of this object relative to where you are standing.

25. During a popular art exhibition, the museum staff finds it necessary to protect the artwork by limiting the total number of viewers in the museum at any particular time. New viewers are admitted at the same rate that others leave. Is this an example of static equilibrium or of dynamic equilibrium? Explain.

26. A favorite object for amateur astronomers is the double star Albireo, with one of its components a golden yellow and the other a bright blue. What do these colors tell you about the relative temperatures of the two stars?

27. The stars you see in the night sky cover a large range of brightness. What does that range tell you about the distances of the various stars? Explain your answer.

28. Why is it not surprising that sunlight peaks in the "visible"?

29. Study Figure 5.24. For which planet is the range of measured temperatures furthest from the predicted value? What accounts for this difference?

30. Suppose you want to find a planet with the same temperature as Earth. What could you say about the size of the orbit of such a planet if it is orbiting a red star? A yellow star? A blue star?

## Applying the Concepts

31. You are tuned to 790 on AM radio. This station is broadcasting at a frequency of 790 kHz ($7.90 \times 10^5$ Hz). You switch to 98.3 on FM radio. This station is broadcasting at a frequency of 98.3 MHz ($9.83 \times 10^7$ Hz).
    a. What are the wavelengths of the AM and FM radio signals?
    b. Which broadcasts at higher frequencies: AM or FM?
    c. What are the photon energies of the two broadcasts?

32. Your microwave oven cooks by vibrating water molecules at a frequency of 2.45 gigahertz (GHz), or $2.45 \times 10^9$ Hz. What is the wavelength, in centimeters, of the microwave's electromagnetic radiation?

33. You observe a spectral line of hydrogen at a wavelength of 502.3 nm in a distant galaxy. The rest wavelength of this line is 486.1 nm. What is the radial velocity of this galaxy? Is it moving toward you or away from you?

34. Assume that an object emitting a pure tone of 440 Hz is on a vehicle approaching you at a speed of 25 m/s. If the speed of sound at this particular atmospheric temperature and pressure is 340 m/s, what will be the frequency of the sound that you hear? (Hint: Keep in mind that frequency is inversely proportional to wavelength.)

35. If half of the phosphorescent atoms in a glow-in-the-dark toy give up a photon every 30 minutes, how bright (relative to its original brightness) will the toy be after 2 hours?

36. How bright would the Sun appear from Neptune, 30 AU from the Sun, compared to its brightness as seen from Earth? The spacecraft *Voyager 1* is now about 130 AU from the Sun and heading out of the Solar System. Compare the brightness of the Sun seen by *Voyager 1* with that seen from Earth.

37. On a dark night you notice that a distant lightbulb happens to have the same brightness as a firefly that is 5 meters away from you. If the lightbulb is a million times more luminous than the firefly, how far away is the lightbulb?

38. Two stars appear to have the same brightness, but one star is 3 times more distant than the other. How much more luminous is the more distant star?

39. A panel with an area of 1 square meter ($m^2$) is heated to a temperature of 500 K. How many watts is it radiating into its surroundings?

40. The Sun has a radius of $6.96 \times 10^5$ km and a blackbody temperature of 5780 K. Calculate the Sun's luminosity.

41. Some of the hottest stars known have a blackbody temperature of 100,000 K. What is the peak wavelength of their radiation? What type of radiation is this?

42. Your body, at a temperature of about 37°C (98.6°F), emits radiation in the infrared region of the spectrum.
    a. What is the peak wavelength, in microns, of your emitted radiation?
    b. Assuming an exposed body surface area of 0.25 m², how many watts of power do you radiate?

43. A planet with no atmosphere at 1 AU from the Sun would have an average blackbody surface temperature of 279 K if it absorbed all the Sun's electromagnetic energy falling on it (albedo = 0).
    a. What would be the average temperature on this planet if its albedo were 0.1, typical of a rock-covered surface?
    b. What would be the average temperature if its albedo were 0.9, typical of a snow-covered surface?

44. The orbit of Eris, a dwarf planet, carries it out to a maximum distance of 97.7 AU from the Sun. Assuming an albedo of 0.8, what is the average temperature of Eris when it is farthest from the Sun?

45. Suppose our Sun had 10 times its current luminosity. What would be the average blackbody surface temperature of Earth, assuming Earth had the same albedo?

## USING THE WEB

46. a. Go to the website for NASA's Astronomy Picture of the Day (http://apod.nasa.gov/apod/ap101027.html) and study the picture of the Andromeda Galaxy in visible light and in ultraviolet light. Which light represents a hotter temperature? What differences do you see in the two images?
    b. Go to the APOD archive (http://apod.nasa.gov/cgi-bin/apod/apod_search) and enter "false color" in the search box. Examine a few images that come up in the search. What does *false color* mean in this context? What wavelength(s) were the pictures exposed in? What is the color coding; that is, what wavelength does each color in the image represent? You can read more about false color here: http://chandra.harvard.edu/photo/false_color.html.

47. Crime scene investigators may use different types of light to examine a crime scene. Search for "forensic lighting" in your browser. What wavelengths of light are used to search for blood and saliva? For fingerprints? Why is it useful for an investigator to have access to different kinds of light? Search on "forensic spectroscopy" and select a recent report. How is spectroscopy being used in crime scene investigations?

48. Using Google Images or an equivalent website, search for "night vision imaging" and "thermal imaging." How do night vision goggles and thermal-imaging devices work differently from regular binoculars or cameras? When are these useful?

49. The Transportation Security Administration (TSA) uses several types of imaging devices to screen passengers in airports. Search for "TSA imaging" in your browser. What wavelengths of light are being used in these devices? What concerns do passengers have about some of these imaging devices?

50. Go to the NASA Earth Observations website (http://neo.sci.gsfc.nasa.gov) and look at the current map of Earth's albedo (click on "albedo" in the menu for "energy" or "land" if it didn't come up). Compare this map with those of 2, 4, 6, 8, and 10 months ago. Which parts of Earth have the lowest and highest albedos? In which parts do the albedos seem to change the most with the time of the year? Would you expect ice, snow, oceans, clouds, forests, and deserts to add or subtract in each case from the total Earth albedo? Which parts of Earth are not showing up on this map?

## smartwork5

If your instructor assigns homework in Smartwork5, access your assignments at digital.wwnorton.com/astro5.

# EXPLORATION

## Light as a Wave, Light as a Photon

digital.wwnorton.com/astro5

Visit the Student Site at the Digital Landing Page, and open the "Light as a Wave, Light as a Photon" AstroTour in Chapter 5. Watch the first section and then click through, using the "Play" button, until you reach "Section 2 of 3."

Here we will explore the following questions: How many properties does a wave have? Are any of these properties related to each other?

Work your way to the experimental section, where you can adjust the properties of the wave. Watch for a moment to see how fast the frequency counter increases.

**1** Increase the wavelength by pressing the arrow key. What happens to the rate of the frequency counter?

_____

_____

_____

**2** Reset the simulation and then decrease the wavelength. What happens to the rate of the frequency counter?

_____

_____

**3** How are the wavelength and frequency related to each other?

_____

_____

**4** Imagine that you increase the frequency instead of the wavelength. How should the wavelength change when you increase the frequency?

_____

_____

**5** Reset the simulation, and increase the frequency. Did the wavelength change in the way you expected?

_____

_____

**6** Reset the simulation, and increase the amplitude. What happens to the wavelength and the frequency counter?

_____

_____

**7** Decrease the amplitude. What happens to the wavelength and the frequency counter?

_____

_____

**8** Is the amplitude related to the wavelength or frequency?

_____

_____

**9** Why can't you change the speed of this wave?

_____

_____

# 6 The Tools of the Astronomer

In the previous chapter, you saw that astronomers learn about the physical and chemical properties of distant planets, stars, and galaxies by studying the light from these objects. This electromagnetic radiation must first be collected and processed before it can be analyzed and converted to useful knowledge. In this chapter, you will learn about the tools that astronomers use to capture and scrutinize that information.

## LEARNING GOALS

By the conclusion of this chapter, you should be able to:

**LG 1** Compare the two main types of optical telescopes and how they gather and focus light.

**LG 2** Summarize the main types of detectors that are used on telescopes.

**LG 3** Explain why some wavelengths of radiation must be observed from space.

**LG 4** Explain the benefits of sending spacecraft to study the planets and moons of our Solar System.

**LG 5** Describe other astronomical tools that contribute to the study of the universe.

The twin 10-meter Keck reflectors on Mauna Kea, Hawaii, have a multiple mirror, compact design. ▶ ▶ ▶

**Why are most
telescopes
on remote
mountaintops?**

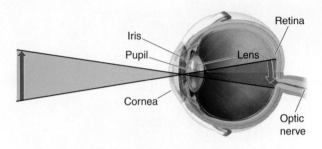

**Figure 6.1** A schematic view of the human eye, creating an image of an object (the blue arrow).

**Nebraska Simulation:** Snell's Law Demonstrator

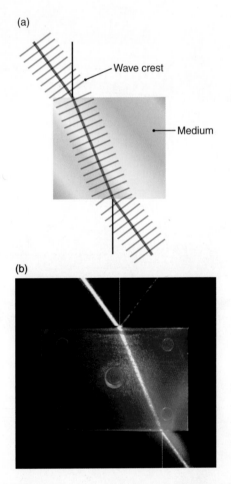

**Figure 6.2** (a) When wavefronts enter a new medium, they bend in a new direction relative to a line perpendicular to the surface (black lines). (b) An actual light ray entering and leaving a medium.

# 6.1 The Optical Telescope Revolutionized Astronomy

Astronomical observations began with the human eye—information about the overall colors of stars and their brightness in the night sky is apparent even to the naked eye, unassisted by binoculars or telescopes or filters. The development of **telescopes**—devices for collecting and focusing light—in the 17th century greatly increased the amount of light that can be collected from astronomical objects. With modern telescopes, astronomers can detect light that has been traveling across space for billions of years—even electromagnetic radiation from soon after the Big Bang, the beginning of the universe itself.

## The Eye

Human eyes are sensitive to light with wavelengths ranging from about 350 nanometers (deep violet) to 750 nanometers (far red). A simplified schematic of the human eye is shown in **Figure 6.1**. The part of the human eye that detects light is called the retina, and the individual receptor cells that respond to light falling on the retina are called rods and cones. The center of the human retina consists solely of cones, which detect color and provide the greatest visual acuity. Away from the center, rods and cones intermingle, with rods dominating far from the center, where they are responsible for peripheral vision.

Our vision is limited by the eye's angular **resolution**, which refers to how close two points of light can be to each other before we can no longer distinguish them. Unaided, the best human eyes can resolve objects separated by 1 arcminute (1/60 of a degree), an angular distance of about 1/30 the diameter of the full Moon. (A more in-depth description of angular units—radians, degrees, arcminutes, and arcseconds—can be found in Appendix 1.4.) This may seem small, but when we look at the sky, thousands of stars and galaxies may reside within a patch of sky with this diameter.

## Refracting Telescopes

Optical telescopes come in two primary types: **refracting telescopes**, which use lenses; and **reflecting telescopes**, which use mirrors. For all telescopes, the "size" of the telescope refers to the diameter of the largest mirror or lens, which determines the light-collecting area. This diameter is called the **aperture**. The light-gathering power of a telescope is proportional to the area of its aperture; that is, to the square of its diameter. The larger the aperture, the more light the telescope can collect. A "1-meter telescope" has a primary mirror (or lens) that is 1 meter in diameter. The aperture of the human eye is about 6–7 millimeters.

In the late 13th century, craftsmen in Venice were making small lentil-shaped disks of glass that could be mounted in frames and worn over the eyes to improve vision. More than 300 years later, Hans Lippershey (1570–1619), a spectacle maker living in the Netherlands, put two of his lenses together in a tube. With this new instrument, he saw distant objects magnified and could see farther. Galileo Galilei heard news of this invention, and he constructed one of his own. Recall from Chapter 3 that by 1610, Galileo had become the first to see the phases of Venus and the moons of Jupiter, and among the first to see craters on the Moon. He was also the first to realize that the Milky Way is made up of large numbers of individual stars. The refracting telescope—one that uses lenses—quickly revolutionized the science of astronomy.

Refraction is the basis for the refracting telescope. Recall from Chapter 5 that the speed of light is constant in a vacuum, but through a medium such as air or glass, the speed of light is always lower. As light enters a new medium, its speed changes. If the light strikes the surface at an angle, some of the crest of the wave arrives at the surface earlier and some arrives later. You can see this in **Figure 6.2a**, a schematic diagram of wave crests (red lines) striking a medium at an angle. Figure 6.2b shows an actual light ray passing into and out of a medium, in this case glass. The ray bends each time the medium changes. This bending of light when it changes the medium through which it travels is called **refraction**.

The amount of refraction is determined by the properties of the medium. A medium's index of refraction ($n$) is equal to the ratio of the speed of light in a vacuum ($c$) to its speed in a medium ($v$). This can be expressed by the equation $n = c/v$. For example, most glass has an index of refraction of approximately 1.5, so the speed of light in glass is 300,000 kilometers per second (km/s) divided by 1.5, or 200,000 km/s. The light bends by an amount that depends on the index of refraction of the materials involved and the angle at which the light strikes.

The primary lens in a refracting telescope is a simple convex lens, called the **objective lens**, shown in **Figure 6.3**, whose curved surfaces refract the light from a distant object. This refracted light forms an image on the telescope's **focal plane**, which is perpendicular to the *optical axis*—the path that light takes through the center of the lens. Because the telescope's glass lens is curved, light at the outer edges of the lens strikes the surface more obliquely than light near the center. Therefore, light at the outer edges of the lens is refracted more than light near its center. The lens concentrates the light rays entering the telescope, bringing them to a sharp focus at a distance called the **focal length**. Sometimes focal length is specified on a telescope as *focal ratio*, which equals focal length divided by aperture size; this term may be familiar to you from lenses used in photography.

**Figure 6.4** illustrates how a telescope uses the light that passes through its lenses. Figure 6.4a shows the light from two stars passing through a lens and converging at the focal plane of the lens. Figure 6.4b shows the same situation for a lens with a longer focal length. Longer focal lengths increase the size and separation of objects in the focal plane. Aperture and focal length are the two most important parameters of a telescope. The image can be viewed with an **eyepiece**—a changeable lens whose focal length determines the magnification (**Working It Out 6.1**). In modern research, however, the images are sent directly to a camera or other detector.

Refracting telescopes have two major shortcomings. First, there are physical limits on the size of refracting telescopes. The larger the area of the objective lens, the more light-gathering power it has and the fainter the stars we can observe. However, as objective lenses get larger, they get heavier, and a massive piece of glass at the end of a very long tube sags too much under the force of gravity. Refracting telescopes grew in size until the 1897 completion of the Yerkes 1-meter refractor (**Figure 6.5**), the world's largest operational refracting telescope. Located in

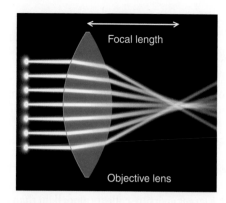

**Figure 6.3** For a curved lens like the one shown, refraction causes the light to focus to a point. This point is in a slightly different location for different wavelengths (colors) of light.

 **Nebraska Simulation:** Telescope Simulator

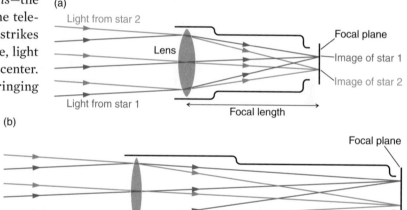

**Figure 6.4** (a) A refracting telescope uses a lens to collect and focus light from two stars, forming images of the stars on its focal plane. (b) Telescopes with longer focal length produce larger images.

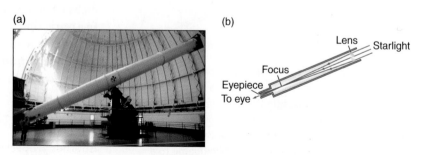

**Figure 6.5** (a) The Yerkes 1-meter refractor is the world's largest refracting telescope. (b) This sketch shows the parts of a refractor.

## 6.1    Working It Out    Telescope Aperture and Magnification

If you were shopping for a telescope, you would likely be told to consider aperture and magnification. Here we will look briefly at each.

### Aperture

The light-gathering power is proportional to the area of the mirror or lens, and thus to the square of the aperture, $\pi \times (D/2)^2$. A telescope with a larger aperture collects more light than does one with a smaller aperture. We can compare a 200-millimeter (mm), or 8-inch, diameter telescope with the light-gathering power of the pupil of your eye, which is about 6 mm in the dark:

$$\text{Light-gathering power of telescope} = \frac{\pi}{4} \times (200 \text{ mm})^2$$

and

$$\text{Light-gathering power of eye} = \frac{\pi}{4} \times (6 \text{ mm})^2$$

So, to compare:

$$\frac{\text{Light-gathering power of telescope}}{\text{Light-gathering power of eye}} = \frac{\frac{\pi}{4}(200 \text{ mm})^2}{\frac{\pi}{4}(6 \text{ mm})^2} = \left(\frac{200}{6}\right)^2 = 1{,}000$$

A typical 8-inch telescope has more than 1,000 times the light-gathering power of your eye.

Comparing this 8-inch telescope to the Keck 10-meter telescope shows why bigger is better: 200 mm = 0.2 meter, and we cancel out the $\frac{\pi}{4}$ again to obtain

$$\frac{\text{Light-gathering power of Keck}}{\text{Light-gathering power of 8-inch telescope}} = \left(\frac{10 \text{ m}}{0.2 \text{ m}}\right)^2 = 2{,}500$$

Even larger telescopes, 25–40 meters in diameter, are currently under construction.

### Magnification

Most telescopes have a set focal length and come with a collection of eyepieces. The magnification of the image in the telescope is given by

$$\text{Magnification} = \frac{\text{Telescope focal length}}{\text{Eyepiece focal length}}$$

Suppose the focal length of the 200-mm telescope in the preceding example is 2,000 mm. Combined with the focal length of a standard eyepiece, 25 mm, this telescope will give the following magnification:

$$\text{Magnification} = \frac{2{,}000 \text{ mm}}{25 \text{ mm}} = 80$$

This telescope and eyepiece combination has a magnifying power of 80, meaning that a crater on the Moon will appear 80 times (80×) larger in the telescope's eyepiece than it does when viewed by the naked eye. An eyepiece that has a focal length of 8 mm will have about 3 times more magnifying power, or 250.

A higher magnification will not necessarily let you see the object better. A faint and fuzzy image will not look clearer when magnified.

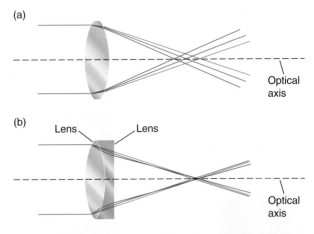

**Figure 6.6** (a) Different wavelengths of light come to different foci along the optical axis of a simple lens, causing chromatic aberration. (b) A compound lens using two types of glass with different indices of refraction can compensate for much of the chromatic aberration, so different colors of light all come to a focus at the same point.

Williams Bay, Wisconsin, the Yerkes telescope carries a 450-kilogram (kg) objective lens mounted at the end of a 19.2-meter tube.

The second major shortcoming of refracting telescopes is **chromatic aberration**. Starlight is made up of all the colors of the rainbow, and each color refracts at a slightly different angle because the index of refraction depends on the wavelength of the light. As seen in **Figure 6.6a**, shorter (bluer) wavelengths are refracted more strongly than longer (redder) wavelengths. This wavelength-dependent difference in refraction, which spreads the white light out into its spectral colors, is called **dispersion**. Dispersion causes bluer light to come to a shorter focus than that of the longer visible wavelengths, creating chromatic aberration. In a refracting telescope with a simple convex lens, chromatic aberration produces haloed images around the star. Manufacturers of quality cameras and telescopes use a **compound lens** composed of two types of glass to correct for chromatic aberration (Figure 6.6b).

### Reflecting Telescopes

Another property of light is **reflection**, the basis for reflecting telescopes. When light encounters a different medium—in this case going from air to glass—there

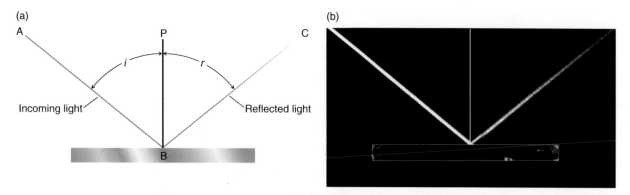

**Figure 6.7** (a) When a ray of incoming light (AB) shines on a flat surface, it reflects from the surface, becoming the reflected ray BC. The angle between AB and PB, the perpendicular to the surface, is the angle of incidence (*i*). The angle between BC and PB is the angle of reflection (*r*). The angles of incidence and reflection are always equal. (b) Light from a laser beam is reflected from a flat glass surface.

will be an amount of light reflected from the surface of the new medium. In other words, some of the light will reverse its direction of travel. The most common example occurs when light encounters an ordinary flat mirror. As shown in **Figure 6.7**, the angle of the incoming light and the angle of outgoing light are always equal. A reflected image from a mirror is a good representation of what falls on it, although left and right are interchanged.

In 1668, Isaac Newton designed a reflecting telescope, which uses mirrors instead of lenses (**Figure 6.8a**). The direction of reflected light does not depend on the wavelength of light; therefore, chromatic aberration is not a problem in reflecting telescopes. A sketch of parts of Newton's reflecting telescope is presented in Figure 6.8b. To make this reflecting telescope, Newton cast a 2-inch primary mirror made of copper and tin and polished it to a special curvature. He then placed this primary mirror at the bottom of a tube with a secondary flat mirror mounted above it at a 45° angle. The second mirror directed the focused light to an eyepiece on the outside of the tube.

Astronomers use mirrors with a surface that curves inward toward the incoming light, called concave mirrors. The same rules of incidence and reflection hold here for each ray of light, but in this case the reflected rays do not maintain the same angle with respect to each other as they do with a flat mirror. Concave mirrors will reflect the rays so that they converge to form an image, as shown in **Figure 6.9**. If the incoming light rays are parallel, as from a distant source like a star, the reflected light rays cross at the focal length of the mirror. The light path from the primary mirror to the focal plane can be "folded" by using a secondary mirror, which enables a significant reduction in the length and weight of the telescope. In many modern telescopes, the primary mirror has a hole so that light can pass back through it; the eyepiece is on the back of the tube of the telescope, and the tube can be shortened.

Large reflecting telescopes did not become common until the latter half of the 18th century. But then the size of the primary mirrors in reflecting telescopes continued to grow; and they became larger every decade. Primary mirrors can be supported from the back, and they can be made thinner and therefore less massive than the objective lenses found in refracting telescopes. The limitation on the size of reflecting telescopes is the cost of their fabrication and support structure.

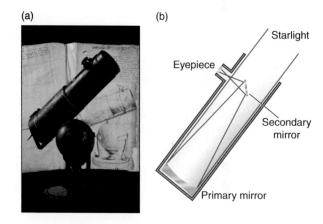

**Figure 6.8** Newton's reflecting telescope (a) has parts shown in the sketch (b).

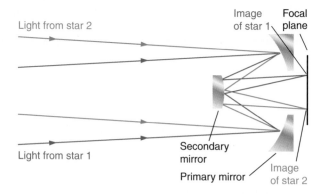

**Figure 6.9** Large reflecting telescopes often use a secondary mirror that directs the light back through a hole in the primary mirror to an accessible focal plane behind the primary mirror. Parallel rays of light that strike a concave parabolic mirror are brought to a focus in the mirror's focal plane.

| TABLE 6.1 | The World's Largest Optical Telescopes |

| Mirror Diameter (meters) | Telescope | Sponsor(s) | Location | Operational Date |
|---|---|---|---|---|
| 39.3 | European Extremely Large Telescope (E-ELT) | European Southern Observatory (Europe, Chile, Brazil) | Cerro Armazones, Chile | Under construction |
| 30.0 | Thirty Meter Telescope (TMT) | International collaboration led by Caltech, U. of California, U. of Hawaii, China, Japan, India, and Canada | Mauna Kea, Hawaii | Under construction |
| 24.5 | Giant Magellan Telescope (GMT) | Carnegie Institution, Harvard U., Smithsonian Institution, U. of Arizona, U. of Texas, Texas A&M, U. of Chicago, Australian National U., Astronomy Australia Ltd., Korea Astronomy and Space Science Institute | Cerro Las Campanas, Chile | Under construction |
| 11.0 | South African Large Telescope (SALT) | South Africa, USA, UK, Germany, Poland, New Zealand, India | Sutherland, South Africa | 2005 |
| 10.4 | Gran Telescopio CANARIAS (GTC) | Spain, Mexico, U. of Florida | Canary Islands | 2007 |
| 10 | Keck I | Caltech, U. of California, NASA | Mauna Kea, Hawaii | 1993 |
| 10 | Keck II | Caltech, U. of California, NASA | Mauna Kea, Hawaii | 1996 |
| 9.2 | Hobby-Eberly Telescope (HET) | U. of Texas, Penn State U., Stanford U., Germany | Mount Fowlkes, Texas | 1999 |
| 8.4 × 2 | Large Binocular Telescope (LBT) | U. of Arizona, Ohio State U., Italy, Germany, Arizona State, and others | Mount Graham, Arizona | 2008 |
| 8.4 | Large Synoptic Survey Telescope (LSST) | Many partners | Cerro Pachón | Under construction |
| 8.3 | Subaru Telescope | Japan | Mauna Kea, Hawaii | 1999 |
| 8.2 × 4 | Very Large Telescope (VLT) | European Southern Observatory | Cerro Paranal, Chile | 2000 |
| 8.1 | Gemini North | USA, UK, Canada, Chile, Brazil, Argentina, Australia | Mauna Kea, Hawaii | 1999 |
| 8.1 | Gemini South | USA, UK, Canada, Chile, Brazil, Argentina, Australia | Cerro Pachón, Chile | 2000 |
| 6.5 | MMT | Smithsonian Institution, U. of Arizona | Tucson, Arizona | 2000 |
| 6.5 | Magellan I | Carnegie Institution, U. of Arizona, Harvard U., U. of Michigan, MIT | Cerro Las Campanas, Chile | 2000 |
| 6.5 | Magellan II | Carnegie Institution, U. of Arizona, Harvard U., U. of Michigan, MIT | Cerro Las Campanas, Chile | 2002 |

▶II **AstroTour:** Geometric Optics and Lenses

**Table 6.1** lists the world's largest optical telescopes. All are reflecting telescopes. The largest single mirrors constructed today are 8 meters in diameter, but reflecting telescopes even bigger than this are designed to make use of an array of smaller segments. The primary mirror of each of the 10-meter, twin Keck telescopes is made up of 36 hexagon-shaped segments that are 1.8 meters in diameter (**Figure 6.10**). Located on 4,100-meter-high Mauna Kea in Hawaii, the Keck telescopes are among the world's largest reflecting telescopes. Each one has 4 million times the light-gathering power of the human eye.

## CHECK YOUR UNDERSTANDING 6.1

Which of the following is a reason that all large astronomical telescopes are reflectors (choose all that apply): (a) chromatic aberration is minimized; (b) they are not as heavy; (c) they can be shorter; (d) the glass doesn't need to be curved.

## Optical and Atmospheric Limitations

Another important characteristic of a telescope is its *resolution*—how close two points of light can be to each other before they are indistinguishable. The concept of resolution is illustrated in **Figure 6.11**. Review Figure 6.4a to see the path followed by rays of light from two distant stars as they pass through the lens of a refracting telescope. Figure 6.4b illustrated that increasing the focal length increases the size of and separation between the images that a telescope produces. This is one reason why telescopes provide a much clearer view of the stars than that obtained with the naked eye. The focal length of a human eye is typically about 20 mm, whereas telescopes used by professional astronomers often have focal lengths of tens or even hundreds of meters. Such telescopes make images that are far larger than those formed by the human eye, and consequently they contain far more detail.

Focal length explains only one difference between the resolution of telescopes and that of the unaided eye. The other difference results from the wave nature of light. **Figure 6.12** shows what happens when light waves pass through the aperture of a telescope: they spread out from the edges of the lens or mirror. The distortion that occurs as light passes the edge of an opaque object is called **diffraction**. Diffraction "diverts" some of the light from its path, slightly blurring the image made by the telescope. The degree of blurring depends on the wavelength of the light and the telescope's aperture. The larger the aperture, the smaller the problem posed by diffraction. The best resolution that a given telescope can achieve is known as the **diffraction limit** (**Working It Out 6.2**).

Larger telescopes have better resolution and can distinguish objects that appear closer together. Theoretically, the 10-meter Keck telescopes have a diffraction-limited resolution of 0.0113 arcseconds (arcsec) in visible light, which would be good enough for you to read newspaper headlines 60 km away. But for telescopes with apertures larger than about a meter, Earth's atmosphere stands in the way of better resolution. If you have ever looked out across a large asphalt parking lot on a summer day, you have seen the distant horizon shimmer as light is bent this way and that by turbulent bubbles of warm air rising off the hot

Segmented primary mirror

**Figure 6.10** Each of the Keck 10-meter reflectors uses an aligned group of 36 hexagonal mirrors to collect light.

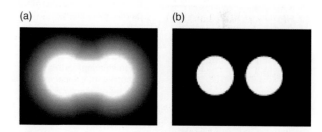

(a)　　　　　　　(b)

**Figure 6.11** Resolution is the ability to separate two images that appear close together. When resolution is lower (a), the two images blend together. When resolution is higher (b), individual images can be seen.

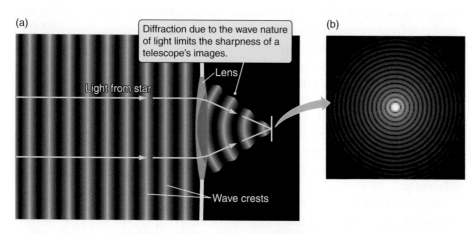

(a)

Diffraction due to the wave nature of light limits the sharpness of a telescope's images.

Light from star

Lens

(b)

Wave crests

**Figure 6.12** (a) Light waves from a star are diffracted by the edges of a telescope's lens or mirror. (b) This diffraction causes the stellar image to be blurred, limiting a telescope's ability to resolve objects.

## 6.2 Working It Out Diffraction Limit

The practical limit on the angular resolution, $\theta$, of a telescope is called the diffraction limit. This limit is determined by the ratio of the wavelength of light, $\lambda$, passing through the telescope to the diameter of the aperture, $D$:

$$\theta = 2.06 \times 10^5 \left(\frac{\lambda}{D}\right) \text{arcsec}$$

With the constant, $2.06 \times 10^5$, the units are arcseconds (arcsec). An **arcsecond** is a tiny angular measure found by first dividing the sky into 360 degrees, and then dividing a degree by 60 to get arcminutes, and then by 60 again to get arcseconds. An arcsecond is 1/1,800 of the size of the Moon in the sky, or about the size of a tennis ball if you could see it from 8 miles away.

Both $\lambda$ and $D$ must be expressed in the same units, usually meters. The smaller the ratio of $\lambda/D$, the better the resolution. For example, the size of the human pupil (see Figure 6.1) ranges from about 2 mm in bright light to 8 mm in the dark. A typical pupil size in the dark is about 6 mm, or 0.006 meter. Visible (green) light has a wavelength ($\lambda$)

of 550 nanometers (nm); that is, $550 \times 10^{-9}$ meter, or $5.5 \times 10^{-7}$ meter. Using these values for the aperture and the wavelength gives

$$\theta = 2.06 \times 10^5 \left(\frac{5.5 \times 10^{-7}\,\text{m}}{0.006\,\text{m}}\right) \text{arcsec} = 19 \,\text{arcsec}$$

or about 0.5 arcmin. The typical resolution of the human eye is 2 arcmin. We do not achieve the theoretical resolution with our eyes because the physical properties of our eyes are not perfect.

How does the resolution of the human eye compare to that of a telescope? Consider the Hubble Space Telescope, when operating in the visible part of the spectrum. Its primary mirror has a diameter of 2.4 meters. Substituting this value for $D$ and again using visible (green) light gives

$$\theta = 2.06 \times 10^5 \left(\frac{5.5 \times 10^{-7}\,\text{m}}{2.4\,\text{m}}\right) \text{arcsec} = 0.047 \,\text{arcsec}$$

or about 600 times better than the theoretical resolving power of the human eye.

---

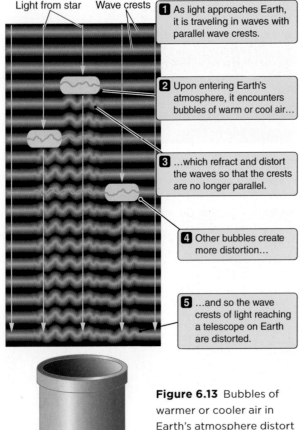

Light from star    Wave crests

1 As light approaches Earth, it is traveling in waves with parallel wave crests.

2 Upon entering Earth's atmosphere, it encounters bubbles of warm or cool air…

3 …which refract and distort the waves so that the crests are no longer parallel.

4 Other bubbles create more distortion…

5 …and so the wave crests of light reaching a telescope on Earth are distorted.

Telescope

**Figure 6.13** Bubbles of warmer or cooler air in Earth's atmosphere distort the wavefront of light from a distant object.

pavement. The problem of the shimmering atmosphere is less pronounced when we look overhead, but the twinkling of stars in the night sky is caused by the same phenomenon. As telescopes magnify the angular diameter of an object, they also magnify the shimmering effects of the atmosphere. The limit on the resolution of a telescope on the surface of Earth caused by this atmospheric distortion is called **astronomical seeing**. One advantage of launching telescopes such as the Hubble Space Telescope into orbit around Earth above the atmosphere is that they are not hindered by astronomical seeing.

Modern technology has improved ground-based telescopes with computer-controlled **adaptive optics** that compensate for much of the atmosphere's distortion. To understand how adaptive optics work, we need to look more closely at how Earth's atmosphere smears out an otherwise perfect stellar image. Light from a distant star arrives at the top of Earth's atmosphere with flat, parallel wave crests. If Earth's atmosphere were perfectly uniform, the crests would remain flat as they reached the objective lens or primary mirror of a ground-based telescope. After making its way through the telescope's optical system, the crests would produce a tiny diffraction disk in the focal plane, as shown in Figure 6.12b. But Earth's atmosphere is not uniform. It is filled with bubbles of air that have slightly different temperatures than those of their surroundings. Different temperatures mean different densities, and different densities mean different refractive properties, so each bubble bends light differently.

These air bubbles act as weak lenses, and by the time the waves reach the telescope they are far from flat, as shown in **Figure 6.13**. Instead of a tiny diffraction disk, the image in the telescope's focal plane is distorted and swollen, degrading the resolution. Adaptive optics flatten out this distortion. First, an optical device within the telescope constantly measures the wave crests. Then, before reaching the telescope's focal plane, the light is reflected off yet another mirror, which has a flexible surface. A computer analyzes the light and bends the flexible mirror so that it accurately corrects for the distortion caused by the air bubbles. **Figure 6.14** shows an example of an image corrected by adaptive optics. The widespread use of adaptive optics has made the image quality of ground-based telescopes competitive with the quality of Hubble images from space at some wavelengths.

## Observatory Locations

What makes a good location for a telescope on Earth? Look back at Table 6.1—what do these locations have in common? Astronomers look for sites that are high, dry, and dark. The best sites are far away from the lights of cities, in locations with little moisture, humidity, or rain, and where the atmosphere is relatively still. Telescopes are located as high as possible so that they get above a significant part of Earth's atmosphere, which distorts images and blocks infrared light. Many telescopes are situated on remote, high mountaintops surrounded by desert or ocean. Recall from Chapter 2 that the stars that can be seen throughout the year depend on latitude, and only at the equator would a telescope have access to all of the stars in the sky. But equatorial latitudes have tropical weather—wet, humid, and stormy—and thus are poor locations for a telescope. So, to cover the entire sky, astronomers have built telescopes in both northern and southern locations. In the United States, large telescopes are located in California, Arizona, New Mexico, Texas, and Hawaii. The largest southern-sky observatories are found in Chile, South Africa, and Australia. The twin Gemini telescopes, designed to be a matched pair, are located in Hawaii in the Northern Hemisphere and in Chile in the Southern Hemisphere.

Newer and larger telescopes are planned for many of the same locations that are listed in Table 6.1. The 8-meter Large Synoptic Survey Telescope (LSST) is headed for Cerro Pachón in Chile, current site of the Gemini South telescope. The Giant Magellan Telescope (GMT), consisting of seven 8-meter mirrors in a pattern equivalent to a 24.5-meter mirror, will be constructed at Cerro Las Campanas in Chile. The Thirty Meter Telescope (TMT) (**Figure 6.15**) is planned for Mauna Kea in Hawaii, current site of the twin Keck telescopes; and the European Southern Observatory (ESO) is building the 39-meter European Extremely Large Telescope (E-ELT) at Cerro Armazones in Chile. As telescopes get larger—and more expensive—international collaboration becomes even more important.

Today's professional astronomers rarely look through the eyepiece of a telescope because they learn much more and make better use of observing time by permanently recording an object's image at a variety of wavelengths or seeing its light spread out into a revealing spectrum. Some astronomers no longer travel to telescopes at all, instead observing remotely from the base of the mountain or far away at their own institutions.

Professional and amateur astronomers alike are concerned about loss of the dark sky. As cities and suburbs grow and expand around the world, the use of outdoor artificial light becomes more widespread. Pictures from space show how bright many areas of Earth are at night. In the United States, two-thirds of the population resides in an area that is too bright to see the Milky Way in the sky at night (**Figure 6.16**), and it has been estimated that by 2025 there will be almost no dark skies in the continental United States. Increased air pollution also dims the

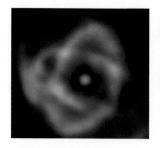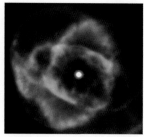

**Figure 6.14** These images of the Cat's Eye Nebula from the Palomar Observatory telescope without (left) and with (right) adaptive optics show the benefit of the technique.

**Figure 6.15** This is an artist's rendering of the Thirty Meter Telescope, a planned reflecting telescope.

**Figure 6.16** This satellite image of the United States at night shows that few populated areas are free from light pollution.

view of the night sky in many locations. The U.S. National Park Service now advertises evening astronomy programs in natural, unpolluted dark skies as one of the reasons to visit some parks. Several international astronomy associations are working with UNESCO (the United Nations Educational, Scientific and Cultural Organization) to promote the "right to starlight," arguing that for historical, cultural, and scientific reasons, it would be a huge loss if humanity could no longer view the stars. These organizations are encouraging countries to create starlight reserves and starlight parks where people can experience increasingly rare dark skies and a natural nocturnal environment.

### CHECK YOUR UNDERSTANDING 6.2

In practice, the smallest angular size that one can resolve with a 10-inch telescope is governed by the: (a) blurring caused by Earth's atmosphere; (b) diffraction limit of the telescope; (c) size of the primary mirror; (d) magnification of the telescope.

## 6.2 Optical Detectors and Instruments Used with Telescopes

Beginning in the 1800s, the development of film photography, and later digital photography, revolutionized astronomy, allowing astronomers to detect fainter and more distant objects than possible to detect with the eye alone. In this section, we will examine some of the more common types of detectors.

### Integration Time and Quantum Efficiency

Originally, the retina of the human eye was the only astronomical detector. The limit of the faintest stars we can see with our unaided eyes is determined in part by two factors that are characteristic of all detectors: *integration time* and *quantum efficiency.*

**Integration time** is the limited time interval during which the eye can add up photons—this is analogous to leaving the shutter open on a camera. The brain "reads out" the information gathered by the eye about every 100 milliseconds (ms). Anything that happens faster than that appears to happen all at once. If two images on a computer screen appear 30 ms apart, you will see them as a single image because your eyes will add up (or integrate) whatever they see over an interval of 100 ms or less. However, if the images occur 200 ms apart, you will see them as separate images. This relatively brief integration time is the most important factor limiting our nighttime vision. Stars too faint to be seen with the unaided eye are those from which you receive too few photons for your eyes to process in 100 ms.

**Quantum efficiency** determines how many responses occur for each photon received. For the human eye, 10 photons must strike a cone within 100 ms to activate a single response. So the quantum efficiency of our eyes is about 10 percent: for every 10 events, the eye sends one signal to the brain. Together, integration time and quantum efficiency determine the rate at which photons must arrive at the retina before the brain says, "Aha, I see something." Astronomers seek to use detectors with longer integration times and higher quantum efficiency than those of our eyes.

## From Photographic Plates to Charge-Coupled Devices

For more than two centuries after the invention of the telescope, astronomers struggled with the problem of surface brightness. Only *point sources* such as stars appear brighter in a telescope; extended astronomical objects like the Moon appear bigger in the eyepiece, but their surfaces are no brighter than they appear to the unaided eye. Even when astronomers built larger telescopes, nebulae and galaxies appeared larger, but the details of these faint objects remained elusive. The problem was not with the telescopes but with the limitations of optics and the human eye. Only with the longer exposure times made possible by the invention of photography and the later development of electronic cameras were astronomers finally able to discern intricate details in faint objects.

In 1840, John W. Draper (1811–1882), a New York chemistry professor, created the earliest known astronomical photograph (**Figure 6.17**). By the late 1800s, astronomers had created thousands of photographic plates with permanent images of planets, nebulae, and galaxies. The quantum efficiency of most photographic systems used in astronomy was poorer than that of the human eye—typically 1–3 percent. But unlike the eye, photography can overcome poor quantum efficiency by leaving the shutter open on the camera, increasing the integration time to many hours of exposure. Photography made it possible for astronomers to record and study objects that were invisible to the human eye. However, one problem is that the response of photography to light is not linear, especially at long exposures, so if you doubled the exposure time, you did not get twice as much light on your image. By the middle of the 20th century, the search was on for electronic detectors that would overcome the sensitivity, spectral range, and nonlinearity problems of photography.

In 1969, scientists at Bell Laboratories invented a detector called a **charge-coupled device**, or **CCD**. By the late 1970s, the CCD had become the detector of choice in almost all astronomical-imaging applications. CCDs are linear, so doubling the exposure means you record twice as much light. Therefore, they are good for measuring objects that vary in brightness, as well as for faint objects that require long exposures. CCDs have a quantum efficiency far superior to that of photography or the eye, up to 80 percent at some wavelengths. This improvement dramatically increases the ability to view faint objects with short exposure times.

A CCD is an ultrathin wafer of silicon—less than the thickness of a human hair—that is divided into a two-dimensional array of picture elements, or **pixels**, as seen in **Figure 6.18a**. When a photon strikes a pixel, it creates a small electric charge within the silicon. As each CCD pixel is read out, the digital signal that flows to the computer is nearly proportional to the accumulated charge. This is what we mean when we say that the CCD is a linear device. However, if a CCD is exposed to too much light, it can lose its linearity.

Liquid nitrogen or helium is used to cool the CCDs down to very low temperatures to reduce noise caused by the movement of the charge-carrying atoms within the silicon wafer. The first astronomical CCDs were small arrays containing a few hundred thousand pixels. The larger CCDs used in astronomy today may contain more than 100 million pixels (Figure 6.18b). Still larger arrays are under development as ever-faster computing power keeps up with image-processing demands.

**Figure 6.17** A photograph of the Moon taken by John W. Draper in 1840.

(a)

(b)

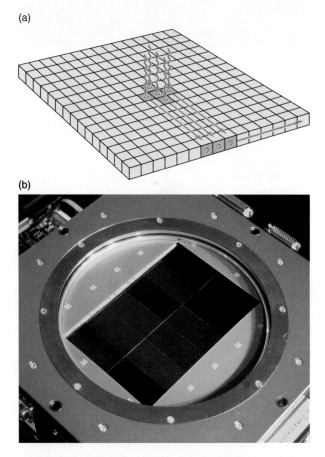

**Figure 6.18** (a) In this simplified diagram of a charge-coupled device (CCD), photons from a star land on pixels (represented by gray squares) and produce free electrons within the silicon. The electron charges are electronically moved sequentially to the collecting register at the bottom. Each row is then moved out to the right to an electronic amplifier, which converts the electric charge of each pixel into a digital signal. (b) This large CCD (about 6 inches across) contains 12,288 × 8,192 pixels.

**Nebraska Simulation:** CCD Simulator

(a)

(b)

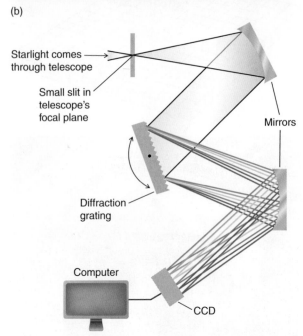

Starlight comes through telescope

Small slit in telescope's focal plane

Mirrors

Diffraction grating

Computer

CCD

**Figure 6.19** (a) A spectrum is created by the reflection of light from the closely spaced tracks of a CD. (b) In a grating spectrograph, light goes through the telescope and then a slit, where it is reflected to the diffraction grating and split into components. The spectrum is recorded on the CCD.

The output from a CCD is a digital signal that can be sent directly from the telescope to image-processing software or stored electronically for later analysis. Nearly every spectacular astronomical image in ultraviolet, visible, or infrared wavelength that you find online was recorded by a CCD in a telescope either on the ground or in space. CCDs are found in many common devices such as digital cameras, digital video cameras, and camera phones.

Your cell phone takes color pictures by using a grid of CCD pixels arranged in groups of three. Each pixel in a group is constructed to respond to only a particular range of colors—only to red light, for example. This is also true for digital image displays. You can see this for yourself if you place a small drop of water on the screen of your smartphone or tablet and turn it on. The water magnifies the grid of pixels so that you can see them individually. This grid degrades the angular resolution of the camera because each spot in the final image requires three pixels of information. Astronomers choose instead to use all the pixels on the camera to measure the number of photons that fall on each pixel, without regard to color. They put filters in front of the camera to allow light of only particular wavelengths to pass through, such as the light of a specific spectral line. Color pictures like those from the Hubble Space Telescope are constructed by taking multiple pictures, coloring each one, and then carefully aligning and overlapping them to produce beautiful and informative images. Sometimes the colors are "true"; that is, they are close to the colors you would see if you were actually looking at the object with your eyes. Other times "false" colors represent different portions of the electromagnetic spectrum and tell you the temperature or composition of different parts of the object. Using changeable filters instead of designated color pixels gives astronomers greater flexibility and greater resolution.

## Spectrographs

**Spectroscopy** is the study of an object's *spectrum* (plural: *spectra*)—its electromagnetic radiation split into component wavelengths. **Spectrographs** (sometimes called **spectrometers**) are instruments that take the spectrum of an object and then record it. The first spectrographs used prisms to disperse the light. Modern spectrographs use a **diffraction grating**, which is made by engraving closely spaced lines on glass to disperse incoming light into its constituent wavelengths. **Figure 6.19a** shows the light reflected from a CD or DVD; the closely spaced tracks act as a grating and create a spectrum. Figure 6.19b shows a grating spectrograph: light from an astronomical object enters a telescope and passes through a slit. The light is reflected onto the diffraction grating, which creates a spectrum like the one shown in Figure 5.14. The spectra are recorded on a CCD and then analyzed. Some modern spectrographs use bundles of optical fibers, or masks with multiple slits, to obtain spectra simultaneously from multiple objects in the field of view of the telescope.

---

### CHECK YOUR UNDERSTANDING 6.3

CCD cameras have much higher quantum efficiency than other detectors. This means that CCD cameras: (a) can collect photons for longer times; (b) can collect photons of different energies; (c) can generate a signal from fewer photons; (d) can split light into different colors.

# 6.3 Astronomers Observe in Wavelengths Beyond the Visible

Recall from Chapter 5 that an object's temperature can be found from the peak wavelength of its continuous spectrum. Extending beyond visible light, radio or infrared telescopes are used to study cool objects, like clouds of dust, whereas X-ray or gamma-ray telescopes are used to study violently hot gas. Therefore, astronomers must utilize telescopes that observe at all the wavelengths of the electromagnetic spectrum. However, not all of these wavelengths reach Earth, so some telescopes must be put into space. **Figure 6.20** shows that Earth has a few **atmospheric windows** that let in parts of the spectrum. The largest window is in radio wavelengths, including microwaves at the short-wavelength end of the radio window. These telescopes can be built on the ground. However, gamma-ray, X-ray, ultraviolet, and most of the infrared light arriving at Earth fails to reach the ground because it is partially or completely absorbed by ozone, water vapor, carbon dioxide, and other molecules in Earth's atmosphere. Light at these wavelengths has to be observed from space.

 **Nebraska Simulation:** EM Spectrum Module

## Radio Telescopes

Karl Jansky (1905–1950), a young physicist working for Bell Laboratories in the early 1930s, identified a radio source in the Milky Way in the direction of the galactic center, in the constellation Sagittarius. Jansky's discovery marked the birth

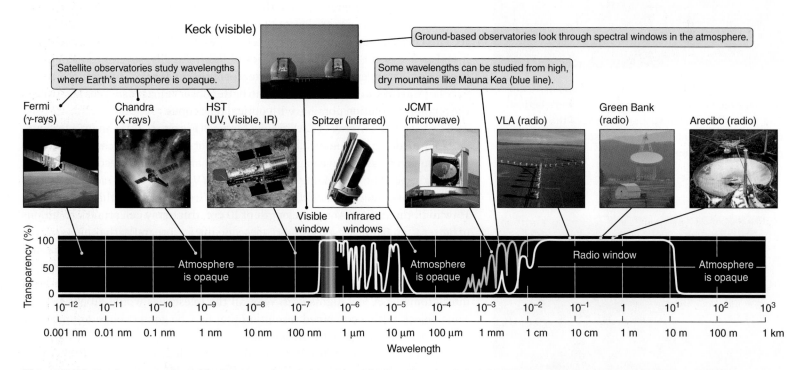

**Figure 6.20** Earth's atmosphere blocks most electromagnetic radiation. Fermi = Fermi Gamma-ray Space Telescope (orbiting); Chandra = Chandra X-ray Observatory (orbiting); HST = Hubble Space Telescope (orbiting); Keck = Keck Observatory (Hawaii); Spitzer = Spitzer Space Telescope (orbiting); JCMT = James Clerk Maxwell Telescope (Hawaii); VLA = Very Large Array (New Mexico); Green Bank = Robert C. Byrd Green Bank Telescope (West Virginia); Arecibo = Arecibo Observatory (Puerto Rico).

(a)

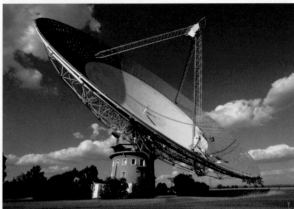

(b)

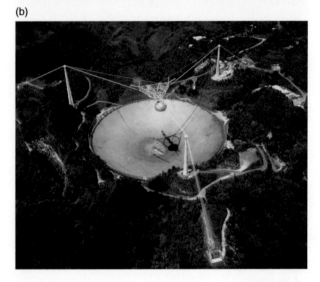

**Figure 6.21** (a) The Parkes radio telescope in Australia. (b) The Arecibo radio telescope is the world's largest. The steerable receiver suspended above the dish permits limited pointing toward celestial targets as they pass close to the zenith.

**Figure 6.22** The VLA in New Mexico combines signals from 27 different telescopes so that they act as one "very large" telescope.

of radio astronomy, and in his honor, the basic unit for the strength of a radio source is called the **jansky (Jy)**. A few years later, Grote Reber (1911–2002), a radio engineer and ham radio operator, built his own radio telescope and conducted the first survey of the sky at radio frequencies. Reber was largely responsible for the rapid advancement in radio astronomy that blossomed in the post–World War II era.

Most radio telescopes are large, steerable dishes, typically tens of meters in diameter, like the one shown in **Figure 6.21a**. The world's largest single-dish radio telescope is the 305-meter Arecibo dish built into a natural bowl-shaped depression in Puerto Rico (Figure 6.21b). (China is constructing a 500-meter single-dish radio telescope with a similar design.) The Arecibo telescope is not steerable, so it can only observe sources that pass within 20° of the zenith as Earth's rotation carries them overhead.

As large as radio telescopes are, they have relatively poor angular resolution. Recall that a telescope's angular resolution is determined by the ratio $\lambda/D$, so a larger ratio means poorer resolution. Radio telescopes have diameters much larger than the apertures of most optical telescopes. However, the wavelengths of radio waves range from about 1 centimeter (cm) to 10 meters, or up to several hundred thousand times greater than the wavelengths of visible light, which makes the ratio larger. Radio telescopes are thus limited by the very long wavelengths they are designed to receive. For example, the resolution of the huge Arecibo dish in Figure 6.21b is typically about 1 arcmin, little better than the unaided human eye.

Radio astronomers have had to develop ways to improve resolution. Mathematically combining the signals from two radio telescopes turns them into a telescope with a diameter equal to the separation between them. For example, if two 10-meter telescopes are located 1,000 meters apart, the $D$ in $\lambda/D$ is 1,000, not 10. This combination of two (or more) telescopes is called an **interferometer**, and it makes use of the wavelike properties of light. Usually, several telescopes are used in an arrangement called an **interferometric array**. Through the use of very large arrays, radio astronomers can better observe bright sources and exceed the angular resolution possible with optical telescopes.

The Very Large Array (VLA) in New Mexico (**Figure 6.22**) is an interferometric array made up of 27 movable dishes spread out in a Y-shaped configuration up to 36 km across. At a wavelength of 10 cm, this array reaches resolutions of less than 1 arcsec. The Very Long Baseline Array (VLBA) uses 10 radio telescopes spread out over more than 8,000 km from the Virgin Islands in the Caribbean to Hawaii in the Pacific. At a wavelength of 10 cm, this array can attain resolutions of better than 0.003 arcsec. A radio telescope put into near-Earth orbit as part of a Space Very Long Baseline Interferometer (SVLBI) overcomes even this limit. The new Event Horizon Telescope will combine many of the most advanced existing radio telescopes, from Greenland to the South Pole, to make an *Earth-sized* interferometer. A few nights each year, all the telescopes would all observe the same object, with a combined resolution that may be good enough to image the objects near the center of the Milky Way.

Some radio telescopes use large numbers of small dishes. The Atacama Large Millimeter/submillimeter Array (ALMA; **Figure 6.23**), located at an elevation of 5,000 meters in the Atacama Desert in Chile, was completed in 2013. This project, an international collaboration of astronomers from Europe, North America, East Asia, and Chile, consists of sixty-six 12- and 7-meter dishes for observations in the 0.3- to 9.6-mm wavelength range. The Square Kilometre Array (SKA) is designed to have *thousands* of small radio dishes, which together will act as one dish with a collecting area of 1 square kilometer ($km^2$). Twenty countries are supporting this telescope, which will be located in Australia and South Africa and is scheduled to be built by 2024.

Optical telescopes can also be combined in an array to yield resolutions greater than those of single telescopes, although for technical reasons the individual units cannot be spread as far apart as radio telescopes. The Very Large Telescope Interferometer (VLTI) in Chile, operated by ESO, combines the four Very Large Telescope (VLT) 8-meter telescopes with four movable 1.8-meter auxiliary telescopes. It has a baseline of up to 200 meters, yielding angular resolution of about 0.001 arcsec. The six-telescope Center for High Angular Resolution Astronomy (CHARA) array in California works in visible and near-infrared and has a baseline of 330 meters with similar resolution.

## Infrared Telescopes

Molecules such as water vapor in Earth's atmosphere block infrared (IR) photons from reaching astronomical telescopes on the ground, so telescopes that observe in the infrared (0.75–30 microns; $\mu$m) are at the highest locations. Mauna Kea, a dormant volcano and home of the Mauna Kea Observatories (MKO), rises 4,200 meters above the Pacific Ocean. At this altitude, the MKO telescopes sit above 40 percent of Earth's atmosphere; but more important, 90 percent of Earth's atmospheric water vapor lies below. Still, for the infrared astronomer the remaining 10 percent is troublesome.

Airborne observatories overcome atmospheric absorption of infrared light by placing telescopes above most of the water vapor in the atmosphere. NASA's Stratospheric Observatory for Infrared Astronomy (SOFIA) (**Figure 6.24**), a joint project with the German Aerospace Center (DLR), is a modified 747 airplane that carries a 2.5-meter telescope and works in the far-infrared region of the spectrum, from 1 to 650 $\mu$m. It flies in the stratosphere at an altitude of about 12 km, above 99 percent of the water vapor in Earth's lower atmosphere. Because airplanes are highly mobile, SOFIA can observe in both the Northern and Southern hemispheres. Other infrared wavelengths must be observed from space.

## Orbiting Observatories

Gaining full access to the complete electromagnetic spectrum requires getting completely above Earth's atmosphere. The first astronomical satellite was the British Ariel 1, launched in 1962 to study solar UV and X-ray radiation. Today, a multitude of orbiting astronomical telescopes cover the electromagnetic spectrum from gamma rays to microwaves, with more in the planning stage (**Table 6.2**). Optical telescopes, such as the 2.4-meter Hubble Space Telescope (HST), operate successfully at low Earth orbit, 600 km above Earth's surface. Launched in 1990, HST has been the workhorse for UV, visible, and IR space astronomy for more than 25 years. Low Earth orbit is also the region where the International Space Station (ISS) and many scientific satellites orbit. For certain other satellites and space telescopes, 600 km is not high enough.

The Chandra X-ray Observatory, NASA's X-ray telescope, cannot see through even the tiniest traces of atmosphere and therefore orbits more than 16,000 km above Earth's surface. NASA's Spitzer Space Telescope, an infrared telescope, is so sensitive that it needs to be completely free from Earth's own infrared

**Figure 6.23** The new Atacama Large Millimeter/submillimeter Array (ALMA) telescope in the Atacama Desert in northern Chile has many international partners.

**Figure 6.24** SOFIA is a 2.5-meter infrared telescope that is mounted in a Boeing 747 aircraft.

**TABLE 6.2** Selected Current and Future Space Observatories

| Telescope | Sponsor(s) | Description | Launch Year |
|---|---|---|---|
| Hubble Space Telescope (HST) | NASA, ESA | Optical, infrared, ultraviolet observations | 1990 |
| Chandra X-ray Observatory | NASA | X-ray imaging and spectroscopy | 1999 |
| X-ray Multi-Mirror Mission (XMM-Newton) | ESA | X-ray spectroscopy | 1999 |
| Galaxy Evolution Explorer (GALEX) | NASA | Ultraviolet observations | 2003 |
| Spitzer Space Telescope | NASA | Infrared observations | 2004 |
| Swift Gamma-Ray Burst Mission | NASA | Gamma-ray bursts | 2004 |
| Convection Rotation and Planetary Transits (COROT) space telescope | CNES (France) | Planet finder | 2006 |
| Fermi Gamma-ray Space Telescope | NASA, European partners | Gamma-ray imaging and gamma-ray bursts | 2008 |
| Planck telescope | ESA | Cosmic microwave background radiation | 2009 |
| Herschel Space Observatory | ESA | Far-infrared and submillimeter observations | 2009 |
| Kepler telescope | NASA | Planet finder | 2009 |
| Solar Dynamics Observatory (SDO) | NASA | Sun, solar weather | 2010 |
| RadioAstron | Russia | Very-long-baseline interferometry in space | 2011 |
| Nuclear Spectroscopic Telescopic Array (NuSTAR) | NASA | High-energy X-ray | 2012 |
| Gaia | ESA | Optical, digital 3D space camera | 2013 |
| James Webb Space Telescope (JWST) | NASA, ESA, Canadian Space Agency | Optical and infrared; replacement for HST | 2018 |

radiation. The solution was to put it into a *solar* orbit, trailing tens of millions of kilometers behind Earth. The James Webb Space Telescope, scheduled to replace the HST, will observe primarily in infrared wavelengths. It will be located 1.5 million miles away from Earth, orbiting the Sun at a fixed distance from the Sun and Earth.

Orbiting telescopes located above the atmosphere are not affected by atmospheric image distortions, weather, or brightening night skies. But space observatories are much more expensive than ground-based observatories and can be difficult or impossible to repair. The HST required several servicing missions, but such missions are not possible for the observatories in more distant Earth orbits. Ground-based telescopes at even the most remote mountaintop locations can receive shipments of replacement parts in a few days; space telescopes cannot. Of course, some wavelengths can be observed from space only. But issues of cost and repair are the reason why ground-based telescopes are much more prevalent.

## CHECK YOUR UNDERSTANDING 6.4

Which of the following is the biggest disadvantage of putting a telescope in space? (a) Astronomers don't have as much control in choosing what to observe. (b) Astronomers have to wait until the telescopes come back to Earth to get their images. (c) Space telescopes can only observe in certain parts of the electromagnetic spectrum. (d) Space telescopes are much more expensive than similar ground-based telescopes.

# 6.4 Planetary Spacecraft Explore the Solar System

Recall from Chapter 2 that everyone always sees the same face of the Moon from Earth because the Moon's orbital and rotational periods are equal. The first view of the "far" side was in 1959, when the Soviet flyby mission *Luna 3* sent back pictures showing that the far side of the Moon was very different from its Earth-facing half. No matter how powerful our ground-based or Earth-orbiting telescopes, sometimes we need to send a spacecraft for a different view.

Only in the past half century has the technology existed to explore the Solar System. Spacecraft have now visited all of the planets and some of their moons, as well as a few comets and asteroids, providing the first close-up views of these distant worlds. The study of the Solar System from space is an international collaboration involving NASA, the European Space Agency (ESA), the Russian Federal Space Agency (Roscosmos), the Japan Aerospace Exploration Agency (JAXA), the China National Space Administration (CNSA), and the Indian Space Research Organisation (ISRO). Other countries may soon join the endeavor. In this section, we will look at the different types of spacecraft used to explore our Solar System.

## Flybys and Orbiters

Exploration of the Solar System began with a reconnaissance phase, using spacecraft to fly by or orbit a planet or other body. A **flyby** is a spacecraft that first approaches and then continues flying past the target. As these spacecraft speed by, instruments aboard them briefly probe the physical and chemical properties of the target and its environment.

Flyby missions are the most common first phase of exploration. They cost less than orbiters or landers and are easier to design and execute. Flyby spacecraft such as *Voyager* are sometimes able to visit several different worlds during their travels (**Figure 6.25**). The downside of flyby missions is that because of the physics of orbits, these spacecraft must move by very swiftly. They are limited to just a few hours or at most a few days in which to conduct close-up studies of their targets. Flyby spacecraft provide astronomers with their first close-up views of Solar System objects, and sometimes the data obtained are then used for planning follow-up studies.

More detailed reconnaissance work is done by spacecraft known as **orbiters** because they orbit around their target. These missions are intrinsically more difficult than flyby missions because they have to make risky maneuvers and use up fuel to change their speed to enter an orbit. But orbiters can linger, looking in detail at more of the surfaces of the objects they are orbiting and studying things that change with time, like planetary weather.

Orbiters use remote-sensing instrumentation like that used by Earth-orbiting satellites to study our own planet. These instruments include tools such as cameras that take images at different wavelength ranges, radar that can map surfaces hidden beneath obscuring layers of clouds, and spectrographs that analyze the electromagnetic spectrum. These instruments enable planetary scientists to map other worlds, measure the heights of mountains, identify geological features and rock types, watch weather patterns develop, measure the composition of atmospheres, and get a general sense of the place. Additional instruments make

(a)

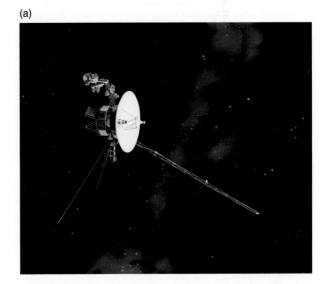

(b)

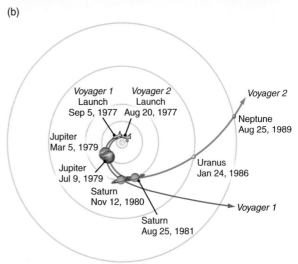

**Figure 6.25** The *Voyager* spacecraft (a) flew past the outer planets and (b) are now near the boundary of our Solar System.

**Figure 6.26** The robotic rover *Curiosity* took this photograph of itself on the surface of Mars.

measurements of the extended atmospheres and space environment through which they travel.

## Landers, Rovers, and Atmospheric Probes

Reconnaissance spacecraft provide a wealth of information about a planet, but there is no better way of exploring a planet than within a planet's atmosphere or on solid ground. Spacecraft have landed on the Moon, Mars, Venus, Saturn's large moon Titan, several asteroids, and a comet. These spacecraft have taken pictures of planetary surfaces, measured surface chemistry, and conducted experiments to determine the physical properties of the surface rocks and soils.

There are several disadvantages of using **landers**—spacecraft that touch down and remain on the surface. Because of the expense, only a few landings in limited areas are practical. Given this limitation, the results may apply only to the small area around the landing site. Imagine, for example, what a different picture of Earth you might get from a spacecraft that landed in Antarctica, as opposed to a spacecraft that landed in a volcano or on the floor of a dry riverbed. Sites to be explored with landed spacecraft must be very carefully chosen on the basis of reconnaissance data. Some landers have wheels and can explore the vicinity of the landing site. Such remote-controlled vehicles, called **rovers**, were used first by the Soviet Union on the Moon four decades ago and more recently by the United States on Mars. **Figure 6.26** shows a self-portrait of the *Curiosity* rover on Mars.

**Atmospheric probes** descend into the atmospheres of planets and continually measure and send back data on temperature, pressure, and wind speed, along with other properties, such as chemical composition. Atmospheric probes have survived all the way to the solid surfaces of Venus and Saturn's moon Titan, sending back streams of data during their descent. An atmospheric probe sent into Jupiter's atmosphere never reached that planet's surface because, as we will discuss later in the book, Jupiter does not have a solid surface in the same sense that terrestrial planets and moons do. After sending back its data, the Jupiter probe eventually melted and vaporized as it descended into the hotter layers of the planet's atmosphere.

## Sample Returns

If you pick up a rock from the side of a mountain road, you might learn a lot from the rock using tools that you could carry in your pocket or your car. But it would be much better to pick up a few samples and carry them back to a laboratory equipped with a full range of state-of-the-art instruments capable of measuring chemical composition, mineral type, age, and other information needed to reconstruct the story of your rock sample's origin and evolution. The same is true of Solar System exploration. One of the most powerful methods for investigating remote objects is to collect samples of the objects and bring them back to Earth for detailed study. So far, only samples of the Moon, a comet, and streams of charged particles from the Sun have been collected and returned to Earth. Scientists have found meteorites on Earth that are likely pieces of Mars that were blasted loose by objects that crashed into Mars. Someday, there may be unmanned "sample and return" missions to Mars.

The missions discussed so far in this section have all been conducted with robotic spacecraft. The only spacecraft that took people to another world were the *Apollo* missions to the Moon. This program ran from 1961 to 1972 and included several missions before the actual Moon landings. The *Apollo 8* astronauts brought back the famous picture of Earth viewed over the surface of the Moon (see the opening figure of Chapter 1). Each mission from *Apollo 11* through *Apollo 17* had three astronauts—two to land on the Moon and one to remain in orbit. One mission (*Apollo 13*) did not reach the Moon but returned to Earth safely. Twelve American astronauts walked on the Moon between 1969 and 1972 and brought back a total of 382 kg of rocks and other material.

The return of extraterrestrial samples to Earth is governed by international treaties and standards to ensure that these samples do not contaminate Earth. For example, before the lunar samples brought back by the *Apollo* missions could be studied, they (and the astronauts) had to be placed in quarantine and tested for alien life-forms. The same international standards apply to spacecraft landing on other planets. The goal of these standards is to avoid transporting life-forms from Earth to another planet. If there is life on other planets, there is concern about introducing potential harm, and we do not want to "discover" life that we, in fact, introduced.

With numerous missions under way and others on the horizon, unmanned exploration of the Solar System is an ongoing, dynamic activity. Appendix 5 summarizes some recent and current missions. Information on the latest discoveries can be found on mission websites and in science news sources.

### CHECK YOUR UNDERSTANDING 6.5

Spacecraft are the most effective way to study planets in our Solar System because: (a) planets move too fast across the sky for us to image them well from Earth; (b) planets cannot be imaged from Earth; (c) they can collect more information than is available just from images from Earth; (d) space missions are easier than long observing campaigns.

## 6.5 Other Tools Contribute to the Study of the Universe

High-profile space missions have sent back stunning images and data from across the electromagnetic spectrum, but astronomers use other tools as well, including particle accelerators and colliders, neutrino and gravitational-wave detectors, and high-speed computers.

### Particle Accelerators

Ever since the early years of the 20th century, physicists have been peering into the structure of the atom by observing what happens when small particles collide. By the 1930s, physicists had developed the technology to accelerate charged subatomic particles such as protons to very high speeds and then observe what happens when they slam into a target. From such experiments, physicists have discovered many kinds of subatomic particles and learned about their physical properties. High-energy particle colliders have proved to be an essential tool for physicists studying the basic building blocks of matter.

Astronomers have realized that to understand the very largest structures seen in the universe, it is important to understand the physics that took place during

**Figure 6.27** The ATLAS particle detector at CERN's Large Hadron Collider near Geneva, Switzerland. The enormous size of this instrument is evident from the person standing near the bottom center of the picture.

the earliest moments in the universe, when everything was extremely hot and dense. High-energy particle colliders that physicists use today are designed to approach the energies of the early universe. The effectiveness of particle accelerators is determined by the energy they can achieve and the number of particles they can accelerate. Modern particle colliders such as the Large Hadron Collider near Geneva, Switzerland (**Figure 6.27**), reach very high energies. Particles can also be studied from space. The Alpha Magnetic Spectrometer, installed on the International Space Station in 2011, searches for some of the most exotic forms of matter, such as dark matter, antimatter, and high-energy particles called cosmic rays.

## Neutrinos and Gravitational Waves

The **neutrino** is an elusive elementary particle that plays a major role in the physics of the interiors of stars. Neutrinos are extremely difficult to detect. In less time than it takes you to read this sentence, a thousand trillion ($10^{15}$) solar neutrinos from the Sun are passing through your body, even during the night. Neutrinos are so nonreactive with matter that they can pass right through Earth (and you) as though it (or you) weren't there at all. A neutrino has to interact with a detector to be observed. Neutrino detectors typically record only one out of every $10^{22}$ (10 billion trillion) neutrinos passing though them, but that's enough to reveal processes deep within the Sun or the violent death of a star 160,000 light-years away.

Experiments designed to look for neutrinos originating outside of Earth are buried deep underground in mines or caverns or under the ocean or ice to ensure that only neutrinos are detected. For example, the ANTARES experiment uses the Mediterranean Sea as a neutrino telescope. Detectors located 2.5 km under the sea, off the coast of France, observe neutrinos that originated in objects visible in southern skies and passed through Earth. In the IceCube neutrino observatory located at the South Pole in Antarctica, the neutrino detectors are 1.5–2.5 km under the ice, and they observe neutrinos that originated in objects visible in northern skies (**Figure 6.28**).

Another elusive phenomenon is the **gravitational wave**. Gravitational waves are disturbances in a gravitational field, similar to the waves that spread out from the disturbance you create when you toss a pebble onto the quiet surface of a pond. There is strong, although indirect, observational evidence for the existence of gravitational waves, but they are so elusive that they have not yet actually been detected (**Process of Science Figure**). Several facilities, including the Laser Interferometer Gravitational-Wave Observatory (LIGO), have been constructed to detect gravitational waves. Scientists are eager to detect gravitational waves—to confirm their existence and to study the physical phenomena they are likely to reveal, such as the birth and evolution of the universe, stellar evolution, or the very force of gravity itself.

## Computers

Astronomers use powerful computers for data gathering, analysis, and interpretation. A single CCD image may contain as many as 100 million pixels, with each pixel displaying roughly 30,000 levels of brightness. That adds up to several trillion pieces of information in each image. To analyze their data, astronomers typically do calculations on *every single pixel* of an image in order to remove unwanted contributions from Earth's atmosphere or to correct for instrumental effects. Astronomers conduct many different types of sky surveys—in which one or more

**Figure 6.28** The IceCube neutrino telescope at the South Pole, Antarctica.

# Process of Science

## TECHNOLOGY AND SCIENCE ARE SYMBIOTIC

**Scientists have been searching for waves that carry gravitational information for nearly 100 years, but the accuracy of their measurements is limited by the available technology.**

## Take 1

### Weber Bar

Precision-machined bars of metal that should "ring" as a gravitational wave passes by.

Sensitive only to extremely powerful gravitational waves.

No detection.

## Take 2

### LIGO

**New Technology: Lasers**

Lasers should interfere as gravitational waves pass by.

Roughly 100 times more sensitive than Weber bar measurements.

No detection (yet).

## Take 3

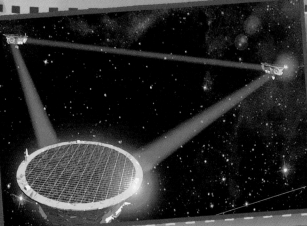

### Future Science Mission

Lasers will interfere as gravitational waves pass by.

Sensitive to more types of objects than LIGO is.

Technology and science develop together. New technologies enable humans to ask new scientific questions. Asking scientific questions pushes the development of better instrumentation. Deeper scientific understanding leads to new technologies.

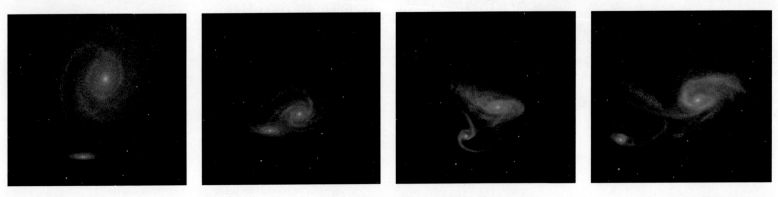

**Figure 6.29** These images show supercomputer simulations of the collision of two galaxies. Astronomers compare simulations like these with telescopic observations.

telescopes survey a specific part of the sky—yielding thousands of images that need to be analyzed.

High-speed computers also play an essential role in generating and testing theoretical models of astronomical objects. Even when we completely understand the underlying physical laws that govern the behavior of a particular object, often the object is so complex that it would be impossible to calculate its properties and behavior without the assistance of high-speed computers. For example, as you learned in Chapter 4, you can use Newton's laws to compute the orbits of two stars that are gravitationally bound to one another, because their orbits take the form of simple ellipses. However, it is not so easy to understand the orbits of the several hundred billion stars that make up the Milky Way Galaxy, even though the underlying physical laws are the same.

Computer modeling is used to determine the interior properties of stars and planets, including Earth. Although astronomers cannot see beneath the surfaces of these bodies, they have a surprisingly good understanding of the interiors of these bodies, as we will describe in later chapters. Astronomers start a model by assigning well-understood physical properties to tiny volumes within a planet or star. The computer assembles an enormous number of these individual elements into an overall representation. When it is all put together, the result is a rather good picture of what the interior of the star or planet is like.

Astronomers also use supercomputers to study the evolution of astronomical objects, systems of objects, or the universe as a whole over time. For example, astronomers create models of galaxies, and then run computer simulations to study how those galaxies might change over billions of years. **Figure 6.29** shows a simulation of the collision of two galaxies. The results of the computer simulations are then compared with telescopic observations. If the simulations do not match the observations, then adjustments are made to the model and the simulations are run again until there is general agreement between them.

## CHECK YOUR UNDERSTANDING 6.6

High-speed computers have become one of an astronomer's most important tools. Which of the following require the use of a high-speed computer? (Choose all that apply.) (a) analyzing images taken with very large CCDs; (b) generating and testing theoretical models; (c) pointing a telescope from object to object; (d) studying the evolution of astronomical objects or systems over time.

# Origins

## Microwave Telescopes Detect Radiation from the Big Bang

In this chapter, we explored the tools of the astronomer, from basic optical telescopes to instruments that observe in different wavelengths. Now let's examine in more detail one type of telescope that has aided the study of the origin of the universe. Recall from Chapter 1 that astronomers think the universe originated with a hot Big Bang. The multiple strands of evidence for this conclusion will be discussed in Chapter 21. Here, we look at one piece: the observation of faint microwave radiation left over from the early hot universe. Two Bell Laboratories physicists, Arno Penzias (1933–) and Robert Wilson (1936–), were working on satellite communications when they first accidentally detected this radiation in 1964 with a microwave dish antenna in New Jersey. Today, we routinely use cell phones and handheld GPS systems that communicate directly with satellites, but at the time, this capability was at the limit of technology.

Penzias and Wilson needed a very sensitive microwave telescope for the work they were doing for Bell Labs, because any spurious signals coming from the telescope itself might wash out the faint signals bounced off a satellite. To that end, they were working very hard to eliminate all possible sources of interference originating from within their instrument, including keeping the telescope free of bird droppings. They found that no matter how carefully they tried to eliminate sources of extraneous noise, they always still detected a faint signal at microwave wavelengths. This faint signal was the same in every direction and

turned out to be from the Big Bang. Penzias and Wilson shared the 1978 Nobel Prize in Physics for the discovery of this **cosmic microwave background radiation (CMB)** left over from the Big Bang itself.

Since 1964, astronomers from around the world have designed increasingly precise instruments to measure this radiation from the ground, from high-altitude balloons, from rockets, and from satellites. The Russian experiment RELIKT-1, launched in 1983, found some limits on the variation of the CMB. The COBE (Cosmic Background Explorer) satellite, launched in 1989, showed that the spectrum of this radiation precisely matched that of a blackbody with a temperature of 2.73 K—exactly what was predicted for the radiation left over from the Big Bang. (Compare **Figure 6.30** with the curves in Figure 5.22.) The data also showed some slight differences in temperature—small fractions of a degree—over the map of the sky. These slight variations tell us about how the universe evolved from one that was dominated by radiation to one that contains structures such as galaxies, stars, planets, and us. John Mather and George Smoot shared the 2006 Nobel Prize in Physics for this work.

In 1998 and 2003, a high-altitude balloon experiment called BOOMERANG (short for "balloon observations of millimetric extragalactic radiation and geophysics") flew over Antarctica at an altitude of 42 km to study CMB variations and estimate the overall geometry of the universe. The *WMAP* (Wilkinson Microwave Anisotropy

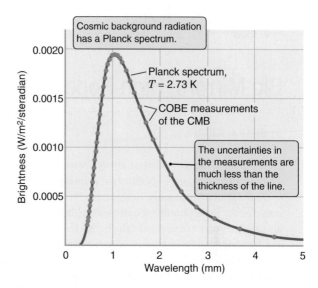

**Figure 6.30** This graph shows the spectrum of the cosmic microwave background radiation (CMB) as measured by the COBE satellite (red dots). A steradian is a unit of solid angle. The uncertainty in the measurement at each wavelength is much less than the size of a dot. The line running through the data is a Planck blackbody spectrum with a temperature of 2.73 K.

Probe) satellite, launched in 2001, created an even more detailed map of the temperature variations in this radiation, yielding more precise values for the age and shape of the universe and the presence of dark matter and dark energy. The Atacama Cosmology Telescope (Chile) and the South Pole Telescope (Antarctica) study this radiation to look for evidence of when galaxy clusters formed. The newest microwave observatory in space, the Planck telescope, was launched in 2009 by the European Space Agency. Planck has much greater sensitivity than *WMAP* and studied these CMB variations in even more precise detail. These experiments and observations have opened up the current era of precision cosmology, in which astronomers can make detailed models of how the universe was born, eventually leading to stars, planets, and us.

# Big Mirrors, High Hopes: Extremely Large Telescope Is a Go

By **CALEB A. SCHARF**

In astronomy, bigger is almost always better. The size of a telescope's aperture (or primary optical element) not only determines how many pesky little photons it can capture, but also the ultimate resolution of the image that can be formed. The challenge is to fabricate optics on large scales, find somewhere really good to put them, and to build the massive structure to house them and the sensitive instruments to analyze the photons that come out of the pipeline.

Now one of the world's next generation of huge telescopes has been given the green light to push ahead with construction and operation. At an astonishing 39.3 meters in diameter, the European Extremely Large Telescope, or E-ELT (one can only hope a more poetic name is eventually chosen), will hold the crown for sheer girth among optical and infrared sensitive telescopes. With an authorized spending of about a billion euros (about $1.24B) the project can move ahead, with an anticipated "first light" sometime in 2024.

Although E-ELT is not alone in the race for so-called "30-meter class" telescopes (with the Thirty Meter Telescope and Giant Magellan observatories also on track, and also extraordinarily powerful), it's definitely the Hulk of the bunch.

Observatories on this scale will have profound impact on how we study the universe around us, from cosmology to planets.

A telescope like E-ELT will peer at the faintest, most distant objects in the young cosmos, and it'll be capable of sensing the atmospheric signatures of life in a few nearby terrestrial-type exoplanets (including the presence of oxygen). These tasks require an enormous "light bucket" to catch enough photons, but perhaps one of the most vivid examples of the gain of a big telescope comes from considering the resolving power.

Equipped with adaptive optics, E-ELT should, for example, be able to routinely study Jupiter down to scales of about 20 kilometers—by comparison the Great Red Spot is at present about 20,000 kilometers across. Mars can be imaged to roughly 5-kilometer resolution (depending of course on the relative separation of Earth). In other words, on a nightly basis we will be able to monitor the worlds in our Solar System with a fidelity comparable to flyby missions of yore.

Preparations have already begun. In June 2014, explosives were used to help level the mountaintop where E-ELT will sit—Cerro Armazones in northern Chile, a dry peak at about 3,000 meters altitude where the night skies are cloudless 89 percent of the time.

What rises there over the next 10 years will help take our understanding of the universe to a whole new level.

1. Why do astronomers want to build at this particular location?
2. What are the advantages of larger telescopes?
3. How can adaptive optics yield images as good as those from old space missions?
4. Why will this telescope observe only in optical and infrared wavelengths?
5. Do a Web search to see the status of this project. What countries are partners? Is there any local opposition to the project in the host country?

# Summary

Earth's atmosphere blocks many spectral regions and distorts telescopic images. Telescopes are sited to be above as much of the atmosphere as possible. Telescopes are matched to the wavelengths of observation, with different technologies required for each region of the spectrum. The aperture of a telescope both determines its light-gathering power and limits its resolution; larger telescopes are better in both measures. Modern CCD cameras have improved quantum efficiency and longer integration times, which allow astronomers to study fainter and more distant objects than were observable with prior detectors. Telescopes observing at microwave wavelengths have detected radiation left over from the Big Bang.

**LG 1** **Compare the two main types of optical telescopes and how they gather and focus light.** The telescope is the astronomer's most important tool. Ground-based telescopes that observe in visible wavelengths come in two basic types: refractors (lenses) and reflectors (mirrors). All large astronomical telescopes are reflectors. Large telescopes collect more light and have greater resolution. The diffraction limit is the limiting resolution of a telescope.

**LG 2** **Summarize the main types of detectors that are used on telescopes.** Photography improved the ability of astronomers to record details of faint objects seen in telescopes. CCDs are today's astronomical detector of choice because

they are much more linear, have a broader spectral response, and can send electronic images directly to a computer. Spectrographs are specialized instruments that take the spectrum of an object to reveal what the object is made of and many other physical properties.

**LG 3** **Explain why some wavelengths of radiation must be observed from space.** Radio telescopes are able to see through our atmosphere. Radio, near-infrared, and optical telescopes can be arrayed to greatly increase angular resolution. Putting telescopes in space solves problems created by Earth's atmosphere.

**LG 4** **Explain the benefits of sending spacecraft to study the planets and moons of our Solar System.** Most of what is known about the planets and moons comes from observations by spacecraft. Flyby and orbiting missions obtain data from space, and landers and rovers collect data from the ground.

**LG 5** **Describe other astronomical tools that contribute to the study of the universe.** Astronomers also use particle accelerators, neutrino detectors, and gravitational-wave detectors to study the universe. High-speed computers are essential to the acquisition, analysis, and interpretation of astronomical data.

## ? UNANSWERED QUESTIONS

- Will telescopes be placed on the Moon? The Moon has no atmosphere to make stars twinkle, cause weather, or block certain wavelengths of light from reaching its surface. The far side of the Moon faces away from the light and radio radiation of Earth, and all parts of the Moon have nights that last for two Earth weeks. One proposal calls for a Lunar Array for Radio Cosmology (LARC), an array of hundreds of radio telescopes that would be deployed on the Moon—after the year 2025—to study the earliest formation of stars and galaxies. Another proposal is for the Lunar Liquid Mirror Telescope (LLMT), with a diameter of 20–100 meters, to be located at one of the Moon's poles. Gravity would settle the rotating liquid into the necessary parabolic shape, and these liquid mirror telescopes are much simpler than are arrays of telescopes with large glass mirrors. The LLMT would observe extremely distant protostars and protogalaxies in infrared wavelengths. Astronomers debate whether telescopes on the Moon would be easier to service and repair than those

in space and whether problems caused by lunar dust would outweigh any advantages.

- Will there be human exploration of the Solar System within your lifetime? Since the *Apollo* program, humans have not returned to the Moon or traveled to other planets or moons in the Solar System. Sending humans to the worlds of the Solar System is much more complicated, risky, and more expensive than sending robotic spacecraft. Humans need life support such as air, water, and food. Radiation in space can be dangerous. Furthermore, human explorers would expect to return to Earth, whereas most spacecraft do not come back. Astronomers and space scientists have heated debates about human spaceflight versus robotic exploration. Some argue that true exploration requires that human eyes and brains actually go there; others argue that the costs and risks are too high for the potential additional scientific knowledge. Beyond basic exploration, we also do not know whether humans will ever permanently colonize space.

# Questions and Problems

## Test Your Understanding

1. You are shopping for telescopes online. You find two in your price range. One of these has an aperture of 20 cm, and the other has an aperture of 30 cm. If aperture size is the only difference, which should you choose, and why?
   a. The 20 cm, because the light-gathering power will be better.
   b. The 20 cm, because the image size will be larger.
   c. The 30 cm, because the light-gathering power will be better.
   d. The 30 cm, because the image size will be larger.

2. Which of the following can be observed from Earth's surface? (Choose all that apply.)
   a. radio waves
   b. gamma radiation
   c. far UV light
   d. X-ray light
   e. visible light

3. Match the following properties of telescopes (lettered) with their corresponding definitions (numbered).
   a. aperture
   b. resolution
   c. focal length
   d. chromatic aberration
   e. diffraction
   f. interferometer
   g. adaptive optics

   (1) two or more telescopes connected to act as one
   (2) distance from lens to focal plane
   (3) diameter
   (4) ability to distinguish close objects
   (5) computer-controlled atmospheric distortion correction
   (6) color-separating effect
   (7) smearing effect due to sharp edge

4. The two Keck 10-meter telescopes, separated by a distance of 85 meters, can operate as an optical interferometer. What is its resolution when it observes in the infrared at a wavelength of 2 microns?
   a. 0.01 arcsec
   b. 0.005 arcsec
   c. 0.2 arsec
   d. 0.05 arcsec

5. Arrays of radio telescopes can produce much better resolution than single-dish telescopes because they work based on the principle of
   a. reflection.
   b. refraction.
   c. diffraction.
   d. interference.

6. Refraction is caused by
   a. light bouncing off a surface.
   b. light changing colors as it enters a new medium.
   c. light changing speed as it enters a new medium.
   d. two light beams interfering.

7. The light-gathering power of a 4-meter telescope is _____ than that of a 2-meter telescope.
   a. 8 times larger
   b. 4 times larger
   c. 16 times smaller
   d. 2 times smaller

8. Improved resolution is helpful to astronomers because
   a. they often want to look in detail at small features of an object.
   b. they often want to look at very distant objects.
   c. they often want to look at many objects close together.
   d. all of the above

9. The part of the human eye that acts as the detector is the
   a. retina.
   b. pupil.
   c. lens.
   d. iris.

10. Cameras that use adaptive optics provide higher-spatial-resolution images primarily because
   a. they operate above Earth's atmosphere.
   b. deformable mirrors are used to correct the blurring due to Earth's atmosphere.
   c. composite lenses correct for chromatic aberration.
   d. they simulate a much larger telescope.

11. The advantage of an interferometer is that
   a. the resolution is dramatically improved.
   b. the focal length is dramatically increased.
   c. the light-gathering power is dramatically increased.
   d. diffraction effects are dramatically decreased.
   e. chromatic aberration is dramatically decreased.

12. The angular resolution of a ground-based telescope is usually determined by
   a. diffraction.
   b. the focal length.
   c. refraction.
   d. atmospheric seeing.

13. A grating is able to spread white light out into a spectrum of colors because of the property of
   a. reflection.
   b. diffraction.
   c. dispersion.
   d. interference.

14. Why would astronomers put telescopes in airplanes?
   a. to get the telescopes closer to the stars
   b. to get the telescopes above the majority of the water vapor in Earth's atmosphere
   c. to be able to observe one object for more than 24 hours without stopping
   d. to allow the telescopes to observe the full spectrum of light

15. If we could increase the quantum efficiency of the human eye, it would
    a. allow humans to see a larger range of wavelengths.
    b. allow humans to see better at night or in other low-light conditions.
    c. increase the resolution of the human eye.
    d. decrease the resolution of the human eye.

## Thinking about the Concepts

16. Galileo's telescope used simple lenses. What is the primary disadvantage of using a simple lens in a refracting telescope?

17. The largest astronomical refractor has an aperture of 1 meter. List several reasons why it would be impractical to build a larger refractor with twice this aperture.

18. Your camera may have a zoom lens, ranging between wide angle (short focal length) and telephoto (long focal length). How does the size of an object in the camera's focal plane differ between wide angle and telephoto?

19. Optical telescopes reveal much about the nature of astronomical objects. Why do astronomers also need information provided by gamma-ray, X-ray, infrared, and radio telescopes?

20. For light reflecting from a flat surface, the angles of incidence and reflection are the same. This is also true for light reflecting from the curved surface of a reflecting telescope's primary mirror. Sketch a curved mirror and several of these reflecting rays.

21. Consider two optically perfect telescopes having different diameters but the same focal length. Is the image of a star larger or smaller in the focal plane of the larger telescope? Explain your answer.

22. Study the Process of Science Figure. Make a flowchart for the symbiosis between technology and science that led to the development of the CCD camera as discussed in Section 6.2.

23. Explain adaptive optics and how they improve a telescope's image quality.

24. Explain integration time and quantum efficiency and how each contributes to the detection of faint astronomical objects.

25. Some people believe that we put astronomical telescopes on high mountaintops or in orbit because doing so gets them closer to the objects they are observing. Explain what is wrong with this popular misconception, and give the actual reason telescopes are located in these places.

26. Humans have sent various kinds of spacecraft—including flybys, orbiters, and landers—to all of the planets in our Solar System. Explain the advantages and disadvantages of each of these types of spacecraft.

27. If there are meteorites that are pieces of Mars on Earth, why is it so important to go to Mars and bring back samples of the martian surface?

28. Humans had a first look at the far side of the Moon as recently as 1959. Why had we not seen it earlier—when Galileo first observed the Moon with his telescope in 1610?

29. Where are neutrino detectors located? Why are neutrinos so difficult to detect?

30. Why do telescopes in space give a better picture of the leftover radiation from the Big Bang?

## Applying the Concepts

31. Compare the light-gathering power of the Thirty Meter Telescope with that of the dark-adapted human eye (aperture 8 mm) and with that of one of the 10-meter Keck telescopes.

32. Study the photograph of light entering and leaving a block of refractive material in Figure 6.2b. Use a protractor to measure the angles of the green light as it enters the block and as it leaves the block. How are these angles related?

33. Many amateur astronomers start out with a 4-inch (aperture) telescope and then graduate to a 16-inch telescope. By what factor does the light-gathering power of the telescope increase with this upgrade? How much fainter are the faintest stars that can be seen in the larger telescope?

34. The resolution of the human eye is about 1.5 arcmin. What would the aperture of a radio telescope (observing at 21 cm) have to be to have this resolution? Even though the atmosphere is transparent at radio wavelengths, humans do not see light in the radio range. Using your calculations and logic, explain why.

35. Assume that you have a telescope with an aperture of 1 meter. Compare the telescope's theoretical resolution when you are observing in the near-infrared region of the spectrum ($\lambda = 1{,}000$ nm) with that when you are observing in the violet region of the spectrum ($\lambda = 400$ nm).

36. Assume that the maximum aperture of the human eye, $D$, is approximately 8 mm and the average wavelength of visible light, $\lambda$, is $5.5 \times 10^{-4}$ mm.
    a. Calculate the diffraction limit of the human eye in visible light.
    b. How does the diffraction limit compare with the actual resolution of 1–2 arcmin (60–120 arcsec)?
    c. To what do you attribute the difference?

37. The diameter of the full Moon in the focal plane of an average amateur's telescope (focal length 1.5 meters) is 13.8 mm. How big would the Moon be in the focal plane of a very large astronomical telescope (focal length 250 meters)?

38. One of the earliest astronomical CCDs had 160,000 pixels, each recording 8 bits (256 levels of brightness). A new generation of astronomical CCDs may contain a billion pixels, each recording 15 bits (32,768 levels of brightness). Compare the number of bits of data that each of these two CCD types produces in a single image.

39. Consider a CCD with a quantum efficiency of 80 percent and a photographic plate with a quantum efficiency of 1 percent. If an exposure time of 1 hour is required to photograph a celestial object with a given telescope, how much observing time would be saved by substituting a CCD for the photographic plate?

40. The VLBA uses an array of radio telescopes ranging across 8,000 km of Earth's surface from the Virgin Islands to Hawaii.
    a. Calculate the angular resolution of the array when radio astronomers are observing interstellar water molecules at a microwave wavelength of 1.35 cm.
    b. How does this resolution compare with the angular resolution of two large optical telescopes separated by 100 meters and operating as an interferometer at a visible wavelength of 550 nm?

41. When operational, the SVLBI may have a baseline of 100,000 km. What will be the angular resolution when studying interstellar molecules emitting at a wavelength of 17 mm from a distant galaxy?

42. The *Mars Reconnaissance Orbiter* (*MRO*) flies at an average altitude of 280 km above the martian surface. If its cameras have an angular resolution of 0.2 arcsec, what is the size of the smallest objects that the *MRO* can detect on the martian surface?

43. *Voyager 1* is now about 125 astronomical units (AU) from Earth, continuing to record its environment as it approaches the limits of our Solar System.
    a. How far away is *Voyager 1*, in kilometers?
    b. How long does it take observational data to come back to us from *Voyager 1*?
    c. How does *Voyager 1*'s distance from Earth compare with that of the nearest star (other than the Sun)?

44. Gravitational waves travel at the speed of light. Their speed, wavelength, and frequency are related as $c = \lambda \times f$. If we were to observe a gravitational wave from a distant cosmic event with a frequency of 10 hertz (Hz), what would be the wavelength of the gravitational wave?

45. Compute the peak of the blackbody spectrum with a temperature of 2.73 K. What region of the spectrum is this?

## USING THE WEB

46. A webcast for the International Year of Astronomy 2009 called "Around the World in 80 Telescopes" can be accessed at http://eso.org/public/events/special-evt/100ha.html. The 80 telescopes are situated all over, including Antarctica and space. Pick two of the telescopes and watch the videos. Do you think these videos are effective for public outreach for the observatory in question or for astronomy in general? For each telescope you choose, answer the following questions: Does the telescope observe in the Northern Hemisphere or the Southern Hemisphere? What wavelengths does the telescope observe? What are some of the key science projects at the telescope?

47. Most major observatories have their own websites. Use the link in question 46 to find a master list of telescopes, and click on a telescope name to link to an observatory website (or run a search on names from Tables 6.1 and 6.2). For the telescope you choose, answer the following questions: (a) What is this telescope's "claim to fame"—is it the largest? at the highest altitude? at the driest location? with the darkest skies? the newest? (b) Does the observatory website have news releases? What is a recent discovery from this telescope?

48. Go to the website for the International Dark Sky Association (http://www.darksky.org/). Click on "Night Sky Conservation" and then "Do you live under light pollution?" Is your location dark? From the menu on the left, is there a "dark sky park" near you? What are the ecological arguments against too much light at night?

49. What is the current status of the James Webb Space Telescope (http://jwst.nasa.gov)? How will this telescope be different from the Hubble Space Telescope? What are some of the instruments for the JWST and its planned projects? What is the current estimated cost of the JWST?

50. Pick a mission from Appendix 5, go to its website, and see what's new. For the mission you choose, answer the following questions: Is the spacecraft still active? Is it sending images? What new science is coming from this mission?

# smartwork5

If your instructor assigns homework in Smartwork5, access your assignments at digital.wwnorton.com/astro5.

# EXPLORATION

digital.wwnorton.com/astro5

Visit the Student Site at the Digital Landing Page, and open the "Geometric Optics and Lenses" animation in Chapter 6. Read through the animation until you reach the optics simulation, pictured in **Figure 6.31**. The simulator shows a converging lens and a pencil. Rays come from the pencil on the left of the converging lens, pass through the lens, and make an image to the right of the lens. The view that would be seen by an observer at the position of the eye is shown in the circle at upper right. Initially, when the pencil is at position 2.3 and the eye is at position 2.0, the pencil is out of focus and blurry.

**1** Is the eraser at the top or the bottom of the actual pencil? (This becomes important later.)

_____

Using the red slider in the upper left of the window, try moving the pencil to the right. Pause when the observer's eye sees a recognizable pencil (even if it's still blurry).

**2** Does the eye see the pencil right side up or upside down?

_____

**3** This is somewhat analogous to the view through a telescope. The objects are very far from the lenses, and the observer sees things upside down in the telescope. If an object in your field of view is at the top of the field and you want it in the center, which way should you move the telescope: up or down?

_____

Now return the pencil to position 2.3. Use the red slider in the lower right of the window to move the eye closer to the lens (to the left).

**4** At what distance does the image of the pencil first become crisp and clear?

_____

**5** Is the pencil right side up or upside down?

_____

**6** In practice at the telescope, we do not move the observer back (away from the eyepiece) to bring the image into focus. Why not?

_____

**7** Instead of moving the observer, we use a focusing knob to move the lens in the eyepiece, which brings the image into focus. Imagine that you are looking through the eyepiece of a telescope and the image is blurry. You turn the focusing knob and things get blurrier! What should you try next?

_____

_____

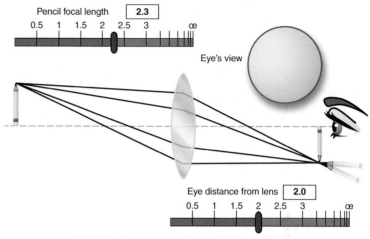

**Figure 6.31** Use this simulation to change the position of the object of the eye to explore what will be seen for various configurations.

**8** Now imagine that you get the image focused just right, so it is crisp and sharp. The next person to use the telescope wears glasses and insists that the image is blurry. But when you look through the telescope again, the image is still crisp. Explain why your experiences differ.

_____

_____

_____

Step through the animation to the next picture. Carefully study the two telescopes shown and the path the light takes through them.

**9** Which telescope has a longer focal length: the top one or the bottom one?

_____

**10** Which telescope produces an image with the red and the blue stars more separated: the top one or the bottom one?

_____

**11** A longer focal length is an advantage in one sense, but it's not the entire story. What are some disadvantages of a telescope with a very long focal length?

_____

_____

_____

# 7 The Birth and Evolution of Planetary Systems

The planetary system containing Earth—our Solar System—is a by-product of the birth of the Sun. But the physical processes that shaped the formation of the Solar System are not unique to it. The same processes have formed numerous multiplanet systems. In this chapter, we will examine how planetary systems are born and evolve.

## LEARNING GOALS

By the conclusion of this chapter, you should be able to:

**LG 1** Describe how our understanding of planetary system formation developed from the work of both planetary and stellar scientists.

**LG 2** Discuss the role of gravity and angular momentum in explaining why planets orbit the Sun in a plane and why they revolve in the same direction that the Sun rotates.

**LG 3** Explain how temperature at different locations in the protoplanetary disk affects the composition of planets, moons, and other bodies.

**LG 4** Discuss the processes that resulted in the formation of planets and other objects in our Solar System.

**LG 5** List how astronomers find planets around other stars, and explain how we know that planetary systems around other stars are common.

From clouds of gas and dust, planetary systems are born. ►►►

How was our
Solar System
born?

# 7.1 Planetary Systems Form around a Star

Earth is part of a collection of **planets**—large, round bodies that orbit a star in individual orbits. Astronomers call a system of planets surrounding a star a **planetary system**. The Solar System, shown in **Figure 7.1**, is the planetary system that includes Earth, other planets, and the Sun. It also includes moons that orbit planets and small bodies that occupy particular regions of the Solar System; for example, in the asteroid belt or in the Kuiper Belt. Our Solar System is a tiny part of our galaxy, which is a tiny part of the universe. Review Figure 1.3 in Chapter 1 to remind yourself of the size scales involved. Light takes about 4 hours to travel to Earth from Neptune, the outermost planet in the Solar System, but light from the most distant galaxies takes nearly *14 billion years* to reach Earth.

Until the latter part of the 20th century, the origin of the Solar System remained speculative. Over the past century, with the aid of spectroscopy, astronomers have determined that the Sun is a typical star, one of hundreds of billions in its galaxy, the Milky Way, and that the Milky Way is a typical galaxy, one of hundreds of billions in the universe. In the past few decades, stellar astronomers studying the formation of stars and planetary scientists analyzing clues about the history of the Solar System have found themselves arriving at the same picture of the early Solar System—but from two very different directions. This unified understanding provides the foundation for the way astronomers now think about the Sun and the myriad objects that orbit it. In this section, we will look at how the work of stellar and planetary scientists converged to inform our understanding of planetary system formation.

## The Nebular Hypothesis

The first plausible theory for the formation of the Solar System, the **nebular hypothesis**, was proposed in 1734 by the German philosopher Immanuel Kant (1724–1804) and conceived independently a few years later by the French astronomer Pierre-Simon Laplace (1749–1827). Kant and Laplace argued that a rotating cloud of interstellar gas, or **nebula** (Latin for "cloud"), gradually collapsed and flattened to form a disk with the Sun at its center. Surrounding the Sun were rings of material from which the planets formed. This configuration would explain why the planets orbit the Sun in the same direction in the same plane. The nebular hypothesis remained popular throughout the 19th century, and these basic principles of the hypothesis are still retained today.

Our modern theory of planetary system formation calculates under what conditions clouds of interstellar gas collapse under the force of their own self-gravity to form stars. Recall from Chapter 4 that self-gravity is the gravitational attraction between the parts of an object such as a planet or star that pulls all the parts toward the object's center. This inward force is opposed by either structural strength (in the case of

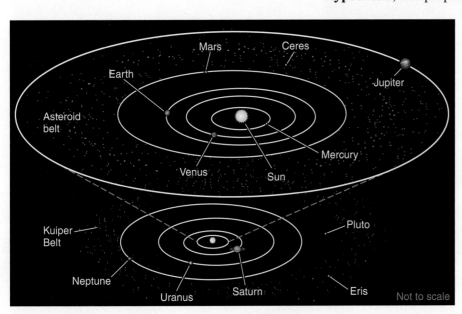

**Figure 7.1** Our Solar System includes planets, moons, and other small bodies. Sizes and distances are not to scale in this sketch.

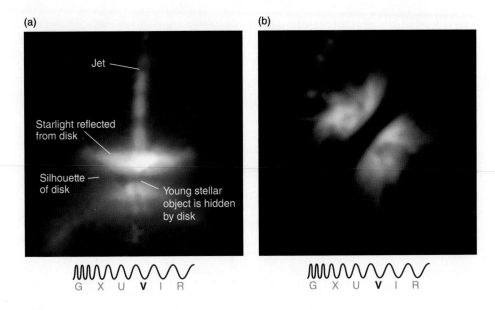

(a)

Jet

Starlight reflected
from disk

Silhouette
of disk

Young stellar
object is hidden
by disk

(b)

G X U V I R

G X U V I R

**Figure 7.2** Hubble Space Telescope images show disks around newly formed stars. (a) The dark band is the silhouette of the disk seen more or less edge-on. Bright regions are dust illuminated by starlight. Some disk material may be expelled in a direction perpendicular to the plane of the disk in the form of violent jets. (b) In this image, the disk is seen in silhouette. Planets may be forming or have already formed in this disk.

rocks that make up terrestrial planets) or the outward force resulting from gas pressure within a star. If the outward force is less than self-gravity, the object contracts; if it is greater, the object expands. In a stable object, the inward and outward forces are balanced.

In support of the nebular hypothesis, disks of gas and dust have been observed surrounding young stellar objects (**Figure 7.2**). From this observational evidence, stellar astronomers have shown that, much like a spinning ball of pizza dough spreads out to form a flat crust, the cloud that produces a star—the Sun, for example—collapses first into a rotating disk. Material in the disk eventually suffers one of three fates: it travels inward onto the forming star at its center, it remains in the disk itself to form planets and other objects, or it is ejected back into interstellar space.

▶❙❙ **AstroTour:** Solar System Formation

## Planetary Scientists and the Convergence of Evidence

While astronomers were working to understand star formation, other groups of scientists with very different backgrounds were piecing together the history of the Solar System. Planetary scientists, geochemists, and geologists looking at the current structure of the Solar System inferred what some of its early characteristics must have been. The orbits of all the planets lie very close to a single plane, so the early Solar System must have been flat. Additionally, all the planets orbit the Sun in the same direction, so the material from which the planets formed must have been orbiting the Sun in the same direction as well.

To find out more, scientists study samples of the very early Solar System. Rocks that fall to Earth from space, known as **meteorites**, include pieces of material that are left over from the Solar System's youth. Many meteorites, such as the one in **Figure 7.3**, resemble a piece of concrete in which pebbles and sand are mixed with a much finer filler, suggesting that the larger bodies in the Solar System must have grown from the aggregation of smaller bodies. This chain of thought suggests an early Solar System in which the young Sun was surrounded by a flattened

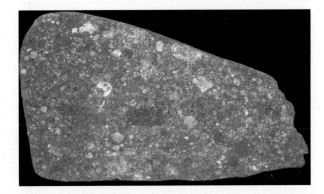

**Figure 7.3** Meteorites are the surviving pieces of Solar System fragments that land on the surfaces of planets. This meteorite formed from many smaller components that stuck together.

# Process of Science

**Astronomers asked: Why is the Solar System a disk, with all planets orbiting in the same direction?**

**Stellar astronomers find dust and gas around young stars.**

**Stellar astronomers observe this gas and dust to be in the shape of disks.**

**Stellar astronomers test the nebular hypothesis, seeking evidence for or against.**

**Mathematicians suggest the nebular hypothesis: a collapsing rotating cloud formed the Solar System.**

**Planetary scientists test the nebular hypothesis, seeking evidence for or against.**

**Planetary scientists study meteorites that show the Solar System bodies formed from many smaller bodies.**

Beginning from the same fundamental observations about the shape of the Solar System, theorists, planetary scientists, and stellar astronomers converge in the nebular theory that stars and planets form together from a collapsing cloud of gas and dust.

disk of both gaseous and solid material. Our Solar System formed from this swirling disk of gas and dust.

As astronomers and planetary scientists compared notes, they realized they had arrived at the same picture of the early Solar System from two completely different directions. The rotating disk from which the planets formed was the remains of the disk that had accompanied the formation of the Sun. Earth, along with all the other orbiting bodies that make up the Solar System, formed from the remnants of an *interstellar cloud* that collapsed to form the local star, the Sun. The connection between the formation of stars and the origin and subsequent evolution of the Solar System is one of the cornerstones of both astronomy and planetary science—a central theme of our understanding of our Solar System (**Process of Science Figure**).

---

## CHECK YOUR UNDERSTANDING 7.1

Which of the following pieces of evidence supports the nebular hypothesis? (Choose all that apply.) (a) Planets orbit the Sun in the same direction. (b) The Solar System is relatively flat. (c) Earth has a large Moon. (d) We observe disks of gas and dust around other stars.

## 7.2 The Solar System Began with a Disk

Now that we've noted the general idea that planets formed in a disk around young stars, let's look at some of the specifics. **Figure 7.4** illustrates the young Solar System as it appeared roughly 5 billion years ago. At that time, the Sun was still a **protostar**—a large ball of gas but not yet hot enough in its center to be a star. As the cloud of interstellar gas collapsed to form the protostar, its gravitational energy was converted into heat energy and radiation. Surrounding the protostellar Sun was a flat, orbiting disk of gas and dust. Each bit of the material in this thin disk orbited the Sun according to the same laws of motion and gravitation that govern the orbits of the planets. The disk around the Sun, like the disks that astronomers see today surrounding protostars elsewhere in our galaxy, is called a **protoplanetary disk**. The disk probably contained less than 1 percent of the mass of the star forming at its center, but this amount was more than enough to account for the bodies that make up the Solar System today.

### The Collapsing Cloud and Angular Momentum

The Solar System formed from a protoplanetary disk, and similar disks are seen around newly formed stars. *Angular momentum* causes these disks to form. **Angular momentum** is a conserved property of a revolving or rotating system with a value that depends on both the velocity and distribution of the system's mass. The angular momentum of an isolated object is always conserved; that is, it remains unchanged unless acted on by an external force. You have likely seen

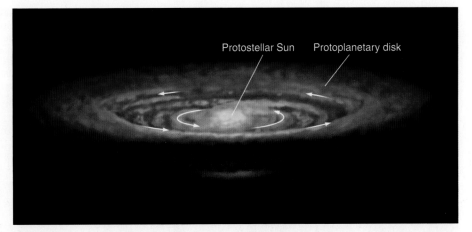

**Figure 7.4** Think of the young Sun as being surrounded by a flat, rotating disk of gas and dust that was flared at its outer edge.

**Figure 7.5** A figure-skater relies on the principle of conservation of angular momentum to change the speed with which she spins.

**Astronomy in Action:** Angular Momentum

a figure-skater spinning on the ice like the one shown in **Figure 7.5**. Like any other rotating object, the spinning ice-skater has some amount of angular momentum. Unless an external force acts on her, such as the ice pushing on her skates, she will always have the same amount of angular momentum.

The amount of angular momentum depends on three factors:

1. How fast the object is rotating. The faster an object is rotating, the more angular momentum it has.

2. The mass of the object. If a bowling ball and a basketball are spinning at the same speed, the bowling ball has more angular momentum because it has more mass.

3. How the mass of the object is distributed relative to the spin axis; that is, how far the object is from the spin axis, or how spread out the object is. For an object of a given mass and rate of rotation, the more spread out it is, the more angular momentum it has. A spread-out object that is rotating slowly might have the same angular momentum as a compact object rotating rapidly.

Both an ice-skater and a collapsing interstellar cloud are affected by **conservation of angular momentum**: the angular momentum must remain the same in the absence of an external force. In order for angular momentum to be conserved, a change in one of the above quantities (the rate of spin, mass, or distribution of mass) must be accompanied by a change of another quantity. For example, an ice-skater can control how rapidly she spins by pulling in or extending her arms or legs. As she pulls in her arms to become more compact, she decreases her distribution of mass and must spin faster to maintain the same angular momentum. When her arms are held tightly in front of her and one leg is wrapped around the other, the skater's spin becomes a blur. She finishes with a flourish by throwing her arms and leg out—an action that abruptly slows her spin by spreading out her mass. Despite the changes in her spin, the skater's angular momentum remains constant throughout the entire maneuver. Similarly—as shown in **Figure 7.6**—the cloud that formed our Sun rotated faster and faster as it collapsed, just as the ice-skater speeds up when she pulls in her arms.

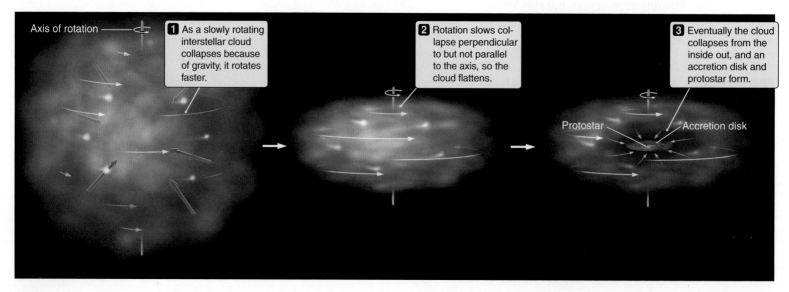

**Figure 7.6** A rotating interstellar cloud collapses in a direction parallel to its axis of rotation, thus forming an accretion disk.

However, this description presents a puzzle. Suppose the Sun formed from a typical cloud—one that was about a light-year across and was rotating so slowly that it took a million years to complete one rotation. By the time such a cloud collapsed to the size of the Sun today, it would have been spinning so fast that one rotation would occur every 0.6 second. This is more than 3 million times faster than our Sun actually spins. At this rate of rotation, the Sun would tear itself apart. It appears that angular momentum was not conserved in the actual formation of the Sun—but that can't be right, because angular momentum must be conserved. We must be missing something. Where did the angular momentum go?

## The Formation of an Accretion Disk

To understand how angular momentum is conserved in disk formation, we must think in three dimensions. Imagine that the ice-skater bends her knees, compressing herself downward instead of bringing her arms toward her body. As she does this, she again makes herself less spread out, but her rate of spin does not change because no part of her body has become any closer to the axis of spin. Similarly, as shown in Figure 7.6, a clump of a molecular cloud can flatten out without speeding up by collapsing parallel to its axis of rotation. Instead of collapsing into a ball, the interstellar cloud flattens into a disk. As the cloud collapses, its self-gravity increases, and the inner parts begin to fall freely inward, raining down on the growing object at the center. The outer portions of the cloud lose the support of the collapsed inner portion, and they start falling inward, too. As this material makes its final inward plunge, it lands on a thin, rotating disk that forms from the accretion of material around a massive object, called an **accretion disk**. The formation of accretion disks is common in the universe.

A visual analogy might be helpful for understanding how interstellar material collects on the accretion disk. Imagine a huge traffic circle, or roundabout, with multiple entrances but with all exits blocked by incoming traffic, as shown in **Figure 7.7a**. As traffic flows into the traffic circle, it has nowhere else to go, resulting in a continual, growing line of traffic driving around and around in an increasingly crowded circle. Eventually, as more and more cars try to pack in, the traffic piles up. This situation is roughly analogous to an accretion disk, shown in Figure 7.7b. Of course, traffic in a roundabout moves on a flat surface, whereas the accretion disk around a protostar forms from material coming in from all directions in three-dimensional space.

As material falls onto the disk, its motion perpendicular to the disk stops abruptly, but its mass motion *parallel* to the surface of the disk adds to the disk's total angular momentum. In this way, the angular momentum of the infalling material is transferred to the accretion disk. The rotating accretion disk has a radius of hundreds of astronomical units and is thousands of times greater than the radius of the star that will eventually form at its center. Therefore, most of the angular momentum in the original interstellar cloud ends up in the accretion disk rather than in the central protostar (see **Working It Out 7.1** for an example of the relevant calculation).

Now we can explain why the Sun does not have the same angular momentum that was present in the original clump of cloud. The radius of a rotating accretion disk is thousands of times greater than the radius of the star that will form at its center. Much of the angular momentum in the original interstellar clump is conserved in its accretion disk rather than in its central protostar.

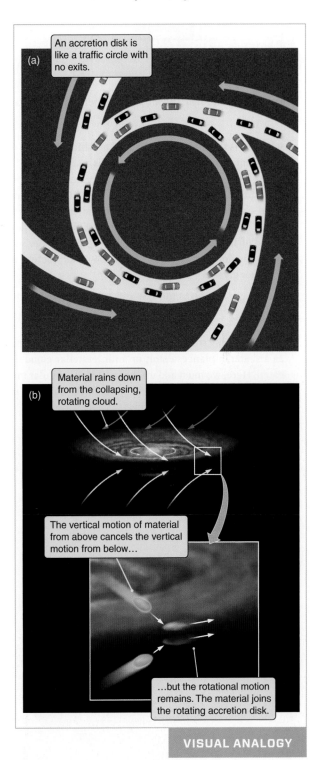

(a) An accretion disk is like a traffic circle with no exits.

(b) Material rains down from the collapsing, rotating cloud.

The vertical motion of material from above cancels the vertical motion from below…

…but the rotational motion remains. The material joins the rotating accretion disk.

**VISUAL ANALOGY**

**Figure 7.7** (a) Traffic piles up in a traffic circle with entrances but no exits. (b) Similarly, gas from a rotating cloud falls inward from opposite sides, piling up onto a rotating disk.

▶‖ **AstroTour:** Traffic Circle Analogy

## 7.1 Working It Out Angular Momentum

In its simplest form, the angular momentum of a system is given by

$$L = m \times v \times r$$

where $m$ is the mass, $v$ is the speed at which the mass is moving, and $r$ represents how spread out the mass is.

As an example, we can apply this relationship to the angular momentum of Jupiter in its orbit about the Sun. The angular momentum from one body orbiting another is called *orbital* angular momentum, $L_{orbital}$. The mass ($m$) of Jupiter is $1.90 \times 10^{27}$ kilograms (kg), the speed of Jupiter in orbit ($v$) is $1.31 \times 10^4$ meters per second (m/s), and the radius of Jupiter's orbit ($r$) is $7.79 \times 10^{11}$ meters. Putting all this together gives

$$L_{orbital} = (1.90 \times 10^{27} \text{ kg}) \times (1.31 \times 10^4 \text{ m/s}) \times (7.79 \times 10^{11} \text{ m})$$

$$L_{orbital} = 1.94 \times 10^{43} \text{ kg m}^2/\text{s}$$

Calculating the *spin* angular momentum of a spinning object, such as a skater, a planet, a star, or an interstellar cloud, is more complicated. Here, we must add up the individual angular momenta of *every tiny mass element* within the object. In the case of a uniform sphere, the spin angular momentum is

$$L_{spin} = \frac{4\pi m R^2}{5P}$$

where $R$ is the radius of the sphere, and $P$ is the rotation period of its spin.

Let's compare Jupiter's orbital angular momentum with the Sun's spin angular momentum to investigate the distribution of angular momentum in the Solar System. Appendix 2 provides the Sun's radius ($6.96 \times 10^8$ meters), mass ($1.99 \times 10^{30}$ kg), and rotation period (24.5 days = $2.12 \times 10^6$ seconds). Assuming that the Sun is a uniform sphere, the spin angular momentum of the Sun is

$$L_{spin} = \frac{4 \times \pi \times (1.99 \times 10^{30} \text{ kg}) \times (6.96 \times 10^8 \text{ m})^2}{5 \times (2.12 \times 10^6 \text{ s})}$$

$$L_{spin} = 1.14 \times 10^{42} \text{ kg m}^2/\text{s}$$

Jupiter's orbital angular momentum is about 17 times greater than the spin angular momentum of the Sun. Most of the angular momentum of the Solar System now resides in the orbits of its major planets.

For a collapsing sphere to conserve spin angular momentum, its rotation period $P$ must be proportional to $R^2$. Like with the skater, when a sphere decreases in radius, its rotation period decreases; that is, it spins faster.

Most of the matter that lands on the accretion disk either becomes part of the star or is ejected back into interstellar space, sometimes in the form of jets or other outflows, as seen in Figure 7.2a. Material swirling in the bipolar jets carries angular momentum away from the accretion disk in the general direction of the poles of the rotation axis. However, a small amount of material is left behind in the disk. It is the objects in this leftover disk—the dregs of the process of star formation—that form planets and other objects that orbit the star. Look again at Figure 7.2 showing images of edge-on accretion disks around young stars. The dark bands are the shadows of the edge-on disks, the top and the bottom of which are illuminated by light from the forming star. Our Sun and Solar System formed from a protostar and disk much like those in these pictures.

### Formation of Large Objects

The chain of events that connects the accretion disk around a young star to a planetary system such as the Solar System begins with random motions of the gas within the protoplanetary disk. As shown in **Figure 7.8**, these motions push the smaller grains of solid material back and forth past larger grains, and as this happens, the smaller grains stick to the larger grains. The "sticking" process among smaller grains is due to the same static electricity that causes dust and hair to cling to plastic surfaces. Starting out at only a few microns ($\mu$m) across—about the size of particles in smoke—the slightly larger bits of dust grow to the size of pebbles and then to clumps the size of boulders, which are not as easily pushed around by gas. When clumps grow to about 100 meters across, the objects are so few and far between that they collide less frequently, and their growth rate slows down but does not stop. Within a protoplanetary disk, the larger dust grains become larger at the expense of the smaller grains.

For two large clumps to stick together rather than explode into many small pieces, they must bump into each other very gently: collision speeds must be about 0.1 m/s or less for colliding boulders to stick together. Your stride is probably about a meter, so to walk as slowly as the collision speed of 0.1 m/s, you would take one step every 10 seconds. The process is not a uniform movement toward larger and larger bodies. Violent collisions do occur in an accretion disk, and larger clumps break back into smaller pieces. But over a long period, large bodies do form.

Objects continue to grow by "sweeping up" smaller objects that get in their way. These objects can eventually measure up to a couple hundred meters across. As the clumps reach the size of about a kilometer, they are massive enough that their gravity begins to pull on nearby bodies, as shown in **Figure 7.9**. These bodies of rock and ice, 100 meters or more in diameter, are known as **planetesimals** ("tiny planets") and eventually combine with each other to form planets. The growth of planetesimals is not fed only by chance collisions with other objects: a planetesimal's gravity can now pull in and capture small objects outside its direct path. The growth of planetesimals speeds up, and larger planetesimals quickly consume most of the remaining bodies in the vicinity of their orbits to become small planets.

---

### CHECK YOUR UNDERSTANDING 7.2

Where does the majority of the angular momentum of the original cloud go?
(a) into the orbital angular momentum of planets; (b) into the star; (c) into the spin of the planets; (d) lost along the jets from the star

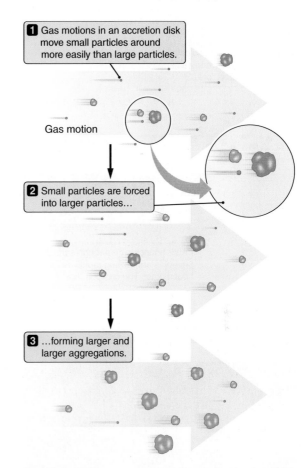

Figure 7.8 Motions of gas in a protoplanetary disk blow smaller particles of dust into larger particles, making the larger particles larger still. This process continues, eventually creating objects many meters in size.

## 7.3 The Inner Disk and Outer Disk Formed at Different Temperatures

In the Solar System, the inner planets are small and mostly rocky, while the outer planets are very large and mostly gaseous. This distinct difference between the inner and the outer Solar System can be explained by how the local disk environment affects the formation process. In this section, we will examine these differences.

### Energy in the Disk

The accretion disks surrounding young stars form from interstellar material that may have a temperature of only a few kelvins, but the disks themselves reach temperatures of hundreds of kelvins or more. Astronomers want to understand what heats up the disk around a forming star so that we can calculate how hot these disks get.

Imagine dumping a box of marbles from the top of a tall ladder onto a rough, hard floor below. The marbles fall, picking up speed as they go. Even though the falling marbles are speeding up, they are all speeding up *together*. If you were riding on one of the marbles, the other marbles would not appear to you to be moving very much; it would be the rest of the room that was whizzing

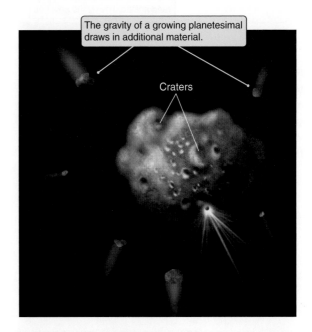

Figure 7.9 The gravity of a planetesimal is strong enough to attract surrounding material, which causes the planetesimal to grow more rapidly.

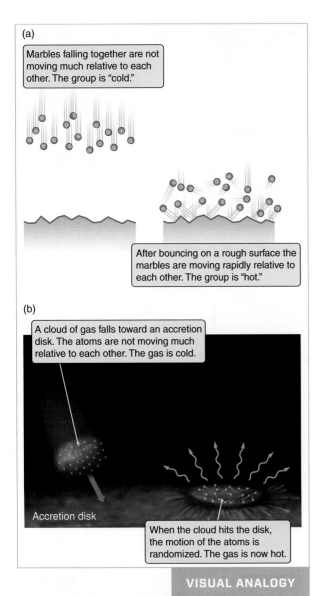

(a)

Marbles falling together are not moving much relative to each other. The group is "cold."

After bouncing on a rough surface the marbles are moving rapidly relative to each other. The group is "hot."

(b)

A cloud of gas falls toward an accretion disk. The atoms are not moving much relative to each other. The gas is cold.

Accretion disk

When the cloud hits the disk, the motion of the atoms is randomized. The gas is now hot.

**VISUAL ANALOGY**

**Figure 7.10** (a) Marbles dropped as a group fall together until they hit a rough floor, at which point their motions become randomized. (b) Similarly, atoms in a gas fall together until they hit the accretion disk, at which point their motions become randomized, raising the temperature of the gas.

by (**Figure 7.10a**). The atoms and molecules in the gas falling toward the protostar are like these marbles. They are picking up speed as they fall as a group toward the protostar, but the gas is still cold because the random thermal velocities of atoms and molecules with respect to each other are still low. Now imagine what happens when the marbles hit the rough floor. They bounce every which way. They are still moving rapidly, but they are no longer moving together. A change has taken place from the ordered motion of marbles falling together to the random motions of marbles traveling in all directions.

Like the falling marble, the atoms and molecules in the gas falling toward the central star scatter in the same fashion when they encounter the uneven gravitational field of the dusty accretion disk. As shown in Figure 7.10b, they are no longer moving as a group, but their random thermal velocities are now very large. The gas is now hot. Similarly, material from the collapsing interstellar cloud falls inward toward the protostar, but because of its angular momentum it misses the protostar and instead falls onto the disk. When this material reaches the disk, its infalling motion comes to an abrupt halt, and the velocity that the atoms and molecules in the gas had before hitting the disk is suddenly converted into random *thermal* velocities instead. The cold gas that was falling toward the disk heats up when it lands on the disk.

Another way to think about why the gas falling on the disk makes the disk hot is to apply another conservation law. The law of **conservation of energy** states that unless energy is added to or taken away from a system from the outside, the total amount of energy in the system must remain constant. But the form the energy takes can change. Imagine you are working against gravity by lifting a heavy object; for example, a brick. It takes energy to lift the brick, and the law of conservation of energy states that energy is never lost. Where does that energy go? The energy is stored and changed into a form called **gravitational potential energy**. If you drop the brick it falls, and as it falls it speeds up. The gravitational potential energy that was stored is converted to energy of motion, which is called **kinetic energy**. When the brick hits the floor, it stops suddenly. The brick loses its energy of motion, so what form does this energy take now?

If the brick cracks, part of the energy goes into breaking the chemical bonds that hold it together. Some of the energy is converted into the sound the brick makes when it hits the floor. Some goes into heating and distorting the floor. But much of the energy is converted into thermal energy. The atoms and molecules that make up the brick are moving about within the brick a bit faster than they were before the brick hit, so the brick and its surroundings, including the floor, grow a tiny bit warmer. Similarly, as gas falls toward the disk surrounding a protostar, gravitational potential energy is converted first to kinetic energy, causing the gas to pick up speed. When the gas hits the disk and stops suddenly, that kinetic energy is turned into thermal energy.

Similarly, material falling onto the accretion disk around a forming star causes the disk to heat up. The amount of heating depends on *where* the material hits the disk. Material hitting the inner part of the disk (the *inner disk*) has fallen farther and picked up greater speed within the gravitational field of the forming star than has material hitting the disk farther out. Like a brick dropped from a tall building, material striking the inner disk is moving quite rapidly when it hits, so it heats the inner disk to high temperatures. In contrast, material falling onto the outer part of the disk (the *outer disk*) is moving much more slowly, like a brick dropped from just a foot or so above the ground. So the temperature at the outermost parts of the disk is not much higher than that of the original interstellar cloud. Stated another way, material falling onto the inner disk converts more gravitational potential energy into thermal energy than does material falling onto the outer disk.

The energy released as material falls onto the disk is not the only source of thermal energy in the disk. Even before the nuclear reactions that will one day power the new star have ignited, conversion of gravitational energy into thermal energy drives the temperature at the surface of the protostar to several thousand kelvins, and it also drives the luminosity of the huge ball of glowing gas to many times the luminosity of the present-day Sun. For the same reasons that Mercury is hot while Pluto is not (see Chapter 5), the radiation streaming outward from the protostar at the center of the disk drives the temperature in the inner parts of the disk even higher, increasing the difference in temperature between the inner and outer parts of the disk.

## The Compositions of Planets

Temperature affects whether or not a material exists in a solid form. On a hot summer day, ice melts and water quickly evaporates; on a cold winter night, water in your breath freezes into tiny ice crystals. Some materials remain solid even at higher temperatures. These include metals and rocky materials, such as iron, **silicates**—which are minerals containing silicon and oxygen—and carbon. Substances that are capable of withstanding high temperatures without melting or being vaporized are called **refractory materials**. Other materials, such as water, ammonia, and methane, remain in a solid form only if their temperature is very low. These materials, which become gases at moderate temperatures, are called **volatile materials** (or *volatiles* for short). Astronomers generally call the solid form of any volatile material an **ice**.

Differences in temperature from place to place within the protoplanetary disk have a significant effect on the makeup of the dust grains in the disk. As **Figure 7.11** illustrates, in the hottest parts of the disk—closest to the protostar—only the most refractory substances can exist in solid form. In the inner disk, dust grains are composed almost entirely of refractory materials. Some substances can survive in solid form somewhat farther out, including some hardier volatiles, such as water ice and certain chemical compounds that are **organic**, meaning that they contain carbon. These solids add to the materials that make up dust grains. In the coldest, outermost parts of the accretion disk, far from the central protostar, highly volatile components such as methane, ammonia, and carbon monoxide ices and other organic molecules survive only in solid form. The different composition of dust grains within the disk determines the composition of the planetesimals formed from the dust. Planets that form closer to the central star tend to be made up mostly of refractory materials such as rock and metals. Planets that form farther from the central star contain refractory materials, but they also contain large quantities of ices and organic materials.

In the Solar System, the inner planets are composed of rocky material surrounding metallic cores of iron and nickel. Objects in the outer Solar System, including moons, giant planets, and comets, are composed largely of ices of various types. But not all planetary systems are so neatly organized as our Solar System. When planets around other stars were first discovered, they appeared to be very different, with large planets close in to their respective stars. Astronomers now think that **chaotic** encounters, in which a small change in the initial state of a system can lead to a large change in the final state of the system, may change the organization of planetary compositions. In a process called **planet migration**, the

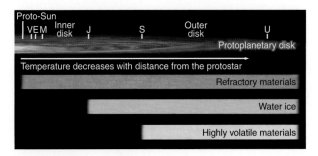

**Figure 7.11** Differences in temperature within a protoplanetary disk determine the composition of dust grains that then evolve into planetesimals and planets. The colored bars show that refractory materials are found throughout the disk, while water ice is found only outside Jupiter's orbit, and highly volatile materials are found only outside Saturn's orbit. Shown here are the proto-Sun (PS) and the orbits of Venus (V), Earth (E), Mars (M), Jupiter (J), Saturn (S), and Uranus (U).

force of gravity from all of the nearby objects can move some planets so that they end up far from the place of their birth. For example, in our Solar System, Uranus and Neptune originally may have formed nearer to the orbits of Jupiter and Saturn, but were then driven outward to their current locations by gravitational encounters with Jupiter and Saturn. A planet can also migrate when it gives up some of its orbital angular momentum to the disk material that surrounds it. Such a loss of angular momentum causes the planet slowly to spiral inward toward the central star. Thus, the order of planets in a system can change over time.

## Formation of an Atmosphere

Once a solid planet has formed, it may continue growing by capturing gas from the protoplanetary disk. To do so, it must act quickly. Young stars and protostars emit fast-moving particles and intense radiation that can quickly disperse the gaseous remains of the accretion disk. Gaseous planets such as Jupiter probably have only about 10 million years or so to form and to grab whatever gas they can. Because of their strong gravitational fields, more massive young planets can capture more of the hydrogen and helium gas that makes up the bulk of the disk. What follows is much like the formation of a star and protoplanetary disk, but on a smaller scale. Just as happens in the accretion disk around the star, gas from a mini accretion disk moves inward and falls onto the planet.

The gas that is captured by a planet at the time of its formation is called the planet's **primary atmosphere**. The primary atmosphere of a large planet can be more massive than the solid body, as in the case of Jupiter. Some of the solid material in the mini accretion disk might stay behind to coalesce into larger bodies in much the same way that particles of dust in the protoplanetary disk came together to form planets. The result is a mini "solar system"—a group of moons that orbit about the planet.

A less massive planet may also capture some gas from the protoplanetary disk, only to lose it later. The gravity of small planets may be too weak to hold low-mass gases such as hydrogen or helium. Even if a small planet is able to gather some hydrogen and helium from its surroundings, this primary atmosphere will not last long. In the inner solar system, the temperatures are higher, so the hydrogen and helium atoms are moving faster than in the outer solar system and will escape from a small planet. The atmosphere that remains around a small planet like Earth is a **secondary atmosphere**, which forms later in the life of a planet. Volcanism is one important source of a secondary atmosphere because it releases heavier and thus slower-moving gases such as carbon dioxide, water vapor, and other gases from the planet's interior. In addition, volatile-rich comets that formed in the outer parts of the disk continue to fall inward toward the new star long after its planets have formed, and they sometimes collide with planets. **Comets** are icy planetesimals that survive planetary accretion. They may provide a significant source of water, organic compounds, and other volatile materials on planets close to the central star.

### CHECK YOUR UNDERSTANDING 7.3

In our Solar System, the inner planets are rocky because: (a) the original cloud had more rocky material near the center; (b) warm temperatures in the inner disk caused the inner planetesimals to be formed of only rocky material; (c) the inner disk filled a smaller volume so it was denser; (d) the hydrogen and helium atoms were too low mass to remain in the inner disk.

# 7.4 The Formation of Our Solar System

We have seen that nearly 5 billion years ago, the Sun was still a protostar surrounded by a protoplanetary disk of gas and dust. During the next few hundred thousand years, much of the dust in the disk had collected into planetesimals—clumps of rock and metal near the emerging Sun and aggregates of rock, metal, ice, and organic materials farther from the Sun. In this section, we will look at the formation of the different types of planets in our own Solar System.

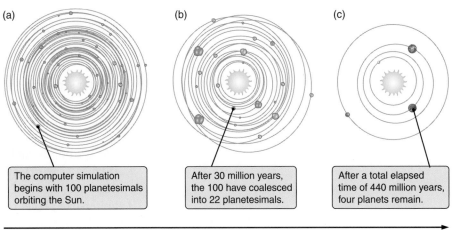

(a) The computer simulation begins with 100 planetesimals orbiting the Sun.

(b) After 30 million years, the 100 have coalesced into 22 planetesimals.

(c) After a total elapsed time of 440 million years, four planets remain.

Time

**Figure 7.12** Computer models simulate how material in the protoplanetary disk became clumped into the planets over time. Only a few planets remain at the end.

## The Terrestrial Planets

Within the inner 5 astronomical units (AU) of the disk, several rock and metal planetesimals quickly grew in size to become the dominant masses in their orbits. With their ever-strengthening gravitational fields, they either captured most of the remaining planetesimals or ejected them from the inner part of the disk. **Figure 7.12** shows some results from a computer simulation of how this might have happened. The dominant planetesimals became planet-sized bodies with masses ranging between 5 percent and 100 percent of Earth's mass. These dominant planetesimals evolved into the **terrestrial planets**, which are planets that are Earth-like, or rocky. Today, the surviving terrestrial planets are Mercury, Venus, Earth, and Mars. Earth's Moon is often grouped with these terrestrial planets because of its similar physical and geological properties, even though it is not a planet itself and formed in a different way. One or two other planets or large moons may have formed in the young Solar System but were later destroyed.

For several hundred million years after the formation of the four surviving terrestrial planets, leftover pieces of debris still in orbit around the Sun continued to rain down on the surfaces of these planets. Today, we can still see the scars of these early impacts on the cratered surfaces of all the terrestrial planets (**Figure 7.13**). This rain of debris continues even today, but at a much lower rate.

Before the proto-Sun became a true star, gas in the inner part of the protoplanetary disk was still plentiful. During this early period the two larger terrestrial planets, Earth and Venus, may have held on to weak primary atmospheres of hydrogen and helium, but these thin atmospheres were soon lost to space. The terrestrial planets did not develop thick atmospheres until the formation of the secondary atmospheres that now surround Venus, Earth, and Mars. Mercury's proximity to the Sun and the Moon's small mass prevented these bodies from retaining significant secondary atmospheres.

## The Giant Planets

Beyond 5 AU from the Sun, in a much colder part of the accretion disk, planetesimals combined to form a number of bodies with masses about 5–20 times that of Earth. These planet-sized objects formed from planetesimals containing volatile ices and organic compounds in addition to rock and metal. In a process astronomers call *core accretion–gas capture*, mini accretion disks formed around

**Figure 7.13** Large impact craters on Mercury (and on other solid bodies throughout the Solar System) record the final days of the Solar System's youth, when planets and planetesimals grew as smaller planetesimals rained down on their surfaces.

these planetary cores, capturing massive amounts of hydrogen and helium and funneling this material onto the planets. Four such massive bodies became the cores of the **giant planets**—Jupiter, Saturn, Uranus, and Neptune. These giant planets are many times the mass of any terrestrial planet.

Jupiter's massive solid core captured and retained the most gas—roughly 300 times the mass of Earth (300 $M_{Earth}$). The other outer planetary cores captured less hydrogen and helium, perhaps because their cores were less massive or because there was less gas available to them. Saturn ended up with less than 100 $M_{Earth}$ of gas, and Uranus and Neptune were able to grab less than twenty Earth masses worth of gas.

The core accretion model indicates that it could take up to 10 million years for a Jupiter-like planet to accumulate. Some planetary scientists do not think that our protoplanetary disk could have survived long enough to form gas giants such as Jupiter through the general process of core accretion. All the gas may have dispersed in roughly half that time, cutting off Jupiter's supply of hydrogen and helium. An alternative explanation is a process called *disk instability*, in which the protoplanetary disk suddenly and quickly fragments into massive clumps equivalent to those of a large planet. It is possible that both core accretion and disk instability played a role in the formation of our own and other planetary systems.

During the formation of the planets, gravitational energy was converted into thermal energy as the individual atoms and molecules moved faster. This conversion warmed the gas surrounding the cores of the giant planets. Proto-Jupiter and proto-Saturn probably became so hot that they actually glowed a deep red color, similar to the heating element on an electric stove. Their internal temperatures may have been even higher.

Some of the material remaining in the mini accretion disks surrounding the giant planets combined into small bodies, which became moons. A **moon** is any natural satellite in orbit about a planet or asteroid. The composition of the moons that formed around the giant planets followed the same trend as that of the planets that formed around the Sun: the innermost moons formed under the hottest conditions and therefore contained the smallest amounts of volatile material. For example, the closest of Jupiter's many moons may have experienced high temperatures from nearby Jupiter glowing so intensely that it would have evaporated most of the volatile substances in the inner part of its mini accretion disk.

## Remaining Planetesimals

Not all planetesimals in the disk went on to become planets. For example, dwarf planets orbit the Sun but have not cleared other, smaller bodies from their orbits. Ceres and Pluto, which are shown in Figure 7.1, are dwarf planets. More dwarf planets, along with a large number of smaller bodies, are found in the Kuiper Belt, beyond Pluto's orbit. Asteroids are small bodies found interior to Jupiter's orbit; most are located in the main asteroid belt between Mars and Jupiter. Jupiter's gravity kept the region between Jupiter and Mars so stirred up that most planetesimals there never formed a large planet.

Planetesimals persist to this day in the outermost part of the Solar System as well. Formed in a deep freeze, these objects have retained most of the highly volatile materials found in the grains present at the formation of the accretion disk. Unlike the crowded inner part of the disk, the outermost parts of the disk had planetesimals that were too sparsely distributed for large planets to grow. Icy planetesimals in the outer Solar System that survived planetary accretion remain today as **comet nuclei**. The frozen, distant dwarf planets Pluto and Eris are especially large examples of these residents of the outer Solar System.

Many Solar System objects show evidence of cataclysmic impacts that reshaped worlds, suggesting that the early Solar System must have been a remarkably violent and chaotic place. The dramatic difference in the terrain of the northern and southern hemispheres on Mars, for example, has been interpreted as the result of one or more colossal collisions. The leading theory for the origin of our Moon is that it resulted from the collision of an object with Earth. Mercury has a crater on its surface from an impact so devastating that it caused the crust to buckle on the opposite side of the planet. In the outer Solar System, one of Saturn's moons, Mimas, has a crater roughly one-third the diameter of the moon itself. Uranus suffered one or more collisions that were violent enough literally to knock the planet on its side. Today, as a result, its equatorial plane is tilted at almost a right angle to its orbital plane. We will see other examples in subsequent chapters.

### CHECK YOUR UNDERSTANDING 7.4

Suppose that astronomers found a rocky, terrestrial planet beyond the orbit of Neptune. What is the most likely explanation for its origin? (a) It formed close to the Sun and migrated outward. (b) It formed in that location and was not disturbed by migration. (c) It formed later in the Sun's history than other planets. (d) It is a captured planet that formed around another star.

## 7.5 Planetary Systems Are Common

When astronomers turn their telescopes to young nearby stars, they see disks of the same type from which the Solar System formed. As illustrated in **Figure 7.14**, when the light from the central star is blocked, evidence of the planetary disk is observed. The physical processes that led to the formation of the Solar System should be commonplace wherever new stars are being born. Compared to stars, however, planets are small and dim objects. They shine primarily by reflection and therefore are millions to billions of times fainter than their host stars. Thus, they were difficult to identify until advances in telescope detector technology enabled astronomers to discover them in the 1990s through indirect methods. In 1995, astronomers announced the first confirmed **extrasolar planet**, also called an **exoplanet**—a planet orbiting around a star other than the Sun. Today, the number of known extrasolar planets has grown to the thousands, and new discoveries are occurring almost daily.

The discovery of extrasolar planets raised the question of what we mean by the term *planet*. The full International Astronomical Union (IAU) definitions for planets and dwarf planets within the Solar System are provided in Appendix 9. The IAU defines an extrasolar planet as an object that orbits a star and has a mass less than 13 Jupiter masses (13 $M_{Jup}$). Objects more massive than 13 $M_{Jup}$ but less massive than 0.08 solar masses (0.8 $M_{Sun}$; about 80 $M_{Jup}$) are **brown dwarfs**. Objects more massive than 0.08 $M_{Sun}$ are defined as stars. **Figure 7.15** compares the diameters of these different objects.

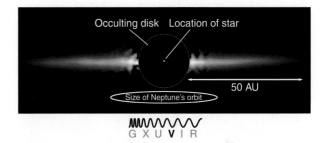

Occulting disk    Location of star

50 AU

Size of Neptune's orbit

G X U **V** I R

**Figure 7.14** An edge-on circumstellar dust disk is seen extending outward to 60 AU from the young (12-million-year-old) star AU Microscopii. The star itself, whose brilliance would otherwise overpower the circumstellar disk, is hidden behind an occulting disk (opaque mask) placed in the telescope's focal plane. Its position is represented by the dot.

Sun

Low-mass star

Brown dwarf

Jupiter

Earth

**Figure 7.15** A comparison of the diameters of the Sun, a low-mass star, a brown dwarf, Jupiter, and Earth.

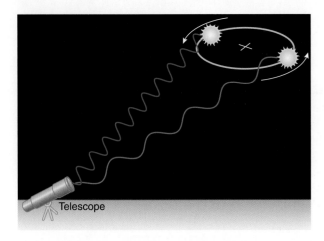

**Figure 7.16** Doppler shifts observed in the spectrum of a star are due to the wobble of the star caused by its planet.

 **Nebraska Simulation:** Influence of Planets on the Sun

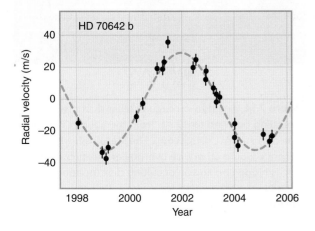

**Figure 7.17** Radial velocity data for a star with a planet. A positive number is motion away from the observer; a negative number is motion toward the observer. The plot repeats as the planet completes another orbit around the star.

 **Nebraska Simulation:** Radial Velocity Graph

**Astronomy in Action:** Doppler Shift

## The Search for Extrasolar Planets

Currently, more than 100 projects are focused on searching for extrasolar planets from the ground and from space. The first planets were discovered indirectly, by observing their gravitational tug on the central star. As technology has improved, other methods have become more productive. Astronomers now have direct images of planets orbiting stars and have also been able to take spectra of planets to observe the composition of their atmospheres. Almost certainly, between the time we write this and the time you read it, there will be new discoveries. The field is advancing extremely quickly. We will now look at each discovery method.

**The Radial Velocity Method**  As a planet orbits a star, the planet's gravity tugs the star around ever so slightly. This motion toward or away from us, its radial velocity, creates an observable Doppler shift in the spectrum of the star. **Figure 7.16** illustrates this method. When the star is moving toward us (negative radial velocity), the light is blueshifted; when the star is moving away from us (positive radial velocity), the light is redshifted. This pattern of radial velocity repeats over time. After detecting this wobble (**Figure 7.17**), astronomers can infer the planet's mass and its distance from the star.

We can see how this works by using the Solar System as an example. Jupiter's mass is greater than the mass of all the other planets, asteroids, and comets combined, so this is the planet most likely to be detected. Both the Sun and Jupiter orbit a common center of gravity (sometimes called center of mass; this is the location where the effect of one mass balances the other) that lies just outside the surface of the Sun, as shown in **Figure 7.18**. Alien astronomers would find that the Sun's radial velocity varies by ±12 m/s, with a period equal to Jupiter's orbital period of 11.86 years. From this information, the astronomers would rightly conclude that the Sun has at least one planet with a mass comparable to Jupiter's but, without greater precision, would be unaware of the other less massive major planets. If the alien astronomers could improve the sensitivity of their instruments to measure radial velocities as small as 2.7 m/s, Saturn would be detectable, and if the precision of their spectrograph extended to radial velocities as small as 0.09 m/s, Earth would be detectable.

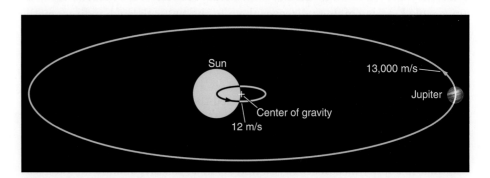

**Figure 7.18** The Sun and Jupiter orbit around a common center of gravity, which lies just outside the Sun's surface. Spectroscopic measurements made by an extrasolar astronomer would reveal the Sun's radial velocity varying by ±12 m/s over 11.86 years, which is Jupiter's orbital period. Jupiter travels around its orbit at a speed of 13,000 m/s. (The orbit is shown in perspective and is not actually very elliptical.)

## 7.2   Working It Out   Estimating the Size of the Orbit of a Planet

In the spectroscopic radial velocity method, the star is moving about its center of mass, and its spectral lines are Doppler-shifted accordingly. Recall from Figure 7.18 that the alien astronomer looking toward the Solar System would observe a shift in the wavelengths of the Sun's spectral lines—caused by the presence of Jupiter—of about 12 m/s.

Figure 7.17 showed the radial velocity data for a star with a planet discovered by this method. How do astronomers use this method to estimate the distance ($A$) of the planet from the star? Recall from Chapter 4 that Newton generalized Kepler's law relating the period of an object's orbit to the orbital semimajor axis:

$$P^2 = \frac{4\pi^2}{G} \times \frac{A^3}{M}$$

where $A$ is the semimajor axis of the orbit, $P$ is its period, and $M$ is the combined mass of the two objects. To find $A$, we rearrange the equation as follows:

$$A^3 = \frac{G}{4\pi^2} \times M \times P^2$$

From the graph of radial velocity observations in Figure 7.17, we can determine that the period of the orbit is 5.7 years. There are $3.16 \times 10^7$ seconds in a year, so

$$P = 5.7 \text{ yr} \times (3.16 \times 10^7 \text{ s/yr})$$

$$P = 1.8 \times 10^8 \text{ s}$$

The mass of the star is much greater than the mass of the planet, so the combined masses of the star and the planet can be approximated as the mass of the star, which in this case is about equal to the mass of the Sun, $2 \times 10^{30}$ kg. (Stellar masses can be estimated from their spectra.) The gravitational constant $G = 6.67 \times 10^{-20}$ km³/(kg s²). Putting in the numbers gives

$$A^3 = \frac{6.67 \times 10^{-20} \dfrac{\text{km}^3}{\text{kg s}^2}}{4\pi^2} \times (2 \times 10^{30} \text{ kg}) \times (1.8 \times 10^8 \text{ s})^2$$

$$A^3 = 1.1 \times 10^{26} \text{ km}^3$$

Taking the cube root of $1.1 \times 10^{26}$ km³ solves for $A$, which is equal to $4.8 \times 10^8$ km. To get a better feel for this number, we might put it into astronomical units (where 1 AU = $1.5 \times 10^8$ km). The semimajor axis of the orbit of this planet is given by

$$A = \frac{4.8 \times 10^8 \text{ km}}{1.5 \times 10^8 \text{ km/AU}} = 3.2 \text{ AU}$$

This planet is more than 3 times farther from its star than Earth is from the Sun.

 **Nebraska Simulation:** Exoplanet Radial Velocity Simulator

---

Current technology limits the precision of radial velocity instruments to about 0.3 m/s, but to date it has been the most successful ground-based approach to finding extrasolar planets. This technique enables astronomers to detect giant planets around solar-type stars, but not yet to find planets with masses similar to Earth's. Finding the signal of the Doppler shift in the noise of the observation requires the star to be quite bright in our sky. So this method is limited to nearby stars, within about 160 light-years from Earth. **Working It Out 7.2** provides additional explanation of the spectroscopic radial velocity method.

 **Nebraska Simulation:** Exoplanet Transit Simulator

**The Transit Method**   Another technique for finding extrasolar planets is the **transit method**, in which we observe the effect of a planet passing in front of its parent star. From Earth it is sometimes possible to see the inner planets Mercury and Venus transit, or pass in front of, the Sun. An alien located somewhere in the plane of Earth's orbit would see Earth pass in front of the Sun and could infer the existence of Earth by detecting the 0.009 percent drop in the Sun's brightness during the transit. Similarly, for astronomers on Earth to observe a planet passing in front of a star, Earth must lie nearly in the orbital plane of that planet. When an extrasolar planet passes in front of its parent star, the light from the star diminishes by a tiny amount, as seen in **Figure 7.19**. Whereas the radial velocity method gives us the mass of the planet and its orbital distance from a star, the transit method provides the size of a planet. **Working It Out 7.3** demonstrates how the radii are estimated.

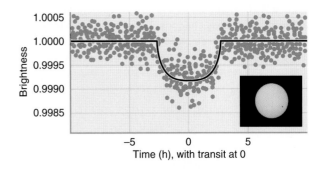

**Figure 7.19** The data show the light curve for Kepler-11c. The inset photograph shows Venus passing in front of the Sun in June 2012, similar to this transit of Kepler-11c.

## 7.3  Working It Out  Estimating the Radius of an Extrasolar Planet

The masses of extrasolar planets can often be estimated using Kepler's laws and the conservation of angular momentum. When planets are detected by the transit method, astronomers can estimate the radius of an extrasolar planet. In this method, astronomers look for planets that eclipse their stars and observe how much the star's light decreases during this eclipse (see Figure 7.19). In the Solar System when Venus or Mercury transits the Sun, a black circular disk is visible on the face of the circular Sun. During the transit, the amount of light from the transited star is reduced by the area of the circular disk of the planet divided by the area of the circular disk of the star:

$$\text{Percentage reduction in light} = \frac{\text{Area of disk of planet}}{\text{Area of disk of star}}$$

$$= \frac{\pi R_{\text{planet}}^2}{\pi R_{\text{star}}^2} = \frac{R_{\text{planet}}^2}{R_{\text{star}}^2}$$

Then, to solve for the radius of the planet, astronomers need an estimate of the radius of the star and a measurement of the percentage reduction in light during the transit. The radius of a star is estimated from the surface temperature and the luminosity of the star.

Let's consider an example. Kepler-11 is a system of at least six planets that transit a star. The radius of the star, $R_{\text{star}}$, is estimated to be 1.1 times the radius of the Sun, or $1.1 \times (7.0 \times 10^5 \text{ km}) = 7.7 \times 10^5 \text{ km}$. The light from planet Kepler-11c is observed to decrease by 0.077 percent, or 0.00077 (see Figure 7.19). What is Kepler-11c's size?

$$0.00077 = \frac{R_{\text{Kepler-11c}}^2}{R_{\text{star}}^2} = \frac{R_{\text{Kepler-11c}}^2}{(7.7 \times 10^5 \text{ km})^2}$$

$$R_{\text{Kepler-11c}}^2 = 4.5 \times 10^8 \text{ km}^2$$

$$R_{\text{Kepler-11c}} = 2.1 \times 10^4 \text{ km}$$

Dividing Kepler-11c's radius by the radius of Earth (6,400 km) shows that the planet Kepler-11c has a radius of $3.3\,R_{\text{Earth}}$.

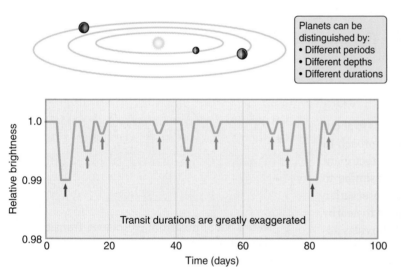

**Figure 7.20** Multiple planets can be detected by multiple transits with different brightness changes. The arrows point to the changes in the total light as the three planets transit the star.

More than a thousand extrasolar planets have been detected from ground-based and space telescopes using the transit method. Current ground-based technology limits the sensitivity of the transit method to about 0.1 percent of a star's brightness. Amateur astronomers have confirmed the existence of several extrasolar planets by observing transits using charge-coupled device (CCD) cameras mounted on telescopes with apertures as small as 20 centimeters (cm). Telescopes in space improve the sensitivity because smaller dips in brightness can be measured. The small French COROT telescope (27 cm) discovered 32 planets during its 6 years of operation (2007–2013). NASA's 0.95-meter Kepler telescope has discovered many planets and has found thousands more candidates that are being investigated further. **Figure 7.20** illustrates how multiplanet systems are identified with this method: if one planet is found, then observations of the variations in timing of the transit can indicate that there are other planets orbiting the same star.

**Microlensing** The gravitational field of an unseen planet can act like a lens, bending the light from a distant star in such a way that it causes the star to brighten temporarily while the planet is passing in front of it. Because the effect is small, it is usually called microlensing. Like the radial velocity method, microlensing provides an estimate of the mass of the planet. To date, several dozen extrasolar planets have been found with this technique.

**The Astrometric Method** Planets may also be detected by astrometry—precisely measuring the position of a star in the sky. If the system is viewed from "above," the star moves in a mini-orbit as the planet pulls it around. This motion is generally tiny and therefore very difficult to measure. However, for systems viewed from above the plane of the planet's orbit, none of the prior methods will work because the planet neither passes in front of the star nor causes a shift in its speed along the line of sight. Space missions such as the Gaia observatory, launched in 2013 by the European Space Agency, is conducting observations of this kind.

**Direct Imaging** Direct imaging means taking a picture of the planet directly. This technique is conceptually straightforward but is technically difficult because it involves searching for a relatively faint planet in the overpowering glare of a bright star—a challenge far more difficult than looking for a star in a clear, bright daytime sky. Even when an object is detected by direct imaging, an astronomer must still determine whether the observed object is actually a planet. Suppose we detect a faint object near a bright star. Could it be a more distant star that just happens to be in the line of sight? Future observations could tell if the object shares the bright star's motion through space. But it could also be a brown dwarf rather than a true planet. An astronomer would need to make further observations to determine the object's mass.

Some planets have been discovered by this method with large ground-based telescopes operating in the infrared region of the spectrum using adaptive optics. **Figure 7.21** is an infrared image of Beta Pictoris b. The first visible-light discovery was made from space while the Hubble Space Telescope was observing Fomalhaut, a bright naked-eye star only 25 light-years away. The planet Fomalhaut b is shown in **Figure 7.22**. It has a mass no more than 3 times that of Jupiter and orbits within a dusty debris ring about 17 billion km from the central star. A related form of direct observation involves separating the spectrum of a planet from the spectrum of its star to obtain information about the planet directly. Large ground-based telescopes have been able to obtain spectra of the atmospheres of some extrasolar planets and have found, for example, carbon monoxide and water in these atmospheres.

## The Discovery of Extrasolar Planets

Searches for extrasolar planets have been remarkably successful. Between the discovery of the first (in 1995) and this writing, nearly 2,000 more have been confirmed, and thousands more candidates are under investigation. As the number of observed systems with single and multiple planets increases, astronomers can compare them with those of the Solar System, and they have found more variation than they expected. The field is changing so fast that the most up-to-date information can only be found online or through mobile applications such as the Kepler App.

The first discoveries included many **hot Jupiters**, which are Jupiter-sized planets orbiting solar-type stars in circular or highly eccentric orbits that bring them closer to their parent stars than Mercury is to our own Sun. These planets were among the first to be detected because they are relatively easy targets for the spectroscopic radial velocity method. The large mass of a nearby hot Jupiter tugs the star very hard, creating large radial velocity variations in the star. In addition, these large planets orbiting close to their parent stars are more likely to pass in front of the star periodically and reveal themselves via the transit method. Therefore, these hot Jupiter systems are easier to find than smaller, more distant planets. Astronomers realized that these hot Jupiter systems are not representative of most planetary systems; they were just easier to find. Scientists call this bias a *selection effect.*

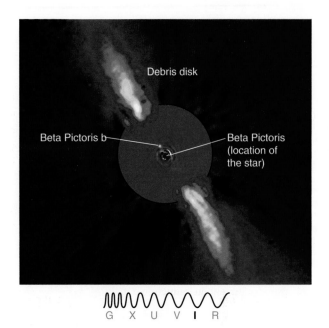

**Figure 7.21** Beta Pictoris b is seen orbiting within a dusty debris disk that surrounds the bright naked-eye star Beta Pictoris. The planet's estimated mass is 8 times that of Jupiter. The star is hidden behind an opaque mask, and the planet appears through a semitransparent mask used to subdue the brightness of the dusty disk.

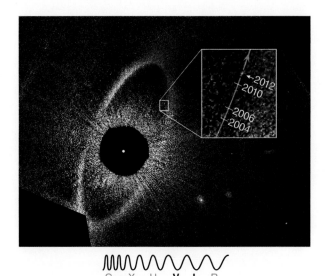

**Figure 7.22** A Hubble Space Telescope image of Fomalhaut b, seen here moving in its orbit around Fomalhaut, a nearby star easily visible to the naked eye. The parent star, hidden by an obscuring mask, is about a billion times brighter than the planet, which is located within a dusty debris ring that surrounds the star.

Astronomers were surprised by the hot Jupiters because, according to the planetary system formation theory available at the time (based only on the Solar System), these giant, volatile-rich planets should not have been able to form so close to their parent stars. The expectation was that Jupiter-type planets should form in the more distant, cooler regions of the protoplanetary disk, where the volatiles that make up much of their composition are able to survive. So astronomers suggested that perhaps hot Jupiters formed much farther away from their parent stars and subsequently migrated inward to a closer orbit. The mechanism by which a planet could migrate over such a distance must involve an interaction with gas or planetesimals in which orbital angular momentum is somehow transferred from the planet to its surroundings, allowing it to spiral inward.

Many of the new planets being discovered by Kepler are mini-Neptunes (gaseous planets with masses of 2–10 $M_{Earth}$) or super-Earths (rocky planets more massive than Earth), but each month brings an announcement of the discovery of smaller planets. Planets with longer orbital periods, and therefore larger orbits, can be discovered only when the observations have gone on long enough to observe more than one complete orbit. Some of the extrasolar planets have highly elliptical orbits compared with those in the Solar System. Planets have been found with orbits that are highly tilted compared with the plane of the rotation of their star, and some planets move in orbits whose direction is opposite that of their star's rotation. Multiple-planet systems have been observed in which the larger mini-Neptunes alternate with smaller super-Earths. The multiple-planet systems that have been found by the transit method reside in flat systems like our own, offering further evidence that the planets formed in a flat protoplanetary disk around a young star. But the current hypothesis to explain the Solar System's inner, small rocky planets and outer, large gaseous planets may not be applicable in these other planetary systems.

In addition, some planets detected by microlensing seem to be wandering freely through the Milky Way. These planets may have been ejected from their solar systems after their formation and are no longer in gravitationally bound orbits around their stars. The frequent new discoveries requiring revisions of existing theories make extrasolar planets one of the most exciting topics in astronomy today.

### CHECK YOUR UNDERSTANDING 7.5

Suppose you hear of the discovery of an Earth-mass planet around a star. This planet was most likely discovered through the _____ method. (a) Doppler spectroscopy; (b) direct imaging; (c) transit; (d) astrometric

# Origins

## Kepler's Search for Earth-Sized Planets

The discovery of planetary systems, many different from the Solar System, shows us that the formation of planets frequently, and perhaps always, accompanies the formation of stars. The implications of this conclusion are profound. Planets are a common by-product of star formation. In a galaxy of 200 billion stars and a universe of hundreds of billions of galaxies, how many planets (and also moons) might exist? And with all of these planets in the universe, how many might have suitable conditions for the particular category of chemical reactions that we refer to as "life"? (We will return to this point in Chapter 24.)

The Kepler Mission was developed by NASA to find Earth-sized and larger planets in orbit about a variety of stars. Kepler is a 1-meter telescope with 42 CCD detectors and is designed to observe approximately 150,000 stars in 100 square degrees of sky and look for planetary transits. To confirm a planetary detection, the transits need to be observed three times with repeatable changes in brightness, duration of transit times, and computed orbital period.

Kepler can detect a dip in the brightness of a star of 0.01 percent—which is sensitive enough to detect an Earth-sized planet. Kepler identified the first Earth-sized planets in 2011. Stars with transiting planets detected by Kepler are also observed spectroscopically to obtain radial velocity measurements that can lead to an estimate of a planet's mass. If a planet's radius and mass are known, the planet's density (mass per volume) can be estimated, too. From the density, astronomers can get a sense of whether the planet is composed primarily of gas, rock, ice, water, or a mixture of some of these.

On Earth, liquid water was essential for the formation and evolution of life. Because life on Earth is the only example of life for which we have evidence, we do not know whether liquid water is a cosmic requirement, but it is a place to start. The primary scientific goal of the Kepler Mission is to look for rocky planets at the right distance from their stars to permit the existence of liquid water, a distance known as the **habitable zone**. If a planet is too close to its star, water will exist only as

a vapor; if it is too far, water will be frozen as ice. In the Solar System, Earth is the only planet in the habitable zone. Although announcements of new planets often state whether the planet is in the habitable zone, just being in the zone doesn't guarantee that the planet actually has liquid water—or that the planet is inhabited! An example of an Earth-sized planet in a habitable zone is shown in **Figure 7.23**.

In 2013, Kepler suffered a mechanical failure that stopped observations, but a work-around was approved, and observations resumed in 2014. Kepler has identified thousands of planet candidates, some in the habitable zones of their respective stars. The candidates must be confirmed by follow-up observations of more transits or of radial velocities before they are officially announced as planet detections. Amateur astronomers can access the candidate lists online (at the "Exoplanet Transit Database") and conduct their own observations. Anyone with Internet access can go to planethunters.org, examine some Kepler data, and contribute to the search.

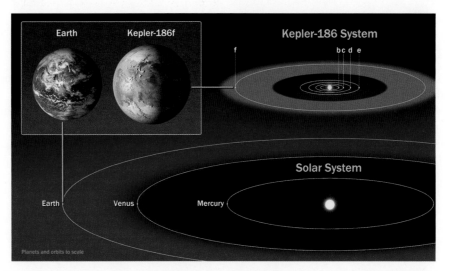

**Figure 7.23** An artist's conception of the Kepler-186 system. Located about 500 light-years from Earth, this system has a planet in its habitable zone. The Solar System is shown for comparison.

# Earth-Size Planet Found in the "Habitable Zone" of Another Star

By **Science@NASA**

Using NASA's Kepler space telescope, astronomers have discovered the first Earth-size planet orbiting in the "habitable zone" of another star (see Figure 7.23). The planet, named "Kepler-186f," orbits an M dwarf, or red dwarf, a class of stars that makes up 70 percent of the stars in the Milky Way Galaxy. The discovery of Kepler-186f confirms that planets the size of Earth exist in the habitable zone of stars other than our Sun.

The "habitable zone" is defined as the range of distances from a star where liquid water might pool on the surface of an orbiting planet. While planets have previously been found in the habitable zone, the previous finds are all at least 40 percent larger in size than Earth, and understanding their makeup is challenging. Kepler-186f is more reminiscent of Earth.

Kepler-186f orbits its parent M dwarf star once every 130 days and receives one-third the energy that Earth gets from the Sun, placing it nearer the outer edge of the habitable zone. On the surface of Kepler-186f, the brightness of its star at high noon is only as bright as our Sun appears to us about an hour before sunset.

"M dwarfs are the most numerous stars," said Elisa Quintana, research scientist at the SETI Institute at NASA's Ames Research Center in Moffett Field, California, and lead author of the paper published today in the journal *Science*. "The first signs of other life in the galaxy may well come from planets orbiting an M dwarf."

However, "being in the habitable zone does not mean we know this planet is habitable," cautions Thomas Barclay, a research scientist at the Bay Area Environmental Research Institute at Ames, and coauthor of the paper. "The temperature on the planet is strongly dependent on what kind of atmosphere the planet has. Kepler-186f can be thought of as an Earth-cousin rather than an Earth-twin. It has many properties that resemble Earth."

Kepler-186f resides in the Kepler-186 system, about 500 light-years from Earth in the constellation Cygnus. The system is also home to four companion planets: Kepler-186b, Kepler-186c, Kepler-186d, and Kepler-186e, whiz around their sun every four, seven, 13, and 22 days, respectively, making them too hot for life as we know it. These four inner planets all measure less than 1.5 times the size of Earth.

Although the size of Kepler-186f is known, its mass and composition are not. Previous research, however, suggests that a planet the size of Kepler-186f is likely to be rocky.

"The discovery of Kepler-186f is a significant step toward finding worlds like our planet Earth," said Paul Hertz, NASA's Astrophysics Division director at the agency's headquarters in Washington.

The next steps in the search for distant life include looking for true Earth-twins—Earth-size planets orbiting within the habitable zone of a Sun-like star—and measuring their chemical compositions. The Kepler space telescope, which simultaneously and continuously measured the brightness of more than 150,000 stars, is NASA's first mission capable of detecting Earth-size planets around stars like our Sun.

Looking ahead, Hertz said, "future NASA missions, like the Transiting Exoplanet Survey Satellite and the James Webb Space Telescope, will discover the nearest rocky exoplanets and determine their composition and atmospheric conditions, continuing humankind's quest to find truly Earth-like worlds."

1. This NASA press release was picked up by business and international news feeds. Why do you think coverage of this discovery was so widespread?
2. The planet is closer to its star than Earth is to the Sun yet receives much less energy. What does that imply about the temperature of the star?
3. Why is the mass of this planet not yet known? What method will be used to find its mass?
4. How will astronomers estimate the planet's composition?
5. Why is this planet called a "cousin" of Earth?

# Summary

Stars and their planetary systems form from collapsing interstellar clouds of gas and dust, following the laws of gravity and conservation of angular momentum. Conservation of angular momentum produces an accretion disk around a protostar that often fragments to form multiple planets, as well as smaller objects such as asteroids and dwarf planets, through the gradual accumulation of material into larger and larger objects. There are multiple methods for finding planets around other stars, and these planets are now thought to be very common. This field of study is evolving very quickly as technology advances.

**LG 1** **Describe how our understanding of planetary system formation developed from the work of both planetary and stellar scientists.** Planets are a common by-product of star formation, and many stars are surrounded by planetary systems. Gravity pulls clumps of gas and dust together, causing them to shrink and heat up. Angular momentum must be conserved, leading to both a spinning central star and an accretion disk that rotates and revolves in the same direction as the central star. Solar System meteorites show that larger objects build up from smaller objects.

**LG 2** **Discuss the role of gravity and angular momentum in explaining why planets orbit the Sun in a plane and why they revolve in the same direction that the Sun rotates.** As particles orbit the forming star, the cloud of dust and gas flattens into a plane. Conservation of angular momentum determines both the speed and the direction of the revolution of the objects in the forming system. Dust grains in the protoplanetary disk first stick together because of collisions and static electricity. As these objects grow, they eventually have enough mass to attract other objects gravitationally. Once this occurs, they begin emptying the space around them. Collisions of planetesimals lead to the formation of planets.

**LG 3** **Explain how temperature at different locations in the protoplanetary disk affects the composition of planets, moons, and other bodies.** Near the central protostar, the temperature is higher. This forces volatile elements, such as water, to evaporate and leave the inner part of the disk. Planets in the inner part of the disk will have fewer volatiles than those in the outer part of the disk. The gas that is captured by a planet at the time of its formation is the planet's primary atmosphere. Less massive planets lose their primary atmospheres and then form secondary atmospheres.

**LG 4** **Discuss the processes that resulted in the formation of planets and other objects in our Solar System.** In the current model of the formation of the Solar System, solid terrestrial planets formed in the inner disk, where temperatures were high, and giant gaseous planets formed in the outer disk, where temperatures were low. Dwarf planets such as Pluto formed in the asteroid belt and in the region beyond the orbit of Neptune. Asteroids and comet nuclei remain today as leftover debris.

**LG 5** **List how astronomers find planets around other stars, and explain how we know that planetary systems around other stars are common.** Astronomers find planets around other stars using a variety of methods: the radial velocity method, the transit method, microlensing, astrometry, and direct imaging. As technology has improved, the number and variety of known extrasolar planets has increased dramatically, with thousands of planets and planet candidates discovered orbiting other stars near the Sun within the Milky Way Galaxy in just the past few years.

## ? UNANSWERED QUESTIONS

- How typical is the Solar System? Only within the past few years have astronomers found other systems containing four or more planets, and so far the observed distributions of large and small planets in these multiplanet systems have looked different from those of the Solar System. Computer simulations of planetary system formation suggest that a system with an orbital stability and a planetary distribution like those of the Solar System may develop only rarely. Improved supercomputers can run more complex simulations, which can be compared with the observations to understand better how solar systems are configured.

- How Earth-like must a planet be before scientists declare it to be "another Earth"? An editorial in the science journal *Nature* cautioned that scientists should define "Earth-like" in advance—before multiple discoveries of planets "similar" to Earth are announced and a media frenzy ensues. Must a planet be of similar size and mass, be located in the habitable zone, and have spectroscopic evidence of liquid water before we call it "Earth 2.0"?

# Questions and Problems

## Test Your Understanding

1. Place the following events in the order that corresponds to the formation of a planetary system.
   a. Gravity collapses a cloud of interstellar gas.
   b. A rotating disk forms.
   c. Small bodies collide to form larger bodies.
   d. A stellar wind "turns on" and sweeps away gas and dust.
   e. Primary atmospheres form.
   f. Primary atmospheres are lost.
   g. Secondary atmospheres form.
   h. Dust grains stick together by static electricity.

2. If the radius of an object's orbit is halved, and angular momentum is conserved, what must happen to the object's speed?
   a. It must be halved.
   b. It must stay the same.
   c. It must be doubled.
   d. It must be squared.

3. Unlike the giant planets, the terrestrial planets formed when
   a. the inner Solar System was richer in heavy elements than the outer Solar System.
   b. the inner Solar System was hotter than the outer Solar System.
   c. the outer Solar System took up more volume than the inner Solar System, so there was more material to form planets.
   d. the inner Solar System was moving faster than the outer Solar System.

4. The terrestrial planets and the giant planets have different compositions because
   a. the giant planets are much larger.
   b. the terrestrial planets formed closer to the Sun.
   c. the giant planets are made mostly of solids.
   d. the terrestrial planets have few moons.

5. The spectroscopic radial velocity method preferentially detects
   a. large planets close to the central star.
   b. small planets close to the central star.
   c. large planets far from the central star.
   d. small planets far from the central star.
   e. the method detects all of these equally well

6. The concept of disk instability was developed to solve the problem that
   a. Jupiter-like planets migrate after formation.
   b. there was not enough gas in the Solar System to form Jupiter.
   c. the early solar nebula likely dispersed too soon to form Jupiter.
   d. Jupiter consists mostly of volatiles.

7. Because angular momentum is conserved, an ice-skater who throws her arms out will
   a. rotate more slowly.
   b. rotate more quickly.
   c. rotate at the same rate.
   d. stop rotating entirely.

8. Clumps grow into planetesimals by
   a. gravitationally pulling in other clumps.
   b. colliding with other clumps.
   c. attracting other clumps with opposite charge.
   d. conserving angular momentum.

9. The transit method preferentially detects
   a. large planets close to the central star.
   b. small planets close to the central star.
   c. large planets far from the central star.
   d. small planets far from the central star.
   e. the method detects all of these equally well

10. If the radius of a spherical object is halved, what must happen to the period so that the spin angular momentum is conserved?
    a. It must be divided by 4.
    b. It must be halved.
    c. It must stay the same.
    d. It must double.
    e. It must be multiplied by 4.

11. The amount of angular momentum in an object does *not* depend on
    a. its radius.
    b. its mass.
    c. its rotation speed.
    d. its temperature.

12. The planets in the inner part of the Solar System are made primarily of refractory materials; the planets in the outer Solar System are made primarily of volatiles. The difference occurred because
    a. refractory materials are heavier than volatiles, so they sank farther into the nebula.
    b. there were no volatiles in the inner part of the accretion disk.
    c. the volatiles on the inner planets were lost soon after the planet formed.
    d. the outer Solar System has gained more volatiles from space since formation.

13. If scientists want to find out about the composition of the early Solar System, the best objects to study are
    a. the terrestrial planets.
    b. the giant planets.
    c. the Sun.
    d. asteroids and comets.

14. The direction of revolution in the plane of the Solar System was determined by
    a. the plane of the galaxy in which the Solar System sits.
    b. the direction of the gravitational force within the original cloud.
    c. the direction of rotation of the original cloud.
    d. the amount of material in the original cloud.

15. A planet in the "habitable zone"
    a. is close to the central star.
    b. is far from the central star.
    c. is the same distance from its star as Earth is from the Sun.
    d. is at a distance where liquid water can exist on the surface.

## Thinking about the Concepts

16. What is the source of the material that now makes up the Sun and the rest of the Solar System?

17. Describe the different ways by which stellar astronomers and planetary scientists each came to the same conclusion about how planetary systems form.

18. What is a protoplanetary disk? What are two reasons that the inner part of the disk is hotter than the outer part?

19. Physicists describe certain properties, such as angular momentum and energy, as being *conserved*. What does this mean? Do conservation laws imply that an individual object can never lose or gain angular momentum or energy? Explain your reasoning.

20. The Process of Science Figure in this chapter makes the point that different areas of science must agree with one another. Suppose that a handful of new exoplanets are discovered that appear not to have formed from the collapse of a stellar nebula (for example, the planetary orbits might be in random orientations). What will scientists do with this new information?

21. How does the law of conservation of angular momentum control a figure-skater's rate of spin?

22. What is an accretion disk?

23. Describe the process by which tiny grains of dust grow to become massive planets.

24. Look under your bed, the refrigerator, or any similar place for dust bunnies. Once you find them, blow one toward another. Watch carefully and describe what happens as they meet. What happens if you repeat this action with additional dust bunnies? Will these dust bunnies ever have enough gravity to begin pulling themselves together? If they were in space instead of on the floor, might that happen? What force prevents their mutual gravity from drawing them together into a "bunny-tesimal" under your bed?

25. Why do we find rocky material everywhere in the Solar System but large amounts of volatile material only in the outer regions?

26. Why were the four giant planets able to collect massive gaseous atmospheres, whereas the terrestrial planets could not? Explain the source of the secondary atmospheres surrounding the terrestrial planets.

27. Describe four methods that astronomers use to search for extrasolar planets. What are the limitations of each method; that is, what circumstances are necessary to detect a planet by each method?

28. Why is it so difficult for astronomers to obtain an image of an extrasolar planet?

29. Many of the first exoplanets that astronomers found orbiting other stars are giant planets with Jupiter-like masses and with orbits located very close to their parent stars. Explain why these characteristics are a selection effect of the discovery method.

30. How does Kepler find Earth-like planets, and what do astronomers mean by "Earth-like"?

## Applying the Concepts

31. Study Figure 7.17. What is the maximum radial velocity of HD 70642b in meters per second? Convert this number to miles per hour (mph). How does this compare to the speed at which Earth orbits the Sun (67,000 mph)?

32. Use Appendix 4 to answer the following:
    a. What is the total mass of all the planets in the Solar System, expressed in Earth masses ($M_{\text{Earth}}$)?
    b. What fraction of this total planetary mass is Jupiter?
    c. What fraction does Earth represent?

33. Compare Earth's orbital angular momentum with its spin angular momentum using the following values: $m = 5.97 \times 10^{24}$ kg, $v = 29.8$ kilometers per second (km/s), $r = 1$ AU, $R = 6,378$ km, and $P = 1$ day. Assume Earth to be a uniform body. What fraction does each component (orbital and spin) contribute to Earth's total angular momentum? Refer to Working It Out 7.1.

34. Venus has a radius 0.949 times that of Earth and a mass 0.815 times that of Earth. Its rotation period is 243 days. What is the ratio of Venus's spin angular momentum to that of Earth? Assume that Venus and Earth are uniform spheres.

35. Jupiter has a mass equal to 318 times Earth's mass, an orbital radius of 5.2 AU, and an orbital velocity of 13.1 km/s. Earth's orbital velocity is 29.8 km/s. What is the ratio of Jupiter's orbital angular momentum to that of Earth?

36. In the text, we give an example of an interstellar cloud having a diameter of $10^{13}$ km and a rotation period of $10^6$ years collapsing to a sphere the size of the Sun ($1.4 \times 10^6$ km in diameter). We point out that if all the cloud's angular momentum went into that sphere, the sphere would have a rotation period of only 0.6 second. Do the calculation to confirm this result.

37. The asteroid Vesta has a diameter of 530 km and a mass of $2.7 \times 10^{20}$ kg.
    a. Calculate the density (mass/volume) of Vesta.
    b. The density of water is 1,000 kg/m³, and that of rock is about 2,500 kg/m³. What does this difference tell you about the composition of this primitive body?

38. Study Figure 7.20.
    a. Recalling Kepler's Laws, put the three planets in order, from fastest to slowest.
    b. Compare the duration of the transits. Why does the outermost planet have the longest duration?

39. The best current technology can measure radial velocities of about 0.3 m/s. Suppose you are observing a spectral line with a wavelength of 575 nanometers (nm). How large a shift in wavelength would a radial velocity of 0.3 m/s produce?

40. Earth tugs the Sun around as it orbits, but it has a much smaller effect (only 0.09 m/s) than that of any known extrasolar planet. How large a shift in wavelength does this effect cause in the Sun's spectrum at 500 nm?

41. If an alien astronomer observed a plot of the light curve as Jupiter passed in front of the Sun, by how much would the Sun's brightness drop during the transit?

42. A planet has been found to orbit a 1-$M_{Sun}$ in 200 days.
    a. What is the orbital radius of this extrasolar planet?
    b. Compare its orbit with that of the planets around our own Sun. What environmental conditions must this planet experience?

43. One of the planets orbiting the star Kepler-11 with an orbital radius of radius 1.1 solar radii, or $R_{Sun}$ has a radius of 4.5 Earth radii ($R_{Earth}$). By how much does the brightness of Kepler-11 decrease when this planet transits the star?

44. Kepler detected a planet with a diameter of 1.7 Earth ($D_{Earth}$).
    a. How much larger is the volume of this planet than Earth's?
    b. Assume that the density of the planet is the same as Earth's. How much more massive is this planet than Earth?

45. The planet COROT-11b was discovered using the transit method, and astronomers have followed up with radial velocity measurements, so both its size (radius 1.43 $R_{Jup}$) and its mass (2.33 $M_{Jup}$) are known. The density provides a clue about whether the object is gaseous or rocky.
    a. What is the mass of this planet in kilograms?
    b. What is the planet's radius in meters?
    c. What is the planet's volume?
    d. What is the planet's density? How does this density compare to the density of water (1,000 kg/m³)? Is the planet likely to be rocky or gaseous?

## USING THE WEB

46. Go to the "Extrasolar Planets Global Searches" Web page (http://exoplanet.eu/searches.php) of the Extrasolar Planets Encyclopedia. Click on one ongoing project under "Ground" and one ongoing project under "Space." What method is used to detect planets in each case? Has the selected project found any planets, and if so, what type are they? Now click on one of the future projects. When will the one you chose be ready to begin? What will be the method of detection?

47. Using the exoplanet catalogs:
    a. Go to the "Catalog" Web page (http://exoplanet.eu/catalog) of the Extrasolar Planets Encyclopedia and set to "All Planets detected." Look for a star that has multiple planets. Make a graph showing the distances of the planets from that star, and note the masses and sizes of the planets. Put the Solar System planets on the same axis. How does this extrasolar planet system compare with the Solar System?
    b. Go to the "Exoplanets Data Explorer" website (http://exoplanets.org) and click on "Table." This website lists planets that have detailed orbital data published in scientific journals, and it may have a smaller total count than the website in part (a). Pick a planet that was discovered this year or last, as specified in the "First Reference" column. What is the planet's minimum mass? What is its semimajor axis and the period of its orbit? What is the eccentricity of its orbit? Click on the star name in the first column to get more

information. Is there a radial velocity curve for this planet? Was it observed in transit, and if so, what is the planet's radius and density? Is it more like Jupiter or more like Earth?

48. Space missions:
    a. Go to the website for the Kepler Mission (http://kepler.nasa.gov). How many confirmed planets has Kepler discovered? Mouse over "confirmed planets": How many planet candidates are there? What kinds of follow-up observations are being done to verify whether the candidates are planets? What is new?
    b. Search for the latest version of the "Kepler Orrery," an animation that shows multiplanet systems discovered by Kepler. Do most of these systems look like our own?
    c. Go to the website for the European Space Agency (ESA) mission Gaia (http://sci.esa.int/gaia). This mission was launched in 2013. Click on the "Exoplanets" link on the left-hand side. What method(s) will *GAIA* use to look for planets? What are the science goals? Have some planets been found?

49. Citizen science projects:
    a. Go to the "PlanetHunters" website at http://planethunters.org. PlanetHunters is part of the Zooniverse, a citizen science project that invites individuals to participate in a major science project using their own computers. To participate in this or any of the other Zooniverse projects mentioned in later chapters, you will need to sign up for an account. Read through the sections under "About," including the FAQ. What are some of the advantages to crowdsourcing Kepler data analysis? Back on the PlanetHunters home page, click on "Tutorial" and watch the "Introduction" and "Tutorial Video." When you're ready to try looking for planets, click on "Classify" and begin. Save a copy of your stars for your homework.
    b. Go to the "Disk Detective" website at http://www.diskdetective.org/, another Zooniverse project for which you will need to make an account as in part (a). In this project, you will look at observations of young stars to see if there is evidence for a planetary disk. Under "Menu," read "Science" and "About," and then "Classify." Work through an example, and then classify a few images.

50. Go to the "Super Planet Crash" Web page (http://www.stefanom.org/spc/ or http://apod.nasa.gov/apod/ap150112.html). Read "Help" to see the rules. First build a system like ours with four Earth-sized planets in the inner 2 AU—is this stable? What happens if you add in super-Earths or "ice giants"? Build up a few completely different planetary systems and see what happens. What types of situations cause instability in the inner 2 AU of these systems?

# smartwork5

If your instructor assigns homework in Smartwork5, access your assignments at digital.wwnorton.com/astro5.

digital.wwnorton.com/astro5

Visit the Student Site at the Digital Landing Page, and open the Exoplanet Radial Velocity Simulator in Chapter 7. This applet has a number of different panels that allow you to experiment with the variables that are important for measurement of radial velocities. First, in the window labeled "Visualization Controls," check the box to show multiple views. Compare the views shown in panels 1–3 with the colored arrows in the last panel to see where an observer would stand to see the view shown. Start the animation (in the "Animation Controls" panel), and allow it to run while you watch the planet orbit its star from each of the views shown. Stop the animation, and in the "Presets" panel, select "Option A" and then click "set."

**1** Is Earth's view of this system most nearly like the "side view" or most nearly like the "orbit view"?

_____

**2** Is the orbit of this planet circular or elongated?

_____

**3** Study the radial velocity graph in the upper right panel. The blue curve shows the radial velocity of the star over a full period. What is the maximum radial velocity of the star?

_____

**4** The horizontal axis of the graph shows the "phase," or fraction of the period. A phase of 0.5 is halfway through a period. The vertical red line indicates the phase shown in views in the upper left panel. Start the animation to see how the red line sweeps across the graph as the planet orbits the star. The period of this planet is 365 days. How many days pass between the minimum radial velocity and the maximum radial velocity?

_____

**5** When the planet moves away from Earth, the star moves toward Earth. The sign of the radial velocity tells the direction of the motion (toward or away). Is the radial velocity of the star positive or negative at this time in the orbit? If you could graph the radial velocity of the planet at this point in the orbit, would it be positive or negative?

_____

In the "Presets" window, select "Option B" and then click "set."

**6** What has changed about the orbit of the planet as shown in the views in the upper left panel?

_____

_____

**7** When is the planet moving fastest: when it is close to the star or when it is far from the star?

_____

_____

**8** When is the star moving fastest: when the planet is close to it or when the planet is far away?

_____

**9** Explain how an astronomer would determine, from a radial velocity graph of the star's motion, whether the orbit of the planet was in a circular or elongated orbit.

_____

_____

**10** Study the "Earth view" panel at the top of the window. Would this planet be a good candidate for a transit observation? Why or why not?

_____

_____

In the "System Orientation" panel, change the inclination to 0.0.

**11** Now is Earth's view of this system most nearly like the "side view" or most nearly like the "orbit view"?

_____

**12** How does the radial velocity of the star change as the planet orbits?

_____

_____

**13** Click the box that says "show simulated measurements," and change the "noise" to 1.0 m/s. The gray dots are simulated data, and the blue line is the theoretical curve. Use the slider bar to change the inclination. What happens to the radial velocity as the inclination increases? (Hint: Pay attention to the vertical axis as you move the slider, not just the blue line.)

_____

_____

_____

**14** What is the smallest inclination for which you would find the data convincing? That is, what is the smallest inclination for which the theoretical curve is in good agreement with the data?

_____

_____

_____

# 8

# The Terrestrial Planets and Earth's Moon

The objects that formed in the inner part of the protoplanetary disk around the Sun are relatively small, rocky worlds, one of which is Earth. A comparison of these worlds reveals the forces that shape a planet. The past six decades have been an exciting time for exploration of and discovery about Earth and the other planets in the Solar System. Robotic probes have visited every planet, and astronauts walked on the surface of the Moon. In addition to discoveries from new space missions and telescopes, improved analytical techniques applied to the rocks and soil brought back from the Moon more than 40 years ago have led to some surprising new results. The information from these missions has revolutionized the understanding of the Solar System, offering insights into the current state of each of the neighboring planets and clues about their histories.

## LEARNING GOALS

By the conclusion of this chapter, you should be able to look at an image of a planet and identify which geological features occurred early in the history of that world and which occurred late. You should also be able to:

LG 1  Describe how impacts have affected the evolution of the terrestrial planets.

LG 2  Explain how radiometric dating is used to measure the ages of rocks and terrestrial planetary surfaces.

LG 3  Explain how scientists use both theory and observation to determine the structure of terrestrial planetary interiors.

LG 4  Describe tectonism and volcanism and the forms they take on different terrestrial planets.

LG 5  Summarize the knowledge of water on the terrestrial planets.

*Mars Reconnaissance Orbiter* image of Newton Crater on Mars. The dark streaks may be indications of flowing water. ▶ ▶ ▶

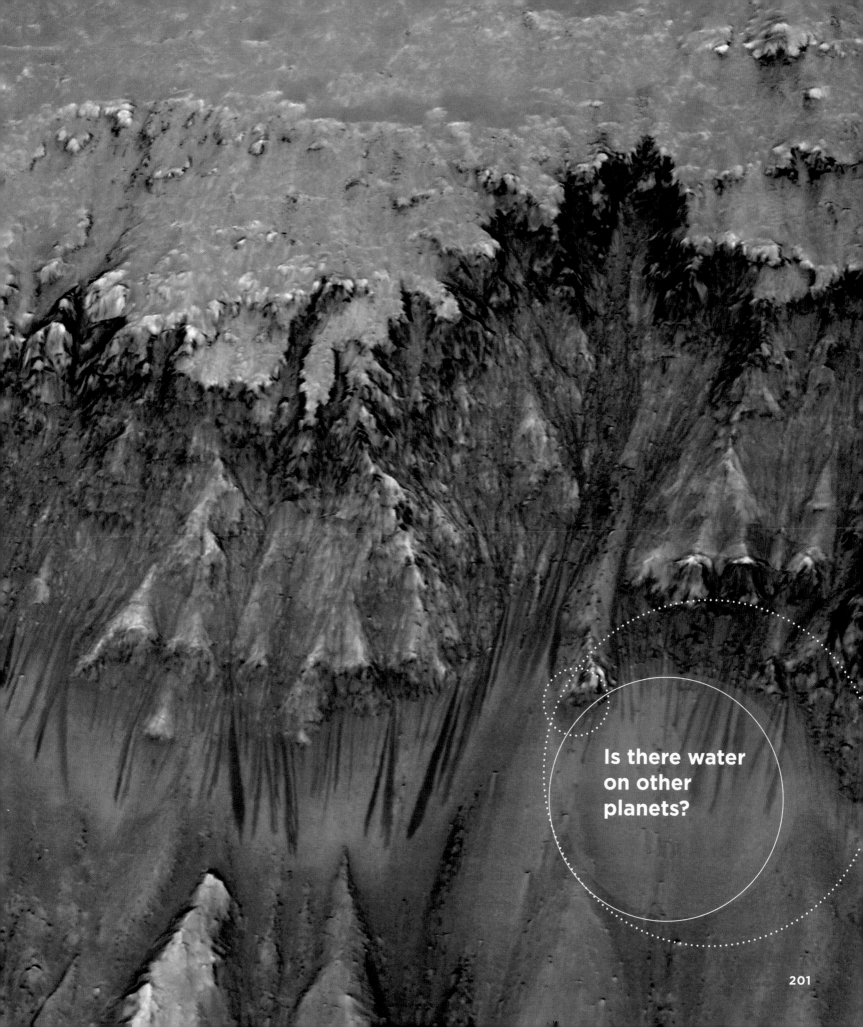

Is there water
on other
planets?

201

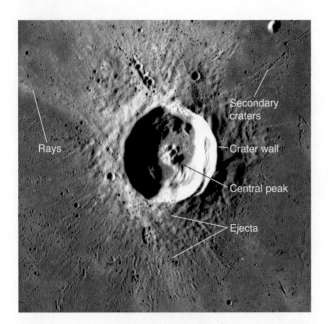

**Figure 8.1** A lunar crater showing the crater wall and central peak surrounded by ejected material (*ejecta*), rays, and secondary craters—all typical features associated with impact craters.

# 8.1 Impacts Help Shape the Evolution of the Planets

Four principal geological processes constantly reshape the planets: *impact cratering, tectonism, volcanism,* and *erosion.* Some geological processes originate in a planet's interior, and other processes are external. The relative importance of each of these processes to each planet varies. Planetary scientists can learn about the evolutionary history of the Solar System by comparing these geological processes on the terrestrial planets and the Moon. In this section, we will examine **impact cratering**, which occurs when large collisions by other Solar System objects leave distinctive scars in the outer layer of a planet.

## Comparative Planetology

The four innermost planets in the Solar System are Mercury, Venus, Earth, and Mars, collectively known as the *terrestrial planets.* Although the Moon is Earth's lone natural satellite, we include it in this chapter because of its close similarity to the terrestrial planets.

When comparing planets, we first compare the basic physical characteristics, such as distance from the Sun, size and density, and gravitational pull at the surface. These characteristics reveal what a planet is made of, what its surface temperature is likely to be, and how well it can hold an atmosphere (planetary atmospheres will be discussed in Chapter 9). By comparing the different planets, scientists can sort out the vast quantity of information returned by space probes. The correct explanation for a particular aspect of one planet must be consistent with what is known about the other planets. For example, an analysis of why the Moon is covered with craters must allow for the fact that preserved craters are rare on Earth. An explanation for why Venus has such a massive atmosphere should point to reasons that Earth and Mars do not. Such comparisons are key to an approach called **comparative planetology**. Some of the basic physical properties of the terrestrial planets are compared in **Table 8.1**.

## Impacts and Craters

Of the four geological processes, impact cratering causes the most concentrated and sudden release of energy. Planets and other objects orbit the Sun at very high speeds. For example, as seen in Table 8.1, Earth orbits the Sun at an average speed of around 30 kilometers per second (km/s), equivalent to 67,000 miles per hour (mph). Collisions between orbiting bodies can release huge amounts of energy and produce craters like the one in **Figure 8.1**. **Figure 8.2** shows the process

| **TABLE 8.1** Comparison of Physical Properties of the Terrestrial Planets and the Moon | | | | | |
|---|---|---|---|---|---|
| | **Mercury** | **Venus** | **Earth** | **Mars** | **Moon** |
| Orbital radius | 0.387 AU | 0.723 AU | 1.000 AU | 1.524 AU | 384,000 km |
| Orbital period | 0.241 yr | 0.615 yr | 1.000 yr | 1.881 yr | 27.32 days |
| Orbital velocity (km/s) | 47.9 | 35.0 | 29.8 | 24.1 | 1.02 |
| Mass ($M_{Earth}$ = 1) | 0.055 | 0.815 | 1.000 | 0.107 | 0.012 |
| Equatorial diameter (km) | 4,880 | 12,104 | 12,756 | 6,794 | 3,476 |
| Equatorial diameter ($D_{Earth}$ = 1) | 0.383 | 0.949 | 1.000 | 0.533 | 0.272 |
| Density (water = 1) | 5.43 | 5.24 | 5.52 | 3.93 | 3.34 |
| Sidereal rotation period* | $58.65^d$ | $243.02^d$ | $23^h56^m$ | $24^h37^m$ | $27.32^d$ |
| Obliquity (degrees)[†] | 0.04 | 177.36 | 23.45 | 25.19 | 6.68 |
| Surface gravity (m/s²) | 3.70 | 8.87 | 9.78 | 3.71 | 1.62 |
| Escape velocity (km/s) | 4.25 | 10.36 | 11.18 | 5.03 | 2.38 |

*The superscript letters *d, h,* and *m* stand for days, hours, and minutes of time, respectively.
[†]An obliquity greater than 90° indicates that the planet rotates in a retrograde, or backward, direction.

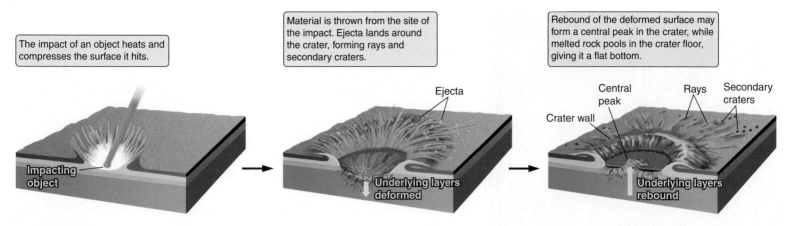

The impact of an object heats and compresses the surface it hits.

Material is thrown from the site of the impact. Ejecta lands around the crater, forming rays and secondary craters.

Rebound of the deformed surface may form a central peak in the crater, while melted rock pools in the crater floor, giving it a flat bottom.

Ejecta

Impacting object

Underlying layers deformed

Central peak

Rays

Secondary craters

Crater wall

Underlying layers rebound

**Figure 8.2** Stages in the formation of an impact crater.

of impact cratering. When an object hits a planet, its kinetic energy heats and compresses the surface that it strikes and throws material far from the resulting impact crater. Sometimes, material thrown from the crater, called **ejecta**, falls back to the surface of the planet with enough energy to cause **secondary craters**. The rebound of heated and compressed material can also lead to the formation of a central peak or a ring of mountains on the crater floor as shown in the lunar crater in Figure 8.1. These processes are similar to what happens when a drop lands in milk, as shown in **Figure 8.3**.

The energy of an impact can be great enough to melt or even vaporize rock. The floors of some craters are the cooled surfaces of melted rock that flowed as lava. The energy released in an impact can also lead to the formation of new minerals. Because some minerals form only during an impact, they are evidence of ancient impacts on Earth's surface. The space rocks that cause these impacts are defined by three closely related terms: **meteoroids** are small (less than 100 meters in diameter) cometary or asteroid fragments in space. A meteoroid that enters and burns up in a planetary atmosphere is called a **meteor**. Any meteoroids that survive to hit the ground are known as **meteorites**.

One of the best-preserved impact structures on Earth is Meteor Crater in Arizona. This impact occurred about 50,000 years ago. From the crater's size and shape and from the remaining pieces of the impacting body, we know that the nickel-iron asteroid fragment was about 50 meters across, had a mass of about 300 million kilograms (kg), and was traveling at 13 km/s relative to Earth, when it hit Earth's upper atmosphere. Approximately half of the original mass was vaporized in the atmosphere before the remainder hit the ground. This collision released about 300 times as much total energy as the first atom bomb. At only 1.2 km in diameter, Meteor Crater is

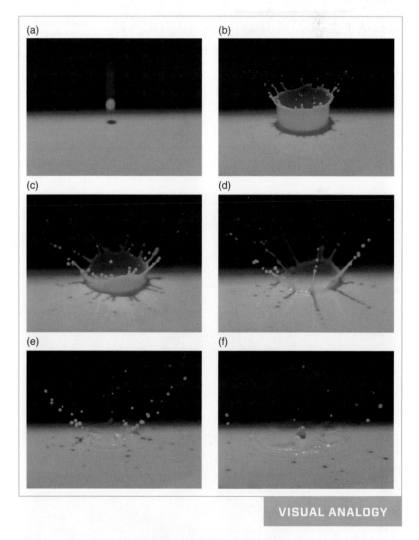

(a) (b) (c) (d) (e) (f)

**VISUAL ANALOGY**

**Figure 8.3** A drop (a) hitting a pool of milk illustrates the formation of features in an impact crater, including crater walls (b, c), secondary craters (d, e), and a central peak (f).

**Figure 8.4** Meteor Crater (also known as Barringer Crater), located in northern Arizona, is an impact crater 1.2 km in diameter that formed some 50,000 years ago by a nickel-iron meteoroid's collision with Earth.

tiny compared to impact craters seen on the Moon or ancient impact scars on Earth (**Figure 8.4**).

Impact craters cover the surfaces of Mercury, Mars, and the Moon. For example, the Moon has millions of craters of all different sizes, one on top of another as seen outlined by yellow circles in **Figure 8.5**. Nearly all of these craters are the result of impacts. On Earth and Venus, by comparison, most impact craters have been destroyed. Fewer than 200 impact craters have been identified on Earth, and about 1,000 have been found on Venus. Earth's crater shortage is primarily due to two processes we will discuss later in the chapter—plate tectonics in Earth's ocean basins and erosion on land. On Venus, lava flows have destroyed most of the craters.

The atmospheres of Earth and Venus provide another explanation for their low number of small craters. The surfaces of the Moon and Mercury are directly exposed to bombardment from space, whereas the surfaces of Earth and Venus are partly protected by their atmospheres. Rock samples from the Moon show craters smaller than a pinhead, formed by micrometeoroids. In contrast, most meteoroids smaller than 100 meters in diameter that enter Earth's atmosphere are either burned up or broken up by friction before they reach the surface. Small meteorites found on the ground on Earth are probably pieces of much larger bodies that broke apart on entering the atmosphere. With an atmosphere far thicker than that of Earth, Venus is even better protected.

Planetary scientists can tell a lot about the surface of a planet by studying its craters because the characteristics of a crater depend on the properties of the planetary surface. An impact in a deep ocean on Earth might create an impressive wave but leave no lasting crater. In contrast, an impact scar formed in an ancient rocky area can be preserved for billions of years. For example, craters on the Moon's pristine surface are often surrounded by strings of smaller secondary craters formed from material thrown out by the impact, like those shown in Figure 8.2.

Some craters on Mars have a very different appearance. They are surrounded by structures that look much like the pattern you might see if you threw a rock into mud (**Figure 8.6**). The flows appear to indicate that the martian surface rocks contained water or ice at the time of the impact. At the time these craters formed, there may have been liquid water on the surface of Mars. Features resembling canyons and dry riverbeds are further evidence of this hypothesis. Not all martian craters have this feature, so the water or ice must have been concentrated in only some areas, and these icy locations might have changed with time.

Another explanation for the appearance of these craters is that the impact heated the surface enough to liquefy temporarily the frozen water in the ground. Today, the surface of Mars is dry in some regions and frozen in others, which suggests that water once on the surface has evaporated, or soaked into the ground, much like water frozen in the ground in Earth's polar regions. The energy released by an impact would have melted this ice, turning the surface material into a slurry with a consistency much like wet concrete. When thrown from the crater by the

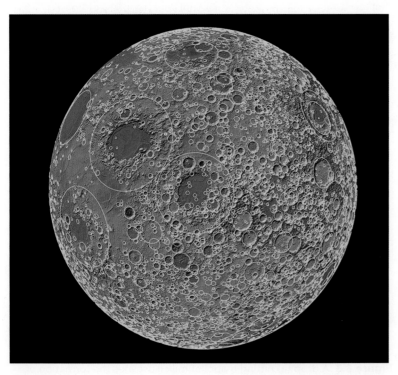

**Figure 8.5** *Lunar Reconnaissance Orbiter* false-color image of craters on the Moon.

force of the impact, this slurry would hit the surrounding terrain and slide across the surface, forming the mudlike craters we see today.

## Giant Impacts Reshape Planets

Because many planetesimals were roaming around the early Solar System, every young planet experienced heavy bombardment early in its history. The last major bombardment is called *the late heavy bombardment*, and it took place from 4.2 billion to 3.8 billion years ago. Observations of uneven thicknesses of the crusts of inner planets have led to theories that some or all of the terrestrial planets were disrupted by at least one giant impact—a collision with an object the size of a large asteroid. A giant impact with Mercury could have removed some of the lighter material of its outside layer, leaving behind an overall denser planet. A giant impact with Venus may have led to its retrograde (backward) rotation. A giant impact probably explains the large differences between the northern and southern hemispheres of Mars. The southern highlands have a thicker crust and formed early in martian history. The northern lowlands have a thinner and younger crust and may have formed when there was melting after the impact. Early giant impacts on Mars also may be responsible for its loss of a magnetic field. Impacts with smaller icy comets from the outer parts of the Solar System brought water, atmospheric gases, and possibly organic molecules to the inner planets. As we'll discuss at the end of this chapter, an impact with Earth about 65 million years ago might have been a crucial event that paved the way for the evolution of *Homo sapiens*.

Although there are several different theories about how the Moon formed, a leading theory involves a giant impact. According to this impact theory, about 4.5 billion years ago, a Mars-sized protoplanet collided with Earth, blasting off and vaporizing parts of Earth's outer layers. The debris from the impact condensed into orbit around Earth and evolved into the Moon. This theory accounts for the similarities in composition between the Moon and Earth's outer layers. It also explains the lower amounts of volatiles on the Moon: during the vaporization stage of the collision, most gases were lost to space, leaving primarily the nonvolatiles to condense as the Moon. Earth, in contrast, was large enough to keep more of its volatiles, which continued to be released from its interior after the collision. Because of the stronger gravity of Earth, these gases were retained as part of Earth's atmosphere. If this account of the Moon's formation is correct, scientists expect to observe chemical evidence of material from the colliding protoplanet that was incorporated into the Moon's composition along with material from Earth. Currently, there is debate about whether chemical evidence of the colliding object has been found (**Process of Science Figure**).

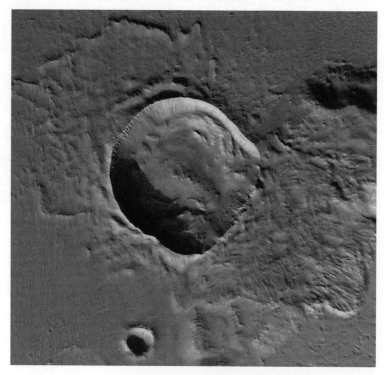

**Figure 8.6** Some craters on Mars look like those formed by rocks thrown into mud, suggesting that material ejected from the crater contained large amounts of water. This crater is about 20 km across.

▶‖ **AstroTour:** Processes That Shape the Planets

## CHECK YOUR UNDERSTANDING 8.1

Geologists can find the relative age of impact craters on a world because: (a) the ones on top must be older; (b) the ones on top must be younger; (c) the larger ones must be older; (d) the larger ones must be younger.

## CERTAINTY IS SOMETIMES OUT OF REACH

There are several hypotheses for how the Moon formed. One of these fits the data better than the others, but none have been absolutely ruled out.

**Was the Moon captured?**

**Did the Moon split-off from Earth?**

**Did the Moon and Earth form together?**

**Did the Moon form from an impact of another object with Earth?**

In some cases, hypotheses cannot be definitely falsified. The working hypothesis is the one that best fits the data, but other ideas are kept in mind.

# 8.2 Radioactive Dating Tells Us the Age of the Moon and the Solar System

The number of visible craters on a planet is determined by the rate at which those craters are destroyed. Geological activity on Earth, Mars, and Venus erased most evidence of early impacts. By contrast, the Moon's surface still preserves the scars of craters dating from about 4 billion years ago. The lunar surface has remained essentially unchanged for more than a billion years because the Moon has no atmosphere or surface water and a cold, geologically dead interior. Mercury also has well-preserved craters, although recent evidence from the *Messenger* mission shows tilted crater floors that are higher on one side than the other—evidence that internal forces lifted the floors unevenly after the craters formed.

Planetary scientists use this cratering record to estimate the ages and geological histories of planetary surfaces: extensive cratering means an older planetary surface that remains relatively unchanged because of minimal geological activity. The amount of cratering can be used as a clock to measure the relative ages of surfaces. But to determine the exact age of a surface based on the number of craters, we need to know how fast the clock runs. In other words, we need to "calibrate the cratering clock."

To assign real dates to these different layers, scientists use a technique called radiometric dating. A geologist can find the age of a rock by measuring the relative amounts of a radioactive element, known as a **radioisotope**, and the decay products it turns into. An isotope is an atom with the same number of protons but a different number of neutrons as other atoms of the same chemical element. The radioactive element is known as the **parent element**, and the decay products are called **daughter products**. Chemical analysis of a rock containing radioactive elements immediately after its formation would reveal the presence of the radioactive parents, but the daughter products of the radioactive decay would be absent because they would not have formed yet. As radioactive atoms decay over time, however, the amount of parent elements decreases and the amount of daughter products builds up. Chemical analysis reveals both the remaining radioactive parent atoms and the daughter products trapped within the structure of the mineral.

The time interval over which a radioactive isotope decays to half its original amount is called its **half-life**. With every half-life that passes, the remaining amount of the radioisotope decreases by a factor of 2. For example, after 3 half-lives the remaining amount of a parent radioisotope will be $\frac{1}{2} \times \frac{1}{2} \times \frac{1}{2} = \frac{1}{8}$ of its original amount. This is illustrated in **Figure 8.7**. At formation, there is 100 percent of the radioactive parent isotope (in red) and no daughter isotope (in blue). After 1 half-life has passed, half of the parent isotope has decayed, and there are equal numbers of parent and daughter. After another half-life has passed, the sample is now only $\frac{1}{4}$ parent and $\frac{3}{4}$ daughter isotopes, and so on. By comparing the percentages of parent and daughter isotopes in a mineral, one can figure out how many half-lives have passed, and thus the age of the mineral. Some numerical examples are discussed in **Working It Out 8.1**.

The age of the Solar System is estimated from radioactive dating of meteorites found on Earth that are 4.5 billion to 4.6 billion years old. Earth may be as young as 4.4 billion years old. The age of the Moon comes from radioactive dating of lunar rocks. Between 1969 and 1976, *Apollo* astronauts and Soviet unmanned probes visited the Moon and brought back samples taken from nine different locations on

(a)

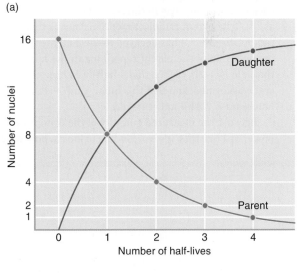

(b)

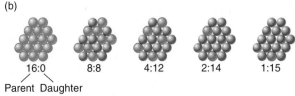

Parent  Daughter

**Figure 8.7** The concept of half-life. A parent population of 16 radioactive nuclei decays over a number of half-lives. This information can be presented (a) graphically or (b) as a collections of particles.

## 8.1  Working It Out  Computing the Ages of Rocks

With every half-life that passes, the remaining amount will decrease by a factor of 2. If we express the number of half-lives more generally as $n$, then we can translate this relationship into math:

$$\frac{P_F}{P_O} = \left(\frac{1}{2}\right)^n$$

where $P_O$ and $P_F$ are the original and final amounts, respectively, of a parent radioisotope; and $n$ is the number of half-lives that have gone by, which equals the time interval of decay (its age) divided by the half-life of the isotope.

For example, the most abundant isotope of the element uranium (uranium-238, or $^{238}U$—the parent) decays through a series of intermediate daughters to an isotope of the element lead (lead-206, or $^{206}Pb$—its final daughter). The half-life of $^{238}U$ is 4.5 billion years. This means that in 4.5 billion years, a sample that originally contained the uranium isotope (the parent) but no lead (its final daughter) would be found instead to contain equal amounts of uranium and lead. If we were to find a mineral with such composition, we would know that half the uranium atoms had turned to lead and that the mineral formed 4.5 billion years ago.

Let's look at another example, this time with a different isotope of uranium ($^{235}U$) that decays to a different lead isotope ($^{207}Pb$) with a half-life of 700 million years. Suppose that a lunar mineral brought back by astronauts has 15 times as much $^{207}Pb$ (the daughter product) as $^{235}U$ (the parent radioisotope). This means that 15/16 of the parent radioisotope ($^{235}U$) has decayed to the daughter product ($^{207}Pb$), leaving only 1/16 of the parent remaining in the mineral sample. Noting that 1/16 is $(1/2)^4$, we see that 4 half-lives have elapsed since the mineral was formed, and that this lunar sample is therefore 4 × 700 million years = 2.8 billion years old.

Because the measured quantity of the isotope is not always a neat power of 2 like this, it's worthwhile to look at how we would solve the equation mathematically. We do this by taking the logarithm on both sides:

$$\log_{10}\frac{P_F}{P_O} = \log_{10}\left(\frac{1}{2}\right)^n$$

$$\log_{10}\frac{P_F}{P_O} = n\log_{10}\frac{1}{2}$$

$$\log_{10}\frac{P_F}{P_O} = -0.3n$$

Putting this back into words, we can write this relationship as

$$\log_{10}\left(\frac{\text{Actual measured quantity of isotope}}{\text{Original quantity of isotope}}\right) = -0.3 \times \frac{\text{Time it has been decaying (age)}}{\text{Half-life}}$$

Solving for age,

$$\text{Age} = -3.3 \times \text{Half-life} \times \log_{10}\left(\frac{\text{Actual measured quantity of isotope}}{\text{Original quantity of isotope}}\right)$$

(Most calculators have a button called "log" or "$\log_{10}$" for calculating such numbers.)

In the second example introduced earlier, $^{235}U$ decays to $^{207}Pb$ with a half-life of 700 million years. The lunar mineral is measured to have 15 times as much lead as uranium, so the mineral currently contains only 1/16 of the original quantity of uranium:

$$\text{Age of mineral} = -3.3 \times (700 \times 10^6\,\text{yr}) \times \log_{10}\left(\frac{1}{16}\right) = 2.8 \times 10^9\,\text{yr}$$

The mineral is 2.8 billion years old.

the lunar surface. By measuring relative amounts of various radioactive elements and the elements into which they decay, scientists were able to assign ages to these different lunar regions. The oldest, most heavily cratered regions on the Moon date back to about 4.4 billion years ago, whereas most of the smoother parts of the lunar surface are typically 3.1 billion to 3.9 billion years old. This suggests the Moon formed after Earth, and the heavy cratering suggests there was heavy bombardment at that time. As you can see in **Figure 8.8**, almost all of the major cratering in the Solar System took place within its first billion years.

### CHECK YOUR UNDERSTANDING 8.2

If radioactive element A decays into radioactive element B with a half-life of 20 seconds, then after 40 seconds: (a) none of element A will remain; (b) none of element B will remain; (c) half of element A will remain; (d) one-quarter of element A will remain.

# 8.3 The Surface of a Terrestrial Planet Is Affected by Processes in the Interior

While impact cratering is driven by forces external to a planet, two other important processes, tectonism and volcanism, are determined by conditions in the interior of the planet. To understand these processes, we must understand the structure and composition of the interiors of planets. But how do we know what the interiors of planets are like? On Earth, the deepest holes ever drilled are about 12 km deep; tiny when compared to Earth's radius of 6,378 km. It is impossible to drill down into Earth's core to observe Earth's interior structure directly. Scientists have determined a lot about the interior of Earth but less about the interiors of the other terrestrial planets.

## Probing the Interior of Earth

The composition of Earth's interior can be determined in two different ways. In one approach, Kepler's or Newton's laws are used to find the mass of Earth; for example, from applying Kepler's third law to a satellite orbiting Earth. Dividing the mass by the volume of Earth gives an average density of 5,500 kilograms per cubic meter ($kg/m^3$), or 5.5 times the density of water. But rocky surface material averages only 2,900 $kg/m^3$. Because the density of the whole planet is greater than the density of the surface, the interior must contain material denser than surface rocks. Another approach to determine the composition of Earth's interior comes from studies of meteorites. Because meteorites are left over from a time when the Solar System was young and Earth was forming from similar materials, the overall composition of Earth should resemble the composition of meteorite material. This material includes minerals with large amounts of iron, which has a density of nearly 8,000 $kg/m^3$. From these considerations, planetary scientists can determine the composition of Earth's interior.

The most important source of information about the structure of Earth's interior comes from monitoring the vibrations from earthquakes. When an earthquake occurs, vibrations spread out through and across the planet as **seismic waves**. There are different classes of seismic waves—those that travel across the surface of a planet and those that travel through a planet. **Surface waves** travel across the surface of a planet, much like waves on the ocean. If conditions are right, surface waves from earthquakes can be seen rolling across the countryside like ripples on water. These waves are responsible for much of the heaving of Earth's surface during an earthquake, causing damage such as the buckling of roadways.

The other types of seismic waves travel *through* Earth, probing the interior of the planet, at a higher speed than surface waves travel. **Primary waves** (P waves) are a type of **longitudinal wave** resulting from alternating compression and decompression of a material. Imagine a stretched-out spring, as illustrated in **Figure 8.9a**. A quick push along its length will make a longitudinal wave. P waves distort the material they travel through, much as compression waves do when they move along the length of a spring. **Secondary waves** (S waves) are a type of **transverse wave** resulting from the sideways motion of material (Figure 8.9b).

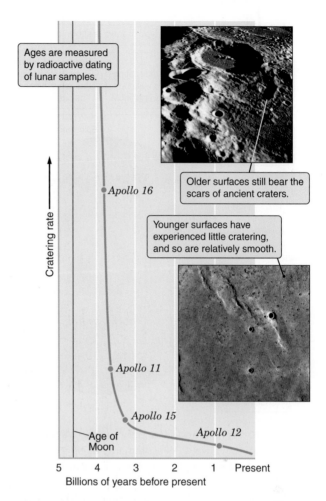

Ages are measured by radioactive dating of lunar samples.

Older surfaces still bear the scars of ancient craters.

Younger surfaces have experienced little cratering, and so are relatively smooth.

Cratering rate

*Apollo 16*

*Apollo 11*

*Apollo 15*

*Apollo 12*

Age of Moon

5    4    3    2    1    Present
Billions of years before present

**Figure 8.8** Radiometric dating of lunar samples returned from specific sites by *Apollo* astronauts was used to determine how the cratering rate has changed over time. Cratering records can then be used to establish the age of other parts of the lunar surface.

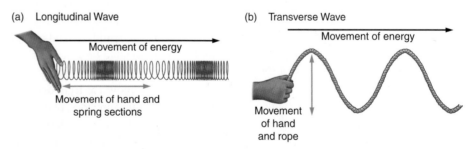

(a) Longitudinal Wave

Movement of energy

Movement of hand and spring sections

(b) Transverse Wave

Movement of energy

Movement of hand and rope

**Figure 8.9** (a) A longitudinal wave involves oscillations along the direction of travel of the wave. (b) A transverse wave involves oscillations that are perpendicular to the direction in which the wave travels. Primary seismic waves are longitudinal; secondary seismic waves are transverse.

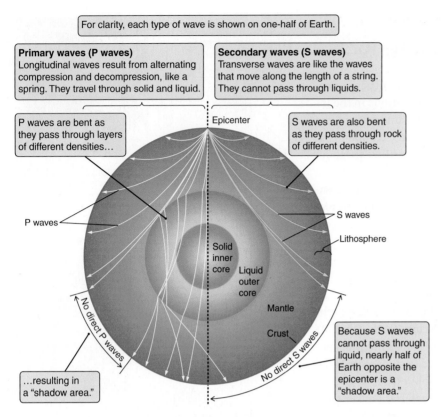

For clarity, each type of wave is shown on one-half of Earth.

**Primary waves (P waves)**
Longitudinal waves result from alternating compression and decompression, like a spring. They travel through solid and liquid.

**Secondary waves (S waves)**
Transverse waves are like the waves that move along the length of a string. They cannot pass through liquids.

P waves are bent as they pass through layers of different densities…

S waves are also bent as they pass through rock of different densities.

Epicenter

P waves

S waves

Lithosphere

Solid inner core

Liquid outer core

Mantle

Crust

No direct P waves

No direct S waves

…resulting in a "shadow area."

Because S waves cannot pass through liquid, nearly half of Earth opposite the epicenter is a "shadow area."

**Figure 8.10** Primary and secondary seismic waves move through the interior of Earth in distinctive ways. Measurements of when and where different types of seismic waves arrive after an earthquake enable scientists to test predictions from detailed models of Earth's interior. Note the "shadow areas" caused by the refraction of primary waves (yellow) at the outer boundary of the liquid outer core and the inability of secondary waves (blue) to pass through the liquid outer core.

The progress of seismic waves through Earth's interior depends on the characteristics of the material through which they are moving. Primary waves (white) can travel through either solids or liquid, but secondary waves (blue) cannot travel through liquids, as shown in **Figure 8.10**. Seismic waves travel at different speeds, depending on the density and composition of the rocks they encounter. As a result, seismic waves moving through rocks of varying densities or composition are bent in much the same way that waves of light are bent when they enter or leave glass. Their speed provides additional information about Earth's interior. The refraction of primary waves at the outer edge of Earth's liquid outer core and the inability of secondary waves to penetrate the liquid outer core create "shadows" of the liquid core on the side of Earth opposite an earthquake's epicenter, as shown in Figure 8.10. Much of scientists' knowledge of Earth's liquid outer core is due to studies of these waves.

Scientists use instruments called seismometers to measure the distinctive patterns of seismic waves. For more than 100 years, thousands of seismometers scattered around the globe have measured the vibrations from countless earthquakes and other seismic events, such as volcanic eruptions and nuclear explosions. A single seismometer can record ground motion at only one place on Earth, but when combined with the recordings of many other seismometers placed all over Earth, scientists can use the data to get a comprehensive picture of the planet's interior.

## Building a Model of Earth's Interior

Geologists use the laws of physics and the properties of materials and how these behave at different temperatures and pressures to model the structure of Earth's interior. The pressure at any point in Earth's interior must be just high enough that the outward forces balance the inward force of the weight of all the material above that point. If the outward pressure at some point within a planet were *less* than the weight per unit area of the overlying material, then that material would fall inward, crushing what was underneath it. If the pressure at some point within a planet were *greater* than the weight per unit area of the overlying material, then the material would expand and push outward, lifting the overlying material. The situation is stable only when the weight of matter above is just balanced by the pressure within the whole interior of the planet. The balance between pressure and weight is known as **hydrostatic equilibrium**, and it is important to the structure of planetary interiors, planetary atmospheres, and the structure and evolution of stars.

From consideration of hydrostatic equilibrium and seismic wave measurements, scientists construct a layered model of Earth's interior. They then test their model by comparing its predictions of how seismic waves would propagate through Earth with actual observations of seismic waves from real earthquakes. The extent to which the predictions agree with observations indicates both strengths and weaknesses of the model. Geologists adjust the model—always remaining consistent with the known physical properties of materials—until a good match is found between prediction and observation.

This is the method geologists used to arrive at the current picture of the interior of Earth shown in Figure 8.10. The innermost region of Earth's interior consists of a **core**. Earth's solid inner core is at a temperature of about 6000 K and is primarily composed of iron, nickel, and other dense metals. The liquid outer core is cooler, at about 4000 K, and is composed of liquid metals. Outside of the outer core is Earth's **mantle**, a rocky shell made of solid, medium-density materials such as silicates. Covering the mantle, the **crust** is a thin, hard layer of lower-density materials that is chemically distinct from the interior.

The cross sections in **Figure 8.11** show the interior structures of each of the terrestrial planets and the Earth's Moon. As you can see, Earth's interior is not uniform. The materials have been separated by density, a process known as **differentiation**. When rocks of different types are mixed together, they tend to stay mixed. Once this rock melts, however, the denser materials sink to the center and the less dense materials float toward the surface. Today, little of Earth's interior is molten, but the differentiated structure shows that Earth was once much hotter, and its interior was liquid throughout. The cores of all the terrestrial planets and the core of the Moon were once molten. When planetary scientists reanalyzed 8 years of data from seismometers left on the Moon by the *Apollo* astronauts using new, improved methods, they found that the Moon has a solid inner core, a liquid outer core, and a partially melted layer between the core and the mantle.

## The Evolution of Planetary Interiors

The balance between energy received and energy produced and emitted governs the temperature within a planet. The interiors of planets evolve as their temperatures change over time. Factors that influence how the temperature changes include the size of the planet, the composition of the material, and heating from various sources. Here, we are concerned with thermal energy—the kinetic energy of particles within a substance that determines the temperature. In general, the interior of a planet cools down over time as heat is emitted from the surface. Because it takes time for heat to travel through rock, the deeper we go within a planet, the higher the temperature. This is similar to the effect of taking a hot pie out of the oven. Over time, the pie radiates heat from the surface and cools down, but the filling takes much longer to cool than the crust.

Planets lose thermal energy from their surfaces primarily through radiation. Recall from Chapter 5 that when objects radiate energy, the hotter they are, the more energy they radiate. The type of energy radiated (infrared, optical, ultraviolet, and so forth) depends on the temperature of the object. The rate at which a planet cools depends on its size. A larger planet has a larger volume of matter and more thermal energy trapped inside. Thermal energy has to escape through the planet's surface, so the planet's surface area determines the rate at which energy is lost. Smaller planets have more surface area in comparison with their small volumes, so they cool off faster, whereas larger planets have a smaller surface area to volume ratio and cool off more slowly (**Working It Out 8.2**). Because geological activity is powered by heat, smaller objects become geologically inactive sooner. Major geological activity ended on Mercury and the Moon first, but the larger terrestrial planets—Venus, Earth, and Mars—continued to have geological activity.

Some of the thermal energy in the interior of Earth is left over from when Earth formed. The tremendous energy of collisions and the energy from

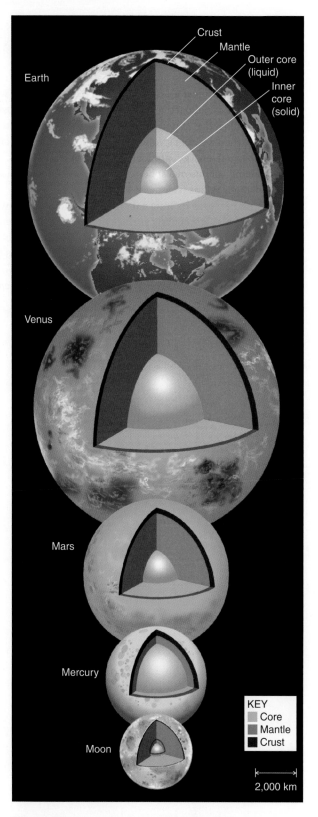

**Figure 8.11** A comparison of the interiors of the terrestrial planets and Earth's Moon. Some fractions of the cores of Mercury, Venus, the Moon, and Mars are probably liquid.

## 8.2 )) Working It Out  How Planets Cool Off

If we assumed that all the terrestrial planets formed with the same percentage of radioactive materials in their bulk composition and that these radioactive materials are their sole source of internal thermal energy, then a planet's volume would determine the total amount of the thermal energy–producing material it contains. The energy-producing volume of a spherical planet is proportional to the cube of the planet's radius (volume = $\frac{4}{3} \times \pi R^3$).

A planet loses its internal energy by radiating it away at its surface, so a planet's cooling surface area determines the rate at which it can get rid of its thermal energy. The cooling surface area of the planet is proportional to only the square of the radius (surface area = $4\pi R^2$). The ratio of the two—the energy-producing volume divided by the surface area through which thermal energy can escape—is

given by

$$\frac{\text{Amount of energy to lose}}{\text{Surface area of energy escape}} = \frac{\text{Volume}}{\text{Surface area}} = \frac{\frac{4}{3} \times \pi R^3}{4\pi R^2} = \frac{R}{3}$$

A planet's ability to transfer internal energy from its hot core to its cooling surface also depends on some of its own internal properties. Nevertheless, all things being equal, planets with larger radii retain their internal energy longer than smaller planets do. For example, Mars has a radius about half that of Earth, so it has been losing its internal thermal energy to space about twice as fast as Earth has. This is one reason that Mars is less geologically active than Earth.

short-lived radioactive elements melted the planet, leading to the differentiated structure. As the surface of Earth radiated energy into space, it cooled rapidly. A solid crust formed above a molten interior. Because a solid crust does not conduct thermal energy well, it helped to retain the remaining heat. Over a long time, energy from the interior of the planet continued to leak through the crust and radiate into space. As a result, the interior of the planet slowly cooled, and the mantle and the inner core solidified.

If the thermal energy from Earth's formation were the only source of heating in Earth's interior, Earth would have long ago solidified completely. Most of the rest of the thermal energy in Earth's interior comes from long-lived radioactive elements trapped in the mantle. As these radioactive elements decay, they release energy, which heats the planet's interior. Today, the temperature of Earth's interior is determined by dynamic equilibrium between the radioactive heating of the interior and the loss of energy to space. As the radioactive elements decay, the amount of thermal energy generated declines, and Earth's interior cools as it ages. A small amount of additional heating of Earth's interior is friction generated by tidal effects of the Moon and Sun.

Although temperature plays an important role in a planet's interior structure, whether a material is solid or liquid also depends on pressure. Higher pressure forces atoms and molecules closer together and makes the material more likely to become a solid. Toward the center of Earth, the effects of temperature and pressure oppose each other: the higher temperatures make it more likely that material will melt, but the higher pressure favors a solid form. In the outer core of Earth, the high temperature wins, allowing the material to exist in a molten state. At the center of Earth, even though the temperature is higher, the pressure is so great that the inner core of Earth is solid.

### CHECK YOUR UNDERSTANDING 8.3

*Differentiation* refers to materials that are separated based on their: (a) weight; (b) mass; (c) volume; (d) density.

## Magnetic Fields

A magnetic field is created by moving charges and exerts a force on magnetically reactive objects, such as iron and on charged particles. A navigation compass is a familiar example on Earth. A compass needle lines up with Earth's magnetic field and points "north" and "south," as shown in **Figure 8.12a**. In the north, a compass needle points to a location in the Arctic Ocean off the coast of northern Canada, near to but not at the geographic North Pole (about which Earth spins). In the south, a compass needle points to a location off the coast of Antarctica, 2,800 km from the geographic South Pole. Earth behaves as if it contained a giant bar magnet that was slightly tilted with respect to the planet's rotation axis and had its two endpoints near the two magnetic poles, as shown in Figure 8.12b. Earth also has a **magnetosphere**, which is the region surrounding a planet that is filled with relatively intense magnetic fields and charged particles.

Earth's magnetic field is not actually due to a bar magnet buried within the planet. A magnetic field is the result of moving electric charges. Earth's magnetic field is created by the combination of Earth's rotation about its axis and a liquid, electrically conducting, circulating outer core. From this combination, Earth converts mechanical energy into magnetic energy. The magnetic field of a planet is an important probe into its internal structure.

Earth's magnetic field is constantly changing. At the moment, the north magnetic pole is traveling several tens of kilometers per year toward the northwest. If this rate and direction continue, the north magnetic pole could be in Siberia before the end of the century. The magnetic pole tends to wander, constantly changing direction as a result of changes in the core.

The geological record shows that much more dramatic changes in the magnetic field have occurred over the history of our planet. When a magnet made of material such as iron gets hot enough, it loses its magnetization. As the material cools, it again becomes magnetized by any magnetic field surrounding it. Thus, iron-bearing minerals record the direction of Earth's magnetic field at the time that they cooled. In this way, a memory of that magnetic field becomes "frozen" into the material. For example, lava extruded from a volcano carries a record of Earth's magnetic field at the moment the lava cooled. By using radiometric techniques to date these materials, geologists obtain a record of how Earth's magnetic field has changed over time. Although Earth's magnetic field has probably existed for at least 3.5 billion years, the north and south magnetic poles switch from time to time. On average, these reversals in Earth's magnetic field take place about every half-million years.

The general idea of how Earth's magnetic field (and those of other planets) originates is called the dynamo theory. In general, magnetic fields result from electric currents, which are moving electric charges. Earth's magnetic field is thought to be a side effect of three factors: Earth's rotation about its axis; an electrically conducting, liquid outer core; and fluid motions within the outer core. This model has been tested with computer simulations, which also produce the pole reversals. The theory suggests that any rotating planet will have a magnetic field if it has an internal heat source.

During the *Apollo* program, astronauts measured the Moon's local magnetic fields, and small satellites have searched for global magnetism. The Moon has a very weak field, possibly none at all, because the Moon is very small and therefore has a solid (not liquid and rotating) inner core. The Moon also has a very small core. However, remnant magnetism is preserved in lunar rocks from an earlier

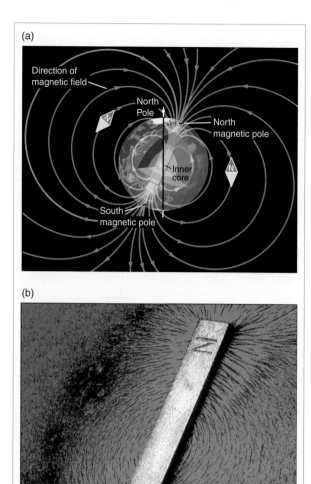

VISUAL ANALOGY

**Figure 8.12** (a) Earth's magnetic field can be visualized as though it were a giant bar magnet tilted relative to Earth's axis of rotation. Compass needles line up along magnetic-field lines and point toward Earth's north magnetic pole. (b) Iron filings sprinkled around a bar magnet help us visualize such a magnetic field.

time when the lunar surface and rocks solidified. Recent analysis of the oldest of the lunar rocks brought back on *Apollo 17* suggests that 4.2 billion years ago, the Moon had a liquid core with a generated magnetic field that lasted at least a few million years. Because the Moon has solidified, it no longer generates a global magnetic field, so what is detected is the "fossil" remains of its early field. Data from India's *Chandrayaan-1* spacecraft suggest that the Moon has a very weak and localized magnetosphere.

Other than Earth, Mercury is the only terrestrial planet with a significant global magnetic field today—although its field is only about 1 percent as strong as Earth's. Slow rotation and a large iron core, parts of which are molten and circulating, cause Mercury's magnetic field. Because the field is so weak, Mercury's magnetosphere is small, meeting the solar wind about 1,700 km above its surface. At this boundary, twisted bundles of magnetic fields transfer magnetic energy from the planet to space.

Planetary scientists expected that Venus would have a magnetic field because its mass and distance from the Sun imply an iron-rich core and partly molten interior like Earth's. Its lack of a magnetic field might be attributed to its extremely slow rotation (see Table 8.1), but this explanation is still uncertain. Or perhaps Venus's magnetic field is temporarily dormant—a condition that Earth is believed to have experienced at times of magnetic field reversals.

Mars has a weak magnetic field, presumably frozen in place early in its history. The magnetic signature occurs only in the ancient crustal rocks, showing that early in the history of Mars, some sort of an internally generated magnetic field must have existed. Geologically younger rocks lack this residual magnetism, so the planet's original magnetic field has long since disappeared. The lack of a strong magnetic field today on Mars might be the result of its small core. Or Mars might have lost its ability to generate a magnetic field after a series of giant impacts early in the planet's history, which could have heated the mantle of Mars enough to reduce the flow of heat out of the core to the mantle. The oldest large impact basins on Mars appear to be magnetized; newer ones are not.

---

### CHECK YOUR UNDERSTANDING 8.4

The dynamo theory says that a planet will have a strong magnetic field if it has: (a) fast rotation and a solid core; (b) slow rotation and a liquid core; (c) fast rotation and a liquid core; (d) slow rotation and a solid core; (e) fast rotation and a gaseous core.

**Figure 8.13** Tectonic processes fold and warp Earth's crust, as seen in these rocks along a roadside in Israel.

## 8.4 Planetary Surfaces Evolve through Tectonism

Now that we have looked at planetary interiors, we can connect the interior conditions to the processes that shape the surface. The crust and part of the upper mantle form the **lithosphere** of a planet. **Tectonism**, the deformation of a planet's lithosphere, warps, twists, and shifts the lithosphere to form visible surface features. If you have driven through mountainous or hilly terrain, you may have seen places like the one shown in **Figure 8.13**, where the roadway has been cut through rock. The exposed layers tell the story of Earth through the vast expanse of geological time. In this section, we will look at tectonic processes that create these layers and play an important part in shaping the surface of a planet.

## The Theory of Plate Tectonics

Early in the 20th century, some scientists recognized that Earth's continents could be fit together like pieces of a giant jigsaw puzzle. In addition, the layers in the rock and the fossil records they hold on the east coast of South America match those on the west coast of Africa. Based on this evidence, Alfred Wegener (1880–1930) proposed a hypothesis that the continents were originally joined in one large landmass that broke apart as the continents began to "drift" away from each other over millions of years. This hypothesis was further developed into the theory known today as **plate tectonics**. Geologists now recognize that Earth's outer shell is composed of a number of relatively brittle segments, or **lithospheric plates**. There are about seven major plates and about a half dozen smaller plates floating on top of the mantle. The motion of these plates is constantly changing the surface of Earth.

Originally, the idea of plate tectonics was met with great skepticism among geologists because they could not imagine a mechanism that could move such huge landmasses. In the late 1950s and early 1960s, however, studies of the ocean floor provided compelling evidence for plate tectonics. These surveys showed surprising characteristics in bands of basalt—a type of rock formed from cooled lava—that were found on both sides of the ocean rifts. Ocean floor rifts such as the Mid-Atlantic Ridge are **spreading centers**. As **Figure 8.14** shows, hot material in these rifts rises toward Earth's surface, becoming new ocean floor. When this hot material cools, it becomes magnetized along the direction of Earth's magnetic field, thus recording the direction of Earth's magnetic field at that time. Greater distance from the rift indicates the ocean floor is older and formed at an earlier time. Combined with radiometric dates for the rocks, this magnetic record proved that the spreading of the seafloor and the motions of the plates have continued over long geological time spans.

Precise surveying techniques and global positioning systems (GPSs) can now determine locations on Earth to within a few centimeters. These measurements confirm that Earth's lithosphere is moving. Some areas are being pulled apart by more than 15 centimeters (cm) each year. Over millions of years, such motions add up. Over 10 million years—a short time by geological standards—15 cm/yr becomes 1,500 km, and maps definitely need to be redrawn.

The theory of plate tectonics is perhaps the greatest advance in 20th century geology. Plate tectonics is responsible for a wide variety of geological features on our planet, including the continental drift that Wegener hypothesized.

## The Role of Convection

The movement of lithospheric plates requires immense forces. These forces are the result of thermal energy escaping from the interior of Earth. The transport of thermal energy by the movement of packets of gas or liquid is known as

▶❙❙ **AstroTour:** Continental Drift

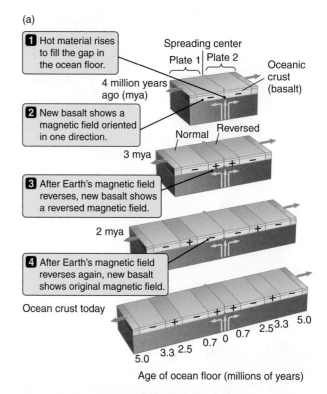

(a)

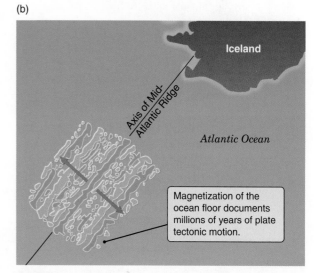

(b)

**Figure 8.14** (a) New seafloor is formed at a spreading center, the cooling rock becomes magnetized, and is then carried away by tectonic motions. (b) Maps like this one of banded magnetic structure in the seafloor near Iceland provide support for the theory of plate tectonics.

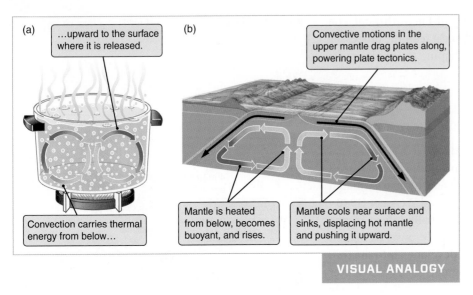

(a) ...upward to the surface where it is released.

Convection carries thermal energy from below...

(b) Convective motions in the upper mantle drag plates along, powering plate tectonics.

Mantle is heated from below, becomes buoyant, and rises.

Mantle cools near surface and sinks, displacing hot mantle and pushing it upward.

**VISUAL ANALOGY**

**Figure 8.15** (a) Convection occurs when a fluid is heated from below. (b) Convection in Earth's mantle drives plate tectonics.

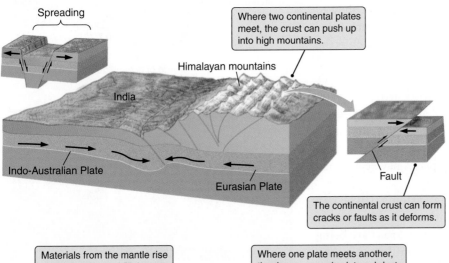

Spreading

Where two continental plates meet, the crust can push up into high mountains.

Himalayan mountains

India

Indo-Australian Plate

Eurasian Plate

Fault

The continental crust can form cracks or faults as it deforms.

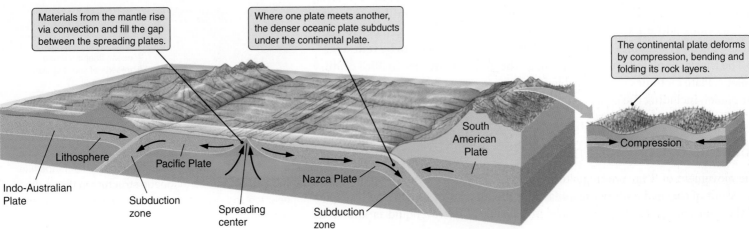

Materials from the mantle rise via convection and fill the gap between the spreading plates.

Where one plate meets another, the denser oceanic plate subducts under the continental plate.

The continental plate deforms by compression, bending and folding its rock layers.

South American Plate

Compression

Lithosphere

Pacific Plate

Nazca Plate

Indo-Australian Plate

Subduction zone

Spreading center

Subduction zone

**Figure 8.16** Divergence and collision of tectonic plates create a wide variety of geological features.

**convection**. **Figure 8.15a** illustrates the process. If you have ever watched water in a heated pot on a stovetop, you have observed convection. Thermal energy from the stove warms water at the bottom of the pot. The warm water expands slightly, becoming less dense than the cooler water above it, and the cooler water with higher density sinks, displacing the warmer water upward. When the lower-density water reaches the surface, it gives up part of its energy to the air and cools; as the water cools, it then becomes denser and sinks back toward the bottom of the pot. Water rises in some locations and sinks in others, forming convection cells. As we will see in later chapters, convection also plays an important role in planetary atmospheres and in the structure of the Sun and stars.

Figure 8.15b shows how convection works in Earth's mantle. Radioactive decay provides the heat source to drive convection in Earth's mantle. We know the mantle is not molten because secondary seismic waves would not be able to travel through it, but the mantle is somewhat mobile. Think of the mantle as having the consistency of hot molten glass. This consistency allows convection to take place very slowly. Convection cells in Earth's mantle drive the plates, carrying both continents and ocean crust along with them. Convection also creates new crust along rift zones in the ocean basins, where mantle material rises up, cools, and slowly spreads out.

**Figure 8.16** illustrates plate tectonics and some its consequences. If material rises and spreads in one location, then it must converge and sink in another. Locations where plates converge and convection currents turn downward are called **subduction zones**.

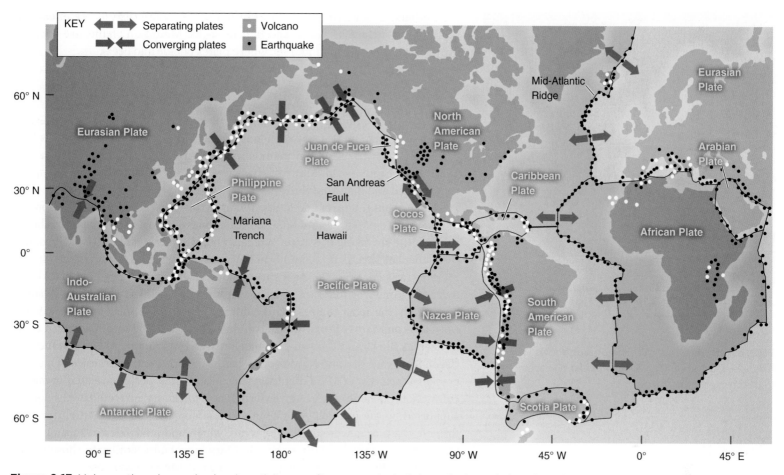

**Figure 8.17** Major earthquakes and volcanic activity are often concentrated along the boundaries of Earth's principal tectonic plates.

In a subduction zone, one plate slides beneath the other, and convection drags the submerged lithospheric material back down into the mantle. The Mariana Trench—the deepest part (11 km) of Earth's ocean floor—is such a subduction zone. Much of the ocean floor lies between spreading centers and subduction zones, and so the ocean floor is the youngest portion of Earth's crust. In fact, the *oldest* seafloor rocks are less than 200 million years old.

In some places, the plates are not sinking but colliding and, consequently, shoved upward. The highest mountains on Earth, the Himalayas, grow a half-meter per century as the Indo-Australian subcontinental plate collides with the Eurasian Plate. In other places, plates meet at oblique angles and slide along past each other. One such place is the San Andreas Fault in California, where the Pacific Plate slides past the North American Plate. A **fault** is a fracture in a planet's crust along which material can slide.

Locations where plates meet tend to be very active geologically. One of the best ways to see the outline of Earth's plates is to look at a map of where earthquakes and volcanism occur, such as the map in **Figure 8.17**. Where plates meet, enormous stresses build up. Earthquakes occur when a portion of the boundary between two plates suddenly slips, relieving the stress. Volcanoes are created when friction between plates melts rock, which is then pushed up through cracks to the surface. Earth also has numerous **hot spots**, where hot deep-mantle material rises, releasing thermal energy. As plates shift, some parts move more rapidly

Figure 8.18 This *Apollo 10* photograph shows Rima Ariadaeus, a 2-km-wide valley between two tectonic faults on the Moon.

than others, causing the plates to stretch, buckle, or fracture. These effects are seen on the surface as folded and faulted rocks. Mountain chains are common near converging plate boundaries, where plates buckle and break.

## Tectonism on Other Planets

We have observed plate tectonics only on Earth. However, all of the terrestrial planets and some moons show evidence of tectonic disruptions. Fractures have cut the crust of the Moon in many areas, leaving fault valleys such as the one pictured in **Figure 8.18**. Many of these features are the result of large impacts that cracked and distorted the lunar crust.

Mercury has fractures and faults similar to those on the Moon. In addition, numerous cliffs on Mercury are hundreds of kilometers long. These appear to be the result of the shrinking of Mercury; recent observations by *Messenger* suggest that the planet has shrunk by about 10–14 km across. Like the other terrestrial planets, Mercury was once molten. As it shrank, Mercury's crust cracked and buckled in much the same way that a grape skin wrinkles as it shrinks to become a raisin.

Possibly the most impressive tectonic feature in the Solar System is Valles Marineris on Mars (**Figure 8.19**). Stretching nearly 4,000 km, and nearly 4 times as deep as the Grand Canyon, this canyon system is as long as the distance between San Francisco and New York. Valles Marineris includes a series of massive cracks in the crust of Mars that formed as local forces, perhaps related to mantle convection, pushed the crust upward from below. The surface could not be equally supported by the interior everywhere, and unsupported segments fell in. Once formed, the cracks were eroded by wind, water, and landslides, resulting in the structure we see today. Other parts of Mars have faults similar to those on the Moon, but cliffs as high and long as those seen on Mercury are absent.

The mass of Venus is only 20 percent less than that of Earth, and its radius is just 5 percent smaller than Earth's, leading to a surface gravity 90 percent that

Figure 8.19 (a) A mosaic of *Viking Orbiter* images shows Valles Marineris, the major tectonic feature on Mars, stretching across the center of the image from left to right. This canyon system is more than 4,000 km long. The dark spots on the left are huge shield volcanoes. (b) This close-up perspective view of the canyon wall was photographed by the European Space Agency's *Mars Express* orbiting spacecraft.

(a)

(b)

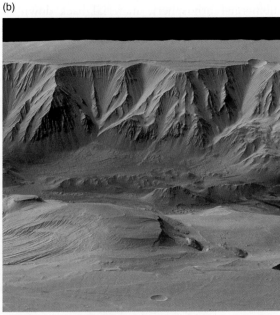

of Earth. Because of the similarities between the two planets, many scientists predicted that Venus might also show evidence of plate tectonics. The NASA *Magellan* mission orbited Venus from 1990 to 1994, and *Magellan* mapped nearly the entire surface of Venus, providing the first high-resolution radar views of the planet's surface (**Figure 8.20**).

The European Space Agency's (ESA's) *Venus Express*, in orbit from 2006 to 2014, mapped Venus in the infrared, which can penetrate through the clouds to enable a view of the surface. The impact craters on Venus seem to be evenly distributed, suggesting that the surface is all about the same age, about a billion years. Venus is mostly covered with smooth lava, but there are two highland regions: Ishtar Terra in the north and Aphrodite Terra in the south. The highland rocks are less smooth and older than those on the rest of Venus and may be similar to granite rocks on Earth. Because granite results from plate tectonics and water, the data hint at the possibility that these highlands on Venus are ancient continents, created by volcanic activity, on a planet with oceans.

Because of the similarities between Venus and Earth, the interior of Venus should be very much like the interior of Earth, and convection should be occurring in its mantle. On Earth, mantle convection and plate tectonism release the most thermal energy from the interior. By contrast, on Venus, hot spots may be the principal way that thermal energy escapes from the planet's interior. Circular fractures called coronae on the surface of Venus, ranging from a few hundred kilometers to more than 2,500 km across, may be the result of upwelling plumes of hot mantle that have fractured Venus's lithosphere. Alternatively, energy may build up in the interior until large chunks of the lithosphere melt and overturn, releasing an enormous amount of energy. Then, the surface cools and solidifies. It is uncertain why Venus and Earth are so different with regard to plate tectonics.

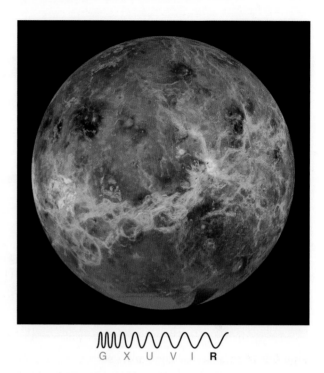

G  X  U  V  I  **R**

**Figure 8.20** The atmosphere of Venus blocks our view of the surface in visible light. This false-color view of Venus is a radar image made by the *Magellan* spacecraft. Bright yellow and white areas are mostly fractures and ridges in the crust. Some circular features seen in the image may be regions of mantle upwelling, or *hot spots*. Most of the surface is formed by lava flows, shown in orange.

### CHECK YOUR UNDERSTANDING 8.5

On which of the following does plate tectonics occur now? (Select all that apply.) (a) Mercury; (b) Venus; (c) Earth; (d) the Moon; (e) Mars

## 8.5 Volcanism Signifies a Geologically Active Planet

You are probably familiar with the image of a volcano spewing molten rock onto the surface of Earth. This molten rock, known as **magma**, originates deep in the crust and in the upper mantle, where sources of thermal energy combine. These sources include rising convection cells in the mantle, heating by friction from movement in the crust, and concentrations of radioactive elements. In this section, we will look at the occurrence of volcanic activity on a planet or moon, which is called **volcanism**. Volcanism not only shapes planetary surfaces but also is a key indicator of a geologically active planet.

### Terrestrial Volcanism Is Related to Tectonism

Volcanoes are usually located along plate boundaries and over hot spots. Maps of geological activity such as the one in Figure 8.17 leave little doubt that most terrestrial volcanism is linked to the same forces responsible for plate motions. A tremendous amount of friction is generated as plates slide under each other. This

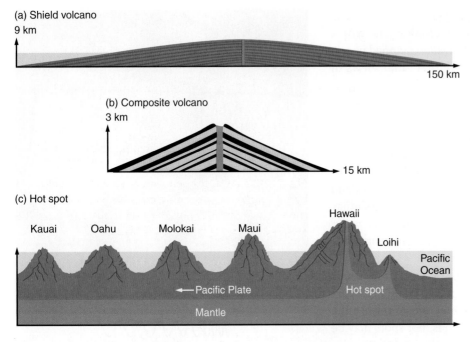

(a) Shield volcano
9 km

150 km

(b) Composite volcano
3 km

15 km

(c) Hot spot

Kauai    Oahu    Molokai    Maui    Hawaii

Loihi

Pacific Ocean

←— Pacific Plate    Hot spot

Mantle

**Figure 8.21** Magma reaching Earth's surface commonly forms (a) shield volcanoes, such as Mauna Loa, which have gently sloped sides built up by fluid lava flows; and (b) composite volcanoes, such as Vesuvius, which have steeply symmetric sides built up by viscous lava flows. (c) Hot spots are convective plumes of lava that can form a successive series of volcanoes as the plate above them slides by.

▶❙❙ **AstroTour:** Hot Spot Creating a Chain of Islands

friction raises the temperature of rock toward its melting point.

Material at the base of a lithospheric plate is under a great deal of pressure because of the weight of the plate pushing down on it. This pressure increases the melting temperature of the material, forcing it to remain solid even at high temperature. As this material is forced up through the crust, its pressure drops, and therefore the material's melting temperature drops too. Material that was solid at the base of a plate becomes molten as it nears the surface. Places where convection carries hot mantle material toward the surface are frequent sites of eruptions. Iceland, which is one of the most volcanically active regions in the world, sits astride one such spreading center—the Mid-Atlantic Ridge (see Figure 8.17). In recent years, volcano eruptions in Iceland have led to travel disruptions for thousands as the airborne volcanic ash made it unsafe for airplanes to fly near the eruption.

Once lava reaches the surface of Earth, it can form many types of structures. Flows often form vast sheets, especially if the eruptions come from long fractures called fissures. If very fluid lava flows from a single *point source*, it spreads out over the surrounding terrain or ocean floor, forming a **shield volcano**, shown in **Figure 8.21a**. A **composite volcano** forms when thick lava flows alternating with explosively generated rock deposits build a steep-sided structure, shown in Figure 8.21b.

Terrestrial volcanism also occurs where convective plumes rise toward the surface in the interiors of lithospheric plates, creating local hot spots. Volcanism over hot spots works much like volcanism elsewhere at a spreading center, except that the convective upwelling occurs at a single spot rather than along the edge of a plate. These hot spots force mantle and lithospheric material toward the surface, where the material emerges as liquid lava.

Earth has numerous hot spots, including the regions around Yellowstone Park and the Hawaiian Islands (Figure 8.21c). The Hawaiian Islands are a chain of shield volcanoes that formed as the lithospheric plate moved across a hot spot. Volcanoes erupt over a hot spot, building an island. The island ceases to grow as the plate motion carries the island away from the hot spot. Meanwhile, a new island grows over the hot spot. Today, the Hawaiian hot spot is located off the southeast coast of the Big Island of Hawaii, where it continues to power the active volcanoes. On top of the hot spot, the newest Hawaiian island, Loihi, is forming. Loihi is already a massive shield volcano, rising more than 3 km above the ocean floor. Loihi will eventually break the surface of the ocean and merge with the Big Island of Hawaii—but not for another 100,000 years.

## Volcanism in the Solar System

**The Moon** Although Earth is the only planet on which plate tectonics is an important process, evidence of volcanism is found throughout the Solar System, including several moons of the outer planets. Some of the first observers to use telescopes to view the Moon noted dark areas that looked like bodies of water—thus they were named **maria** (singular: *mare*), Latin for "seas." Early photographs showed flowlike features in the dark regions of the Moon. We now know that the maria are actually vast, hardened lava flows, similar to volcanic rocks known as basalts on Earth. Because the maria contain relatively few craters, these volcanic flows must have occurred after the period of heavy bombardment ceased.

Many of the rock samples that the *Apollo* astronauts brought back from the lunar maria were found to contain gas bubbles typical of volcanic materials (**Figure 8.22**). The lava that flowed across the lunar surface must have been relatively fluid. This fluidity, due partly to the lava's chemical composition, explains why lunar basalts form vast sheets that fill low-lying areas such as impact basins (**Figure 8.23**). It also partly explains the Moon's lack of classic volcanoes: the lava was too fluid to pile up, like motor oil poured from a container spreading out.

The lunar rock samples also showed that most of the lunar lava flows are older than 3 billion years. Samples from the heavily cratered terrain of the Moon also originated from magma, indicating that the young Moon went through a molten stage. These rocks cooled from a "magma ocean" and are more than 4 billion years old, preserving the early history of the Solar System. Most of the sources of heating and volcanic activity on the Moon must have shut down some 3 billion years ago—unlike on Earth, where volcanism continues. This conclusion is consistent with the idea that smaller objects and planets cool more efficiently and thus are less active than larger planets.

Only in a few limited areas of the Moon are younger lavas thought to exist; most of these have not been sampled directly. The *Lunar Reconnaissance Orbiter* observed volcanic cones that were likely built up from volcanic rocks erupting from the surface. These volcanic rocks are far different from the mare basalt rocks and contain silica and thorium. These domes could have been formed as recently as 800 million years ago, which would make them the result of the most recent volcanic activity found on the Moon.

**Mercury**  Mercury also shows evidence of past volcanism. The *Mariner 10* and *Messenger* missions revealed smooth plains on Mercury similar in appearance to the lunar maria. These sparsely cratered plains are the youngest areas on Mercury and, like those on the Moon, are almost certainly volcanic in origin, created when fluid lavas flowed into and filled huge impact basins. Many of the volcanic plains on Mercury are also associated with impact scars. The volcanic activity that created the plains likely ceased 3.8 billion years ago, possibly from the shrinking of the planet as it cooled. The ending of the late heavy bombardment might also have been a factor. High-resolution imaging by *Messenger* has also identified a number of volcanoes. Vents that could be from explosive volcanism have been found around the large, old impact basin Caloris (**Figure 8.24**) and may be as young as 1 billion to 2 billion years old.

**Mars**  Mars has also been volcanically active. More than half the surface of Mars is covered with volcanic rocks. Lavas covered huge regions of Mars, flooding the older, cratered terrain. Most of the vents or long cracks that created these flows are buried under the lava that poured forth from them. Among the most impressive features on Mars are its enormous shield volcanoes. These volcanoes are the largest mountains in the Solar System. Olympus Mons,

**Figure 8.22** This rock sample from the Moon, collected by the *Apollo 15* astronauts from a lunar lava flow, shows gas bubbles typical of gas-rich volcanic materials. The rock is about 6 × 12 cm.

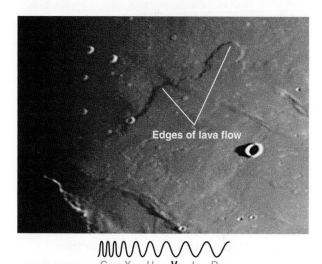

G X U **V** I R

**Figure 8.23** The lava flowing across the surface of Mare Imbrium on the Moon must have been extremely fluid to spread out for hundreds of kilometers in sheets that are only tens of meters thick.

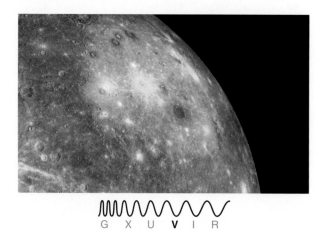

G X U **V** I R

**Figure 8.24** The Caloris Basin on Mercury (yellow) is one of the largest impact basins in the Solar System, with a span of about 1,500 km. The orange regions may be volcanic vents. The false color is enhanced to show more detail.

G X U V I R

**Figure 8.25** The largest known volcano in the Solar System, Olympus Mons is a 27-km-high shield-type volcano on Mars, similar to but much larger than Hawaii's Mauna Loa. This partial view of Olympus Mons was taken by the *Mars Global Surveyor*.

standing 27 km high at its peak and 550 km wide at its base (**Figure 8.25**), would tower over Earth's largest mountains. Olympus Mons and its neighbors grew as the result of hundreds of thousands of individual eruptions. In the absence of plate tectonics on Mars, its volcanoes have remained over their hot spots for billions of years, growing ever taller and broader in the lower surface gravity with each successive eruption.

Lava flows and other volcanic landforms span nearly the entire history of Mars, estimated to extend from the formation of crust some 4.4 billion years ago to geologically recent times and to cover more than half of the red planet's surface. Although some "fresh-appearing" lava flows have been identified on Mars, until rock samples are radiometrically dated we will not know the age of these latest eruptions. Mars could, in principle, experience eruptions today.

**Venus** Of the terrestrial planets, Venus has the most volcanoes. Radar images reveal a wide variety of volcanic landforms. These include highly fluid flood lavas covering thousands of square kilometers, shield volcanoes approaching those of Mars in terms of size and complexity, and lava channels thousands of kilometers long. These lavas must have been extremely hot and fluid to flow for such long distances. Some of the volcanic eruptions on Venus are thought to have been associated with deformation of Venus's lithosphere above hot spots such as the circular features mentioned earlier.

Lavas on Venus are basalts, much like the lavas on Earth, the Moon, Mars, and possibly Mercury. The *Venus Express* spacecraft imaged three of the nine Hawaii-like hot spots in infrared wavelengths. Each hot spot has several volcanoes, with altitudes of 500–1,200 meters above the nearby plains. It was found that some of the volcanic regions were radiating heat more efficiently than the regions nearby. This suggests that these volcanic regions have younger material and that volcanic activity had taken place within the past 2.5 million years, and perhaps as recently as a few thousand years ago. As we'll describe in the next chapter, Venus has some of the volcanic gas sulfur dioxide in its atmosphere, so Venus may still be cooling its interior through volcanic activity.

A geological timescale for Venus has not yet been devised, but from its relative lack of impact craters, most of the surface is considered to be less than 1 billion years old, and some of it may be much more recent. When volcanism began on Venus and how much active volcanism exists today remain unanswered questions.

### CHECK YOUR UNDERSTANDING 8.6

Which is *not* a reason for the large size of volcanoes on Mars compared to Earth's smaller volcanoes? (a) absence of plate tectonics; (b) distance from the Sun; (c) lower surface gravity than Earth's; (d) many repeated eruptions

## 8.6 The Geological Evidence for Water

Today, Earth is the only planet in the Solar System where the temperature and atmospheric conditions allow extensive liquid surface water to exist. The other inner planets do not have extensive liquid surface water, but there is evidence for water ice in deep craters or in the polar regions and for water below the surface in permafrost, subsurface glaciers, or possibly as liquid in the core.

Life, as we know it on Earth, requires water as a solvent and as a delivery mechanism for essential chemistry. Because of this, the search for water is central to the search for life in the Solar System. Additionally, if humans are ever going to live on another terrestrial planet, they will need a source of water. In this section, we will look at how water modifies the surface of a planet and then discuss the search for water in the Solar System.

## Water and Erosion

Tectonism, volcanism, and impact cratering affect Earth's surface by creating variations in the height of the surface. **Erosion** is the wearing away of a planet's surface by mechanical action. The term *erosion* covers a wide variety of processes. Erosion by running water, but also by wind and by the actions of living organisms, wears down hills, mountains, and craters; the resulting debris fills in valleys, lakes, and canyons. If erosion were the only geological process operating, it would eventually smooth out the surface of the planet completely. Because Earth is a geologically and biologically active world, however, its surface is an ever-changing battleground between processes that build up topography and those that tear it down.

Weathering is the first step in the process of erosion. During weathering, rocks are broken into smaller pieces and may be chemically altered. For example, rocks on Earth are physically weathered along shorelines, where the pounding waves break them into beach sand. Other weathering processes include chemical reactions, such as when oxygen in the air combines with iron in rocks to form a type of rust. One of the most efficient forms of weathering is caused by water: liquid water runs into crevices and then freezes. As the water freezes, it expands and shatters the rock.

After weathering, the resulting debris can be carried away by flowing water, glacial ice, or blowing wind and deposited in other areas as sediment. Where material is eroded, we can see features such as river valleys, wind-sculpted hills, or mountains carved by glaciers. Where eroded material is deposited, we see features such as river deltas, sand dunes, or piles of rock at the bases of mountains and cliffs. Erosion is most efficient on planets with water and wind. On Earth, where water and wind are prevalent, most impact craters on land are worn down and filled in.

Even though the Moon and Mercury have almost no atmosphere and no running water, a type of slow erosion is still at work. Radiation from the Sun and from deep space slowly decomposes some types of minerals, effectively weathering the rock. Such effects are only a few millimeters deep at most. Impacts of micrometeoroids also chip away at rocks. In addition, landslides can occur wherever gravity and differences in elevation are present. Although water enhances landslide activity, landslides are also seen on dry bodies like Mercury and the Moon.

As we will discuss in the next chapter, Earth, Mars, and Venus have atmospheres, and all three planets show the effects of windstorms. Images of Mars and Venus returned by spacecraft landers show surfaces that have been subjected to the forces of wind. Sand dunes are common on Earth and Mars (**Figure 8.26**), and some have been identified on Venus. Orbiting spacecraft have also found wind-eroded hills and surface patterns called wind streaks. These surface patterns appear, disappear, and change in response to winds blowing sediments around hills, craters, and cliffs. They serve as local weather vanes, telling plane-

G X U **V** I R

**Figure 8.26** A *Mars Reconnaissance Orbiter* image of the Nili Patera dune field on Mars. The dunes change over months because of winds on Mars.

tary scientists about the direction of local prevailing surface winds. Planet-encompassing dust storms have been seen on Mars.

## The Search for Water in the Solar System

A priority of recent planetary exploration missions is the search for water on the terrestrial planets and the Moon. Some of the evidence for past water comes from the geological processes discussed earlier in the chapter. Water was brought in by impacts in the early Solar System, and it is affected by geological and atmospheric activity on a planet. The search for water includes examination of images of the terrain obtained by flybys, orbiters, and landers. For the Moon, the search has included reanalyzing 40-year-old lunar rocks and soil brought back to Earth in the *Apollo* missions and crashing spacecraft into the surface in order to analyze the debris that is kicked up.

**Mars**  The search for water on Mars goes back almost a century and a half. Even small telescopes show polar ice caps, which change with the martian seasons. In 1877, Italian astronomer Giovanni Schiaparelli (1835–1910) observed what appeared to be linear features on Mars and dubbed them *canali* ("channels" in Italian). Unfortunately, some other observers, including the American observer of Mars, Percival Lowell (1855–1916), incorrectly translated Schiaparelli's *canali* as "canals," implying that they were artificially constructed by intelligent life, rather than naturally formed by geology. Lowell strongly advocated the theory that these "canals" were built to move water around a drying planet Mars. Other observers of the time disputed this idea, arguing that these were optical illusions seen in telescopes. Larger telescopes and astrophotographs did not show canals, astronomical spectroscopy did not find water vapor, and this idea of artificial canals went out of favor after Lowell died.

Scientists debate about how recently there has been significant liquid water flow on the surface of Mars. The geological evidence suggests that at one time, water flowed across the surface of Mars in vast quantities. Canyons and huge, dry riverbeds attest to tremendous floods that poured across the martian surface. In addition, many regions on Mars show small networks of valleys that probably were carved by flowing water (**Figure 8.27**). Large deposits of subsurface ice have been detected under the surface. Mars may have contained oceans at one time, including one ocean that might have covered a third of the planet's surface.

In 2004, NASA sent two instrument-equipped roving vehicles, *Opportunity* and *Spirit*, to search for evidence of water on Mars. *Opportunity* landed inside a crater. For the first time, martian rocks were available for study in the original order in which they had been laid down. Previously, the only rocks that landers and rovers had come across were those that had been dislodged from their original settings by either impacts or river floods. The layered rocks at the *Opportunity* site revealed that they had once been soaked in or transported by water. The form of the layers was typical of layered sandy deposits laid down by gentle currents of water. Magnified images of the rocks showed "blueberries," small, bluish spheres a few millimeters across that probably formed in place among the layered rocks. Analysis of the spheres revealed abundant hematite, an iron-rich mineral that forms in the presence of water.

Observations by ESA's *Mars Express* and NASA's *Mars Odyssey* and *Mars Reconnaissance Orbiter* have shown the hematite signature and the presence of

G X U V I R

**Figure 8.27** A photograph of gully channels in a crater on Mars taken by the *Mars Reconnaissance Orbiter.* The gullies coming from the rocky cliffs near the crater's rim (out of the image, to the upper left) show meandering and braided patterns similar to those of water-carved channels on Earth.

sulfur-rich compounds in a vast area surrounding the *Opportunity* landing site. These observations suggest the existence of an ancient martian sea larger than the combined area of the Great Lakes and as much as 500 meters deep.

In August 2012, NASA's Mars rover *Curiosity* landed in Gale Crater, a large (150 km) crater just south of the equator of Mars. *Curiosity* found evidence of a stream that flowed at a rate of about 1 meter per second and was as much as 2 feet deep. The streambed is identified by water-worn gravel (**Figure 8.28**). The rover, which is about the size of a car, includes cameras, a drill, and an instrument to measure chemical composition. When the rover drilled into a rock, it found sulfur, nitrogen, hydrogen, oxygen, phosphorus, and carbon, together with clay minerals that formed in a water-rich environment that was not very salty. Taken together, these pieces of evidence indicate that Mars may have had conditions suitable to support Earth-like microbial life in the distant past.

**Figure 8.28** This image compares a photograph taken by NASA's *Curiosity* rover (left) with a photograph of a streambed on Earth (right). The Mars image shows water-worn gravel embedded in sand, sure evidence of an ancient streambed.

Where did the water go? Some escaped into the thin atmosphere of Mars, and some is locked up as ice in the polar regions, just as the ice caps on Earth hold much of its water. Unlike Earth's polar caps of frozen water, those on Mars are a mixture of frozen carbon dioxide and frozen water. Water must be hiding elsewhere on Mars. Small amounts of water can be found on the surface, and in 2008 NASA's *Phoenix* lander found water ice just a centimeter or so beneath surface soils at high northern latitudes (**Figure 8.29**). However, most of the water on Mars appears to be trapped well below the surface. Radar imaging by *Mars Express* and the *Mars Reconnaissance Orbiter* (*MRO*) indicates huge quantities of subsurface water ice, not only in the polar areas as expected but also at lower latitudes under craters. In addition, *MRO* images suggest that there might be seasonal salt water that flows on the surface far from the poles. Salt water freezes at a lower temperature, so some sites could be warm enough to have temporary liquid salt water. Another location for liquid water may be in martian volcanoes.

**Venus** Evidence for liquid water on Venus comes primarily from water vapor in its atmosphere, but there are some geological indications of past water, such as color differences between the highland and lowland regions. As noted earlier, on Earth such a difference indicates the presence of granite, which requires water for its formation. We will return to the subject of what happened to the water on Venus and Mars in the next chapter, when we discuss their atmospheres.

**The Moon** Infrared measurements of the Moon by the U.S. *Clementine* mission in 1994 returned information supporting the possibility of ice at the lunar poles. In 1998, NASA's *Lunar Prospector* observations suggested subsurface water ice in

**Figure 8.29** Water ice appears a few centimeters below the surface in this trench dug by a robotic arm on the *Phoenix* lander. The trench measures about 20 × 30 cm.

the polar regions. When its primary mission was completed, NASA crashed *Lunar Prospector* into a crater near the Moon's south pole while ground-based telescopes searched for evidence of water vapor above the impact site, but none was seen. Since 2007, several new missions have been sent to the Moon, including Japan's *Kaguya*, India's *Chandrayaan-1*, and NASA's *Lunar Reconnaissance Orbiter* (*LRO*) and its companion *Lunar Crater Observation and Sensing Satellite* (*LCROSS*).

In 2009, NASA crashed the *LCROSS* launch vehicle into the Cabeus Crater at the lunar south pole, which sent up dust and vapor that was analyzed by the *LCROSS* and *LRO* spacecraft. More than 5 percent of the resulting plume was water, which makes this part of the Moon wetter than many Earth deserts. Other volatiles were also detected. Measurements of hydrogen by the *LRO* suggest that there is a fair amount of buried water ice in the cold southern polar region.

These space observations of lunar ice sent planetary scientists back to the collections of lunar rocks and soil returned to Earth decades ago by the *Apollo* mission astronauts. New analysis of volcanic glass beads in lunar soil suggests that the interior of the Moon may have a much larger amount of volatiles than previously believed. Reanalysis of the *Apollo* lunar rocks also found evidence of water. One way to distinguish whether water on the Moon originated from its interior or from impacts is to look at the ratio of water molecules composed of regular hydrogen (with one proton) and oxygen to water molecules composed of oxygen, hydrogen, and an isotope of hydrogen called deuterium (hydrogen with one proton and one neutron). The ratio is higher in the water in lunar rocks than in the water on Earth. Water that originated in comets or in meteorites rich in water has a different ratio. Alternatively protons from the solar wind or from high-energy cosmic rays in space could have combined with oxygen on the lunar surface, yielding a different ratio. A recent study suggests that the Moon's water came from the solar wind.

These results on lunar water are preliminary. There is still considerable debate among scientists about exactly how much water ice exists on the lunar surface and how much liquid water is in the interior, how the Moon acquired this water, and how the presence and origin of water affect the current theories of lunar formation. Several countries—and some private companies—are considering proposals to send robotic spacecraft to the Moon to collect additional lunar material and bring it back to Earth for analysis.

**Mercury** Water ice has also been detected in the polar regions of Mercury. Some deep craters in the polar regions of Mercury have floors that are in perpetual shadow, and thus receive no sunlight. Temperatures in these permanently shadowed areas remain very cold, below 180 K. For many years, planetary scientists had speculated that ice could be found in these polar craters, and there was a possible detection by radar in the early 1990s. The *Messenger* spacecraft, which orbited Mercury from 2011–2015, found deposits of ice at craters at the planet's north pole. Other frozen volatiles were also seen in the polar craters. The icy areas have sharp boundaries, which indicates they are relatively recent, either from comet impacts or from some ongoing process on the planet.

## CHECK YOUR UNDERSTANDING 8.7

Which of the following worlds show evidence of the current presence of liquid or frozen water? (Choose all that apply.) (a) Mercury; (b) Venus; (c) Earth; (d) the Moon; (e) Mars

# Origins

## The Death of the Dinosaurs

When large impacts happen on Earth, they can have far-reaching consequences for Earth's climate and for terrestrial life. One of the biggest and most significant impacts happened at the end of the Cretaceous Period, which lasted from 146 million years ago to 65 million years ago. At the end of the Cretaceous Period, more than 50 percent of all living species, including the dinosaurs, became extinct. This mass extinction is marked in Earth's fossil record by the Cretaceous-Tertiary boundary, or *K-T boundary* (the *K* comes from *Kreide*, German for "Cretaceous"). Fossils of dinosaurs and other now-extinct life-forms are found in older layers below the K-T boundary. Fossils in the newer rocks above the K-T boundary lack more than half of all previous species but contain a record of many other newly evolving species. Big winners in the new order were the mammals—distant ancestors of humans—that moved into ecological niches vacated by extinct species.

How do scientists know that an impact was involved? The K-T boundary is marked in the fossil record in many areas by a layer of clay. Studies at more than 100 locations around the world have found that this layer contains large amounts of the element iridium, as well as traces of soot. Iridium is very rare in Earth's crust but is common in meteorites. The soot at the K-T boundary possibly indicates that widespread fires burned the world over. The thickness of the layer of clay at the K-T boundary and the concentration of iridium increases toward what is today the Yucatán Peninsula in Mexico. Although the original crater has largely been erased by erosion, geophysical

**Figure 8.30** This artist's rendition depicts an asteroid or comet, perhaps 10 km across, striking Earth 65 million years ago in what is now the Yucatán Peninsula in Mexico. The lasting effects of the impact might have killed off most forms of terrestrial life, including the dinosaurs.

surveys and rocks from drill holes in this area show a deeply deformed subsurface rock structure, similar to that seen at known impact sites. These results provide compelling evidence that 65 million years ago, an asteroid about 10 km in diameter struck the area, throwing great clouds of red-hot dust and other debris into the atmosphere (**Figure 8.30**) and possibly igniting a worldwide conflagration. The energy of the impact is estimated to have been more than that released by 5 billion nuclear bombs.

An impact of this energy clearly would have had a devastating effect on terrestrial life. In addition to the possible firestorm ignited by the impact, computer models suggest there would have been earthquakes and tsunamis. Dust from the collision and soot from the firestorms thrown into Earth's upper atmosphere would have remained there for years, blocking out sunlight and plunging Earth into decades of a

cold and dark "impact winter." Recent measurements of ancient microbes in ocean sediments suggest that Earth may have cooled by 7°C. The firestorms, temperature changes, and decreased food supplies could have led to a mass starvation that would have been especially hard on large animals such as the dinosaurs.

Not all paleontologists believe that this mass extinction was the result of an impact; some think volcanic activity was important as well. However, the evidence is compelling that a great impact did occur at the end of the Cretaceous Period. Life on our planet has had its course altered by sudden and cataclysmic events when asteroids and comets have slammed into Earth. It seems very possible that we owe our existence to the luck of our remote ancestors—small rodent-like mammals—that could live amid the destruction after such an impact 65 million years ago.

Scientists report that the Moon may have had volcanic activity more recently than expected.

# Did Volcanoes Erupt on the Moon while Dinosaurs Roamed Earth?

By **AMINA KHAN, Los Angeles Times**

Ever looked up into the night sky and seen the ancient face of the "man on the Moon"? Well, turns out he may have had some recent work done.

Scientists thought the Moon has been cold and dead for roughly a billion years. But strange small features on the surface discovered by NASA's *Lunar Reconnaissance Orbiter* reveal that there could have been volcanic activity during the time of the dinosaurs. That's practically just last week, by geological timescales.

The findings, described in the journal *Nature Geoscience*, could force researchers to reconsider established theories on the Moon's evolution.

The large dark patches on the lunar surface that give shape to the Moon's "face" are called *maria*, and they're thought to be the remains of volcanic activity on the Moon that started 3.5 billion years ago. Scientists thought the Moon cooled quickly, and this period of volcanism ended abruptly, around a billion years back.

There were some anomalies—for example a strange feature called Ina, imaged from orbit by *Apollo 15* astronauts in 1971, that seemed to be very young. But Ina was thought to be an exception, and it wasn't clear what its existence meant.

Now, using the NASA orbiter, a team led by Arizona State University scientists picked out 70 of these strange features—round, smooth areas, surrounded by rough, choppy terrain. They're called "irregular mare patches," and

they're too small to be distinguished with the naked eye from Earth. Though they range in size from 328 feet to 3.1 miles, the average is 1,591 feet, or 0.3 miles (**Figure 8.31**).

Since the scientists have no way to bring this rock back to the lab to study, they relied on a commonly used method to date these strange features. Essentially, the more craters that pock their surfaces, the older they are (because older features have had more time to be smashed by space debris). These crater counts have been calibrated using the laboratory-measured ages of Moon rock samples brought back by *Apollo* astronauts.

The researchers found that scores of these features were less than 100 million years old—which would put them in range of the Cretaceous Period on Earth, which was the dinosaurs' heyday. Some were even younger—Ina could be 33 million years old

and another patch, Sosigenes, could be just 18 million years old.

That means there could have been regular volcanic activity all around the Moon in very recent times—not like the dramatic volcanism that produced the enormous *maria*, but still significant and widespread. It also means that the Moon cooled more gradually than scientists thought, and that we may not really understand how much heat still remains inside of it. Theories about the Moon's thermal evolution might need a serious rethink.

The best way to know for sure? Return to the lunar surface and bring back rock samples that can be analyzed in the lab, the authors say. (The last time that happened was in the 1970s.)

"Sample return will be required for radiometric age dating to confirm the relatively young ages implied by remote sensing observations," the study authors wrote.

**Figure 8.31** This feature on the lunar surface, called Maskelyne, is one of many newly-discovered young volcanic deposits on the moon. These "irregular mare patches" are thought to be remnants of small eruptions that occurred just a few tens of millions of years ago.

1. Why did scientists expect that the Moon cooled quickly?
2. Why are the features called "maria" distributed in patches?
3. What was the evidence that indicated volcanic activity ended a billion years ago?
4. Why would bringing a sample back to Earth yield a more accurate age estimate for more recent volcanic activity?
5. Do a search for "Moon volcanic activity." Are there any new findings about either more recent or billion-year-old volcanic activity?

# Summary

The terrestrial planets in the Solar System are Mercury, Venus, Earth, and Mars, all of which have evidence of past or present water. The Moon is usually included in discussions of the terrestrial worlds because it is similar to them in many ways. Comparative planetology is the key to understanding the planets. Four geological processes—impacts, volcanism, tectonism, and erosion—are responsible for topography on the terrestrial planets. Active volcanism and tectonics are the results of a "living" planetary interior: one that is still hot inside. Over time, the interiors cool, and tectonics and volcanism weaken. On Earth, radioactive decay and tidal effects from the Moon contribute to heat in the interior. Erosion is a surface phenomenon that results from weathering by wind or water. Surface features on the terrestrial planets, such as tectonic plates, volcanoes, mountain ranges, or canyons, are the result of the interplay between these four processes. Evolution on Earth may have been affected by impacts, such as the impact of an asteroid 65 million years ago that might have led to the death of the dinosaurs.

**LG 1** **Describe how impacts have affected the evolution of the terrestrial planets.** Impact cratering is the result of a direct interaction of an astronomical object with the surface of a planet. The layering of craters gives their relative ages, with more recent craters found superimposed on older ones. Crater densities can be used to find the relative ages of regions on a surface; more heavily cratered regions are older than less cratered ones. Planets protected by atmospheres, like Earth and Venus, have fewer small impact craters. The Moon was probably created when a Mars-sized protoplanet collided with Earth.

**LG 2** **Explain how radiometric dating is used to measure the ages of rocks and planetary surfaces.** Radioactive isotopes found in rocks can be used to measure their age. The oldest rocks measured, from the Moon and from meteorites, give the age of the Solar System of 4.5 billion to 4.6 billion years.

**LG 3** **Explain how scientists use both theory and observation to determine the structure of planetary interiors.** Models of Earth's interior are used to predict how seismic waves should propagate through the interior, and these predictions are compared to observations of seismic waves. The interiors of other planets are modeled using physical principles, along with observational data on their magnetic fields. Earth has a strong magnetic field, but Venus and Mars do not. The cause for this difference between the terrestrial planets is uncertain.

**LG 4** **Describe tectonism and volcanism and the forms they take on different planets.** Tectonism folds, twists, and cracks the outer surface of a planet. Plate tectonics is unique to Earth, although other types of tectonic disruptions are observed on the other terrestrial planets, such as cracking and buckling on the surface. Smooth areas on the Moon and Mercury are ancient lava flows. While Venus has the most volcanoes, the largest mountains in the Solar System are volcanoes on Mars. Earth's surface is still changing as volcanic hot-spot activity forms new members of island chains.

**LG 5** **Summarize the knowledge of water on the terrestrial planets.** Geological features observed in orbiting and surface missions to Mars suggest there once was liquid water on the surface. Mars today has large amounts of subsurface water ice. Venus might have had liquid oceans early in its history. Space mission data indicate that water ice exists near the poles of the Moon and Mercury. This search for water is important to both the search for extraterrestrial life and the possibilities of human colonization of space.

## ? UNANSWERED QUESTIONS

- Why is Venus so different from Earth? These two planets of similar size, mass, and composition are very different geologically, with respect to magnetic fields, plate tectonics, and recent activity, and it is not yet known why. In addition, how did Venus end up rotating in the direction opposite that of its revolution around the Sun? Did it form with a different orbit or rotation? Was it the result of an impact early in its history? Did it change very slowly over time because of tidal effects from other planets?

- Will humans someday "live off the land" on the Moon? Recent space missions have provided evidence that there is some water ice on the Moon. If there is water on the Moon, that would certainly make living there more practical than if water had to be brought from Earth or synthesized from hydrogen and oxygen extracted from the lunar soil. Scientists and engineers have been studying several methods to see whether oxygen can be extracted from lunar rocks to make breathable air for people. Others have looked at using lunar rock as a building material; for example, to make concrete. But the most valuable material on the Moon might turn out to be an isotope of helium, helium-3 ($^3$He, which is helium with two protons and one neutron). On Earth, this isotope exists only as a by-product of nuclear weapons, but there may be up to a million tons of it on the Moon. Some scientists and engineers think that helium-3 could be used in a "clean" type of nuclear energy. This helium-3 could be brought back from the Moon for use on Earth or possibly even used in a power plant on the Moon to create energy for a lunar colony.

# Questions and Problems

## Test Your Understanding

1. _____, _____, and _____ build up structures on the terrestrial planets, while _____ tears them down.
   a. impacts, erosion, volcanism; tectonism
   b. impacts, tectonism, volcanism; erosion
   c. tectonism, volcanism, erosion; impacts
   d. tectonism, impacts, erosion; volcanism

2. Geologists can determine the relative age of features on a planet because
   a. the ones on top must be older.
   b. the ones on top must be younger.
   c. the larger ones must be older.
   d. the larger ones must be younger.

3. Scientists can learn about the interiors of the terrestrial planets from
   a. seismic waves.
   b. satellite observations of gravitational fields.
   c. physical arguments about cooling.
   d. satellite observations of magnetic fields.
   e. all of the above

4. Earth's interior is heated by
   a. angular momentum and gravity.
   b. radioactive decay and gravity.
   c. radioactive decay and tidal effects.
   d. angular momentum and tidal effects.
   e. gravity and tidal effects.

5. If a radioactive element has a half-life of 10,000 years, what fraction of it is left in a rock after 40,000 years?
   a. 1/2         c. 1/8         e. 1/32
   b. 1/4         d. 1/16

6. Lava flows on the Moon and Mercury created large, smooth plains. We don't see similar features on Earth because
   a. Earth has less lava.
   b. Earth had fewer large impacts in the past.
   c. Earth has plate tectonics that recycle the surface.
   d. Earth is large compared to the size of these plains, so they are not as noticeable.
   e. Earth rotates much faster than either of these other worlds.

7. Scientists know the history of Earth's magnetic field because
   a. the magnetic field hasn't changed since Earth formed.
   b. they see today's changes and project backward in time.
   c. the magnetic field becomes frozen into rocks, and plate tectonics spreads those rocks apart.
   d. they compare the magnetic fields on other planets to Earth's.

8. Suppose an earthquake occurs on an imaginary planet. Scientists on the other side of the planet detect primary waves but not secondary waves after the quake. This suggests that
   a. part of the planet's interior is liquid.
   b. all of the planet's interior is solid.
   c. the planet has an iron core.
   d. the planet's interior consists entirely of rocky materials.
   e. the planet's mantle is liquid.

9. Geologists can determine the actual age of features on a planet by
   a. radiometric dating of rocks retrieved from the planet.
   b. comparing cratering rates on one planet to those on another.
   c. assuming that all features on a planetary surface are the same age.
   d. both a and b
   e. both b and c

10. Impacts on the terrestrial planets and the Moon
    a. are more common than they used to be.
    b. have occurred at approximately the same rate since the Solar System formed.
    c. are less common than they used to be.
    d. periodically become more common and then less common.
    e. never occur anymore.

11. Earth has fewer craters than Venus. Why?
    a. Earth's atmosphere protects better than Venus's.
    b. Earth is a smaller target than Venus.
    c. Earth is closer to the asteroid belt.
    d. Earth's surface experiences more erosion.

12. Scientists propose an early period of heavy bombardment in the Solar System because
    a. the Moon is heavily cratered.
    b. all the craters on the Moon are old.
    c. the smooth part of the Moon is nearly as old as the heavily cratered part.
    d. all the craters on the Moon are young.

13. Scientists know that Earth was once completely molten because
    a. the surface is smooth.
    b. the interior layers are denser.
    c. the chemical composition indicates this.
    d. volcanoes exist today.

14. What is the main reason that Earth's interior is liquid today?
    a. tidal force of the Moon on Earth
    b. seismic waves that travel through Earth's interior
    c. decay of radioactive elements
    d. convective motions in the mantle
    e. pressure on the core from Earth's outer layers

15. Mars has a diameter that is approximately half that of Earth. If the interiors of these planets are heated by radioactive decays, how does the heating rate of the interior of Mars compare to that of Earth?
    a. The heating rate of Mars is 0.125 times that of Earth.
    b. The heating rate of Mars is 8 times that of Earth.
    c. The heating rate of Mars is 0.5 times that of Earth.
    d. The heating rate of Mars is 4 times that of Earth.
    e. The heating rates are about the same.

## Thinking about the Concepts

16. In discussing the terrestrial planets, why do we include Earth's Moon?

17. Can all rocks be dated with radiometric methods? Explain.

18. Explain how scientists know that rock layers at the bottom of the Grand Canyon are older than those found on the rim.

19. Describe the sources of heating that are responsible for generating Earth's magma.

20. Explain why the Moon's core is cooler than Earth's.

21. Explain the difference between longitudinal waves and transverse waves.

22. How do we know that Earth's core includes a liquid zone?

23. Study the Process of Science Figure. What evidence makes the impactor theory the currently preferred favorite explanation for the origin of the Moon? What evidence remains to be found to rule out the competing theories?

24. Compare and contrast tectonism on Venus, Earth, and Mercury.

25. Explain plate tectonics and identify the only planet on which this process has been observed.

26. Volcanoes have been found on all of the terrestrial planets. Where are the largest volcanoes in the inner Solar System?

27. Explain the criteria you would apply to images (assume adequate resolution) in order to distinguish between a crater formed by an impact and one formed by a volcanic eruption.

28. What are the primary reasons that the surfaces of Venus, Earth, and Mars have been determined to be younger than those of Mercury and the Moon?

29. Explain some of the geological evidence suggesting that Mars once had liquid water on its surface.

30. What evidence supports the theory suggesting that a mass extinction occurred as a consequence of an enormous impact on Earth 65 million years ago?

## Applying the Concepts

31. Study Figure 8.8.
    a. How has the cratering rate changed over time? Has it fallen off gradually or abruptly?
    b. At present, what is the cratering rate compared to that about 4 billion years ago?
    c. Explain why this falloff in cratering rate fits nicely in the theory of planet formation.

32. Study Figure 8.7. Are the vertical and horizontal axes linear or logarithmic? After how many half-lives will the number of parent isotopes equal the number of daughter isotopes? Is this result unique to this example? Why or why not?

33. Study Figure 8.7. The destruction of the parent isotope is an example of exponential decay. Is the growth of the daughter isotope an example of exponential growth? How can you tell?

34. Compare Figures 8.18 and 8.23. Which of these regions is older? How do you know?

35. Assume that Earth and Mars are perfect spheres with radii of 6,371 km and 3,390 km, respectively.
    a. Calculate the surface area of Earth.
    b. Calculate the surface area of Mars.
    c. If 0.72 (72 percent) of Earth's surface is covered with water, compare the amount of Earth's land area to the total surface area of Mars.

36. Compare the kinetic energy ($= \frac{1}{2}mv^2$) of a 1-gram piece of ice (about half the mass of a dime) entering Earth's atmosphere at a speed of 50 km/s to that of a 2-metric-ton SUV (mass $= 2 \times 10^3$ kg) speeding down the highway at 90 km/h.

37. The object that created Arizona's Meteor Crater was estimated to have a radius of 25 meters and a mass of 300 million kg. Calculate the density of the impacting object, and explain what that may tell you about its composition.

38. Using the information in Table 8.1 and Working It Out 8.2, determine the relative rates of internal energy loss experienced by Earth and the Moon.

39. Earth's mean radius is 6,371 km, and its mass is $6.0 \times 10^{24}$ kg. The Moon's mean radius is 1,738 km, and its mass is $7.2 \times 10^{22}$ kg.
    a. Calculate Earth's average density. Show your work; do not look this value up.
    b. The average density of Earth's crust is 2,600 kg/m$^3$. What does this value tell you about Earth's interior?
    c. Compute the Moon's average density. Show your work.
    d. Compare the average densities of the Moon, Earth, and Earth's crust. What do these values tell you about the Moon's composition compared to that of Earth and of Earth's crust?

40. Suppose you find a piece of ancient pottery and find that the glaze contains radium, a radioactive element that decays to radon and has a half-life of 1,620 years. There could not have been any radon in the glaze when the pottery was being fired, but now it contains three atoms of radon for each atom of radium. How old is the pottery?

41. Archaeological samples are often dated by radiocarbon dating. The half-life of carbon-14 is 5,700 years.
    a. After how many half-lives will the sample have only 1/64 as much carbon-14 as it originally contained?
    b. How much time will have passed?
    c. If the daughter product of carbon-14 is present in the sample when it forms (even before any radioactive decay happens), you cannot assume that every daughter you see is the result of carbon-14 decay. If you did make this assumption, would you overestimate or underestimate the age of a sample?

42. Different radioisotopes have different half-lives. For example, the half-life of carbon-14 is 5,700 years, the half-life of uranium-235 is 704 million years, the half-life of potassium-40 is 1.3 billion years, and the half-life of rubidium-87 is 49 billion years.
    a. Why wouldn't you use an isotope with a half-life similar to that of carbon-14 to determine the age of the Solar System?
    b. The age of the universe is approximately 14 billion years. Does that mean that no rubidium-87 has decayed yet?

43. Assume that the east coast of South America and the west coast of Africa are separated by an average distance of 4,500 km. Assume also that GPS measurements indicate that these continents are now moving apart at a rate of 3.75 cm/yr. If this rate has been constant over geological time, how long ago were these two continents joined together as part of a supercontinent?

44. Shield volcanoes are shaped something like flattened cones. The volume of a cone is equal to the area of its base multiplied by one-third of its height. The largest volcano on Mars, Olympus Mons, is 27 km high and has a base diameter of 550 km. Compare its volume with that of Earth's largest volcano, Mauna Loa, which is 9 km high and has a base diameter of 120 km.

45. Using the data in Table 8.1, compare the surface gravity on Mars with that on Earth. How does this help explain why the volcanoes on Mars can grow so high?

## USING THE WEB

46. Go to the U.S. Geological Survey's "Earthquake" website (http://earthquake.usgs.gov/earthquakes/map). Set "Zoom" to "World," set the "Settings" icon in the upper right to "Seven Days, Magnitude 2.5+," and look at the earthquakes for the past week. Were there any really large ones? Compare the map of recent earthquakes with Figure 8.17 in the text. Are any of the quakes in surprising locations? Where was the most recent one? Now change the "Zoom" to the United States (or to "Your location") and change the settings to "30 days, Magnitude 2.5+." Has there been seismic activity, and if so where?

47. Use Google Earth to explore the Moon, Mercury, and Mars.
    a. View all sides of the Moon. Does one hemisphere look more heavily cratered than others, and if so, why?
    b. View the planet Mercury. In what ways is Mercury similar to and in what ways different from the Moon? (You might need to get the Mercury KMZ file: http://messenger.jhuapl.edu/the_mission/google.html.)
    c. View all sides of the planet Mars. What differences can you see between the northern and southern hemispheres?

48. Citizen science:
    a. Go to the website for "Moon Zoo" (http://moonzoo.org), a project that lets everyone participate in the analysis of images from NASA's *Lunar Reconnaissance Orbiter*. Read through the FAQ, then click on "Tutorials" and select "How to Take Part." (You will need to create an account if you haven't already done so for another Zooniverse project.) In this project you count craters on the Moon, noting where there are boulders, classifying some of these features, and looking for hardware left over from exploration missions.
    b. Go to the website for Cosmoquest (http://cosmoquest.org) and click on "Mercury Mappers." You will need to create an account for the Cosmoquest projects. Click on the circled question mark under the blue check box, and read the FAQ and watch the tutorial. What is the goal of this project? Where did the data come from? Classify some images.
    c. Go to the website for Cosmoquest (http://cosmoquest.org) and click on "Moon Mappers." As in part (b), you will need an account. Click on the circled question mark under the blue check box and read the FAQ and watch the four tutorials. What are some of the basic features? How does the angle of the sunlight and the direction of illumination affect what you see? Now classify a few craters.

49. Space missions:
    a. Go to the website for NASA's *Messenger* mission to Mercury (http://messenger.jhuapl.edu). Click on "Gallery" and then "Science Images," and look at a few of the pictures. Are the color images using real or false colors? Click on "News Center." Describe a result.
    b. Go to the website for the *Mars Science Laboratory Curiosity* (http://mars.jpl.nasa.gov/msl), which landed in 2012. What are the latest science results?
    c. The Google Lunar X Prize (http://googlelunarxprize.org) goes to the first privately funded team to send a robot to the Moon. The winning robot must travel some distance on the Moon's surface and send back pictures. On the website, click on "Teams" and read about a few that are still competing. What kind of people and companies are on the team? What is their plan to go to the Moon? Aside from this prize, why do they want to go to the Moon: what commercial opportunities on the Moon do they anticipate?

50. Video:
    a. Watch one of the available documentaries about the *Apollo* missions to the Moon (for example, *In the Shadow of the Moon*, 2008). Why did the United States decide to send astronauts to the Moon? Why did the *Apollo* program end? Are there current plans to send people to the Moon?
    b. The first science fiction film was the short *Voyage to the Moon* (Georges Méliès, 1902). A version with an English narration can be viewed at https://archive.org/details/Levoyagedanslalune. A restored digitized and colorized version was released in 2011 and can be found at http://vimeo.com/39275260. Where do the "Selenians" live on the Moon? In this first cinematic depiction of contact with life from outside of Earth, what do the human astronomers do to the Selenians? Contrast what the astronomers in the film find on the Moon with what the *Apollo* astronauts actually saw.

## smartw⦿rk**5**

If your instructor assigns homework in Smartwork5, access your assignments at digital.wwnorton.com/astro5.

digital.wwnorton.com/astro5

Knowledge of the exponential function is critical to understanding life in modern times because this function shows up in many contexts, from economics to population studies to climate change.

Radioactive decay exhibits exponential decay, where the amount of a radioactive material that remains after an elapsed time is proportional to the amount that was present at the beginning of the time period. In this Exploration, you will use a small (1.69-ounce) bag of plain M&M's to investigate this behavior.

This size bag of M&M's contains approximately 56 pieces of candy. Each piece has an *M* stamped on one side. Thus, there are two ways for each piece to fall when dropped: *M* side up or *M* side down. This is much like what happens to a radioactive nucleus: for any given period of time, either it decays or it doesn't.

Rather than acquiring 10 bags of M&M's, with all the mess that would make, you will use the same bag 10 times, and then add your results together. This approach makes the sample large enough that the exponential behavior becomes apparent.

**Step 1** For the first trial, shake all the M&M's into your hands and then pour them onto the table.

**Step 2** Count how many land *M* side up, and record that number in the provided table under Trial 1, Time Step 1.

**Step 3** Set the *M*-side-up candies to one side.

**Step 4** Repeat steps 1–3 for time steps 2, 3, 4, and so on, until the last candy lands *M* side up.

**Step 5** Repeat steps 1–4 for 9 more trials (making a total of 10 trials).

**Step 6** Add together the results of the trials for each time step, and record them in the Sum column of the table.

**Step 7** It's very difficult to make sense of numbers in tabular form like this, so plot the results on a graph with the sum on the *y*-axis and the time step on the *x*-axis.

| Time Step | Trial 1 | Trial 2 | Trial 3 | Trial 4 | Trial 5 | Trial 6 | Trial 7 | Trial 8 | Trial 9 | Trial 10 | Sum |
|---|---|---|---|---|---|---|---|---|---|---|---|
| 1 | | | | | | | | | | | |
| 2 | | | | | | | | | | | |
| 3 | | | | | | | | | | | |
| 4 | | | | | | | | | | | |
| 5 | | | | | | | | | | | |
| 6 | | | | | | | | | | | |
| 7 | | | | | | | | | | | |
| 8 | | | | | | | | | | | |
| 9 | | | | | | | | | | | |
| 10 | | | | | | | | | | | |
| 11 | | | | | | | | | | | |
| 12 | | | | | | | | | | | |

**1** Study your graph. At first, does the number of M&M's landing *M* side up decrease slowly or quickly? In later time steps (such as 8 or 9), does the number decrease slowly or quickly?

_____

**2** Generalize your answer to question 1 to the behavior of radioactive sources. Does the radioactivity fall off slowly or quickly at the beginning of an observation? How about at later times?

_____

**3** As time goes by, what happens to the number of M&M's that remain? What happens to the number of M&M's in the pile that is set aside?

_____

**4** Generalize your answer to question 3 to the behavior of radioactive sources. What happens to the number of radioactive isotopes over time? What happens to the number of daughter products?

_____

**5** Imagine that you walk into a room and observe another student performing this experiment. She pours the M&M's onto the table and counts 10 candies that landed *M* side up. About how many time steps have passed since this student started the experiment?

_____

_____

**6** Apply your answer to question 5 to the behavior of radioactive sources. When scientists study radioactive sources, they generally study the ratio of the number of radioactive isotopes to the number of daughter products. Explain how this method, while different from what you did in step 5, contains the same information about the amount of time that has passed.

_____

_____

_____

# 9 Atmospheres of the Terrestrial Planets

E arth's atmosphere surrounds its inhabitants like an ocean of air. It is evident in the blueness of the sky and in the breezes in the air. Without Earth's atmosphere, there would be neither clouds nor oceans. Without an atmosphere, Earth would look something like the Moon, and life would not exist on our planet. Among the five terrestrial bodies that we discussed in Chapter 8, only Venus and Earth have dense atmospheres. Mars has a very low-density atmosphere, and the atmospheres of Mercury and the Moon are so sparse that they can hardly be detected. To understand the origins of the atmospheres of Venus, Earth, and Mars, how they have changed over time, how they compare to one another, and how they are likely to evolve in the future requires us to look back nearly 5 billion years to a time when the planets were just completing their growth.

## LEARNING GOALS

In this chapter, we will compare the atmospheres of the terrestrial planets. By the conclusion of this chapter, you should be able to:

LG 1  Identify the processes that cause primary and secondary atmospheres to be formed, retained, and lost.

LG 2  Compare the strength of the greenhouse effect and differences in the atmospheres of Earth, Venus, and Mars.

LG 3  Describe the layers of the atmospheres on Earth, Venus, and Mars, and explain how Earth's atmosphere has been reshaped by the presence of life.

LG 4  Compare the atmospheres of Venus and Mars with the atmosphere of Earth.

LG 5  Describe how comparative planetology contributes to a better understanding of the changes in Earth's climate.

Earth from space. ▶ ▶ ▶

Why can you breathe only on Earth?

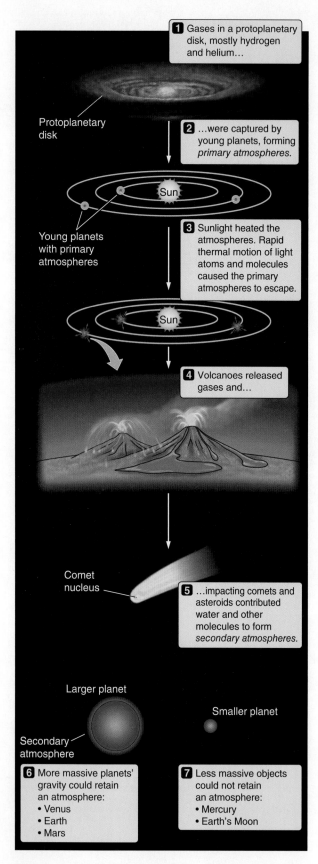

**1** Gases in a protoplanetary disk, mostly hydrogen and helium…

Protoplanetary disk

**2** …were captured by young planets, forming *primary atmospheres.*

Sun

Young planets with primary atmospheres

**3** Sunlight heated the atmospheres. Rapid thermal motion of light atoms and molecules caused the primary atmospheres to escape.

Sun

**4** Volcanoes released gases and…

Comet nucleus

**5** …impacting comets and asteroids contributed water and other molecules to form *secondary atmospheres.*

Larger planet

Smaller planet

Secondary atmosphere

**6** More massive planets' gravity could retain an atmosphere:
• Venus
• Earth
• Mars

**7** Less massive objects could not retain an atmosphere:
• Mercury
• Earth's Moon

**Figure 9.1** Planetary atmospheres form and evolve in phases.

# 9.1 Atmospheres Change over Time

An atmosphere is a layer of gas that sits above the surface of a solid body such as a terrestrial planet. A blanket of atmosphere warms and sustains Earth's temperate climate. On Venus, by contrast, a thick, carbon dioxide atmosphere pushes the planet's surface temperature very high. A thin atmosphere leaves the surface of Mars unprotected and frozen. Mercury and the Moon have essentially no atmosphere. Why should some of the terrestrial planets have dense atmospheres while others have little or none? In this section, we will look at the formation of planetary atmospheres.

## Formation and Loss of Primary Atmospheres

Planetary atmospheres formed in phases, which are shown in **Figure 9.1**. At the time of formation, the young planets were initially enveloped by the remaining hydrogen and helium that filled the protoplanetary disk surrounding the Sun, and they captured some of this surrounding gas. Gas capture continued until soon after formation of the planets when the gaseous disk dissipated and the supply of gas ran out. The gaseous atmosphere collected by a newly formed planet is called its primary atmosphere. This primary atmosphere was lost from the terrestrial planets as these lightweight atoms and molecules escaped from the planet's gravity. To understand this, we must consider how particles move within a planetary atmosphere.

In Chapter 7, you saw that the terrestrial planets are less massive than the giant planets and therefore have weaker gravitational attraction. These planets lack the ability to hold light gases such as hydrogen and helium. Giant impacts by large planetesimals early in the history of the Solar System may have blasted away some of their primary atmospheres. When the supply of gas in the protoplanetary disk ran out, the primary atmospheres of the terrestrial planets began leaking back into space. How can gas molecules escape from a planet? Recall from Chapter 4 that any object—from a molecule to a spacecraft—can escape a planet if the object reaches a speed greater than the escape velocity and is pointed in the right direction. Intense radiation from the Sun—which is the primary source of thermal, or kinetic, energy in the atmospheres of the terrestrial planets—raises the temperature and thus the speed of atmospheric molecules enough for some to escape.

Let's look more closely at how molecules move within a planetary atmosphere. Imagine a large box that contains air. In thermal equilibrium, each type of molecule in the box, from the lightest to the most massive, will have the same average kinetic energy. Because the kinetic energy of a molecule (or any object) is determined by its mass and its speed, if each type has the same average energy, then the lightest molecules must be moving faster than the more massive ones. This average kinetic energy of the gas molecules is directly proportional to the temperature of the gas. So for a gas at a given temperature, if each type of molecule has the same average energy, then the less massive molecules must be moving faster than the more massive ones. For example, in a mixture of hydrogen and oxygen at room temperature, hydrogen molecules will be rushing around the box at about 2,000 meters per second (m/s) on average, while the much more massive oxygen molecules will be moving at a slower 500 m/s. Remember, though, that these are the average speeds. A few of the molecules will always be moving much faster or slower than average.

Deep within a planet's atmosphere, fast molecules near the ground will almost certainly collide with other molecules before the fast molecules have a chance to escape. Higher regions of the atmosphere contain fewer molecules. Therefore, fast molecules in the upper atmosphere are less likely to collide with other molecules and have a better chance of escaping as long as they are heading more or less upward. At a given temperature, lighter molecules such as hydrogen and helium move faster and are more quickly lost to space than more massive molecules such as nitrogen or carbon dioxide.

Solar heating caused the molecules to move quickly on the young terrestrial planets. In addition, small planets, like the terrestrial planets, have only a weak gravitational grasp. These conditions caused the terrestrial planets to lose the hydrogen and helium they had acquired as a primary atmosphere. This process was likely assisted by collisions with other planetesimals. Because the giant planets were farther from the Sun, they were far more massive and also cooler: stronger gravity and lower temperatures enabled them to retain nearly all of their massive primary atmospheres.

▶❚❚ **AstroTour:** Atmospheres: Formation and Escape

## The Formation of Secondary Atmospheres

Although their primary atmospheres were lost, some of the terrestrial planets do have an atmosphere today, known as a secondary atmosphere. Where did this secondary atmosphere come from? Accretion, volcanism, and impacts are responsible for the atmospheres of Earth, Venus, and Mars today. During the planetary accretion process, minerals containing water, carbon dioxide, and other volatile matter collected in the planetary interiors. Later, as an interior heated up, these gases were released from the minerals that had held them. Volcanism then brought the gases to the surface, where they accumulated and created a secondary atmosphere, as shown in step 4 of Figure 9.1.

Impacts by comets and asteroids were another important source of gases. Huge numbers of comets formed in the outer parts of the Solar System and were therefore rich in volatiles. As the giant planets of the outer Solar System grew to maturity or migrated their orbits, their gravitational perturbations stirred up the comets and asteroids that orbited relatively nearby. Many of these icy bodies were flung outward by the giant planets to join other existing planetesimals in the Kuiper Belt, 30–50 astronomical units (AU) from the Sun. Other bodies joined the part of the Solar System known as the Oort Cloud, a spherical cloud of icy planetesimals that surrounds the Sun at a distance ranging from the Kuiper Belt to about 50,000 AU from the Sun—nearly one-quarter of the way to the nearest star. Other comets were scattered into the inner parts of the Solar System. Upon impact with the terrestrial planets, these objects brought ices such as water, carbon monoxide, methane, and ammonia. On the terrestrial planets, cometary water mixed with the water that had been released into the atmosphere by volcanism, as shown in step 5 of Figure 9.1. On Earth, and perhaps Mars and Venus as well, most of the water vapor then condensed as rain and flowed into the lower areas to form the earliest oceans.

Sunlight also influenced the composition of secondary atmospheres. Ultraviolet (UV) light from the Sun easily fragments cometary molecules such as ammonia ($NH_3$) and methane ($CH_4$). Ammonia, for example, is broken down into hydrogen and nitrogen. When this happens, the lighter hydrogen atoms quickly escape to space, leaving behind the much heavier nitrogen atoms. Pairs of nitrogen atoms then combine to form more massive nitrogen molecules ($N_2$), and these molecules are even less likely to escape into space. Decomposition of ammonia by sunlight became the primary source of molecular nitrogen in the atmospheres of the terrestrial planets. Molecular nitrogen makes up the bulk of Earth's atmosphere.

Among the terrestrial planets, today only Venus, Earth, and Mars have significant secondary atmospheres. What happened in the case of Mercury and the Moon? Even if these two bodies experienced less volcanism than the other terrestrial planets (see Chapter 8), they would have had the same early bombardment of comet nuclei from the outer Solar System. Some carbon dioxide and water must have accumulated during volcanic eruptions and comet impacts. A secondary atmosphere can be lost through the same processes that cause the loss of a primary atmosphere. Large impacts were less frequent as the Solar System aged, but atmospheric escape continued over the past 4 billion years. In addition, decreases in the magnetic field as Mercury and the Moon cooled might have contributed to atmospheric escape. With a weaker magnetic field, the planet became less protected from the **solar wind**—a constant stream of charged particles from the Sun. The solar wind can accelerate atmospheric particles to escape velocity.

Both the Moon and Mercury have virtually no atmosphere today. Mercury lost nearly its entire secondary atmosphere to space, just as it had previously lost its primary atmosphere. Even molecules as massive as carbon dioxide can escape from a small planet if the temperature is high enough, as it is on Mercury's sunlit side. Furthermore, intense UV radiation from the Sun can break molecules into less massive fragments, which are lost to space even more quickly. Because the distance from the Sun to the Moon is much farther than the distance from the Sun to Mercury, the Moon is much cooler than Mercury, but its mass is so small that molecules easily escaped even at relatively low temperatures. The ability of planets to hold on to their atmospheres is explored further in **Working It Out 9.1**.

**Nebraska Simulation:** Gas Retention Simulator

---

### CHECK YOUR UNDERSTANDING 9.1

Which are reasons Mercury has so little gas in its atmosphere? (Choose all that apply.) (a) Its mass is small. (b) It has a high temperature. (c) It is close to the Sun. (d) Its escape velocity is low. (e) It has no moons.

............................................................................................................................

## 9.2 Secondary Atmospheres Evolve

Although Venus, Earth, and Mars most likely started out with atmospheres of similar composition, they ended up being very different from one another. Earth is volcanically active, Venus might still be volcanically active, and Mars has been volcanically active in the recent past. All three planets must have shared the intense cometary showers of the early Solar System. Their similar geological histories suggest that their early secondary atmospheres might also have been quite similar. However, Earth's secondary atmosphere has changed significantly since it formed—the development of life increased the amount of oxygen. Earth's atmosphere is made up primarily of nitrogen and oxygen, with only a trace of carbon dioxide. In contrast, the composition of the atmospheres of Venus and Mars today are nearly identical—mostly carbon dioxide, with much smaller amounts of nitrogen. The atmospheres of these planets differ for two reasons we will explore: planetary mass and the greenhouse effect.

To estimate a planet's ability to retain its atmosphere, we compare the escape velocity from the planet (which depends on the planet's gravity, determined by the mass and radius) with the average speed of the molecules in a gas (which depends on the temperature of the gas and the mass of the molecules that make up the gas). The escape velocity is defined as

$$v_{esc} = \sqrt{\frac{2GM}{R}}$$

and the values of $v_{esc}$, in kilometers per second (km/s), for the inner planets are given at the bottom of Table 8.1.

We said in Chapter 4 that the temperature, $T$, of a gas is proportional to the kinetic energy of the particles, $\frac{1}{2}mv^2$. We can rearrange that relationship to solve for $v$ and insert the constants of proportionality to get the average speed of a molecule in a gas:

$$v_{molecule} = \sqrt{\frac{3kT}{m}}$$

where $T$ is the temperature of the gas in kelvins, $m$ is the mass of the molecule in kilograms (kg), and $k$ is the Boltzmann constant. The atomic mass of a molecule is found by adding up the atomic masses of its composite atoms as specified in the periodic table. (Atomic masses of atoms come from the total number of neutrons and protons; the electron weighs little in comparison). Oxygen molecules ($O_2$), for example, are 16 times as massive as hydrogen molecules ($H_2$).

If we put in the value of the Boltzmann constant ($k = 1.38 \times 10^{-23}$ joule per kelvin, or J/K) and the mass of the hydrogen atom ($m = 1.67 \times 10^{-27}$ kg), then $v_{molecule}$, in kilometers per second, is given by

$$v_{molecule} = 0.157 \text{ km/s} \times \sqrt{\frac{\text{Temperature of gas}}{\text{Atomic weight of molecule}}}$$

The higher the temperature, the higher the average kinetic energy of the individual molecules, and the faster the average speed of the particles. This difference explains why Earth can hold onto the oxygen in its atmosphere but loses hydrogen to space.

To use our example of hydrogen and oxygen molecules, $H_2$ has an atomic weight of 2 (1 for each hydrogen atom), and $O_2$ has an atomic weight of 32 (16 for each oxygen atom, from 8 protons and 8 neutrons). Earth has an average temperature of 288 K. The average speeds of the molecules are thus

$$\text{For } H_2: v_{molecule} = 0.157\sqrt{\frac{288}{2}} = 1.88 \text{ km/s}$$

$$\text{For } O_2: v_{molecule} = 0.157\sqrt{\frac{288}{32}} = 0.47 \text{ km/s}$$

Thus, in a gas containing both hydrogen and oxygen molecules, the average hydrogen molecule will be moving 4 times faster than the average oxygen molecule. At any given temperature, the lighter molecules will be moving faster. Not all molecules in a gas are moving at the average speed: some move faster and some slower (**Figure 9.2**). The general rule is that over the age of the Solar System, a planet can keep its atmosphere if, for that type of gas molecule:

$$v_{molecule} \leq \frac{1}{6}v_{esc}$$

The escape velocity from Earth is 11.2 km/s, and one-sixth of this is 1.87 km/s. These numbers explain why Earth has been able to keep its $O_2$ but not its $H_2$. A similar analysis shows that on the Moon, with its lower $v_{esc}$ value, both the $H_2$ and the $O_2$ molecules escape. On Jupiter, with its much colder temperatures and higher $v_{esc}$ value, both the $H_2$ and the $O_2$ molecules are retained.

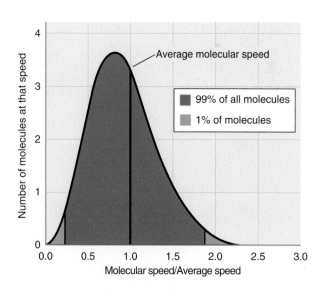

**Figure 9.2** This plot shows the distribution of the speeds of molecules in a gas. The shape of the curve and the exact numbers will depend on the temperature of the gas and the masses of the molecules. In all cases, some of the speedier molecules may be able to escape if they are faster than the escape velocity.

## TABLE 9.1 Atmospheres of the Terrestrial Planets

*Physical Properties and Composition*

| | PLANET | | |
|---|---|---|---|
| | Venus | Earth | Mars |
| Surface pressure (bars) | 92 | 1.0 | 0.006 |
| Atmospheric mass (kg) | $4.8 \times 10^{20}$ | $5.1 \times 10^{18}$ | $2.5 \times 10^{16}$ |
| Surface temperature (K) | 740 | 288 | 210 |
| Carbon dioxide (%) | 96.5 | 0.039 | 95.3 |
| Nitrogen (%) | 3.5 | 78.1 | 2.7 |
| Oxygen (%) | 0.00 | 20.9 | 0.13 |
| Water (%) | 0.002 | 0.1 to 3 | 0.02 |
| Argon (%) | 0.007 | 0.93 | 1.6 |
| Sulfur dioxide (%) | 0.015 | 0.02 | 0.00 |

## The Effect of Planetary Mass on a Planet's Atmosphere

**Table 9.1** shows that the atmospheres of Venus and Mars today are nearly identical in composition. They are both composed mostly of carbon dioxide, with much smaller amounts of nitrogen. Carbon dioxide and water vapor came from volcanic gases; and nitrogen came from decomposed cometary ammonia. However, the total *amount* of atmosphere is very different among the three planets. The atmospheric pressure on the surface of Venus is nearly 100 times greater than Earth's. By contrast, the average surface pressure on Mars is less than a hundredth that on Earth. Venus is nearly 8 times as massive as Mars, so we assume it probably had about 8 times as much carbon within its interior to produce carbon dioxide, the principal secondary-atmosphere component of both planets. Even allowing for the differences in planetary mass, however, Venus today has greater than 2,500 times more atmospheric mass than Mars.

The large difference in atmospheric mass comes from the relative strengths of each planet's surface gravity, which involves both the mass and the radius of a planet. Venus has the gravitational pull necessary to hang on to its atmosphere; Mars has less gravitational attraction to keep its atmosphere (see Working It Out 9.1). Furthermore, when a planet such as Mars began to lose its atmosphere to space, the process began to take on a *runaway* behavior. With a thinner atmosphere, there were fewer slow molecules to keep fast molecules from escaping, and the rate of escape increased. This process in turn led to even less atmosphere and still greater escape rates.

Mars might also have lost atmosphere in a giant impact. In addition, scientists debate how much atmospheric loss arises from the effects of the solar wind on planetary atmospheres, especially in the absence of a planetary magnetic field. All three planets are currently losing some atmosphere to space, even though Earth has a magnetic field and Venus and Mars do not. The extent to which the lack of a magnetic field on Mars played a role in its atmospheric loss is being studied by NASA's *Mars Atmosphere and Volatile EvolutioN* (MAVEN) mission, which arrived at Mars in September 2014.

## The Atmospheric Greenhouse Effect

Differences in the present-day masses of the atmospheres of Venus, Earth, and Mars have a large effect on their surface temperatures. Recall from Chapter 5 that the temperature of a planet is determined by a balance between the amount of sunlight being absorbed and the amount of energy being radiated back into space. When we calculated the temperature of a planet by finding the equilibrium between the amount of energy it receives and the amount of energy it radiates, we found that this calculation gives a good result for planets without atmospheres. But Earth is somewhat warmer than expected, and Venus is very much hotter than this simple model predicted. When the predictions of a model fail, the implication is that something was left out of the model. In this case, that something was the *atmospheric greenhouse effect*, which traps solar radiation.

The atmospheric greenhouse effect in planetary atmospheres and the conventional **greenhouse effect** operate in different ways, although the end results are much the same. Planetary atmospheres and the interiors of greenhouses are both heated by trapping the Sun's energy, but here the similarities end. The conventional greenhouse effect is the rise in temperature in a car on a sunny day when you leave the windows closed, or what allows plants to grow in the winter in a greenhouse. Sunlight pours through the glass, heating the interior and raising the internal air temperature. With the windows closed, hot air is trapped, and temperatures can climb as high as 80°C (about 180°F). Heating by solar radiation is most efficient when an enclosure is transparent, which is why the walls and roofs of real greenhouses, which allow plants to grow in the winter, are made mostly of glass.

The atmospheric greenhouse effect is illustrated in **Figure 9.3**. Atmospheric gases freely transmit visible light, allowing the Sun to warm the planet's surface. The warmed surface radiates the energy in the infrared region of the spectrum. Some of the atmospheric gases strongly absorb this infrared radiation and convert it to thermal energy, which is released in random directions. Some of the thermal energy continues into space, but much of it goes back to the ground, which causes a planet's surface temperature to rise. As a result of this radiation, the planet's surface receives thermal energy from both the Sun and the atmosphere. Gases that transmit visible radiation but absorb infrared radiation are known as **greenhouse gases**. Examples of atmospheric greenhouse gases include water vapor, carbon dioxide, methane, and nitrous oxide, as well as industrial chemicals such as halogens. The presence of greenhouse gases in a planet's atmosphere will cause its surface temperature to rise.

This rise in temperature continues until the surface becomes sufficiently hot—and therefore radiates enough energy—that the fraction of infrared radiation leaking out through the atmosphere balances the absorbed sunlight, and equilibrium is reached. Convection also helps maintain equilibrium by transporting thermal energy to the top of the atmosphere, where it can be more easily radiated to space. In short, the temperature rises until an equilibrium between absorbed sunlight and thermal energy radiated away by the planet is reached. If the amount of greenhouse gases increases in the atmosphere, the trapping effect increases, and the temperature at which energy input and output balances also increases. Even though the mechanisms are somewhat different, the conventional greenhouse effect and the atmospheric greenhouse effect produce the same net result: the local environment is heated by trapped solar radiation.

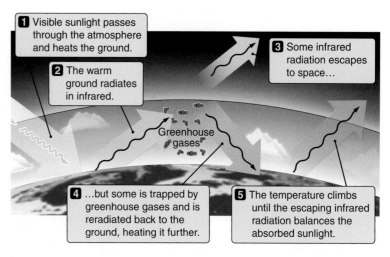

1 Visible sunlight passes through the atmosphere and heats the ground.

2 The warm ground radiates in infrared.

3 Some infrared radiation escapes to space…

Greenhouse gases

4 …but some is trapped by greenhouse gases and is reradiated back to the ground, heating it further.

5 The temperature climbs until the escaping infrared radiation balances the absorbed sunlight.

**Figure 9.3** In the atmospheric greenhouse effect, greenhouse gases such as water vapor and carbon dioxide trap infrared radiation, increasing a planet's temperature.

**Astronomy in Action:** Changing Equilibrium

**AstroTour:** Greenhouse Effect

## Similarities and Differences among the Terrestrial Planets

Let's look more closely at how the atmospheric greenhouse effect operates on Mars, Earth, and Venus. What really matters is the actual number of greenhouse molecules in a planet's atmosphere, not the fraction they represent. For example, even though the atmosphere of Mars is composed almost entirely of carbon dioxide (see Table 9.1)—an effective greenhouse molecule—the atmosphere is very thin and contains relatively few greenhouse molecules compared to the atmospheres of Venus or Earth. As a result, the atmospheric greenhouse effect is relatively weak on Mars and raises the average surface temperature by only about 5 K (9°F). At the other extreme, Venus's massive atmosphere of carbon dioxide and sulfur compounds raises its average surface temperature by more than 400 K, to about 740 K (467°C, or 870°F). At such high temperatures, any remaining water and most carbon dioxide locked up in surface rocks are driven into the atmosphere, further enhancing the atmospheric greenhouse effect.

The atmospheric greenhouse effect on Earth is not as severe as it is on Venus—the average global temperature near Earth's surface is about 288 K (15°C, or 59°F). Temperatures on Earth's surface are about 35 K (63°F) warmer than they would be in the absence of an atmospheric greenhouse effect, mainly because of water vapor and carbon dioxide. Yet this comparatively small difference has been crucial in shaping the Earth we know. Without this greenhouse effect, Earth's average global temperature would be −18°C (0°F), well below the freezing point of water, leaving us with a world of frozen oceans and ice-covered continents.

How has the atmospheric greenhouse effect made the composition of Earth's atmosphere so different from the high-carbon-dioxide atmospheres of Venus and Mars? The answer lies in Earth's location in the Solar System. Earth and Venus have about the same mass, but Venus orbits the Sun somewhat closer than Earth, at 0.7 AU. Volcanism and cometary impacts produced large amounts of carbon dioxide and water vapor to form early secondary atmospheres on both planets. Most of Earth's water quickly rained out of the atmosphere to fill vast ocean basins. But because Venus was closer to the Sun, its surface temperatures were higher than those of Earth. As the Sun itself aged and brightened, Venus got warmer, and most of the rainwater on Venus immediately reevaporated, much as water does in Earth's desert regions. Venus was left with a surface that contained very little liquid water and an atmosphere filled with water vapor. The water vapor caused even higher temperatures, which led to the release of more carbon dioxide from the rocks to the atmosphere. The continuing buildup of both water vapor and carbon dioxide in the atmosphere of Venus led to a runaway atmospheric greenhouse effect that drove up the surface temperature of the planet even more. Ultimately, the surface of Venus became so hot that no liquid water could exist on it.

This early difference between a watery Earth and an arid Venus forever changed the ways that their atmospheres and surfaces evolved. On Earth, water erosion caused by rain and rivers continually exposed fresh minerals, which then reacted chemically with atmospheric carbon dioxide to form solid carbonates. This reaction removed some of the atmospheric carbon dioxide, burying it within Earth's crust as a component of a rock called limestone. Later, the development of life in Earth's oceans accelerated the removal of atmospheric carbon dioxide. Tiny sea creatures built their protective shells of carbonates, and as they died they built up massive beds of limestone on the ocean floors. Water erosion and the chemistry of life tied up all but a trace of Earth's total inventory of carbon dioxide in limestone beds. Earth's particular location in the Solar System seems to have spared it from the runaway atmospheric greenhouse effect. If all the carbon dioxide now in limestone beds had not been locked up by these reactions, Earth's atmosphere would be composed of about 98 percent carbon dioxide, similar to that of Venus or Mars. The atmospheric greenhouse effect would be much stronger, and Earth's temperature would be much higher.

The details of the differences in the amount of water on Venus, Earth, and Mars are not well understood. Geological evidence indicates that liquid water was once plentiful on the surface of Mars. Several of the spacecraft orbiting Mars have found evidence that significant amounts of water still exist on Mars in the form of subsurface ice—far more than the atmospheric abundance indicated in Table 9.1. Earth's liquid and solid water supply is even greater, about 0.02 percent of its total mass. More than 97 percent of Earth's water is in the oceans, which have an average depth of about 4 km. Earth today has 100,000 times more water than Venus.

Some scientists think that Venus once had as much water as Earth—as liquid oceans or as more water vapor than is measured today. As the Sun aged and became brighter, and the planets received more solar energy, water molecules high in the atmosphere of Venus were broken apart into hydrogen and oxygen by solar UV radiation. The low-mass hydrogen atoms were quickly lost to space. Oxygen escaped more slowly, so some eventually migrated downward to the planet's surface, where it was removed from the atmosphere by bonding with minerals on the surface. The *Venus Express* spacecraft has measured hydrogen and some oxygen escaping from the upper levels of Venus's atmosphere in support of this theory.

---

### CHECK YOUR UNDERSTANDING 9.2

The main greenhouse gases in the atmospheres of the terrestrial planets are: (a) oxygen and nitrogen; (b) methane and ammonia; (c) carbon dioxide and water vapor; (d) hydrogen and helium.

.................................................................

# 9.3 Earth's Atmosphere Has Detailed Structure

Now that we have considered some of the overall processes that have influenced the evolution of the terrestrial planet atmospheres, we will look in depth at each of them. We begin with the composition and structure of Earth's atmosphere, not only because we know it best, but also because it will help us better understand the atmospheres of other worlds.

## Life and the Composition of Earth's Atmosphere

Two principal gases make up Earth's atmosphere: about four-fifths is nitrogen ($N_2$) and one-fifth is oxygen ($O_2$) (see Table 9.1). There are also many important minor constituents, such as water vapor and carbon dioxide ($CO_2$), the amounts of which vary depending on global location and season. The composition of Earth's atmosphere is relatively uniform on a global scale, but temperatures can vary widely. Atmospheric temperatures near Earth's surface can range from as high as 60°C (140°F) in the deserts to as low as −90°C (−130°F) in the polar regions. The mean global temperature is about 15°C.

Oxygen   Table 9.1 shows that Earth's atmosphere contains abundant amounts of oxygen ($O_2$) while the atmospheres of other planets do not. Oxygen is a highly reactive gas: it chemically combines with, or oxidizes, almost any material it touches. The rust (iron oxide) that forms on steel is an example. The reddish surface of Mars is coated with oxidized iron-bearing minerals, and this is one reason the martian atmosphere is almost completely free of oxygen. A planet with significant amounts of oxygen in its atmosphere requires a means of replacing oxygen lost through oxidation. On Earth, plants perform this role.

The oxygen concentration in Earth's atmosphere has changed over the history of the planet, as shown in **Figure 9.4**. When Earth's secondary atmosphere first appeared about 4 billion years ago, it had very little oxygen because $O_2$ is not found in volcanic gases or comets. Studies of ancient sediments show that about 2.8 billion years ago, an ancestral form of cyanobacteria—single-celled organisms that contain chlorophyll, which enables them to obtain energy from

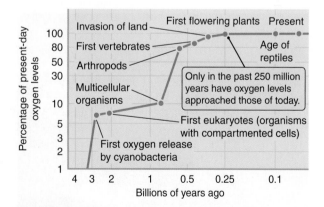

**Figure 9.4** The amount of oxygen in Earth's atmosphere has built up over time as a result of plant life on the planet.

sunlight—began releasing oxygen into Earth's atmosphere as a waste product of their metabolism. At first, this biologically generated oxygen combined with exposed metals and minerals in surface rocks and soils and so was removed from the atmosphere as quickly as it formed. Ultimately, the explosive growth of cyanobacteria and then plant life accelerated the production of oxygen, building up atmospheric concentrations that approached today's levels only about 250 million years ago.

All true plants, from tiny green algae to giant redwoods, use the energy of sunlight to build carbon compounds out of carbon dioxide and produce oxygen as a metabolic waste product in a process called photosynthesis. In this way, emerging life dramatically changed the very composition and appearance of Earth's surface—the first of many such widespread modifications imposed on Earth by living organisms. Earth's atmospheric oxygen content is held in a delicate balance primarily by plants. If plant life on the planet were to disappear, so, too, would nearly all of Earth's atmospheric oxygen, and therefore all animal life—including us.

Ozone Ozone ($O_3$) is another constituent in Earth's atmosphere. Ozone is formed when UV light from the Sun breaks molecular oxygen ($O_2$) into its individual atoms. These oxygen atoms can then recombine with other oxygen molecules to form ozone (the net reaction is $O_2 + O \rightarrow O_3$). Most of Earth's natural ozone is concentrated in the upper atmosphere at altitudes between 20 and 50 km. There it acts as a very strong absorber of UV sunlight. Without the ozone layer, this radiation would reach all the way to Earth's surface, where it would be lethal to nearly all forms of life. However, ozone in the lower atmosphere occurs primarily as a by-product of power plants, factories, and automobiles. This human-made pollutant is a health hazard, which raises the risk of respiratory and heart problems.

In the mid-1980s, scientists began noticing that the measured amount of ozone in Earth's upper atmosphere had been decreasing seasonally since the 1970s, primarily over the polar latitudes during springtime in both the Northern and Southern hemispheres. They called these depleted regions "ozone holes," although they are more like depressions than real holes in the ozone layer. Ozone depletion is caused by a seasonal buildup of atmospheric halogens—mostly chlorine, fluorine, and bromine—such as those found in industrial refrigerants, especially chlorofluorocarbons (CFCs). Halogens diffuse upward into the stratosphere, where they destroy ozone without themselves being consumed. Such agents are called catalysts—materials that participate in and accelerate chemical reactions but are not themselves modified in the process. Because they are not modified or used up, halogens may remain in Earth's upper atmosphere for decades or even centuries. Even though more of the chemicals originated in the north, the depletions are greater in the Southern Hemisphere because the colder temperatures in the southern polar regions produce a type of cloud that provides a surface on which the ozone-destroying chemical reactions can take place (**Figure 9.5**).

Scientists predicted that the continuing removal of ozone from the high atmosphere could cause trouble for terrestrial life as more and more UV radiation reached the ground. Measured increases in the levels of UV radiation appear to be related to increases in skin cancer in humans, and the mutating effects it may have on other life-forms are not yet understood. By the late 1980s, international agreements on phasing out the production of CFCs and other ozone-depleting chemicals were signed, and world consumption has steadily declined. The largest

**Figure 9.5** Polar stratospheric clouds form in the polar springtime and provide the surface upon which ozone destruction takes place.

Southern Hemisphere ozone hole occurred over Antarctica in 2006. The largest Arctic hole to date occurred in 2011, but this may have been due to unusually cold temperatures. Full recovery to 1980 levels is not expected until the late 21st century at the earliest, but scientists are hopeful that the international agreements have solved the problem.

Carbon Dioxide Carbon dioxide ($CO_2$) is another variable component of Earth's atmosphere. A complex pattern of carbon dioxide *sources* (places where it originates) and *sinks* (places where it goes) determines how much carbon dioxide will be present in the atmosphere at any one time. Plants consume carbon dioxide in great quantities as part of their metabolic process. Coral reefs are colonies of tiny ocean organisms that build their protective shells with carbonates produced from dissolved carbon dioxide. Fires, decaying vegetation, and human burning of fossil fuels all release carbon dioxide back into the atmosphere. This balance between carbon dioxide sources and sinks changes with time. As we'll describe later, the amount of carbon dioxide in the atmosphere has varied historically but has been increasing more rapidly for almost two centuries—since the industrial revolution. This recent increase in carbon dioxide in turn has had a direct effect on global temperature because carbon dioxide is a powerful greenhouse gas.

Water Vapor Water vapor ($H_2O$) in Earth's atmosphere also affects daily life and is a powerful greenhouse gas. Over the range of temperatures on Earth, the amount of water in the atmosphere varies from time to time and from place to place. In warm, moist climates, water vapor may account for as much as 3 percent of the total atmospheric composition. In cold, arid climates, it may be less than 0.1 percent. The continual process of condensation and evaporation of water involves the exchange of thermal and other forms of energy, making water vapor a major contributor to Earth's weather.

## The Layers of Earth's Atmosphere

Earth's atmosphere is a blanket of gas that is several hundred kilometers thick. It has a total mass of approximately $5 \times 10^{18}$ kg, which is less than one-millionth of Earth's total mass. The weight of Earth's atmosphere creates a force of approximately 100,000 newtons (N) acting on each square meter of the planet's surface, equivalent to about 14.7 pounds pressing on every square inch. This amount of pressure is called a **bar** (from the Greek *baros*, meaning "weight" or "heavy"). Earth's average atmospheric pressure at sea level is approximately 1 bar. A millibar (mb) is one-thousandth of 1 bar and is more commonly used in meteorology and in weather reports. One bar of pressure is equivalent to what you would experience underwater at a depth of 10 meters, or 33 feet. We are largely unaware of Earth's atmospheric pressure because the same pressure exists both inside and outside our bodies, so the force pushing out precisely balances the force pushing in.

Recall from Chapter 8 that the pressure at any point within a planet's interior must be great enough to balance the weight of the overlying layers. The same principle holds true in a planetary atmosphere. The atmospheric pressure on a planet's surface must be great enough to support the weight of the overlying atmosphere. Different forms of matter provide the pressure within a planet's interior and in its atmosphere. In the interior of a solid planet, solid materials exert pressure as they resist being compressed. In a planetary atmosphere, the motions of gas molecules exert sufficient pressure to support the atmosphere.

**Figure 9.6** These graphs show (a) temperature and (b) pressure plotted for Earth's atmospheric layers as a function of height. Most human activities are confined to the bottom layers of Earth's atmosphere.

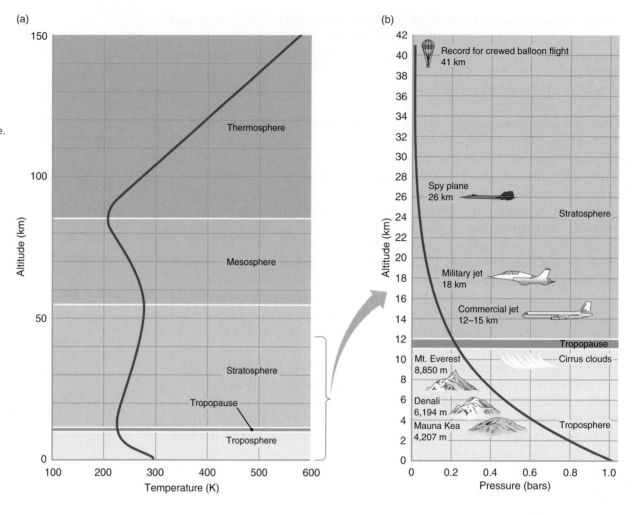

(a)

(b)

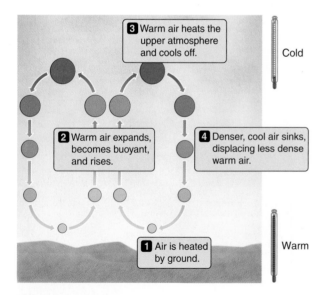

**Figure 9.7** Atmospheric convection carries thermal energy from the Sun-heated surface of Earth upward through the atmosphere.

The atmosphere of Earth is made up of several distinct layers, shown in **Figure 9.6**. These layers are distinguished by the changes in temperature and pressure through the atmosphere. The lowest layer, the one in which humans live and breathe, is called the **troposphere**. It contains 90 percent of Earth's atmospheric mass and is the source of all our weather. At Earth's surface, usually called sea level, the troposphere has an average temperature of 15°C (288 K). Within the troposphere, atmospheric pressure, density, and temperature all decrease as altitude increases. For example, at an altitude of 5.5 km (18,000 feet, which is a few thousand feet below the summit of Denali in Alaska), the atmospheric pressure and density are only 50 percent of their sea-level values, and the average temperature has dropped to −20°C (253 K). Still higher, at an altitude of 12 km, where commercial jets cruise, the temperature is −60°C (213 K), and the density and pressure are less than one-fifth what they are at sea level.

The atmosphere is warmer near Earth's surface because the air is closer to the sunlight-heated ground, which warms the air by infrared radiation. The atmosphere is cooler at very high altitudes because there the atmosphere freely radiates its thermal energy into space. In fact, it would get colder with increasing altitude even faster if not for convection. **Figure 9.7** illustrates how convection carries thermal energy upward through Earth's atmosphere. At a given pressure, cold air is denser than warm air. So when cold air encounters warm air, the denser cold air slips under the less dense warm air, pushing the warm air upward. This convection sets up air circulation between the lower and upper levels of the atmosphere and tends to diminish the temperature extremes caused by heating at the bottom and cooling at the top.

Convection also affects the vertical distribution of atmospheric water vapor. The ability of air to hold water in the form of vapor depends very strongly on the air temperature: the warmer the air, the more water vapor it can hold. The amount of water vapor in the air relative to what the air could hold at a particular temperature is called the relative humidity. Air that is saturated with water vapor has a relative humidity of 100 percent. As air is convected upward, it cools, limiting its capacity to hold water vapor. When the air temperature decreases to the point at which the air can no longer hold all its water vapor, water begins to condense to tiny droplets or ice crystals. In large numbers these become visible as clouds. When these droplets combine to form large drops, convective updrafts can no longer support them, and they fall as rain or snow. For this reason, most of the water vapor in Earth's atmosphere stays within 2 km of the surface. At an altitude of 4 km, the Mauna Kea Observatories (see the Chapter 6 opening figure) are higher than approximately one-third of Earth's atmosphere, but they lie above nine-tenths of the atmospheric water vapor. This is important for astronomers who observe in the infrared region of the spectrum, because water vapor strongly absorbs infrared light. The water in the atmosphere is more often visible as condensed water in the form of clouds and ice.

Returning to Figure 9.6, you can see that above the troposphere and extending upward to an altitude of 50 km above sea level is the **stratosphere**. The boundary between the troposphere and stratosphere is called the **tropopause**, the height at which temperature no longer decreases with increasing altitude. This change in atmospheric behavior is caused by heating from absorbed sunlight within the atmospheric layers that lie above the tropopause. The tropopause varies between 10 and 15 km above sea level, depending on latitude, and is highest at the equator. In this region, little convection takes place, because the temperature-altitude relationship reverses at the tropopause, and the temperature begins to increase with altitude. This temperature reversal is caused by the ozone layer, which warms the stratosphere by absorbing UV radiation from the Sun.

The region above the stratosphere is the **mesosphere**, which extends from an altitude of 50 km to about 90 km. In the mesosphere there is no ozone to absorb sunlight, so temperatures once again decrease with altitude. The base of the stratosphere and the upper boundary of the mesosphere are two of the coldest levels in Earth's atmosphere. Higher in Earth's atmosphere, interactions with space begin to be important. The solar wind is a flow of high-energy particles that stream continually from the Sun. At altitudes above 90 km, solar UV radiation and high-energy particles from the solar wind strip electrons from, or **ionize**, atmospheric molecules, causing the temperature once again to increase with altitude. This region, called the **thermosphere**, is the hottest part of the atmosphere. The temperature can reach 1000 K near the top of the thermosphere, at an altitude of 600 km.

The atoms and molecules in the gases within and beyond the thermosphere are ionized by UV photons and high-energy particles from the Sun. This region of ionized atmosphere is called the **ionosphere**, and it overlaps the thermosphere but also extends farther into space. The ionosphere reflects certain frequencies of radio waves back to the ground. For example, the frequencies used by AM radio bounce back and forth between the ionosphere and the surface, enabling radio receivers to pick up stations at great distances from the transmitters. Amateur radio operators are able to communicate with each other around the world by bouncing their signals off the ionosphere.

## Earth's Magnetosphere

Even farther out than the ionosphere is Earth's magnetosphere, which surrounds Earth and its atmosphere. The magnetosphere is a large region filled with electrons, protons, and other charged particles from the Sun that have been captured by the planet's magnetic field. This region has a radius approximately 10 times that of Earth and fills a volume more than 1,000 times as large as the volume of the planet itself. Magnetic fields only affect moving charges. Charged particles are free to move along the direction of the magnetic field but cannot cross magnetic

**Figure 9.8** (a) Charged particles, in this case electrons, spiral in a uniform magnetic field. (b) When the field is pinched, charged particles can be trapped in a "magnetic bottle." (c) Earth's magnetic field acts like a bundle of magnetic bottles, trapping particles in Earth's magnetosphere. In all these images, the radius of the helix that the charged particle follows is greatly exaggerated.

(a)

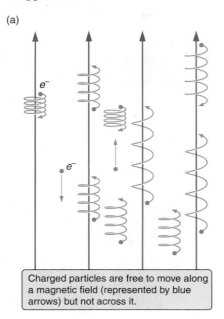

(b)

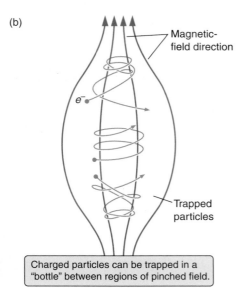

(c)

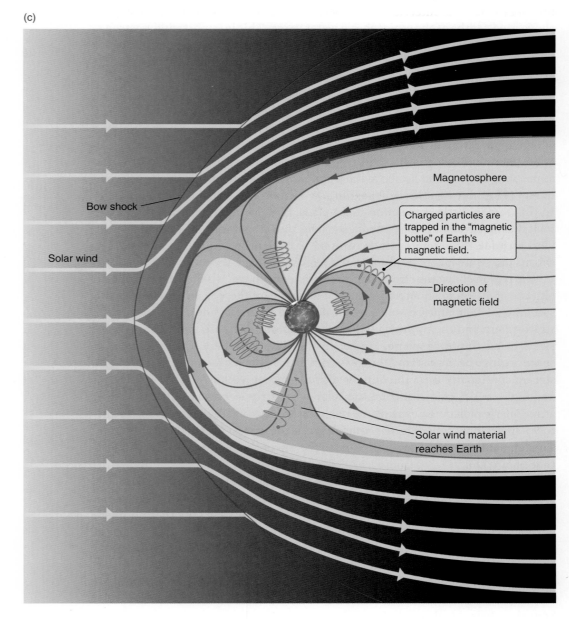

field lines. If they try to move across the direction of the field, they experience a force that is perpendicular both to the attempted motion of the particles across the field and to the direction of the magnetic field, as illustrated in **Figure 9.8a**. This force causes them to spiral around the direction of the magnetic field. Charged particles act like beads on a string, free to slide along the direction of the magnetic field but unable to cross it.

If the magnetic field is pinched together at some point, particles moving into the pinch will feel a magnetic force that reflects them back along the direction they came from, creating a sort of "magnetic bottle" that contains the charged particles. If charged particles are located in a region where the field is pinched on both ends, as shown in Figure 9.8b, then they may bounce back and forth many times. Earth's magnetic field is pinched together at the two magnetic poles and spreads out around the planet.

Earth and its magnetic field are immersed in the solar wind. When the charged particles of the solar wind first encounter Earth's magnetic field, the smooth flow is interrupted and their speed suddenly drops—they are diverted by Earth's magnetic field like a river is diverted around a boulder. As they flow past, some of these charged particles become trapped by Earth's magnetic field, where they bounce back and forth between Earth's magnetic poles as illustrated in Figure 9.8c.

An understanding of Earth's magnetosphere is of great practical importance. Regions in the magnetosphere that contain especially strong concentrations of energetic charged particles, called **radiation belts**, can be very damaging to both electronic equipment and astronauts. Yet it is not necessary to leave the surface of the planet to witness the dramatic effects of the magnetosphere. Disturbances in Earth's magnetosphere caused by changes in the solar wind can lead to changes in Earth's magnetic field that are large enough to trip power grids, cause blackouts, and disrupt communications.

Earth's magnetic field also funnels energetic charged particles down into the ionosphere in two rings located around the magnetic poles. These charged particles (mostly electrons) collide with atoms and molecules such as oxygen, nitrogen, and hydrogen in the upper atmosphere, causing them to glow like the gas in a neon sign. Interactions with different atoms cause different colors. These glowing rings, called **auroras**, can be seen from space (**Figure 9.9a**). When viewed from the ground (Figure 9.9b), auroras appear as eerie, shifting curtains of multicolored light. People living far from the equator are often treated to spectacular displays of the aurora borealis ("northern lights") in the Northern Hemisphere or the aurora australis in the Southern Hemisphere. When the solar wind is particularly strong, auroras can be seen at lower latitudes far from their usual zone. Auroras have also been seen on Venus, Mars, all of the giant planets, and some moons.

The general structure we have described here is not limited to Earth's atmosphere. The major vertical structural components—troposphere, tropopause, stratosphere, and ionosphere—also exist in the atmospheres of Venus and Mars, as well as in the atmospheres of Titan and the giant planets. The magnetospheres of the giant planets are among the largest structures in the Solar System.

## Wind and Weather

**Weather** is the local day-to-day state of the atmosphere. Local weather is caused by winds and convection. Recall from Chapter 5 that heating a gas increases its

**(a)**

G X U V I R

**(b)**

G X U V I R

**Figure 9.9** Auroras result when particles trapped in Earth's magnetosphere collide with molecules in the upper atmosphere. (a) An auroral ring around Earth's south magnetic pole, as seen from space. (b) Aurora borealis—the "northern lights"—viewed from the ground in Alaska.

**Astronomy in Action:** Charged Particles and Magnetic Fields

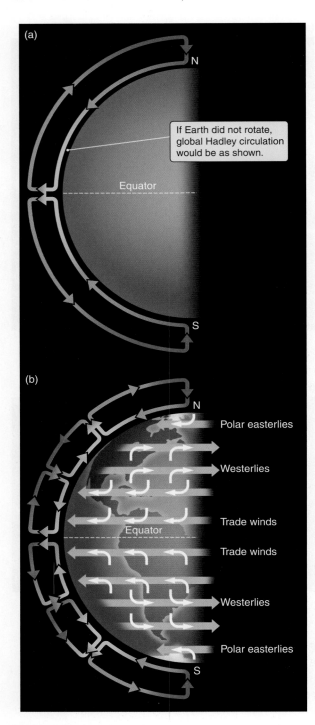

(a)

N

If Earth did not rotate, global Hadley circulation would be as shown.

Equator

S

(b)

N

Polar easterlies

Westerlies

Trade winds

Equator

Trade winds

Westerlies

Polar easterlies

S

**Figure 9.10** (a) Hadley circulation covers an entire hemisphere. (b) On Earth, Hadley circulation breaks up into smaller circulation cells due to the Coriolis effect which diverts the north–south flow into east–west zonal flow.

pressure, which causes it to push into its surroundings. These pressure differences cause winds. Winds are the natural movement of air, both locally and on a global scale, in response to variations in temperature from place to place. The air is usually warmer in the daytime than at night, warmer in the summer than in winter, and warmer at the equator than in the polar regions. Large bodies of water, such as oceans, also affect atmospheric temperatures. The strength of the winds is governed by the size of the temperature difference from place to place.

Recall from Chapter 2 that the effect of Earth's rotation on winds—and on the motion of any object—is called the Coriolis effect (see Figure 2.11). As air in Earth's equatorial regions is heated by the warm surface, convection causes it to rise. The warmed surface air displaces the air above it, which then has nowhere to go but toward the poles. This air becomes cooler and denser as it moves toward the poles, and so it sinks back down through the atmosphere. It displaces the surface polar air, which is forced back toward the equator, completing the circulation. As a result, the equatorial regions remain cooler and the polar regions remain warmer than they otherwise would be. Air moves between the equator and poles of a planet in a pattern known as **Hadley circulation**, which is shown in **Figure 9.10a**.

On Earth, other factors break up the planetwide flow into a series of smaller Hadley cells. Most planets and their atmospheres rotate rapidly, and the Coriolis effect strongly interferes with Hadley circulation by redirecting the horizontal flow, shown in Figure 9.10b. The Coriolis effect creates winds that blow predominantly in an east–west direction and are often confined to relatively narrow bands of latitude. Meteorologists call these **zonal winds**. More rapidly rotating planets have a stronger Coriolis effect and stronger zonal winds. Between the equator and the poles in most planetary atmospheres, the zonal winds alternate between winds blowing from the east toward the west (easterlies) and winds blowing from the west toward the east (westerlies).

In Earth's atmosphere, several bands of alternating zonal winds lie between the equator and each hemisphere's pole. This zonal pattern is called Earth's **global circulation** because its extent is planetwide. The best-known zonal currents are the subtropical trade winds—more or less easterly winds that once carried sailing ships from Europe westward to the Americas—and the midlatitude prevailing westerlies that carried them home again.

Embedded within Earth's global circulation pattern are systems of winds associated with large high-pressure and low-pressure regions. A combination of a low-pressure region and the Coriolis effect produces a circulating pattern called **cyclonic motion** (**Figure 9.11**). Cyclonic motion is associated with stormy weather, including hurricanes. Similarly, high-pressure systems are localized regions where the air pressure is higher than average. We think of these regions of greater-than-average air concentration as "mountains" of air. Owing to the Coriolis effect, high-pressure regions rotate in a direction opposite to that of low-pressure regions. These high-pressure circulating systems experience **anticyclonic motion** and are generally associated with fair weather.

Earth has a water cycle: water from the oceans enters the air and later returns to the oceans. When liquid water in Earth's oceans, lakes, and rivers absorbs enough thermal energy from sunlight, it turns to water vapor. The water vapor carries this thermal energy as it circulates throughout the atmosphere, releasing the energy to its surroundings when the water vapor recondenses. This process powers rainstorms, thunderstorms, hurricanes, and other dramatic weather. For example, a thunderstorm begins when moist air close to the ground is warmed by the Sun and is convected upward, cooling as it gains altitude, until it condenses as rain. With strong solar heating and an adequate supply of moist air, this self-feeding process can grow within minutes to become a violent thunderstorm. Water falls as rain, eventually returning to lakes and oceans, wearing down mountains and eroding the soil as it flows.

Coriolis forces acting on air rushing into regions of low atmospheric pressure create huge circulating systems that result in hurricanes. The conditions must be just right: warm tropical seawater, light winds, and a region of low pressure in which air spirals inward. Warm seawater evaporates; then the moisture-laden air rises and releases energy as it condenses at cooler levels, similar to the process that leads to thunderstorms. Sustained winds near the center of the storm can reach speeds of greater than 300 kilometers per hour (km/h), causing widespread damage and fatalities. Tornadoes are small but violent circulations of air associated with storm systems. Dust devils are similar in structure to tornadoes, but they are generally smaller and less intense and usually occur in fair weather; for example, in the deserts of the American Southwest. Diameters of dust devils range from a few meters to a few dozen meters, with average heights of several hundred meters. The lifetime of a typical tornado or dust devil is brief, limited to a dozen or so minutes.

**CHECK YOUR UNDERSTANDING 9.4**

All weather and wind on Earth are a result of convection in the: (a) troposphere; (b) stratosphere; (c) mesosphere; (d) ionosphere; (e) thermosphere.

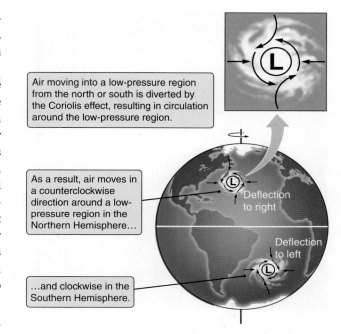

**Figure 9.11** As a result of the Coriolis effect, air circulates around regions of low pressure on the rotating Earth.

## 9.4 The Atmospheres of Venus and Mars Differ from Earth's

As shown in Table 9.1, the atmospheres of Venus, Earth, and Mars are very different. The atmosphere of Venus is very hot and dense compared to that of Earth, while the atmosphere of Mars is very cold and thin. The greenhouse effect has turned Venus hellish, with extremely high temperatures and choking amounts of sulfurous gases. Compared to Venus, the surface of Mars is almost hospitable. Understanding why and how these atmospheres are so different helps us understand how Earth's atmosphere may evolve in the future.

### Venus

Venus and Earth are similar enough in size and mass that they were once thought of as sister planets. Indeed, when we used the laws of radiation in Chapter 5 to predict temperatures for the two planets, we concluded that they should be very similar. However, spacecraft visits to Venus in the 1960s revealed that the temperature, density, and pressure of Venus's atmosphere were all much higher than for Earth's atmosphere. Ninety-six percent of Venus's massive atmosphere is carbon dioxide, with only 3.5 percent of nitrogen and lesser amounts of other gases. These atmospheric properties are due to the greenhouse effect and the role of carbon dioxide in blocking the infrared radiation typically emitted by a planetary surface. This thick blanket of carbon dioxide effectively traps the infrared radiation, raising the temperature at the surface of the planet to a sizzling 740 K, which is hot enough to melt lead (**Figure 9.12**). The atmospheric pressure at the

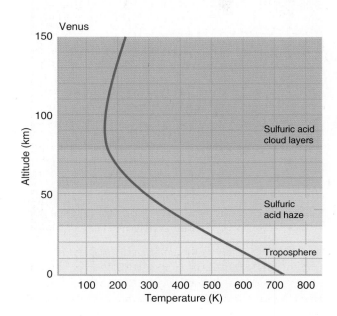

**Figure 9.12** The temperature of the atmosphere on Venus primarily drops with increase in altitude, unlike temperatures in Earth's atmosphere, which fall and rise and fall again through the troposphere, stratosphere, and mesosphere, respectively.

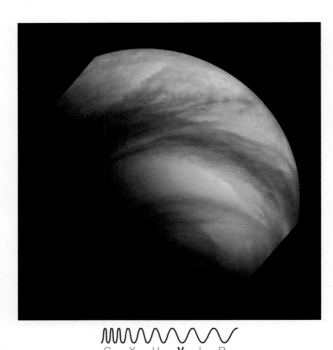

Figure 9.13 This image from *Venus Express* shows the thick clouds obscuring the view of the surface of Venus.

**Figure 9.14** This image of Venus is from the 1982 Soviet *Venera 14* mission. The spacecraft is in the foreground. Note the rocky ground and the orange sky.

**Figure 9.15** This true-color image of the surface of Mars was taken by the rover *Spirit*. In the absence of dust, the sky's thin atmosphere would appear deep blue. In this image, windblown dust turns the sky pinkish.

surface of Venus is 92 times greater than that at Earth's surface: this is equal to the water pressure at an ocean depth of 900 meters, which is more than enough pressure to crush the hull of a submarine.

As you can see in Figure 9.12, the atmospheric temperature of Venus decreases continuously with altitude throughout the planet's troposphere—similar to Earth, dropping to a low of about 160 K at the tropopause. At an altitude of approximately 50 km, Venus's atmosphere has an average temperature and pressure similar to Earth's atmosphere at sea level. At altitudes between 50 and 80 km, the atmosphere is cool enough for sulfurous oxide vapors to react with water vapor to form dense clouds of concentrated sulfuric acid droplets ($H_2SO_4$). These dense clouds completely block Earth's view of the surface of Venus, as **Figure 9.13** shows. Large variations in the observed amounts of sulfurous compounds in the high atmosphere of Venus suggest that the sulfur arises from sporadic episodes of volcanic activity. This along with some bright spots seen near a large shield volcano strengthens the possibility that Venus is currently volcanically active.

In the 1960s, radio telescopes and spacecraft with cloud-penetrating radar provided low-resolution views of the surface of Venus. It was not until 1975, when the Soviet Union succeeded in landing cameras there, that scientists got a clear picture of the surface. These images showed fields of rocks 30–40 centimeters (cm) across and basalt-like slabs surrounded by weathered material. A series of Soviet landers in the 1980s revealed similar landscapes (**Figure 9.14**). Radar images taken by the *Magellan* spacecraft in the early 1990s (see Figure 8.20) produced a global map of the surface of Venus. The high atmospheric temperatures on Venus also mean that neither liquid water nor liquid sulfurous compounds can exist on its surface, leaving an extremely dry lower atmosphere with only 0.01 percent water and sulfur dioxide vapor.

Imagine yourself standing on the surface of Venus. Because sunlight cannot easily penetrate the dense clouds above you, noontime on the surface of Venus is no brighter than a very cloudy day on Earth. High temperatures and very light winds keep the lower atmosphere of Venus free of clouds and hazes. The local horizon can be seen clearly, but strong scattering of light by molecules in the dense atmosphere of Venus would greatly soften any view you might have of distant mountains.

Unlike the other Solar System planets, Venus rotates on its axis in a direction opposite to its motion around the Sun. Astronomers call this *retrograde rotation*. Relative to the stars, Venus rotates on its axis once every 243 Earth days. However, a solar day on Venus—the time it takes for the Sun to return to the same place in the sky—is only 117 Earth days. The slow rotation means that Coriolis effects on the atmosphere are small. Global Hadley circulation seldom occurs in planetary atmospheres, because other factors, such as planet rotation, break up the planetwide flow into a series of smaller Hadley cells. However, because of its slow rotation, Venus is an exception and as a result is the only planet with global circulation that is quite close to a classic Hadley pattern (see Figure 9.10a).

The massive atmosphere on Venus is highly efficient in transporting thermal energy around the planet, so the polar regions are only a few degrees cooler than the equatorial regions, and there is almost no temperature difference between day and night. Because Venus's equator is nearly in the plane of the planet's orbit, seasonal effects are quite small, producing only negligible changes in surface temperature. Such small temperature variations also mean that wind speeds near the surface of Venus are quite low, typically about a meter per second, so wind erosion is weak compared to that on Earth and Mars. High in the atmosphere, 70 km up, temperature differences are larger, contributing to super hurricane force winds that reach speeds of 110 m/s (400 km/h), circling the planet in only 4 days. The variation of this high-altitude wind speed with latitude can be seen in the chevron, or V-shaped, cloud patterns.

When the *Pioneer Venus* spacecraft was orbiting Venus during the 1980s, its radio receiver picked up many bursts of lightning static—so many that Venus appears to have a rate of lightning activity comparable to that of Earth. On Venus, as on Earth, lightning is created in the clouds; but Venus's clouds are so high—typically 55 km above the surface of the planet—that the lightning bolts never hit the ground. More recently, *Venus Express* observed magnetic signatures of lightning on Venus.

## Mars

Mars has a stark landscape, colored reddish by the oxidation of iron-bearing surface minerals. The sky is sometimes dark blue but more often a pinkish color caused by windblown dust (**Figure 9.15**). The lower density of the martian atmosphere makes it more responsive than Earth's atmosphere to heating and cooling, so its temperature extremes are greater. The surface near the equator at noontime is a comfortable 20°C—a cool room temperature (68°F) on Earth. However, nighttime temperatures typically drop to a frigid −100°C, and during the polar night the air temperature can reach −150°C—cold enough to freeze carbon dioxide out of the air in the form of a dry-ice frost. The temperature profile of the atmosphere of Mars shown in **Figure 9.16** has a range of only 100 degrees up to about 125 km. Above that. the temperature rises because of absorption of sunlight in the upper atmosphere. The temperature profile of Mars is more similar to that of Earth than that of Venus.

The average atmospheric surface pressure of Mars is equivalent to the pressure at an altitude of 35 km above sea level on Earth, far higher than Earth's highest mountain. There is no "sea level" on Mars because there are no oceans. Surface pressure varies from 11.5 mb in the lowest impact basins of Mars to 0.3 mb at the summit of Olympus Mons. Recall that Earth's pressure at sea level is about 1 bar, so the highest pressure on Mars is only 1.1 percent of Earth's pressure at sea level. Like Earth, Mars has some water vapor in its atmosphere, but its low temperatures condense much of the water vapor out as clouds of ice crystals. Mars can have early-morning ice fog in the lowlands (**Figure 9.17**) and clouds hanging over the mountains.

In the absence of plants, Mars has only a tiny trace of oxygen, which is so important to life on Earth. Like Venus, the atmosphere of Mars is composed almost entirely of carbon dioxide (95 percent) and a lesser amount of nitrogen (2.7 percent). The near absence of oxygen means that Mars has very little ozone. Without ozone, solar UV radiation reaches all the way to the surface. These UV rays could be lethal to any surface life-forms, so any life would either need to develop protective layers or be located away from direct exposure on the surface of the planet; for example, in caves or below the surface.

The tilt of the rotation axis of Mars is similar to Earth's at present, so both planets have similar seasons. But seasonal effects on Mars are larger for two reasons: Mars varies more in its annual orbital distance from the Sun than Earth does, and the low density of the martian atmosphere makes it more responsive to

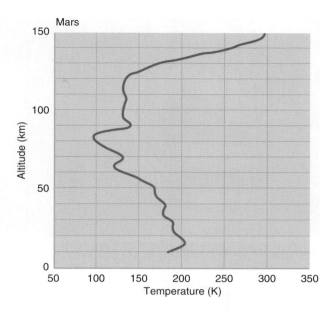

**Figure 9.16** The temperature profile of the atmosphere of Mars. Note the differences in temperature and structure between this profile and the profile of the atmosphere of Venus (see Figure 9.12).

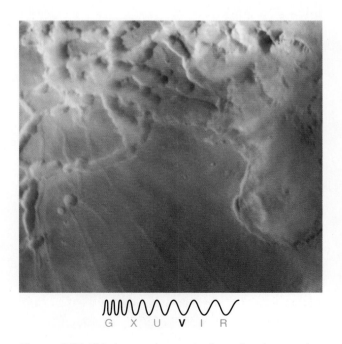

**Figure 9.17** This image shows patches of early-morning water vapor fog forming in canyons on Mars.

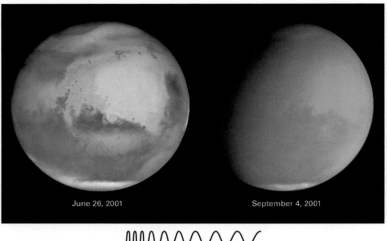

June 26, 2001          September 4, 2001

G  X  U  V  I  R

**Figure 9.18** Hubble Space Telescope images show the development of a global dust storm that enshrouded Mars in September 2001. The same region of the planet is shown in both images; surface features are obscured by the thick layer of dust.

**Figure 9.19** This dust devil on Mars was imaged by the *Mars Reconnaissance Orbiter*.

seasonal change. The large daily, seasonal, and latitudinal surface temperature differences on Mars often create locally strong winds, some estimated to be higher than 100 m/s (360 km/h). High winds can stir up huge quantities of dust and distribute it around the planet's surface. For more than a century, astronomers have watched the seasonal development of springtime dust storms on Mars. The stronger storms spread quickly and can envelop the entire planet in a shroud of dust within a few weeks (**Figure 9.18**). Such large amounts of windblown dust can take many months to settle out of the atmosphere. Seasonal movement of dust from one area to another alternately exposes and covers large areas of dark, rocky surface. This phenomenon led some astronomers of the late 19th and early 20th centuries to believe that they were witnessing the seasonal growth and decay of vegetation on Mars.

The *Viking* landers first noticed dust devils on Mars in 1976. More recently, *Mars Reconnaissance Orbiter* spotted a large number of dust devils, visible because of the shadows they cast on the martian surface. **Figure 9.19** shows a 20-km-high, 70-m-wide dust devil. Most martian dust devils leave dark meandering trails behind them where they have lifted bright surface dust, revealing the dark surface rock that lies beneath the dust. Dust devils on Mars— typically higher, wider, and stronger than dust devils on Earth—reach heights of up to 8 km and have diameters ranging from a few dozen to a few hundred meters.

Mars likely had a more massive secondary atmosphere in the distant past. As discussed in Chapter 8, geological evidence strongly suggests that liquid water once flowed across its surface. But the low incidence (possibly cessation) of volcanism and the planet's low gravity, and perhaps the decrease of its magnetic field, was responsible for the loss of much of this earlier atmosphere. Scientists have not yet reached a consensus on how massive the martian atmosphere was in the past.

## Mercury and the Moon

The ultrathin atmospheres of Mercury and the Moon are known as **exospheres**, and they are less than a million-billionth ($10^{-15}$) as dense as Earth's atmosphere. The recent NASA *Lunar Atmosphere and Dust Environment Explorer* (*LADEE*) mission found helium, argon, and dust in the Moon's exosphere. Other atoms, such as sodium, calcium, and even water-related ions, were seen in Mercury's exosphere by the *Messenger* spacecraft, and they may have been blasted loose from Mercury's surface by the solar wind or micrometeoroids. The exospheres of Mercury and the Moon probably vary somewhat with the strength of the solar wind and the atoms of hydrogen and helium they capture from it. Exospheres have no effect on local surface temperatures, but astronomers study their interactions with the solar wind.

Mercury and Venus are too hot for people to visit or live on their surfaces. If people ever create settlements on the Moon or Mars, they will need to take along or find local materials to produce a pressurized environment with oxygen to breathe and to construct protection from solar radiation.

## CHECK YOUR UNDERSTANDING 9.5

Rank, from greatest to smallest, the seasonal variations on the following planets: (a) Mercury; (b) Venus; (c) Earth; (d) Mars

# 9.5 Greenhouse Gases Affect Global Climates

**Climate** is the *average* state of an atmosphere, including temperature, humidity, winds, and so on. Climate describes the planet as a whole. This is an important distinction from weather, which is the state of an atmosphere at any given time and place. The study of climate change on Earth and Mars is not new to the 21st century. Nineteenth-century scientists found evidence of past ice ages and knew that Earth's climate had been very different at earlier times in its history. Observations of changes in the martian ice caps led to speculation about whether Mars also had ice ages in its history. In this section, we will look at the natural factors that can cause climates to change on planets, and then we will examine the additional factors that affect Earth.

## Factors That Can Cause Climate Change on a Planet

Scientists study the astronomical, geological, and, on Earth, biological mechanisms controlling climate on the planets. Astronomical mechanisms that influence changes in planetary temperature include changes in the Sun's energy output, which has increased slowly as the Sun ages, and possibly changes in the galactic environment as the Sun travels in its orbit around the center of the Milky Way. Scientists have suggested that sporadic bursts of gamma rays or cosmic rays (fast-moving protons) from distant exploding stars could interact with planetary atmospheres. These mechanisms would affect *all* of the planets in the Solar System at the same time.

Other astronomical mechanisms relevant to climate change and specific to each planet are the Milankovitch cycles, named for geophysicist Milutin Milanković (1879–1958). For example, a planet's energy balance can be affected by periodic changes in its orbital eccentricity, or its precession, or the tilt of its rotational axis. Recall from Working It Out 5.4 that numerous factors affect a planet's energy balance and therefore its temperature. If a planet's orbit becomes more eccentric, the amount of energy it receives from the Sun will vary more over its year. If the tilt of a planet increases, its seasons will become more extreme, and its temperature variation during the year will increase. The precession cycle affects which hemisphere is pointed to the Sun at different times of the elliptical orbit, so that one hemisphere may have longer winters and the other longer summers.

The tilt (obliquity) of Earth's axis varies from 22.1° to 24.5°, and Earth's relatively large Moon keeps that tilt from changing more than this. In contrast, the moons of Mars are small, and the gravitational influence of Jupiter is a greater factor on Mars. The tilt of Mars is thought to vary from 13° to 40°, or possibly even more. Given the precession of Mars and its more eccentric orbit, which creates differences in season length between its northern and southern hemispheres, Mars may have had very large swings in its climate throughout its history as the obliquity changed.

Another set of factors that can affect climate is the geological activity of a planet. Volcanic eruptions can produce dust, aerosol particles, clouds, or hazes that block sunlight over the entire globe and lower the temperature. Impacts by large objects can kick up sunlight-blocking particles. Tectonic activity can also affect climate. For example, on Earth the shifting of the plates has led to different configurations of the oceans and the shifting continents, affecting global atmospheric and oceanic circulatory patterns. The albedo of a planet can increase if there are more clouds and ice or decrease if ice melts or is covered by volcanic ash. Changes may also arise from variations in carbon cycles. On Earth, long-term interactions of the oceans, the land, and the atmosphere can affect the levels of greenhouse gases such as carbon dioxide and water vapor.

A third set of mechanisms that trigger climate changes is biological. Recall from Section 9.3 that over billions of years on Earth, photosynthesis by bacteria and later by plants removed carbon dioxide from the atmosphere and replaced it with oxygen. Biological (and geological) activity on Earth can produce methane, a strong greenhouse gas. Certain microorganisms produce methane as a metabolic by-product. For example, those bubbles you see rising to the surface of a stagnant pond—swamp gas—contain biologically produced methane. Methane is also emitted from the guts of grain-fed livestock (and, in the past, from some of the large dinosaurs) as well as from termites. Another biological effect on climate could come from phytoplankton. If the oceans get more solar energy and warm up from one of the astronomical mechanisms, then the phytoplankton in the ocean may grow faster, leading to more aerosol release and more cloud formation, which increases albedo. Finally, human activities are triggering some major changes, as discussed in the next subsection.

In short, many factors affect the temperature of a planet. Earth's climate is the most complicated of those of the terrestrial planets because Earth is the most geologically and biologically active. How do scientists sort out all of these factors? They use the scientific method. Scientists create mathematical models to simulate the general circulation and energy balance of a planet, incorporating all of the appropriate factors. The goal is to create a global climate model that reproduces the empirical data from observations of a planet. Once the model correctly predicts past and present climate, then it can be used to predict future climate. The first simple climate models for Earth were run on the earliest computers in the 1950s and 1960s. One set of models developed at NASA's Goddard Institute for Space Studies in the 1970s was a spinoff of a program originally designed for the study of Venus. The insights from comparative planetology are very important for producing better models that will aid in scientific predictions of climate change on Earth.

## Climate Change on Earth

**Paleoclimatology** is the study of changes in Earth's climate throughout its history. Earth's atmosphere is so sensitive to even small temperature changes that it takes a drop of only a few degrees in the mean global temperature to plunge the planet's climate into an ice age. Scientists use evidence from geology and paleontology such as sediments, ice sheets, rocks, tree rings, coral, shells, and fossils to get data on Earth's past climate. They have found that Earth's climate has lengthy temperature cycles, some lasting hundreds of thousands of years and some tens of thousands of years. As you can see in the middle plot in **Figure 9.20**, there have been periods of

**Figure 9.20** This graph shows global variations in carbon dioxide ($CO_2$), temperature, and methane ($CH_4$) concentrations over the past 800,000 years of Earth's history. Notice the multiple *y*-axes on this graph. The axis on the right relates to the temperature data (black); the axes on the left relates to the $CO_2$ (blue) and $CH_4$ (red) data. These data sets have been plotted on the same graph to make the similarities and differences easier to see. The low points correspond to ice ages. ppb = parts per billion; ppm = parts per million. 2015 values for methane (red) and $CO_2$ (blue) are indicated by the large arrows.

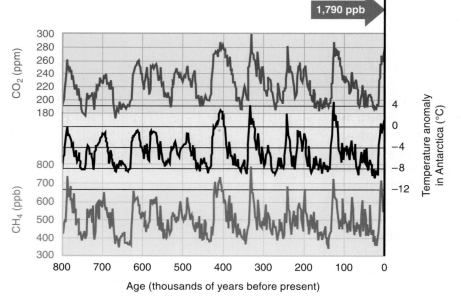

colder temperatures on Earth, known as ice ages. These oscillations in the mean global temperature are far smaller than typical geographic or seasonal temperature variations.

Periodic Milankovitch cycle changes in Earth's orbit correspond to some temperature cycles. As shown in **Figure 9.21**, Earth's tilt varies in cycles of 41,000 years; the eccentricity of Earth's orbit varies with two cycles of about 100,000 and 413,000 years, respectively; and the time of the year when Earth is closest to the Sun varies in cycles of about 21,000 years. Global climate models using these Milankovitch cycles have successfully replicated much of the observed paleo-climatology data. Temperature changes that are not periodic may have been triggered internally by volcanic eruptions or long-term interactions between Earth's oceans and its atmosphere or by other factors mentioned already.

If Earth's climate has been changing naturally for most of its history, then why are scientists especially concerned about the current trend in global climate? Figure 9.20 shows the carbon dioxide levels, methane levels, and temperature of Earth's atmosphere over the past 800,000 years, obtained from measuring deep ice-cores in the polar regions. Notice that these three factors are correlated. When one rises, so do the others. These data show the naturally occurring ranges since before the existence of the first humans. The temperature difference between ice ages and interglacial periods is only 10°C–15°C. Note that these changes are gradual, occurring over tens of thousands of years.

Two major changes have taken place on Earth during the past 150 years. First, the industrial revolution led to an increase in the production of greenhouse gases, especially from the burning of fossil fuels, which releases carbon dioxide into the atmosphere. In 1896, Svante Arrhenius (1859–1927), a Nobel Prize–winning chemist, produced calculations showing that $CO_2$ released from burning fossil fuels could increase the greenhouse effect and raise Earth's surface temperature. The second change has been the rapid growth in human population. When populations increase, people burn forests to clear land for agriculture and industry, removing photosynthesizing plants, increasing $CO_2$ emissions, and locally changing Earth's albedo. A higher population means more agricultural soil that releases nitrous oxide, more livestock that releases methane, and more people who release carbon dioxide by burning fossil fuels. The data in **Figure 9.22** show that concentrations of these greenhouse gases in the atmosphere have been increasing since

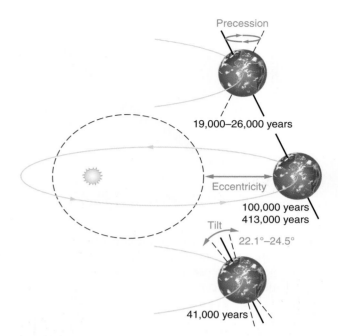

**Figure 9.21** An illustration of the Milankovitch cycles for Earth. The precession of Earth and the rotation of its elliptical orbit combine to a cyclic variation of about 21,000 years; its eccentricity varies with two cycles of about 100,000 and 413,000 years, and its tilt varies in cycles of 41,000 years. (Not to scale.)

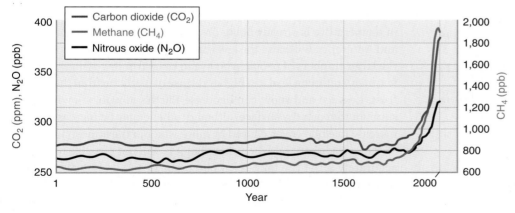

**Figure 9.22** Concentrations of greenhouse gases from the year 1 to 2005, showing the increases beginning at the time of industrialization.

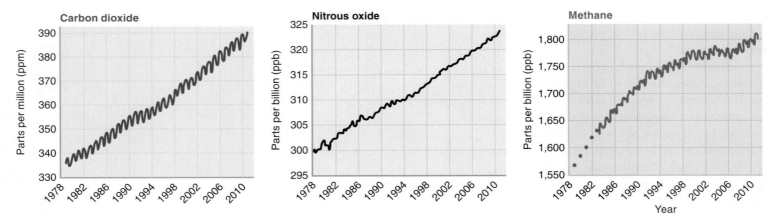

**Figure 9.23** The National Oceanic and Atmospheric Administration (NOAA) plots of global average amounts of the major greenhouse gases. $CO_2$ reached 400 ppm in some months in 2014 and 2015.

**Figure 9.24** Earth's global average temperature and $CO_2$ concentrations since 1880. This graph shows that global temperatures are climbing along with concentrations of carbon dioxide. Annual variations in atmospheric $CO_2$ can be attributed to seasonal variations in plant life and fossil fuel use, while the overall steady climb is due to human activities.

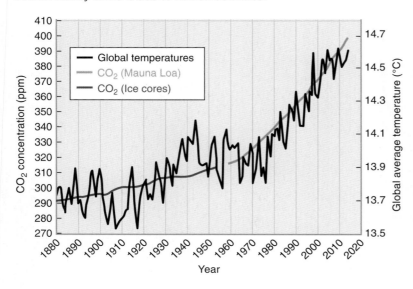

the industrial revolution. The $CO_2$ level has risen and is higher than any of the levels seen in Figure 9.20. **Figures 9.23** and **9.24** show that there has been a steady increase in nitrous oxide, methane, and the mean global temperature.

The vast majority of climatologists accept the computer models indicating that this trend represents the beginning of a long-term change in temperature caused by the buildup of human-produced greenhouse gases. This anthropogenic (human-caused) change is happening much faster than the changes seen in Figure 9.20. Even an average increase of a few degrees can greatly affect Earth. The planet's atmosphere is a delicately balanced mechanism. Earth's climate is a complex, chaotic system within which tiny changes can produce enormous and often unexpected results. To add to the complexity, Earth's climate is intimately tied to ocean temperatures and currents. Ocean currents are critical in transporting energy from one part of Earth to another, and it is uncertain how increased temperatures may affect those systems. Warmer oceans evaporate, leading to wetter air, which can mean more intense summer and winter storms (including more snow). We see examples of this connection in the periodic El Niño and La Niña conditions, small shifts in ocean temperature that cause much larger global changes in air temperature and rainfall. The **Process of Science Figure** discusses how scientists think about complex issues like this.

Changes in climate affect where plants and animals can live, and thus dates and locations of breeding, migration, hibernation, and so on. Agricultural growing seasons and pollination are also affected, as is the availability of freshwater. The melting of mountain glaciers and polar sea ice from the increase in temperature is already being observed. Melted ice from Greenland and Antarctica will raise the level of the oceans—a serious problem for the large numbers of people who live in coastal or low-lying regions. A warmer Arctic can lead to a higher release of methane from the permafrost. Less ice and snow can decrease Earth's albedo, allowing more sunlight to reach the surface, although albedo may be increased by an increase in cloud cover, caused by more water in the atmosphere.

The processes are so complex that it is still not possible to predict accurately all of the long-term outcomes of small changes that humans are now making to the composition of Earth's atmosphere and to Earth's albedo. In a real sense, we are *experimenting* with Earth. We are asking the question, What happens to Earth's climate if we steadily increase the number of greenhouse molecules in its atmosphere? We do not yet know the full answer, but we are already seeing some of the consequences.

**Climate change is an example of a complex scientific issue. When confronted with complex science, there are several questions you should ask.**

## What is the quality of the evidence?

**How many studies have been done?**

**How many kinds of studies have been done?**

## Does the basic physics make sense?

**Is it logical?**

**Is the explanation natural or supernatural?**

**If this claim were not true, what else would that imply?**

**If this scientific claim were true, what else would that imply?**

## Is the claim falsifiable?

**If I hold this in my head as if it were true, what would convince me it was false?**

**If I hold this in my head as if it were false, what would convince me it was true?**

Scientific issues become remarkably complex when they relate to policy decisions with a broad reach. Keeping an open mind in such cases means thinking carefully about the quality of the evidence, as well as what data would make you change your mind. If your mind cannot be changed, you are not participating in science.

# Origins

## Our Special Planet

We might ask the question, Why are we here on Earth, instead of elsewhere in the Solar System? In Chapter 7, we defined the habitable zone of a planetary system as the distance from its star where a planet could have a surface temperature such that large amounts of water could exist in liquid form. In this chapter, you have learned that the temperature of a planet can depend on more than just distance from the star, and it can change over time because of numerous factors, such as the planet's ability to hold on to its atmosphere, the atmospheric greenhouse effect, and orbital variations. Mercury and the Moon were too small to hold on to either their primary or their secondary atmospheres, leaving them as airless rocks. Venus, Earth, and Mars all had some liquid water at one time: early in the history of the Solar System all three might have been classified as habitable. Their atmospheres were likely similar then, too, before the atmosphere of Mars escaped, the atmosphere of Venus was heated by the greenhouse effect, and the atmosphere of Earth was oxygenated. It is even possible that primitive life developed on Venus and Mars at the same time it developed on Earth, about a billion years after the Solar System formed.

Ultimately, the three planets evolved differently, and currently only Earth has the liquid water that is vital for our biology. Venus receives more energy from the Sun and was slightly warmer than the young Earth. The young Venus might have been too hot for liquid water to form oceans or any liquid water might have quickly evaporated. Because Venus had a lot of water vapor in its atmos-phere and lacked a liquid ocean to store carbon dioxide, the greenhouse effect created a thicker atmosphere. This thicker atmosphere fed back to a stronger greenhouse effect, which created an even thicker atmosphere. With this resulting runaway greenhouse effect, Venus became just too hot: the evaporated water molecules broke apart, the hydrogen escaped to space, and any water cycle was completely destroyed. It is possible that primitive bacteria lingered in water vapor in the clouds on Venus, but life did not evolve into anything more complex.

Mars is smaller and less massive than Venus or Earth, so it has a weaker gravity. It has a larger orbit than Venus and Earth, and so receives less energy from the Sun. Over time, much of its atmosphere escaped and was not replaced by volcanic emissions, and the atmospheric pressure became too low to maintain liquid water. But whereas Venus was too hot, when Mars was young it was too cold. Back then it had a thicker atmosphere and liquid rain. Images of the surface of Mars show flood basins indicating huge rivers. Liquid water on the surface of Mars was too effective at scrubbing the planet's atmosphere of carbon dioxide. The process that prevented Earth from becoming a Venus-like hothouse continued further on Mars, and the temperature fell until the water froze. There are hints of subsurface water on Mars, and perhaps some form of martian bacteria will be found beneath the ground.

Thus, our Solar System contains astronomical evidence of how the greenhouse effect can influence planetary atmospheres, including Earth's. Planetary scientists view these different planets as a cautionary tale, showing the results of varied "doses" of greenhouse gases. Only Earth stayed "just right" and was able to retain the liquid oceans in which more complex life evolved. We owe our lives to the blanket of atmosphere that covers the planet. The study of the Solar System reveals that Earth is maintained by the most delicate of balances. Over billions of years, life has shaped Earth's atmosphere, and today, through the activities of humans, life is reshaping the planet's atmosphere once again. Human civilization is very young compared to Earth, and it has been brief compared to the cycles of climate on the planet. The past 10,000 years have been a relatively stable period of Earth's climate: many argue that this stability is what enabled agriculture and civilization to develop. It is uncertain what will happen to agriculture, civilization, or the planet itself if the climate undergoes fast, intense changes. The other planets in the Solar System are not places where billions of people from our planet can live. Earth is the only planet suitable for human life.

Planetary scientists conclude that Mars had more liquid water than expected.

(X)

# Mars Once Had an Entire Ocean—and then Lost It, Scientists Say

By **AMINA KHAN, Los Angeles Times**

Dry, dusty Mars once had an ocean that held as much water as the Arctic Ocean and covered a larger share of the red planet's surface than the Atlantic Ocean does on Earth, according to a surprising new study.

The findings, described online in the journal *Science*, examined the patterns in the martian atmosphere to try to understand how much water it has lost in the last few billion years—and finds that the planet may have been wetter and for longer than scientists may have thought.

As the scientists examined their surprising findings, "the story started to make sense," said lead author Geronimo Villanueva, a planetary scientist at NASA's Goddard Space Flight Center in Greenbelt, Maryland.

Researchers have gone back and forth on whether Mars held enough water for long enough to have given microbial life a sporting chance to emerge on the red planet. NASA's *Curiosity* rover has tasted the air and found that the martian atmosphere may have been stripped so long ago that there was a slim chance for life; but studies of rocks that the rover has drilled in Gale Crater have revealed signs of a series of lakes that lasted for many millions of years.

To get at this question, an international team of researchers used ground-based telescopes to study the composition of the traces of water in the atmosphere over almost six Earth years (which are roughly three Mars years). Thus, they were able to map the atmospheres, and witness seasonal and microclimate changes over the entire planet.

They specifically looked at two isotopes of water left in the atmosphere: regular water, made of an oxygen and two hydrogens, and semi-heavy water, where one of the hydrogens has an extra neutron in its nucleus. Regular water, which is lighter, tends to rise up and escape the atmosphere at a faster rate, while the heavier water stays put. So over time, the share of heavy water grows—and the greater the share of heavy water today, the more water must have been lost over time.

The scientists took a particular interest in the atmosphere near the polar regions, because much of the red planet's water is stored in its polar ice caps. Based on their calculations, the scientists found that the share of heavy water in the atmosphere near the polar areas was about 7 times as high as in the water on Earth.

At one point, the water reserves must have been about 6.5 times larger than the reserves mostly stored in the martian polar ice caps today. An early Mars would have held about 20 million cubic kilometers (4.8 million cubic miles) of water.

Where did all this water lie? While it probably could have covered the entire planet with a 450-foot-deep layer, it was probably mostly contained on the low-lying northern plains, and in some places could have gone about a mile deep.

"If you drop all that water on the planet, it will accumulate in the northern part of the planet," Villanueva said. "So that's [where] we think it formed an ocean."

NASA's MAVEN spacecraft is studying what remains of Mars' now-thin atmosphere to see if scientists can learn how much of it escaped—data that will be of great use to planetary scientists looking to solve the mystery of the Martian water. The European Space Agency's first ExoMars mission is also scheduled to arrive at Mars in 2016.

"In the next five years we're going to probably change our perception of what Mars was in the past," Villanueva said.

1. Sketch the heavy water molecule described in the article. How much more massive is it than a "regular" water molecule?
2. Explain why regular water escapes faster than heavy water.
3. Why did the scientists want to study the atmosphere over the polar regions of Mars?
4. Why did Mars lose its ocean of water?
5. Do a news search for "Mars water." What are the most recent discoveries? What is the evidence being reported, and when was the water thought to have existed? What was different on Mars at that time so that it was able to have liquid water?

# Summary

Earth's atmosphere is thick enough to warm the surface to life-sustaining temperatures, but not so thick that Earth becomes overheated. Earth, Venus, and Mars all have significant atmospheres that are different from the original atmospheres they captured when they formed. These atmospheres are complex, both in chemical composition and in physical characteristics such as temperature and pressure. The climates of Earth, Venus, and Mars are all determined by their individual atmospheres. Venus, Earth, and Mars are warmer than they would be from solar illumination alone. The atmospheres of Earth, Venus, and Mars have different chemical compositions. These, in turn, lead to dramatic differences in temperature and pressure. Life has altered Earth's atmosphere several times, most notably in the distant past from an increase in the amount of oxygen in the atmosphere and in modern times from an increase in greenhouse gases. Mars and Venus might have been habitable early in the history of the Solar System, but now only Earth has liquid water.

**LG 1** **Identify the processes that cause primary and secondary atmospheres to be formed, retained, and lost.** Planetary atmospheres evolve over time. The primary atmospheres consisted mainly of hydrogen and helium captured from the protoplanetary disk. The terrestrial planets lost their primary atmospheres soon after the planets formed. Secondary atmospheres were created by volcanic gases and from volatiles brought in by impacting comets and asteroids. Planetary bodies must have sufficient mass to hold on to their atmospheres.

**LG 2** **Compare the strength of the greenhouse effect and differences in the atmospheres of Earth, Venus, and Mars.** Naturally occurring greenhouse gases exist on Venus, Earth, and Mars, which increase the average surface temperature of each planet. The amount by which these greenhouse gases raise the temperature of a planet depends on the number of greenhouse gas molecules in the atmosphere. The differences in global temperatures among these planets can be explained in part by their distances from the Sun. However, the different compositions and densities of their atmospheres is a far more significant factor in global temperature. The atmospheric greenhouse effect keeps Earth from freezing, but it turns the atmosphere of Venus into an inferno.

**LG 3** **Describe the layers of the atmospheres on Earth, Venus, and Mars, and explain how Earth's atmosphere has been reshaped by the presence of life.** The atmospheres of Venus, Earth, and Mars have different temperatures, pressures, and compositions. Earth's atmosphere, in particular, has many layers. The layers are determined by the variations in temperature and absorption of solar radiation vertically throughout the atmosphere. Temperature and pressure decrease with altitude in the troposphere of Venus, Earth, and Mars. Earth's magnetosphere shields the planet from the solar wind. The oxygen levels in Earth's atmosphere have been enhanced through photosynthesis by bacteria and then by plants. Increased oxygen allowed for the development of more advanced forms of life.

**LG 4** **Compare the atmospheres of Venus and Mars with the atmosphere of Earth.** Venus has a massive, hot atmosphere of carbon dioxide and sulfur compounds. It has surprisingly fast winds for a slowly rotating planet. Mars has a thin, cold, carbon dioxide atmosphere that may have been much thicker in the past. Earth's atmosphere is thicker than Mars', but thinner than Venus', and is composed primarily of nitrogen.

**LG 5** **Describe how comparative planetology contributes to a better understanding of the changes in Earth's climate.** Astronomical, geological, and biological processes can lead to large changes in the climate of planets. The study of climate on the terrestrial planets expands scientists' knowledge of Earth's past, present, and future conditions. Large variations in global temperature over the past 800,000 years correlate strongly with the number of greenhouse molecules in the atmosphere. The current level of greenhouse gases in Earth's atmosphere is higher than any seen during this period and correlates with a subsequent rise in temperature.

## ? UNANSWERED QUESTIONS

- What is the role of magnetic fields in helping a planet retain its atmosphere? In the absence of a magnetic field, it has been assumed that the solar wind is more likely to sweep away an atmosphere, and this idea has been invoked to explain what happened on Mars. But Venus doesn't have a magnetic field either, and recent observations have shown that Earth leaks as much atmosphere to space as do the planets that lack magnetic fields.

- Will there be a way to slow or stop the rise in greenhouse gases on Earth? Climate scientists worry that current changes are abrupt compared with the natural cycles of changes in climate that take place gradually over thousands of years. Will there be an international agreement to reduce the production of these gases, as there was to reduce the use of chemicals that created the ozone hole?

# Questions and Problems

## Test Your Understanding

1. Place in chronological order the following steps in the formation and evolution of Earth's atmosphere.
   a. Plant life converts carbon dioxide ($CO_2$) to oxygen.
   b. Hydrogen and helium are lost from the atmosphere.
   c. Volcanoes, comets, and asteroids increase the inventory of volatile matter.
   d. Hydrogen and helium are captured from the protoplanetary disk.
   e. Oxygen enables the growth of new life-forms.
   f. Life releases $CO_2$ from the subsurface into the atmosphere.

2. On which of these planets is the atmospheric greenhouse effect strongest?
   a. Venus
   b. Earth
   c. Mars
   d. Mercury

3. The oxygen molecules in Earth's atmosphere
   a. were part of the primary atmosphere.
   b. arose when the secondary atmosphere formed.
   c. are the result of life.
   d. are being rapidly depleted by the burning of fossil fuels.

4. The differences in the climates of Venus, Earth, and Mars are caused primarily by
   a. the composition of their atmospheres.
   b. their relative distances from the Sun.
   c. the thickness of their atmospheres.
   d. the time at which their atmospheres formed.

5. The words *weather* and *climate*
   a. mean essentially the same thing.
   b. refer to very different timescales.
   c. refer to very different size scales.
   d. both b and c

6. Less massive molecules tend to escape from an atmosphere more often than more massive molecules because
   a. the gravitational force on them is less.
   b. they are moving faster.
   c. they are more buoyant.
   d. they are smaller and so experience fewer collisions on their way out.

7. Venus is hot and Mars is cold primarily because
   a. Venus is closer to the Sun.
   b. Venus has a much thicker atmosphere.
   c. the atmosphere of Venus is dominated by $CO_2$, but the atmosphere of Mars is not.
   d. Venus has stronger winds.

8. Studying climate on other planets is important to understanding climate on Earth because (select all that apply)
   a. underlying physical processes are the same on every planet.
   b. other planets offer a range of extremes to which Earth can be compared.
   c. comparing climates on other planets helps scientists understand which factors are important.
   d. other planets can be used to test atmospheric models.

9. The atmosphere of Mars is often pink-orange because
   a. it is dominated by carbon dioxide.
   b. the Sun is at a low angle in the sky.
   c. Mars has no oceans to reflect blue light to the sky.
   d. winds lift dust into the atmosphere.

10. Auroras are the result of
    a. the interaction of particles from the Sun and Earth's atmosphere and magnetic field.
    b. upper-atmosphere lightning strikes.
    c. the destruction of stratospheric ozone, which leaves a hole.
    d. the interaction of Earth's magnetic field with Earth's atmosphere.

11. The ozone layer protects life on Earth from
    a. high-energy particles from the solar wind.
    b. micrometeorites.
    c. ultraviolet radiation.
    d. charged particles trapped in Earth's magnetic field.

12. Hadley circulation is broken into zonal winds by
    a. convection from solar heating.
    b. hurricanes and other storms.
    c. interactions with the solar wind.
    d. the planet's rapid rotation.

13. The _____ of greenhouse gas molecules affects the temperature of an atmosphere.
    a. percentage
    b. fraction
    c. number
    d. mass

14. Over the past 800,000 years, Earth's temperature has closely tracked
    a. solar luminosity.
    b. oxygen levels in the atmosphere.
    c. the size of the ozone hole.
    d. carbon dioxide levels in the atmosphere.

15. Convection in the _____ causes weather on Earth.
    a. stratosphere
    b. mesosphere
    c. troposphere
    d. ionosphere

## Thinking about the Concepts

16. Primary atmospheres of the terrestrial planets were composed almost entirely of hydrogen and helium. Explain why they contained only these gases and not others.

17. How were the secondary atmospheres of the terrestrial planets created?

18. Nitrogen, the principal gas in Earth's atmosphere, was not a significant component of the protostellar disk from which the Sun and planets formed. Where did Earth's nitrogen come from?

19. What are the likely sources of Earth's water?

20. In what way is the atmospheric greenhouse effect beneficial to terrestrial life?

21. In what ways does plant life affect the composition of Earth's atmosphere?

22. What is the difference between ozone in the stratosphere and ozone in the troposphere? Which is a pollutant, and which protects terrestrial life?

23. What is the principal cause of winds in the atmospheres of the terrestrial planets?

24. Global warming appears to be responsible for increased melting of the ice in Earth's polar regions.
    a. Why does the melting of Arctic ice, which floats on the surface of the Arctic Ocean, *not* affect the level of the oceans?
    b. How is the melting of glaciers in Greenland and Antarctica affecting the level of the oceans?

25. Why are we unable to get a clear view of the surface of Venus, as we have so successfully done with the surface of Mars?

26. What is the evidence that the greenhouse effect exists on Earth, Venus, and Mars?

27. Explain why surface temperatures on Venus hardly vary between day and night and between the equator and the poles.

28. Why do scientists think that Mars and Venus were once more habitable, but no longer are?

29. Examine the Process of Science Figure. The last step is one that anyone can carry out about any complex issue. Write down your current take on the issue of anthropogenic climate change: do you accept the evidence or not? Then write down a piece of scientific evidence that would convince you to change your mind. This exercise may help you to think critically about any issue.

30. Given the current conditions on Venus and Mars, which planet might be easier to engineer to make it habitable to humans? Explain.

## Applying the Concepts

31. Study Figure 9.21.
    a. Are the axes linear or logarithmic?
    b. Compare the $CO_2$ levels (top) with the temperature relative to present (middle). How would you describe the relationship between the two graphs?
    c. How much higher is the current $CO_2$ level than the previous highest value?

32. Study Figure 9.23.
    a. When (approximately) did greenhouse gases begin rising exponentially?
    b. These graphs show that several greenhouse gases have behaved similarly in recent times. Was this true in the past (in general)?
    c. Speculate on possible causes for the common behavior of greenhouse gases in modern times.

33. Study Figure 9.6. Why do commercial jet planes fly at those altitudes?

34. Commercial jets are pressurized. If you take a bag of chips on a commercial jet airplane, the bag puffs up as you travel to the cruising altitude of 14 km.
    a. Is the pressure in the cabin higher or lower than the pressure on the ground?
    b. If there were a second bag attached to the outside of the plane, which bag would puff up more? About how much more (assume an unbreakable bag)? (See Figure 9.6.)

35. Atmospheric pressure is caused by the weight of a column of air above you pushing down. At sea level on Earth, this pressure is equal to $10^5$ newtons per square meter ($N/m^2$).
    a. Estimate the total force on the top of your head from this pressure.
    b. Recall that the acceleration due to gravity is 9.8 $m/s^2$. If the force in part (a) were caused by a kangaroo sitting on your head, what would the mass of the kangaroo be?
    c. Assume a typical kangaroo has a mass of 60 kg. How many kangaroos would have to be sitting on your head to be equal to the extremely massive kangaroo in part (b)?
    d. Why are you not crushed by this astonishing force on your head?

36. Repeat the calculations in question 35 for Venus.

37. Repeat the calculations in question 35 for Mars.

38. Increasing the temperature of a gas inside a closed, rigid box increases the pressure. (This is why you do not put an unopened can of soup directly on the stove!) Explain how Figure 9.6 shows this phenomenon at work in Earth's atmosphere.

39. The total mass of Earth's atmosphere is $5 \times 10^{18}$ kg. Carbon dioxide ($CO_2$) makes up about 0.06 percent of Earth's atmospheric mass.
    a. What is the mass of $CO_2$ (in kilograms) in Earth's atmosphere?
    b. The annual global production of $CO_2$ is now estimated to be $3 \times 10^{13}$ kg. What annual fractional increase does this represent?
    c. The mass of a molecule of $CO_2$ is $7.31 \times 10^{-26}$ kg. How many molecules of $CO_2$ are added to the atmosphere each year?
    d. Why does an increase in $CO_2$ have such a big effect, even though it represents a small fraction of the atmosphere?

40. The ability of wind to erode the surface of a planet is related in part to the wind's kinetic energy.
    a. Compare the kinetic energy of a cubic meter of air at sea level on Earth (mass 1.23 kg) moving at a speed of 10 m/s with a cubic meter of air at the surface of Venus (mass 64.8 kg) moving at 1 m/s.
    b. Compare the kinetic-energy value you determined for Earth in part (a) with that of a cubic meter of air at the surface of Mars (mass 0.015 kg) moving at a speed of 50 m/s.
    c. Why do you think there is not more evidence of wind erosion on Earth?

41. Suppose you seal a rigid container that has been open to air at sea level when the temperature is 0°C (273 K). The pressure inside the sealed container is now exactly equal to the outside air pressure: $10^5$ N/m². 
    a. What would be the pressure inside the container if it were left sitting in the desert shade where the surrounding air temperature was 50°C (323 K)?
    b. What would be the pressure inside the container if it were left sitting out in an Antarctic night where the surrounding air temperature was −70°C (203 K)?
    c. What would you observe in each case if the walls of the container were not rigid?

42. Oxygen molecules ($O_2$) are 16 times as massive as hydrogen molecules ($H_2$). Carbon dioxide molecules ($CO_2$) are 22 times as massive as $H_2$.
    a. Compare the average speed of $O_2$ and $CO_2$ molecules in a volume of air.
    b. Does this ratio of the speeds in part (a) depend on air temperature?

43. Calculate the average speed of a carbon dioxide molecule in the atmospheres of Earth and Mars. Compare these speeds with their respective escape velocities. What does this tell you about each planet's hold on its atmosphere?

44. The average surface pressure on Mars is 6.4 mb. Using Figure 9.6, estimate how high you would have to go in Earth's atmosphere to experience the same atmospheric pressure that you would experience if you were standing on Mars.

45. Water pressure in Earth's oceans increases by 1 bar for every 10 meters of depth. Compute how deep you would have to go to experience pressure equal to the atmospheric surface pressure on Venus.

## USING THE WEB

46. Look up the data on this year's ozone hole. NASA's "Ozone Hole Watch" website (http://ozonewatch.gsfc.nasa.gov) shows a daily image of southern ozone, as well as animations for current and previous years and some comparative plots. Other comparative plots are available on NOAA's "Southern Hemisphere Ozone Hole Area" Web page (http://www.cpc.ncep.noaa.gov/products/stratosphere/sbuv2to/gif_files/ozone_hole_plot.png). At what time of year is the hole the largest, and why? How do the most recent ozone holes compare to previous ones in size and minima? Do they seem to be getting smaller?

47. Mars:
    a. Go to http://www.planetfour.org, a Zooniverse Citizen Science Project in which people examine images of the surface of Mars. Log in or create a Zooniverse account if you don't have one. Read through "About": Where did these data come from? What are the goals of this project? Why is it useful to have many people look at the data? Read through "Classify": "Show Tutorial" and "See Examples" and "FAQs." Now classify some images.
    b. Go to the website for the *MAVEN* mission, which entered the orbit of Mars in 2014. (http://lasp.colorado.edu/home/maven). What are the scientific goals of the mission? Is this mission a lander, an orbiter, or a flyby? What instruments are on this mission? How will this mission contribute to the understanding of climate change on Mars? Go to the NASA Web page for *MAVEN* (http://www.nasa.gov/mission_pages/maven/main/index.html). Are there any results?

48. Earth:
    a. Go to the National Snow & Ice Data Center (NSIDC) websites (http://nsidc.org/data/seaice_index/ and http://nsidc.org/arcticseaicenews/). What are the current status and the trend of the Arctic sea ice? How does it compare with previous years and with the median shown? Is anything new reported about Antarctic ice? Qualitatively, how might a change in the amount of ice at Earth's poles affect the albedo of Earth, and how does the albedo affect Earth's temperature?

b. Go to the website for NASA's Goddard Institute for Space Studies (http://www.giss.nasa.gov), click on "Datasets & Images," and select "Surface Temperatures." The graphs are updated every year. Note that the temperature is compared to a baseline of the average temperature in the period 1951–1980. What has happened with the temperature in the past few years? If the annual mean decreased, does that change the trend? What does the 5-year running mean show? How much warmer is it on average now than in 1880?

c. Go to NOAA's "Trend in Atmospheric Carbon Dioxide" Web page on carbon dioxide levels at the observatory on Mauna Loa (http://esrl.noaa.gov/gmd/ccgg/trends/mlo.html). What is the current level of $CO_2$? How does this compare with the level from 1 year ago? Scroll down the page and click on "A description of how we make measurements at Mauna Loa." Why is this a good site for measuring $CO_2$? What exactly is measured? Are the numbers cross-checked with other measurements?

49. Climate change:
a. Go to the timeline on the "Discovery of Global Warming" Web page of the American Institute of Physics (http://aip.org/history/climate/timeline.htm). When did scientists first suspect that $CO_2$ produced by humans might affect Earth's temperature? When were other anthropogenic greenhouse gases identified? When did scientific opinion about global warming start to converge? Click on "Venus & Mars": How did observations of these planets add to an understanding of global climate change? Click on "Aerosols": How do these contribute to "global dimming"?

b. The Fifth Assessment report from the Intergovernmental Panel on Climate Change (IPCC) was released in October 2014. Go to the IPCC website section on the 2014 Synthesis report (http://ipcc.ch/report/ar5/syr/) and watch the 16-minute video. What are some of the causes of the increase in warming? What are some of the effects of warming seen in the polar regions? How are measurements from the past and present used to predict the climate in the future?

c. Advanced: Go to the website for "Educational Global Climate Modeling," or EdGCM (http://edgcm.columbia.edu). This is a version of the NASA GISS modeling software that will enable students to run a functional three-dimensional global climate model on their laptop computers. Download the trial version and install it on your computer. What can you study with this program? What factors that contribute to global warming or to global cooling on Earth can you adjust in the model? Your instructor may give you an assignment using this program and the Earth Exploration Toolbook (http://serc.carleton.edu/eet/envisioningclimatechange/index.html).

50. Mars movies:
a. Watch a science fiction film about people going to Mars. How does the film handle the science? Can people breathe the atmosphere? Are the low surface gravity and atmospheric pressure correctly portrayed? Do the astronauts have access to water?

b. At the end of the film *Total Recall* (1990), Arnold Schwarzenegger's character presses an alien button, the martian volcanoes start spewing, and within a few minutes the martian sky is blue, the atmospheric pressure is Earth-like, and the atmosphere is totally breathable. (Probably you can find the scene online.) What, scientifically, is wrong with this scene? That is, why would volcanic gases *not* quickly create a breathable atmosphere on Mars?

## smartwork5

If your instructor assigns homework in Smartwork5, access your assignments at digital.wwnorton.com/astro5.

digital.wwnorton.com/astro5

One prediction about climate change is that as the planet warms, ice in the polar caps and in glaciers will melt. Such melting certainly seems to be occurring in the vast majority of glaciers and ice sheets around the planet. It is reasonable to ask whether this actually matters, and why. In this Exploration, we will explore several consequences of the melting ice on Earth.

## Experiment 1: Floating Ice

For this experiment, you will need a permanent marker, a translucent plastic cup, water, and ice cubes. Place a few ice cubes in the cup and add water until the cubes float (that is, they don't touch the bottom). Mark the water level on the outside of the cup with the marker, and label this mark "Initial water level."

**1** As the ice melts, what do you expect to happen to the water level in the cup?

_____

_____

Wait for the ice to melt completely, then mark the cup again.

**2** What happened to the water level in the cup when the ice melted?

_____

_____

**3** Given the results of your experiment, what can you predict will happen to global sea levels when the Arctic ice sheet, which floats on the ocean, melts?

_____

_____

## Experiment 2: Ice on Land

For this experiment, you will need the same materials as in experiment 1, plus a paper or plastic bowl. Fill the cup about halfway with water and then mark the water level, labeling the mark "Initial water level." Poke a hole in the bottom of the bowl and set the bowl over the cup. Add some ice cubes to the bowl.

**4** As the ice melts, what do you expect will happen to the water level in the cup?

_____

_____

Wait for the ice to melt completely, then mark the cup again.

**5** What happened to the water level in the cup when the ice melted?

_____

_____

**6** In this experiment, the water in the cup is analogous to the ocean, and the ice in the bowl is analogous to ice on land. Given the results of your experiment, what can you predict will happen to global sea levels when the Antarctic ice sheet, which sits on land, melts?

_____

_____

## Experiment 3: Why Does It Matter?

Search online using the phrase "Earth at night" to find a satellite picture of Earth taken at night. The bright spots on the image trace out population centers. In general, the brighter a spot is, the more populous the area is (although there is a confounding factor relating to technological advancement).

**7** Where do humans tend to live—near coasts or inland? Coastal regions are, by definition, near sea level. If both the Greenland and Antarctic ice sheets melted completely, sea levels would rise by as much as 80 meters. How would a sea-level rise of a few meters (in the range of reasonable predictions) over the next few decades affect the global population? (Note that one story of a building is about 3 meters.)

_____

_____

# 10

# Worlds of Gas and Liquid—The Giant Planets

**U**nlike the solid planets of the inner Solar System, the four worlds in the outer Solar System were able to capture and retain gases and volatile materials from the Sun's protoplanetary disk and swell to enormous size and mass. These planets have dense cores, very large atmospheres, and rotate faster than Earth.

## LEARNING GOALS

By the conclusion of this chapter, you should be able to:

**LG 1**   Differentiate the giant planets from each other and from the terrestrial planets.

**LG 2**   Describe the atmosphere of each giant planet.

**LG 3**   Explain the extreme conditions deep within the interiors of the giant planets.

**LG 4**   Describe the magnetosphere of each of the giant planets.

**LG 5**   Compare the planets of our Solar System to those in extrasolar planetary systems.

This *Cassini* image was taken at the end of a massive storm in 2010–2011. The red, orange, and green clouds are in false color to highlight more details.  ▶ ▶ ▶

What is a
storm like
on Saturn?

| TABLE 10.1 Physical Properties of the Giant Planets | Jupiter | Saturn | Uranus | Neptune |
|---|---|---|---|---|
| Orbital semimajor axis (AU) | 5.20 | 9.6 | 19.2 | 30 |
| Orbital period (Earth years) | 11.9 | 29.5 | 84.0 | 164.8 |
| Orbital velocity (km/s) | 13.1 | 9.7 | 6.8 | 5.4 |
| Mass ($M_{Earth}$ = 1) | 317.8 | 95 | 14.5 | 17.1 |
| Equatorial radius (km) | 71,490 | 60,270 | 25,560 | 24,300 |
| Equatorial radius ($R_{Earth}$ = 1) | 11.2 | 9.5 | 4.0 | 3.8 |
| Oblateness | 0.065 | 0.098 | 0.023 | 0.017 |
| Density (water = 1) | 1.33 | 0.69 | 1.27 | 1.64 |
| Rotation period (hours) | 9.9 | 10.7 | 17.2 | 16.0 |
| Tilt (degrees) | 3.13 | 26.7 | 97.8 | 28.3 |
| Surface gravity (relative to Earth's) | 2.53 | 1.07 | 0.89 | 1.14 |
| Escape speed (km/s) | 59.5 | 35.5 | 21.3 | 23.5 |

## 10.1 The Giant Planets Are Large, Cold, and Massive

Collectively, Jupiter, Saturn, Uranus, and Neptune are known as the *giant planets*. They are sometimes referred to as the *Jovian planets*, after Jupiter, the largest of the giant planets. (*Jove* is another name for Jupiter, the highest-ranking deity of ancient Rome.) As with the terrestrial planets, we learn much about the giant planets by comparing them to each other. We begin our discussion of the giant planets by comparing their physical properties and their compositions, shown in **Table 10.1**. Comparative planetology is useful both within planetary groups and between groups. Throughout most of the chapter, we will be comparing giant planets with giant planets. But it is useful to fix in your mind as a reference point a comparison of at least one giant planet with Earth. For example, to understand the size of the giant planets, it is helpful to know that Jupiter is 11 times larger than Earth, and its mass is 318 times greater than Earth's. In this section, we examine the physical properties of the giant planets.

### Characteristics of the Giant Planets

The giant planets orbit the Sun far beyond the orbits of Earth and Mars. Jupiter, the closest giant planet, is more than 5 astronomical units (AU) from the Sun. At this distance, the Sun is very faint and provides very little warmth. From Jupiter, the Sun appears to be a tiny disk, 1/27 as bright as it appears from Earth. At the distance of Neptune, the Sun no longer looks like a disk at all: it appears as a brilliant star about 500 times brighter than the full Moon is in Earth's sky. Daytime on Neptune is only as bright as twilight on Earth. With so little sunlight available for warmth, daytime temperatures hover around 123 kelvins (K) at the cloud tops on Jupiter, and they can dip to just 58 K on Neptune.

Jupiter, Saturn, Uranus, and Neptune are enormous compared to the rocky terrestrial planets. Jupiter is the largest of the eight planets and is more than 1/10 the diameter of the Sun. Saturn is slightly smaller than Jupiter, with a diameter of

**Figure 10.1** (a) Images of the giant planets, shown to the same physical scale. (b) The same images, scaled according to how the planets would appear as seen from Earth.

(a)

Jupiter          Saturn          Uranus          Neptune

G  X  U  V  I  R

9.5 Earths. Uranus and Neptune are each about 4 Earth diameters across. **Figure 10.1** compares the actual relative diameters of the giant planets with their apparent relative diameters as seen from Earth.

The most accurate method of finding the diameter of a giant planet is to observe a **stellar occultation**, which occurs when the planet eclipses a star in the sky. As shown in **Figure 10.2**, the star disappears behind the planet and then reappears a short while later. Because we know the relative orbital speeds of Earth and the giant planets, we can calculate the size of the eclipsing giant planet from the length of time the star is eclipsed. Occultations of the radio signals transmitted from orbiting spacecraft and images taken by spacecraft cameras also provide accurate measures of the diameters and shapes of planets and their moons.

Although far away, the giant planets are so large that all but Neptune can be seen with the unaided eye. Jupiter and Saturn were known to the ancients, but Uranus was not discovered until 1781, when William Herschel accidently noticed a tiny disk in the eyepiece of his 6-inch telescope. At first he thought he had found a comet, but the object's slow nightly motion soon convinced him that it was a planet beyond the orbit of Saturn. During the decades that followed Herschel's discovery, astronomers found that Uranus's position differed from the path predicted by Newton's laws of motion and suggested that the gravitational pull of an unknown planet caused Uranus's surprising behavior. Using the astronomers' measured positions of Uranus, two young mathematicians—Urbain-Jean-Joseph Le Verrier (1811–1877) in France and John Couch Adams (1819–1892) in England—independently predicted the location of the hypothetical planet. The German astronomer Johann Gottfried Galle (1812–1910) found the planet on his first observing night, just where Le Verrier and Adams had predicted it would be. Thus, Galle's discovery of Neptune in 1846 became a triumph for mathematical prediction based on physical law—and for the subsequent confirmation of theory by observation (**Process of Science Figure**). Neptune remains

**Figure 10.2** (a) Occultations occur when a planet, moon, or ring passes in front of a star. (b) As the planet moves (from right to left as seen from Earth), the starlight is blocked. (c) The amount of time that the star is hidden combined with information about how fast the planet is moving gives the size of the planet.

**Process of Science**

SCIENTIFIC LAWS MAKE TESTABLE PREDICTIONS

Newton's laws of motion and gravity do more than describe what we see. They also enable us to predict the existence of things as yet unseen.

Uranus is discovered in 1781.

That's odd. Uranus's orbit does not match predictions.

Could another planet's gravitational pull be acting on Uranus?

Mathematicians use observations of Uranus and Newton's laws to predict the location of another planet.

Neptune is discovered in 1846.

Newton's laws pass another test!

Laws make predictions that can be tested, in order to verify or falsify them. Each test that does not falsify a scientific law strengthens confidence in its predictions.

**Figure 10.3** (a) The gas giants: Saturn, seen via the Hubble Space Telescope (HST), and Jupiter, imaged by *Cassini*. (b) The ice giants: Uranus and Neptune, imaged in visible light by *Voyager*.

(a) Saturn

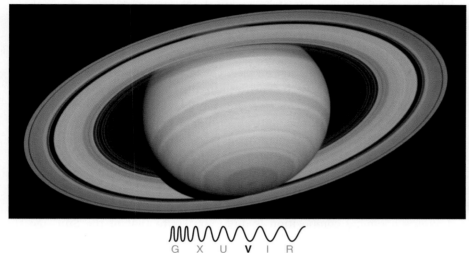

Jupiter

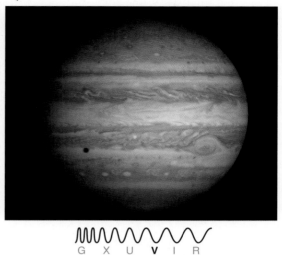

G   X   U   **V**   I   R

(b) Uranus

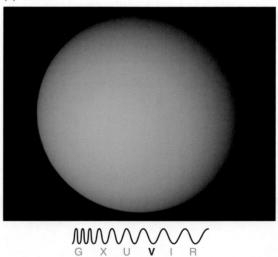

the outermost planet in the Solar System and cannot be seen without the aid of binoculars.

The giant planets contain 99.5 percent of all the nonsolar mass in the Solar System. All other Solar System objects—terrestrial planets, dwarf planets, moons, asteroids, and comets—make up the remaining 0.5 percent. Even though Jupiter is only about a thousandth as massive as the Sun, it contains more than twice the mass of all the other planets combined. Jupiter is 318 times as massive as Earth, 3.3 times as massive as Saturn, and about 20 times as massive as either Uranus or Neptune.

The mass of a planet can be calculated by observing the motions and orbital size of a planet's moon. In Chapter 4, you saw that Newton's law of gravitation and Kepler's third law together show a relation between the motion of an orbiting object and the mass of the body it is orbiting. Planetary spacecraft now make it possible to measure the masses of planets even more accurately. As a spacecraft flies by a planet, the planet's gravity deflects it. By tracking and comparing the spacecraft's radio signals using several antennae here on Earth, astronomers can detect tiny changes in the spacecraft's path and accurately measure the planet's mass.

G   X   U   **V**   I   R

Neptune

## Composition of the Giant Planets

The giant planets are made up primarily of gases and liquids. Jupiter and Saturn are composed of hydrogen and helium and are therefore known as **gas giants**. Uranus and Neptune are known as **ice giants** because they both contain much larger amounts of water and other ices than Jupiter and Saturn. On a giant planet, a relatively shallow atmosphere merges seamlessly into a deep liquid ocean, which in turn merges smoothly into a denser liquid or solid core. There is no abrupt transition from atmosphere to solid ground, as is found on the terrestrial planets. Although the atmospheres of the giant planets are shallow compared with the depth of the liquid layers below, they are still much thicker than the atmospheres of the terrestrial planets—thousands of kilometers rather than hundreds. As with Venus, only the very highest levels of the atmospheres of the gas giants are visible to us. In the case of Jupiter and Saturn, we see the top of a layer of thick clouds that obscures deeper layers (**Figure 10.3a**). There are only a few

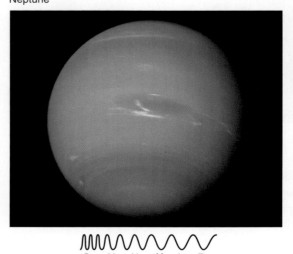

G   X   U   **V**   I   R

thin clouds visible on Uranus, although atmospheric models suggest that thick cloud layers must lie below. Neptune displays a few high clouds with a deep, clear atmosphere showing between them (Figure 10.3b).

As discussed in Chapter 8, the terrestrial planets are composed mostly of rocky minerals, such as silicates, along with various amounts of iron and other metals. While the atmospheres of the terrestrial planets contain lighter materials, the masses of these atmospheres—and even of Earth's oceans—are insignificant compared with the total planetary masses. The terrestrial planets are the densest objects in the Solar System, with densities ranging from 3.9 (Mars) to 5.5 (Earth) times the density of water. In contrast, the giant planets have lower densities because they are composed almost entirely of lighter materials, such as hydrogen, helium, and water. Among the giant planets, Neptune has the highest density, about 1.6 times that of water. Saturn has the lowest density, only 0.7 times the density of water. This means that Saturn would float—if you had an immobile and deep enough body of water—with 70 percent of its volume submerged. Jupiter and Uranus have densities between those of Neptune and Saturn.

Jupiter's chemical composition is quite similar to that of the Sun—mostly hydrogen and helium. (Recall Figure 5.16, the astronomer's periodic table.) Only 2 percent of Jupiter's mass is made up of **heavy elements**, which astronomers define as all elements more massive than helium. Many of these heavy elements combine chemically with hydrogen (H). For example, atoms of oxygen (O), carbon (C), nitrogen (N), and sulfur (S) combine with hydrogen to form molecules of water ($H_2O$), methane ($CH_4$), ammonia ($NH_3$), and hydrogen sulfide ($H_2S$). More complex combinations produce materials such as ammonium hydrosulfide ($NH_4HS$). Jupiter's liquid core, which contains most of the planet's iron and silicate and much of its water, is leftover from the original rocky planetesimals around which Jupiter grew. Computer models of the density are required for an understanding of compositions deep in the cores of the giant planets.

The principal compositional differences among the four giant planets lie in the amounts of hydrogen and helium that each of them contains. Because of its larger mass, Jupiter accumulated more hydrogen and helium when it formed than the other planets did. Saturn is more abundant in heavy elements than Jupiter and therefore less abundant in hydrogen and helium. Heavy elements are significant components of Uranus and Neptune. Methane is a particularly important molecule in the atmospheres of these two planets, giving them their characteristic blue-green color. These differences in composition are important clues to understanding how the giant planets formed.

## Days and Seasons on the Giant Planets

As shown in Table 10.1, giant planets rotate rapidly so their days are short, ranging from 10 to 17 hours. The rapid rotation of the giant planets distorts their shapes—if they did not rotate, they would be perfectly spherical. Instead, the rapidly rotating planets are oblate—they bulge at their equators and have an overall flattened appearance. Saturn's appearance is very oblate: its equatorial diameter is almost 10 percent greater than its polar diameter (**Figure 10.4** and Table 10.1). In comparison, the oblateness of Earth is only 0.3 percent.

Recall from Chapter 2 that the intensity of a planet's seasons is determined by the tilt of its axis. Earth's tilt of 23.5° causes our distinct seasons. The tilts of the giant planets are shown in Table 10.1. With a tilt of only 3°, Jupiter has almost no seasons at all. The tilts of Saturn and Neptune are slightly larger than those of

G  X  U  V  I  R

**Figure 10.4** This Hubble Space Telescope image of Saturn was taken in 1999. The oblateness of the planet is apparent. The large orange moon Titan appears near the top of the disk of Saturn, along with its black shadow.

Earth or Mars, which causes moderate but well-defined seasons. Curiously, Uranus spins on an axis that lies nearly in the plane of its orbit—its tilt is about 98°. Uranus's high tilt causes its seasons to be extreme, with each polar region alternately experiencing 42 years of continual sunshine followed by 42 years of total darkness. Averaged over an entire orbit, the poles receive more sunlight than the equator—a situation quite different from that of any other Solar System planet.

Viewed from Earth, Uranus appears to be either spinning face-on to Earth or rolling along on its side (or something in between), depending on where Uranus happens to be in its orbit. A tilt greater than 90° indicates that the planet rotates in a clockwise direction when seen from above its orbital plane. Why is Uranus tilted so differently from most other planets? One possible explanation is that Uranus was "knocked over" by the impact of one huge or several large planetesimals near the end of its accretion phase. Venus, Pluto, Pluto's moon Charon, and Neptune's moon Triton also have tilts greater than 90°.

---

### CHECK YOUR UNDERSTANDING 10.1

Uranus and Neptune are different from Jupiter and Saturn in that: (a) Uranus and Neptune have a higher percentage of ices in their interiors; (b) Uranus and Neptune have more hydrogen; (c) Uranus and Neptune have no storms; (d) Uranus and Neptune are closer to the Sun.

## 10.2 The Giant Planets Have Clouds and Weather

When we observe the giant planets through a telescope or in visible images from a spacecraft, we are seeing only the top layers of the atmosphere. In some cases, we can see a bit deeper into the clouds, but in essence, we are seeing a two-dimensional view of the cloud tops. The deeper cloud layers on these giant planets are inferred from physical models of temperature as a function of depth. In this section, we explore the atmospheres of the giant planets.

### Viewing the Cloud Tops

Even when viewed through small telescopes, Jupiter is very colorful. Parallel bands, ranging in hue from bluish gray to various shades of orange, reddish brown, and pink, stretch out across its large, pale yellow disk. Astronomers call the darker bands *belts* and the lighter bands *zones*. Many clouds—some dark and some bright, some circular and others more oval—appear along the edges of, or within, the belts. The most prominent of these is a large, red, oval feature in Jupiter's southern hemisphere known as the **Great Red Spot**.

The Great Red Spot (**Figure 10.5**) was first observed more than three centuries ago, shortly after the telescope was invented. Since then, it has varied unpredictably in size, shape, color, and motion as it drifts among Jupiter's clouds. In the 1800s, the Great Red Spot was so large that three Earths could fit inside of it, but now it has shrunk to the size of one. Observations of small clouds circling the perimeter of the Great Red Spot show that it is an enormous atmospheric whirlpool, swirling in a counterclockwise direction with a period of about a week. Its cloud pattern looks a lot like that of a terrestrial hurricane, but it rotates in the

**Figure 10.5** This is a digital enhancement of an image of Jupiter taken in 1979 by the *Voyager 1* spacecraft as it flew by Jupiter. The Great Red Spot is a hurricane larger than Earth.

(a)

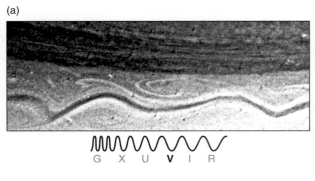

(b)

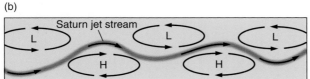

**Figure 10.6** (a) This *Voyager* image of a jet stream in Saturn's northern hemisphere is similar to jet streams in Earth's atmosphere. (b) The jet stream dips equatorward below regions of low pressure and is forced poleward above regions of high pressure.

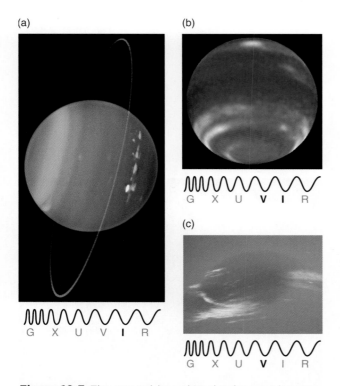

**Figure 10.7** The ground-based Keck telescope image of Uranus (a) and Hubble Space Telescope (HST) image of Neptune (b) were taken at wavelengths of light that are strongly absorbed by methane. The visible clouds are high in the atmosphere. (c) The Great Dark Spot on Neptune disappeared between the time *Voyager 2* flew by Neptune in 1989 and the time HST images were obtained in 1994.

opposite direction, exhibiting *anticyclonic* rather than cyclonic flow, an indication of a high-pressure system. Comparable whirlpool-like behavior is observed in many of the smaller oval-shaped clouds found elsewhere in Jupiter's atmosphere and in similar clouds observed in the atmospheres of Saturn and Neptune. Whirlpool-like, swirling features are known as vortices (the singular is vortex). These vortices are familiar to us on Earth as high- and low-pressure systems, hurricanes, and supercell thunderstorms.

Jupiter is a turbulent, swirling giant with atmospheric currents and vortices so complex that scientists still do not fully understand the details of how they interact with one another, even after decades of analysis. In a series of time-lapse images, *Voyager 2* observed a number of Alaska-sized clouds being swept into the Great Red Spot. Some of these clouds were carried around the vortex a few times and then ejected, while others were swallowed up and never seen again. Other smaller clouds with structure and behavior similar to that of the Great Red Spot are seen in Jupiter's middle latitudes.

Because Saturn is farther away than Jupiter and somewhat smaller in radius, from Earth it appears less than half as large as Jupiter (see Figure 10.1b). Saturn also displays atmospheric bands, but they tend to be wider than those on Jupiter, and their colors and contrasts are much more subdued. A relatively narrow, meandering band in the mid-northern latitudes encircles the planet in a manner similar to Earth's jet stream (**Figure 10.6**). The largest atmospheric features on Saturn are about the size of the continental United States, but many are smaller than terrestrial hurricanes. Close-up views from the *Cassini* spacecraft show immense lightning-producing storms in a region of Saturn's southern hemisphere known as "storm alley." Individual clouds on Saturn are not seen often from Earth, but in December 2010 a large storm appeared (see the chapter-opening figure) that was visible in even small amateur telescopes. This large storm eventually wrapped itself around the planet.

From Earth, in most telescopes, Uranus and Neptune look like tiny, featureless, pale bluish green disks. But with the largest ground-based telescopes or the telescopes in space, optical and infrared imaging reveals a number of individual clouds and belts. Images show atmospheric bands and small clouds suggestive of those seen on Jupiter and Saturn, but more subdued (**Figure 10.7a**). The strong absorption of reflected sunlight by methane causes the atmospheres of Uranus and Neptune to appear dark in the near infrared, allowing the highest clouds and bands to stand out in contrast against the dark background.

A number of bright cloud bands appear in the Hubble Space Telescope (HST) image of Neptune's atmosphere (Figure 10.7b). Located near the planet's tropopause, these cloud bands cast their shadows downward through the clear upper atmosphere onto a dense cloud layer 50 km below. A large, dark, oval feature seen in the southern hemisphere first observed in images taken by *Voyager 2* in 1989 reminded astronomers of Jupiter's Great Red Spot, so they called it the Great Dark Spot (Figure 10.7c). However, the Neptune feature was gray rather than red, and it changed in length and shape more rapidly than the Great Red Spot. When HST observed Neptune in 1994, the Great Dark Spot had disappeared, but a different dark spot of comparable size had appeared briefly in Neptune's northern hemisphere.

## The Structure Below the Cloud Tops

Although our visual impression of the giant planets from Earth is based on a two-dimensional view of their cloud tops, atmospheres are three-dimensional structures. As we saw when we discussed the terrestrial planets, atmospheric temperature, density, pressure, and even chemical composition vary with height and over horizontal distances. As a rule, atmospheric temperature, density, and pressure all decrease with increasing altitude, although temperature is sometimes higher at very high altitudes, as in Earth's thermosphere. The stratospheres above the cloud tops of the giant planets appear relatively clear, but closer inspection shows that they contain layers of thin haze that show up best when seen in profile above the edges of the planets. The composition of the haze particles remains unknown, but they may be smoglike products created when ultraviolet sunlight acts on hydrocarbon gases such as methane.

Water is the only substance in Earth's lower atmosphere that can condense into clouds, but the atmospheres of the giant planets have a much larger range of temperatures and pressures, so more kinds of volatiles can condense and form clouds. (Recall that volatiles are materials that become gases at moderate temperatures.) **Figure 10.8** shows how the ice layers are stacked in the tropospheres of the giant planets. Because each kind of volatile, such as water or ammonia, condenses at a particular temperature and pressure, each forms clouds at a different altitude. Convection carries volatile materials upward along with other atmospheric gases, and when each particular volatile reaches an altitude with its condensation temperature, most of that volatile condenses and separates from the other gases, so very little of it is carried higher aloft. These volatiles form dense layers of cloud separated by regions of relatively clear atmosphere.

The farther the planet is from the Sun, the colder its troposphere will be. Therefore, the distance from the Sun determines the altitude at which a particular volatile, such as ammonia or water, will condense to form a cloud layer on each of the planets. If temperatures are too high, some volatiles may not condense at all. The highest clouds in the frigid atmospheres of Uranus and Neptune are crystals of methane ice. The highest clouds on Jupiter and Saturn are made up of ammonia ice. Methane never freezes to ice in the warmer atmospheres of Jupiter and Saturn.

In 1995, an atmospheric probe on the *Galileo* spacecraft descended slowly via parachute into the atmosphere of Jupiter. Near the top of Jupiter's troposphere at a temperature of about 130 K (about –140°C), it found that ammonia had condensed. Next it found a layer of ammonium hydrosulfide clouds at a temperature of about 190 K (about –80°C). Soon after descending to an atmospheric pressure of 22 bars and a temperature of about 373 K (100°C), the *Galileo* probe failed, presumably because its transmitter got too hot. The *Juno* mission that is expected to enter an orbit around Jupiter in 2016 will use infrared and microwave instruments to further analyze the atmosphere.

Why are some clouds so colorful, especially Jupiter's? In their purest form, the ices that make up the clouds of the giant planets are all white, similar to snow on Earth. The colorful tints and hues must come from impurities in the ice crystals, similar to how syrups color snow cones. These impurities are elemental sulfur and phosphorus, as well as various organic materials produced when ultraviolet sunlight breaks up hydrocarbons such as methane, acetylene, and ethane. The molecular fragments can then recombine to form complex organic compounds that condense into solid particles, many of which are quite colorful. These reactions also occur in Earth's atmosphere. Some of the photochemical products produced close to the ground on Earth are called *smog*.

The atmospheric composition of Uranus and Neptune give these planets their bluish green color. The upper troposphere of Uranus and Neptune are relatively

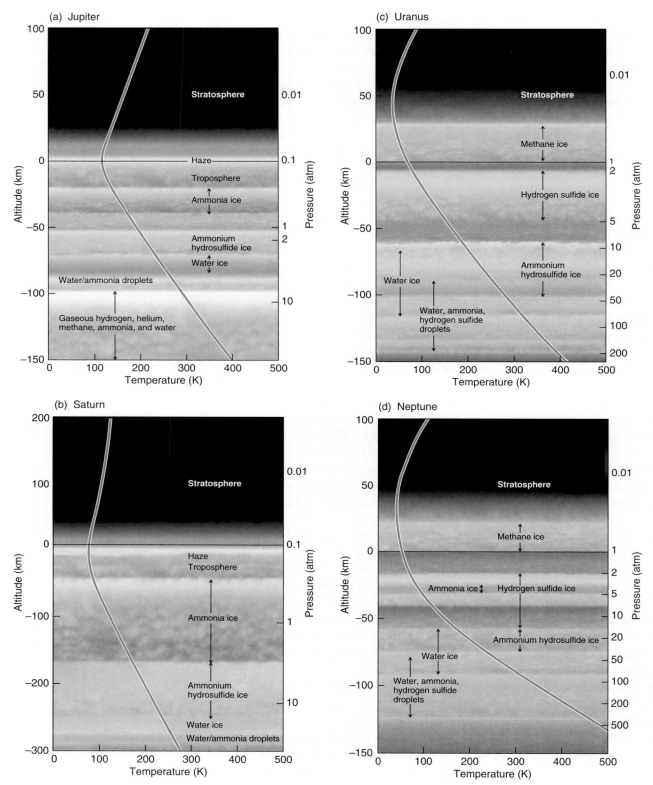

**Figure 10.8** Volatile materials condense at different levels in the atmospheres of the giant planets, leading to chemically different types of clouds at different depths in the atmospheres. The red line in each diagram shows how atmospheric temperature changes with height. Because these planets have no solid surface, the zero point of altitude is arbitrary. In these figures, the arbitrary zero points of altitude are at 0.1 atmosphere (atm) for Jupiter (a) and Saturn (b) and at 1.0 atm for Uranus (c) and Neptune (d). The value of 1.0 atm corresponds to the atmospheric pressure at sea level on Earth. Note that Saturn's altitude scale is compressed to show the layered structure better.

clear, with only a few white clouds that are probably composed of methane ice crystals. Methane gas is much more abundant in the atmospheres of Uranus and Neptune than in those of Jupiter and Saturn. Like water, methane gas tends to selectively absorb the longer wavelengths of light—yellow, orange, and red. Absorption of the longer wavelengths leaves only the shorter wavelengths—green and blue—to be scattered from the atmospheres of Uranus and Neptune.

## Winds and Weather

On the giant planets, the thermal energy that drives convection comes both from the Sun and from the hot interiors of the planets themselves. Recall from Chapter 9 that convection results from vertical temperature differences. As heating drives air up and down, the Coriolis effect shapes that convection into atmospheric vortices, visible as isolated circular or oval cloud structures, such as the Great Red Spot on Jupiter and the Great Dark Spot on Neptune. As the atmosphere rises near the center of a vortex, it expands and cools. Cooling condenses certain volatile materials into liquid droplets, which then fall as rain. As they fall, the raindrops collide with surrounding molecules, stripping electrons from the molecules and thereby developing tiny electric charges in the air. The cumulative effect of countless falling raindrops can be an electric field so great that it ionizes the molecules in the atmosphere and creates a surge of current and a flash of lightning. A single observation of Jupiter's night side by *Voyager 1* revealed several dozen lightning bolts within an interval of 3 minutes. *Cassini* has also imaged lightning flashes in Saturn's atmosphere, and radio receivers on *Voyager 2* picked up lightning static in the atmospheres of both Uranus and Neptune.

The giant planets have much stronger zonal winds than the terrestrial planets. Because they are farther from the Sun, less thermal energy is available. However, they rotate rapidly, which makes the Coriolis effect very strong. In fact, the Coriolis effect is more important than atmospheric temperature patterns in determining the structure of the global winds. If we know the radius of the planet, we can find out how fast the features are moving, as shown in **Working It Out 10.1**, and find rotation speeds and wind speeds. **Figure 10.9** shows the wind speeds at various latitudes on the different planets.

**Jupiter** On Jupiter, the strongest winds are equatorial, blowing from the west, at speeds up to 550 kilometers per hour (km/h), as seen in Figure 10.9a. At higher latitudes, the winds alternate between blowing from the west or east in a pattern that might be related to Jupiter's banded structure. Near a latitude of 20° south, the Great Red Spot vortex appears to be caught between a pair of winds blowing from the west and east with opposing speeds of more than 200 km/h. This indicates a relationship between zonal flow and vortices.

**Saturn** The equatorial winds on Saturn also blow from the west but are stronger than those on Jupiter. The maximum wind speeds at any given time vary between 990 km/h and 1,650 km/h. Saturn's winds appear to decrease with height in the atmosphere, so the apparent time variability of Saturn's equatorial winds may be nothing more than changes in the height of the cloud tops. Alternating winds blowing from the east or west also occur at higher latitudes; but unlike on Jupiter, this alternation seems to bear no clear association with Saturn's atmospheric bands, shown in Figure 10.9b. This is just one example of the many unexplained differences among the giant planets.

(a) Jupiter
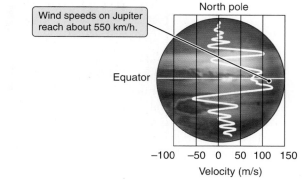

(b) Saturn
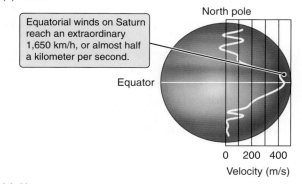

(c) Uranus
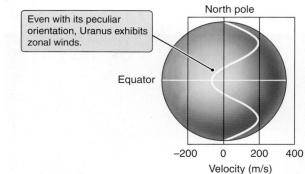

(d) Neptune
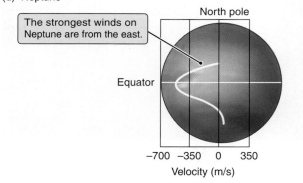

**Figure 10.9** The speed of the wind at various latitudes on the giant planets is shown by the white line. When the speed is positive (to the right of zero), the wind is from the west. When the speed is negative (to the left of zero), the wind is from the east.

## 10.1 ) Working It Out   Measuring Wind Speeds on Different Planets

How do astronomers measure wind speeds on planets that are so far away? As on Earth, clouds are carried by local winds. The local wind speed can be calculated by measuring the positions of individual clouds and noting how much they move during a time interval. To find the speed of the wind relative to the planet's rotating surface, the rotation speed of the planet must also be known. In the case of the giant planets, there is no solid surface against which to measure the winds. Scientists must instead assume a hypothetical surface—one that rotates as though it were somehow "connected" to the planet's deep interior. This is usually found from observing periodic bursts of radio waves that are generated as the planet's magnetic field rotates.

Let's look at an example of how this works using a small white cloud in Neptune's atmosphere. The cloud, on Neptune's equator, is observed to be at longitude 73.0° west on a given day. (This longitude system is anchored in the planet's deep interior.) The spot is then seen at longitude 153.0° west exactly 24 hours later. Neptune's equatorial winds have carried the white spot 80.0° in longitude in 24 hours.

The circumference, $C$, of a planet is given by $2\pi r$, where $r$ is the equatorial radius. The equatorial radius of Neptune is 24,760 km. So Neptune's circumference is

$$C = 2\pi r = 2\pi \times 24,760 \text{ km} = 155,600 \text{ km}$$

There are 360° of longitude in the full circle represented by the circumference. The spot has moved 80°/360° of the circumference and thus has traveled

$$\frac{80}{360} \times 155,600 \text{ km} = 34,580 \text{ km}$$

in 24 hours. This means that the wind speed is given by

$$\text{Speed} = \frac{\text{Distance}}{\text{Time}} = \frac{34,580 \text{ km}}{24 \text{ h}} = 1,440 \text{ km/h} = 400 \text{ m/s}$$

The equatorial winds are very strong and are blowing in a direction opposite to the planet's rotation. On Earth, the much slower equivalents of these winds are called *trade winds*.

---

Saturn's jet stream, at latitude 45° north, is a narrow meandering river of atmosphere with alternating crests and troughs, curving around regions of high and low pressure to create a wavelike structure. It is similar to Earth's jet streams, where high-speed winds blow generally from west to east but wander toward and away from the poles. Nested within the crests and troughs of Saturn's jet stream are anticyclonic and cyclonic vortices. They appear remarkably similar in both form and size to terrestrial high- and low-pressure systems, which bring alternating periods of fair and stormy weather.

**Uranus**   Less is known about global winds on Uranus, illustrated in Figure 10.9c, than about those on the other giant planets. When *Voyager 2* flew by Uranus in 1986, the few visible clouds were in the southern hemisphere because the northern hemisphere was in complete darkness at the time. The strongest winds observed were 650 km/h from the west in the middle to high southern latitudes, as shown in Figure 10.9c, and no winds from the east were seen. Because Uranus's peculiar orientation makes its poles warmer than its equator, some astronomers had predicted that the global wind system of Uranus might be very different from that of the other giant planets. But *Voyager 2* observed that the Coriolis force dominate on Uranus as they do on other planets, so the dominant winds on Uranus are zonal, just as they are on the other giant planets.

As Uranus moves along in its orbit, regions previously unseen by modern telescopes have become visible (**Figure 10.10**). Observations by HST and ground-based telescopes showed bright cloud bands in the far north extending more than 18,000 km in length and revealed wind speeds of up to 900 km/h. As Uranus approaches northern summer solstice in the year 2027, much more about its northern hemisphere will be learned.

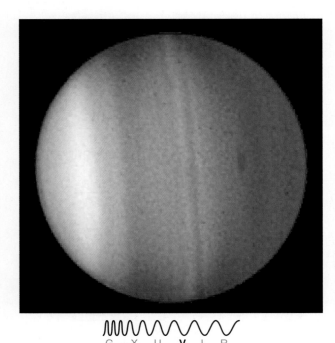

G  X  U  V  I  R

**Figure 10.10** Uranus is approaching equinox in this 2006 HST image. Much of its northern hemisphere is becoming visible. The dark spot in the northern hemisphere (to the right) is similar to but smaller than the Great Dark Spot seen on Neptune in 1989.

**Neptune**  On Neptune, the southern hemisphere's summer solstice occurred in 2005, so much of the north is still in darkness. Observers will have to wait until Neptune's equinox in 2045 to see its northern hemisphere. As shown in Figure 10.9d, the strongest winds on Neptune occur in the tropics, similar to winds on Jupiter and Saturn. However, on Neptune the winds are from the east rather than from the west, with speeds in excess of 2,000 km/h. Winds from the west with speeds higher than 900 km/h have been seen in Neptune's south polar regions. With wind speeds 5 times greater than those of the fiercest hurricanes on Earth, Neptune and Saturn are the windiest planets known.

### CHECK YOUR UNDERSTANDING 10.2

Why is Jupiter reddish in color? (a) because it is very hot; (b) because of the composition of its atmosphere; (c) because it is moving very quickly; (d) because it is rusty, like Mars.

## 10.3  The Interiors of the Giant Planets Are Hot and Dense

At the center of each giant planet is a dense, liquid core consisting of a very hot mixture of heavier materials such as water, molten rock, and metals. **Figure 10.11** illustrates the interior structure of the giant planets. As you can see, the gas giants differ from the ice giants in their amounts of hydrogen and ices. We will look at each in turn.

### The Cores of Jupiter and Saturn

The overlying layers of Jupiter and Saturn press down upon the liquid core, which raises the temperature. For example, the pressure at Jupiter's core is about 45 million bars, and this high pressure heats the fluid to 35,000 K. Central temperatures and pressures of the other, less massive giant planets are correspondingly lower than those of Jupiter. Water is still liquid at these temperatures of tens of thousands of degrees because the extremely high pressures at the centers of the giant planets prevent water from turning to steam.

The internal energy that lies deep within the giant planets is a leftover from their formation. The giant planets are still contracting and converting their gravitational potential energy into thermal energy today as they did when they first formed, but they are doing it more slowly. The annual amount of contraction necessary to sustain their internal temperature is only a tiny fraction of their radius. Jupiter, for example, is contracting by about 2 centimeters per year (cm/yr). The thermal energy from the core drives convection in the atmosphere and eventually escapes to space as radiation (**Working It Out 10.2**). In addition, in Saturn's case, and perhaps Jupiter's, under the right conditions, liquid helium separates from a hydrogen-helium mixture and rains downward toward the core. As the droplets of liquid helium sink, they release their gravitational potential energy as thermal energy. Planetary physicists think that most of Saturn's internal energy and perhaps some of Jupiter's internal energy come from this separation of liquid helium. The continual production of thermal energy is sufficient to replace the energy that is escaping from their interiors.

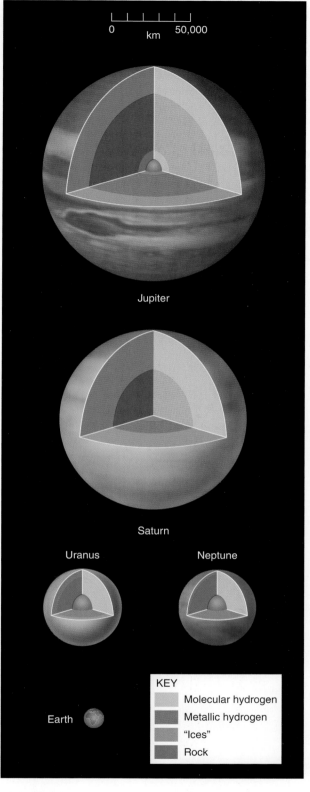

**Figure 10.11** The interiors of the giant planets have central cores and outer liquid shells. Only Jupiter and Saturn have significant amounts of the molecular and metallic forms of liquid hydrogen surrounding their cores.

## 10.2 Working It Out Internal Thermal Energy Heats the Giant Plants

Chapter 5 described the equilibrium between the absorption of sunlight and the radiation of infrared light into space, and in Chapter 9 we explained how the resulting equilibrium temperature is modified by the greenhouse effect on Venus, Earth, and Mars. When we calculate this equilibrium for the giant planets, it doesn't match the measurements. According to these calculations, the equilibrium temperature for Jupiter should be 109 K, but when it is measured, scientists find instead an average temperature of about 124 K. A difference of 15 K might not seem like much, but remember that according to the Stefan-Boltzmann law, the energy radiated by an object depends on its temperature raised to the fourth power. Applying this relationship to Jupiter, we get:

$$\left(\frac{T_{\text{actual}}}{T_{\text{expected}}}\right)^4 = \left(\frac{124\,\text{K}}{109\,\text{K}}\right)^4 = 1.67$$

This implies that Jupiter is radiating roughly two-thirds more energy into space than it absorbs from sunlight. Similarly, the internal energy escaping from Saturn is observed to be about 1.8 times greater than the sunlight that it absorbs. Neptune emits 2.6 times more energy than it absorbs from the Sun. Therefore, these planets are not in equilibrium; they are slowly contracting and thus generating more heat. However, the internal energy escaping from Uranus is small compared to the absorbed solar energy.

The pressure within the atmospheres of Jupiter and Saturn increases with depth because overlying layers of atmosphere press down on lower layers. At depths of a few thousand kilometers, the atmospheric gases of Jupiter and Saturn are so compressed by the weight of the overlying atmosphere that they turn to liquid. This roughly marks the lower boundary of the atmosphere. The difference between a liquid and a highly compressed, very dense gas is subtle, so on Jupiter and Saturn there is no clear boundary between the atmosphere and the ocean of liquid that lies below. Jupiter's atmosphere is about 20,000 km deep, and Saturn's atmosphere is about 30,000 km deep; at these depths, the pressure climbs to 2 million bars and the temperature reaches 10,000 K (hotter than the surface of the Sun). Under these conditions, hydrogen molecules are battered so violently that their electrons are stripped free, and the hydrogen acts like a liquid metal. In this state, it is called metallic hydrogen. These oceans of hydrogen and helium are tens of thousands of kilometers deep. Uranus and Neptune are less massive than Jupiter and Saturn, have lower interior pressures, and contain a smaller fraction of hydrogen—their interiors probably contain only a small amount of liquid hydrogen, with little or none of it in a metallic state.

Differentiation has occurred and is still occurring in Saturn, and perhaps in Jupiter, too. On Saturn, helium condenses out of the hydrogen-helium oceans. Helium can also be compressed to a metal, but it does not reach this metallic state under the physical conditions existing in the interiors of the giant planets. Because these droplets of helium are denser than the hydrogen-helium liquid in which they condense, they sink toward the center of the planet, converting gravitational energy to thermal energy. This process heats the planet and enriches the helium concentration in the core while depleting it in the upper layers. In Jupiter's hotter interior, by contrast, the liquid helium is mostly dissolved together with the liquid hydrogen.

The heavy-element components of the cores of Jupiter and Saturn have masses of about 5–20 Earth masses. Jupiter and Saturn have total masses of 318 and 95 Earth masses, respectively. The heavy materials in their cores contribute little to their average chemical composition. This means Jupiter and Saturn have approximately the same composition as the Sun and the rest of the universe: about 98 percent hydrogen and helium, leaving only 2 percent for everything else.

## The Cores of Uranus and Neptune

Uranus and Neptune are less massive than Jupiter and Saturn, have lower interior pressures, and contain smaller fractions of hydrogen—their interiors probably contain only a small amount of liquid hydrogen, with little or none of it in a metallic state. Uranus and Neptune are made of denser material than Saturn and Jupiter. Neptune, the densest of the giant planets, is about 1.6 times denser than water and only about half as dense as rock. Uranus is less dense than Neptune. These observations tell us that water and other low-density ices, such as ammonia and methane, must be the major compositional components of Uranus and Neptune, along with lesser amounts of silicates and metals. The total amount of hydrogen and helium in these planets is probably limited to no more than 1 or 2 Earth masses, and most of these gases reside in the relatively shallow atmospheres of the planets. Computer simulations suggest that under their conditions of high pressure and temperature, the water that makes up so much of Uranus and Neptune might be *super-ionic*, a state in-between a liquid and a solid that behaves like both.

Why do Jupiter and Saturn have so much hydrogen and helium compared with Uranus and Neptune? We have seen that although each of the giant planets formed around cores of rock and metal, they turned out differently. These differences are an important clue to their origins. The variation may be due to the time that it took for these planets to form and to the distribution of material from which they formed. The cores of Uranus and Neptune were smaller and formed much later than those of Jupiter and Saturn, at a time when most of the gas in the protoplanetary disk had been blown away by the emerging Sun. The icy planetesimals from which they formed were more widely dispersed at their greater distances from the Sun. With more space between planetesimals, their cores would have taken longer to build up. Saturn may have captured less gas than Jupiter both because its core formed somewhat later and because less gas was available at its greater distance from the Sun. As we will discuss in the Origins section later, some astronomers hypothesize that Uranus and Neptune might have formed in a location different from where they are now.

### CHECK YOUR UNDERSTANDING 10.3

The interiors of the giant planets are heated by gravitational contraction. We know this because: (a) the cores are very hot; (b) the giant planets radiate more energy than they receive from the Sun; (c) the giant planets have strong winds; (d) the giant planets are mostly atmosphere.

# 10.4 The Giant Planets Are Magnetic Powerhouses

All of the giant planets have magnetic fields that are much stronger than Earth's: their field strengths range from 50 to 20,000 times stronger. However, because field strength falls off with distance, fields at the cloud tops of Saturn, Uranus, and Neptune are comparable in strength to Earth's surface field. Even in the case of Jupiter's exceptionally strong field, the field strength at the cloud tops is only about 15 times that of Earth's surface field. In Jupiter and Saturn, circulating currents within deep layers of metallic hydrogen generate magnetic fields. In Uranus and Neptune, magnetic fields arise within deep oceans of water and ammonia made electrically conductive by dissolved salts or ionized molecules. The

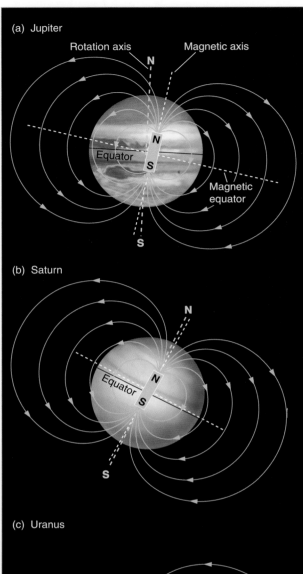

(a) Jupiter

Rotation axis

Magnetic axis

N

N

S

Equator

Magnetic
equator

S

(b) Saturn

N

Equator

N

S

S

magnetospheres of the giant planets are very large and interact with both the solar wind (as Earth's does) and the rings and moons that orbit the giant planets.

## The Size and Shape of the Magnetospheres

Just as Earth's magnetic field traps energetic charged particles to form Earth's magnetosphere, the magnetic fields of the giant planets also trap energetic particles to form magnetospheres of their own. By far, the most colossal of these is Jupiter's magnetosphere. Its radius is 100 times that of the planet itself, roughly 10 times the radius of the Sun. Even the relatively weak magnetic fields of Uranus and Neptune form magnetospheres that are comparable in size to the Sun. Evidence of the giant planets' magnetospheres comes from spacecraft in the outer Solar System, from telescopes orbiting Earth, and from radio emissions received on Earth.

**Figure 10.12** illustrates the geometry of the magnetic fields of the giant planets as if they came from bar magnets. The differences in the orientations of the magnetic field axes are not well understood. Jupiter's magnetic axis is inclined 10° to its rotation axis—an orientation similar to Earth's—but it is offset about a tenth of a radius from the planet's center. Saturn's magnetic axis is located almost precisely at the planet's center and is almost perfectly aligned with the rotation axis. The magnetic axis of Uranus is inclined nearly 60° to its rotation axis and is offset by a third of a radius from the planet's center. The orientation of Neptune's rotation axis is similar to that of Earth, Mars, and Saturn. But Neptune's magnetic-field axis is inclined 47° to its rotation axis, and the center of this magnetic field is displaced from the planet's center by more than half the radius—an offset even greater than that of Uranus. The field is displaced primarily toward Neptune's southern hemisphere, thereby creating a field 20 times stronger at the southern

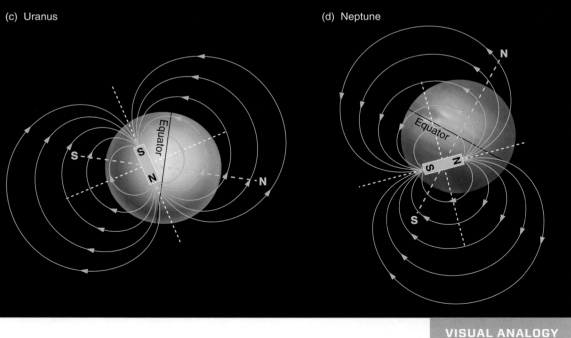

(c) Uranus

Equator

S

S

N

N

(d) Neptune

N

Equator

S

N

S

**Figure 10.12** The magnetic fields of the giant planets can be approximated by the fields from bar magnets offset and tilted with respect to the planets' axes. Compare these with Earth's magnetic field, shown in Figure 8.12.

**VISUAL ANALOGY**

cloud tops than at the northern cloud tops. The reason for the unusual geometries of the magnetic fields of Uranus and Neptune remains uncertain, but it is not related to the orientations of their rotation axes.

The magnetosphere is also influenced by solar wind. Recall from Chapter 9 that the solar wind supplies some of the particles for a magnetosphere. In addition, the pressure of the solar wind also pushes on and compresses a magnetosphere, so the size and shape of a planet's magnetosphere depends on how the solar wind is blowing at any particular time. Planetary magnetic fields also divert the solar wind, which flows around magnetospheres the way a stream flows around boulders. Just as a rock in a river creates a wake that extends downstream as illustrated in **Figure 10.13a**, the magnetosphere of a planet produces a wake that can extend for great distances.

Figure 10.13b shows that the wake of Jupiter's magnetosphere extends well past the orbit of Saturn. Jupiter's magnetosphere is the largest permanent "object" in the Solar System, surpassed in size only by the tail of an occasional comet. If your eyes were sensitive to radio waves, then the second-brightest object in the sky would be Jupiter's magnetosphere. The Sun would still be brighter, but even at a distance from Earth of 4.2–6.2 AU, Jupiter's magnetosphere would appear roughly twice as large as the Sun in the sky.

Saturn's magnetosphere would also be large enough to see, if we could see radio waves, but it would be much fainter than Jupiter's. Even though Saturn has a strong magnetic field, pieces of rock, ice, and dust in Saturn's spectacular rings act like sponges, soaking up magnetospheric particles soon after those particles enter the magnetosphere. With far fewer magnetospheric electrons, there is much less radio emission from Saturn. The magnetic tails of Uranus and Neptune have a curious structure. For both Uranus and Neptune, the tilt and the large displacement of the magnetic field from the center of each planet cause the magnetosphere to wobble as the planet rotates. This wobble causes the tail of the magnetosphere to twist like a corkscrew as it stretches away.

Rapidly moving electrons in planetary magnetospheres spiral around the magnetic field lines, and as they do so they emit a type of radiation, known as

**Figure 10.13** (a) Water flowing past a rock sweeps the algae against the rock and into a "tail" pointing in the direction of the water's flow. (b) The solar wind compresses Jupiter's (or any other) magnetosphere on the side toward the Sun and draws it out into a magnetic tail away from the Sun. Jupiter's tail stretches beyond the orbit of Saturn. Note that this drawing is not to scale.

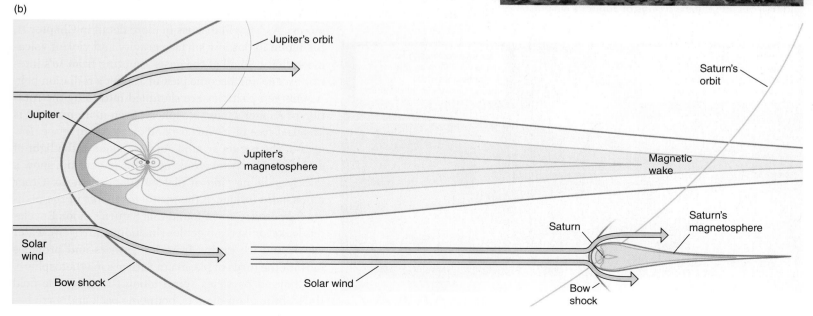

**synchrotron radiation**, concentrated in the low-energy radio part of the spectrum. Precise measurement of periodic variations in the radio signals "broadcast" by the giant planets indicates the planets' true rotation periods. The magnetic field of each planet is locked to the conducting liquid layers deep within the planet's interior, so the magnetic field rotates with exactly the same period as the deep interior of the planet. Given the fast and highly variable winds that push around clouds in the atmospheres of the giant planets, measurement of radio emission is the only way to determine the true rotation periods of the giant planets.

## Radiation Belts and Auroras of the Giant Planets

As a planet rotates on its axis, it drags its magnetosphere around with it, and charged particles are swept around at high speeds. These fast-moving charged particles slam into neutral atoms, and the energy released in the resulting high-speed collisions heats the **plasma** to extreme temperatures. (A plasma is a gas consisting of electrically charged particles.) In 1979, while passing through Jupiter's magnetosphere, *Voyager 1* encountered a region of tenuous plasma with a temperature of more than 300 million K, 20 times the temperature at the center of the Sun. *Voyager 1* did not melt when passing through this region because the plasma is so tenuous that the plasma's particles were very far apart in space. Although each particle was extraordinarily energetic, there were so few of them that the probe passed unscathed through the plasma.

Charged particles trapped in planetary magnetospheres are concentrated in *radiation belts*. Although Earth's radiation belts are severe enough to worry astronauts, the radiation belts that surround Jupiter are searing in comparison. In 1974, the *Pioneer 11* spacecraft passed through the radiation belts of Jupiter. Several of the instruments onboard were permanently damaged as a result, and the spacecraft itself barely survived to continue its journey to Saturn.

In addition to protons and electrons from the solar wind, the magnetospheres of the giant planets contain large amounts of various elements, some ionized, including sodium, sulfur, oxygen, nitrogen, and carbon. These elements come from several sources, including the planets' extended atmospheres and the moons that orbit within them. The most intense radiation belt in the Solar System is a doughnut-shaped ring of plasma associated with Io, the innermost of Jupiter's four Galilean moons. As we will discuss in more detail in Chapter 11, the moon Io has low surface gravity and violent volcanic activity. Some of the gases erupting from Io's interior escape and become part of Jupiter's radiation belt. As charged particles are slammed into Io by the rotation of Jupiter's magnetosphere, even more material is knocked free of its surface and ejected into space. Images of the region around Jupiter, taken in the light of emission lines from atoms of sulfur or sodium, show a faintly glowing ring of plasma supplied by Io. Other moons also influence the magnetospheres of the planets they orbit. *Cassini* found that Saturn's moon Enceladus leaks ionized molecules (including nitrogen), water vapor, and ice grains from icy geysers and provides most of the torus of plasma in Saturn's magnetosphere.

Charged particles spiral along the magnetic-field lines of the giant planets, bouncing back and forth between each planet's two magnetic poles, just as they do

**Figure 10.14** The Hubble Space Telescope took images of auroral rings around the poles of Jupiter (a) and Saturn (b). The auroral images (bright rings near the poles) were taken in ultraviolet light and then superimposed on visible-light images. (High-level haze obscures the ultraviolet views of the underlying cloud layers, as the insets show.) The bright spot and trail outside the main rings of Jupiter's auroras are the footprint and tail of Io's flux tube.

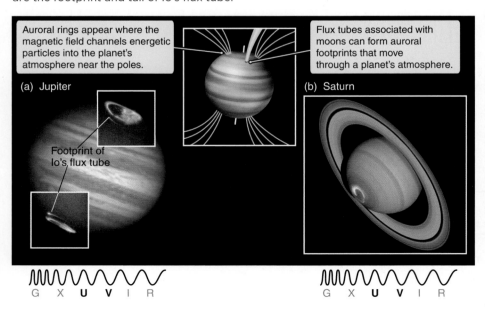

Auroral rings appear where the magnetic field channels energetic particles into the planet's atmosphere near the poles.

Flux tubes associated with moons can form auroral footprints that move through a planet's atmosphere.

(a) Jupiter

Footprint of Io's flux tube

(b) Saturn

G X U V I R

G X U V I R

around Earth. As is the case with Earth, these energetic particles collide with atoms and molecules in a planet's atmosphere, knocking them into excited energy states that decay and emit light. The results are bright auroral rings, shown in **Figure 10.14**. These auroral rings surround the magnetic poles of the giant planets, just as the aurora borealis and aurora australis ring the north and south magnetic poles of Earth.

Jupiter's auroras have an added twist that is not seen on Earth. As Jupiter's magnetic field sweeps past Io, electrons spiral along Jupiter's magnetic-field lines. The result is a magnetic channel, called a **flux tube**, that connects Io with Jupiter's atmosphere near the planet's magnetic poles (**Figure 10.15**). Io's flux tube carries power roughly equivalent to the total power produced by all electrical generating stations on Earth. Much of the power generated within the flux tube is radiated away as radio energy. These radio signals are received at Earth as intense bursts. However, a substantial fraction of the energy of particles in the flux tube is also deposited into Jupiter's atmosphere. At the very location where Io's flux tube intercepts Jupiter's atmosphere, there is a spot of intense auroral activity. As Jupiter rotates, this spot leaves behind an auroral trail in Jupiter's atmosphere. The footprint of Io's plasma torus, along with its wake, can be seen outside the main auroral ring in Figure 10.14a.

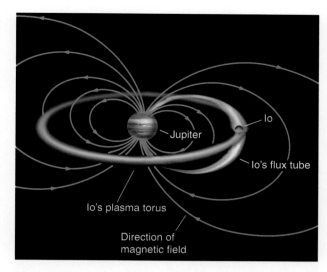

**Figure 10.15** This illustration shows the geometry of Io's plasma torus and flux tube.

---

### CHECK YOUR UNDERSTANDING 10.4

The radiation belts around Jupiter are much stronger than those found around Earth because: (a) Jupiter has larger storms than Earth; (b) Jupiter is colder than Earth; (c) Jupiter rotates faster than Earth; (d) Jupiter has a stronger magnetic field than Earth.

## 10.5 The Planets of Our Solar System Might Not Be Typical

In the past three chapters, we have discussed the planets of our Solar System in detail. The categories of inner terrestrial rocky planets versus outer gas and ice giants was based on our eight planets—but how typical are these categories when compared with the numerous extrasolar planets that have been detected? Do other multiplanet systems resemble our own? As the number of confirmed extrasolar planets increases, astronomers can compare them statistically with those of our own system. Planetary scientists were surprised to find that extrasolar planets differ from those of our Sun. In this section, we will examine some of these differences.

### Different Types of "Jupiters"

As noted in Chapter 7, the first exoplanets were discovered using the radial velocity method, which measures the wobble of a star caused by the gravity of its planet. This method is most successful at finding a massive planet located close to its star, where the planet's gravitational tug on the star is stronger than if the planet were of smaller mass or far away. These *hot Jupiters* are gas giants, within a few percent of an AU to half an AU from their respective stars, with correspondingly short and sometimes highly elliptical orbits. Current models of planetary formation suggest that there would not have been enough excess hydrogen that close to their stars for these planets to form there. Some of these planets may be

**Figure 10.16** This image is an artist's depiction of the super-Jupiter planet Kappa Andromedae b.

*puffy Jupiters,* with a larger radius and a lower density than Jupiter, the density closer to that of Saturn. The larger radius is thought to come from a heated and thus expanding gaseous atmosphere.

Astronomers have identified several hundred extrasolar planets with masses of about 2–13 times the mass of Jupiter, known as *super-Jupiters.* Some are also hot Jupiters, but most of them are not. Their higher mass gives them stronger gravitational contraction than that in planets with Jupiter's mass, so they shrink over time. As a result, most of them have a higher density than Jupiter. The gases can be compressed by self-gravity so much that the super-Jupiter could be denser than Earth. The stronger gravitational contraction would create hotter cores, so they might have more intense winds and weather than Jupiter. An artist's depiction of the super-Jupiter planet Kappa Andromedae b, with about 13 times the mass of Jupiter, is shown in **Figure 10.16**.

## Super-Earths to Mini-Neptunes

Currently, observations suggest that the most common size of planet is one with a radius between that of Earth and that of Neptune (4 times larger than Earth). However, no planet in our Solar System falls in this range. Planets with about 1.5–10 $M_{\text{Earth}}$ are called *super-Earths.* Up to about 2 $R_{\text{Earth}}$, these planets get denser as they get larger, as expected for rocky planets. But above 2 $R_{\text{Earth}}$, most planets are puffier—a gaseous envelope surrounds the rocky core. **Figure 10.17** shows a plot of radii versus density for some of these observations. These gaseous planets at the higher end of this range are sometimes called *mini-Neptunes* or *gas dwarfs.* Some planets don't quite fit these rules; for example, KOI-314c has Earth's mass but 1.6 times Earth's radius, which gives it a density more like Neptune than like Earth (that is, a mini-Neptune). Kepler 10c has 2.3 $R_{\text{Earth}}$ and a mass as high as 17 $M_{\text{Earth}}$—making it a rocky planet with a surprisingly high mass.

In Chapter 7, we presented a scenario in which the gaseous giant planets formed in the cold outer Solar System and the small rocky terrestrial planets formed in the warm inner Solar System. This model seemed to make sense chemically and physically, so astronomers were quite shocked when the hot Jupiters were first discovered in the mid-1990s. To explain hot Jupiters, astronomers worked with computer models, which showed that gravitational interactions between a planet and the protoplanetary disk or among the planets could cause planets to migrate to different orbital distances from where they had formed. As more exoplanets that were not hot Jupiters were discovered, computer models suggested that migration might explain their locations, too, especially for super-Earths close in to their respective stars.

To date, the mix of planet types in our Solar System—outer giant gaseous planets and inner, small rocky planets, all with nearly circular orbits—has not been seen in several hundred extrasolar multiplanet systems. Observations and computer models show that many combinations of planetary sizes, masses, and compositions can exist within planetary systems. It is not yet known if planetary systems like ours are rare or if they just haven't yet been discovered. Current observations favor the detection of large, short-period planets. More time is needed to find out the answer to this question.

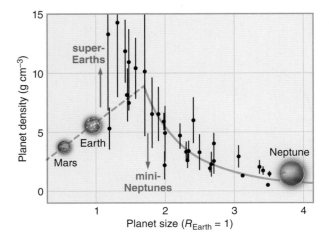

**Figure 10.17** Data on nearby super-Earths suggest that at about 2 $R_{\text{Earth}}$, planet density decreases with size because the planet has more gas accumulated on its rocky core.

## CHECK YOUR UNDERSTANDING 10.5

Place in order of increasing diameter the following types of extrasolar planets: (a) super-Earths; (b) puffy Jupiters; (c) super-Jupiters; (d) mini-Neptunes.

# Origins

## Giant Planet Migration and the Inner Solar System

Newton's law of gravitation is not complicated: as long as only two objects are involved, the resulting motions are simple. Kepler's laws describe the regular, repeating elliptical orbits of planets around the Sun. When more than two objects are involved, however, the resulting motions may be anything but simple and regular. Each planet in the Solar System moves under the gravitational influence of the Sun combined with that of all the other planets. Although these extra influences are small, they are not negligible. Over millions of years, they lead to significant differences in the locations of planets in their orbits. In many possible extrasolar planetary systems, such interactions among planets might cause planets to dramatically change their orbits or even be ejected from the system entirely.

Computer models developed to understand the extrasolar planet systems have been applied to the Solar System as well, and the results are intriguing. Computer models of the formation of the Solar System show that the giant planets may not have formed in their current locations and could have migrated substantially in the early Solar System. A key point seems to be the gravitational influence of Saturn on Jupiter, especially the ratio of their orbital periods. **Figure 10.18** shows one type of migration model. When Saturn's orbital period became twice that of Jupiter, their respective orbits became more elongated. The result was an outward migration of Uranus and Neptune, whose orbits grew larger. Uranus and Neptune may actually have switched places with each other. This shuffling cleared away nearby planetesimals, sending some to the inner Solar System, and made the planetary orbits more stable. In another set of models, Jupiter migrated inward to 1.5 AU—the current orbital distance of Mars. Then Saturn migrated inward even faster to a point at which its orbital period was 1.5 times that of Jupiter, and then they both migrated outward, pushing Uranus and Neptune into larger orbits. Some of these computer models could explain the lower mass of Mars (at 1.5 AU) and the distribution of material in the small bodies of the Solar System.

In some extrasolar planetary systems, gas giants are observed at the right distance from their respective stars to be in the habitable zone (see Chapter 7), but it is not known whether gas giants can support life. Super-Earths are also sometimes found in the habitable zone of their stars. In the Solar System, migrations of the giant planets could have helped confine and stabilize the orbits of the inner planets, so that they reside in or near the habitable zone. Migration of Jupiter may have scattered nearby planetesimals and kept Mars small, thereby reducing the ability of Mars to hold on to its atmosphere and its liquid water, and making it a less likely starting point for life. The shuffling about of the outer planets may be responsible for the period of heavy bombardment, which brought at least some of the atmospheric gases, water, and possibly organic molecules that were needed for life to form on Earth.

(a)                                    (b)                                    (c)

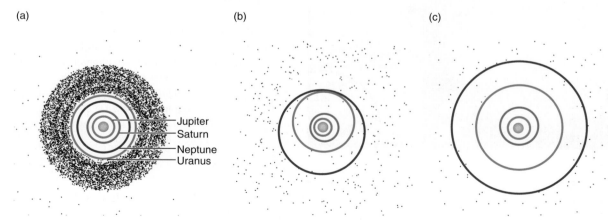

**Figure 10.18** These three snapshots from computer simulations of planet migration show the effects of Jupiter and Saturn on the outer Solar System. The four circles are the orbits of the Jovian planets, and the black dots are planetesimals. (a) The orbital period of Saturn becomes twice that of Jupiter. (b) Planetesimals are scattered as the giant planets change orbits. (c) In some models, Neptune and Uranus change places. The inner Solar System is left more stable.

# Hubble Sees Jupiter's Red Spot Shrink to Smallest Size Ever

By **BOB KING**, UniverseToday.com

Earlier this year we reported that amateur astronomers had observed and photographed the recent shrinking of Jupiter's iconic Great Red Spot. Now, astronomers using the Hubble Space Telescope concur:

"Recent Hubble Space Telescope observations confirm that the spot is now just under 10,250 miles (16,500 km) across, the smallest diameter we've ever measured," said Amy Simon of NASA's Goddard Space Flight Center in Maryland, USA (**Figure 10.19a**).

Using historic sketches and photos from the late 1800s, astronomers determined the spot's diameter then at 25,475 miles (41,000 km) across (Figure 10.19b). Even the smallest telescope would have shown it as a huge red hot dog. Amateur observations starting in 2012 revealed a noticeable increase in the spot's shrinkage rate.

The spot's "waistline" is getting smaller by just under 620 miles (1,000 km) per year while its north-south extent has changed little. In a word, the spot has downsized and become more circular in shape. Many who've attempted to see Jupiter's signature feature have been frustrated in recent years not only because the spot's pale color makes it hard to see against adjacent cloud features, but because it's physically getting smaller.

Jupiter's Great Red Spot or GRS is located in a "bay" or hollow south of the swirly South Equatorial Belt. A titanic storm that has raged hurricane-like for at least 400 years, the top of the spot's cloud deck rises 5 miles (8 km) above the planet's clouds and rotates in an anticlockwise direction about once every 4 days.

As to what is causing the drastic downsizing, there are no firm answers yet:

"In our new observations it is apparent that very small eddies are feeding into the storm," said Simon. "We hypothesized that these may be responsible for the accelerated change by altering the internal dynamics of the Great Red Spot."

The Great Red Spot has been a trademark of the planet for at least 400 years—a giant hurricane-like storm whirling in the planet's upper cloud tops with a period of 6 days. But as it has shrunk, its period has likewise grown shorter and now clocks in at about 4 days.

The storm appears to be conserving angular momentum by spinning faster the same way an ice-skater spins up when she pulls in her arms. Wind speeds are increasing too, making one wonder whether they'll ultimately shrink the spot further or bring about its rejuvenation.

Definitely worth keeping an eye on.

1995

2009

2014

**Figure 10.19** In this comparison of the Great Red Spot (insets) as seen by the Hubble Space Telescope, the top photograph was taken in 1995 and shows the spot at a diameter of just under 13,050 miles (21,000 km); the middle photograph was taken in 2009 and shows the spot at a diameter of just under 11,180 miles (18,000 km); and the bottom photograph was taken in 2014 and shows the spot at its smallest size yet, with a diameter of just 9,940 miles (16,000 km).

1. Why has the Great Red Spot been seen for only 400 years?
2. Explain how astronomers measure the size of the Great Red Spot.
3. Why is it difficult for astronomers to understand what is happening below the cloud tops?
4. Why might it be difficult to drop a probe into the Great Red Spot?
5. Search for news on Jupiter's Great Red Spot. Is it still shrinking?

# Summary

The giant planets are much larger and less dense than the terrestrial planets, consist primarily of light elements rather than rock, and their outer atmospheres are much colder. Because of their rapid rotation and the Coriolis effect, zonal winds are very strong on these planets. Storms, such as those on Saturn, tend to be larger and longer-lived than storms on Earth. Volatiles become ices at various heights in these atmospheres, leading to a layered cloud structure. Jupiter, Saturn, and Neptune are still shrinking, and their gravitational energy is being converted to thermal energy, heating both the cores and the atmospheres from the inside. Uranus does not seem to have as large a heat source inside. Temperatures and pressures in the cores of the giant planets are very high, leading to unusual states of matter, such as metallic hydrogen and super-ionic water. The current locations of these planets might be very different from where they formed: models suggest that their positions may have migrated.

**LG1 Differentiate the giant planets from each other and from the terrestrial planets.** Uranus and Neptune were discovered by telescope, unlike all the other planets, which have been known since ancient times. The giant planets can be divided into two classes: Jupiter and Saturn are gas giants, and Uranus and Neptune are ice giants. Jupiter and Saturn are made up mostly of hydrogen and helium—a composition similar to that of the Sun. Uranus and Neptune contain larger amounts of ices such as water, ammonia, and methane than that found in Jupiter and Saturn. These compositions set them apart from the terrestrial planets. In addition, all four giant planets are much larger than Earth.

**LG 2 Describe the atmosphere of each giant planet.** We see only atmospheres on the giant planets because solid or liquid surfaces, if they exist, are deep below the cloud layers. Clouds on Jupiter and Saturn are composed of various kinds of ice crystals colored by impurities. Uranus and Neptune have relatively few clouds, and so their atmospheres appear more uniform. The most prominent atmospheric feature is the Great Red Spot in Jupiter's southern hemisphere.

**LG 3 Explain the extreme conditions deep within the interiors of the giant planets.** The ongoing collapse of the giant planets converts gravitational energy to thermal energy. This process heats most of the giant planets from within, producing convection. Powerful convection and the Coriolis effect drive high-speed winds in the upper atmospheres of all of the giant planets. The interiors of the giant planets are very hot and very dense because of the high pressures exerted by their overlying atmospheres.

**LG 4 Describe the magnetosphere of each of the giant planets.** The giant planets have enormous magnetospheres that emit synchrotron radiation and interact with their moons. The rotation speed of a gas giant is found from periodic bursts of radio waves that are generated as the planet's magnetic field rotates.

**LG 5 Compare the planets of our Solar System to those in extrasolar planetary systems.** Systems of extrasolar planets found to date do not contain our Solar System's division of small, rocky inner planets and large, giant outer planets. Most of the planets found to date fall between Earth and Neptune in size. Our understanding of how solar systems form is incomplete.

## ? UNANSWERED QUESTIONS

- Did the Solar System start with more planets? In the same types of computer models of the early Solar System that we discussed in Origins, astronomers are able to run simulations with different types of initial configurations of planets to see which configurations evolve over time and then compare these to what is observed today. In one set of models, astronomers found that starting with five giant planets best reproduced the current outer Solar System. The fifth planet would have been kicked out of the Solar System after a close encounter with Jupiter, and it may still be wandering through the Milky Way.

- What are the mass and size of the core of each giant planet? Is there a rocky core underneath the thick atmosphere? Is the core the size of a terrestrial planet or larger? There may be an answer in a few years for Jupiter. The NASA *Juno* mission en route to Jupiter, with a scheduled arrival in 2016, will measure Jupiter's gravitational and magnetic fields and map the amount and distribution of mass in its core and atmosphere.

# Questions and Problems

## Test Your Understanding

1. The following steps lead to convection in the atmospheres of giant planets. After (a), place (b)–(f) in order.
   a. Gravity pulls particles toward the center.
   b. Warm material rises and expands.
   c. Particles fall toward the center, converting gravitational energy to kinetic energy.
   d. Expanding material cools.
   e. Thermal energy heats the material.
   f. Friction converts kinetic energy to thermal energy.

2. Deep in the interiors of the giant planets, water is still a liquid even though the temperatures are tens of thousands of degrees above the boiling point of water. This can happen because
   a. the density inside the giant planets is so high.
   b. the pressure inside the giant planets is so high.
   c. the outer Solar System is so cold.
   d. space has very low pressure.

3. Assume you want to deduce the radius of a planet in our Solar System as it occults a background star when the relative velocity between the planet and Earth is 30 km/s. If the star crosses through the middle of the planet and disappears for a total of 26 minutes, what is the planet's radius?
   a. 3,000 km
   b. 23,000 km
   c. 15,000 km
   d. 5,000 km

4. Neptune's existence was predicted because
   a. Uranus did not seem to obey Newton's laws of motion.
   b. Uranus wobbled on its axis.
   c. Uranus became brighter and fainter in an unusual way.
   d. some of the solar nebula's mass was unaccounted for.

5. Which of the giant planets has the most extreme seasons?
   a. Jupiter
   b. Saturn
   c. Uranus
   d. Neptune

6. The magnetic fields of the giant planets
   a. align closely with the rotation axis.
   b. extend far into space.
   c. are thousands of times stronger at the cloud tops than at Earth's surface field.
   d. have an axis that passes through the planet's center.

7. An occultation occurs when
   a. a star passes between Earth and a planet.
   b. a planet passes between Earth and a star.
   c. a planet passes between Earth and the Sun.
   d. Earth passes between the Sun and a planet.

8. Occultations directly determine a planet's
   a. diameter.
   b. mass.
   c. density.
   d. orbital speed.

9. The chemical compositions of Jupiter and Saturn are most similar to those of
   a. Uranus and Neptune.
   b. the terrestrial planets.
   c. their moons.
   d. the Sun.

10. Individual cloud layers in the giant planets have different compositions. This happens because
   a. the winds are all in the outermost layer.
   b. the Coriolis effect only occurs close to the "surface" of the inner core.
   c. there is no convection on the giant planets.
   d. different volatiles freeze out at different temperatures.

11. The Great Red Spot on Jupiter is
   a. a surface feature.
   b. a storm that has been raging for more than 300 years.
   c. caused by the interaction between the magnetosphere and Io.
   d. about the size of North America.

12. Uranus and Neptune are different from Jupiter and Saturn in that
   a. Uranus and Neptune have a higher percentage of ices in their interiors.
   b. Uranus and Neptune have no rings.
   c. Uranus and Neptune have no magnetic field.
   d. Uranus and Neptune are closer to the Sun.

13. What could have caused the planets to migrate through the Solar System?
   a. gravitational pull from the Sun
   b. interaction with the solar wind
   c. accreting gas from the solar nebula
   d. gravitational pull from other planets

14. Zonal winds on the giant planets are stronger than those on the terrestrial planets because
   a. they have more thermal energy.
   b. the giant planets rotate faster.
   c. the moons of giant planets provide additional pull.
   d. the moons feed energy to the planet through the magnetosphere.

15. A "hot Jupiter" gets its name from the fact that
   a. its temperature has been measured to be higher than Jupiter's.
   b. it is located around a much hotter star than the Sun.
   c. it has very high density, and therefore its temperature is high.
   d. it orbits very close to its central star.

# Thinking about the Concepts

16. Describe how the giant planets differ from the terrestrial planets.

17. Jupiter's chemical composition is more like that of the Sun than Earth's is. Yet both planets formed from the same protoplanetary disk. Explain why they are different today.

18. What can be learned about a Solar System object when it occults a star?

19. What drives the zonal winds in the atmospheres of the giant planets?

20. Compare the sequence of events in the Process of Science Figure in this chapter with the flowchart of the Process of Science Figure in Chapter 1. Redraw the flowchart, incorporating each of the events leading to the discovery of Uranus as examples in the appropriate boxes.

21. None of the giant planets are truly round. Explain why they have a flattened appearance.

22. What is the source of color in Jupiter's clouds? Uranus and Neptune, when viewed through a telescope, appear distinctly bluish green in color. What are the two reasons for their striking appearance?

23. Which of the giant planets have seasons similar to Earth's, and which one experiences extreme seasons?

24. Jupiter's core is thought to consist of rocky material and ices, all in a liquid state at a temperature of 35,000 K. How can materials such as water be liquid at such high temperatures?

25. Explain how astronomers measure wind speeds in the atmospheres of the giant planets.

26. What is the Great Red Spot?

27. Jupiter, Saturn, and Neptune radiate more energy into space than they receive from the Sun. What is the source of the additional energy?

28. When viewed by radio telescopes, Jupiter is the second-brightest object in the sky. What is the source of its radiation?

29. What creates auroras in the polar regions of Jupiter and Saturn?

30. How might migration of the outer giant planets affect the sizes and orbits of the inner planets?

# Applying the Concepts

31. Figure 10.1 shows two different sets of pictures of the outer planets. What is the difference between Figure 10.1a and Figure 10.1b?

32. Figure 10.9d shows the winds on Neptune. The graph, however, does not cover the full planet. Is this likely to mean that the wind speed is zero where there is no white line or that the wind speed is unknown where there is no white line? Explain your reasoning.

33. What creates metallic hydrogen in the interiors of Jupiter and Saturn, and why do we call it metallic?

34. Use Figure 10.15 to estimate the radius of Io's plasma torus in terms of the radius of Jupiter. Convert this value to kilometers, and then look up the answer on the Internet. How close did you get with your simple measurement?

35. The Sun appears 400,000 times brighter than the full Moon in Earth's sky. How far from the Sun (in astronomical units) would you have to go for the Sun to appear only as bright as the full Moon appears in Earth's nighttime sky? How does the distance you would have to travel compare to the semimajor axis of Neptune's orbit?

36. Uranus occults a star at a time when the relative motion between Uranus and Earth is 23.0 km/s. An observer on Earth sees the star disappear for 37 minutes 2 seconds and notes that the center of Uranus passed directly in front of the star.
    a. On the basis of these observations, what value would the observer calculate for the diameter of Uranus?
    b. What could you conclude about the planet's diameter if its center did not pass directly in front of the star?

37. Jupiter's equatorial radius ($R_{\text{Jup}}$) is 71,500 km, and its oblateness is 0.065. What is Jupiter's polar radius ($R_{\text{Polar}}$)? (Oblateness is given by $[R_{\text{Jup}} - R_{\text{Polar}}]/R_{\text{Jup}}$.)

38. Ammonium hydrosulfide ($NH_4HS$) is a molecule in Jupiter's atmosphere responsible for many of its clouds. Using the periodic table in Appendix 3, calculate the molecular weight of an ammonium hydrosulfide molecule, where the atomic weight of a hydrogen atom is 1. (Recall from Working It Out 9.1 that the weight of a molecule is equal to the sum of the weights of its component atoms.)

39. Jupiter is an oblate planet with an average radius of 69,900 km, compared to Earth's average radius of 6,370 km.
    a. Given that volume is proportional to the cube of the radius, how many Earth volumes could fit inside Jupiter?
    b. Jupiter is 318 times as massive as Earth. Show that Jupiter's average density is about one-fourth that of Earth's.

40. The tilt of Uranus is 98°. From one of the planet's poles, how far from the zenith would the Sun appear on summer solstice?

41. A small cloud in Jupiter's equatorial region is observed to be at a longitude of 122.0° west in a coordinate system rotating at the same rate as the deep interior of the planet. (West longitude is measured along a planet's equator toward the west.) Another observation, made exactly 10 Earth hours later, finds the cloud at a longitude of 118.0° west. Jupiter's equatorial radius is 71,500 km. What is the observed equatorial wind speed, in kilometers per hour? Is this wind from the east or west?

42. The equilibrium temperature for Saturn should be 82 K, but the observed temperature is 95 K. How much more energy does Saturn radiate than it absorbs?

43. Neptune radiates 2.6 times as much energy into space as it absorbs from the Sun. Its equilibrium temperature (see Chapter 5) is 47 K. What is its true temperature?

44. Compare the graphs in Figures 10.8a and b. Does atmospheric pressure increase more rapidly with depth on Jupiter or on Saturn? Compare the graphs in Figures 10.8c and d. Does pressure increase more rapidly with depth on Uranus or on Neptune? Of the four giant planets, which has the fastest pressure rise with depth? Which has the slowest?

45. Using Figure 10.8, find the temperature at an altitude of 100 km on each of the four giant planets.

## USING THE WEB

46. Go to the *Cassini* website (http://saturn.jpl.nasa.gov). Its final mission is scheduled for late 2016 to 2017. Click on "News." What discovery was reported in a recent news release about Saturn (not about the rings or moons)? Why is this discovery important?

47. Another website for *Cassini* images is found at http://ciclops .org. What do the most recent images of Saturn show? What wavelengths were observed? Are the pictures shown in false color, and if so, why? Why are these images important?

48. a. Go to websites for the NASA *Juno* mission (http://www .nasa.gov/mission_pages/juno and http://missionjuno.swri .edu), a spacecraft that was launched in 2011 and is scheduled to arrive at Jupiter in 2016. What are the science goals of the mission? Examine the mission's trajectory. Why did it loop around the Sun and pass Earth again in 2013 before heading to Jupiter? Why is there a plaque dedicated to Galileo Galilei on the spacecraft?

    b. What are the main instruments for this mission? Are there any data yet? Have any discoveries been reported?

49. Go to the website for the *Voyager 1* and *2* missions (http:// voyager.jpl.nasa.gov), which collected data on all four of the giant planets more than two decades ago.

    a. Where are the spacecraft now? Click on "Images & Video." These are still the only close-up images of Uranus and Neptune. What was learned about these planets?

    b. Click on the icon of "The Golden Record," and then on the right, look at scenes, greetings, music, and sounds from Earth. Suppose you were asked to make a new version of the Golden Record, a playlist to send on an upcoming space mission to outside of the Solar System. What would you include in one or more of those categories?

50. Go to the Extrasolar Planets encyclopedia (http://exoplanet .eu/catalog/).

    a. Under "Mass," look for a super-Jupiter planet with a mass significantly larger than that of Jupiter. How far is it from its star—is it a hot Jupiter? Click on the planet name—how was it discovered? If a radius is given, is it more or less dense than Jupiter? Click twice on "Mass" to get a list in descending order—what is the most massive super-Jupiter in the catalog?

    b. Under "Mass," click on "$M_{Jup}$" so it changes to "$M_{Earth}$"; do the same under "Radius" so it shows "$R_{Earth}$." Look for a "Super-Earth." What is its radius? How was it detected? Is there an estimated Mass—if so, what is its density compared with that of Earth? Is it a hot or cold super-Earth?

## smartw⊛rk**5**

If your instructor assigns homework in Smartwork5, access your assignments at digital.wwnorton.com/astro5.

# EXPLORATION

digital.wwnorton.com/astro5

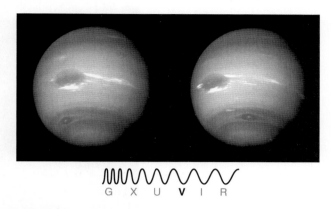

Figure 10.20 These two images of Neptune were taken 17.6 hours apart by *Voyager 2*.

Study the two images of Neptune in **Figure 10.20**. The image on the left was taken first, and the image on the right was taken 17.6 hours later; during this time, the Great Dark Spot completed nearly one full rotation. The small storm at the bottom of the image completed slightly more than one rotation. You would be very surprised to see this result for locations on Earth.

**1** What do these observations tell you about the rotation of visible cloud tops of Neptune?

_____

You can find the rotation period of the smaller storm by equating two ratios. First, use a ruler to find the distance (in millimeters) from the left edge of the planet to the small storm in each image (**Figure 10.21a**). The right edge of the planet is not illuminated, so you will have to estimate the radius of the circles traveled by the storms. You can do this by measuring from the edge of the planet to a line through the planet's center (Figure 10.21b). Because the small storm travels along a line of latitude close to a pole, the distance it travels is significantly less than the circumference of the planet.

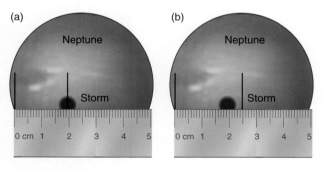

Figure 10.21 (a) How to measure the position of the storm. (b) How to measure the radius of the circle that the storm traveled.

**2** Estimate the radius of the circle that this small storm makes around Neptune (in millimeters) by measuring from the edge of the disk to the line through the center of the planet.

_____

**3** Find the circumference of this circle (in millimeters).

_____

Because the small storm rotated *more than* one time, the total distance it traveled is the circumference of the circle plus the distance between its locations in the two images.

**4** Add the numbers you obtained in steps 2 and 3 to get the total distance traveled (in millimeters) between these images.

_____

Now take a ratio and find the rotation period. The ratio of the rotation period, *T*, to the time elapsed, *t*, must be equal to the ratio of the circumference of the circle around which it travels, *C* (in millimeters), to the total distance traveled, *D* (in millimeters): $T/t = C/D$.

**5** You have all the numbers you need to solve for *T*. What value do you calculate for the small storm's rotation period? (To check your work, note that your answer should be less than 17.6 hours. Why?)

_____

You may be wondering why this calculation works at all. Clearly, the actual distance the small storm traveled is not a small number of millimeters, nor is the circumference of the circle around which it travels. To find the actual distance or circumference, you would multiply both values by the same constant of proportionality. Because you are taking a ratio, however, that constant cancels out, so you might as well leave it out from the beginning.

**6** Perform the corresponding measurements and calculations for the Great Dark Spot. What is its rotation period? Think carefully about how to find the total distance traveled, as the Great Dark Spot has rotated *less than* one time around the planet. (To check your work, note that your answer should be more than 17.6 hours. Why?)

_____

**7** How similar are the rotation periods for these two storms?

_____

**8** What does this comparison tell you about determining the rotation periods of the giant planets using this method?

_____

**9** What method do astronomers use instead?

_____

# 11

# Planetary Moons and Rings

For centuries, Saturn's rings and the Galilean moons of Jupiter delighted those who looked through telescopes. Since the dawn of the space age, robotic explorers traveling through the Solar System have revealed even more of the diverse collection of moons and rings orbiting other planets.

## LEARNING GOALS

By the conclusion of this chapter, you should be able to:

**LG 1** Compare and contrast the orbits and formation of regular and irregular moons.

**LG 2** Describe the evidence for geological activity and liquid oceans on some of the moons.

**LG 3** Describe the composition, origin, and general structure of the rings of the giant planets.

**LG 4** Explain the role gravity plays in the structure of the rings and the behavior of ring particles.

Saturn and its rings as viewed by the *Cassini* spacecraft. ▶ ▶ ▶

**Why do some planets have rings?**

# 11.1 Many Solar System Planets Have Moons

Most of the planets and some of the dwarf planets in our Solar System have moons. As of 2015, the planets of the Solar System have nearly 150 confirmed moons and a few dozen "provisional" moons that need further confirmation. (Moons around asteroids will be discussed in the next chapter.) New moons in the outer Solar System are still being discovered. Some of these planetary moons are listed in Appendix 4, and an updated list can be found through the "Using the Web" problems at the end of this chapter. Many of these moons are unique worlds of their own, exhibiting geological processes similar to those on the terrestrial planets. Some moons have volcanic activity and atmospheres, and some are likely to contain liquid water under their icy surfaces. Recent discoveries suggest that a few of these moons could have conditions suitable to some forms of life. In this section, we will discuss the orbits and the formation of the moons.

## The Distribution of the Moons

The planetary moons of the Solar System are not distributed equally; most are among the giant planets. In the inner part of the Solar System there are only three moons: Earth has one, and Mars has two. Among the dwarf planets, Pluto has five known moons, Haumea has two, and Eris has one. All of the remaining planetary

**Figure 11.1** This figure shows the major moons of the Solar System, as imaged by various spacecraft. The images are shown to scale. The planet Mercury and dwarf planet Pluto are shown for comparison. The martian moons, Phobos and Deimos, are too small to be shown.

moons belong to the giant planets. Mercury and Venus failed to capture or keep any moons of their own. Earth likely has a moon because of a cataclysmic collision when the planet was young. While the larger planets were forming, they had greater attracting mass and greater amounts of debris around them; consequently, they have more moons.

**Figure 11.1** shows the major moons in the Solar System. Some, like Earth's Moon, are made of rock. Others, especially in the outer Solar System, are mixtures of rock and water ice. A few are made almost entirely of ice. Only two moons, Jupiter's Ganymede and Saturn's Titan, are larger in diameter than Mercury, and the smallest known moons are only a kilometer in diameter. Although most moons have no atmosphere, Titan has an atmosphere denser than Earth's, and several have very low-density atmospheres. Scientists suspect that moons accreted from smaller bodies in much the same way that planets accreted from planetesimals, although some may be the product of collisions.

## The Orbits of the Moons

Moons can be classified according to their orbits into one of two categories: *regular moons* and *irregular moons*. A **regular moon** lies in its planet's equatorial plane, is close to its planet, and has a nearly circular orbit in the same direction in which its planet rotates (**Figure 11.2a**). About one-third of the moons in the Solar System are regular. These are moons that likely formed from an accretion disk around a host planet at around the time the planet was forming. Our Moon, the Galilean moons of Jupiter, and Saturn's Titan are large, regular moons. With few exceptions, regular moons are tidally locked to their parent planets. Recall from Chapter 4 that tidal locking causes a body to rotate synchronously with respect to its orbit, as Earth's Moon does. When a moon is in synchronous rotation around its planet, the leading hemisphere permanently faces the direction in which the moon is traveling in its orbit around the planet. The trailing hemisphere faces backward. The leading hemisphere is always flying directly into any local debris surrounding the planet, so it may have more impact craters on its surface than the trailing hemisphere.

An **irregular moon** has a more elliptical and more inclined orbit than a regular moon and generally is farther away from its planet than regular moons are from their planets, as shown in Figure 11.2b. Most irregular moons orbit in a direction that is opposite to the rotation of their respective planets; that is, in retrograde orbits. You may recall from Chapter 3 that apparent backward motion of a planet in the sky is called retrograde motion. The largest irregular moons are Neptune's Triton and Saturn's Phoebe. Most of the recently discovered moons of the outer planets are irregular, and many are only a few kilometers across. These are almost certainly bodies that formed elsewhere and were later captured by the planets.

Some of the regular moons have strange orbital characteristics. For example, the moon that is closest to a planet is Phobos, one of the two small moons of Mars (**Figure 11.3**). Phobos is so close that it actually orbits Mars faster than Mars rotates: as seen from Mars, Phobos rises in the west and sets in the east twice a day. It is not known if Phobos and the other moon of Mars, Deimos, were captured from the nearby asteroid belt or if they evolved together with Mars, possibly after a collision early in the history of Mars. Another strange regular moon is Saturn's Hyperion. Hyperion's rotation is chaotic, meaning that it tumbles in its orbit with a rotation period and a spin-axis orientation that are constantly and unpredictably changing. (A *chaotic* system is one in which the final state is exquisitely sensitive to small variations in the initial state. Typically, the result is unpredictable behavior.) No other known moon in the Solar System tumbles like this.

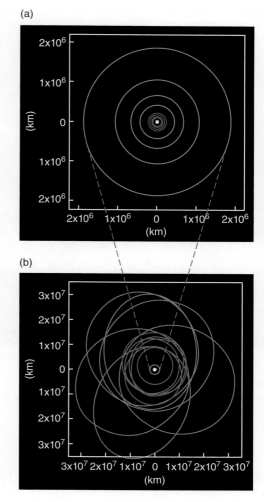

**Figure 11.2** These diagrams illustrate the view from "above" the orbits of some of Jupiter's moons. (a) Closer moons, including the Galilean moons, are regular, with nearly circular orbits in Jupiter's equatorial plane. (b) Most of the more distant moons are irregular, with more elliptical, retrograde orbits that are not in the equatorial plane.

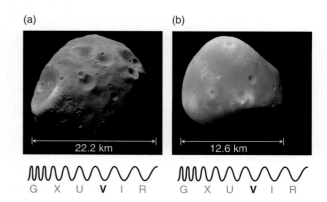

**Figure 11.3** These photographs from the *Mars Reconnaissance Orbiter* show the two tiny moons of Mars: (a) Phobos and (b) Deimos.

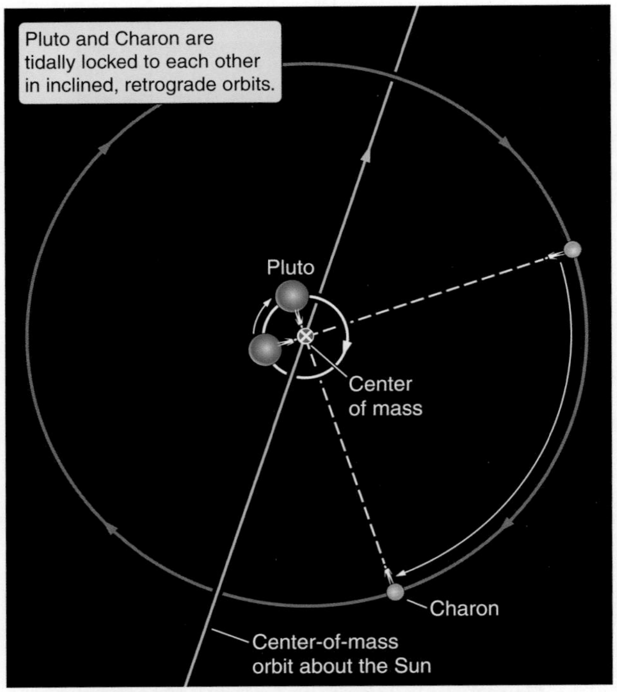

Pluto and Charon are tidally locked to each other in inclined, retrograde orbits.

Pluto

Center of mass

Charon

Center-of-mass orbit about the Sun

**Figure 11.4** This diagram shows the doubly synchronous rotation and revolution in the Pluto-Charon system. The two bodies permanently face one another.

Yet another example of a strange regular moon is Pluto's moon Charon, which is about half as big as Pluto. Pluto and Charon are the only known pair in the Solar System in which both objects are tidally locked to each other. The two are in synchronous rotation, so each has one hemisphere that always faces the other body and another hemisphere that never faces the other body, as shown in **Figure 11.4**. It seems likely that Pluto's highly compact moon system was created by a massive collision between Pluto and another planetesimal, producing a cloud of debris that coalesced to form Charon and the four smaller moons of Pluto, perhaps similar to the way Earth's Moon formed.

Some sets of moons are in synchronized orbits, called **orbital resonances**, where the orbital period of one is a multiple of the orbital period of another. For example, Jupiter's moons Ganymede, Europa, and Io are in a resonance of 1:2:4; that is, for every one orbit of Ganymede, there are two orbits of Europa and four of Io. When the moons are aligned, gravitational effects elongate Io's orbit, which creates variability in the tidal forces of Jupiter on Io. Pluto's five moons are in what appears to be a 1:3:4:5:6 sequence of near resonances. Pairs of some of Saturn's moons are in resonance, and there are also resonances between its moons and gaps in its rings.

The orbits of the moons not only indicate something about their origin. They can also be used to find the masses of their host planets using Kepler's laws, just as you can find the mass of the Sun from the orbital properties of the planets. An example of this is shown in **Working It Out 11.1**.

## CHECK YOUR UNDERSTANDING 11.1

Which of the following are characteristics of regular moons? (Choose all that apply.) (a) They revolve around their planets in the same direction as the planets rotate. (b) They have orbits that lie nearly in the equatorial planes of their planets. (c) They are usually tidally locked to their parent planets. (d) They are much smaller than all of the known planets.

## 11.1 Working It Out Using Moons to Compute the Mass of a Planet

Recall from Working It Out 4.3 that Newton's version of Kepler's law for planets orbiting the Sun could be used to estimate the mass of the Sun. In Chapter 4, we used the following equation to calculate the mass ($M$) of the Sun:

$$M = \frac{4\pi^2}{G} \times \frac{A^3}{P^2}$$

where $A$ is the semimajor axis of the orbit, and $P$ is the orbital period of any planet.

For moons orbiting a planet, the same equation applies, as long as the moon is much less massive than the planet. Thus, we can use the orbital motion of the moons to estimate the mass of the planet. For example, let's use Jupiter's moon Io, which has an orbital semimajor axis of 422,000 kilometers (km) and an orbital period of $P = 1.77$ days. To match the units in $G$, we need to put $P$ into seconds:

$$1.77\,\text{days} = 1.77\,\text{days} \times 24\frac{\text{h}}{\text{day}} \times 60\frac{\text{min}}{\text{h}} \times 60\frac{\text{s}}{\text{min}} = 152,928\,\text{s}$$

The universal gravitational constant $G$ is equal to $6.67 \times 10^{-20}$ km³/(kg s²). Then, the mass of Jupiter is given by

$$M_{\text{Jup}} = \frac{4\pi^2}{G} \times \frac{A^3}{P^2} = \frac{4\pi^2}{6.67 \times 10^{-20}\,\text{km}^3/(\text{kg}\,\text{s}^2)} \times \frac{(422,000\,\text{km})^3}{(152,928\,\text{s})^2}$$

$$M_{\text{Jup}} = 1.90 \times 10^{27}\,\text{kg}$$

You would get the same answer using any other moon of Jupiter.

Back before Newton published his law of gravity and before any measured value of the gravitational constant $G$ was possible, Galileo and Kepler showed that $P^2/A^3$ was the same for each of the four Galilean moons of Jupiter. This demonstrated that Kepler's law applied to systems other than planets orbiting the Sun.

# 11.2 Some Moons Have Geological Activity and Water

There are several ways to group the moons of the Solar System. Some groupings are based on the sequence of the moons in their orbits around their parent planets; others are based on the sizes or compositions of the moons. In this section, we will organize our discussion of the moons by considering some of the same properties we discussed for the terrestrial planets: the history of the moons' geological activity, and the presence of water and an atmosphere.

Some moons in the Solar System have been frozen in time since their formation during the early history of the Solar System, while others are even more geologically active than Earth. As with the terrestrial planets and Earth's Moon, surface features provide critical clues to a moon's geological history. For example, water ice is a common surface material among the moons of the outer Solar System, and the freshness of that ice indicates the age of those surfaces. Meteorite dust darkens the icy surfaces of moons just as dirt darkens snow late in the season in urban areas. A bright surface often means a fresh surface. The size and number of impact craters indicate the relative timing of events such as volcanism, and this timing enables scientists to gauge whether and when a moon may have been active in the past. Older surfaces have more craters. Observations of erupting volcanoes, which are found on Io and Enceladus, for example, are direct evidence that some moons are geologically active today.

## Io, the Most Geologically Active Moon

One of the more spectacular surprises in Solar System exploration was the discovery of active volcanoes on Io, the innermost of the four large moons of Jupiter. Yet in one of those rare coincidences that happen in science, the changing direction and strength of tidal forces from Jupiter and the nearby moons enabled planetary scientists to predict Io's volcanism just 2 weeks before the moon's volcanic activity was discovered. Why is Io so active? Did you ever take a piece of metal and bend it back and forth, eventually breaking it in half? Touch the crease line, and you can burn your fingers. Just as bending metal in your hands creates heat, the continual flexing of Io generates enough energy to melt parts of its mantle. In this way, Jupiter's gravitational energy is converted into thermal energy, powering the most active volcanism in the Solar System.

Io is just slightly larger than our Moon. Its surface is covered with volcanic features, including vast lava flows, volcanoes, and volcanic craters (**Figure 11.5a**). Lava flows and volcanic ash bury impact craters as quickly as they form, so no impact craters have been observed on the surface. The *Voyager*, *Galileo*, and *New Horizons* spacecraft and the Keck telescope have observed hundreds of volcanic vents and active volcanoes on Io. The most vigorous eruptions spray sulfurous gases and solids hundreds of kilometers above the surface. Some of this material escapes entirely from Io. Ash and other particles rain onto the surface as far as 600 km from the vents. The moon is so active that several huge eruptions often occur at the same time. Figure 11.5b shows the volcanic activity on Io—the source of the material supplying Io's plasma torus and flux tube discussed in Chapter 10.

The surface of Io displays a wide variety of colors—pale shades of red, yellow, orange, and brown. Mixtures of sulfur, sulfur dioxide frost, and sulfurous salts of sodium and potassium likely cause the wide variety of colors on Io's surface. Bright patches may be fields of sulfur dioxide snow. Liquid sulfur dioxide flows

(a)

G X U **V** I R

(b)

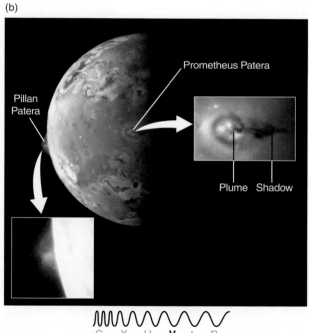

G X U **V** I R

**Figure 11.5** (a) This composite image of Jupiter's volcanically active moon Io was constructed from pictures obtained by *Galileo*. (b) The plume from the crater Pillan Patera rises 140 km above the limb of the moon on the left, while the shadow of a 75-km-high plume can be seen to the right of the vent of Prometheus Patera, a volcanic crater near the moon's center.

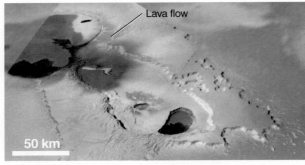

Figure 11.6 This *Galileo* image of Io shows regions where lava has erupted within a caldera. The molten lava flow is shown in false color to make it more visible.

Figure 11.7 This high-resolution *Galileo* image of Jupiter's moon Europa shows where the icy crust has been broken into slabs that, in turn, have been rafted into new positions. These areas of chaotic terrain are characteristic of a thin, brittle crust of ice floating atop a liquid or slushy ocean.

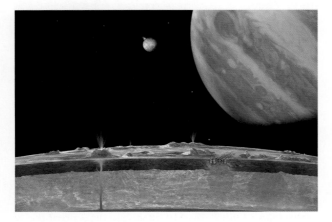

Figure 11.8 This artist's conception of Europa shows liquid bubbling up from the liquid ocean underneath the icy surface. Jupiter and Io are visible in the Europan sky.

beneath Io's surface, held at high pressure by the weight of overlying material. Like water from a spring, this pressurized sulfur dioxide is pushed out though fractures in the crust, producing sprays of sulfur dioxide snow crystals that travel for up to hundreds of kilometers before settling back to the moon's surface. A similar process takes place with a carbon dioxide fire extinguisher. These fire extinguishers contain liquid carbon dioxide at high pressure that immediately turns to "dry ice" snow as it leaves the nozzle.

Spacecraft images reveal the plains, irregular volcanic craters, and flows, all related to the eruption of mostly silicate magmas onto the surface of Io. They also show tall mountains, some nearly twice the height of Mount Everest. Huge structures have multiple summit craters showing a long history of repeated eruptions followed by the collapse of partially emptied magma chambers. Many of the chamber floors are very hot (**Figure 11.6**) and might still contain molten material similar to the magnesium-rich lavas that erupted on Earth more than 1.5 billion years ago. Volcanoes on Io are spread much more randomly than those on Earth, implying a lack of plate tectonics.

Because of its active volcanism, Io's mantle has turned inside-out more than once in the past, leading to chemical differentiation. Volatiles such as water and carbon dioxide probably escaped into space long ago, while most of the heavier materials sank to the interior to form a core. Sulfur and various sulfur compounds, as well as silicate magmas, are constantly being recycled to form the complex surface we see today.

## Evidence of Liquid Oceans on Europa and Enceladus

Jupiter's moon Europa is slightly smaller than our Moon and is made of rock and ice and has an iron core. *Voyager* observed an outer shell of water ice with surface cracks and creases. There are few impact craters, so the surface must be young. Regions of chaotic terrain, as shown in **Figure 11.7**, are places where the icy crust has been broken into slabs that have shifted into new positions. In other areas, the crust has split apart, and the gaps have filled in with new dark material rising from the interior. The young surface implies activity, likely powered by continually changing tidal stresses from Jupiter similar to what happens on Io. This activity varies as Europa orbits Jupiter. However, the forces are not as strong on Europa as they are on Io because Europa is farther from the planet (**Working It Out 11.2**).

The *Galileo* spacecraft measured Europa's magnetic field and found that it is variable, indicating an internal electrically conducting fluid. Detailed computer models of the interior of Europa suggest that Europa has a global ocean 100 km deep that contains more water than all of Earth's oceans. This ocean might be salty with dissolved minerals. The lightly cratered and thus geologically young surface indicates that there is energy exchange between the icy crust and liquid water, but it is not known if the icy crust is tens or hundreds of kilometers thick. It is also not yet known if Europa has volcanic activity at the seafloor.

The Hubble Space Telescope may have detected two transient plumes of water vapor erupting from the icy surface, but they were not seen in subsequent observations. The surface is too cold for liquid water, but there might be lakes a few kilometers underneath the surface. Scientists have reanalyzed observations of Europa in light of what has been learned about the numerous subsurface lakes in Antarctica, subglacial volcanoes in Iceland, and ice sheets at both poles of Earth. There may be many shallow "great lakes" underneath the ice, and these would be prime targets for future exploration. **Figure 11.8** is an

## 11.2 Working it Out Tidal Forces on the Moons

Recall from Chapter 4 that the tidal force between a planet and its moon depends on the masses of the planet and the moon, and the size of the moon, divided by the cube of the distances between them:

$$F_{tidal} = \frac{2GM_{Jup}M_{moon}R_{moon}}{d_{Jup\text{-}moon}^3}$$

We can use this equation to compare the tidal forces between Jupiter and two of its moons by taking a ratio. For Io and Europa:

$$\frac{F_{tidal\text{-}Io}}{F_{tidal\text{-}Europa}} = \frac{\dfrac{2GM_{Jup}M_{Io}R_{Io}}{d_{Io}^3}}{\dfrac{2GM_{Jup}M_{Europa}R_{Europa}}{d_{Europa}^3}} = \frac{\dfrac{M_{Io}R_{Io}}{d_{Io}^3}}{\dfrac{M_{Europa}R_{Europa}}{d_{Europa}^3}}$$

After canceling out the mass of Jupiter $M_{Jup}$ and the constants $2G$, and using data on Io and Europa from Appendix 4—$M_{Io} = 8.9 \times 10^{22}$ kg, $M_{Europa} = 4.8 \times 10^{22}$ kg, $R_{Io} = 1,820$ km, $R_{Europa} = 1,560$ km, $d_{Io} = 422,000$ km, and $d_{Europa} = 671,000$ km—we have

$$\frac{F_{tidal\text{-}Io}}{F_{tidal\text{-}Europa}} = \frac{\dfrac{(8.9 \times 10^{22}) \times 1,820}{(422,000)^3}}{\dfrac{(4.8 \times 10^{22}) \times 1,560}{(671,000)^3}} = 8.7$$

This comparison shows that the tidal forces on Io are much stronger than those on Europa.

artist's schematic of Europa showing some of the water from the ocean leaking out at the surface, where it would be easier to study. Jupiter and Io are seen in Europa's sky.

Enceladus, one of Saturn's icy moons, shows a wide variety of ridges, faults, and smooth plains. This evidence of tectonic processes is unexpected for a small (500 km) body. The activity on Enceladus is an example of **cryovolcanism**, which is similar to terrestrial volcanism but is driven by subsurface low-temperature liquids such as water and hydrogen rather than molten rock. Some impact craters appear softened, perhaps by the viscous flow of ice, like the flow that occurs in the bottom layers of glaciers on Earth. Parts of the moon have no craters, indicating recent resurfacing. Terrain near the south pole of Enceladus is cracked and twisted (**Figure 11.9a**). The cracks are warmer than their surroundings, suggesting that tidal heating and radioactive decay within the moon's rocky core heat the surrounding ice and drive it to the surface.

Enceladus has a liquid ocean buried beneath 30–40 km of ice crust and is 10 km deep, as an artist has illustrated in Figure 1.10 in Chapter 1. The cracks are warmer than their surroundings, implying that tidal heating and radioactive decay within the moon's rocky core heat ice and drive it to the surface. Active cryovolcanic plumes, like those seen in Figure 11.9b, expel water vapor, tiny ice crystals, and salts. Some of the crystals fall back onto the surface as an extremely fine, powdery snow. The rate of accumulation is very low—a fraction of a millimeter per year—but over time the snow builds up. *Cassini* scientists estimate that the snow may be 100 meters thick in one area near the south pole of Enceladus, indicating that the plume activity has continued on and off for at least tens of millions of years.

Tidal flexing is the likely source of the heat energy coming from Enceladus, as it is on Io and Europa. Enceladus has an orbital resonance with Saturn's moon Dione. A moon made completely of ice would be too stiff for tidal heating to be effective; tidal heating works more effectively with ice over liquid water or cracked ice with some liquid. It remains a mystery why Enceladus is so active while Mimas, a neighboring moon of about the same size—but closer to Saturn and also subject to tidal heating—appears to be geologically dead.

(a)

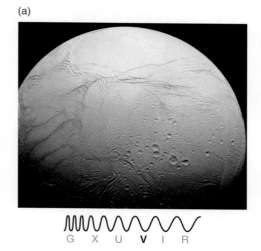

G X U V I R

(b)

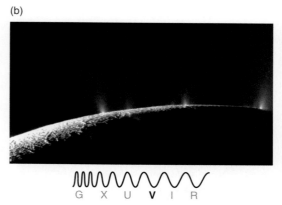

G X U V I R

**Figure 11.9** These images of Enceladus were taken by *Cassini*. (a) The deformed ice cracks (shown blue in false color) were found to be the sources of cryovolcanism. (b) Cryovolcanic plumes in the south polar region are seen spewing ice particles into space.

(a)

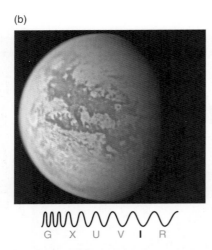

G X U V I R

(b)

G X U V I R

**Figure 11.10** These images of Saturn's largest moon, Titan, were taken by *Cassini*. (a) Titan's orange atmosphere is caused by organic, smoglike particles. (b) Infrared-light imaging penetrates Titan's smoggy atmosphere and reveals surface features.

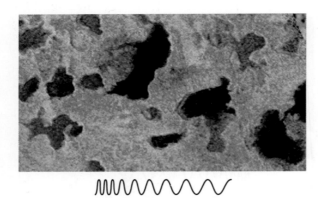

G X U V I **R**

**Figure 11.11** Radar imaging (false color) near Titan's north pole shows lakes of liquid hydrocarbons covering 100,000 square kilometers ($km^2$) of the moon's surface. Features such as islands, bays, and inlets, are visible in many of these radar images.

## Titan's Atmosphere and Ocean

Saturn's moon Titan is slightly larger than Mercury and has a composition of about 45 percent water ice and 55 percent rocky material. What makes Titan especially remarkable is its thick atmosphere. Whereas Mercury's secondary atmosphere has been lost to space, Titan's greater mass and distance from the Sun have allowed it to retain an atmosphere that is 30 percent denser than that of Earth. Titan's atmosphere, like Earth's, is mostly nitrogen. As Titan differentiated, various ices, including methane ($CH_4$) and ammonia ($NH_3$), emerged from the interior to form an early atmosphere. Ultraviolet photons from the Sun have enough energy to break apart ammonia and methane molecules—a process called **photodissociation**. Photodissociation of ammonia is the likely source of Titan's atmospheric nitrogen. Methane breaks into fragments that recombine to form organic compounds including complex hydrocarbons such as ethane. These compounds tend to cluster in tiny particles, creating organic smog much like the air over Los Angeles on a bad day; this gives Titan's atmosphere its characteristic orange hue (**Figure 11.10a**).

Close-up views of Titan's surface were obtained by the *Cassini* spacecraft. Haze-penetrating infrared imaging showed broad regions of dark and bright terrain (Figure 11.10b). Radar imaging of Titan revealed irregularly shaped features in its northern hemisphere that appear to be widespread lakes and seas of methane, ethane, and other hydrocarbons (**Figure 11.11**). The photodissociative process by sunlight should have destroyed all atmospheric methane within a geologically brief period of about 50 million years, so there must be a process for renewing the methane that is being destroyed by solar radiation. This along with the near absence of impact craters on the surface of Titan suggests recent methane-producing activity. Radar views also indicate an active surface, showing features that resemble terrestrial sand dunes and channels. Heat supplied by radioactive decay could cause cryovolcanism that releases "new" methane from underground. The evidence of active cryovolcanism on Titan is indirect—the presence of abundant atmospheric methane and of methane lakes strongly suggests that Titan has some geological activity.

Titan has terrains reminiscent of those on Earth, with networks of channels, ridges, hills, and flat areas that may be dry lake basins. These terrains suggest a sort of methane cycle (analogous to Earth's water cycle) in which methane rain falls to the surface, washes the ridges free of the dark hydrocarbons, and then collects into drainage systems that empty into low-lying, liquid methane pools. Stubby, dark channels appear to be springs where liquid methane emerges from the subsurface; bright, curving streaks could be water ice that has oozed to the surface to feed glaciers. An infrared camera photographed a reflection of the Sun from such a lake surface. The type of reflection observed proves that the lake contains a liquid and is not frozen or dry. Recent observations might indicate waves on one of these lakes.

Titan is the only moon (aside from Earth's) that has been landed upon. In 2005, *Cassini* released a probe, *Huygens*, which plunged through Titan's atmosphere measuring the moon's composition, temperature, pressure, and wind speeds, and taking pictures as it descended. *Huygens* confirmed the presence of nitrogen-bearing organic compounds in the clouds. During its descent, *Huygens* encountered 120-meter-per-second (m/s) winds and temperatures as low as 88 K. As it reached the surface, though, winds died down to less than 1 m/s and the temperature warmed to 112 K. The pictures taken by *Huygens* showed that the surface was wet with liquid methane, which evaporated as the probe—heated during its passage through the atmosphere—landed in the frigid soil. The surface was also rich with other organic (carbon-bearing) compounds, such as cyanogen and ethane. As shown in **Figure 11.12**, the surface around the landing site is relatively flat and littered with rounded "rocks" of water ice. The dark "soil" is probably a mixture of water and hydrocarbon ices.

As is the case for Saturn's moon Enceladus and Jupiter's moon Europa, gravitational mapping provides indirect evidence that Titan also has an ocean buried beneath its surface. Some of Titan's surface features move by as much as 35 km, which suggests that the crust is sliding on an underlying liquid layer. The current model of Titan is that its rigid ice shell varies in thickness and surrounds an ocean 100 km below the surface, shown in **Figure 11.13**. This ocean would be made of water mixed with dissolved salts—possibly saltier than Earth's Dead Sea. In this model, methane outgassing would occur in hot spots.

Titan is the only moon with a significant atmosphere and the only Solar System body besides Earth that has standing liquid on the surface and a cycle of liquid rain and evaporation. In many ways, Titan resembles a primordial Earth, albeit at much lower temperatures. The presence of liquids and of organic compounds that could be biological precursors for life in the right environment makes Titan another high-priority target for continued exploration.

## CHECK YOUR UNDERSTANDING 11.2

Which of the following moons is *not* thought to have an ocean of water beneath its surface? (a) Io; (b) Europa; (c) Enceladus; (d) Titan

## Cryovolcanism on Triton

Cryovolcanism also occurs on Triton, Neptune's largest moon. Triton is an irregular moon with a retrograde orbit, which suggests that Triton was captured by Neptune after the planet's formation. As Triton achieved its current circular, synchronous orbit, it experienced extreme tidal stresses from Neptune, generating large amounts of thermal energy. The interior may have melted, allowing Triton to become chemically differentiated.

Triton has a thin atmosphere and a surface composed mostly of ices and frosts of methane and nitrogen at a temperature of about 38 K. The relative absence of craters tells us the surface is geologically young. Part of Triton is covered with terrain that looks like the skin of a cantaloupe (**Figure 11.14**), with irregular pits and hills that may be caused by slushy ice emerging onto the surface from the interior. Veinlike features include grooves and ridges that could result from ice oozing out along fractures. The rest of Triton is covered with smooth volcanic plains. Irregularly shaped depressions as wide as 200 km formed when mixtures of water, methane, and nitrogen ice melted in the interior of Triton and erupted onto the surface, much as rocky magmas erupted onto the lunar surface and filled impact basins on Earth's Moon.

Clear nitrogen ice creates a localized greenhouse effect, in which solar energy trapped beneath the ice raises the temperature at the base of the ice layer. A temperature increase of only 4 K vaporizes the nitrogen ice. As this gas is formed, the expanding vapor exerts very high pressures beneath the ice cap. Eventually, the ice ruptures and vents the gas explosively into the low-density atmosphere. *Voyager 2* found four of these active geyserlike cryovolcanoes on Triton. Each consisted of a plume of gas and dust as much as 1 km wide rising 8 km above the surface, where the plume was caught by upper atmospheric winds and carried for

**Figure 11.12** The two water-ice "rocks" just below the center of this Huygens image are about 85 centimeters (cm) from the camera and roughly 15 and 4 cm across, respectively.

G X U V I R

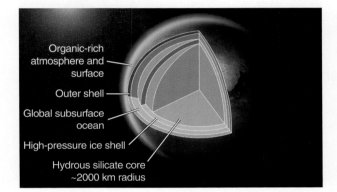

Organic-rich atmosphere and surface

Outer shell

Global subsurface ocean

High-pressure ice shell

Hydrous silicate core ~2000 km radius

**Figure 11.13** This artist's conception of Titan's internal shows how Titan is differentiated, with a core of water-bearing rocks and a subsurface ocean of liquid water. A layer of high-pressure ice surrounds the core, and an outer ice shell is on top of the subsurface ocean.

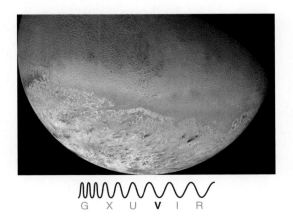

G X U V I R

**Figure 11.14** This *Voyager 2* mosaic shows various terrains on the Neptune-facing hemisphere of Triton. The lack of impact craters in the "cantaloupe terrain," visible at the top, indicates a geologically younger age than that of the bright, cratered terrain at the bottom.

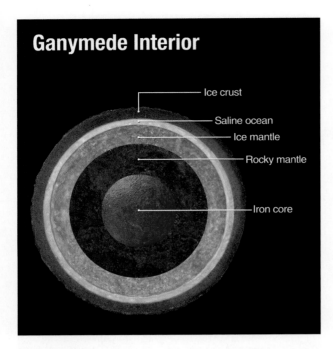

**Figure 11.15** This artist's conception of Jupiter's moon Ganymede illustrates that the ocean and ice may be stacked up in multiple layers.

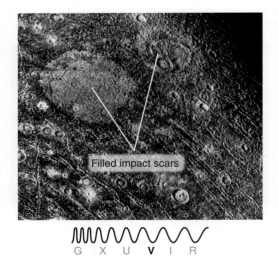

**Figure 11.16** This *Voyager* image shows filled impact scars on Jupiter's moon Ganymede.

hundreds of kilometers downwind. Dark material, perhaps silicate dust or radiation-darkened methane ice grains, is carried along with the expanding vapor into the atmosphere, from which it subsequently settles to the surface, forming dark patches streaked out by local winds, as seen near the lower right of Figure 11.14.

## Formerly Active Moons

Some moons show clear evidence of past ice volcanism and tectonic deformation, but no current geological activity. For example, Jupiter's moon Ganymede, the largest moon in the Solar System, is larger than the planet Mercury. When Jupiter was forming, low temperatures enabled grains of water ice to survive and coalesce along with dust grains into larger bodies at the distance of Ganymede's orbit. In less than half a million years, these bodies accreted to form Ganymede, Jupiter's largest moon. Heating from accretion melted parts of Ganymede so that it is fully differentiated, with outer water layers, an inner silicate zone, and an iron-rich liquid core. As the moon cooled, much of the outer water layer froze, forming a dirty ice crust. Most of the denser materials sank to the central core, leaving an intermediate ice-silicate zone. Ganymede might also have a large, salty ocean underneath its icy surface, maybe 800 km deep, containing 25 times the volume of Earth's oceans (**Figure 11.15**).

Its surface is composed of two prominent terrains: a dark, heavily cratered (and therefore ancient) terrain, and a bright terrain characterized by ridges and grooves. The abundance of impact craters on Ganymede's dark terrain reflects the period of intense bombardment during the early history of the Solar System. The largest region of ancient dark terrain includes a semicircular area more than 3,200 km across on the leading hemisphere. Furrowlike depressions occurring in many dark areas are among Ganymede's oldest surface features. They may represent surface deformation from internal processes or they may be relics of impact-cratering processes.

Impact craters on Ganymede range up to hundreds of kilometers in diameter, and the larger craters are proportionately shallower. The icy crater rims slowly slump, like a lump of soft clay. They are seen as bright, flat, circular patches found principally in the moon's dark terrain (**Figure 11.16**) and are thought to be scars left by early impacts onto a thin, icy crust overlying water or slush (**Figure 11.17**). In Chapter 8, we discussed how planetary surfaces can be fractured by faults or folded by compression resulting from movements initiated in the mantle. On Ganymede, the tectonic processes have been so intense that the fracturing and faulting have completely deformed the icy crust, destroying all signs of older features, such as impact craters, and creating the bright terrain. The energy that powered Ganymede's early activity was liberated during a period of differentiation when the moon was very young. After differentiation was complete, that source of internal energy ran out, and geological activity ceased.

Many other moons show evidence that they experienced an early period of geological activity that resulted in a dazzling array of terrains. A 400-km impact crater scars Saturn's moon Tethys, covering 40 percent of its diameter, and an enormous canyonland wraps at least three-fourths of the way around the moon's equator. Saturn's moon Dione shows bright ice cliffs up to several hundred meters high, created by tectonic fracturing. The trailing hemisphere of Saturn's Iapetus is bright, reflecting half the light that falls on it, while much of the leading

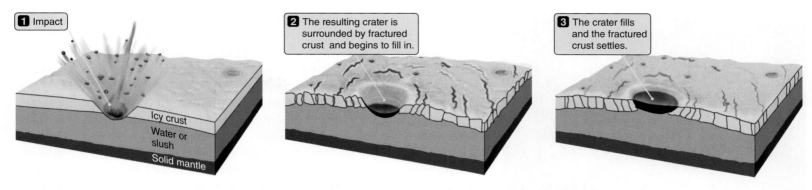

**1** Impact

icy crust
Water or
slush
Solid mantle

**2** The resulting crater is
surrounded by fractured
crust and begins to fill in.

**3** The crater fills
and the fractured
crust settles.

**Figure 11.17** Filled impact scars form as viscous flow smooths out structures left by impacts on icy surfaces.

hemisphere is as black as tar. These dark deposits appear *only* in the leading hemisphere of Iapetus, suggesting that they might be debris that was blasted off small retrograde moons of Saturn by micrometeoritic impacts and swept up by Iapetus as it moved along in its prograde orbit around Saturn.

Saturn's moon Mimas, no larger than the state of Ohio, is heavily cratered with deep, bowl-shaped depressions. The most striking feature on Mimas is a huge impact crater in the leading hemisphere. Named "Herschel" after astronomer Sir William Herschel, who discovered many of Saturn's moons, the crater is 130 km across, a third the size of Mimas itself (**Figure 11.18**). It is doubtful that Mimas could have survived the impact of a body much larger than the one that created Herschel. Some astronomers think that Mimas (and perhaps other small, icy moons as well) was hit many times in the past by objects so large as to fragment the moon into many small pieces. Each time this happened, the individual pieces still in Mimas's orbit would coalesce to re-form the moon, perhaps in much the same way that Earth's Moon coalesced from fragments that remained in orbit around Earth after a large planetesimal impacted Earth early in its history.

Areas on Uranus's small moon Miranda have been resurfaced by eruptions of icy slush or glacierlike flows. Other moons of Uranus—Oberon, Titania, and Ariel—are covered with faults and additional signs of early tectonism. On Ariel, in particular, very old, large craters appear to be missing, perhaps obliterated by earlier volcanism.

## Geologically Dead Moons

Geologically dead moons, such as Jupiter's Callisto, Saturn's **Hyperion,** Uranus's Umbriel, and a large assortment of irregular moons, are moons for which there is little or no evidence of internal activity having occurred at any time since their formation. The surfaces of these moons are heavily cratered and show no modification other than the cumulative degradation caused by a long history of impacts.

Callisto is the third largest moon in the Solar System, just slightly smaller than Mercury. It is also the darkest of the Galilean moons of Jupiter, yet it is still twice as reflective as Earth's Moon. This brightness indicates that Callisto is rich in water ice, but with a mixture of dark, rocky materials. Except in areas that experienced large impact events, the surface is essentially uniform, consisting of

G  X  U  **V**  I  R

**Figure 11.18** This *Cassini* image shows Saturn's moon Mimas and the crater Herschel.

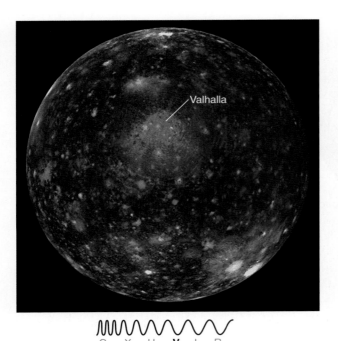

G X U V I R

**Figure 11.19** This *Galileo* image shows Jupiter's second-largest moon, Callisto. Its ancient surface is dominated by impact craters and shows no sign of early internal activity.

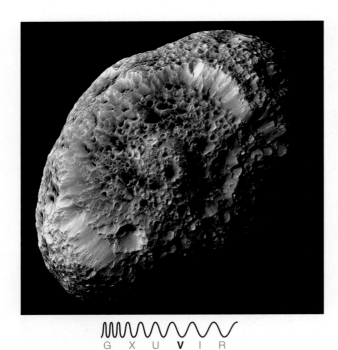

G X U V I R

**Figure 11.20** Saturn's moon Hyperion rotates chaotically, with its rotation period and spin axis constantly changing. The 250-km moon's low density and spongelike texture, seen in this *Cassini* image, suggest that its interior houses a vast system of caverns.

relatively dark, heavily cratered terrain. Callisto's most prominent feature is a 2,000-km, multiringed structure of impact origin named Valhalla (the largest bright feature visible on Callisto's face in **Figure 11.19**). *Galileo* results suggest that a liquid ocean containing water or water mixed with ammonia could exist beneath the heavily cratered surface. Callisto may have partially differentiated, with rocky material separating from ices and sinking deeper into the interior.

Saturn's Hyperion is one of the largest irregularly shaped moons and could be the remnant of an impact. The extensive craters look almost like sponges (**Figure 11.20**). Hyperion crosses Saturn's magnetosphere in its chaotic orbit, which seems to have left the moon with some electric charge. Umbriel, the darkest and third largest of Uranus's moons, appears uniform in color, reflectivity, and general surface features, indicative of an ancient surface. The real puzzle posed by Umbriel is why it is geologically dead, while the surrounding large moons of Uranus have been active at least at some time in their past.

### CHECK YOUR UNDERSTANDING 11.3

Rank these moons in terms of the density of impact craters you would expect to observe on the surface. Rank them from most to least. (a) Callisto; (b) Titan; (c) Io; (d) Ganymede

## 11.3  Rings Surround the Giant Planets

A planetary ring is a collection of particles—varying in size from tiny grains to house-sized boulders—that orbit individually around a planet, forming a flat disk. Ring systems do not occur in the terrestrial planets but are found around each of the giant planets. **Figure 11.21** shows how the ring system of each giant planet varies in size and complexity: some systems extend for hundreds of thousands of kilometers, and some systems have detailed structure that includes numerous small rings. In this section, we discuss ring formation, composition, and evolution.

### The Discovery of Planetary Rings

Saturn's rings have been observed for centuries. In 1610, Galileo observed two small objects next to Saturn and thought they might be similar to the four moons orbiting Jupiter. But Saturn's "moons" did not move, and 2 years later they disappeared. In 1655, Dutch instrument maker Christiaan Huygens (1629–1695) pointed a superior telescope of his own design at Saturn. Huygens observed that an apparently continuous flat ring surrounds the planet and that the ring's visibility changes with its apparent tilt as Saturn orbits the Sun. Over the next three centuries, astronomers discovered more rings around Saturn, but searches failed to detect rings around any other planet.

Most Solar System rings were more recently discovered. In 1977, a team of astronomers studying the atmosphere of Uranus during stellar occultations saw brief, minute changes in the brightness of a star as it first approached and then receded from the planet. The astronomers realized this meant that Uranus has rings. Over the next several years, stellar occultations revealed a total of nine rings surrounding the planet. In 1986, *Voyager 2* imaged two additional rings of

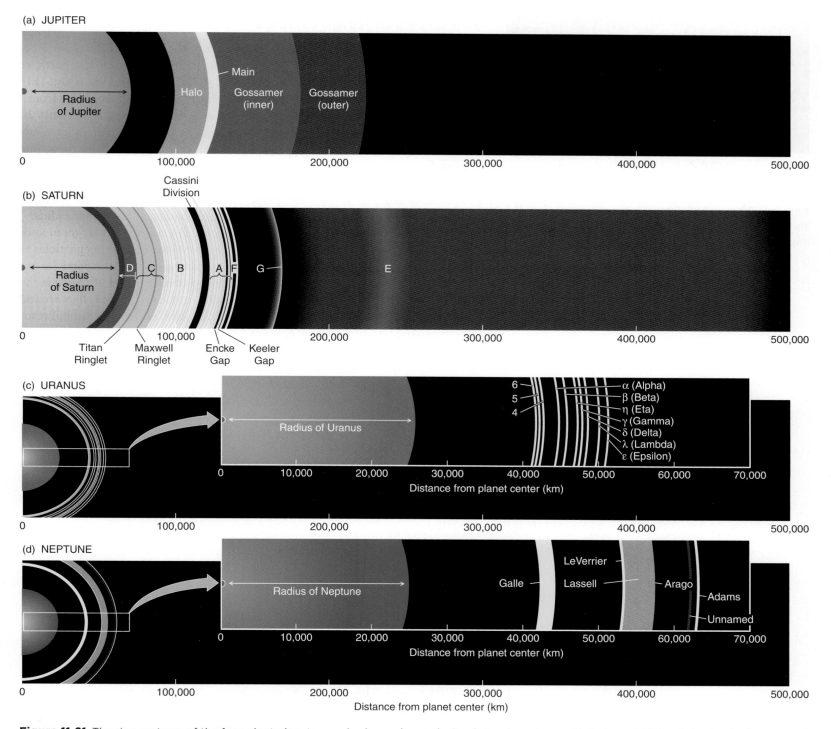

**Figure 11.21** The ring systems of the four giant planets vary in size and complexity. Saturn's system, with its broad E Ring, is by far the largest and has the most complex structure in the inner rings.

Uranus, and in 2005 the Hubble Space Telescope recorded two more, bringing the total to 13. In 1979, cameras on *Voyager 1* recorded a faint ring around Jupiter. The occultation technique also revealed arclike ring segments around Neptune, which were determined to be complete rings when *Voyager 2* reached Neptune in 1989.

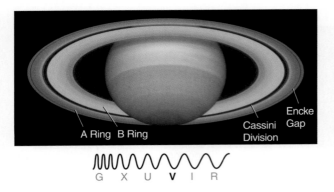

**Figure 11.22** A Hubble Space Telescope image showing Saturn and its A Ring, B Ring, Cassini Division, and Encke Gap. The C Ring is too dim to be seen clearly.

## The Orbits of Ring Particles

Ring particles follow Kepler's laws, and therefore the speed and orbital period of each particle must vary with its distance from the planet. The closest particles move the fastest and have the shortest orbital periods (see Working It Out 11.1). The orbital periods of particles in Saturn's bright rings, for example, range from 5 hours 45 minutes at the inner edge of the innermost bright ring to 14 hours 20 minutes at the outer edge of the outermost bright ring. Ring particles have low speeds relative to one another because they are all orbiting in the same direction. A particle moving on an upward trajectory will bump into another particle on a downward trajectory and the upward and downward motion will cancel, leaving the particles moving in the same plane. A similar process occurs for particles moving inward and outward, leaving the particles moving at a constant radius.

The orbits of ring particles can also be influenced by the planet's larger moons. If the moon is massive enough, it exerts a gravitational tug on the ring particles as it passes by. If this happens over and over through many orbits, the particles are pulled out of the area, leaving a lower-density gap (see Figure 11.21). Such is the case with Saturn's moon Mimas, which causes the famous gap in the rings around Saturn called the **Cassini Division** (**Figure 11.22**). Mimas is in a 1:2 orbital resonance with the Cassini Division, giving a ring particle located in the Cassini Division an orbital period about Saturn that is equal to half the orbital period of Mimas. Such resonances with other moons are known to produce some of the gaps that appear in Saturn's bright rings. One of the gaps is caused by a 4:1 resonance between the ring particles and Mimas.

Other kinds of orbital resonances are also possible. For example, most narrow rings are caught up in a periodic gravitational tug-of-war with nearby moons, known as shepherd moons because of the way they "herd" the flock of ring particles. Shepherd moons are usually small and often come in pairs, with one orbiting just inside and the other just outside a narrow ring. A shepherd moon just outside a ring robs orbital energy from any particles that drift outward beyond the edge of the ring, causing the particles to move back inward. A shepherd moon just inside a ring gives up orbital energy to a ring particle that has drifted too far in, nudging it back in line with the rest of the ring. In some cases, narrow rings are trapped between two shepherd moons in slightly different orbits.

## Ring Formation and Evolution

Much of the material found in planetary rings is thought to be the result of tidal stresses. If a moon (or other planetesimal) orbits a large planet, the force of gravity will be stronger on the side of the moon close to the planet and weaker on the side farther away. This difference in gravitational force stretches out the moon, as you saw in the discussion of tidal forces in Chapter 4. If the tidal stresses are greater than the self-gravity that holds the moon together, the moon will be torn apart. The distance at which the tidal stresses exactly equal the self-gravity is known as the Roche limit. The Roche limit applies only to objects that are held together by their own gravity; it does not apply to objects held together by other forces—like people or cars. If a moon or planetesimal comes within the Roche limit of a planet, it is pulled apart by tidal stresses, leaving many small pieces orbiting the planet. These pieces gradually spread out, and their orbits are circularized and flattened out by collisions. The fragmented pieces of the disrupted body are then distributed around the planet in the form of a ring.

Planetary rings do not have the long-term stability of most Solar System objects. Ring particles are constantly colliding with one another in their tightly packed environment, either gaining or losing orbital energy. This redistribution of orbital energy can cause particles at the ring edges to leave the rings and drift away, aided by nongravitational influences such as the pressure of sunlight. Although moons may help guide the orbits of ring particles and delay the dissipation of the rings themselves, at best this condition can be only temporary. Saturn's brightest rings might be nearly as old as Saturn, but most planetary rings eventually disperse.

Even Earth may have had several short-lived rings at various times during its long history. Any number of comets or asteroids must have passed within Earth's Roche limit (about 25,000 km for rocky bodies and more than twice that for icy bodies) to disintegrate into a swarm of small fragments to create a temporary ring. However, unlike the giant planets, Earth lacks shepherd moons to provide orbital stability to rings.

## The Composition of Ring Material

Because much of the material in the rings of the giant planets comes from their moons, the composition of the rings is similar to the composition of the moons. Saturn's bright rings probably formed when a moon or planetesimal came within the Roche limit of Saturn. These rings reflect about 60 percent of the sunlight falling on them. They are made of water ice, though a slight reddish tint tells us they are not made of pure ice but must contain small amounts of other materials, such as silicates. The icy moons around Saturn or the frozen comets of the outer Solar System could easily provide this material.

Saturn's rings are the brightest in the Solar System and are the only ones that we know are composed of water ice. In stark contrast, the rings of Uranus and Neptune are among the darkest objects known in the Solar System. Only 2 percent of the sunlight falling on them is reflected back into space, which makes the ring particles blacker than coal or soot. No silicates or similar rocky materials are this dark, so these rings are likely composed of organic materials and ices that have been radiation darkened by high-energy, charged particles in the magnetospheres of these planets. (Radiation blackens organic ices such as methane by releasing carbon from the ice molecules.) Jupiter's rings are of intermediate brightness, suggesting that they may be rich in silicate materials, like the innermost of Jupiter's small moons.

The jumble of fragments that make up Saturn's rings is understood to be a product of tidal disruption of a moon or planetesimal, but moons can contribute material to rings in other ways. The brightest of Jupiter's rings is a relatively narrow strand only 6,500 km across, consisting of material from the moons Metis and Adrastea. These two moons orbit in Jupiter's equatorial plane, and the ring they form is narrow. Beyond this main ring, however, are the very different wispy rings called gossamer rings. The gossamer rings are supplied with dust by the moons Amalthea and Thebe. The innermost ring in Jupiter's system, called the halo ring, consists mostly of material from the main ring. As the dust particles in the main ring drift slowly inward toward the planet, they pick up an electric charge and are pulled into this rather thick torus by electromagnetic forces associated with Jupiter's powerful magnetic field.

Finally, moons may contribute ring material through volcanism. Volcanoes on Jupiter's moon Io continually eject sulfur particles into space, many of which are

## 11.3 Working It Out Feeding the Rings

The moons of the giant planets have a low surface gravity and a much lower escape velocity than that of Earth, which is 11.2 km/s. Thus, volcanic emissions from some of these small moons can escape and supply material to a ring. Recall the equation from Working It Out 4.2 for escape velocity from a spherical object of mass $M$ and radius $R$:

$$v_{esc} = \sqrt{\frac{2GM}{R}}$$

Saturn's moon Enceladus has a mass of $1.08 \times 10^{20}$ kg and a radius of 250 km. The escape velocity from Enceladus is given by

$$v_{esc} = \sqrt{\frac{2 \times [6.67 \times 10^{-20}\,\text{km}^3/(\text{kg s}^2)] \times (1.08 \times 10^{20}\,\text{kg})}{250\,\text{km}}}$$

$v_{esc} = 0.24$ km/s; or multiply by 3,600 s/h to get 864 km/h

This escape velocity is lower than the speed of the volcanic plumes on Enceladus, which is nearly 2,200 km/h. The icy particles from the plumes supply particles to Saturn's E Ring.

G   X   U   **V**   I   R

**Figure 11.23** Saturn's moon Enceladus (the large bright spot appearing to be on the ring) is the source of material in Saturn's E Ring. Note the distortion in the distribution of ring material in the immediate vicinity of the moon caused by its gravitational influence on the orbits of ring particles. Other bright objects in this *Cassini* image are stars.

pushed inward by sunlight and find their way into a ring. The particles in Saturn's E Ring are ice crystals ejected from icy geysers on the moon Enceladus, which is located in the very densest part of the E Ring (**Working It Out 11.3**). Ice particles ejected into space replace particles continually lost from Saturn's E Ring (**Figure 11.23**). The E Ring will survive for as long as Enceladus remains geologically active.

### CHECK YOUR UNDERSTANDING 11.4

If rings are observed around a planet, this indicates that: (a) there is a recent source of ring material; (b) the planet is newly formed; (c) the rings formed with the planet; (d) the rings are made of fine dust.

## 11.4 Ring Systems Have a Complex Structure

Huygens, with his mid-17th-century understanding of physics, thought that Saturn's ring was a solid disk surrounding the planet. It was not until the middle of the 19th century that the brilliant Scottish mathematical physicist James Clerk Maxwell showed that solid rings would be unstable and would quickly break apart. In this section, we will examine the details of the rings of each planet of the outer planets. You may want to refer back to Figure 11.21.

### Saturn's Magnificent Rings—A Closer Look

Saturn is adorned by a magnificent and complex system of rings, unmatched by any other planet in the Solar System. Figure 11.21b shows the individual components of Saturn's ring system and its major divisions and gaps. Among the four giant planets, Saturn's rings are the widest and brightest. The outermost bright ring, the A Ring, is the narrowest of the three bright rings. It has a sharp outer edge and contains several narrow gaps.

In 1675, the Italian-French astronomer Jean-Dominique Cassini (1625–1712) found a gap in the planet's seemingly solid ring. Saturn appeared to have two rings rather than one, and the gap that separated them became known as the *Cassini Division*. The Cassini Division is so wide (4,700 km) that the planet Mercury would almost fit within it. Astronomers once thought that it was completely empty, but images taken by *Voyager 1* show the Cassini Division is filled with material, although it is less dense than the material in the bright rings.

The B Ring, whose width is roughly twice Earth's diameter, is the brightest of Saturn's rings and has no internal gaps on the scale of those seen in the other bright rings. The C Ring is so much fainter than neighboring rings that it often fails to show up in normally exposed photographs. Through the eyepiece of a telescope, this ring appears like delicate gauze. There is no known gap between the C Ring and either of the adjacent rings; only an abrupt change in brightness marks the boundary between them. The cause of this sharp change in the amount of ring material remains an unanswered question. Too dim to be seen next to Saturn's bright disk, the D Ring is a fourth wide ring that was unknown until it was imaged by *Voyager 1*. It shows less structure than any of the bright rings, and it does not appear to have a definable inner edge. The D Ring may extend all the way down to the top of Saturn's atmosphere, where its ring particles would burn up as meteors.

Saturn's bright rings are not uniform. The A and C rings contain hundreds, and the B Ring thousands, of individual ringlets, some only a few kilometers wide (**Figure 11.24**). Each of these ringlets is a narrowly confined concentration of ring particles bounded on both sides by regions of relatively little material. About every 15 years, the plane of Saturn's rings lines up with Earth, and we view them edge on. The rings are so thin that they all but vanish for a day or so in even the largest telescopes. While the glare of the rings is absent, astronomers search for undiscovered moons or other faint objects close to Saturn. In 1966, an astronomer was looking for moons when he found weak but compelling evidence for a faint ring near the orbit of Saturn's moon Enceladus. In 1980, *Voyager 1* confirmed the existence of this faint ring, now called the E Ring, and found another closer one known as the G Ring.

The E and G rings are examples of diffuse rings. In a diffuse ring, particles are far apart, and rare collisions between them can cause their individual orbits to become eccentric, inclined, or both. Because collisions are rare, the particles tend to remain in these disturbed orbits. Diffuse rings spread out horizontally and thicken vertically, sometimes without any obvious boundaries.

Diffuse rings contain tiny particles that show up best when the viewer is looking at these rings in the direction of the Sun. In contrast, larger objects such as pebbles and boulders are easiest to see when the light illuminating them is coming from behind the viewer. Dust and other small particles, however, stand out most strongly when you look *into* the light. For example, dust particles on your windshield appear brightest when you are driving toward the Sun. Photographers call this effect backlighting and often place their subjects in front of a bright light to highlight hair (**Figure 11.25a**). Backlighting happens when light falls on very small objects—those with dimensions a few times to several dozen times the wavelength of light. Cat fur and human hair is near the upper end of this range. Light falling on strands of hair tends to continue in the direction away from the source of illumination. Very little of the light is scattered off to the side, and almost none is scattered back toward the source.

Some of the dustier planetary rings are filled with particles that are just a few times larger than the wavelength of visible light. To a spacecraft approaching from the direction of the Sun, such rings may be difficult or even impossible to

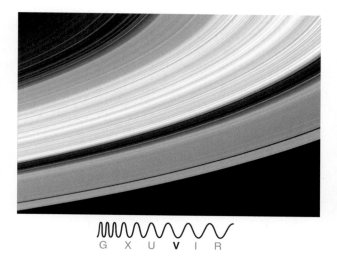

G X U V I R

**Figure 11.24** This *Cassini* image of the rings of Saturn shows so many ringlets and minigaps that it looks like a close-up of an old-fashioned phonograph record. The cause of most of this structure has yet to be explained in detail.

(a)

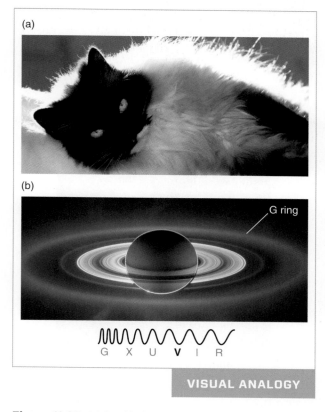

(b)

G ring

G X U V I R

**VISUAL ANALOGY**

**Figure 11.25** (a) Backlighting of hair creates a halo effect. (b) This backlit image of Saturn shows the G Ring.

**Figure 11.26** This artist's conception shows the highly inclined giant dust ring recently discovered around Saturn. This ring is so large that the rest of Saturn appears as a speck in the center (magnified in the inset).

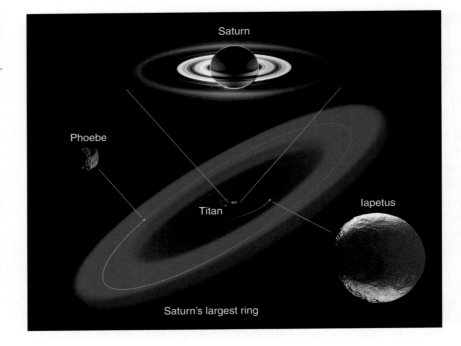

see. These tiny ring particles scatter very little sunlight back toward the Sun and the approaching spacecraft. However, when the spacecraft passes by the planet and looks backward in the general direction of the Sun, these same dusty rings suddenly appear as a circular blaze of light, much like a halo surrounding the nighttime hemisphere of the planet. Many planetary rings are best seen with backlighting, and some, such as Saturn's G Ring (see Figure 11.25b), have been observed only under these conditions. In 2009, astronomers using the infrared Spitzer Space Telescope discovered another diffuse ring around Saturn. This dusty ring is thicker than other rings, about 20 times larger than Saturn from top to bottom, and is tilted 27° with respect to the plane of the rest of the rings (**Figure 11.26**).

Although Saturn's bright rings are very wide—more than 62,000 km from the inner edge of the C Ring to the outer edge of the A Ring—they are extremely thin. Saturn's bright rings are no more than 100 meters thick and probably only a few tens of meters from their lower to upper surfaces. The diameter of Saturn's bright ring system is 10 million times the thickness of the rings. If the bright rings of Saturn were the thickness of a page in a book, six football fields laid end to end would stretch across them.

*Voyager 1* images showed that Saturn's F Ring is separated into several strands that appear to be intertwined and also display what appear to be a number of knots and kinks. Saturn's F Ring is now understood to be a dramatic example of the action of a pair of shepherd moons. The F Ring is flanked by Prometheus, a moon that orbits 860 km inside the ring, and Pandora, a moon that orbits 1,490 km on the outside, as seen in a more recent *Cassini* image (**Figure 11.27**). Both moons are irregular in shape, with average diameters of 85 and 80 km, respectively. Because of their relatively large size and proximity, the moons exert significant gravitational forces on nearby ring particles. The resulting tug-of-war between Prometheus pulling ring particles in its vicinity into larger orbits and Pandora drawing its neighboring particles into smaller orbits is the cause of the bizarre structure in the F Ring (**Process of Science Figure**).

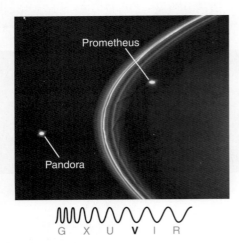

**Figure 11.27** This *Cassini* image shows Saturn's F Ring and its shepherd moons Pandora and Prometheus. You can also see some "kinks" in the inner ring.

# Process of Science

**Scientists expected to find the dust particles in the rings of Saturn moving on undisturbed orbits. Instead, the F Ring particles seemed to disobey the laws of physics!**

**The *Voyager 1* spacecraft discovers multiple intertwining strands, knots, and kinks in Saturn's F Ring.**

**The media proclaim: "The laws of physics are wrong!"**

**Scientists proclaim: "We must not have accounted for everything."**

**Further observations reveal previously unobserved moons that produce the unexpected behavior of F Ring particles.**

**Observers find more examples of deformed rings and the "shepherd moons" that cause them.**

**Theorists verify gravitational effects with simulations.**

**The theory that rings can be distorted by the gravitational influence of nearby moons becomes widely accepted as it repeatedly passes tests of observation and simulation.**

Scientists are excited by apparent violations of well-supported theories because it may lead to a new discovery. One consequence is that unexpected or contradictory results often receive more attention than confirming results.

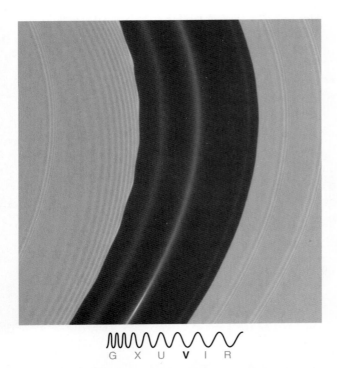

**Figure 11.28** In this *Cassini* high-resolution view, Saturn's Encke Gap reveals a scalloped pattern along its inner edge that is caused by the moon Pan.

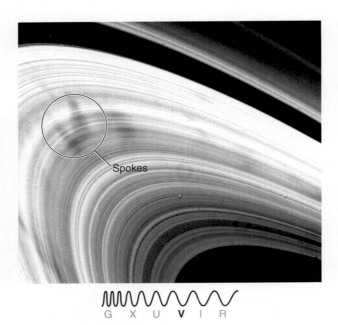

**Figure 11.29** Spokes in Saturn's B Ring appear dark in normal viewing but bright with backlighting.

The F Ring is not an isolated case. The 360-km-wide Encke Gap in the outer part of Saturn's A Ring contains two narrow rings that show bright knots and dark gaps—a structure that must be related to a 20-km-diameter moon named Pan that orbits within the gap. Small moons orbiting within ring gaps can also disturb ring particles along the edges of the gaps. **Figure 11.28** shows the scalloped pattern caused by Pan that is found along the inner edge of the Encke Gap. Similarly, the 7-km-diameter moon Daphnis disrupts the inner and outer edges of Saturn's 35-km-wide Keeler Gap, located near the outer edge of the A Ring.

*Voyager 1* and then *Cassini* observed dozens of dark, spokelike features in the outer part of Saturn's B Ring (**Figure 11.29**). These temporary features grow in a radial direction and last for less than half an orbit around Saturn, indicating that the particles in the spokes must be suspended above the ring plane, probably by electrostatic forces. One explanation is that when the charged particles interact with Saturn's magnetic field, the spokes rotate as the planet spins.

## Rings around the Other Outer Planets

Ring structure among the other giant planets is not as diverse as Saturn's. Most rings other than Saturn's are quite narrow, although a few are diffuse. When *Voyager 1* scientists looked at Jupiter's ring system with the Sun behind the camera, all they saw was a narrow, faint strand. But when *Voyager 2* looked back toward the Sun while in the shadow of the planet, Jupiter's rings suddenly blazed into prominence. **Figure 11.30a** shows a nearly edge-on, backlit view of Jupiter's rings taken by the *Galileo* spacecraft. Most of the material in Jupiter's rings is made up of fine dust dislodged by meteoritic impacts on the surfaces of Jupiter's small inner moons. The moons are shown orbiting among the rings in Figure 11.30b.

Of the 13 rings of Uranus, 9 are very narrow and widely spaced relative to their widths (see Figure 11.21c). Most are only a few kilometers wide, but they are many hundreds of kilometers apart. The two rings discovered by the Hubble Space Telescope in 2005 (**Figure 11.31**) are much wider and more distant than the narrow rings. The most prominent ring of Uranus, the Epsilon Ring, is eccentric and the widest of the planet's inner narrow rings, varying in width between 20 and 100 km. The innermost ring is wide and diffuse, with an undefined inner edge. As with Saturn's D Ring, material in this ring may be spiraling into the top of the planetary atmosphere. When viewed under backlit conditions by *Voyager 2*, the space *between* the rings of Uranus turned out to be filled with dust, much as in Jupiter's ring system.

The rings in the Encke Gap are unusual but not unique. If shepherd moons are in eccentric or inclined orbits, they cause the confined ring also to be eccentric or inclined. This is the case for the Epsilon Ring of Uranus. Because shepherd moons can be so small, they often escape detection. According to current theories of ring dynamics, a number of still-unknown shepherd moons must be interspersed among the ring systems of the outer Solar System.

For a while, Neptune seemed to be the only giant planet devoid of rings. Then, in the early to mid 1980s, occultation searches by teams of astronomers began yielding confusing results. Several occultation events that appeared to be due to rings were seen on only one side of the planet. The astronomers concluded that Neptune was surrounded not by complete rings but rather by several arclike ring

(a)

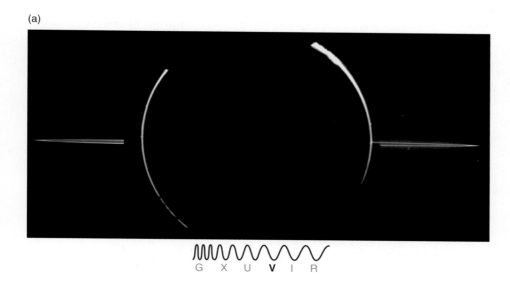

(b)

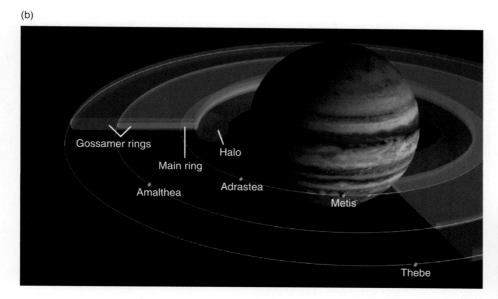

**Figure 11.30** (a) This backlit *Galileo* image of Jupiter's rings also shows the forward scattering of sunlight by tiny particles in the upper layers of Jupiter's atmosphere. (b) A diagram of the Jupiter ring system and the small moons that form the rings.

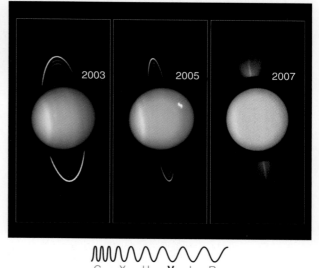

**Figure 11.31** The appearance of rings depends dramatically on lighting conditions and the angle from which they are seen. Earth's view of the rings of Uranus changes over the course of several years, as shown here.

segments. When *Voyager 2* reached Neptune in 1989 it was determined that Neptune's rings are complete. The **ring arcs** are high-density segments within one of its narrow rings. All of Neptune's rings are faint and, with the exception of the ring arcs, they contain too little material to be detected by the stellar occultation technique.

Four of Neptune's six rings are very narrow, similar to the 13 narrow rings surrounding Uranus. The other two have widths of a few thousand kilometers (see Figure 11.21d). Neptune's rings are named for 19th century astronomers who made major contributions to Neptune's discovery. Of these, the Adams Ring attracts the greatest attention. Much of the material in the Adams Ring is clumped together into several ring arcs. These high-density ring segments extend over lengths of 4,000–10,000 km, yet are only about 15 km wide. When first discovered, the ring arcs were a puzzle, because mutual collisions among their particles

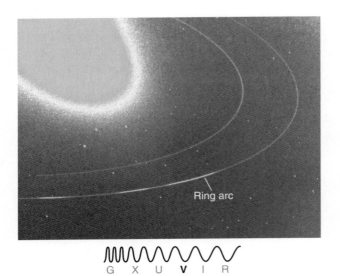

**Figure 11.32** This *Voyager 2* image shows the three brightest arcs in Neptune's Adams Ring. Neptune itself is very much overexposed in this image.

**Figure 11.33** This artist's conception shows a hypothetical Earth-like moon around a Saturn-like exoplanet.

should cause the particles to be spread more or less uniformly around their orbits. Most astronomers now attribute this clumping to orbital resonances with the moon Galatea that orbits just inside the Adams Ring (**Figure 11.32**). Images obtained by the Hubble Space Telescope in 2004 and 2005 compared with those taken by *Voyager 2* in 1989 show that some parts of the Neptune ring arcs are unstable. Slow decay is seen in two of the arcs, Liberté and Courage, suggesting that they may disappear entirely before the end of the century. Uranus's Lambda Ring and Saturn's G Ring also show ring arcs.

## Moons and Rings around Extrasolar Planets

In previous chapters, we have shown extrasolar planets are common. Large and small planets, many in multiplanet systems, have been detected within our galaxy. If our Solar System is typical, then we might expect that other planetary systems contain planets with rings and planets with large moons, perhaps with geological activity and water. However, the identification of extrasolar moons and rings is at the limit of current astronomical instrumentation. As of this writing, there have not been confirmed detections.

The proposed methods for detecting *exomoons* are similar to those for detecting exoplanets. Recall from Chapter 7 that *Kepler* detects planets through the transit method (see Figures 7.19 and 7.20). The presence of a moon near the planet can slightly alter the depth and duration of the change in the light curve. The moon is likely in a different place in its orbit each time the planet orbits the star, so these alterations in the light curve will be different for each cycle. Over the course of several cycles, the signature of a moon might be detected. It is also possible that astronomers could detect a large exomoon as the moon itself transits its star. Or a large moon could make its planet "wobble" in its orbit, and this wobble could be detected in the transit signal. NASA supercomputers are being used to analyze the large database from *Kepler* to look for signatures of exomoons.

Similarly, a large extrasolar planet with an extensive ring system, especially if the ring system has gaps, might be detectable through changes in the transit signal. The depth and length of the changes in the star's light curve could indicate the presence of such a system. **Figure 11.33** is an artist's conception of an extrasolar planet with rings and a large exomoon.

### CHECK YOUR UNDERSTANDING 11.5

If you wanted to search for faint rings around a giant planet by sending a spacecraft on a flyby, it would be best to make your observations: (a) as the spacecraft approached the planet; (b) after the spacecraft passed the planet; (c) while orbiting the planet; (d) during the closest flyby; (e) while orbiting one of its moons.

# Origins

## Extreme Environments

During the 1980s and 1990s, as the *Voyager* spacecraft were exploring the outer Solar System, biologists back on Earth were identifying strange forms of life. Off the coast of the Galápagos Islands, 2,500 meters beneath the ocean's surface, plates grind against one another, creating friction, high temperatures, and seafloor volcanism. Mineral-rich, superheated water pours out of hydrothermal vents. The surrounding water contains very little dissolved oxygen. No sunlight reaches these depths, yet in the total darkness of the ocean bottom, life abounds. From tiny bacteria to shrimp to giant clams and tube worms, sea life thrives in this severe environment. In the complete absence of sunlight, the small, single-celled organisms at the bottom of the local food chain get their energy from *chemosynthesis*, a process by which inorganic materials are converted into food through the use of chemical energy. Biologists call these life-forms *extremophiles.*

Similarly, robust types of bacteria are found flourishing in the scalding waters of Yellowstone's hot springs; in the bone-dry oxidizing environment of Chile's Atacama Desert; and in the Dead Sea, where salt concentrations run as high as 33 percent. Bacteria have even been found in core samples of ancient ice 3,600 meters below the surface of the East Antarctic ice sheet. When it comes to harsh habitats, life is amazingly adaptable. If life can exist under such extreme conditions on Earth, might it also exist on those moons of the giant planets that have the ingredients necessary for life on Earth: liquid water, an energy source, and the presence of organic compounds?

When scientists realized that Mars and Venus were not as Earth-like as once imagined, prospects for finding life elsewhere in the Solar System seemed dim. Now astrobiologists are turning their attention to some of the small worlds that circle the giant planets far from the Sun. These moons may supply clues about the history of life in the Solar System. Their environments may be similar to some of the ecological niches on Earth that support extremophiles. The conditions necessary to create and support life on Earth—liquid water, heat, and organic material—could all be present in oceans on some of these moons.

Enceladus, Saturn's geologically active ice moon, spews salty water-ice grains, indicating liquid water below the surface. Perhaps its south polar region is a habitable zone. The fractured ice floes that make up the surface of Jupiter's Europa may cover an ocean warmed and enriched by geothermal vents not unlike those that dot the floors of Earth's oceans. Like Europa, Callisto also shows magnetic variability, a possible signature of a salty ocean.

Saturn's Titan has an atmosphere and a possible subsurface ocean, and it is similar in several ways to a much earlier Earth.

The presence of comet-borne organic material in large bodies of water on Europa, Callisto, Enceladus, or Titan cannot yet be confirmed. Methane can arise from biological processes or it can come from chemical or geochemical processes, so its detection on Enceladus and Titan is tantalizing but is not evidence of life. In addition to the presence of methane, spectroscopy reveals organic gases in Titan's massive atmosphere. Titan's nitrogen atmosphere contains compounds of biological interest. For example, five molecules of hydrogen cyanide (HCN) will spontaneously combine to form adenine, one of the four primary components of DNA and RNA. HCN is also a building block of amino acids, which in turn combine to form proteins. Photodissociation and recombination of these various gases produce complex organic molecules that then rain out onto Titan's surface as a frozen tarry sludge. Biochemists think that many of these substances are biological precursors, similar to the organic molecules that preceded the development of life on Earth.

Astronomers anticipate future exploration of these moons, which may provide fascinating clues to the origins of terrestrial life.

# Possible New Moon Forming around Saturn

By **Science@NASA**

NASA's *Cassini* spacecraft has documented the formation of a small icy object within the rings of Saturn. Informally named "Peggy," the object may be a new moon. Details of the observations were published online today by the journal *Icarus*.

"We have not seen anything like this before," said Carl Murray of Queen Mary University of London, and the report's lead author. "We may be looking at the act of birth, where this object is just leaving the rings and heading off to be a moon in its own right."

Images taken with *Cassini*'s narrow-angle camera on April 15, 2013, show disturbances at the very edge of Saturn's A Ring—the outermost of the planet's large, bright rings (**Figure 11.34**). One of these disturbances is an arc about 20 percent brighter than its surroundings, 750 miles (1,200 kilometers) long and 6 miles (10 kilometers) wide. Scientists also found unusual protuberances in the usually smooth profile at the ring's edge. Scientists believe the arc and protuberances are caused by the gravitational effects of a nearby object.

The object is not expected to grow any larger, and may even be falling apart. But the process of its formation and outward movement aids in our understanding of how Saturn's icy moons, including the cloud-wrapped Titan and ocean-holding Enceladus, may have formed in more massive rings long ago. It also provides insight into how Earth and other planets in our Solar System may have formed and migrated away from our star, the Sun.

"Witnessing the possible birth of a tiny moon is an exciting, unexpected event," said *Cassini* Project Scientist Linda Spilker, of NASA's Jet Propulsion Laboratory (JPL) in

**Figure 11.34** The small object "Peggy" is at the edge of the ring.

Pasadena, California. According to Spilker, *Cassini*'s orbit will move closer to the outer edge of the A Ring in late 2016 and provide an opportunity to study Peggy in more detail and perhaps even image it.

Peggy is too small to see in images so far. Scientists estimate it is probably no more than about a half mile in diameter. Saturn's icy moons range in size depending on their proximity to the planet—the farther from the planet, the larger. And many of Saturn's moons are composed primarily of ice, as are the particles that form Saturn's rings. Based on these facts, and other indicators, researchers recently proposed that the icy moons formed from ring particles and then moved outward, away from the planet, merging with other moons on the way.

"The theory holds that Saturn long ago had a much more massive ring system capable of giving birth to larger moons," Murray said. "As the moons formed near the edge, they depleted the rings."

It is possible the process of moon formation in Saturn's rings has ended with Peggy, as Saturn's rings now are, in all likelihood, too depleted to make more moons. Because they may not observe this process again, Murray and his colleagues are wringing from the observations all they can learn.

1. What does this article suggest about the link between the formation of moons and rings?
2. Is a moon this size likely to be geologically active? Explain.
3. Some, but not all, of Saturn's moons are listed in Appendix 4 (and on the websites in end-of-chapter question 47). Is it correctly stated that the farther the icy moon is from the planet, the larger it is in size?
4. Looking at the names of the other moons of Saturn, is this moon likely to keep the name "Peggy"?
5. Has *Cassini* revisited this area of the rings? Do a search to see what it found.

# Summary

The moons of the outer Solar System are composed of rock and ice. A few moons are geologically active, but most are dead. All four giant planets have ring systems, which are transitory, created from and maintained by moons also in orbit around these planets. Scientists are excited about the evidence for oceans on several of the moons and the possibility that some form of life may exist in these oceans.

**LG 1**  **Compare and contrast the formation and orbits of regular and irregular moons.** Most of the regular moons were formed along with their parent planets and have short and nearly circular orbits. Irregular moons were captured later, have more elongated orbits, and often travel in the opposite direction of the planet's rotation. Observations of the orbits of moons can be used to find the masses of their host planets.

**LG 2**  **Describe the evidence for geological activity and liquid oceans on some of the moons.** Jupiter's Io is the most volcanically active body in the Solar System. Jupiter's moon Europa contains an enormous subsurface ocean and probably has some geological activity. Saturn's moon Titan has lakes of liquid methane and perhaps a deep, salty ocean. Saturn's moon Enceladus and Neptune's moon Triton have cryovolcanoes. The large Galilean moons of Jupiter, Ganymede and Callisto, may also have subsurface oceans. Some moons were geologically active in the past, as indicated by crater scars and areas smoothed by flowing fluids. Moons that have always been geologically dead show nothing but impact craters on their surfaces.

**LG 3**  **Describe the composition, origin, and general structure of the rings of the giant planets.** Rings are formed of countless numbers of particles all in the same plane, held to the host planet by gravity. Some rings form when moons cross a planet's Roche limit. The composition of these moons determines the composition of the rings that form from them. Shepherd moons often maintain and shape rings by gravitationally pulling and pushing these particles as they pass by. Ring particles also interact gravitationally with each other. Some rings may be transient features held in place by moons. Saturn's bright rings and its E Ring are made primarily of water ice: the rings of the other planets are composed of darker materials.

**LG 4**  **Explain the role gravity plays in the structure of the rings and the behavior of ring particles.** Saturn's ring system is the most complex, and it is the best laboratory for understanding gravitational interactions between moons and rings, interactions between rings, and ring formation and dissipation. Gravity holds the ring particles in orbit around the planet, and gravitational interactions with moons determines the size and shape of rings.

## ? UNANSWERED QUESTIONS

- What is the source of Titan's nitrogen atmosphere? One group of experimenters studied this by blasting a laser at water-ammonia ($H_2O$-$NH_3$) ice to simulate cometary impacts and see whether nitrogen gas ($N_2$) forms. They concluded that the observed amount of $N_2$ in Titan's atmosphere could have been created from ammonia ice in this way. Another theory is that these gases were accreted during the formation of the moon. Using data from the *Huygens* probe, other researchers conclude that if Titan had differentiated like Ganymede, then hydrothermal activity released the gases from a hot core. Astronomers want to understand why Titan has an atmosphere and the other larger moons do not.

- What is the origin of Saturn's brightest rings? One early hypothesis is that the rings come from a moon that approached too close to the planet. But Saturn's moons are composed of rock and ice, and the rings are solely ice. Some recent computer models start with a differentiated moon the size of Titan that had a rock and iron core and a large, icy mantle. As the moon in the model slowly migrates toward Saturn and crosses the Roche limit, tidal forces rip away its water ice, but not the rocky core. According to this model, the core might have continued migrating inward until it fell into Saturn, and the ice would have formed Saturn's ring. As time went on, the ring spread, and as material crossed the Roche limit outward, Saturn's small moonlets formed. Computers are only now getting fast enough fully to test these models, which suggest a unified origin for many of the moons and rings.

- When will there be robotic space missions to the outer moons? Several missions have been proposed to study Europa, Ganymede, Titan, and Enceladus. Missions to Europa to investigate the liquid ocean beneath the icy surface have the highest priority. One proposed mission would crash a probe onto the surface of Europa and collect and analyze the debris plume. Another would have a probe land on the ice and drill into it until it reached liquid water. (This has been done at Lake Vostok, an under-ice lake in Antarctica.) Proposed missions to Titan include a balloon that would hover in its atmosphere for many months taking data or a probe that would land and float on one of the lakes. NASA and the European Space Agency are discussing possible missions for launch in the mid-2020s.

# Questions and Problems

## Test Your Understanding

1. Categorizing moons by geological activity is helpful because
   a. comparing them reveals underlying physical processes.
   b. geological activity levels drop with distance from the Sun.
   c. geological activity determines the size and composition of the moons.
   d. most moons are very similar to each other.

2. Why are Ganymede and Callisto geologically dead while the other two Galilean moons of Jupiter are active?
   a. they are larger
   b. they are farther from Jupiter
   c. they are more massive
   d. they have retrograde orbits

3. Moons of outer planets may provide a home for life because
   a. some have liquid water.
   b. some have organic molecules.
   c. some have an interior source of energy.
   d. all of the above

4. Io has the most volcanic activity in the Solar System because
   a. it is continually being bombarded with material in Saturn's E Ring.
   b. it is one of the largest moons and its interior is heated by radioactive decays.
   c. of gravitational friction caused by the moon Enceladus.
   d. its interior is tidally heated as it orbits around Jupiter.
   e. the ice on the surface creates a large pressure on the water below.

5. Gravitational interactions with moons produce
   a. fine structure within rings.
   b. short-lived rings.
   c. smoothed-out rings.
   d. rings with spokes.

6. Saturn's bright rings are located within the Roche limit of Saturn. This fact supports the theory that these rings (select all that apply)
   a. formed of moons torn apart by tidal stresses.
   b. formed at the same time that Saturn formed.
   c. are relatively recent.
   d. are temporary.

7. The story of the F Ring of Saturn is an example of
   a. an unexplained phenomenon.
   b. media bias.
   c. the self-correcting nature of science.
   d. a violation of causality.

8. If a moon revolves opposite to its planet's rotation, it probably
   a. was captured after the planet formed.
   b. had its orbit altered by a collision.
   c. has a different composition from other moons.
   d. formed very recently in the Solar System's history.

9. Under what lighting conditions are the tiny dust particles found in some planetary rings best observed?
   a. viewed from the shadowed side of the planet, looking toward deep space
   b. viewed from the shadowed side of the planet, looking toward the Sun
   c. viewed from the near side of the planet, looking toward deep space
   d. viewed from the near side of the planet, looking toward the Sun

10. Planets in the outer Solar System have more moons than those in the inner Solar System because
    a. the solar wind was weaker there.
    b. there was more debris around the outer planets when they were forming.
    c. the outer planets captured most of their moons.
    d. there were more planetesimal collisions far from the Sun.

11. The difference between a moon and a planet is that
    a. moons orbit planets, whereas planets orbit stars.
    b. moons are smaller than planets.
    c. moons and planets have different compositions.
    d. moons and planets formed in different ways.

12. Scientists determine the geological history of the moons of the outer planets from
    a. seismic probing.
    b. radioactive dating.
    c. surface features.
    d. time-lapse photography.

13. The energy that keeps Io's core molten comes from
    a. the Sun.
    b. radioactivity in the core.
    c. residual heat from the collapse.
    d. Jupiter's gravity.

14. We classify moons as formerly active if they
    a. are completely covered in craters.
    b. have no young craters.
    c. have regions with few craters.
    d. have regions with no craters.

15. Volcanoes on Enceladus affect the E Ring of Saturn by
    a. pushing the ring around.
    b. stirring the ring particles.
    c. supplying ring particles.
    d. dissipating the ring.

## Thinking about the Concepts

16. Explain the process that drives volcanism on Jupiter's moon Io.

17. Describe cryovolcanism and explain its similarities and differences with respect to terrestrial volcanism. Which moons show evidence of cryovolcanism?

18. Discuss evidence supporting the idea that Europa might have a subsurface ocean of liquid water.

19. Titan contains abundant amounts of methane. What process destroys methane in this moon's atmosphere?

20. In certain ways, Titan resembles a frigid version of the early Earth. Explain the similarities.

21. Some moons display signs of geological activity in the past. Identify some of the evidence for past activity.

22. Why do the outer planets but not the inner planets have rings? Describe a ground-based technique that led to the discovery of rings around the outer planets.

23. What are ring arcs and where are they found?

24. Identify and explain two possible mechanisms that can produce planetary ring material.

25. Explain two mechanisms that create gaps in Saturn's bright-ring system.

26. Describe ways in which diffuse rings differ from other planetary rings.

27. In Chapter 1, we stated that "all scientific theories are provisional." Explain how the discovery of the detailed structure of Saturn's F Ring (as described in the Process of Science Figure) challenged a scientific theory, and how this apparent conflict was ultimately resolved.

28. Astronomers think that most planetary rings eventually dissipate. Explain why the rings do not last forever. Describe and explain a mechanism that keeps planetary rings from dissipating.

29. Name one ring that might continue to exist indefinitely, and explain why it could survive when others might not.

30. Make a case for sending a space mission to one of the moons. Which moon would you choose to explore, and what types of observations would you try to obtain?

## Applying the Concepts

31. Io has a mass of $8.9 \times 10^{22}$ kg and a radius of 1,820 km.
    a. Using the formula provided in Working It Out 11.3, calculate Io's escape velocity.
    b. How does Io's escape velocity compare with the vent velocities of 1 km/s from its volcanoes?

32. Use the value of $P^2/A^3$ for Europa, as in Working It Out 11.1, to compute the mass of Jupiter.

33. Follow Working It Out 11.1 to compute the mass of Saturn using one of its moons.

34. Study Figure 11.2.
    a. Are the scales on (a) and (b) linear or logarithmic?
    b. About how much larger is the space shown in (b) than in (a)?

35. Planetary scientists have estimated that Io's extensive volcanism could be covering the moon's surface with lava and ash to an average depth of up to 3 millimeters (mm) per year.
    a. Io's radius is 1,820 km. If you assume Io is a sphere, what are its surface area and volume?
    b. What is the volume of volcanic material deposited on Io's surface each year?
    c. How many years would it take for volcanism to perform the equivalent of depositing Io's entire volume on its surface?
    d. How many times might Io have "turned inside out" over the age of the Solar System?

36. Consider the formula for tidal forces in Working It Out 11.2. If the radius of the moon increases but its mass stays the same, what happens to the tidal force? If the radius of the moon's orbit decreases, what happens to the tidal force? If the mass of the central planet increases, what happens to the tidal force?

37. Follow Working It Out 11.2 to compare the tidal force between Jupiter and Io with the tidal force between Earth and its Moon.

38. Imagine that a 60-kg astronaut is spacewalking outside the International Space Station, 380 km above Earth. Follow Working It Out 11.2 to find the tidal force on the astronaut, assuming she is oriented with her feet toward the center of Earth.

39. Assuming that all other numbers are held constant, make a graph of the tidal force versus the distance between a planet and its moon. On the same graph, plot the gravitational force (which falls off like $1/d^2$). Compare the two graphs to determine the relative importance of tidal and gravitational forces at various distances.

40. Particles at the very outer edge of Saturn's A Ring are in a 7:6 orbital resonance with the moon Janus. If the orbital period of Janus is 16 hours 41 minutes ($16^{\mathrm{h}}\ 41^{\mathrm{m}}$), what is the orbital period of the outer edge of Ring A?

41. Follow Working It Out 11.3 to find the escape velocity from Saturn's moon Janus.

42. The inner and outer diameters of Saturn's B Ring are 184,000 and 235,000 km, respectively. If the average thickness of the ring is 10 meters and the average density is 150 kilograms per cubic meter ($kg/m^3$), what is the mass of Saturn's B Ring?

43. The mass of Saturn's small, icy moon Mimas is $3.8 \times 10^{19}$ kg. How does this mass compare with the mass of Saturn's B Ring, as calculated in question 42? Why is this comparison meaningful?

44. The inner and outer diameters of Saturn's B Ring are 184,000 and 235,000 km, respectively. Use this information to find the ratio of the periods of particles at these two diameters. Does the B Ring orbit like a solid disk or like a collection of separate particles?

45. Consider the escape velocity equation in Working It Out 11.3. For more massive planets, is the escape velocity higher or lower? For larger planets, is the escape velocity higher or lower? If you know the escape velocity of a planet, what other piece of information do you need in order to find the planet's mass?

## USING THE WEB

46. Go to *Sky & Telescope*'s "Jupiter's Moons" Web page (http://www.skyandtelescope.com/wp-content/observing-tools/jupiter_moons/jupiter.html). Enter your date and time. Where are the four Galilean moons? Keep clicking on "[plus]1 hour" to see how their positions change over time. Which moon passes in front of (transits) Jupiter? If possible, observe these moons for a couple of nights through a small telescope, binoculars, or telephoto camera lens. Sketch the positions of the moons.

47. Look at the updated lists of giant planet moons on NASA's "Our Solar System" website (http://solarsystem.nasa.gov/planets; click on the planet's name and "Moons") or on the Carnegie Institution of Washington Department of Terrestrial Magnetism's (DTM) "Jupiter Satellite Page" (http://www.dtm.ciw.edu/users/sheppard/satellites). What are two of the more recently discovered moons on one of the planets? Are the orbits retrograde? What are the eccentricities and inclinations of the orbits? Where would these new moons fit on the graph of orbits on the DTM's website? Where did the names come from? Why are some of the more recent moons labeled "provisional"?

48. Go to the website for the *Cassini* mission (http://saturn.jpl.nasa.gov). Is *Cassini* still making observations? Look at a *Cassini* image of one of Saturn's moons. What does this image reveal about Saturn? Watch the video at http://saturn.jpl.nasa.gov/video/videodetails/?videoID=232 to listen to the "hiss" from the aurora. What moon is the cause of Saturn's aurora?

49. Missions to the moons:
    a. Do a search for the European Space Agency *JUICE* (*JUpiter ICy moons Explorer*) mission, scheduled for launch in the mid-2020s. Which moons will this mission study? What are the goals of the mission? What is the status of this project?
    b. Go to the "Destination Europa" website (http://europa.seti.org/). What are they advocating for? Do a search on "Europa Clipper" and "Europa mission" to see the status of proposed NASA space missions to Europa. Have any of them been approved for funding? What will the mission study?

50. Do a search to see if moons and rings have been confirmed on extrasolar planets. Why is this of interest to astronomers?

## smartwork5

If your instructor assigns homework in Smartwork5, access your assignments at digital.wwnorton.com/astro5.

# EXPLORATION

digital.wwnorton.com/astro5

## Part A: Finding the Image Scale

Finding the scale of an image is like finding the scale on a map. On a map, each inch or centimeter represents miles or kilometers of actual space. The same thing is true in an image. If you take a picture of a meter stick and then measure the meter stick in the picture to be 10 cm long, you know that 10 cm in the picture represents 1 meter of actual space, and 1 cm in the picture represents 10 cm of actual space.

To find the scale, you must compare the size of something in the image with its actual size in space. In **Figure 11.35**, the moon Io is the known object.

**1** Use a ruler to measure the diameter of this image of Io in millimeters.

_____

**2** Estimate the error in your measurement. (How far off could your measurement be?)

_____

**3** Find the radius from this diameter.

_____

**4** Look up the actual radius of Io (in kilometers) in Appendix 4 or online.

_____

**5** Find the image scale, *s*, as follows:

$$s = \frac{\text{Actual size of object}}{\text{Size of object on the image}}$$

_____

**6** What are the units of this image scale?

_____

## Part B: Finding the Sizes of Features on Io

**7** There is a geyser near the center of Io, surrounded by a black circle and a white ring. This is Prometheus Patera. What is the diameter of the black ring around Prometheus in this image (in millimeters)? (Do not use the inset image.)

_____

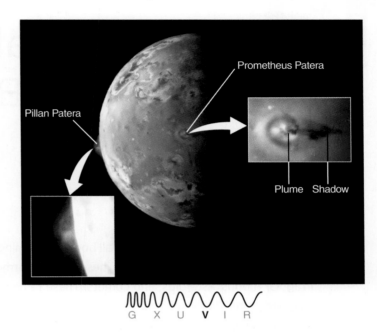

Figure 11.35 A *Galileo* image of Jupiter's moon Io, a very active moon. This image offers an opportunity to make some measurements.

**8** Multiply the measured diameter by the image scale to find the actual size of the circle around Prometheus. Identify something on Earth that is about the same size.

_____

**9** On the limb of the moon, there is a purple plume from the eruption of a sulfur geyser. Follow steps 7 and 8 to find the height of that plume. Identify something on Earth that is about the same height.

_____

There are at least two sources of error in this measurement. One is the error of measurement—how well you use a ruler, and how accurately you determined the top and the bottom of the plume. The other source of error is uncertainty about whether the plume is *exactly* at the limb of the moon. If the eruption occurred on the far side of Io, would your calculation be an overestimate or underestimate of the plume height?

# 12

# Dwarf Planets and Small Solar System Bodies

In this chapter, we explore the small bodies remaining from the formation of the Solar System (see Chapter 7). These bodies formed today's dwarf planets, irregular moons, asteroids, and comets. These remaining planetesimals, and the fragments that some of them continually create, have revealed much about the physical and chemical conditions of the earliest moments in the history of the Solar System and how the Solar System formed and evolved. In addition, these planetesimals are important because some fraction of the water, gases, and organic material found on Earth and in the inner Solar System came from comets and asteroids.

## LEARNING GOALS

By the conclusion of this chapter, you should be able to:

**LG 1** List the categories of small bodies and identify their locations in the Solar System.

**LG 2** Describe the defining characteristics of the dwarf planets in the Solar System.

**LG 3** Describe the origin of the different types of asteroids, comets, and meteorites.

**LG 4** Explain how asteroids, comets, and meteoroids provide important clues about the history and formation of the Solar System.

**LG 5** Describe what has been learned from observations of recent impacts in the Solar System.

The European Space Agency landed a probe on Comet 67P/Churyumov-Gerasimenko in 2014. ▶ ▶ ▶

Why land a
spacecraft on
a comet?

327

# 12.1 Dwarf Planets May Outnumber Planets

Recall from Chapter 7 that very early in the history of the Solar System—at the same time that the Sun was becoming a star—tiny grains of primitive material stuck together to produce swarms of small bodies called *planetesimals*. Those that formed in the hotter, inner part of the Solar System were composed mostly of rock and metal; those in the colder, outer part were composed of ice, organic compounds, and rock. Some of the objects collided to become planets and moons. However, many are still present, and they remain a small but scientifically important component of the present-day Solar System.

Dwarf planets, asteroids, Kuiper Belt objects, comets, and meteoroids are smaller than planets and orbit the Sun. Dwarf planets are found in the asteroid belt and in the **Kuiper Belt**. The asteroid belt in the region between the orbits of Mars and Jupiter contains most of the asteroids in the Solar System. The Kuiper Belt is a disk-shaped population of comet nuclei extending from Neptune's orbit to perhaps thousands of astronomical units from the Sun.

The dwarf planets orbit the Sun and have round shapes, but because they have relatively small mass, they have not cleared the area around their orbits. As of this writing, there are five officially recognized dwarf planets in the Solar System: Pluto, Eris, Haumea, Makemake, and Ceres (their properties are tabulated in Appendix 4). Ceres is a large object in the main asteroid belt; the other dwarf planets are found in the Kuiper Belt. There are many dwarf planet candidates, but their shapes have not yet been measured well enough for certain classification.

## Pluto

Throughout the 19th century, discrepancies were observed between the observed and predicted orbital positions of Uranus and Neptune. Early in the 20th century, astronomers hypothesized that an unseen body was perturbing the orbits of these planets. They called this body Planet X and estimated that it had 6 times Earth's mass and was located beyond Neptune's orbit. Astronomer Clyde W. Tombaugh (1906–1997) discovered Planet X in 1930, not far from its predicted position. It became the Solar System's ninth planet and was named Pluto for the Roman god of the underworld. However, observational evidence soon indicated that the mass of Pluto was far too small to have produced the perturbations in the orbits of Uranus and Neptune. When astronomers reanalyzed the 19th century observations, they found that the orbital "discrepancies" were a mistake. Pluto's discovery thus turned out to be a coincidence.

Pluto's orbit is 248 Earth years long and quite elliptical, and it is tilted with respect to the plane of the Solar System. Its orbit periodically crosses inside of Neptune's nearly circular orbit—from 1979 to 1999, Pluto was closer to the Sun than Neptune. Pluto is only two-thirds as large as our Moon. Pluto has five known moons, the largest of which is Charon, about half the size of Pluto. The total mass of the Pluto-Charon system is 1/400 that of Earth, or 1/5 the mass of the Moon. As with Uranus, the plane of Pluto's equator is nearly perpendicular to its orbital plane. Pluto and Charon are a tidally locked pair: each has one hemisphere that always faces the other (**Figure 12.1a**).

Pluto and Charon were not on *Voyager's* route through the Solar System, so until recently there was limited information about their surface properties or geological history. This changed when NASA's *New Horizons* flyby spacecraft passed within 12,500 km to Pluto in July 2015. Astronomers were surprised by the varied surface features seen in *New Horizons* images of Pluto, including a large bright region primarily composed of carbon monoxide ice (Figure 12.1b). Figure 12.1c shows 3500-meter high ice mountains and icy plains. Pluto's surface contains an icy mixture of frozen water, carbon dioxide, methane, and carbon monoxide, with flowing nitrogen ice. Pluto has a thin atmosphere of nitrogen, methane, ethane, and carbon monoxide: these gases freeze out of the atmosphere when Pluto is more distant from the Sun and therefore cooler. Charon has no atmosphere. Its surface has deep canyons, which might have formed as an ancient ocean froze and pushed surface outwards (Figure 12.1d). Few craters are seen on either body, which suggests the surfaces are relatively young, but it is not yet known if there was recent geological activity or impacts.

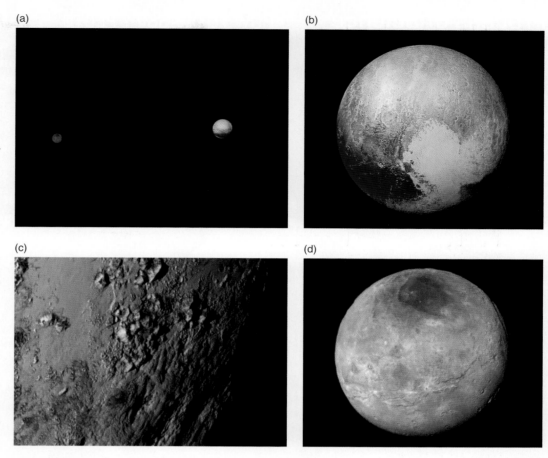

**Figure 12.1** (a) Pluto (right) and Charon (to scale). All images are from the New Horizons spacecraft flyby in 2015. (b) Enhanced color image of Pluto. (c) Young mountains of ice on Pluto rise to 3500m, suggesting recent geological activity. (d) Image of Pluto's largest moon Charon.

As astronomers discovered more about Pluto and other objects beyond Neptune's orbit, some questioned Pluto's classification as a planet, and a debate ensued. In 2005, astronomers identified an object more distant than Pluto, later named Eris, and then Eris's moon, Dysnomia. Observations of Dysnomia's orbit yielded a mass for Eris, which turned out to be about 28 percent greater than Pluto's mass. Pluto and Eris have similar nitrogen and methane abundances and a relatively large moon. At this point, the inevitable question emerged: Should astronomers consider Eris to be the Solar System's tenth planet? Or should neither Pluto nor Eris be called a planet? The International Astronomical Union (IAU) made its decision in August 2006 (**Process of Science Figure**): Pluto is round like the classical planets, but it is not able to clear its neighborhood, so it was reclassified as a dwarf planet (details of the IAU resolution can be found in Appendix 9).

## Eris, Haumea, and Makemake

These three dwarf planets were discovered in the 21st century. Eris is about the same size as Pluto but is more massive. Eris also has a relatively large moon, called Dysnomia. The highly eccentric orbit of Eris shown in **Figure 12.2** carries it from 37.8 astronomical units (AU) out to 97.6 AU away from the Sun, with an orbital

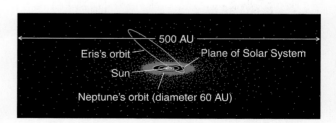

**Figure 12.2** Eris's orbit is both highly eccentric and highly inclined to the rest of the Solar System.

# Process of Science

## HOW TO CLASSIFY PLUTO

Pluto's reclassification from planet to dwarf planet in 2006 got a lot of publicity. The reclassification of Pluto is a clear example of the scientific method in practice.

Pluto orbits the Sun.

### 1930
Pluto is discovered. It is classified as a planet because it orbits the Sun.

Pluto's composition is not like other planets.

Pluto orbits the Sun.

Pluto crosses Neptune's orbit.

Pluto has a moon nearly as large as itself.

### 1978
Pluto's moon Charon is discovered. As information about Pluto accumulates, it becomes clear that Pluto is different from the other planets.

Pluto orbits the Sun.

### 2006
In light of recent discoveries, the International Astronomical Union redefines the word *planet*. Pluto falls outside the new definition and is classified as a "dwarf planet."

Several other Pluto-like objects are discovered. One is larger than Pluto.

Pluto's orbit is inclined significantly.

Scientific decision making must follow the weight of the evidence.

## 12.1 Working It Out Eccentric Orbits

Many of the objects discussed in this chapter are far from the Sun, and their complete orbits take many years. But observing a complete orbit is not necessary for determining an object's semimajor axis and eccentricity: these values can be obtained from watching how the object moves in just a fraction of its orbit. Astronomers can calculate the orbits of distant objects as they approach the Sun in a highly elliptical orbit and determine if they will come near to Earth.

Recall Kepler's law for objects orbiting the Sun: $P^2 = A^3$. **Figure 12.3** shows how eccentricity is defined mathematically as the distance from the center of the orbit to one focus (the Sun) divided by the semimajor axis ($A$). (The orbits of the planets in the Solar System range from nearly circular for Venus to an eccentricity [$e$] of 0.2 for Mercury.) The types of objects discussed in this chapter generally have higher eccentricities.

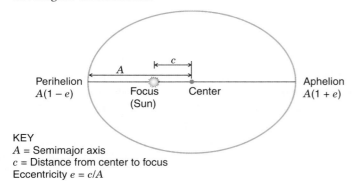

KEY
$A$ = Semimajor axis
$c$ = Distance from center to focus
Eccentricity $e = c/A$

**Figure 12.3** This drawing of an elliptical orbit shows eccentricity, aphelion, and perihelion.

We can relate the eccentricity to the closest approach and the farthest distance in the orbit, as seen in Figure 12.3, so that the object's closest approach to the Sun, its **perihelion**, equals $A(1 - e)$, and the object's farthest distance from the Sun, its **aphelion**, equals $A(1 + e)$. So if we know the semimajor axis and eccentricity of an orbit, we can calculate how close to and how far away from the Sun an object's orbit takes it.

For a first example, consider the orbit of dwarf planet Eris. The eccentricity of Eris's orbit is 0.44, and the semimajor axis of its orbit ($A$) is 67.7 AU. Therefore, we can calculate how close to the Sun and how far away Eris gets as follows:

Perihelion $= A(1 - e) = 67.7(1 - 0.44) = 67.7 \times 0.56 = 37.9$ AU

Aphelion $= A(1 + e) = 67.7(1 + 0.44) = 67.7 \times 1.44 = 97.5$ AU

At this time, Eris is close to aphelion. But when it approaches perihelion, it will cross the orbit of Pluto, whose distance varies from 29.7 to 48.9 AU.

For a second example, let's look at Apollo asteroid 2005 YU55, which has a semimajor axis of 1.14 AU and an orbital eccentricity of 0.43. We can calculate its perihelion and aphelion similarly:

Perihelion $= A(1 - e) = 1.14(1 - 0.43) = 1.14 \times 0.57 = 0.65$ AU

Aphelion $= A(1 + e) = 1.14(1 + 0.43) = 1.14 \times 1.43 = 1.63$ AU

These results indicate that the orbit of 2005 YU55 crosses the orbits of Earth and Mars. In November 2011, this asteroid passed 324,900 kilometers (km) from Earth—which is about 85 percent of the distance to the Moon.

period of 557 years (**Working It Out 12.1**). Eris is near the most distant point in its orbit, making it the most remote known object in the Solar System. At this distance it is about 100 times fainter than Pluto. The eccentric orbits of other Solar System bodies will eventually carry them farther away than Eris, however, so Eris will not always be the most distant object known.

When astronomers combine the observed brightness of Eris with its diameter, they find it has a surprisingly high albedo of 0.96. (Recall from Chapter 5 that albedo is a measure of how much light an object reflects.) The surface of Eris is more highly reflecting than that of any other major Solar System body except Enceladus, so Eris too must have a coating of pristine ice. The surface of Enceladus is water ice, while Eris is covered with methane ice. At its present location, the average surface temperature on Eris is cold enough to freeze out any atmospheric methane, but it will probably develop a methane atmosphere when it comes closest to the Sun in the year 2257.

Haumea and Makemake are both smaller and have slightly larger orbits than that of Pluto (**Figure 12.4**). Haumea has two

**Figure 12.4** This NASA illustration shows the five dwarf planets compared to our Moon (Luna) and Earth.

(a)

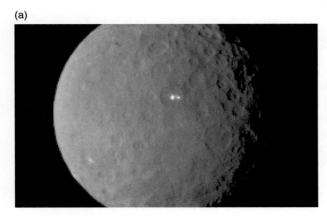

(b)

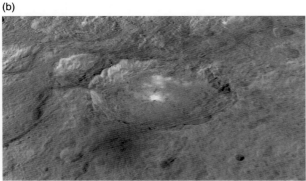

**Figure 12.5** (a) Dwarf planet Ceres photographed by the *Dawn* mission. (b) Haze has been observed above the white spots, which suggests they are made of ice.

moons—Hi'iaka and Namaka—enabling astronomers to calculate the system's mass. Although Haumea has sufficient mass to pull itself into a spherical shape, it spins so rapidly on its axis that its shape is flattened, with an equatorial radius that is approximately twice its polar radius. This difference between the equatorial and polar radii gives Haumea an oblateness (a measure of how far an object is from perfectly round) of 0.5, the most distorted shape of any of the planets or dwarf planets. HST infrared imaging indicates that Haumea and its two moons are covered in water ice. Astronomers think that these three objects and some smaller debris were left over after a larger body broke up following a collision. No moons have been discovered orbiting Makemake, so less is known about this dwarf planet than about Pluto, Haumea, or Eris.

## Ceres

The dwarf planet Ceres was discovered in 1801 when Sicilian astronomer Giuseppe Piazzi found a bright object between the orbits of Mars and Jupiter. Piazzi named the new object Ceres. When Piazzi discovered Ceres, he thought he might have found a hypothetical "missing planet." But as more objects were discovered orbiting the region between Mars and Jupiter, astronomers classified Ceres as belonging to a new category of Solar System objects called asteroids. Ceres is the largest body in the main asteroid belt. It also is now called a dwarf planet, because although it is round (**Figure 12.5a**), it has not cleared its surroundings.

With a diameter of about 940 km, Ceres is larger than most moons but smaller than any planet. It contains about a third of the total mass in the asteroid belt, but only about 1.3 percent of the mass of Earth's Moon. Ceres rotates on its axis with a period of about 9 hours, typical of many asteroids. As a large planetesimal, Ceres seems to have survived largely intact, although it appears to have undergone differentiation at some point in its early history.

About a quarter of its mass exists in the form of a water-ice mantle that surrounds a rocky inner core. Water vapor coming from two locations on Ceres indicates that there is water ice in specific locations on the surface. NASA's *Dawn* mission went into orbit around Ceres in 2015. Observed geological features include a 5-km high mountain and craters 4-5 km deep. One crater has bright spots and haze within the boundaries of its rim, suggesting the spots are made of ice (Figure 12.5b). Dawn will remain in Ceres' orbit for the remainder of its mission.

---

### CHECK YOUR UNDERSTANDING 12.1

Why are these objects called dwarf *planets* even though they are smaller than some moons?

..............................................................................................

## 12.2 Asteroids Are Pieces of the Past

After Piazzi found Ceres in 1801, a number of similar objects were discovered in the region between the orbits of Mars and Jupiter. Because these new objects appeared in astronomers' eyepieces as nothing more than faint points of light, William and Caroline Herschel (the brother-sister pair of astronomers who discovered Uranus) named them *asteroids*, a Greek word meaning "starlike." As the years went by, more asteroids were discovered, and there are now estimated to be 1 million to 2 million asteroids larger than 1 km in size, and many more that are smaller. Because these objects are in the Solar System, many of them move quickly enough across the sky that their motion is noticeable over a few hours. Both professional and amateur astronomers have discovered asteroids.

Recall that an asteroid is a primitive planetesimal that did not become part of the accretion process that formed planets. The planetesimals that formed our Solar System's planets and moons have been so severely modified by planetary processes that nearly all information about their original physical condition and chemical composition has been lost. By contrast, asteroids and comet nuclei constitute an ancient and far more pristine record of what the early Solar System was like. Asteroids are composed of the same type of rocky material that became the inner planets, and comets are composed of the same type of icy material that became the outer planets. Thus, planetary scientists study asteroids in order to learn about the inner planets and their formation. In this section, we will study the orbits and composition of the asteroids.

## The Distribution of Asteroids

Asteroids are found throughout the Solar System. Most orbit the Sun in several distinct zones, with the majority residing between the orbits of Mars and Jupiter in the **main asteroid belt**. The main belt contains at least 1,000 objects larger than 30 km in diameter, of which about 200 are larger than 100 km. Although there are a great number of asteroids, they account for only a tiny fraction of the matter in the Solar System. Some of the asteroids are bound to another asteroid in a double system, and more than 200 asteroids have moons, some similar in size to the asteroids themselves. At least one asteroid has a ring.

Asteroids are not distributed randomly throughout the main asteroid belt: there are several empty regions. **Figure 12.6** shows that there are very few asteroids that orbit at specific distances from the Sun. These "gaps" in the asteroid belt are called Kirkwood gaps after Daniel Kirkwood (1814–1895), the astronomer who first recognized them. Recall the idea of orbital resonances discussed in the past chapter: the orbital periods of some moons around their planets are numerically related. Similarly, all of the Kirkwood gaps in the asteroid belt correspond to resonances: asteroid orbits that are related to the orbital period of Jupiter by the ratio of two small integers. The boundaries of the asteroid belt are set by some of these resonances. The inner boundary of the asteroid belt, at 1.8 AU, corresponds to the 5:1 orbital resonance of Jupiter; the outer boundary, at 3.3 AU, corresponds to the 2:1 orbital resonance.

To understand the Kirkwood gaps, consider the example of an asteroid starting with an orbital period exactly half that of Jupiter, a 2:1 orbital resonance. After two complete asteroid orbits, the asteroid, Jupiter, and the Sun are lined up in the same place where they started. As Jupiter and the asteroid continue in their orbits, they line up at this same location again and again, every 11.86 years (the orbital period of Jupiter). The gravitational force of the Sun on the asteroid is more than 360 times stronger than the gravitational force that Jupiter exerts on this asteroid at its closest approach. A single close pass between Jupiter and the asteroid does very little to the asteroid's orbit. For an asteroid that is *not* in orbital resonance with Jupiter, the tiny gravitational tugs from Jupiter come at a different place in its orbit each time. The effects of these random tugs average out, and as a result even multiple passes close to Jupiter have little overall effect. For an asteroid that has a 2:1 orbital resonance with Jupiter as in this example, the tug from Jupiter comes at the same place in its orbit every time.

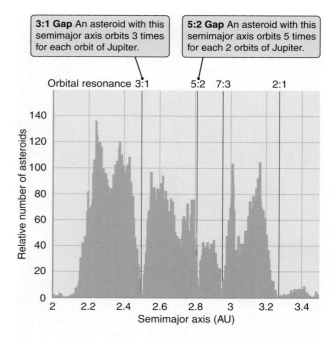

**Figure 12.6** This plot shows the relative number of asteroids in the main belt with a given orbital semimajor axis. The gaps in the distribution of asteroids, called *Kirkwood gaps*, are caused by orbital resonances with Jupiter.

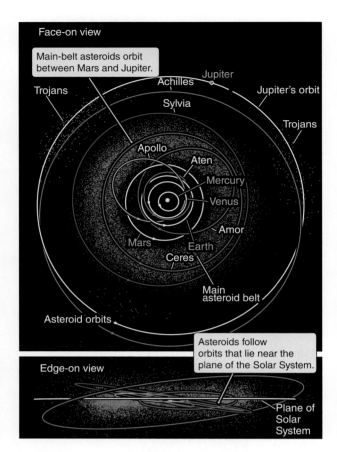

**Figure 12.7** This illustration shows face-on and edge-on views of asteroid orbits. Blue dots show the locations of known asteroids at a single point in time. The orbits of Aten, Amor, and Apollo (prototype members of some groups of asteroids) are shown. Most asteroids, such as Sylvia, are main-belt asteroids. Achilles was the first Trojan asteroid to be discovered.

The repeated tugs from Jupiter at the same place add up, changing the asteroid's orbit. Thus, an asteroid in such an orbit would not stay there long. This is why there are no asteroids with orbital periods equal to half the orbital period of Jupiter. Other orbital resonances, such as a 3:1 resonance, will have a similar effect. The reason asteroids are not found in the Kirkwood gaps is that their gravitational interaction with Jupiter prevents them from staying there.

There are several groups of asteroids not in the main asteroid belt. As shown in **Figure 12.7**, these are divided according to their orbital characteristics. Trojans share Jupiter's orbit and are held in place by interactions with Jupiter's gravitational field. Three other groups are defined by their relationship to the orbits of Earth and Mars: Apollo asteroids cross the orbits of Earth and Mars, Aten asteroids cross Earth's orbit but not that of Mars, and Amor asteroids cross the orbit of Mars but not that of Earth. All three of these groups are named for a prototype that is representative of the group.

Asteroids whose orbits bring them within 1.3 AU of the Sun are called **near-Earth asteroids**. These asteroids, along with a few comet nuclei, are known collectively as **near-Earth objects (NEOs)**. NEOs occasionally collide with Earth or the Moon. Astronomers estimate that between 500 and 1,000 NEOs have diameters larger than a kilometer. Collisions with such objects are geologically important and have dramatically altered the history of Earth and life on Earth as discussed in Chapter 8.

Part of NASA's mission is to identify and track NEOs. NASA's Wide-field Infrared Survey Explorer (WISE), an infrared telescope in space, surveyed the entire sky during 2010. The data suggest there are about 20,000 mid-sized asteroids (100 meters to 1 km in diameter) near Earth. WISE also observed more than 150,000 asteroids in the main belt, including 33,000 new ones, as well as 2,000 Jovian Trojans. The WISE mission was reactivated in late 2013 and is currently searching for NEOs.

## The Composition and Classification of Asteroids

Most asteroids are relics of rocky or metallic planetesimals that originated in the region between the orbits of Mars and Jupiter. Although early collisions between these planetesimals created several bodies large enough to differentiate, Jupiter's tidal disruption and possible orbital migration prevented them from forming a single Moon-sized planet. As they orbit the Sun, asteroids continue to collide with one another, producing small fragments of rock and metal. Most meteorites are pieces of these asteroidal fragments that have found their way to Earth and crashed to its surface.

With a few exceptions, the mass of an asteroid is too small for self-gravity to have pulled it into a spherical shape. Some asteroids have highly elongated irregular shapes, somewhat like potatoes, suggesting objects that either are fragments of larger bodies or were created haphazardly from collisions between smaller bodies. Astronomers have measured the masses of a number of asteroids by noting the effect of their gravity on Mars, on spacecraft passing nearby, on each other, or by the orbits of their moons. The total mass of the asteroids in the main belt is estimated to be about 3 times the mass of Ceres, or 4 percent the mass of the Moon. Their densities can be found from their mass and size and range between 1.3 and 3.5 times the density of water. The lower-density asteroids are shattered heaps of rubble, with large voids between the fragments.

Asteroids rotate just as planets and moons do, although irregularly shaped asteroids can wobble a lot as they spin. The rotation periods of some asteroids range from 2 hours to longer than 40 Earth days. Rotation periods for asteroids are measured by watching changes in their brightness as they alternately present their broad and narrow faces to Earth. Different groups of asteroids have different average rotations.

Asteroids can also be classified by composition. Meteorites found on Earth come from asteroids, which come from planetesimals, as shown in **Figure 12.8**. As larger planetesimals accreted smaller objects, thermal energy from impacts and the decay of radioactive elements heated them. Despite this heating, some planetesimals never reached the high temperatures needed to melt their interiors: they simply cooled off. They look like rubble piles, pretty much as they were when they formed. These planetesimals, the most common type of asteroid in the main belt, are called **C-type** (carbon) asteroids. They are composed of primitive material that has largely been unmodified since the origin of the Solar System almost 4.6 billion years ago.

In contrast, some planetesimals were heated enough by impacts and radioactive decay to cause them to melt and differentiate, with denser matter such as iron sinking to their centers. Lower-density material—such as compounds of calcium, silicon, and oxygen—floated toward the surfaces of these planetesimals and combined to form mantles and crusts of silicate rocks. **S-type** (stony) asteroids may be pieces of the mantles and crusts of such differentiated planetesimals and are

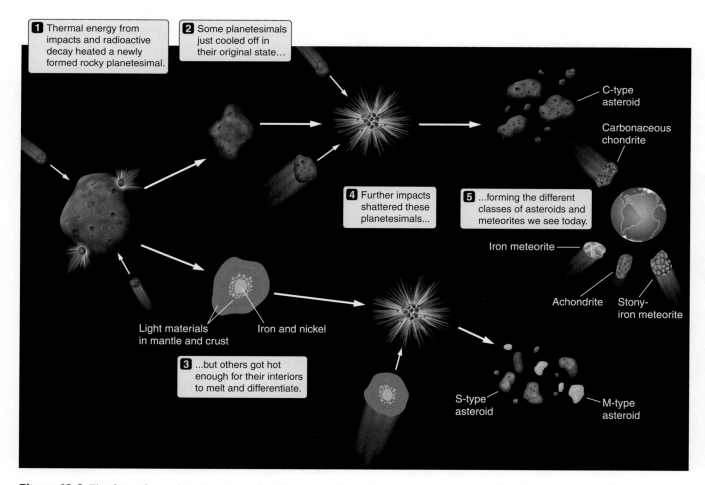

**Figure 12.8** The fate of a rocky planetesimal in the young Solar System depends on whether it gets large and hot enough to melt and differentiate, as well as on the impacts it experiences. Different histories led to the different types of asteroids and meteorites found today. (Images not to scale.)

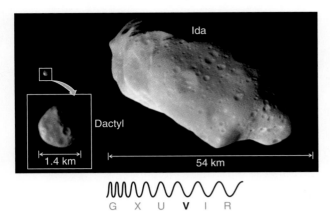

**Figure 12.9** This *Galileo* spacecraft image shows the asteroid Ida with its tiny moon, Dactyl (shown enlarged in the inset).

chemically similar to volcanic rocks found on Earth. They were hot enough at some point to lose their carbon compounds and other volatile materials to space. Similarly, **M-type** (metal) asteroids are fragments of the iron- and nickel-rich cores of one or more differentiated planetesimals that shattered into small pieces during collisions with other planetesimals.

Recently, some asteroids have been shown to have ice on their surface. Using a ground-based infrared telescope, astronomers found ice on 24 Themis: it is one of the largest main-belt asteroids (diameter 200 km) and orbits the Sun at the outer edge of the asteroid belt. Water ice covers its entire surface, and organic molecules were also found there. Hydrated minerals have been found on meteorites thought to have come from outer-main-belt asteroids, but this was the first direct detection of water ice on an asteroid. This discovery may indicate that there is a continuum rather than a strict boundary between icy comets and rocky asteroids. The observations support the idea that both asteroids and comets brought water and organic material to the early Earth.

## Asteroids Viewed Up Close

Several asteroids have been visited by spacecraft. In 1991, the *Galileo* mission passed by two S-type asteroids while on its way to Jupiter. The small asteroid Gaspra is cratered and irregular in shape, about 18 × 11 × 9 km in size. Faint, groovelike patterns may be fractures from the impact that chipped Gaspra from a larger planetesimal. Distinctive colors imply that Gaspra is covered with a variety of rock types. Later, *Galileo* passed close to asteroid Ida in the outer part of the main asteroid belt (**Figure 12.9**). *Galileo* flew so close to Ida that its cameras could see details as small as 10 meters across. Ida is 60 × 19 × 25 km in size, and its surface is about a billion years old, twice the age estimated for Gaspra. Like Gaspra, Ida contains fractures, indicating that these asteroids must be made of relatively solid rock. This supports the idea that some asteroids are chips from larger, solid objects. The *Galileo* images also revealed a tiny moon orbiting Ida, called Dactyl, which is only 1.4 km across and cratered from impacts.

The first spacecraft to land on an asteroid was *NEAR Shoemaker*, which was gently crash-landed into asteroid Eros in 2002 after a year of taking observations. Chemical analyses confirmed that the composition of Eros is like that of primitive meteorites. In November 2005, the Japanese spacecraft *Hayabusa* made contact with the small (less than 0.5 km) S-type asteroid Itokawa. *Hayabusa* collected small samples of dust that were returned to Earth in 2010—the first sample-return mission from an asteroid. Chemical analysis showed that such S-type asteroids are the parents of a type of meteorite found on Earth. They also suggested that Itokawa had been much larger, more than 20 km in diameter, when it formed.

In 2011, NASA's *Dawn* spacecraft went into orbit around Vesta (**Figure 12.10**), the second most massive body in the asteroid belt (after Ceres). Vesta is small (525 km in diameter) compared to the terrestrial planets but large compared to the other visited asteroids. The data from *Dawn* indicate that Vesta is a leftover intact protoplanet that formed within the first 2 million years of the condensation of the first solid bodies in the Solar System. It has an iron core and is differentiated, so it is more like the planets than like other asteroids.

Vesta's spectrum matches the reflection spectrum of a peculiar group of meteorites that look like rocks taken from iron-rich lava flows on Earth and the Moon. A collision—or two—that created the two large impact basins in the south polar

**Figure 12.10** This image of Vesta was taken by *Dawn* in 2012. Its north pole is in the middle of the image.

region of Vesta (**Figure 12.11a**) blasted material into space that then landed on Earth as these meteorites. These basins are only 1 billion to 2 billion years old. The younger basin is 500 km across and 19 km deep—a depth greater than the height of Mauna Kea in Hawaii. Smaller adjacent impact craters in the northern hemisphere (Figure 12.11b) were nicknamed "Snowman."

## CHECK YOUR UNDERSTANDING 12.2

Remnants of volcanic activity on the asteroid Vesta indicate that members of the asteroid belt: (a) were once part of a single protoplanet that was shattered by collisions; (b) have all undergone significant chemical evolution since formation; (c) occasionally grow large enough to become differentiated and geologically active; (d) used to be volcanic moons orbiting other planets.

# 12.3 Comets Are Clumps of Ice

Early cultures viewed the sudden and unexpected appearance of a bright comet as an omen. Comets were often seen as dire warnings of disease, destruction, and death, but sometimes as portents of victory in battle or as heavenly messengers announcing the impending birth of a great leader. The earliest records of comets date from as long ago as the 23rd century BCE. Until the end of the Middle Ages, comets were regarded as mysterious temporary atmospheric phenomena rather than as astronomical objects. In the 16th century, Tycho Brahe reasoned that if comets were atmospheric phenomena like clouds, then their appearance and location in the sky should be very different to observers located many miles apart. But when Tycho compared sightings of comets made by observers at several different sites, he found no evidence of such differences, and he concluded that comets must be at least as far away as the Moon.

Today, we know that comets are icy planetesimals that formed from primordial material. They spend most of their time adrift in the frigid outer reaches of the Solar System. Comet nuclei put on a show only when their orbit brings them deep enough into the inner Solar System to undergo destructive heating from the Sun—they emit streams of dust and gas. In this section, we will examine the orbits and composition of the comets.

## The Homes of the Comets

A *comet* is a complex object consisting of a small, solid, icy nucleus, an atmospheric halo, and a tail of dust and gas: a comet nucleus is the "heart" of the comet and contains most of the comet's mass. When very distant from the Sun, the comet is entirely nucleus—frozen throughout. As it approaches the Sun, the coma forms first, and then the tail forms. When they are near enough to the Sun to show the effects of solar heating, they are called **active comets**, or often simply *comets*. Most of these icy bodies are much too small and far away to be seen, so no one really knows how many there are. Estimates for our Solar System range as high as a trillion ($10^{12}$) comet nuclei—more than the number of stars in the Milky Way Galaxy—but astronomers have seen only several thousand.

We know where comets come from by observing their orbits as they pass through the inner Solar System. Comets fall into two distinct groups named for scientists Gerard Kuiper (1905–1973) and Jan Oort (1900–1992).

(a)

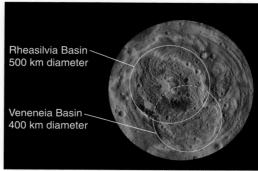

Rheasilvia Basin
500 km diameter

Veneneia Basin
400 km diameter

(b)

G  X  U  V  I  R

**Figure 12.11** (a) This image shows impact basins at the south pole of Vesta. By counting the craters on top of it, astronomers estimate Rheasilvia to be 1 billion years old. Veneneia is partly beneath Rheasilvia and is estimated to be 2 billion years old. Red is higher elevation. (b) This image shows three craters in the northern hemisphere—60, 50, and 22 km across, respectively. The feature was nicknamed "Snowman."

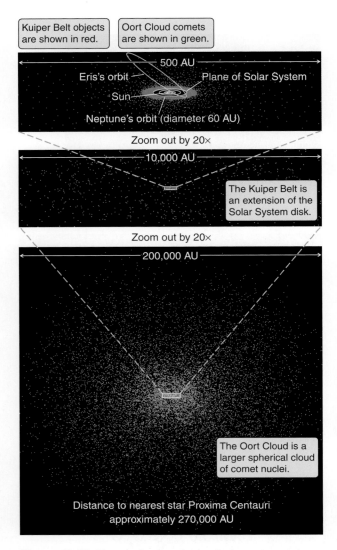

Kuiper Belt objects are shown in red.

Oort Cloud comets are shown in green.

500 AU

Eris's orbit

Plane of Solar System

Sun

Neptune's orbit (diameter 60 AU)

Zoom out by 20×

10,000 AU

The Kuiper Belt is an extension of the Solar System disk.

Zoom out by 20×

200,000 AU

The Oort Cloud is a larger spherical cloud of comet nuclei.

Distance to nearest star Proxima Centauri approximately 270,000 AU

**Figure 12.12** The top image shows that most comets near the inner Solar System populate an extension to the disk of the Solar System called the Kuiper Belt (red). The middle image zooms out to show the expanded Kuiper Belt. The bottom image zooms out to illustrate the spherical Oort Cloud, which is far larger and contains many more comet nuclei (green).

**Kuiper Belt** The Kuiper Belt is a disk-shaped population of comet nuclei that begins at about 30 AU from the Sun, near the orbit of Neptune, and extends outward to about 50 AU (**Figure 12.12**). Comets from the Kuiper Belt orbit the Sun in a disk-shaped region aligned with the Solar System. The innermost part of the Kuiper Belt contains tens of thousands of icy planetesimals known as Kuiper Belt objects (KBOs), or sometimes as *trans-Neptunian objects (TNOs)*. The largest KBOs are similar in size to Pluto and Eris. With a few exceptions, the sizes of KBOs are difficult to determine, because although brightness and approximate distance are known, their albedos are uncertain. Reasonable limits for the albedos can set maximum and minimum values for their size. Like asteroids, some KBOs have moons, and at least one has three moons. We know very little of the chemical and physical properties of KBOs because of their great distance. After it encountered Pluto in 2015, the *New Horizons* spacecraft is scheduled to continue outward into the Kuiper Belt, where it will be maneuvered to fly close to one or more KBOs.

One of the larger known KBOs, Quaoar (pronounced "kwa-whar"), is also one of the few whose size astronomers have independently measured—about 900 km. From its apparent brightness, distance, and size, astronomers calculate Quaoar's albedo to be 0.20, making it more reflective than the nuclei of those comets that have entered the inner Solar System but far less reflective than Pluto. Quaoar's remote location and pristine condition have apparently allowed some volatile ices to survive on its surface, including crystalline water ice, methane, and ethane. Quaoar has a nearly circular orbit about the Sun and has a small moon, which enables astronomers to estimate its mass.

The icy planetesimals in the Kuiper Belt are packed closely enough to interact gravitationally from time to time. In such events, one object gains energy while the other loses it. The "winner" may gain enough energy to be sent into an orbit that reaches far beyond the boundary of the Kuiper Belt. The "loser" may fall inward toward the Sun.

**Oort Cloud** Unlike the flat disk of the Kuiper Belt, the **Oort Cloud** is a spherical distribution of planetesimals (Figure 12.12) that are much too distant to be seen by even the most powerful telescopes. Astronomers determine the size and shape of the Oort Cloud from the orbits of this region's comets, which approach the Sun from all directions and from as far as 100,000 AU away—nearly halfway to the nearest stars.

Sedna is an interesting object in the inner Oort Cloud whose highly elliptical orbit around the Sun takes it from 76 AU out to 937 AU. With such an extended orbit, Sedna requires more than 11,000 years to make a single trip around the Sun. When discovered in 2003, Sedna was about 90 AU from the Sun and getting closer. It will reach its perihelion in 2076. Herschel Space Observatory data suggest an albedo of 0.30 and a size of 1,000 km. Sedna has no known moon, so it is difficult to estimate its mass. Water and methane ices have been detected in its spectrum. Like dwarf planet Eris, Sedna has a highly eccentric orbit. A second object in the inner Oort Cloud, 2012 VP113, was recently detected. Its distance ranges from 80 AU to 452 AU from the Sun, and it is thought to be about half the size of Sedna.

Inner Solar System objects are close enough to the Sun that disturbances external to the Solar System never exert more than a tiny fraction of the gravitational force of the Sun on them. In the distant Oort Cloud, however, comet nuclei are so far from the Sun, and the Sun's gravitational force on them is so weak, that they are barely bound to the Sun at all. The tug of a slowly passing star or interstellar cloud can compete with the Sun's gravity, significantly stirring up the Oort Cloud and changing the orbits of its objects. If the interaction adds to the orbital energy of a comet nucleus, the comet may move outward to an even more distant orbit or perhaps escape from the Sun completely. A comet nucleus that loses orbital energy as a result of this type of interaction will fall inward. Some of these comet nuclei come all the way into the inner Solar System, where they may appear briefly in Earth's skies before returning once again to the Oort Cloud.

## The Orbits of Comets

The lifetime of a comet nucleus depends on how frequently it passes by the Sun and how close it gets. There are about 400 known **short-period comets**, which by definition have periods less than 200 years. Additionally, each year astronomers discover about six new **long-period comets**, whose orbital periods are longer than 200 years. The total number of long-period comets observed to date is about 3,000.

**Figure 12.13** shows the orbits of a number of comets, nearly all highly elliptical, with one end of the orbit close to the Sun and the other in the distant parts of the Solar System. Most comets passing through the inner Solar System have long orbital periods that carry them back to the Oort Cloud or the Kuiper Belt. Long-period comets were scattered to the outer Solar System by gravitational interactions, so they come into the inner Solar System from all directions. Some orbit the Sun in the same direction that the planets orbit (prograde), and some orbit in the opposite direction (retrograde).

▶❚❚ **AstroTour:** Cometary Orbits

**Figure 12.13** This figure illustrates the orbits of a number of comets in face-on and edge-on views of the Solar System. Populations of (a) short-period comets and (b) long-period comets have very different orbital properties. Comet Halley, which appears in both diagrams for comparison, is a short-period comet.

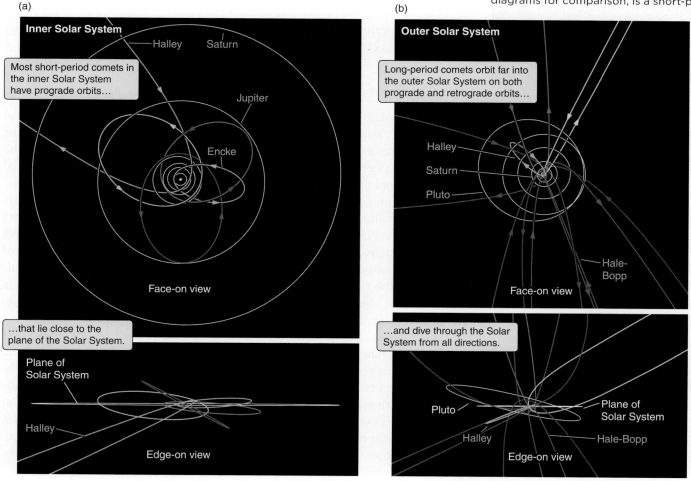

Conversely, short-period comets tend to be prograde and to have orbits in the ecliptic plane, and they frequently pass close enough to a planet for its gravity to change the comet's orbit about the Sun. Short-period comets presumably originated in the Kuiper Belt, but as they fell in toward the Sun, they were forced into their current short-period orbits relatively close to the Sun by gravitational encounters with Jupiter.

Comet Halley is the brightest and most famous of the short-period comets. In 1705, Edmund Halley, using the gravitational laws of his colleague Isaac Newton, noted that a bright comet from 1682 had an orbit remarkably similar to those of comets seen in 1531 and 1607. He concluded that all three were the same comet and predicted that it would return in 1758. When it reappeared, astronomers quickly named it Halley's Comet and heralded it as a triumph for the genius of both Newton and Halley. Comet Halley's highly elongated orbit takes it from perihelion, about halfway between the orbits of Mercury and Venus, out to aphelion beyond the orbit of Neptune. Astronomers and historians have now identified possible sightings of the comet that go back at least to 240 BCE. Comet Halley has an average period of 76 years. Its most recent appearance was in 1986, and it was not especially spectacular compared to 1910 because in 1986, Comet Halley and Earth were on opposite sides of the Sun. Comet Halley will return and become visible to the naked eye once again in summer 2061.

Hundreds of long-period comets have well-determined orbits. Some have orbital periods of hundreds of thousands or even millions of years. Almost all their time is spent in the Oort Cloud in the frigid, outermost regions of the Solar System. Orbits of a few long-period comets are shown in Figure 12.13b. These are the comets that reveal the existence of the Oort Cloud. Because of their very long orbital periods, these comets have made at most one appearance throughout the course of recorded history.

## Anatomy of an Active Comet

Unlike asteroids, which have been through a host of chemical and physical changes as a result of collisions, heating, and differentiation, most comet nuclei have been preserved over the past 4.6 billion years by the "deep freeze" of the outer Solar System. Comet nuclei are made of the most nearly pristine material remaining from the formation of the Solar System.

The comet nucleus at the center is the smallest component of a comet, but it is the source of all the material that we see stretched across the sky as the comet nears the Sun (**Figure 12.14**). Comet nuclei range in size from a few dozen meters to several hundred kilometers across. These "dirty snowballs" are composed of ice, organic compounds, and dust grains. They have been described as being similar to deep-fried ice cream, with a soft and porous interior surrounded by a crunchy crust of hardened water-ice crystals, topped off with sooty dust and organic molecules.

As a comet nucleus nears the Sun, sunlight heats its surface, vaporizing ices that stream away from its nucleus, and these gases carry dust particles along with them. This process of conversion from solid to gas is called **sublimation**. For example, dry ice (frozen carbon dioxide) does not melt like water ice but instead turns directly into carbon dioxide gas. Dry ice sublimates—that is why it is called "dry." Set a piece of dry ice out in the Sun on a summer day, and you will get a pretty good idea of what happens to a comet. The gases and dust driven from the nucleus of an active comet form a nearly spherical atmospheric cloud around the nucleus called the

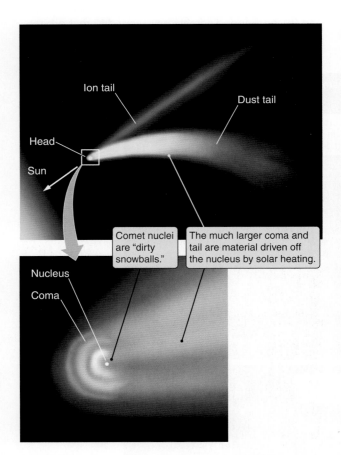

**Figure 12.14** The principal components of a fully developed active comet are the nucleus, the coma, and two types of tails called the dust tail and the ion tail. Together, the nucleus and the coma are called the head.

**coma**. The nucleus and the inner part of the coma are sometimes called the comet's **head**. Pointing from the head of the comet in a direction more or less away from the Sun are long streamers of dust, gas, and ions called **tails**.

The tails are the largest and most spectacular part of a comet. The tails are also the "hair" for which comets are named. (*Comet* comes from the Greek word *kometes*, which means "hairy one.") Active comets have two different types of tails, as shown in Figure 12.14. One is the **ion tail**. Many of the atoms and molecules that make up a comet's coma are ions. Because they are electrically charged, ions in the coma feel the effect of the solar wind—the stream of charged particles that blows continually away from the Sun. The solar wind pushes on these ions, rapidly accelerating them to speeds of more than 100 kilometers per second (km/s)—far greater than the orbital velocity of the comet itself—and sweeps them out into a long wispy structure. Because the particles that make up the ion tail are so quickly picked up by the solar wind, an ion tail is usually very straight: beginning at the head of the comet, an ion tail points directly away from the Sun.

Dust particles in the coma can also have a net electric charge and feel the force of the solar wind. Sunlight also exerts a force on cometary dust. But dust particles are much more massive than individual ions, so they are accelerated more gently and do not reach such high relative speeds as those of the ions. As a result, the dust particles are unable to keep up with the comet, and the **dust tail** often curves away from the head of the comet as the dust particles are gradually pushed from the comet's orbit in the direction away from the Sun (Figure 12.14).

**Figure 12.15** shows the tails of a comet at various points in its orbit. Remember that both types of tails always point away from the Sun, regardless of which direction the comet is moving. As the comet approaches the Sun, its two tails trail behind its nucleus. But the tails extend ahead of the nucleus as the comet moves away from the Sun. Tails vary greatly from one comet to another. Some comets display both types of tails simultaneously; others, for reasons that are not understood, produce no tails at all. A tail often forms as a comet crosses the orbit of Mars, where the increase in solar heating drives gas and dust away from the nucleus.

The gas in a comet's tail is even more tenuous than the gas in its coma, with densities of no more than a few hundred particles per cubic centimeter. This is much, much less than the density of Earth's atmosphere, which at sea level contains more than $10^{19}$ molecules per cubic centimeter. Dust particles in the tail are typically about 1 micron ($\mu$m) in diameter, roughly the size of smoke particles.

The nuclei of short-period comets have been badly worn out by their repeated exposure to heating by the Sun. As the volatile ices are driven from a nucleus, some of the dust and organics are left behind on the surface. The buildup of this covering slows down cometary activity. (Envision how, as a pile of dirty snow melts, the dirt left behind is concentrated on the surface of the snow.) In contrast, long-period comets are usually relatively pristine. More of their supply of volatile ices still remains close to the surface of the nucleus, and they can produce a truly magnificent show.

Most naked-eye comets develop first a coma and then an extended tail as they approach the inner Solar System. Comet McNaught in 2007 was such a comet, and it was the brightest to appear in nearly 50 years. The comet's nucleus and coma were visible in broad daylight as its orbit carried it within 25 million km of the Sun. When Comet McNaught had passed behind the Sun and next appeared in the evening skies to observers in the Southern Hemisphere, its tail had grown

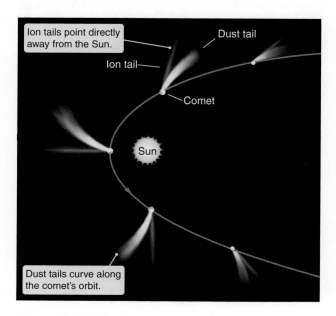

**Figure 12.15** This drawing illustrates the orientation of the dust and ion tails at several points in a comet's orbit. The ion tail points directly away from the Sun, while the dust tail curves along the comet's orbit.

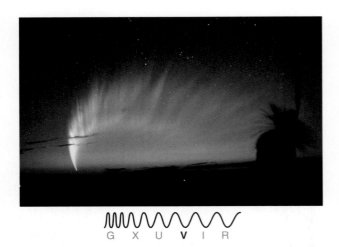

G X U V I R

**Figure 12.16** Comet McNaught in 2007 was the brightest comet to appear in decades, but its true splendor was visible only to observers in the Southern Hemisphere.

G X U **V** I R

**Figure 12.17** Comet Hale-Bopp was a great comet in 1997. The ion tail is blue in this image, and the dust tail is white.

to a length of more than 160 million km and stretched 35° across the sky (**Figure 12.16**). Comet McNaught came into the inner Solar System from the Oort Cloud, but it left on a path that will carry it out of the Solar System.

Comet Hale-Bopp in 1997 was a spectacular long-period comet with a long, beautiful tail (**Figure 12.17**). Hale-Bopp was a large comet, with a nucleus estimated at 60 km in diameter. It was discovered far from the Sun, near Jupiter's orbit, 2 years before its perihelion passage. This early discovery extended the total time available to study its development and plan observations as it approached the Sun. Warmed by the Sun, the nucleus produced large quantities of gas and dust and as much as 300 tons of water per second, with lesser amounts of carbon monoxide, sulfur dioxide, cyanogen, and other gases. Comet Hale-Bopp will continue its outward journey for more than 1,000 years, and it will not return to the inner Solar System until sometime around the year 4530.

Comet Ikeya-Seki is a member of a family of comets called **sungrazers**, comets whose perihelia are located very close to the surface of the Sun. Many sungrazers fail to survive even a single orbit of the Sun. Ikeya-Seki became so bright as it neared perihelion in 1965 that it was visible in broad daylight, close to the Sun in the sky. Sungrazers generally come in groups, with successive comets following in nearly identical orbits. Each member of such a group started as part of a single larger nucleus that broke into pieces during an earlier perihelion passage.

A half dozen or so long-period comets arrive each year. Most pass through the inner Solar System at relatively large distances from Earth or the Sun and never become bright enough to attract much public attention. On average, a spectacular comet appears about once per decade.

## Visits to Comets

Comets provide an engineering challenge for spacecraft designers. There is seldom enough advance knowledge of a comet's visit or its orbit to mount a successful mission to intercept it. The relative speed between an Earth-launched spacecraft and a comet can be extremely high. Observations must be made very quickly, and there is a danger of high-speed collisions with debris from the nucleus. About a dozen spacecraft have been sent to rendezvous with comets, including five spacecraft sent to Comet Halley by the Soviet, European, and Japanese space agencies in 1986. Much of what we know about comet nuclei and the innermost parts of the coma comes from data sent back by these missions. The spacecraft observed gas and dust jets, impacts craters, and ice and dust on the comet nuclei.

Two Soviet *Vega* and the European *Giotto* spacecraft entered the coma of Comet Halley when they were still nearly 300,000 km from its nucleus. We learned that the dust from Comet Halley was a mixture of light organic substances and heavier rocky material, and the gas was about 80 percent water and 10 percent carbon monoxide with smaller amounts of other organic molecules. The surface of Comet Halley's nucleus is among the darkest known objects in the Solar System, which means that it is rich in complex organic matter that must have been present as dust in the disk around the young Sun—perhaps even in the interstellar cloud from which the Solar System formed. As the three spacecraft passed close by Halley's nucleus, they observed jets of gas and dust moving away from its surface at speeds of up to 1 km/s, far above the escape velocity. By observing the jets of material streaming away from the nucleus of Halley, planetary scientists estimated that Comet Halley must have lost one-tenth of 1 percent of its mass as it went around the Sun.

Several space missions have visited short-period comets. In 2004, NASA's *Stardust* spacecraft flew within 235 km of the nucleus of Comet Wild 2. Comet Wild 2 had previously resided in the region between the orbits of Jupiter and Uranus, but a close encounter with Jupiter in 1974 had perturbed its orbit, bringing this relatively pristine body closer to the Sun as it traveled between the orbits of Jupiter and Earth. At the time of *Stardust*'s encounter with Wild 2, the comet had made only five trips around the Sun in its new orbit. Wild 2's nearly spherical nucleus is about 5 km across. At least 10 gas jets were active, some of which carried large chunks of surface material. The surface of Wild 2 is covered with features that may be impact craters modified by ice sublimation, small landslides, and erosion by jetting gas (**Figure 12.18**). Some craters show flat floors, suggesting a relatively solid interior beneath a porous surface layer.

The *Stardust* mission collected dust samples from Wild 2, which were returned to Earth in 2006. It found new kinds of organic materials unlike any seen before in materials from space. They are more primitive than those observed in meteorites and may have formed before the Solar System itself. These grains can be used to investigate the conditions under which the Sun and planets formed. Minerals that form at high temperature have also been found, supporting the idea that the solar wind blew material out of the inner Solar System very early in the system's history. Scientists will be studying the particles from this mission in detail for many years.

In 2005, NASA's *Deep Impact* spacecraft launched a 370-kg impacting projectile into the nucleus of Comet Tempel 1 at a speed of more than 10 km/s. The impact sent 10,000 tons of water and dust flying off into space at speeds of 50 meters per second (m/s—**Figure 12.19**). A camera mounted on the projectile snapped photos of its target until it was vaporized by the impact. Observations of the event were made both locally by *Deep Impact* and back on Earth by orbiting and ground-based telescopes. Water, carbon dioxide, hydrogen cyanide, iron-bearing minerals, and a host of complex organic molecules were identified in the Comet Tempel 1 impact. The comet's outer layer is composed of fine dust with a consistency of talcum powder. Beneath the dust are layers made up of water ice and organic materials. Well-formed impact craters, which had been absent in close-up images of Comet Wild 2, were also seen.

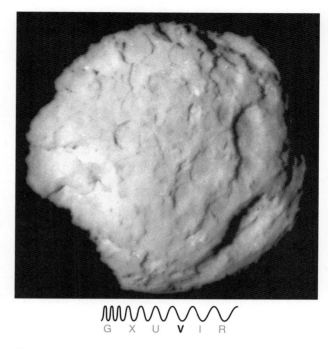

G  X  U  **V**  I  R

**Figure 12.18** The nucleus of Comet Wild 2 was imaged by the *Stardust* spacecraft, which also sampled its tail.

(a)

(b)

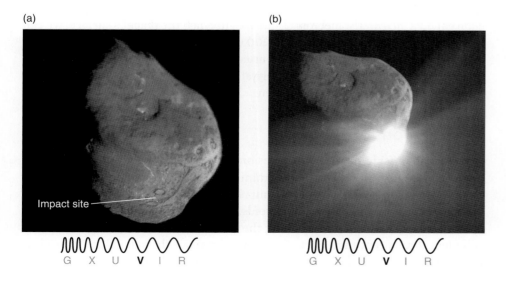

Impact site

G  X  U  **V**  I  R

G  X  U  **V**  I  R

**Figure 12.19** (a) The surface of the nucleus of Comet Tempel 1 is shown just before impact by the *Deep Impact* projectile. The impact occurred between the two 370-meter-diameter craters located near the bottom of the image. The smallest features appearing in this image are about 5 meters across. (b) Sixteen seconds after the impactor struck the comet, the parent spacecraft took this image of the initial ejecta.

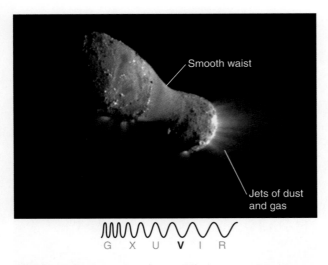

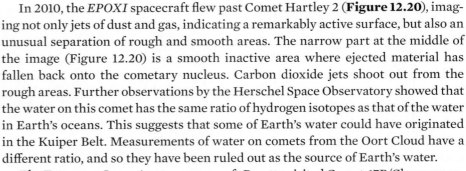

G X U V I R

**Figure 12.20** This image of Comet Hartley 2 taken by the *EPOXI* spacecraft reveals two distinct surface types. Water seeps through the dust at the smooth "waist" of the comet's nucleus, whereas carbon dioxide jets shoot gas, dust, and chunks of ice from the rough areas.

 **Nebraska Simulation:** Driving through Snow

In 2010, the *EPOXI* spacecraft flew past Comet Hartley 2 (**Figure 12.20**), imaging not only jets of dust and gas, indicating a remarkably active surface, but also an unusual separation of rough and smooth areas. The narrow part at the middle of the image (Figure 12.20) is a smooth inactive area where ejected material has fallen back onto the cometary nucleus. Carbon dioxide jets shoot out from the rough areas. Further observations by the Herschel Space Observatory showed that the water on this comet has the same ratio of hydrogen isotopes as that of the water in Earth's oceans. This suggests that some of Earth's water could have originated in the Kuiper Belt. Measurements of water on comets from the Oort Cloud have a different ratio, and so they have been ruled out as the source of Earth's water.

The European Space Agency spacecraft *Rosetta* visited Comet 67P/Churyumov-Gerasimenko in 2014 (see the chapter-opening figure). A small separate spacecraft soft-landed on the comet and sent back data for 2.5 days before running out of power. The landing site was dust-covered solid ice, too thick to drill. *Rosetta* orbited the comet as it approached the Sun and observed changes as it heated up. Dust and gas were released from the comet, including a bright jet.

### CHECK YOUR UNDERSTANDING 12.3

The nucleus of a comet is mostly: (a) solid ice; (b) solid rock; (c) a porous mix of ice and dust; (d) frozen carbon dioxide.

## 12.4 Meteorites Are Remnants of the Early Solar System

Comet nuclei that enter the inner Solar System generally disintegrate within a few hundred thousand years as a result of their repeated passages near the Sun. Asteroids have much longer lives but still are slowly broken into pieces from occasional collisions with each other. The disintegration of comet nuclei and collisions between asteroids create most of the debris that fills the inner part of the Solar System. As Earth and other planets move along in their orbits, they continually sweep up this fine debris. The cometary and asteroidal debris is the source of most of the *meteoroids* that Earth encounters. Meteoroids are small solid bodies ranging in size from 10 $\mu$m to 100 meters. When a meteoroid enters Earth's atmosphere, frictional heat causes the air to glow, producing an atmospheric phenomenon called a *meteor*. If a meteoroid survives to reach the planet's surface, we call it a *meteorite*. Earth sweeps up some 100,000 kg of meteoritic debris every day, and particles smaller than 100 $\mu$m eventually settle to the ground as fine dust. In this section, we will look more closely at meteorites and what can be learned about the early Solar System from them.

### Observations of Meteors

If you stand outside for a few minutes on a moonless, starry night, away from bright city lights, you will almost certainly see a meteor, commonly known as a *shooting star*. The larger pieces that survive the plunge through Earth's atmosphere are usually fragments of asteroids. Most of the smaller pieces that burn up in the atmosphere before reaching the ground are cometary fragments typically less than a centimeter across and having about the same density as cigarette ash.

A 1-gram meteoroid (about half the mass of a dime) entering Earth's atmosphere at 50 km/s has a kinetic energy comparable to that of an automobile cruising along at the fastest highway speeds. Scientists measuring meteor heights with radar find that the altitudes of meteors are between 50 and 150 km. Most meteoroids are so small and fragile that they burn up completely before reaching Earth's surface. A meteor may streak across 100 km of Earth's atmosphere and last at most a few seconds. Meteorites likely litter the surfaces of all solid planets and moons.

Nearly all ancient cultures were fascinated by these rocks from the sky. Iron from meteorites was used to make the earliest tools. Early Egyptians preserved meteorites along with the remains of their pharaohs, Japanese placed them in Shinto shrines, and ancient Greeks worshipped them. Despite numerous eyewitness accounts of meteorite falls, however, many people were slow to accept that these peculiar rocks actually come from far beyond Earth. By the early 1800s, scientists had documented so many meteorite falls that their true origin was indisputable. Today, hardly a year passes without a recorded meteorite fall, including some that have caused damage.

Fragments of asteroids are much denser than cometary meteoroids. If an asteroid fragment is large enough—about the size of your fist—it can survive all the way to the ground to become a meteorite. The fall of a 10-kg meteoroid can produce a fireball so bright that it lights up the night sky more brilliantly than the full Moon. Such a large meteoroid, traveling many times faster than the speed of sound, may create a sonic boom heard hundreds of kilometers away. It may even explode into multiple fragments as it nears the end of its flight. Some fireballs glow with a brilliant green color, caused by elements in the meteoroid that created them.

**Meteor showers** occur when Earth's orbit crosses the orbit of a comet or asteroid and passes through a concentration of cometary or asteroidal debris. During a shower, many meteors can be observed in just a few hours. More than a dozen comets and at least two asteroids have orbits that come close enough to Earth's orbit to produce annual meteor showers, as listed in **Table 12.1**. Because the meteoroids in a shower are all in similar orbits, they all enter Earth's atmosphere moving in the same direction—the paths through the sky are parallel to one another. Therefore, all the meteors appear to originate from the same point in the sky (**Figure 12.21a**),

| TABLE 12.1 | Selected Meteor Showers | |
|---|---|---|
| **Shower** | **Approximate Date** | **Parent Object** |
| Quadrantids | January 3–4 | Asteroid 2003 EH1 |
| Lyrids | April 21–22 | Comet Thatcher |
| Eta Aquariids | May 5–6 | Comet Halley |
| Perseids | August 12–13 | Comet Swift-Tuttle |
| Draconids | October 8–9 | Comet Giacobini-Zinner |
| Orionids | October 21–22 | Comet Halley |
| Taurids | November 5–6 | Comet Encke |
| Leonids | November 17–18 | Comet Tempel-Tuttle |
| Geminids | December 13–14 | Asteroid Phaethon |
| Ursids | December 22–23 | Comet Tuttle |

(a) Meteors in a shower appear to move away from a point called the radiant.

G X U **V** I R

(b) Actually the meteor paths are parallel. The radiant is a vanishing point.

VISUAL ANALOGY

**Figure 12.21** (a) Meteors appear to stream away from the radiant of the Leonid meteor shower. (b) Such streaks are actually parallel paths that appear to emerge from a vanishing point, as in our view of these railroad tracks.

just as the parallel rails of a railroad track appear to vanish to a single point in the distance (Figure 12.21b). This point is called the shower's **radiant**.

For example, the Perseid shower in August is the result of Earth crossing the orbit of Comet Swift-Tuttle. Although spread out along the comet's orbit, the debris is more concentrated in the vicinity of the comet itself. In 1992, Comet Swift-Tuttle returned to the inner Solar System for the first time since its discovery in 1862, resulting in an exceptional Perseid meteor shower with counts of up to 500 meteors per hour.

In mid-November of each year, Earth passes almost directly through the orbit of Comet Tempel-Tuttle, a short-period comet with an orbital period of 33.2 years. This produces the Leonid meteor shower, which is usually weak because in most years, Comet Tempel-Tuttle distributes little of its debris around its orbit. In 1833 and 1866, however, Tempel-Tuttle was not far away when Earth passed through its orbit, and the Leonid showers were so intense that meteors filled the sky with as many as 100,000 meteors per hour. Further perturbations of the comet's orbit caused a spectacular Leonid shower in 1966 that may have produced as many as a half-million meteors per hour. The Leonid shower put on less spectacular but still impressive shows between 1999 and 2003, when several thousand meteors per hour were seen.

## Types of Meteoroids

As asteroids orbit the Sun, they occasionally collide with each other, chipping off smaller rocks and bits of dust. Sometimes, one of these fragments is captured by Earth's gravity and survives its fiery descent through Earth's atmosphere as a meteor. Thousands of meteorites reach the surface of Earth every day, but only a tiny fraction of these are ever found and identified. Antarctica offers the best meteorite hunting in the world because in many places, the only stones to be found on the ice are meteorites. Because Antarctica is actually very dry, Antarctic meteorites also tend to show little weathering or contamination from terrestrial dust or organic compounds, which makes them excellent specimens for study.

Meteorites are grouped into three categories according to their materials and the degree of differentiation they experienced within their parent bodies (**Figure 12.22**). More than 90 percent of meteorites are included in the first category, **stony meteorites**, which are similar to terrestrial silicate rocks. A stony meteorite is characterized by the thin coating of melted rock that forms as it passes through the atmosphere. Many stony meteorites contain small round spherules called **chondrules**, once-molten droplets that rapidly cooled to form crystallized spheres ranging in size from that of sand grains to that of marbles. Stony meteorites containing chondrules are called **chondrites** (Figure 12.22a); those without chondrules are known as **achondrites** (Figure 12.22b). **Carbonaceous chondrites** are chondrites that are rich in carbon: these are the most primitive of the meteorites. Indirect measurements suggest that these meteorites are about 4.56 billion years old—consistent with all other measurements of the time that has passed since the Solar System was formed.

**Figure 12.22** Cross sections of several kinds of meteorites: (a) a chondrite (a stony meteorite with chondrules); (b) an achondrite (a stony meteorite without chondrules); (c) an iron meteorite; (d) a stony-iron meteorite.

The second major category of meteorites, **iron meteorites** (Figure 12.22c), comes from M-type asteroids. Iron meteorites can be recognized by their melted and pitted appearance generated by frictional heating as it streaked through the atmosphere. Many iron meteorites are never found, either because they land in water or because they are not recognized as meteorites. The Mars *Opportunity* rover discovered a few iron meteorites on the martian surface (**Figure 12.23**). Both their appearance—typical of iron meteorites found on Earth—and their position on the smooth, featureless plains made them instantly recognizable.

The third category of meteorites is the **stony-iron meteorites**, which consist of a mixture of rocky material and iron-nickel alloys (Figure 12.22d). Stony-iron meteorites are relatively rare.

**Figure 12.23** This 7-foot-long iron meteorite lying on the surface of Mars was imaged by the Mars exploration rover *Curiosity*.

## Meteorites and the History of the Solar System

Meteorites are extremely valuable because they are samples of the same relatively pristine material that makes up asteroids. Astronomers can take meteorites into the laboratory and study them. Scientists compare meteorites to rocks found on Earth and the Moon and contrast their structure and chemical makeup with rocks studied by spacecraft that have landed on Mars and Venus. Comparing the spectra of meteorites with those of asteroids and planets reveals their origin.

Meteorites come from asteroids, which derive from stony-iron planetesimals. A few planetesimals in the region between the orbits of Mars and Jupiter evolved toward becoming tiny planets before being shattered by collisions. Some became volcanically active, with eruption of lava onto their surfaces. But rather than forming planets, these planetesimals broke into pieces in collisions with other planetesimals.

Some types of meteorites fail to follow the patterns just discussed. Whereas most achondrites have ages in the range 4.5 billion to 4.6 billion years, some are less than 1.3 billion years old. Other achondrites are chemically and physically similar to the soil and the atmospheric gases that NASA's lander instruments have measured on Mars. The similarities are so strong that most planetary scientists think these meteorites are pieces of Mars that were knocked into space by large asteroidal impacts—so that researchers can study pieces of Mars in laboratories here on Earth. In 1996, a NASA research team announced that the meteorite ALH84001, found in Antarctica, showed possible physical and chemical evidence of past life on Mars, but the claim is still debated (see Chapter 24).

Another group of meteorites bear striking similarities to samples returned from the Moon. Like the meteorites from Mars, these are chunks of the Moon that were blasted into space by impacts and later fell to Earth. It is possible, therefore, that meteorites from Earth fell on the Moon. If they are ever collected from the unchanging Moon, they could tell us about what conditions were like on the early Earth.

## Zodiacal Dust

Like meteoroids, **zodiacal dust** is a mixture of cometary debris and ground-up asteroidal material. Just as you can "see" sunlight streaming through an open window by observing its reflection from dust drifting in the air, you can see the sunlight reflected off tiny zodiacal dust particles that fill the inner parts of the Solar System close to the plane of the ecliptic. On a clear, moonless night, not long after the western sky has grown dark, this dust is visible as a faint column of

**Figure 12.24** Zodiacal light shines in the western sky after sunset as seen from the La Silla Observatory in Chile.

light slanting upward from the western horizon along the path of the ecliptic. This band, called the **zodiacal light**, can also be seen in the eastern sky just before dawn (**Figure 12.24**). With good eyes and an especially dark night, you may be able to follow the zodiacal dust band all the way across the sky. In its brightest parts, the zodiacal light can be several times brighter than the Milky Way, for which it is sometimes mistaken.

The dust grains are roughly a millionth of a meter in diameter—the size of smoke particles. In the vicinity of Earth, each cubic kilometer of space contains only a few particles of zodiacal dust. The total amount of zodiacal dust in the entire Solar System is estimated to be $10^{16}$ kg, equivalent to a solid body about 25 km across, or roughly the size of a large comet nucleus. Grains of zodiacal dust are constantly being lost as they are swept up by planets or pushed out of the Solar System by the pressure of sunlight. Such interplanetary dust grains have been recovered from Earth's upper atmosphere by aircraft flying very high. If not replaced by new dust from comets, all zodiacal dust would be gone within the brief span of 50,000 years.

In the infrared region of the spectrum, thermal emission from the band of warm zodiacal dust makes it one of the brightest features in the sky. It is so bright that astronomers wanting to observe faint infrared sources are frequently hindered by its foreground glow.

### CHECK YOUR UNDERSTANDING 12.4

Meteorites contain clues to which of the following? (Choose all that apply.) (a) the age of the Solar System; (b) the temperature in the early solar nebula; (c) changes in the composition of the primitive Solar System; (d) changes in the rate of cratering in the early Solar System; (e) the physical processes that controlled the formation of the Solar System.

## 12.5 Collisions Still Happen Today

Almost all hard-surfaced objects in the Solar System still bear the scars of a time when tremendous impact events were common. Although such impacts are far less frequent today than they once were, they still happen. In this section, we'll examine a few examples of recent collisions.

### Comet Shoemaker-Levy 9 Collided with Jupiter

Early in the 20th century, the orbit of a comet nucleus called Shoemaker-Levy 9 from the Kuiper Belt was perturbed, and the comet's new orbit carried it close to Jupiter. Eventually, it was captured by Jupiter and orbited the planet. In 1992, this comet passed so close to Jupiter that tidal stresses broke it into two dozen major fragments, which subsequently spread out along its orbit. The fragments took one more 2-year orbit around the planet, and throughout a week in 1994, the entire string of fragments crashed into Jupiter. The impacts occurred just behind the limb of the planet, so they were not visible on Earth until Jupiter's rotation put the impact points in view. Astronomers using ground-based telescopes and the Hubble Space Telescope could see immense plumes rising from the impacts to heights of more than 3,000 km above the cloud tops at the limb. The debris in these plumes then rained back onto Jupiter's stratosphere, causing ripples like pebbles thrown into a pond. **Figure 12.25** shows some HST images of the impact features. Sulfur and carbon compounds released by the impacts formed Earth-sized scars in the atmosphere that persisted for months.

## Collisions with Earth

In summer 1908, a remote region of western Siberia was blasted with the energy equivalent of 2,000 times the energy of the atomic bomb dropped on Hiroshima. Eyewitness accounts detailed the destruction of dwellings, the incineration of reindeer (including one herd of 700), and the deaths of at least five people. Although trees were burned or flattened over more than 2,150 square kilometers (km²)—an area greater than metropolitan New York City—no crater was left behind. The Tunguska event (named for the nearby river) was the result of a tremendous high-altitude explosion that occurred when a small body hit Earth's atmosphere, ripped apart, and formed a fireball before reaching Earth's surface. Recent expeditions to the Tunguska area have recovered resin from the trees blasted by the event. Chemical traces in the resin suggest that the impacting object may have been a stony asteroid.

In February 2013, a known near-Earth object about half the size of an American football field passed so close to Earth that it came within the orbit of man-made satellites. This near miss was uneventful, and the object simply continued on its way. However, in an unrelated event on the same day, a previously unknown meteoroid estimated to have a radius of about 20 meters exploded over Chelyabinsk, Russia. The shock waves from this explosion damaged thousands of buildings in six cities and injured more than 1,000 people. This was likely the largest impact on Earth since the Tunguska event, and there were many recorded observations of the effect on this less remote location.

From car dashboard, cell phone, and security camera video and images (**Figure 12.26a**), as well as reports of the time between the brighter-than-the-Sun flash and the sonic boom that followed, scientists determined the trajectory and speed of the incoming object as it traveled through the atmosphere. They estimate a preimpact orbit of the object in the inner asteroid belt and think it originally broke off from a known 2-km-sized asteroid. From small pieces collected over a wide area and from a large, 600-kg chunk found in a frozen-over lake (Figure 12.26b), scientists could analyze the object's composition and density. It seems to

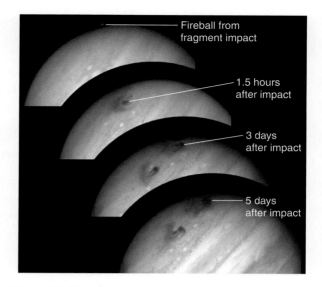

**Figure 12.25** HST images of the evolution of the scar left by one fragment of Comet Shoemaker-Levy 9 when it impacted Jupiter in 2004.

Fireball from fragment impact

1.5 hours after impact

3 days after impact

5 days after impact

(a)

(b)

**Figure 12.26** (a) In February 2013, a meteoroid entered the atmosphere over Russia, creating a fireball that eyewitnesses said was brighter than the Sun. (b) A 600-kg piece of the meteorite on display.

## 12.2 Working It Out Impact Energy

How much energy can be released from the impact of a comet nucleus? The kinetic energy of a moving object is given by

$$E_K = \frac{1}{2}mv^2$$

where $E_K$ is the kinetic energy in joules (J), $m$ is the mass in kilograms, and $v$ is the speed in meters per second.

Suppose an asteroid or comet nucleus that is 10 km in diameter with a mass of $5 \times 10^{14}$ kg hits Earth at a speed of 20 km/s = $20 \times 10^3$ m/s. Putting these values into the preceding equation gives us

$$E_K = \frac{1}{2} \times (5 \times 10^{14}\,\text{kg}) \times (20 \times 10^3\,\text{m/s})^2$$

$$E_K = 1.0 \times 10^{23}\,\text{J}$$

How much energy is this? A 1-megaton hydrogen bomb (67 times as energetic as the Hiroshima atomic bomb) releases $4.2 \times 10^{15}$ J. If we divide the energy from our comet impact by this number, we have

$$\frac{1.0 \times 10^{23}\,\text{J}}{4.2 \times 10^{15}\,\text{J/megaton H-bomb}} = 2.4 \times 10^7$$

$$= 24\,\text{million 1-megaton H-bombs}$$

That's a lot of energy, and it shows why impacts have been so important in the history of the Solar System.

be similar in composition to Itokawa, the asteroid whose dust was collected and returned to Earth. It is estimated that only a few hundredths of 1 percent of the original 10-million-kg mass has been found on the surface of Earth. The energy of the explosion was about 30 times the blast power of a World War II atomic bomb. However, most of that energy went into the atmosphere, heating and breaking up the meteoroid at a much higher altitude than where bombs are detonated, so the effects on the ground were less than that of a bomb.

These impacts are sobering events. The distribution of relatively large asteroids in the inner Solar System indicates it is highly improbable that an asteroid will impact a populated area on Earth within your lifetime. However, there are many comets and smaller asteroids with unknown orbits, and several previously unknown long-period comets enter the inner Solar System each year. If on a collision course with Earth, even a large comet might not be noticed until just a few weeks or months before impact. For example, Comet Hyakutake was discovered only 2 months before it passed near Earth, and a potentially destructive asteroid that just missed Earth in 2002 was not discovered until 3 days *after* its closest approach. The Chelyabinsk meteoroid was in an orbit such that it wasn't detectable at all. Earth's geological and historical record suggests that actual impacts by large bodies are infrequent events.

There may be as many as 10 million asteroids larger than a kilometer across, but only about 130,000 have well-known orbits, and most of the unknown asteroids are too small to see until they come very close to Earth. The U.S. government—along with the governments of several other nations—is aware of the risk posed by Near-Earth Objects. Although the probability of a collision between a small asteroid and Earth is quite small, the consequences could be catastrophic (**Working It Out 12.2**), so NASA has been given a congressional mandate to catalog all NEOs and to scan the skies for those that remain undiscovered.

### CHECK YOUR UNDERSTANDING 12.5

How do astronomers determine the origin of a meteorite that reaches Earth?

# Origins

## Comets, Asteroids, Meteoroids, and Life

Water is essential to life on Earth, so it seems important to know where the water came from. However, the origin of the water is still under debate. Scientists have thought that some of Earth's water was contributed during impacts of icy planetesimals early in the history of the Solar System. The icy planetesimals condensed from the protoplanetary disk surrounding the young Sun and grew to their current size near the orbits of the giant planets. These planetesimals subsequently suffered strong orbital disturbances from the giant planets, which may themselves have been migrating to and from the inner Solar System. In such interactions, some of the planetesimals were flung outward to form the Kuiper Belt and Oort Cloud, and some were thrown inward toward the Sun, possibly hitting Earth. Because much of the mass in comet nuclei and some in asteroids appears to be in the form of water ice, it is possible that some of Earth's current water supply came from this early bombardment. Spacecraft have measured the type of water in several comets and asteroids, and to date the water on Earth best matches that of the primitive carbonaceous chondrites and a few, but not most, comets.

Comets and asteroids can threaten life on Earth. Occasional collisions of comet nuclei and asteroids with Earth have probably resulted in widespread devastation of Earth's ecosystem and in the extinction of many species. Passing stars or the periodic passage of the Sun through giant gas clouds located in denser regions of the Milky Way Galaxy may have resulted in showers of comet nuclei into the inner Solar System, possibly contributing to a change in the climate and mass extinctions. Although such events certainly qualify as global disasters for the plants and animals alive at the time, they also represent global opportunities for new life-forms to evolve and fill the niches left by species that did not survive. As noted in Chapter 8, such a collision with an asteroid or comet likely played a central role in ending the 180-million-year reign of dinosaurs and provided an opportunity for the evolution of mammals.

In studying comets, astronomers may also have found a key to the chemical origins of life on Earth. Comets are rich in complex organic material—the chemical basis for terrestrial life—and cometary impacts on the young Earth may have played a role in chemically seeding the planet. If comets are pristine samples of the material from which the Sun and planets formed, then organic material must be widely distributed throughout interstellar space. Radio telescope observations of vast interstellar clouds throughout the Milky Way confirm the presence of organic material. The fact that asteroid belts and storms of comets have been observed in distant solar systems could have significant implications as astronomers consider the possibility of life elsewhere in the universe.

# READING ASTRONOMY NEWS

ARTICLES    QUESTIONS

Scientists report on the water observed on Comet 67P/Churyumov-Gerasimenko.

# *Rosetta* Spacecraft Finds Water on Earth Didn't Come from Comets

By **REBECCA JACOBSON, PBS NewsHour**

It's a mystery that has baffled scientists for decades: Where did Earth's water come from?

Some scientists believed comets might have been the original source of the Earth's oceans. But a study published this week in the journal *Science* is sending scientists back to the drawing board. In its first published scientific data, the ROSINA mass spectrometer on board the *Rosetta* probe found that water on Comet 67P/Churyumov-Gerasimenko doesn't match the water on Earth.

The result is surprising, says Kathrin Altwegg, principal investigator for ROSINA at the University of Bern and one of the authors of the study. For decades, scientists had ruled out comets from the Oort Cloud at the very edge of our Solar System as the source of Earth's water.

But three years ago, an analysis of water on the Hartley 2 comet near Jupiter found a perfect match to the Earth's oceans. That finding led scientists to believe that Earth's water could have come from much closer comets, either near Jupiter or in the Kuiper Belt just beyond Neptune. Comet 67P/Churyumov-Gerasimenko is one of those Jupiter family comets, which scientists believe originated in the Kuiper Belt.

"That was a big surprise, but now we are back to what I expected," she said. "I think it's very nice to see the diversity we have in Kuiper Belt, to see that not everything is as simple as it seemed."

To find the origin of Earth's water, scientists analyze the water's "fingerprint," says Claudia Alexander, project scientist of the U.S. *Rosetta* Project at NASA's Jet Propulsion Laboratory. Water has a chemical isotopic signature, which works just like a fingerprint. Planets, comets, even minerals all have a fingerprint, Alexander says, and scientists are looking for a match to Earth's.

On Earth, water is mostly two parts hydrogen and one part oxygen—$H_2O$. But there's also "heavy" water, Alexander explained, which is made with deuterium—a hydrogen atom with a neutron. That heavy water is what the *Rosetta* spacecraft found on Comet 67P/Churyumov-Gerasimenko. It's also a closer match to the water scientists have found on other comets, ruling them out as Earth's water source, Alexander said.

"The clues don't quite all add up," she said.

Altwegg agrees, saying that it's not likely the other Kuiper Belt comets have a match to Earth's water, but further studies would be helpful.

"You would have to assume that 67P is the exception in the Kuiper Belt," she said. "We need more missions to Kuiper Belt comets, which would be fabulous."

There are several ideas to explain the origin of Earth's water, Alexander said. Some believe that water has been on Earth since its formation, that it was beaten out of other minerals as the planet formed. Others think "wet planetesimals" near Jupiter—which were like planetary Silly Putty, loose sticky blobs of rock and ice, Alexander said—collided with Earth in the early formation of the Solar System.

Alexander believes the answer could be a combination of any of these ideas. The *Dawn* mission in 2015 will study the water on the asteroid Ceres, near Jupiter. If it's a match for Earth's water, it may be another clue, Alexander said. But the *Rosetta* finding is a huge step in solving the mystery, she said.

"I think this is a big deal. . . . For me, I've not always been a believer in the story that comets brought the water," Alexander said. "In some respects, I'm somewhat relieved (this finding) doesn't confirm it. It's more complicated than that. I think we need more forensic evidence to settle the score."

1. This article uses the word "fingerprint" when discussing isotope ratios. Earlier in this book, we used it in our discussion of spectral lines. How are the two phenomena similar?
2. What might explain why the water on Comet Hartley 2 matched Earth's water but the water on Comet 67P/Churyumov-Gerasimenko did not?
3. Is the water on other comets more like the water on Comet Hartley 2 or on Comet 67P/Churyumov-Gerasimenko?
4. What are other possible sources of Earth's water?
5. Do an internet search to see if there have been any results from the Dawn mission indicating the type of water observed on Ceres.

# Summary

The story of how planetesimals, asteroids, and meteorites are related is a great success of planetary science. Scientists have assembled a wealth of information about this diverse collection of objects to piece together a picture of how planetesimals grow, differentiate, and then shatter in subsequent collisions. This story fits well with the even larger story of how most planetesimals were accreted into the planets and their moons. Comets, asteroids and meteoroids may have supplied Earth with water, volatiles, and organic material in the early history of the Solar System. Impacts with large asteroids or comet nuclei may have led to mass extinctions on Earth that eventually enabled mammals to evolve. Recent spacecraft missions to comets and asteroids have begun to reveal details about the composition of these bodies.

**LG 1** **List the categories of small bodies and identify their locations in the Solar System.** Small bodies in the Solar System that orbit the Sun include dwarf planets, asteroids, comets, Kuiper Belt objects, and meteoroids. Most asteroids are located in the main asteroid belt between the orbits of Mars and Jupiter. Comets are small, icy planetesimals that reside in the frigid regions of the Kuiper Belt and the Oort Cloud, beyond the planets. The orbits of nearly all comets are highly elliptical, with one end of the orbit close to the Sun and the other in the distant parts of the Solar System.

**LG 2** **Describe the defining characteristics of the dwarf planets in the Solar System.** Pluto, Eris, Haumea, Makemake, and Ceres are classified as dwarf planets because, although they are sufficiently massive to have pulled themselves into round shapes, they are not massive enough to have cleared their surroundings of other bodies and are therefore not planets.

**LG 3** **Describe the origin of the different types of asteroids, comets, and meteorites.** Asteroids are small Solar System bodies made of rock and metal. Although early collisions between these planetesimals created several bodies large enough to differentiate, Jupiter's tidal disruption (and possible migration) prevented them from forming a single planet. Comets that venture into the inner Solar System are warmed by the Sun, producing an atmospheric coma and a tail. Meteoroids are small fragments of asteroids and comets. When a meteoroid enters Earth's atmosphere, frictional heat causes the air to glow, producing a meteor. Meteor showers occur when Earth passes through a trail of cometary debris. A meteoroid that survives to a planet's surface is called a meteorite. The various types of meteoroids that are formed depend on the differentiation of the parent body.

**LG 4** **Explain how asteroids, comets, and meteoroids provide important clues about the history and formation of the Solar System.** Asteroids, comets, and meteoroids are leftover debris from the formation of the Solar System. Asteroids are composed of the same type of material that became the inner planets, and comets are composed of the same type of material that became the outer planets. They provide samples of the initial composition and properties of the Solar System and furnish samples of material from its entire history.

**LG 5** **Describe what has been learned from observations of recent impacts in the Solar System.** Impacts by comets and meteoroids have been observed recently on Jupiter and Earth. The debris from these impacts helps astronomers understand the conditions in the early Solar System when impacts were much more frequent.

## ? UNANSWERED QUESTIONS

- What can asteroids and comets reveal about the dynamics of the early Solar System? It is not accidental that the main asteroid belt and the Kuiper Belt straddle the orbits of the giant planets. As noted in Chapter 10, the giant planets may have moved around quite a bit in the early Solar System. This migration of giant planets may explain the spread of orbits of the objects in the main asteroid belt. Migration may also have brought icy objects—such as comet nuclei—out of the Kuiper Belt and into the main asteroid belt. The Kuiper Belt would have been closer to the Sun originally, and Jupiter and Saturn would have pushed it outward. And finally, the migration might have sent objects from both belts into the inner Solar System, creating the heavy bombardment of 4 billion years ago.

- Is there a small, dim star in the neighborhood of the Sun that comes close periodically and stirs up the Oort Cloud, sending a much higher than average number of comets into the inner Solar System? Some scientists think the fossil data show that there have been periodic mass extinctions on Earth, and they have investigated astronomical causes of the extinctions. One hypothesis is that a distant companion to the Sun has a wide orbit and periodically has come closer to the Sun as they both have traveled around the Milky Way Galaxy and has stirred up the Oort Cloud as a result. This Oort Cloud disturbance sent a large number of comets to the inner Solar System, which could have caused many impacts on Earth, leading to a mass extinction similar to the one that wiped out the dinosaurs. But some scientists dispute that Earth's extinctions have occurred at regular intervals, and recent surveys with infrared telescopes have found no evidence of a small companion star.

# Questions and Problems

## Test Your Understanding

1. This chapter deals with leftover planetesimals. What became of most of the others?
   a. They evaporated.
   b. They left the Solar System.
   c. They became part of larger bodies.
   d. They fragmented into smaller pieces.

2. The three types of meteorites come from different parts of their parent bodies. Stony-iron meteorites are rare because
   a. they are hard to find.
   b. the volume of a differentiated body that has both stone and iron is small.
   c. there is very little iron in the Solar System.
   d. the magnetic field of the Sun attracts the iron.

3. As a comet leaves the inner Solar System, the ion tail points
   a. back along the orbit.
   b. forward along the orbit.
   c. toward the Sun.
   d. away from the Sun.

4. Congress tasked NASA with searching for near-Earth objects because
   a. they might impact Earth, as others have in the past.
   b. they are close by and easy to study.
   c. they are moving fast.
   d. they are scientifically interesting.

5. Meteorites can provide information about all of these except
   a. the early composition of the Solar System.
   b. the composition of asteroids.
   c. the composition of comets.
   d. the Oort Cloud.

6. Perihelion is the point in an orbit _____ the Sun; aphelion is the point in an orbit _____ the Sun.
   a. closest to; farthest from
   b. farthest from; closer to
   c. at one focus of; at the other focus of

7. Kuiper Belt objects (KBOs) are actually comet nuclei. Why do they not display comae and tails?
   a. Most of the material has already been stripped from the objects.
   b. They are too far from the Sun.
   c. They are too close to the Sun.
   d. The comae and tails are pointing away from Earth, behind the object.

8. Asteroids are small
   a. rock and metal objects orbiting the Sun.
   b. icy objects orbiting the Sun.
   c. rock and metal objects found only between Mars and Jupiter.
   d. icy bodies found only in the outer Solar System.

9. Aside from their periods, short-period and long-period comets differ because
   a. short-period comets orbit prograde, while long-period comets orbit in either sense.
   b. short-period comets contain less ice, while long-period comets contain more.
   c. short-period comets do not develop ion tails, while long-period comets do.
   d. short-period comets come closer to the Sun at closest approach than long-period comets.

10. On average, a bright comet appears about once each decade. Statistically, this means that
    a. one will definitely be observed every tenth year.
    b. one will definitely be observed in each 10-year period.
    c. exactly 10 comets will be observed in a century.
    d. about 10 comets will be observed in a century.

11. Most asteroids are located between the orbits of
    a. Earth and Mars.
    b. Mars and Jupiter.
    c. Jupiter and Saturn.
    d. the Kuiper Belt and the Oort Cloud.

12. Comets, asteroids, and meteoroids may be responsible for delivering a significant fraction of the current supply of _____ to Earth.
    a. mass
    b. water
    c. oxygen
    d. carbon

13. An iron meteorite most likely came from
    a. an undifferentiated asteroid.
    b. a differentiated asteroid.
    c. a planet.
    d. a comet.

14. Meteor showers occur because Earth passes through the path of
    a. another planet.
    b. a planetesimal.
    c. a comet.
    d. the Moon.

15. Dwarf planets differ from the other planets in that they
    a. have no atmosphere.
    b. have no moons.
    c. are all very far from the Sun.
    d. have lower mass.
    e. are covered in ice.

## Thinking about the Concepts

16. Describe ways in which Pluto differs significantly from the Solar System planets.

17. By what criteria did Pluto fail to be considered a planet under the new IAU definition? Explain how this decision demonstrates the self-correcting nature of science.

18. How does the composition of an asteroid differ from that of a comet nucleus?

19. Define *meteoroid*, *meteor*, and *meteorite*.

20. What are the differences between a comet and a meteor in terms of their size, distance, and how long they remain visible?

21. Most meteorites are 4.54 billion years old. Carbonaceous chondrites, however, are 20 million years older. What determines the time of "birth" of these pieces of rock? What does this information tell you about the history of their parent bodies?

22. Most asteroids are found between the orbits of Mars and Jupiter, but astronomers are especially interested in the relative few whose orbits cross that of Earth. Why?

23. How could you and a friend, armed only with your cell phones and knowledge of the night sky, prove conclusively that meteors are an atmospheric phenomenon?

24. Suppose you find a rock that has all the characteristics of a meteorite. You take it to a physicist friend who confirms that it is a meteorite but says that radioisotope dating indicates an age of only a billion years. What might be the origin of this meteorite?

25. Describe differences between the Kuiper Belt and the Oort Cloud as sources of comets. What is the ultimate fate of a comet from each of these reservoirs?

26. What are the three parts of a comet? Which part is the smallest in radius? Which is the most massive?

27. In 1910, Earth passed directly through the tail of Comet Halley. Among the various gases in the tail was hydrogen cyanide, deadly to humans. Yet nobody became ill from this event. Why?

28. Comets have two types of tails. Describe them and explain why they sometimes point in different directions.

29. What is zodiacal light, and what is its source?

30. How might comets and asteroids have contributed to the origin of life on Earth?

## Applying the Concepts

31. Comet 67P/Churyumov-Gerasimenko has a diameter of 4 km and a mass of $10^{13}$ kg.
    a. What is the density of the comet? How does that compare with the density of water?
    b. What is the escape velocity from the surface of this comet?

32. Ceres has a diameter of 975 km and a period of about 9 hours. What is the rotational speed of a point on the surface of this dwarf planet?

33. Figure 12.12 shows the scale of the Solar System out to the Oort Cloud. Judging from this figure, what fraction of the distance between the Sun and Proxima Centauri is occupied by the Oort Cloud, a part of the Solar System?

34. Follow Working It Out 12.1 (and use the information in Appendix 4) to find the perihelion and aphelion distances for Pluto and Eris.

35. Follow Working It Out 12.2 to find the impact energy (in joules) of an asteroid with a mass of $4.6 \times 10^{11}$ kg traveling at 40 km/s. Does this energy depend on the "target" of impact? What is the equivalent in 1-megaton H bombs?

36. Earth's Moon has a diameter of 3,474 km and orbits at an average distance of 384,400 km. At this distance. it subtends an angle just slightly larger than half a degree in Earth's sky. Pluto's moon Charon has a diameter of 1,186 km and orbits at a distance of 19,600 km from the dwarf planet.
    a. Compare the appearance of Charon in Pluto's skies with the Moon in Earth's skies.
    b. Describe where in the sky Charon would appear as seen from various locations on Pluto.

37. One recent estimate concludes that nearly 800 meteorites with mass greater than 100 grams (massive enough to cause personal injury) strike the surface of Earth each day. Assuming you present a target of 0.25 square meter (m²) to a falling meteorite, what is the probability that you will be struck by a meteorite during your 100-year lifetime? (Note that the surface area of Earth is approximately $5 \times 10^{14}$ m².)

38. Electra is a 182-km-diameter asteroid accompanied by a small moon orbiting at a distance of 1,350 km in a circular orbit with a period of 3.92 days.
    a. What is the mass of Electra?
    b. What is Electra's density?

39. Calculate the orbital radius of the Kirkwood gap that is in a 3:1 orbital resonance with Jupiter.

40. The orbital periods of Comets Encke, Halley, and Hale-Bopp are 3.3 years, 76 years, and 2,530 years, respectively. Their orbital eccentricities are 0.847, 0.967, and 0.995, respectively.
    a. What are the semimajor axes (in astronomical units) of the orbits of these comets?
    b. What are the minimum and maximum distances from the Sun (in astronomical units) reached by Comets Halley and Hale-Bopp in their respective orbits?
    c. Which region of the Solar System did each likely come from?
    d. Which would you guess is the most pristine comet among the three? Which is the least? Explain your reasoning.

41. Comet Halley has a mass of approximately $2.2 \times 10^{14}$ kg. It loses about $3 \times 10^{11}$ kg each time it passes the Sun.
    a. The first confirmed observation of the comet was made in 240 BCE. Assuming a constant period of 76.4 years, how many times has it reappeared since that early sighting?
    b. How much mass has the comet lost since 240 BCE?
    c. What percentage of the comet's total mass today does this amount represent?

42. If Comet Halley is approximated as a sphere 5 km in radius, what is its density if it has a mass of $2.2 \times 10^{14}$ kg? How does that density compare to that of water ($1,000$ kg/m³)?

43. A cubic centimeter of the air you breathe contains about $10^{19}$ molecules. A cubic centimeter of a comet's tail may typically contain 200 molecules. Calculate the cubic volume of comet tail material that would hold $10^{19}$ molecules.

44. Some near-Earth objects are in binary systems, so it is possible to estimate their mass. How much energy would be released if a near-Earth asteroid with mass $m = 4.6 \times 10^{11}$ kg hit Earth at a speed ($v$) of 5 km/s?

45. The estimated amount of zodiacal dust in the Solar System remains constant at approximately $10^{16}$ kg. Yet zodiacal dust is constantly being swept up by planets or removed by the pressure of sunlight.
    a. If all the dust disappeared (at a constant rate) over a span of 30,000 years, what would the average production rate, in kilograms per second, have to be to maintain the current content?
    b. Is this an example of static or dynamic equilibrium? Explain your answer.

## USING THE WEB

46. Dwarf planets:
    a. Go to planetary astronomer Mike Brown's website of dwarf planets (http://gps.caltech.edu/~mbrown/dps.html). How many dwarf planets does he think are in the Solar System? Why is it difficult officially to certify an object as a dwarf planet?
    b. Go to the website for the *New Horizons* mission (http://pluto.jhuapl.edu), which reached Pluto in 2015 and is scheduled to visit Kuiper Belt objects afterward. Click on "Where Is New Horizons?" What is the spacecraft's current location? How far is it from Earth, and how far from Pluto? How long would it take to send a radio signal to the spacecraft? Click on "News Center." What has been learned from this mission?

47. Go to the website for the *Dawn* mission (http://dawn.jpl.nasa.gov).
    a. Read the sections on "Mission" and "Science," and look at the videos and images. What was learned about Vesta on this mission?
    b. What was learned about dwarf planet Ceres during *Dawn*'s visit in 2015?

48. Citizen science projects:
    a. Go to Asteroid Zoo (http://www.asteroidzoo.org/). What are the science goals of this project? Click on "Classify" and read through the Tutorial and Guide. Classify some frames, and save a copy for your homework.
    b. Go to Cosmoquest (https://cosmoquest.org), and click on Asteroid Mappers. What are the science goals? Read through the FAQ and the Tutorial. If you don't already have an account (from Moon Mappers), create one. Log in and get some images, and mark some craters. Do they have data on Ceres? If so, analyze some of those images too.

49. Go to NASA's Asteroid Watch website (http://www.jpl.nasa.gov/asteroidwatch). What is new? Has there been a new discovery or a recent flyby? Was the asteroid studied with a spacecraft, an orbiting telescope, or a ground-based telescope? What has been learned about the object?

50. Go to the Space Weather website (http://spaceweather.com). Are any comets currently visible with the naked eye? Scroll down to "Near Earth Asteroids." Are any "close encounters" coming up in the next few months? Click on a few asteroid names to access the JPL Small-Body Database, where you can view an animation of the orbits. In each case, how close will the NEO be to Earth when it is at its closest? Note the values of $e$ and $a$ in the table under the orbit. Calculate the NEO's closest and farthest distances from the Sun. How large is the object?

# smartw⬡rk**5**

If your instructor assigns homework in Smartwork5, access your assignments at digital.wwnorton.com/astro5.

Astronomers often discover asteroids and other small Solar System objects by comparing two (or more) images of the same star field and looking for bright spots that have moved between the images. The four images in **Figure 12.27** are all "negative images": every dark spot would actually be bright on the sky, and all the white space is dark sky. A negative image sometimes helps the observer pick out faint details, and it is preferable for printing and photocopying. Study these four images. Can you find an asteroid that moves across the field in these

images? That's the hard way to do it. A much easier method is to use a "blink comparison," which lets you look at one image and then another very quickly. Make three photocopies of each of these images, cut out each one, and align all of them carefully on top of one another in sequence, so that the stars overlap. You should have 12 pieces of paper, in this order: Image 1, Image 2, Image 3, Image 4; Image 1, Image 2, . . . and so on. Staple the top edge and flip the pages with your thumb, looking carefully at the images. Can you find the asteroid now?

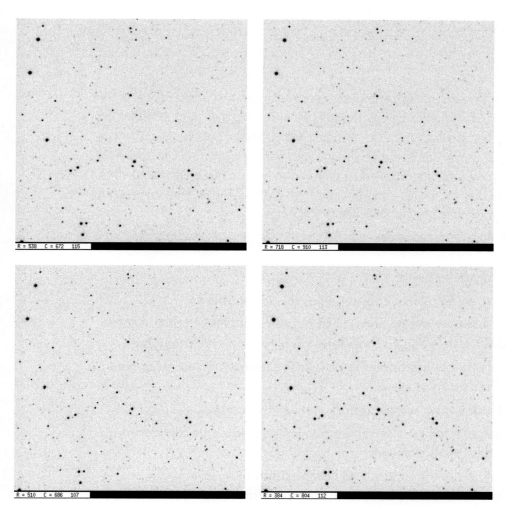

R = 538   C = 672   115

R = 718   C = 910   113

R = 510   C = 686   107

R = 384   C = 804   112

**Figure 12.27**

**1** Circle the asteroid in each image.

The "blink comparison" method takes advantage of a feature of the human eye-brain connection. Humans are much better at noticing things that move than things that do not.

**2** Why might this feature be a helpful evolutionary adaptation?

There's a third method, which sometimes makes things easier to see but requires high-quality digital images. In that method, one image is subtracted from another.

**3** If you used the subtraction method with two of these images, what would you expect to see in the resulting image?

# 13

# Taking the Measure of Stars

To all but the largest telescopes, even nearby stars are just points of light in the night sky. Astronomers study the stars by observing their light, by using the laws of physics discussed in earlier chapters, and by finding patterns in subgroups of stars that are extrapolated to other stars. Astronomers use knowledge of geometry, radiation, and orbits to begin to answer basic questions about stars, such as how they are similar to or different from the Sun, and whether they might have planets orbiting around them as the Sun does.

## LEARNING GOALS

By the conclusion of this chapter, you should be able to:

LG 1  Explain how the brightness of nearby stars and their distances from Earth are used to determine how luminous they are.

LG 2  Explain how astronomers obtain the temperatures, sizes, and composition of stars.

LG 3  Describe how astronomers estimate the masses of stars.

LG 4  Classify stars, and organize this information on a Hertzsprung-Russell (H-R) diagram.

LG 5  Explain how the mass and composition of a main-sequence star determine its luminosity, temperature, and size.

The constellation Orion. The reddish object in the middle of the vertical line of stars is a nebula. ▶ ▶ ▶

Why do stars
have different
colors?

359

# 13.1 Astronomers Measure the Distance, Brightness, and Luminosity of Stars

When looking up at the stars in the sky, it is immediately noticeable that they differ in brightness and color. However, we don't know if one star appears brighter than another because it has a higher luminosity and is emitting more light or because it is closer to us. In this section, you will learn how to find the distances to nearby stars and how to use distance and apparent brightness to find the luminosity of a star.

## Stereoscopic Vision

Your two eyes have different views that depend on the distance to the object you are viewing. Hold up your finger in front of you, quite close to your nose. View it with your right eye only and then with your left eye only. Each eye sends a slightly different image to your brain, so your finger *appears* to move back and forth relative to the background behind it. Now hold up your finger at arm's length, and blink your right eye, then your left. Your finger appears to move much less. The way your brain combines the different information from your eyes to perceive the distances to objects around you is called **stereoscopic vision**. **Figure 13.1a** shows an overhead view of the experiment you just performed with your finger. The left eye sees the blue pencil almost directly between the green balls on the bookshelf. But the right eye sees the blue pencil to the left of both balls. Similarly, the position of the pink pencil appears to vary. Because the pink pencil is closer to the observer, its position appears to change more than the position of the blue pencil—it seems to move from the right of the blue pencil to the left of the blue pencil.

Stereoscopic vision enables you to judge the distances of objects as far away as a few hundred meters, but beyond that it is of little use. Your right eye's view of a mountain several kilometers away is indistinguishable from the view seen by your left eye—all you can determine is that the mountain is too far away for you to judge its distance stereoscopically. The distance over which your stereoscopic vision works is limited by the separation between your eyes, about 6 centimeters (cm). If you could separate your eyes by several meters, the view from each eye would be different enough for you to judge the distances to objects that are kilometers away.

Of course, you cannot separate your eyes, but you can compare pictures taken with a camera from two widely separated locations. The greatest separation we can obtain without leaving Earth is to let Earth's orbital motion carry us from one side of the Sun to the other. If you take a picture of the sky tonight and then wait 6 months and take another picture, the distance between the two locations is the diameter of Earth's orbit (2 astronomical units [AU]), which gives us more powerful stereoscopic vision.

**Astronomy in Action:** Parallax

## Distances to Nearby Stars

Figure 13.1b shows how astronomers apply this concept of stereoscopic vision to measure the distances to stars. This illustration shows Earth's orbit as viewed from far above the Solar System. The change in position of Earth over 6 months is like the distance between the right eye and the left eye in Figure 13.1a. The nearby (pink and blue) stars are like the pink and blue pencils, while the distant yellow stars are like the green balls on the bookcase. Because of the shift in perspective as Earth orbits the Sun, nearby stars appear to shift their positions. The pink star, which is closer, appears to move farther than the more distant blue star. Over the

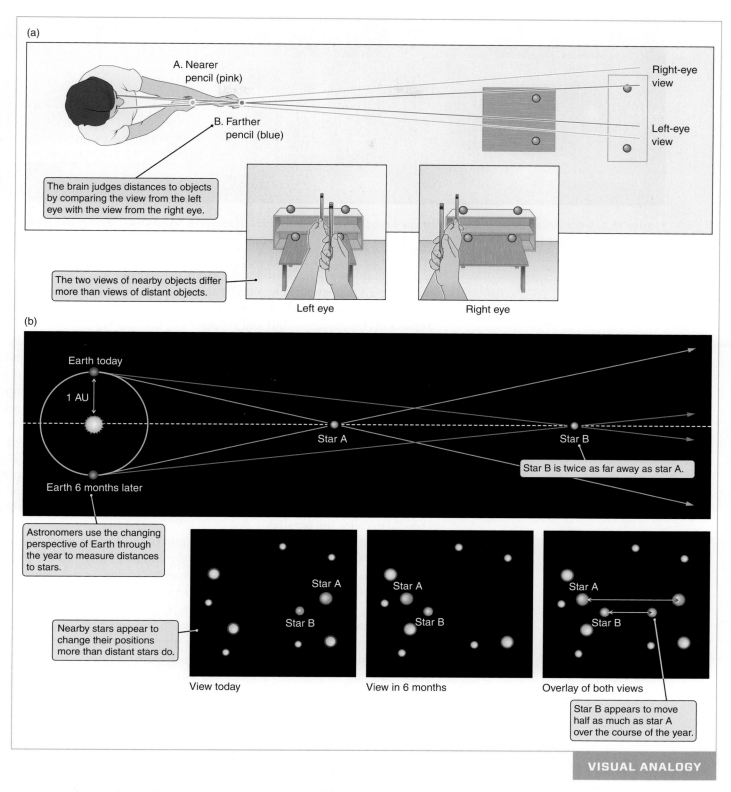

**Figure 13.1** (a) Stereoscopic vision enables you to determine the distance to an object by comparing the view from each eye. (b) Similarly, comparing views from different places in Earth's orbit enables astronomers to determine the distances to stars. As Earth moves around the Sun, the apparent positions of nearby stars change more than the apparent positions of more distant stars. (The diagram is not to scale.)

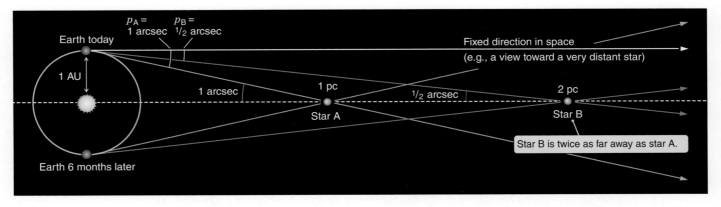

**Figure 13.2** The parallax (*p*) of a star is inversely proportional to its distance. More distant stars have smaller parallaxes. (The diagram is not to scale.)

**Nebraska Simulation:** Parallax Calculator

course of 1 full year, the nearby star appears to move one way and then back again with respect to distant background stars, returning to its original position at the end of that 1-year period. We can determine the distance to the star using the amount of this apparent shift and geometry.

The eye cannot detect the changes in position of a nearby star throughout the year, but telescopes can reveal these small shifts relative to the background stars. **Figure 13.2** shows Earth, the Sun, and two stars in a similar configuration as that in Figure 13.1b. Look first at star A, the closest star. When Earth is at the top of the figure, it forms a right triangle with the Sun and star A at the other corners. (Remember that a right triangle is one with a 90° angle in it.) The short leg of the triangle is the distance from Earth to the Sun, which is 1 AU. The long leg of the triangle is the distance from the Sun to star A. The small angle near star A is called the *parallactic angle*, or simply **parallax**, of the star. As Earth completes an orbit around the Sun, the star's position in the sky appears to shift back and forth, returning to its original position at the end of that year. The amount of this shift is equal to twice the parallax.

More distant stars make longer and skinnier triangles with smaller parallaxes. Star B is twice as far away as star A, and its parallax is only half the parallax of star A. If you were to draw a number of such triangles for different stars, you would find that increasing the distance to the star always reduces the star's parallax. Moving a star 3 times farther away reduces its parallax to $\frac{1}{3}$ of its original value. Moving a star 10 times farther away reduces its parallax to $\frac{1}{10}$ of its original value. The parallax of a star (*p*) is inversely proportional to its distance (*d*).

The parallaxes of real stars are tiny. Recall from Chapter 2 that the full circle of the sky can be divided into 360 degrees. The apparent diameter of the full Moon in the sky averages about half a degree. Just as an hour on the clock is divided into minutes and seconds, a degree of sky can be divided into arcminutes and arcseconds. An **arcminute** (abbreviated **arcmin**) is 1/60 of a degree, and an **arcsecond** (abbreviated **arcsec**) is 1/60 of an arcminute. An arcsecond is about equal to the angle formed by the diameter of a golf ball at a distance of 9 km.

Recall from Chapter 1 that we usually use units of light-years to indicate distances to stars. One light-year is the distance that light travels in 1 year—about 9.5 trillion kilometers (km). We use this unit because it is the unit you are most likely to see online or in a popular book about astronomy. When astronomers discuss distances to stars and galaxies, however, the unit they often use is the **parsec (pc)**, which is equal to 3.26 light-years (or 206,265 AU). The term is short for *parallax second*—a star at a distance of 1 parsec has a parallax of 1 arcsecond.

## 13.1 Working It Out  Parallax and Distance

As seen in Figure 13.2, if star B is twice as far away from us as star A, then star B will have half the parallax of star A. The parallax of a star ($p$) is inversely proportional to its distance ($d$):

$$p \propto \frac{1}{d} \quad \text{or} \quad d \propto \frac{1}{p}$$

If the angle at the apex of a triangle is 1 arcsec, and the base of the triangle is 1 AU, then the length of the triangle is 1 parsec. One parsec equals 206,265 AU (see Appendix 1), or $3.09 \times 10^{16}$ meters, or 3.26 light-years. Astronomers use the unit of parsecs because it makes the relationship between distance and parallax easier than that with the use of light-years.

As illustrated in Figure 13.2, a star with a parallax of 1 arcsec is at a distance of 1 pc. The inverse proportionality between distance and parallax becomes

$$\left( \begin{array}{c} \text{Distance measured} \\ \text{in parsecs} \end{array} \right) = \frac{1}{\left( \begin{array}{c} \text{Parallax measured} \\ \text{in arcseconds} \end{array} \right)}$$

or

$$d(\text{pc}) = \frac{1}{p(\text{arcsec})}$$

Suppose that the parallax of a star is measured to be 0.5 arcsec. The distance can be found by

$$d(\text{pc}) = \frac{1}{0.5} = 2 \,\text{pc}$$

Similarly, a star with a measured parallax of 0.01 arcsec is located at a distance of $1/0.01 = 100$ pc.

After the Sun, the next closest star to Earth is Proxima Centauri. Located at a distance of 4.24 light-years, Proxima Centauri is a faint member of a system of three stars called Alpha Centauri. What is this star's parallax? First, we convert the distance to parsecs:

$$d = 4.24 \,\text{light-years} \times \frac{1 \,\text{parsec}}{3.26 \,\text{light-years}} = 1.30 \,\text{parsecs}$$

Then,

$$p(\text{arcsec}) = \frac{1}{1.30 \,\text{pc}} = 0.77 \,\text{arcsec}$$

Even the closest star to the Sun has a parallax of only about $\frac{3}{4}$ arcsec.

When astronomers began to measure the parallax angles of stars, they discovered that stars are very distant objects (**Working It Out 13.1**). The first successful measurement of the parallax of a star was made by F. W. Bessel (1784–1846), who in 1838 reported a parallax of 0.314 arcsec for the star 61 Cygni. This finding implied that 61 Cygni was 3.2 pc away, or 660,000 times as far away as the Sun. With this one measurement, Bessel increased the known volume of the universe by a factor of 10,000. Today, astronomers know of about 60 stars in 54 single-, double-, or triple-star systems within 5 pc (16.3 light-years) of the Sun. In the neighborhood of the Sun, each star or star system has on average a volume of about 260 cubic light-years of space to itself.

Most stars are so far away that the parallax angle is too small to measure using ground-based telescopes, which are limited by Earth's atmosphere. In the 1990s, the European Space Agency's Hipparcos satellite measured the positions and parallaxes of 120,000 stars, thus greatly improving our picture of the Sun's stellar neighborhood. The accuracy of any given Hipparcos parallax measurement is about $\pm 0.001$ arcsec. Because of this observational uncertainty, measurements of the distances to stars are not perfect. For example, a star with a Hipparcos-measured parallax of $0.004 \pm 0.001$ arcsec really has a parallax between 0.003 and 0.005 arcsec. This gives a corresponding distance range of 200–330 pc from Earth. As an analogy, consider your speed while driving down the road. If your digital speedometer says 10 kilometers per hour (km/h), you might actually be traveling 10.4 km/h or 9.6 km/h. The precision of your speedometer is limited to the nearest 1 km/h.

A successor to Hipparcos is Gaia, a space mission to study stellar parallaxes that was launched at the end of 2013. Gaia is expected to observe 1 billion stars and measure the parallaxes of 20 million of them with a high precision. Other methods of measuring distance to more remote stars will be discussed later.

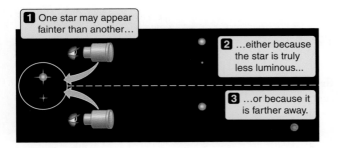

**Figure 13.3** The brightness of a star visible in our sky depends on both its luminosity—how much light it emits—and its distance.

## Luminosity, Brightness, and Distance

The stars in Earth's sky are of different brightnesses. In Chapter 5, you saw that brightness corresponds to the amount of energy falling on a square meter of area each second in the form of electromagnetic radiation. Although the brightness of a star can be measured directly, it does not immediately give much information about the star itself. As illustrated in **Figure 13.3**, a bright star in the night sky may in fact only appear bright because it is nearby. Conversely, a faint star may be a powerful beacon, still visible despite its tremendous distance.

Astronomers measure the brightness of stars by comparing them to one another. The system they use dates back 2,100 years, when the Greek astronomer Hipparchus classified stars according to their brightness. The details of his system, which is still in use today, are discussed in **Working It Out 13.2** and Appendix 7.

To learn about the actual properties of a star, astronomers need to know the total energy radiated by a star each second—the star's luminosity. Recall from

---

## 13.2 Working It Out The Magnitude System

The **magnitude** system of brightness for celestial objects can be traced back 2,100 years to the ancient Greek astronomer Hipparchus, who classified the brightest stars he could see as being "of the first magnitude" and the faintest as being "of the sixth magnitude." Later, astronomers defined Hipparchus's 1st magnitude stars as being exactly 100 times brighter than his 6th magnitude stars. Hipparchus must have had typical eyesight, because an average person under dark skies can see stars only as faint as 6th magnitude. Today, telescopes extend our vision far into space. The Hubble Space Telescope can integrate for long exposures and detect stars as faint as 30th magnitude. The limits have also been extended to stars brighter than 1st magnitude using zero and negative numbers. A negative magnitude signifies that an object is *brighter* than an object at zero magnitude. For example, Sirius, the brightest star in the sky, has a magnitude of −1.46. Venus can be as bright as magnitude −4.4, or about 15 times brighter than Sirius and bright enough to cast a shadow. The magnitude of the full Moon is −12.6, and that of the Sun is −26.7 (**Figure 13.4**).

Looking at this mathematically, with five steps between the 1st and 6th magnitudes, each step is equal to the fifth root of 100, or $100^{1/5}$, which is approximately 2.512. This system is logarithmic, but instead of the usual base 10, it is base 2.512. Thus, a 5-magnitude difference in brightness equals $(2.512)^5$, or 100, times difference in brightness. Fifth magnitude stars are 2.512 times brighter than 6th magnitude stars, and 4th magnitude stars are $2.512 \times 2.512 = 6.310$ times brighter than 6th magnitude stars. A 2.5-magnitude difference equals $(2.512)^{2.5}$, or 10, times difference in brightness. The brightness ratio between any two stars is equal to $(2.512)^N$, where $N$ is the magnitude difference between them.

Since the limit of the Hubble Space Telescope (HST) is 30 and $30 − 6 = 24$, HST can detect stars that are $(2.512)^{24} = 4 \times 10^9$, or 4 billion, times fainter than the magnitude 6 that the naked eye can see. Or if we compare the Sun and the Moon, the Sun is 14 magnitudes, or $(2.512)^{14} = 4 \times 10^5$ (400,000) times brighter than the full Moon. (More detailed calculations and a table of magnitudes and brightness differences are located in Appendix 7.)

The magnitude of a star, as we have discussed it, is called the star's **apparent magnitude** because it is the brightness of the star as it *appears* in Earth's sky. Now imagine that all stars were located at exactly 10 pc (32.6 light-years) away from Earth. The brightness of each star would then reflect its luminosity. If the distance from Earth to a star is known, astronomers compute how bright the star would appear if it were located at 10 pc. The **absolute magnitude** of a star—its apparent magnitude at a distance of 10 pc—measures the star's luminosity.

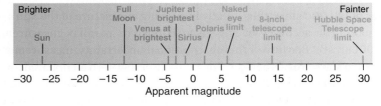

**Figure 13.4** Apparent magnitude indicates the apparent brightness of an object in our sky. The brightest objects have a negative apparent magnitude, while telescopes have extended the observable range to fainter objects with higher magnitudes.

Chapter 5 that the brightness of an object that has a known luminosity and is located at a distance $d$ is given by the following equation:

$$\text{Brightness} = \frac{\text{Total light emitted per second}}{\text{Area of a sphere of radius } d} = \frac{\text{Luminosity}}{4\pi d^2}$$

You can rearrange this equation, moving the quantities you know how to measure (distance and brightness) to the right-hand side and the quantity you would like to know (luminosity) to the left, to get

$$\text{Luminosity} = 4\pi d^2 \times \text{Brightness}$$

This equation is used to find how much total light a star must be giving off in order to appear as bright as it does when seen from Earth.

Different stars have different luminosities. The Sun provides a convenient comparison when measuring the properties of stars, including their luminosity. The luminosity of the Sun is measured at $L_{\text{Sun}} = 3.9 \times 10^{26}$ watts (W). The most luminous stars exceed a million times the luminosity of the Sun ($10^6\ L_{\text{Sun}}$). The least luminous stars have luminosities less than 1/10,000 that of the Sun ($10^{-4}\ L_{\text{Sun}}$). The most luminous stars are therefore more than 10 billion ($10^{10}$) times more luminous than the least luminous stars. Only a very small fraction of stars are near the upper end of this range of luminosities. The vast majority of stars are at the faint end of this distribution, less luminous than the Sun. **Figure 13.5** shows the relative number of stars compared to their luminosities in solar units. (Distances for the nearest stars are obtained from their parallaxes; other methods—to be discussed later—are used for the more distant stars.)

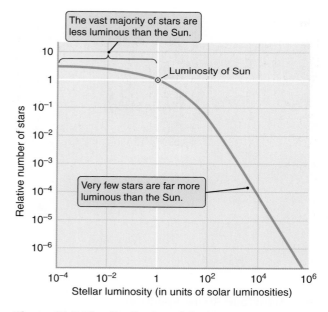

Figure 13.5 The distribution of the luminosities of stars is plotted logarithmically in this graph, so that increments are in powers of 10.

 **Nebraska Simulation:** Stellar Luminosity Calculator

---

### CHECK YOUR UNDERSTANDING 13.1

Stars A and B appear equally bright, but star A is twice as far away from us as star B. Which of the following is true? (a) Star A is twice as luminous as star B. (b) Star A is 4 times as luminous as star B. (c) Star B is twice as luminous as star A. (d) Star B is 4 times as luminous as star B. (e) Star A and star B have the same luminosity because they have the same brightness.

## 13.2 Astronomers Can Determine the Temperature, Size, and Composition of Stars

Two everyday concepts—stereoscopic vision and the fact that objects appear brighter when closer—have provided the tools needed to measure the distance and luminosity of the closest stars. Stars that appear to be faint points of light in the night sky are in fact luminous beacons located at great distances. The laws of radiation that we described in Chapter 5 reveal still more about stars.

Stars are gaseous, but they are dense enough that the radiation from a star comes close to obeying the same laws as the radiation from solid objects like the heating element on an electric stove. We can therefore use our knowledge of Planck blackbody radiation to understand the radiation from stars. Recall both the Stefan-Boltzmann law from Chapter 5, which states that among same-sized objects, the hotter objects are more luminous, and Wien's law, which states that hotter objects are bluer. In this section, we will use these two laws to measure the temperatures and sizes of stars. We will also develop a more detailed understanding of the line emission mentioned in Chapter 5 to obtain information about the composition of stars.

## Wien's Law Revisited: The Color and Surface Temperature of Stars

Wien's law (see Working It Out 5.3) shows that the temperature of an object determines the peak wavelength of its spectrum. The hotter the surface of an object, the bluer the light that it emits. Stars with especially hot surfaces are blue, stars with especially cool surfaces are red, and yellow-white stars such as the Sun are in-between. If you obtain a spectrum of a star and measure the wavelength at which the spectrum peaks, then Wien's law will tell you the temperature of the star's surface. The color of a star tells you about the temperature only at the surface, because this layer is giving off most of the radiation that we see. (Stellar interiors are far hotter than this, as we will discuss in the next chapter.)

▶️II **AstroTour:** Stellar Spectrum

In practice, it is not necessary to obtain a complete spectrum of a star to determine its temperature. Astronomers often measure the colors of stars by comparing the brightness at two different, specific wavelengths. The brightness of a star is usually measured through an optical **filter**—sometimes just a piece of colored glass—that lets through only a small range of wavelengths. Two of the most common filters are a blue filter that allows light with wavelengths of about 440 nanometers (nm) to pass through and a "visual" (yellow-green) filter that allows light with wavelengths of about 550 nm to pass through. The ratio of brightness between the blue and visual filters is called the *color index* of the star (more details are discussed in Appendix 7). From a pair of pictures of a group of stars, each taken through a different filter, we can find an approximate value of the surface temperature of every star in the picture—perhaps hundreds or even thousands—all at once. This type of analysis shows that there are many more cool stars than hot stars, and most stars have surface temperatures lower than that of the Sun.

## Classification of Stars by Surface Temperature

Although the hot "surface" of a star emits radiation with a spectrum very close to a smooth Planck blackbody curve, this light must then escape through the outer layers of the star's atmosphere. The atoms and molecules in the cooler layers of the star's atmosphere leave their absorption line fingerprints in the escaping light, as shown in **Figure 13.6**. Under some circumstances, the atoms and molecules in the star's atmosphere, along with any gas that might be found near the star, can produce emission lines in stellar spectra. Absorption and emission lines complicate how astronomers use the laws of Planck blackbody radiation to interpret light from stars, but spectral lines provide a wealth of information about the state of the gas in a star's atmosphere.

The spectra of stars were first classified during the late 1800s, long before stars, atoms, or radiation were well understood. Stars were classified by the appearance of the dark bands (now known as absorption lines) seen in their spectra. The original ordering of this classification was arbitrarily based on the prominence of particular absorption lines known to be associated with the element hydrogen. Stars with the strongest hydrogen lines were denoted *A stars*, stars with somewhat weaker hydrogen lines were denoted *B stars*, and so on.

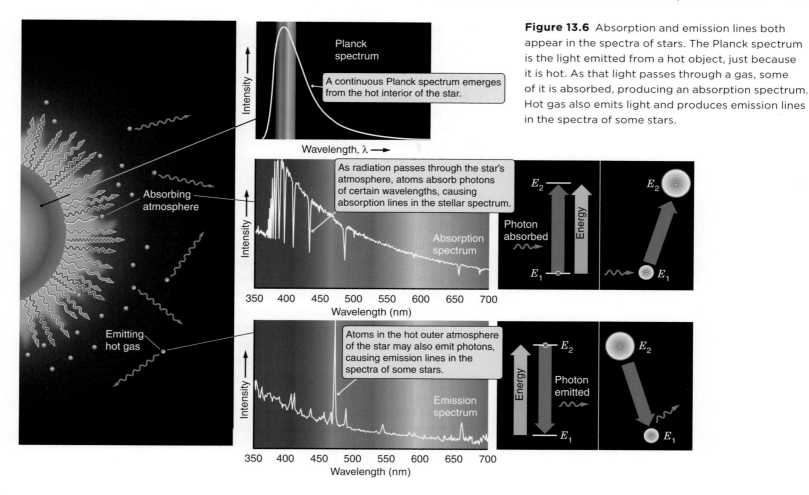

**Figure 13.6** Absorption and emission lines both appear in the spectra of stars. The Planck spectrum is the light emitted from a hot object, just because it is hot. As that light passes through a gas, some of it is absorbed, producing an absorption spectrum. Hot gas also emits light and produces emission lines in the spectra of some stars.

Annie Jump Cannon (1863–1941) led an effort at the Harvard College Observatory to examine and classify the spectra of hundreds of thousands of stars systematically. She dropped many of the earlier spectral types, keeping only seven that were subsequently reordered based on surface temperatures. Spectra of stars of different spectral types are shown between the horizontal bars in **Figure 13.7**. The hottest stars, with surface temperatures above 30,000 K, are denoted *O stars*. O stars have only weak absorption lines from hydrogen and helium. The coolest stars—*M stars*—have temperatures as low as about 2800 K. M stars show absorption lines from many different types of atoms and molecules. The sequence of **spectral types** of stars, from hottest to coolest, is O, B, A, F, G, K, M. This sequence has undergone several modifications over time, most recently to add cooler objects known as brown dwarfs with spectral types L, T, and Y.

Astronomers divide the main spectral types into a finer sequence of subclasses by adding numbers to

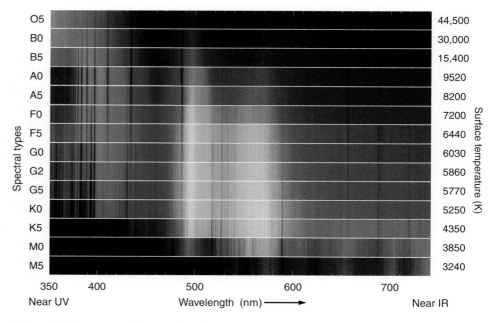

**Figure 13.7** Spectra of stars with different spectral types, ranging from hot, blue O stars to cool, red M stars. Hotter stars are more luminous at shorter wavelengths. The dark lines are absorption lines.

the letter designations. For example, the hottest B stars are B0 stars, slightly cooler B stars are B1 stars, and so on. The coolest B stars are B9 stars, which are only slightly hotter than A0 stars. The Sun is a G2 star. The boundaries between spectral types are not always easy to determine. A hotter-than-average G star is very similar to a cooler-than-average F star.

In Figure 13.7, notice that not only are hot stars bluer than cool stars but also the absorption lines in their spectra are quite different. The temperature of the gas in the atmosphere of a star affects the state of the atoms in that gas, which in turn affects the energy level transitions available to absorb radiation (see Section 5.2 to review the concept of atomic energy levels). In O stars, the temperature is so high that most atoms have had one or more electrons stripped from them by energetic collisions within the gas. Few transitions are available in the visible part of the electromagnetic spectrum, so the visible spectrum of an O star is relatively featureless. At lower temperatures, there are more atoms in energy states that can absorb light in the visible part of the spectrum, so the visible spectra of cooler stars are far more complex than the spectra of O stars.

All absorption lines have a temperature at which they are strongest. For example, absorption lines from hydrogen are most prominent at temperatures of about 10,000 K, which is the surface temperature of an A star. At the very lowest stellar temperatures, atoms in the atmosphere of a star react with each other, forming molecules. Molecules such as titanium oxide (TiO) are responsible for much of the absorption in the atmospheres of cool M stars.

Because different spectral lines are formed at different temperatures, astronomers can use these absorption lines to measure a star's temperature. The surface temperatures of stars measured in this way agree extremely well with the surface temperatures of stars measured using Wien's law, again confirming that the physical laws that apply on Earth apply to stars as well.

## The Composition of Stars

Most of the variations in the lines of a particular chemical element seen in stellar spectra are due to temperature, but the details of the absorption and emission lines found in starlight also carry a wealth of other information. By applying the physics of atoms and molecules to the study of stellar absorption lines, astronomers can accurately determine not only surface temperatures of stars but also pressures, chemical compositions, magnetic-field strengths, and other physical properties. In addition, by making use of the Doppler shift of emission and absorption lines, astronomers can measure rotation rates, atmospheric motions, expansion and contraction, "winds" driven away from stars, and other dynamic properties of stars.

The chemical composition of a star is also found from its spectra. In Chapter 5, you saw that because each type of atom has different energy levels, each type of atom has different spectral lines. The patterns of spectral lines are measured in laboratories on Earth and then used to identify the types of atoms (or molecules) in stars. For example, if a star has absorption lines that correspond to the energy difference between two levels in the calcium atom, then we know that calcium is present in the atmosphere of the star.

The strengths of various absorption lines tell us not only what kinds of atoms are present in the gas but also the amount of each. However, we must take great care in interpreting spectra to account properly for the temperature and density of the gas in the atmosphere of a star. Recall the "astronomer's periodic table of the elements" from Figure 5.16. Typically, hydrogen composes more than 90 percent of the atoms in the atmosphere of a star, while helium accounts for most of what remains. All of the other chemical elements, which collectively are called **heavy elements** or **massive elements**, are present in only very small amounts.

**Table 13.1** shows the chemical composition of the atmosphere of the Sun. The Sun's composition is fairly typical for stars in its vicinity, but the percentages of various heavy elements can vary tremendously from star to star. Some stars have lower amounts of heavier elements than the Sun. The existence of such stars, all but devoid of more massive elements, provides important clues about the origin of chemical elements and the chemical evolution of the universe. Note that many of the atoms that make up Earth and its atmosphere (for example, iron, silicon, nitrogen, oxygen, and carbon) exist as only a small percentage of the Sun.

## The Stefan-Boltzmann Law and Finding the Sizes of Stars

Stars are so far away that only two can be imaged as more than point sources of light. To determine the size of a star, astronomers must use other measurements: the temperature and the luminosity.

The temperature of a star can be found directly, either from its color through Wien's law (**Figure 13.8a**) or from the strength of its spectral lines. The temperature of the surface of a star is one factor that influences its luminosity. If a large star and a small star are the same temperature, they will emit the same energy from every patch of surface, but the large star has more patches, so it is more luminous altogether. Conversely, if two stars are the same size, the hotter one will be more luminous than the cooler one. This is an application of the Stefan-Boltzmann law, shown in Figure 13.8b. A small, hot star might even be more luminous than a larger cool star.

| TABLE 13.1 | The Relative Amounts of Different Chemical Elements in the Atmosphere of the Sun |

| Element | Percentage of Atoms in the Sun This Element Represents | Percentage of Sun's Mass This Element Represents |
|---|---|---|
| Hydrogen | 92.5 | 74.5 |
| Helium | 7.4 | 23.7 |
| Oxygen | 0.064 | 0.82 |
| Carbon | 0.039 | 0.37 |
| Neon | 0.012 | 0.19 |
| Nitrogen | 0.008 | 0.09 |
| Silicon | 0.004 | 0.09 |
| Magnesium | 0.003 | 0.06 |
| Iron | 0.003 | 0.16 |
| Sulfur | 0.001 | 0.04 |
| Total of others | 0.001 | 0.03 |

**Wien's Law:** Blue stars are hot, red stars are cool.

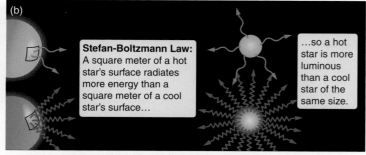

**Stefan-Boltzmann Law:** A square meter of a hot star's surface radiates more energy than a square meter of a cool star's surface…

…so a hot star is more luminous than a cool star of the same size.

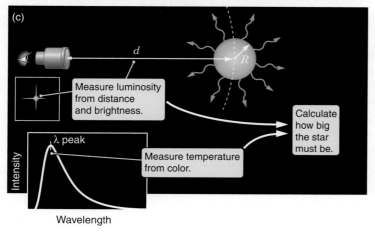

Measure luminosity from distance and brightness.

Measure temperature from color.

Calculate how big the star must be.

**Figure 13.8** (a) The temperature of a star can be found from its color through Wien's law. (b) The luminosity depends on both the temperature and the size of the star. (c) Once the temperature and the luminosity are known, the size of the star can be calculated.

The luminosity of a star can be found from its brightness and its distance. Because the luminosity depends on both the temperature and the size of the star, we can use the luminosity and the temperature to determine the radius of the star, as shown in Figure 13.8c (**Working It Out 13.3**). Astronomers have used the luminosity-temperature-radius relationship to estimate the radii of many thousands of stars. The radius of the Sun, written $R_{Sun}$, is 696,000 km. One of the

## 13.3 Working It Out Estimating the Sizes of Stars

In Chapter 5, you learned that according to the Stefan-Boltzmann law, the amount of energy radiated each second by each square meter of the surface of a star is equal to the constant $\sigma$ multiplied by the surface temperature of the star raised to the fourth power. Written as an equation, this relationship says:

$$\left( \begin{array}{c} \text{Energy radiated each} \\ \text{second by } 1\,\text{m}^2 \text{ of surface} \end{array} \right) = \sigma T^4$$

To find the total amount of light radiated each second by the star, we need to multiply the radiation per second from each square meter by the number of square meters of the star's surface:

$$\left( \begin{array}{c} \text{Energy radiated} \\ \text{each second} \end{array} \right) = \left( \begin{array}{c} \text{Energy radiated each} \\ \text{second by } 1\,\text{m}^2 \text{of surface} \end{array} \right) \times \left( \begin{array}{c} \text{Surface} \\ \text{area} \end{array} \right)$$

The left-hand term in this equation—the total energy emitted by the star per second (in units of joules per second [J/s] = watts [W])—is the star's luminosity, $L$. The middle term—the energy radiated by each square meter of the star per second (in units of joules per square meter per second; $J/m^2/s$)—can be replaced with the $\sigma T^4$ factor from the Stefan-Boltzmann law. The remaining term—the number of square meters covering the surface of the star—is the surface area of a sphere, $A_{sphere} = 4\pi R^2$ (in units of square meters; $m^2$), where $R$ is the radius of the star.

If we replace the words in the equation with the appropriate mathematical expressions for the Stefan-Boltzmann law and the area of a sphere, our equation for the luminosity of a star looks like this:

$$\text{Luminosity} = \sigma T^4 \times 4\pi R^2$$

Combining gives

$$L = 4\pi R^2 \sigma T^4 \text{ J/s (W)}$$

This last equation is called the **luminosity-temperature-radius relationship** for stars. Because the constants (4, $\pi$, and $\sigma$) do not change, the luminosity of a star is proportional only to $R^2 T^4$. Make a star 3 times as large, and its surface area becomes $3^2 = 9$ times as large. There is 9 times as much area to radiate, so there is 9 times as much radiation. Make a star twice as hot, and each square meter of the star's surface radiates $2^4 = 16$ times as much energy. Larger, hotter stars are more luminous than smaller, cooler stars.

Now turn this question around and ask, How large must a star of a given temperature be to have a total luminosity of $L$? The star's luminosity ($L$) and temperature ($T$) are quantities that we can measure, and the star's radius ($R$) is what we want to know. We can rearrange the previous equation, moving the properties that we know how to measure (temperature and luminosity) to the right-hand side of the equation and the property that we would like to know (the radius of the star) to the left-hand side. After a couple of steps of algebra, we find:

$$R = \sqrt{\frac{L}{4\pi\sigma T^4}} = \frac{1}{T^2}\sqrt{\frac{L}{4\pi\sigma}}$$

Again, the right-hand side of the equation contains only things that we know or can measure. The constants 4, $\pi$, and $\sigma$ are always the same. We can find $L$, the luminosity of the star, by use of the measurements of the star's brightness and parallax (although only for nearby stars with known parallax). $T$ is the surface temperature of the star, which can be measured from its color. From the relationship of these measurements, we now know something new: the size of the star.

Often, we compare two stars and the constants all cancel out, leaving $L$, $T$, and $R$:

$$\frac{L_{\text{star1}}}{L_{\text{star2}}} = \frac{R_{\text{star1}}^2}{R_{\text{star2}}^2} \times \frac{T_{\text{star1}}^4}{T_{\text{star2}}^4} \quad \text{or} \quad \frac{R_{\text{star1}}}{R_{\text{star2}}} = \sqrt{\frac{L_{\text{star1}}}{L_{\text{star2}}} \times \frac{T_{\text{star2}}^2}{T_{\text{star1}}^2}}$$

Suppose we compare the Sun to the second brightest star in the constellation Orion, a red star called Betelgeuse. From its spectrum, we know that Betelgeuse's surface temperature $T$ is about 3500 K. Its distance is about 200 pc, and from that and its brightness, its luminosity is estimated to be 140,000 times that of the Sun. What can we say about the size of Betelgeuse? Using the preceding equation, we can determine the following:

$$\frac{R_{\text{Betelgeuse}}}{R_{\text{Sun}}} = \sqrt{\frac{L_{\text{Betelgeuse}}}{L_{\text{Sun}}} \times \frac{T_{\text{Sun}}^2}{T_{\text{Betelgeuse}}^2}}$$

$$\frac{R_{\text{Betelgeuse}}}{R_{\text{Sun}}} = \sqrt{\frac{140,000}{1} \times \frac{5,800^2}{3,500^2}} = 374 \times 2.7 = 1,010$$

Betelgeuse has a radius more than 1,000 times larger than that of the Sun; such stars are called *supergiants*.

smallest types of stars, called white dwarfs, have radii that are only about 1 percent of the Sun's radius—about the size of Earth. The largest stars, called red supergiants, can have radii more than 1,000 times that of the Sun. There are many more stars toward the small end of this range—smaller than the Sun—than there are giant stars.

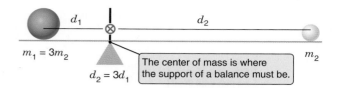

**Figure 13.9** The center of mass of two objects is the "balance" point on a line joining the centers of two masses.

---

## CHECK YOUR UNDERSTANDING 13.2

If star A has twice the surface temperature of the Sun but has the same luminosity as the Sun, the diameter of star A must be _____ the diameter of the Sun.
(a) 16 times; (b) 4 times; (c) $\frac{1}{2}$; (d) $\frac{1}{4}$

..................................................................................

# 13.3 Measuring the Masses of Stars in Binary Systems

Determining the mass of a star is difficult. Astronomers cannot use the amount of light from a star or the star's size as a measure of its mass. Stars can be large or small, faint or luminous. However, more massive stars *always* have stronger gravity. When astronomers are trying to determine the masses of astronomical objects, they almost always wind up looking for the effects of gravity. In Chapter 4, you learned that Kepler's laws of planetary motion are the result of gravity, and that the properties of the orbit of a planet can be used to measure the mass of the Sun. Similarly, astronomers can study two stars that orbit each other to determine their masses.

About half of the higher-mass stars in the sky are actually systems consisting of several stars moving about under the influence of their mutual gravity. Most of these are **binary stars** in which two stars orbit each other in elliptical orbits as predicted by Newton's version of Kepler's laws. This version of Kepler's laws can be used to find the mass of a star, as we will demonstrate. However, most low-mass stars are single, and low-mass stars far outnumber higher-mass stars, so most stars are single and their mass cannot be found this way. In this section, we will look at how astronomers measure the masses of stars in binary systems.

## Binary Star Orbits

The **center of mass** is the balance point of a system. If the two objects were sitting on a seesaw in a gravitational field, the support of the seesaw would have to be directly under the center of mass for the objects to balance, as shown in **Figure 13.9**. When Newton applied his laws of motion to the problem of orbits, he found that two objects must move in elliptical orbits around each other, and that their common center of mass lies at one focus shared by both of the ellipses, as shown in **Figure 13.10**. The center of mass, which lies along the line between the two objects, remains stationary. The two objects will always be found on exactly opposite sides of the center of mass.

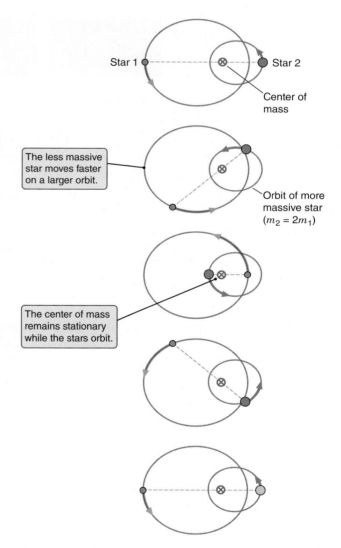

**Figure 13.10** In a binary star system, the two stars orbit on elliptical paths about their common center of mass. In this case, the blue star has twice the mass of the red one. The eccentricity of the orbits is 0.5. There are equal time steps between the frames.

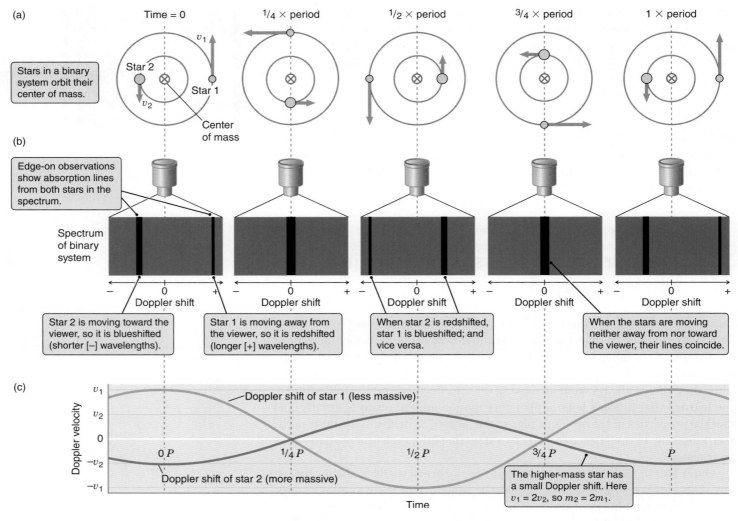

**Figure 13.11** (a) The view from "above" the binary system shows that both stars orbit a common center of mass. (b) The spectrum of the combined system (seen edge-on) shows the spectral lines of each star shift back and forth. (c) Graphing the Doppler shift of star 1 with star 2 versus time reveals that star 2 has half the maximum Doppler shift, so star 2 is twice as massive as star 1. $P$ is the period of the orbit.

**Nebraska Simulation:** Center of Mass Simulator

Imagine that you are watching a binary star as shown in **Figure 13.11a**. As seen from above, two stars orbit the common center of mass. Star 1, which is less massive, must complete its orbit in the same time as star 2, which is more massive. Because the less massive star has farther to go around the center of mass, it must be moving *faster* than the more massive star. In this view, no determination of the Doppler shift (Chapter 5) can be made because all the motion is in the plane of the sky, and none is toward or away from the observer.

When a system is edge-on to the observer, however, the observer can take advantage of the Doppler shift to find out about the motion. Figure 13.11b shows observations of the spectrum of the combined system associated with each position in Figure 13.11a. The spectral lines of the stars shift back and forth as they move toward and away from the observer. Because the two stars are always exactly on opposite sides of their center of mass, they are always moving in opposite directions. When star 2 approaches, star 1 recedes. The light coming from star 2 will be shifted to shorter wavelengths by the Doppler effect as it approaches, so the light will be blueshifted, and the light coming from star 1 will be shifted to longer wavelengths as it recedes, so the light will be redshifted. Half an orbital period later, the situation is reversed: lines from star 1 are blueshifted, and lines from star 2 are redshifted.

The less massive star has a larger orbit—and consequently moves more quickly—than the more massive star. Figure 13.11c compares the velocity obtained from the maximum Doppler shift for star 1 with the velocity obtained from maximum Doppler shift for star 2. This comparison gives the ratio of the masses of the two stars:

$$\frac{v_1}{v_2} = \frac{m_2}{m_1}$$

By observing the spectrum of the system, we can find from this equation the relative masses of the two stars; star 2 is two times as massive as star 1. But we can't find the actual mass of either star from these observations alone.

## The Masses of Binary Stars

In Chapter 4, we ignored the complexity of the motion of two objects around their common center of mass. Now, however, this very complexity enables us to measure the masses of the two stars in a binary system. If we can measure the period of the binary system and the average separation between the two stars, then Kepler's third law gives us the total mass in the system: the sum of the two masses. Because the analysis in the previous subsection gives us the ratio of the two masses, we now have two different relationships between two different unknowns. We have all we need to determine the mass of each star separately. In other words, if we know that star 2 is 2 times as massive as star 1, and we know that star 1 and star 2 together are 3 times as massive as the Sun, then we can calculate separate values for the masses of star 1 and star 2.

Depending on the type of system, there are several ways to measure the average separation and the orbital period. In a **visual binary** system, the system is close enough to Earth, and the stars are far enough from each other, that we can take pictures that show the two stars separately (**Figure 13.12**). Then, astronomers can directly measure the shapes and periods of the orbits of the two stars just by watching them as they orbit each other. These can be used with Doppler measurements of the radial (line-of-sight) velocities of the stars to solve for the ratio of the two masses.

In most binary systems, however, the two stars are so close together and far away from us that we cannot actually see the stars separately. The identification of these stars as binary systems is more indirect and comes from observing periodic variations in the *light* from the star or from observing periodic changes in the *spectrum* of the star. If we view a binary system nearly edge-on, so that one star passes in front of the other, it is called an **eclipsing binary**. An observer will see a repeating dip in brightness as one star passes in front of (eclipses) the other. If the stars are of different temperatures, there will be a repeating pattern of a smaller dip in brightness when the hotter star eclipses the cooler one, followed by a larger dip in brightness when the cooler star eclipses the hotter one, as shown in **Figure 13.13**. The pattern of these dips also gives an estimate of the relative sizes (radii) of the two stars. This procedure for identifying binary systems is similar to the transit method for finding extrasolar planets discussed in Chapter 7, and it works only when the system is viewed nearly edge-on. The Kepler space telescope

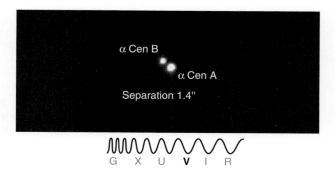

**Figure 13.12** The two stars of a visual binary are resolved. These stars are two components of Alpha Centauri, the nearest star system to the Solar System.

 **Nebraska Simulation:** Eclipsing Binary Simulator

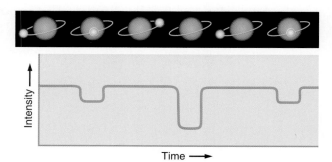

**Figure 13.13** In an eclipsing binary system, the system is viewed nearly edge-on, so that the stars repeatedly pass behind one another, blocking some of the light. When the blue star passes in front of the larger, cooler star, less light is blocked than when it passes behind the red star. The shape of the dips in the light curve of an eclipsing binary can reveal information about the relative size and surface brightness of the two stars.

In Working It Out 4.3, we used Newton's version of Kepler's third law to calculate the mass of the Sun by observing the orbital period of one of its planets. In that special case, the Sun's mass is so much greater than the mass of a planet that the planet's mass is negligible. In the case of two stars, however, the masses are similar, so neither is negligible, and we need to keep both in the equations. Newton showed that if two objects with masses $m_1$ and $m_2$ are in orbit about each other, then the period of the orbit, $P$, is related to the average distance between the two masses, the semimajor axis $A$, by the equation

$$P^2 = \frac{4\pi^2 A^3}{G(m_1 + m_2)}$$

Rearranging this equation a bit turns it into an expression for the sum of the masses of the two objects:

$$m_1 + m_2 = \frac{4\pi^2}{G} \times \frac{A^3}{P^2}$$

We could use the equation this way, with the masses of the two stars in kilograms (kg), the distance between them in kilometers, the period of their orbit in seconds, and the gravitational constant $G$ in kilometers, kilograms, and seconds. However, astronomers often think about stellar masses in units of the Sun's mass. If we divide this equation by the "mass of the Sun" equation in Working It Out 4.3,

$$M_{Sun} = \frac{4\pi^2}{G} \times \frac{A^3}{P^2}$$

(where $M_{Sun}$ = mass of the Sun, $A$ = 1 AU, and $P$ = 1 year), then the constants cancel out and this equation simplifies to

$$\frac{m_1}{M_{Sun}} + \frac{m_2}{M_{Sun}} = \frac{A_{AU}^3}{P_{years}^2}$$

Therefore, if we know both $m_1/m_2$ from measuring velocities by Doppler shifts and $m_1 + m_2$ from the observed orbital properties, we can solve for the separate values of $m_1$ and $m_2$.

Suppose you are an astronomer studying a binary star system. After observing the star for several years, you accumulate the following information about the system:

1. The star is an eclipsing binary.
2. The period of the orbit is 2.63 years.
3. Star 1 has a Doppler velocity that varies between +20.4 and −20.4 km/s.
4. Star 2 has a Doppler velocity that varies between +6.8 and −6.8 km/s.
5. The stars are in circular orbits. You know this because the Doppler velocities about the star are symmetric; the approach and recession speeds of the star are equal.

These data are summarized in **Figure 13.14.** You begin your analysis by noting that the star is an eclipsing binary, which tells you that the orbit of the star is edge-on to your line of sight. The Doppler velocities tell you the total orbital velocity of each star, and you determine the size of the orbits using the relationship

$$\text{Distance} = \text{Speed} \times \text{Time}$$

In one orbital period, star 1 travels around a circle—a distance of

$$d = (20.4 \text{ km/s}) \times (2.63 \text{ yr}) = 53.7 \text{ km} \times \text{yr/s}$$

Multiply by the number of seconds in a year:

$$d = 53.7\frac{\text{km} \times \text{yr}}{\text{s}} \times \frac{3.16 \times 10^7 \text{s}}{\text{yr}} = 1.70 \times 10^9 \text{km}$$

This distance is the circumference of the star's orbit, or $2\pi$ times the radius of the star's orbit, $A_1$. Thus, star 1 is following an orbit with a radius of

$$A_1 = \frac{d}{2\pi} = \frac{1.70 \times 10^9 \text{km}}{2\pi} = 2.7 \times 10^8 \text{km}$$

has been observing and discovering thousands of eclipsing binaries in addition to finding new extrasolar planets.

If a binary system is a **spectroscopic binary**, the spectral lines of the two stars show periodic changes as they are Doppler-shifted away from each other, first in one direction and then in the other, as shown in Figure 13.11. The period of the orbit is determined from the time it takes for a set of spectral lines to go from approaching to receding and back again. The orbital velocities of the stars and the period of the orbit give the size of the orbit because distance equals velocity multiplied by time. Consequently, astronomers can estimate the combined masses of the two stars. To calculate the individual masses, an estimate of the tilt of the orbit is needed. Thus, spectroscopic binary masses are more approximate than those in eclipsing binary systems.

To convert this to astronomical units, use the relation 1 AU = $1.50 \times 10^8$ km:

$$A_1 = 2.7 \times 10^8 \, \text{km} \times \frac{1 \, \text{AU}}{1.50 \times 10^8 \, \text{km}} = 1.8 \, \text{AU}$$

A similar analysis of star 2 shows that its orbit has a radius of $A_2 = 0.6$ AU.

Next, apply Newton's version of Kepler's third law. Because the stars are always on opposite sides of the center of mass, $A_{AU} = 1.8$ AU + 0.6 AU = 2.4 AU. Because you know $A$ and the period $P$ (measured as 2.63 years), you can calculate the total mass of the two stars:

$$\frac{m_1}{M_{Sun}} + \frac{m_2}{M_{Sun}} = \frac{(A_{AU})^3}{(P_{years})^2} = \frac{(2.4)^3}{(2.63)^2} = 2.0$$

So you have learned that the combined mass of the two stars is twice the mass of the Sun. To sort out the individual masses of the stars, use

the measured velocities and the fact that the mass and velocity are inversely proportional:

$$\frac{m_2}{m_1} = \frac{v_1}{v_2} = \frac{20.4 \, \text{km/s}}{6.8 \, \text{km/s}} = 3.0$$

Star 2 is 3 times as massive as star 1. In mathematical terms, $m_2 = 3 \times m_1$. Substituting into the equation

$$m_1 + m_2 = 2.0 \, M_{Sun}$$

gives

$$m_1 + 3m_1 = 2.0 \, M_{Sun}$$

or $4m_1 = 2.0 \, M_{Sun}$, so $m_1 = 0.5 \, M_{Sun}$. Because $m_2 = 3 \times m_1$, then $m_2 = 1.5 \, M_{Sun}$.

Star 1 has a mass of $0.5 \, M_{Sun}$, and star 2 has a mass of $1.5 \, M_{Sun}$. You have just found the masses of two distant stars.

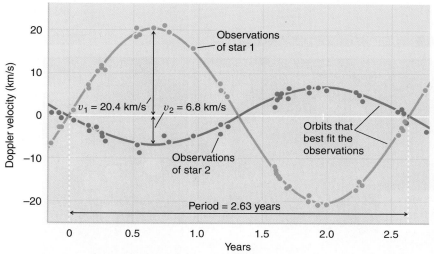
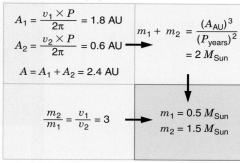

**Figure 13.14** Doppler velocities of the stars in an eclipsing binary are used to measure the masses of the stars.

A binary system can fall into more than one of these three categories, regardless of how it was originally discovered. If a spectroscopic binary system is also a visual or eclipsing binary, then the orbit and masses of the stars can be completely solved (**Working It Out 13.4**). Historically, most stellar masses were measured for stars in eclipsing binary systems, rather than for those in visual or spectroscopic binaries. But new observational capabilities have increased the number of known visual binaries by greatly improving the ability to see the stars in a binary directly. Accurate measurements of masses have been obtained for several hundred binary stars, about half of which are eclipsing binaries. The range of stellar masses found in this way is not nearly as great as the range of stellar luminosities. The least massive stars have masses of about $0.08 \, M_{Sun}$; the most massive stars appear to have masses up to about $200 \, M_{Sun}$.

## 13.4 The Hertzsprung-Russell Diagram Is the Key to Understanding Stars

**Nebraska Simulation:** Hertzsprung-Russell Diagram Explorer

Measuring some basic properties of stars is only the first step in understanding; the next step is to look for patterns in their properties. In the early part of the 20th century, Ejnar Hertzsprung (1873–1967) and Henry Norris Russell (1877–1957) independently studied the properties of stars. Both astronomers plotted the

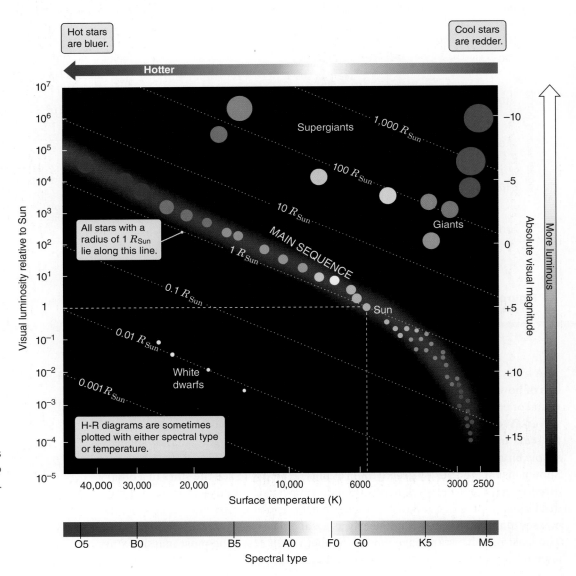

**Figure 13.15** The Hertzsprung-Russell, or H-R, diagram is used to plot the properties of stars. More luminous stars are at the top of the diagram. Hotter stars are on the left. Stars of the same radius (*R*) lie along the dotted lines moving from upper left to lower right. Absolute magnitudes are discussed in Working It Out 13.2 and Appendix 7.

luminosities of stars versus their surface temperatures—a diagram that came to be known as the Hertzsprung-Russell diagram, or simply **H-R diagram**. We will use this diagram often for the study of stars. In this section, we take a first look at this important diagram and the way stars are organized within it.

## The H-R Diagram

We begin with the layout of the H-R diagram, shown in **Figure 13.15**. The spectral type is plotted on the horizontal axis (the *x*-axis), along with the surface temperature plotted backward: temperature is higher on the left and lower on the right. Hot blue stars are on the left side of the H-R diagram; cool red stars are on the right. Temperature is plotted logarithmically. This means that the size of an interval along the axis from a point representing a star with a surface temperature of 40,000 K to one with a surface temperature of 20,000 K—a temperature change by a factor of 2—is the same as the size of an interval between points representing a star with a temperature of 10,000 K and a star with a temperature of 5000 K, which is also a temperature change by a factor of 2. The horizontal axis is sometimes labeled with another characteristic that corresponds to temperature, such as the color.

The luminosity of stars is plotted along the vertical axis (the *y*-axis). More luminous stars are toward the top of the diagram, and less luminous stars are toward the bottom. Sometimes, the luminosity axis is labeled with the absolute visual magnitude instead of luminosity, as shown on the right-hand *y*-axis. As with the temperature axis, luminosities are plotted logarithmically. In this case, each step along the left-hand *y*-axis corresponds to a multiplicative factor of 10 in the luminosity. To understand why the plotting is done this way, recall that the most luminous stars are 10 billion times more luminous than the least luminous stars, yet all of these stars must fit on the same plot.

Each point on the H-R diagram is specified by a surface temperature and luminosity. Therefore, we can use the luminosity-temperature-radius relationship described earlier in the chapter to find the radius of a star at that point as well. A star in the upper right corner of the H-R diagram is very cool, so each square meter of its surface radiates only a small amount of energy. But this star is also extremely luminous. It must be huge to account for its high luminosity, despite the feeble radiation coming from each square meter of its surface. Conversely, a star in the lower left corner of the H-R diagram is very hot, which means that a large amount of energy is coming from each square meter of its surface. However, this star has a very low overall luminosity, so it must be very small. Moving up and to the right takes you to larger and larger stars. Moving down and to the left takes you to smaller and smaller stars. All stars of the same radius lie along slanted lines across the H-R diagram. Astronomers can note the properties of a star—its temperature, color, size, and luminosity—from a glance at its position on the H-R diagram. The discovery and study of these patterns led to an understanding of the astrophysics of stars (**Process of Science Figure**).

## The Main Sequence

**Figure 13.16** shows 16,600 nearby stars plotted on an H-R diagram. The data are based on observations of stars near enough for parallax measurements obtained by the Hipparcos satellite. A quick look at this diagram immediately

▶‖ **AstroTour:** H-R Diagram

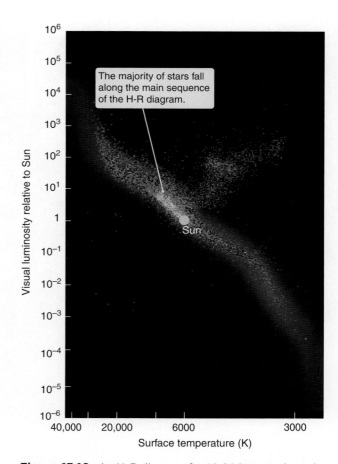

The majority of stars fall along the main sequence of the H-R diagram.

Sun

**Figure 13.16** An H-R diagram for 16,600 stars plotted from data obtained by the Hipparcos satellite clearly shows the main sequence. Most of the stars lie along this band running from the lower right of the diagram toward the upper left.

**An understanding of the meaning behind stellar data took decades, and the contributions of dozens of people, all working toward a common goal.**

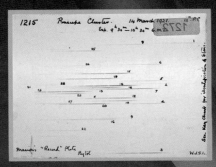

**The Observations (~1880s):**
About 500,000 photographs of stellar spectra are obtained, by many astronomers at many telescopes.

**The Classification (~1900s):**
Annie Jump Cannon leads a team that classifies all the spectra according to the strengths of particular absorption lines at particular wavelengths.

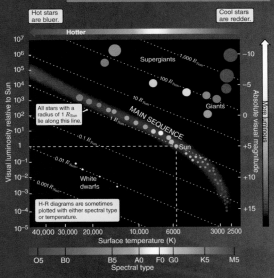

**The Graph (~1910s):**
Hertzsprung and Russell independently develop what will later be called the H-R diagram. They do not understand why the *x*-axis, when ordered O-B-A-F-G-K-M, gives such a nice band across the middle. Russell hypothesizes this must come from a single stellar characteristic.

**The Understanding (~1920s):**
Meghnad Saha shows that the stellar characteristic in question is temperature. Cecilia Payne-Gaposchkin shows that stars are mostly composed of hydrogen and helium. Modern astrophysics is born; others go on to develop the understanding of stellar atmospheres.

Scientific discoveries sometimes seem to occur suddenly. However, new scientific knowledge is usually the effort of many minds working for years to solve a problem.

shows what was first discovered in the original diagrams of Hertzsprung and Russell. About 90 percent of the stars in the sky lie in a well-defined region running across the H-R diagram from lower right to upper left, known as the **main sequence**. On the left end of the main sequence are the O stars: hotter, larger, and more luminous than the Sun. On the right end of the main sequence are the M stars: cooler, smaller, and fainter than the Sun. If you know where a star lies on the main sequence, then you know its approximate luminosity, surface temperature, and size.

The H-R diagram supplies a useful method for finding the distance to main-sequence stars. Astronomers can determine whether a star is on the main sequence by looking at the absorption lines in its spectrum. The spectral type is also determined from the spectral lines, and this spectral type indicates the star's temperature. Once this value on the *x*-axis is known, we can then read up to the main sequence and then across to the *y*-axis to find the star's luminosity. Recall that the luminosity, brightness, and distance are all connected. We can think about how far away a star of a particular luminosity must be to have the observed brightness. So we can find the star's distance by comparing a star's luminosity, obtained from the H-R diagram, with its apparent brightness. This method of determining distances to main-sequence stars from the spectra, luminosity, and brightness of stars is called **spectroscopic parallax**. Details of this method are discussed in Appendix 7. Despite the similarity between the names, this method is very different from the parallax method using trigonometry. Spectroscopic parallax is useful to much larger distances than trigonometric parallax, although it is less precise.

From a combination of observations of binary star masses, parallax, luminosity measurements, and mathematical models, astronomers have determined that stars of different masses lie on different parts of the main sequence. Stellar mass increases smoothly from the lower right to the upper left along the main sequence. If a main-sequence star is less massive than the Sun, it is also smaller, cooler, redder, and less luminous than the Sun, and it is located to the lower right of the Sun on the main sequence. Conversely, if a main-sequence star is more massive than the Sun, it is also larger, hotter, bluer, and more luminous than the Sun, and it is located to the upper left of the Sun on the main sequence, as illustrated in **Figure 13.17**. The mass of a star determines where on the main sequence the star will lie.

**Table 13.2** summarizes the properties of the different spectral classes of main-sequence stars. *All* main-sequence stars with a mass of 1 $M_{Sun}$ are G2 stars like the Sun and have the *same* surface temperature, size, and luminosity as the Sun. Similarly, if a main-sequence star is classified as B0, it has these properties: a surface temperature of about 30,000 K, a luminosity about 32,500 times that of the Sun, a mass of about 17.5 $M_{Sun}$, and a radius of about 6.7 $R_{Sun}$. If a different main-sequence star is classified as M5, it has a surface temperature of 3,170 K, a luminosity of about 0.008 $L_{Sun}$, a mass of about 0.21 $M_{Sun}$, and a radius of about 0.29 $R_{Sun}$.

The relationship between the mass and the luminosity of stars is very sensitive. Relatively small differences in the masses of stars result in large differences in their main-sequence luminosities. From determining the luminosities of binary stars with measured mass, a relationship between the mass and luminosity was found. This

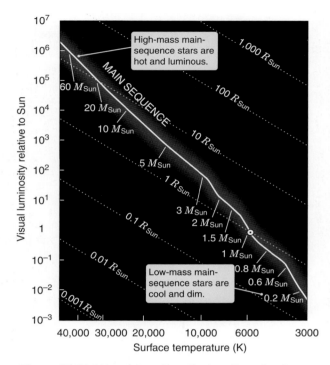

**Figure 13.17** Mass determines the location of a star along the main sequence..

 **Nebraska Simulation:** Spectroscopic Parallax Simulator

| TABLE 13.2 | The Properties of Main Sequence Stars | | | |
|---|---|---|---|---|
| Spectral Type | Temperature (K) | Mass ($M_{Sun}$) | Radius ($R_{Sun}$) | Luminosity ($L_{Sun}$) |
| O5 | 42,000 | 60 | 13 | 500,000 |
| B0 | 30,000 | 17.5 | 6.7 | 32,500 |
| B5 | 15,200 | 5.9 | 3.2 | 480 |
| A0 | 9800 | 2.9 | 2.0 | 39 |
| A5 | 8200 | 2.0 | 1.8 | 12.3 |
| F0 | 7300 | 1.6 | 1.4 | 5.2 |
| F5 | 6650 | 1.4 | 1.2 | 2.6 |
| G0 | 5940 | 1.05 | 1.06 | 1.25 |
| G2 (Sun) | 5780 | 1.00 | 1.00 | 1.0 |
| G5 | 5560 | 0.92 | 0.93 | 0.8 |
| K0 | 5150 | 0.79 | 0.93 | 0.55 |
| K5 | 4410 | 0.67 | 0.80 | 0.32 |
| M0 | 3840 | 0.51 | 0.63 | 0.08 |
| M5 | 3170 | 0.21 | 0.29 | 0.008 |

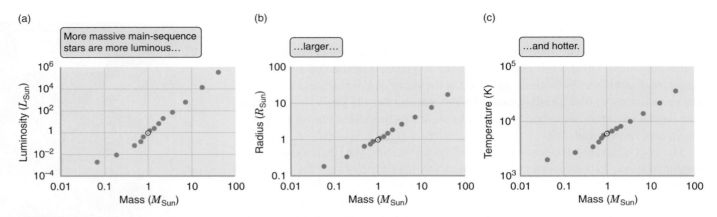

**Figure 13.18** These graphs plot luminosity (a), radius (b), and temperature (c) versus mass for stars along the main sequence. The mass of a main-sequence star determines its other properties.

**mass-luminosity relationship**, usually expressed as $L \propto M^{3.5}$, is shown in **Figure 13.18a**. The exact exponent varies somewhat for different ranges of stellar masses, but this method is useful for estimating masses of single stars. The mass also correlates to the size of a star, shown in Figure 13.18b, and to the temperature of a star, shown in Figure 13.18c.

The mass and chemical composition of a main-sequence star determine its other characteristics: how large it is, what its surface temperature is, how luminous it is, what its internal structure is, how long it will live, how it will evolve, and what its final fate will be. A star must have a balance between gravity trying to hold the star together and the energy released by nuclear reactions in the interior of the star trying to blow it apart. The mass of the star determines the strength of its gravity, which in turn determines how much energy must be generated in its interior to prevent it from collapsing under its own weight. The mass of a star determines where the balance is struck.

## Stars Not on the Main Sequence

Although 90 percent of stars are main-sequence stars, some stars are found in the upper right portion of the H-R diagram, well above the main sequence (see Figure 13.15). So they must be luminous, cool, and large, with radii hundreds or thousands of times the radius of the Sun. These stars are called giants. At the other extreme are stars found in the far lower left corner of the H-R diagram. These stars are the tiny white dwarfs, comparable to the size of Earth. Their small surface areas explain why they have such low luminosities, despite having high temperatures.

Stars that lie off the main sequence on the H-R diagram can be identified by their luminosities (determined by their distance) or by slight differences in their spectral lines. The width of a star's spectral lines is an indicator of the density and surface pressure of gas in the star's atmosphere. In general, denser stars have broader lines. Puffed-up stars above the main sequence have lower densities and lower surface pressure and narrower absorption lines compared to main-sequence stars.

When using the H-R diagram to estimate the distance to a star by the spectroscopic parallax method, astronomers must know whether the star is on, above, or below the main sequence in order to find the star's luminosity. The spectral line widths of stars both on and off the main sequence indicate **luminosity class**, which tells us the relative *size* of the star within each spectral class. Supergiant stars, which are the largest stars that we see, are luminosity class I, bright giants are class II, giants are class III, subgiants are class IV, main-sequence stars are class V, and white dwarfs are class WD. Luminosity classes I–IV lie above the main sequence, while class WD falls below and to the left of the main sequence, as shown in **Figure 13.19**. Thus, the complete spectral classification of a star includes both its spectral type (which indicates temperature and color) and its luminosity class (which indicates relative size).

The existence of the main sequence, together with the fact that the mass of a main-sequence star determines where on the sequence it will lie, is a grand pattern that points to the possibility of a deep understanding of stars. The existence of stars that do *not* follow this pattern raises yet more questions. In the coming chapters, you will learn that the main sequence tells us what stars are and how they work, and that stars off the main sequence reveal how stars form, how they evolve, and how they die. **Table 13.3** summarizes the techniques that astronomers use to determine some of the basic properties of stars. Of the properties listed in the table, only temperature, distance, and composition can be *measured*. Luminosity must be *inferred* from the H-R diagram or calculated from distance and brightness, and size and mass must be *calculated*. Other properties that can be measured include brightness, color, spectral type, and parallax shift.

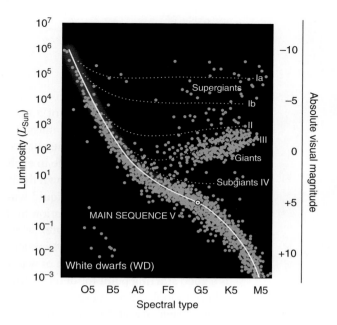

**Figure 13.19** Stellar luminosity classes indicate the size (radius) of a star at each spectral type.

## CHECK YOUR UNDERSTANDING 13.4

Choose the two qualities that describe a star located in the lower right of the H-R diagram: (a) hot; (b) cold; (c) high luminosity; (d) low luminosity.

| TABLE 13.3 | Taking the Measure of Stars |
| --- | --- |

| Property | Methods |
| --- | --- |
| Luminosity | • For a star with a known distance, measure the brightness, then apply the inverse square law of radiation:<br><br>$$\text{Luminosity} = 4\pi \times \text{Distance}^2 \times \text{Brightness}$$<br><br>• For a star without a known distance, take a spectrum of the star to determine its spectral and luminosity classes, plot them on an H-R diagram, and read the luminosity from the diagram. |
| Temperature | • Measure the color index of the star using blue and visual filters. Use Wien's law to relate the color to a temperature.<br><br>• Take a spectrum of the star, and estimate the temperature from its spectral class by noting which spectral lines are present. |
| Distance | • For a relatively nearby star (within a few hundred parsecs), measure the parallax shift of the star over the course of the year.<br><br>• For a more distant star, find the luminosity using the H-R diagram as noted earlier, and then use the spectroscopic parallax method to relate luminosity, distance, and brightness. |
| Size | • For a few of the largest and closest stars, measure the size directly or by the length of eclipse in eclipsing binary stars.<br><br>• From the width of the star's spectral lines, estimate the luminosity class (supergiant, giant, or main sequence).<br><br>• For a star with known luminosity and temperature, use the Stefan-Boltzmann law to calculate the star's radius (the luminosity-temperature-radius relationship). |
| Mass | • Measure the motions of the stars in a binary system, and use these to determine the orbits of the stars, then apply Newton's form of Kepler's third law.<br><br>• For a non-binary star, use the mass-luminosity relationship to estimate the mass from the luminosity. |
| Composition | • Analyze the lines in the star's spectrum to measure chemical composition. |

# Origins

## Habitable Zones

How might a basic property of a star, such as its luminosity, color, mass, or surface temperature, affect the chance of there being a planet with life in orbit around that star? The only known life is that on planet Earth, where liquid water was essential for its formation and evolution. Whether liquid water is an absolute requirement for life elsewhere is not known, but the presence of water is a good starting point for determining where to look. So astronomers look for planets that are at the right distance from their stars to have a planetary temperature that permits water to exist in a liquid state on their surfaces—a range of distances known as the **habitable zone**. On planets that lie inside the habitable zone, water would exist only as a vapor—if at all. On planets that lie outside the habitable zone, water would be permanently frozen as ice.

Recall from the Chapter 5 "Origins" and Working It Out 5.4 that an important factor for estimating the temperature of a planet is the brightness of the sunlight that falls on that planet. This factor depends on the luminosity of the star and the planet's distance from the star. In the Solar System, the habitable zone ranges from about ~0.9 to ~1.4 AU, which includes Earth but just misses Venus and Mars. Main-sequence stars that are less luminous than the Sun are cooler and have narrower habitable zones, minimizing the chance that a planet will form within that slender zone. Main-sequence stars that are more massive than the Sun are hotter and have larger habitable zones. **Figure 13.20** illustrates these zones around Sun-like, hotter, and cooler stars.

In the past few years, astronomers have started to find planets in the habitable zones of their respective stars. Methods of planet detection, as discussed in Chapter 7, work best when the planet is close to its star. Using the transit method, the Kepler Mission has identified and confirmed 1–2 dozen planets in habitable zones as of this writing and has found more candidate planets that need to be confirmed.

The distance from a star at a certain temperature is not the only consideration for whether a planet has water. The presence of a planetary atmosphere is also a factor. More massive planets can retain their atmospheres, which can trap heat and raise the planet's temperature, as we saw in Chapter 5 for Venus and Earth. Smaller planets have a lower gravitational pull and may not be able to keep an atmosphere. Additionally, some habitable zones may be near planets, not stars. Some of the giant planets in the cold outer part of our own Solar System have moons with liquid water. The heat keeping the water liquid comes from the nearby planet, not from the Sun.

Finally, we note that *habitable* does not mean *inhabited*; it only means the planet is at the right distance from its star that it could have liquid water. Identifying planets in their habitable zone is a first step to selecting which planets are most interesting for further study.

**Figure 13.20** The distance and extent of a habitable zone (green) surrounding a star depends on the star's temperature. Regions too close to the star are too hot (red) and those too far away are too cold (blue) to permit the existence of liquid water. The orbits of Mercury, Venus, Earth, and Mars have been drawn around these stars for scale.

*NASA reports on a new technique for using the Hubble Space Telescope to measure parallax angles.* ⓧ

# NASA's Hubble Extends Stellar Tape Measure 10 Times Farther into Space

**NASA Press Release**

Using NASA's Hubble Space Telescope, astronomers now can precisely measure the distance of stars up to 10,000 light-years away—10 times farther than previously possible.

Astronomers have developed yet another novel way to use the 24-year-old space telescope by employing a technique called spatial scanning, which dramatically improves Hubble's accuracy for making angular measurements. The technique, when applied to the age-old method for gauging distances called astronomical parallax, extends Hubble's tape measure 10 times farther into space (**Figure 13.21**).

"This new capability is expected to yield new insight into the nature of dark energy, a mysterious component of space that is pushing the universe apart at an ever-faster rate," said Nobel laureate Adam Riess of the Space Telescope Science Institute (STScI) in Baltimore, Maryland.

Parallax, a trigonometric technique, is the most reliable method for making astronomical distance measurements, and a practice long employed by land surveyors here on Earth. The diameter of Earth's orbit is the base of a triangle, and the star is the apex where the triangle's sides meet. The lengths of the sides are calculated by accurately measuring the three angles of the resulting triangle.

Astronomical parallax works reliably well for stars within a few hundred light-years of Earth. For example, measurements of the distance to Alpha Centauri, the star system closest to our Sun, vary only by 1 arcsecond. This variance in distance is equal to the apparent width of a dime seen from 2 miles away.

Stars farther out have much smaller angles of apparent back-and-forth motion that are extremely difficult to measure. Astronomers have pushed to extend the parallax yardstick ever deeper into our galaxy by measuring smaller angles more accurately.

This new long-range precision was proven when scientists successfully used Hubble to measure the distance of a special class of bright stars called Cepheid variables, approximately 7,500 light-years away in the northern constellation Auriga. The technique worked so well, they are now using Hubble to measure the distances of other far-flung Cepheids.

Such measurements will be used to provide firmer footing for the so-called cosmic "distance ladder." This ladder's "bottom rung" is built on measurements to Cepheid variable stars that, because of their known brightness, have been used for more than a century to gauge the size of the observable universe. They are the first step in calibrating far more distant extra-galactic milepost markers such as Type Ia supernovae.

Riess and the Johns Hopkins University in Baltimore, Maryland, in collaboration with Stefano Casertano of STScI, developed a technique to use Hubble to make measurements as small as five-billionths of a degree.

To make a distance measurement, two exposures of the target Cepheid star were taken 6 months apart, when Earth was on opposite sides of the Sun. A very subtle shift in the star's position was measured to an accuracy of 1/1,000 the width of a single image pixel in Hubble's Wide Field Camera 3, which has 16.8 megapixels total. A third exposure was taken after another 6 months to allow for the team to subtract the effects of the subtle space motion of stars, with additional exposures used to remove other sources of error.

Riess shares the 2011 Nobel Prize in Physics with another team for his leadership in the 1998 discovery that the expansion rate of the universe is accelerating—a phenomenon widely attributed to a mysterious, unexplained dark energy filling the universe. This new high-precision distance measurement technique is enabling Riess to gauge just how much the universe is stretching. His goal is to refine estimates of the universe's expansion rate to the point where dark energy can be better characterized.

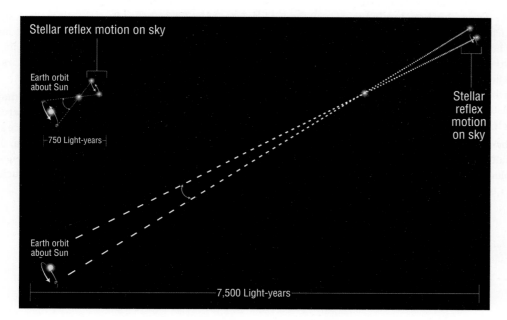

**Figure 13.21** By applying a technique called spatial scanning to an age-old method for gauging distances called astronomical parallax, scientists now can use NASA's Hubble Space Telescope to make precision distance measurements 10 times farther into our galaxy than previously possible.

ARTICLES | QUESTIONS

1. How many parsecs are in 7,500 light-years? in 10,000 light-years?
2. What is the parallax angle of a star that is 7,500 light-years away? 10,000 light-years away?
3. Why were the exposures taken 6 months apart?
4. Why is it better to make parallax measurements from space than from the ground on Earth?
5. How does improving the accuracy of the distances to nearby stars from trigonometric parallax affect astronomers' estimates of the distances to farther stars using spectroscopic parallax?

# Summary

Finding the distances to stars is a difficult but important task for astronomers. Parallax and spectroscopic parallax are two of the methods that astronomers use to determine distances to stars. Brightness and distance can be used to obtain the luminosity. Careful study of the light from a star, including its spectral lines, gives the temperature, size, and composition of the star. Study of binary systems gives the masses of stars of various spectral types, which we can extend to all stars of the same spectral type. The H-R diagram shows the relationship among the various physical properties of stars. The mass of a star will be the major determining factor in its changes over time. The habitable zone is the distance from a star in which a planet could have the right temperature for liquid water to exist on its surface. Stars of different luminosities and temperatures have habitable zones of different widths at different distances from the star.

LG 1 **Explain how the brightness of nearby stars and their distances from Earth are used to determine how luminous they are.** The distance to a nearby star is measured by finding its parallax—by measuring how the star's apparent position changes in the sky over the course of a year. The nearest star (other than the Sun) is about 4 light-years (1.3 parsecs) away. The brightness of a star in the sky can be measured directly, and brightness and distance can be used to obtain the star's luminosity—how much light the star emits.

LG 2 **Explain how astronomers obtain the temperatures, sizes, and composition of stars.** The temperature of a star is determined by its color, with blue stars being hotter and red stars being cooler. The radius can be computed from the temperature and luminosity of the star. Small, cool stars greatly outnumber large, hot stars. Spectral lines carry a great deal of information about a star, including what chemical elements and molecules are present in the star.

LG 3 **Describe how astronomers estimate the masses of stars.** Masses of stars are measured in binary star systems by observing the effects of the gravitational pull between the stars. Newton's universal law of gravitation and Kepler's laws connect the motion of the star to the forces they experience, and thus to their masses.

LG 4 **Classify stars, and organize this information on a Hertzsprung-Russell (H-R) diagram.** The H-R diagram shows the relationship among the various physical properties of stars. Temperature increases to the left, so that hotter stars lie on the left side of the diagram, while cooler stars lie on the right. Luminosity increases vertically, so that the most luminous stars lie near the top of the diagram. A star's luminosity class and temperature indicate its size. The mass and composition of a main-sequence star determine its luminosity, temperature, and size. Ninety percent of stars lie along the main sequence.

LG 5 **Explain how the mass and composition of a main-sequence star determine its luminosity, temperature, and size.** The mass and composition of a main-sequence star determine its original position on the H-R diagram. The main sequence on the H-R diagram is actually a sequence of masses. This position connects its other properties such as its luminosity, temperature, and size.

# Questions and Problems

## Test Your Understanding

1. Star A and star B are nearly the same distance from Earth. Star A is half as bright as star B. Which of the following statements must be true?
   a. Star B is farther away than star A.
   b. Star B is twice as luminous as star A.
   c. Star B is hotter than star A.
   d. Star B is larger than star A.

2. Star A and star B are two nearby stars. Star A is blue, and star B is red. Which of the following statements must be true?
   a. Star A is hotter than star B.
   b. Star A is cooler than star B.
   c. Star A is farther away than star B.
   d. Star A is more luminous than star B.

3. Star A and star B are two stars nearly the same distance from Earth. Star A is blue, and star B is red, but they have equal brightness. Which of the following statements is true?
   a. Star A is more luminous than star B.
   b. Star A is larger than star B.
   c. Star A is smaller than star B.
   d. Star A is less luminous than star B.

4. If a star has very weak hydrogen lines and is blue, what does that most likely mean?
   a. The star is too hot for hydrogen lines to form.
   b. The star has no hydrogen.
   c. The star is too cold for hydrogen lines to form.
   d. The star is moving too fast to measure the lines.

5. Star A and star B are a binary system. The Doppler shift of star A's absorption lines is 3 times the Doppler shift of star B's absorption lines. Which of the following statements is true?
   a. Star A is 3 times as massive as star B.
   b. Star A is one-third as massive as star B.
   c. Star A is closer than star B.
   d. The binary pair is moving toward Earth, but star A is farther away.

6. Star A and star B are two red stars at nearly the same distance from Earth. Star A is many times brighter than star B. Which of the following statements is true?
   a. Star A is a main-sequence star, and star B is a red giant.
   b. Star A is a red giant, and star B is a main-sequence star.
   c. Star A is hotter than star B.
   d. Star A is a white dwarf, and star B is a red giant.

7. Star A and star B are two blue stars at nearly the same distance from Earth. Star A is many times brighter than star B. Which of the following statements is true?
   a. Star A is a main-sequence star, and star B is a red giant.
   b. Star A is a main-sequence star, and star B is a blue giant.
   c. Star A is a white dwarf, and star B is a blue giant.
   d. Star A is a blue giant, and star B is a white dwarf.

8. In which region of an H-R diagram would you find the main-sequence stars with the widest habitable zones?
   a. upper left
   b. upper right
   c. center
   d. lower left
   e. lower right

9. Star A is more massive than star B. Both are main-sequence stars. Therefore, star A is _____ than star B. (Choose all that apply.)
   a. more luminous
   b. less luminous
   c. hotter
   d. colder
   e. larger
   f. smaller

10. A telescope on Mars would be able to measure the distances to more stars than can be measured from Earth because
    a. the resolution of the telescope would be better.
    b. Mars has a thin atmosphere.
    c. it would be closer to the stars.
    d. the parallax "baseline" would be longer.

11. Star A and star B are two nearby stars. Star A has a parallactic angle 4 times as large as star B's. Which of the following statements is true?
    a. Star A is one-quarter as far away as star B.
    b. Star A is 4 times as far away as star B.
    c. Star A has moved through space one-quarter as far as star B.
    d. Star A has moved through space 4 times as far as star B.

12. Star A appears twice as bright as star B, but is also twice as far away. Star A is _____ as luminous as star B.
    a. 8 times
    b. 4 times
    c. twice
    d. half

13. Table 13.1 shows two ways of reporting the amount of an element in the Sun. The percentage of hydrogen drops when changing from percentage by number of atoms to percentage by mass. But the percentage of helium grows. Why?
    a. Hydrogen is more massive than helium.
    b. Helium is more massive than hydrogen.
    c. Hydrogen is located in a different part of the Sun.
    d. It is difficult to measure the mass of hydrogen.

14. Capella (in the constellation Auriga) is the sixth brightest star in the sky. When viewed with a high-power telescope, it is clear that Capella is actually two pairs of binary stars: the first pair are G-type giants; the second pair are M-type main-sequence stars. What color does Capella appear to be?
    a. red
    b. yellow
    c. blue
    d. color cannot be determined from this information

15. An eclipsing binary system has a primary eclipse (star A is eclipsed by star B) that is deeper (more light is removed from the light curve) than the secondary eclipse (star B is eclipsed by star A). What does this tell you about stars A and B?
    a. Star A is hotter than star B.
    b. Star B is hotter than star A.
    c. Star B is larger than star A.
    d. Star B is moving faster than star A.

## Thinking about the Concepts

16. The distances of nearby stars are determined by their parallaxes. Why is there greater uncertainty in the distances of stars that are farther from Earth?

17. To know certain properties of a star, you must first determine the star's distance. For other properties, knowledge of distance is not necessary. Explain why an astronomer does or does not need to know a star's distance to determine each of the following properties: size, mass, temperature, color, spectral type, and chemical composition? In each case, state your reason(s).

18. Albireo, in the constellation Cygnus, is a visual binary system whose two components can easily be seen with even a small, amateur telescope. Viewers describe the brighter star as "golden" and the fainter one as "sapphire blue."
    a. What does this description tell you about the relative temperatures of the two stars?
    b. What does this description tell you about their respective sizes?

19. Very cool stars have temperatures around 2500 K and emit Planck spectra with peak wavelengths in the red part of the spectrum. Do these stars emit any blue light? Explain your answer.

20. The stars Betelgeuse and Rigel are both in the constellation Orion. Betelgeuse appears red, and Rigel is bluish white. To the eye, the two stars seem equally bright. If you can compare the temperature, luminosity, or size from just this information, do so. If not, explain why.

21. Explain why the stellar spectral types (O, B, A, F, G, K, M) are not in alphabetical order. What sequence of temperatures is defined by these spectral types?

22. Other than the Sun, the only stars whose mass astronomers can measure *directly* are those in eclipsing or visual binary systems. Why? How do astronomers estimate the masses of stars that are not in eclipsing or visual binary systems?

23. Once the mass of a certain spectral type of star located in a binary system has been determined, it can be assumed that all other stars of the same spectral type and luminosity class have the same mass. Why is this a reasonable assumption?

24. Explain why the Kepler Mission is finding eclipsing binary stars while it is searching for extrasolar planets using the transit method.

25. Scientific advances often require the participation of scientists from all over the world, working on the same problem over many decades, even centuries. Compare and contrast this mode of "collaboration" with collaborations in your courses (perhaps on final projects or papers). What mechanisms must be in place to allow scientists to collaborate across space and time in this way?

26. What would happen to our ability to measure stellar parallax if we were on the planet Mars? What if we were on Venus or Jupiter?

27. In Figure 13.7, there is an absorption line at about 410 nm that is weak for O stars and weak for G stars but very strong in A stars. This particular line comes from the transition from the second excited state of hydrogen up to the sixth excited state. Why is this line weak in O stars? Why is it weak in G stars? Why is it strongest in the middle of the range of spectral types?

28. Which kinds of binary systems are best observed edge-on? Which kind are best observed face-on?

29. In Figure 13.10, two stars orbit a common center of mass.
    a. Explain why star 2 has a smaller orbit than star 1.
    b. Re-sketch this picture for the case where star 1 has a very low mass, perhaps close to that of a planet.
    c. Re-sketch this picture for the case where star 1 and star 2 have the same mass.

30. If our Sun were a blue main-sequence star, and Earth was still 1 AU from the Sun, would you expect Earth to be in the habitable zone? What about if our Sun were a red main-sequence star?

## Applying the Concepts

31. Look at Figure 13.1b. Suppose the figure included a third star, located 4 times as far away as star A. How much less than star A would it appear to move each year? How much less than star B?

32. Suppose you see an object jump from side to side by half a degree as you blink back and forth between your eyes. How much farther away is an object that moves only one-third of a degree?

33. Logarithmic (log) plots show major steps along an axis scaled to represent equal factors, most often factors of 10. Why do astronomers sometimes use a log plot instead of the more conventional linear plot? Is the horizontal axis of the H-R diagram in Figure 13.15 logarithmic or linear?

34. Examine Figure 13.5. This figure is plotted logarithmically on both axes. The luminosities are in units of solar luminosities.
    a. How much more luminous than the Sun is a star on the far right side of the plot?
    b. How much less luminous than the Sun is a star on the far left side of the plot?

35. Study Figure 13.17. Compared to the Sun, how luminous, large, and hot is a star that has 10 times the mass of the Sun?

36. Sirius, the brightest star in the sky, has a parallax of 0.379 arcsec. What is its distance in parsecs? in light-years? How long does it take light to reach Earth?

37. Sirius is actually a binary pair of two A-type stars. The brighter of the two stars is called the "Dog Star" and the fainter is called the "Pup Star" because Sirius is in the constellation Canis Major (meaning "Big Dog"). The Dog Star appears about 6,800 times brighter than the Pup Star, even though both stars are at the same distance from Earth. Compare the temperatures, luminosities, and sizes of these two stars.

38. Sirius and its companion orbit around a common center of mass with a period of 50 years. The mass of Sirius is 2 times the mass of the Sun.
    a. If the orbital velocity of the companion is 2.35 times greater than that of Sirius, what is the mass of the companion?
    b. What is the semimajor axis of the orbit?

39. Sirius is 25 times more luminous than the Sun, and Polaris (the "North Pole Star") is 2,500 times more luminous than the Sun. Sirius appears 24 times brighter than Polaris. How much farther away is Polaris than Sirius? Use your answer from problem 36 to find the distance of Polaris in light-years.

40. Betelgeuse (in Orion) has a parallax of 0.00763±0.00164 arcsec, as measured by the Hipparcos satellite. What is the distance to Betelgeuse, and what is the uncertainty in that measurement?

41. Rigel (also in Orion) has a Hipparcos parallax of 0.00412 arcsec. Given that Betelgeuse and Rigel appear equally bright in the sky, which star is actually more luminous. Knowing that Betelgeuse appears reddish while Rigel appears bluish white, which star would you say is larger and why?

42. The Sun is about 16 trillion ($1.6 \times 10^{13}$) times brighter than the faintest stars visible to the naked eye.
    a. How far away (in astronomical units) would an identical solar-type star be if it were just barely visible to the naked eye?
    b. What would be its distance in light-years?

43. Study Figure 13.9. If $m_1 = m_2$, where would the center of mass be located? If $m_1 = 2m_2$, where would the center of mass be located?

44. Find the peak wavelength of blackbody emission for a star with a temperature of about 10,000 K. In what region of the spectrum does this wavelength fall? What color is this star?

45. About 1,470 watts (W) of solar energy hits each square meter of Earth's surface. Use this value and the distance to the Sun to calculate the Sun's luminosity.

## USING THE WEB

46. Go to the European Space Agency's Gaia mission website (http://esa.int/science/gaia). How will it help astronomers determine the distances to more stars? Why is it better to make parallax measurements from space than from the ground on Earth? Have any data been released?

47. Go to the "Eclipsing Binary Stars Lab" website (http://astro .unl.edu/naap/ebs/ebs.html). Click on "Eclipsing Binary Simulator." Select preset Example 1, in which the two stars are identical. The animation will run with inclination 90° and show a 50 percent eclipse. What happens when you slowly change your viewing angle to the system—the inclination. How does this change the eclipse? At what value of inclination do you no longer see eclipses? What does the system look like at 0°? Reset the inclination to 90° and adjust the separation of the two stars. How does the light curve change when the separation is larger or smaller? Now make the two stars different. Change star 2 so that its radius is 3.0 $R_{Sun}$ and its temperature is 4000 K. At what value of inclination do you no longer see eclipses? What types of eclipsing binary systems do you think are the easiest to detect?

48. Go to the Kepler home page (http://kepler.nasa.gov) and mouse over "Confirmed Planets" on the upper right. How many eclipsing binary stars has Kepler found? Go to the Kepler Eclipsing Binary Catalog (http://keplerebs.villanova.edu) to see what new observations look like. Pick a few stars to study. What is the inclination ("sin i")? Look at the last 2 columns ("Figures"). The "raw" and "dtr" figures are rough, but the "pf" figure shows a familiar light curve. How deep is the eclipse; that is, how much lower is the "normalized flux" during maximum eclipse?

49. Do a search for a photograph of your favorite constellation (or go outside and take a picture yourself). Can you see different colors in the stars? What do the colors tell you about the surface temperatures of the stars? From your photograph, can you tell which are the three brightest stars in the constellation? These stars will be named "alpha" ($\alpha$), "beta" ($\beta$), and "gamma" ($\gamma$) for that constellation. Look up the constellation online and see if you chose the right stars. What are their temperatures and luminosities? What are their distances?

50. Citizen science: Go to the website for the Stellar Classification Online Public Exploration (SCOPE) project (http://scope.pari .edu/takepart.php). This project uses crowdsourcing to classify stars seen on old photographic plates of photographs taken in the Southern Hemisphere. Create an account, review the science and the FAQ, and then click on "To Take Part" to see some practice examples. Then go to "Classify," choose a photographic plate, and classify a few stars.

# smartw⊛rk5

If your instructor assigns homework in Smartwork5, access your assignments at digital.wwnorton.com/astro5.

# EXPLORATION

## The H-R Diagram

Open the "HR Explorer" interactive simulation for this chapter at the Student Site at the Digital Landing Page. This simulation enables you to compare stars on the H-R diagram in two ways. You can compare an individual star (marked by a red $X$) to the Sun by varying its properties in the box in the left half of the window. Or you can compare groups of the nearest and brightest stars. Play around with the controls for a few minutes to familiarize yourself with the simulation.

Begin by exploring how changes to the properties of the individual star change its location on the H-R diagram. First, press the "Reset" button at the top right of the window.

Decrease the temperature of the star by dragging the temperature slider to the left. Notice that the luminosity remains the same. Because the temperature has decreased, each square meter of star surface must be emitting less light. What other property of the star changes in order to keep the total luminosity of the star constant?

Predict what will happen when you slide the temperature slider all the way to the right. Now do it. Did the star behave as you expected?

**1** As you move to the left across the H-R diagram, what happens to the radius?

**2** What happens as you move to the right?

Press "Reset" and experiment with the luminosity slider.

**3** As you move up on the H-R diagram, what happens to the radius?

**4** What happens as you move down?

Press "Reset" again and predict how you would have to move the slider bars to move your star into the red giant portion of the H-R diagram (upper right). Adjust the slider bars until the star is in that area. Were you correct?

**5** How would you adjust the slider bars to move the star into the white dwarf area of the H-R diagram?

Press the "Reset" button and explore the right-hand side of the window. Add the nearest stars to the graph by clicking their radio button under "Plotted Stars." Using what you learned above, compare the temperatures and luminosities of these stars to the Sun (marked by the $X$).

**6** Are the nearest stars generally hotter or cooler than the Sun?

**7** Are the nearest stars generally more or less luminous than the Sun?

Press the radio button for the brightest stars. This action will remove the nearest stars and add the brightest stars in the sky to the plot. Compare these stars to the Sun.

**8** Are the brightest stars generally hotter or cooler than the Sun?

**9** Are the brightest stars generally more or less luminous than the Sun?

**10** How do the temperatures and luminosities of the brightest stars in the sky compare to the temperatures and luminosities of the nearest stars? Does this information support the claim in the chapter that there are more low-luminosity stars than high-luminosity stars? Explain.

**Student Site :** digital.wwnorton.com/astro5

**389**

# 14 Our Star— The Sun

Because the Sun is the only star close to Earth, much of the detailed knowledge about stars has come from studying the Sun. In Chapter 13, we looked at the physical properties of distant stars, including their mass, luminosity, size, temperature, and chemical composition. In this chapter, we ask fundamental questions about Earth's local star. How does the Sun work? Where does it get its energy? How has its luminosity been able to remain so constant over the billions of years since the Solar System formed?

## LEARNING GOALS

By the conclusion of this chapter, you should be able to:

**LG 1** Describe the balance between the forces that determine the structure of the Sun.

**LG 2** Explain how mass is converted into energy in the Sun's core and how long it will take the Sun to use up its fuel.

**LG 3** Sketch a physical model of the Sun's interior, and list the different ways that energy moves outward from the Sun's core toward its surface.

**LG 4** Describe how observations of solar neutrinos and seismic vibrations on the surface of the Sun test astronomers' models of the Sun.

**LG 5** Describe the solar activity cycles of 11 and 22 years, and explain how these cycles are related to the Sun's changing magnetic field.

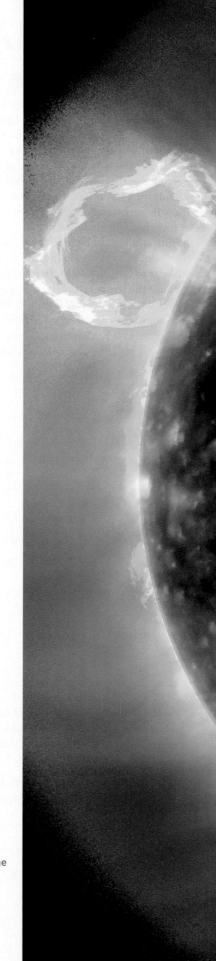

This image is a combination of several extreme ultraviolet images of the Sun from the Solar Dynamics Observatory. ▶ ▶ ▶

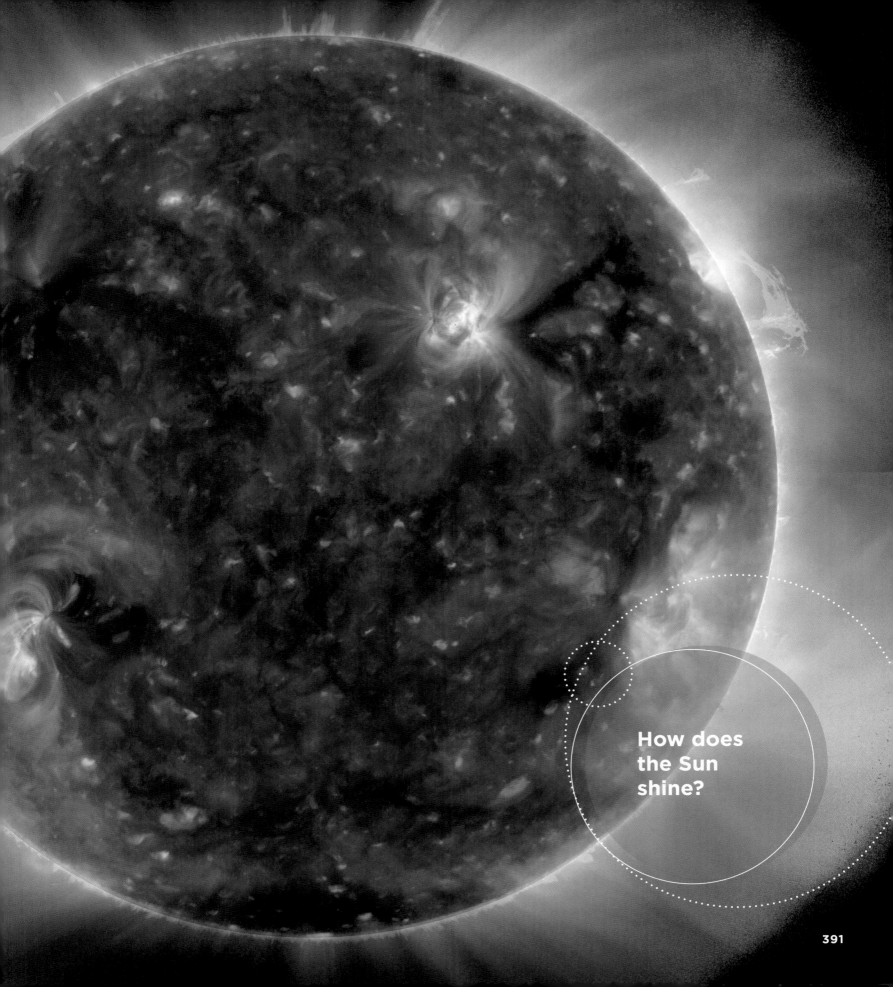

How does the Sun shine?

391

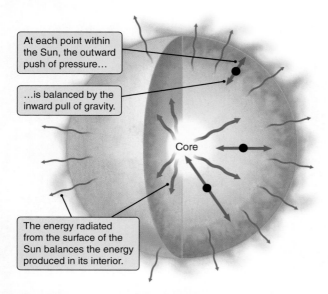

At each point within the Sun, the outward push of pressure…

…is balanced by the inward pull of gravity.

Core

The energy radiated from the surface of the Sun balances the energy produced in its interior.

**Figure 14.1** The structure of the Sun is determined by the balance between the forces of pressure and gravity and the balance between the energy generated in its core and the energy radiated from its surface.

(a)

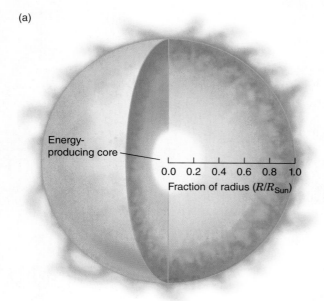

Energy-producing core

0.0 0.2 0.4 0.6 0.8 1.0
Fraction of radius ($R/R_{Sun}$)

**Figure 14.2** (a) This cutaway figure shows how the fraction of radius given in the *x*-axis of the graphs in (b) is measured. The energy produced by the Sun is generated in the Sun's core. (b) Pressure, density, and temperature increase toward the center of the Sun.

# 14.1 The Sun Is Powered by Nuclear Fusion

Energy from the Sun is responsible for daylight, for Earth's weather and seasons, and for terrestrial life itself. At a luminosity of $3.85 \times 10^{26}$ watts (W), the Sun produces more energy in a second than all of the electric power plants on Earth could generate in a half-million years. In this section, we will look at energy production in the Sun.

## Hydrostatic Equilibrium

Geologists learn about the interior of Earth by using a combination of physics, detailed computer models, and experiments that test the predictions of those models. The task of exploring the interior of the Sun is much the same. Like Earth's structure, the structure of the Sun is governed by a number of physical processes and relationships. Using physics, chemistry, and the properties of matter and radiation, astronomers can express these processes and relationships as mathematical equations. They then solve these equations and create a model of the Sun. One of the great successes of 20th century astronomy was the construction of a physical model of the Sun that agrees with observations of the mass, composition, size, temperature, and luminosity of the real thing.

The structure of the Sun is a matter of balance between the pressure outward and the force of gravity inward: this balance is known as **hydrostatic equilibrium**. The pressure results from energy finding its way to the surface of the Sun from deep in its interior. To understand how hydrostatic equilibrium affects the Sun, we need to know how these forces are produced and how they continually change to balance each other.

The balance between the forces due to pressure and gravity is illustrated in **Figure 14.1**. The Sun is a huge ball of hot gas. Deep in the Sun's interior, the outer layers press downward because of gravity, producing a large inward force. To maintain balance, the outward force due to pressure must be equally large. If gravity were not balanced by pressure, the Sun would collapse. If pressure were not balanced by gravity, the Sun would blow itself apart. At every point within the Sun's interior, the pressure must be just enough to hold up the weight of all the layers above that point. If the Sun were not in a stable hydrostatic equilibrium, forces within it would not be in balance, and the size of the Sun would change.

Hydrostatic equilibrium becomes an even more powerful concept when combined with the way gases behave. Deeper in the interior of the Sun, the weight of the material above becomes greater, and hence the pressure must increase. In a gas, higher pressure means higher density and/or higher temperature. **Figure 14.2** shows how conditions vary inside the Sun. As illustrated in the graphs in Figure 14.2b, calculations show that toward the center of the Sun, the pressure, the density, and the temperature of the gas increase.

(b) Pressure (billions of atmospheres)

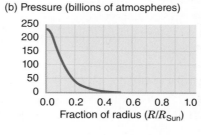

250
200
150
100
50
0

0.0 0.2 0.4 0.6 0.8 1.0
Fraction of radius ($R/R_{Sun}$)

Density (thousands of kg/m³)

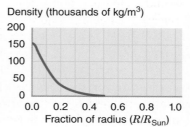

200
150
100
50
0

0.0 0.2 0.4 0.6 0.8 1.0
Fraction of radius ($R/R_{Sun}$)

Temperature (millions of K)

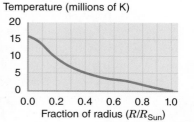

20
15
10
5
0

0.0 0.2 0.4 0.6 0.8 1.0
Fraction of radius ($R/R_{Sun}$)

## CHECK YOUR UNDERSTANDING 14.1

Hydrostatic equilibrium in the Sun means that: (a) the Sun does not change over time; (b) the Sun absorbs and emits equal amounts of energy; (c) pressure balances the weight of overlying layers; (d) energy produced in the core per unit time equals energy emitted at the surface per unit time.

## Nuclear Fusion

A second fundamental balance within the Sun is the balance of energy (see Figure 14.1). Stars like the Sun are remarkably stable objects. Geological records show that the luminosity of the Sun has remained nearly constant for billions of years. To remain in balance, the Sun must produce just enough energy in its interior to replace the energy radiated away from its surface. This energy balance tells us how much energy must be produced in the interior of the Sun and how that energy finds its way from the interior to the Sun's surface, where it is radiated away. Models of stellar evolution indicate that the luminosity of the Sun is increasing with time, but very, very slowly. The Sun's luminosity 4.5 billion years ago was about 70 percent of its current luminosity.

The amount of energy produced by the Sun each second is truly astronomical: $3.85 \times 10^{26}$ W. One of the most basic questions facing the pioneers of stellar astrophysics was how the Sun and other stars get their energy. In the 19th century, physicists proposed that the Sun was slowly shrinking, and that the core was heating up as a result of this gravitational contraction. However, calculations soon showed that this would power the Sun for only millions of years. Geological and biological evidence available at the time suggested that Earth was tens of millions or hundreds of millions of years old. By the early 20th century, radiometric dating suggested that Earth was more than a billion years old, and therefore gravitational contraction could not be the source of the Sun's energy. In the 1930s, using theoretical and laboratory physics, nuclear physicists concluded that the Sun's energy comes from nuclear reactions at its core, capable of powering the star for billions of years.

Recall from Chapter 5 that the nucleus of most hydrogen atoms consists of a single proton. Nuclei of all other atoms are built from a mixture of protons and neutrons. Most helium nuclei, for example, consist of two protons and two neutrons. Protons have a positive electric charge, and neutrons have no electric charge. Because like charges repel, and the closer they are the stronger the force, all of the protons in an atomic nucleus are continually repelling each other with a tremendous force. The nuclei of atoms should fly apart due to electric repulsion—yet atomic nuclei are held together by the **strong nuclear force**, which overcomes this repulsion. However, the strong nuclear force acts only over very short distances, of the order $10^{-15}$ meter, about the size of the atomic nucleus, or about a hundred-thousandth the size of an atom.

Compared to the energy required to free an electron from an atom, the amount of energy required to tear a nucleus apart is enormous. Conversely, when an atomic nucleus (with a mass up to and including the nucleus of iron) is formed from component parts, energy is released. **Nuclear fusion**—the process of combining two less massive atomic nuclei into a single more massive atomic nucleus—occurs when atomic nuclei are brought close enough together for the strong nuclear force to overcome the force of electric repulsion, as illustrated in **Figure 14.3**. Many kinds of nuclear fusion can occur in stars. In main-sequence stars like the Sun, the primary energy generation process is the fusion of hydrogen into helium—a process often called **hydrogen burning** (even though it has nothing to

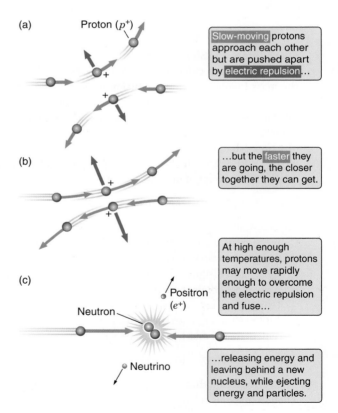

**Figure 14.3** (a) Atomic nuclei are positively charged and electrically repel each other. (b) The faster that two nuclei are moving toward each other, the closer they will get before veering away. (c) At the temperatures and densities found in the centers of stars, nuclei can overcome this electric repulsion, so fusion takes place.

do with fire or other chemical combustion). The fusion of hydrogen into helium always takes several steps, but the net result is that four hydrogen nuclei become the one helium nucleus plus energy.

The energy produced in nuclear reactions comes from the conversion of mass into energy. The exchange rate between mass and energy is given by Einstein's famous equation, $E = mc^2$, in which $E$ is energy, $m$ is mass, and $c^2$ is the speed of light squared. For any nuclear reaction, we can determine the mass that is turned into energy by calculating the mass that is lost. To find this lost mass, we subtract the mass of the outputs from the mass of the inputs. In hydrogen burning, the inputs are four hydrogen nuclei, and the output is a helium nucleus plus energy. The mass of four separate hydrogen nuclei is 1.007 times greater than the mass of a single helium nucleus; so when hydrogen fuses to make helium, 0.7 percent of the mass of the hydrogen is converted to energy.

Although each fusion reaction produces a small amount of energy, the total mass of the Sun is very large, so there is much hydrogen to "burn." When the amount of energy produced by nuclear burning is compared with the luminosity of the Sun, we see that these reactions can power the Sun for 10 billion years, a time frame that is longer than the 4.6-billion-year age of the Solar System measured from radioactive dating. Details of this calculation are provided in **Working It Out 14.1**.

## 14.1 Working It Out The Source of the Sun's Energy

Like all stars, the Sun's lifetime is limited by the amount of fuel available to it. We can calculate how long the Sun will live by comparing the mass involved in nuclear fusion with the amount of mass available. Converting four hydrogen nuclei (protons) into a single helium nucleus results in a loss of mass. The mass of a single hydrogen nucleus is $1.6726 \times 10^{-27}$ kilogram (kg). So, four hydrogen nuclei have a mass of 4 times that, or $6.6904 \times 10^{-27}$ kg. The mass of a helium nucleus is $6.6447 \times 10^{-27}$ kg, which is less than the mass of the four hydrogen nuclei. The amount of mass lost, $m$, is

$$m = 6.6904 \times 10^{-27}\,\text{kg} - 6.6447 \times 10^{-27}\,\text{kg} = 0.0457 \times 10^{-27}\,\text{kg}$$

We can write this as $4.57 \times 10^{-29}$ kg—a mass loss of about 0.7 percent. Conversion of 0.7 percent of the mass of the hydrogen into energy might not seem very efficient—until we compare it with other sources of energy and discover that it is millions of times more efficient than even the most efficient chemical reactions.

Using Einstein's equation $E = mc^2$, where $c$ is the speed of light ($3 \times 10^8$ m/s), along with the definition of a joule (1 J = 1 kg m²/s²), we can calculate the energy released by this mass-to-energy conversion:

$$E = mc^2 = (4.57 \times 10^{-29}\,\text{kg}) \times (3.00 \times 10^8\,\text{m/s})^2 = 4.11 \times 10^{-12}\,\text{J}$$

Each reaction that takes four hydrogen nuclei and turns them into a helium nucleus releases $4.11 \times 10^{-12}$ J of energy, which doesn't seem like very much. But atoms are very small. Fusing a single kilogram of hydrogen into helium releases about $6.3 \times 10^{14}$ J of energy—about the equivalent of the chemical energy released in burning 100,000 barrels of oil. To see how much the Sun must be fusing per second to produce its current luminosity, we divide the luminosity of the Sun by this amount of energy per kilogram:

$$\frac{\text{Luminosity of Sun}}{\text{Energy per kilogram}} = \frac{4 \times 10^{26}\,\text{J/s}}{6.3 \times 10^{14}\,\text{J/kg}} = 6.2 \times 10^{11}\,\text{kg/s}$$

For the Sun to produce as much energy as it does, it must convert roughly 620 billion kg of hydrogen into helium every second (and about 4 billion kg of matter is converted to energy in the process). The Sun has been burning hydrogen at this rate for at least the age of Earth and the Solar System—4.6 billion years. How much longer will the Sun last?

Astronomers estimate that only 10 percent of the Sun's total mass will ever be involved in fusion, because the other 90 percent will never get hot enough or dense enough for the strong nuclear force to make fusion happen. Ten percent of the mass of the Sun is $(0.1) \times (2 \times 10^{30})$ kg, or $2 \times 10^{29}$ kg. That is the amount of "fuel" the Sun has available. The Sun consumes hydrogen at a rate of 620 billion kg/s, so each year the Sun consumes:

$$M_{\text{year}} = (6.2 \times 10^{11}\,\text{kg/s}) \times (3.16 \times 10^7\,\text{s/yr}) = 2 \times 10^{19}\,\text{kg/yr}$$

If we know how much fuel the Sun has ($2 \times 10^{29}$ kg), and we know how much the Sun burns each year ($2 \times 10^{19}$ kg/yr), then we can divide the amount by the rate to find the lifetime of the Sun:

$$\text{Lifetime} = \frac{M_{\text{fuel}}}{M_{\text{year}}} = \frac{2 \times 10^{29}\,\text{kg}}{2 \times 10^{19}\,\text{kg/yr}} = 10^{10}\,\text{yr}$$

When the Sun was formed, it had enough fuel to power it for about 10 billion years. The Sun is nearly halfway through its lifetime of hydrogen burning.

Energy is produced in the Sun's innermost region, the **core**, where the conditions are the most extreme. The density of matter in the core is about 150 times the density of water, and the temperature is about 15 million kelvins (K). Under these conditions, the atomic nuclei have tens of thousands of times more kinetic energy than that of atoms at room temperature and can slam into each other hard enough to overcome the electric repulsion, allowing the strong nuclear force to act (Figure 14.3c). In hotter and denser gases, such collisions happen more frequently. For this reason, the rate of nuclear fusion reactions is extremely sensitive to the temperature and the density of the gas, which is why these energy-producing collisions are concentrated in the Sun's core. Half of the energy produced by the Sun is generated within the inner 9 percent of the Sun's radius, or less than 0.1 percent of the volume of the Sun.

The conversion of four hydrogen nuclei to one helium nucleus is the most significant source of energy in main-sequence stars. Hydrogen is the most abundant element in the universe, so it is the most abundant source of nuclear fuel at the beginning of a star's lifetime. Hydrogen is also the easiest type of atom to fuse. Hydrogen nuclei—protons—have an electric charge of +1. The electric barrier that must be overcome to fuse protons is the repulsion of one proton against another. To fuse 2 carbon nuclei together, for example, the repulsion of the six protons in one carbon nucleus pushing against the six protons in another carbon nucleus must be overcome. The repulsion between two carbon nuclei is 36 times stronger than that between two hydrogen nuclei. Therefore, hydrogen fusion occurs at a much lower temperature than any other type of nuclear fusion.

▶❙❙ **AstroTour:** The Solar Core

## The Proton-Proton Chain

To test the theory that the Sun shines because of nuclear fusion, astronomers can analyze the predicted by-products of the nuclear reactions. In the core of the Sun and in other low-mass stars, hydrogen burning takes place in a series of nuclear reactions called the **proton-proton chain**, which has three different branches. The most important branch, responsible for about 85 percent of the energy generated in the Sun, consists of three steps, illustrated in **Figure 14.4**. Each step produces particles and/or energy in the form of light. We will begin by following the creation of the helium nucleus, and then go back to find out what happens to the other products of the reaction.

Follow along in Figure 14.4 as we step through the proton-proton chain. The nucleus of hydrogen consists of one proton. In the first step, two protons fuse. During this process, one of the protons is transformed into a neutron. To conserve spin and charge, two particles are emitted: a positively charged particle called a **positron** and a neutral particle called a **neutrino**. Energy is also emitted in the form of photons carrying electromagnetic radiation. The new atomic nucleus formed by the first

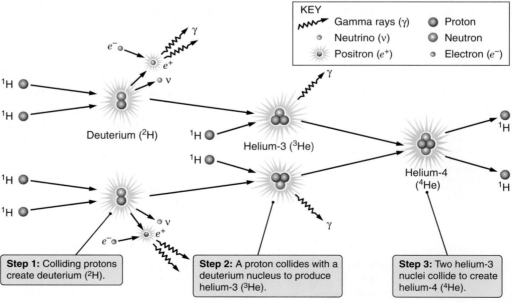

KEY
〰➤ Gamma rays (γ)   ● Proton
∘ Neutrino (ν)   ● Neutron
☀ Positron ($e^+$)   ∘ Electron ($e^-$)

Deuterium ($^2$H)

Helium-3 ($^3$He)

Helium-4 ($^4$He)

**Step 1:** Colliding protons create deuterium ($^2$H).

**Step 2:** A proton collides with a deuterium nucleus to produce helium-3 ($^3$He).

**Step 3:** Two helium-3 nuclei collide to create helium-4 ($^4$He).

**Figure 14.4** The Sun and all other main-sequence stars get their energy by fusing the nuclei of four hydrogen atoms together to make a single helium atom. In the Sun, about 85 percent of the energy produced comes from the branch of the proton-proton chain illustrated here.

step in the chain consists of a proton and a neutron. Recall from Chapter 5 that an *isotope* of an element has the same number of protons and a different number of neutrons. Thus, the new atomic nucleus is still hydrogen because it has only one proton. This particular isotope of hydrogen is common enough that it has its own name—deuterium, written as $^2$H (H is the element symbol for hydrogen, which always has one proton, and 2 is the *atomic mass number*—the total number of protons and neutrons in the nucleus).

In the second step of the proton-proton chain, another proton slams into the deuterium nucleus, forming the nucleus of an isotope of helium, $^3$He, consisting of two protons and a neutron. The energy released in this step is carried away as a highly energetic gamma-ray photon. Notice that the first two steps are shown twice, along the top of the figure and the bottom, because these steps must occur twice to produce a single $^4$He nucleus.

In the third and final step of the proton-proton chain, two $^3$He nuclei collide and fuse, producing an ordinary $^4$He nucleus and ejecting two protons in the process. The energy released in this step is the kinetic energy of the helium nucleus and two ejected protons. Overall, four hydrogen nuclei have combined to form one helium nucleus.

Now let's go back and look at what happens to the other products of the reaction. In step 1, a positron—a particle of antimatter—is produced. **Antimatter** particles have the same mass as a corresponding matter particle but have opposite values of other properties, such as charge. The positron ($e^+$) is the antimatter counterpart of an electron ($e^-$). When matter (electrons) and antimatter (positrons) meet, they annihilate each other, and their total mass is converted to energy in the form of gamma-ray photons ($\gamma$). This happens to the emitted positrons inside the Sun, and the emitted photons from the annihilation carry away part of the energy released when the two protons fused. These photons heat the surrounding gas. The gamma rays emitted in step 2 similarly heat the gas. The thermal energy produced in the core of the Sun takes 100,000 years to find its way to the Sun's surface, and so the light we see from the Sun indicates what the Sun was doing 100,000 years ago.

The neutrino emitted in step 1 has a very different fate. Neutrinos are particles that have no charge, very little mass, and travel at nearly the speed of light. They interact weakly with ordinary matter, so weakly that the neutrino escapes from the Sun without further interactions with any other particles. The core of the Sun lies buried beneath 700,000 kilometers (km) of dense, hot matter, yet the Sun is transparent to neutrinos—essentially all of them travel into space as if the outer layers of the Sun were not there. Because they travel at nearly the speed of light, neutrinos from the center of the Sun arrive at Earth after only $8\frac{1}{3}$ minutes. Therefore, we can use them to probe what the Sun is doing today.

This dominant branch of the proton-proton chain can be written symbolically as follows:

**Step 1:** $^1$H + $^1$H → $^2$H + $e^+$ + $\nu$ and then $e^+$ + $e^-$ → $\gamma$ + $\gamma$

**Step 2:** $^2$H + $^1$H → $^3$He + $\gamma$

**Step 3:** $^3$He + $^3$He → $^4$He + $^1$H + $^1$H

 **Nebraska Simulation:** Proton-Proton Animation

The rate of the proton-proton chain reaction depends on both temperature and density. At the temperature and pressure that exist within the Sun's core, the reaction rate is relatively slow—in fact, extremely slow compared to a nuclear bomb

explosion. The Sun's slow nuclear fusion is fortunate for life on Earth: if its hydrogen burned quickly, the Sun would have exhausted its supply long ago, and life might not have had time to evolve.

The other 15 percent of the Sun's energy is generated by variations of the proton-proton chain. The most common variation happens in step 3, where $^3$He fuses with an existing $^4$He to create beryllium ($^7$Be), which decays to lithium ($^7$Li) and energy, and then the $^7$Li plus one $^1$H become two $^4$He. In a less common variation, the beryllium combines with hydrogen to become boron ($^8$B), which then decays to beryllium and then to two $^4$He. In both of these variations, ultimately four hydrogen nuclei become one helium nucleus.

---

### CHECK YOUR UNDERSTANDING 14.2

When hydrogen is fused into helium, energy is released from: (a) gravitational collapse; (b) conversion of mass to energy; (c) the increase in pressure; (d) the decrease in the gravitational field.

---

## 14.2 Energy Is Transferred from the Interior of the Sun

Although geologists cannot travel deep inside Earth to find out how it is structured, they are able to build a model of its interior using data on how seismic waves travel during earthquakes. Similarly, astronomers can create a model of the Sun's interior using their knowledge of the balance of forces and energy within the Sun and an understanding of how energy moves from one place to another. These models can be tested by observations of waves traveling through the Sun and by studying neutrinos from the Sun.

### Energy Transport

Some of the energy released by hydrogen burning in the core of the Sun escapes directly into space in the form of neutrinos. However, most of the energy heats the solar interior and then moves outward through the Sun to the surface, a process known as **energy transport**. Energy transport, a key determinant of the Sun's structure, can occur by *conduction*, *convection*, or *radiation*.

**Conduction** is important primarily in solids. For example, when you pick up a hot object, your fingers are heated by conduction. This happens because energetic thermal vibrations of atoms and molecules cause neighboring atoms and molecules to vibrate more rapidly as well. Conduction is typically ineffective in a gas because the atoms and molecules are too far apart to transmit vibrations to one another efficiently. Conduction does not play a key role in the transport of energy from the core of the Sun to its surface, but it will be relevant later when we discuss dying stars.

In the Sun, energy is transported by convection and radiation through different zones, as shown in **Figure 14.5**. The mechanism of energy transport from the center of the Sun outward depends on the decreasing temperature and density as the radius increases. First, energy moves outward through the inner layers of the

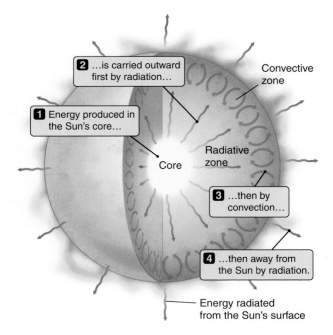

**Figure 14.5** The interior structure of the Sun is divided into zones on the basis of where energy is produced and how it is transported outward.

Sun as radiation in the form of photons. Next, energy moves by convection in parcels of gas. Finally, energy radiates from the Sun's surface as light. We will look at each process in turn.

Near the core, **radiation** transfers energy from hotter to cooler regions via photons, which carry the energy with them. Consider a hotter region of the Sun located next to a cooler region, as shown in **Figure 14.6**. Recall from your study of radiation in Chapter 5 that the hotter region contains more (and more energetic) photons than the cooler region. More photons move from the hotter, very crowded region to the cooler, less crowded region than in the reverse direction. A net transfer of photons and photon energy occurs from the hotter region to the cooler region, and radiation carries energy outward from the Sun's core.

The transfer of energy from one point to another by radiation also depends on how freely radiation can move from one point to another within a star. The degree to which matter blocks the flow of photons through it is called **opacity**. The opacity of a material depends on many things, including the density of the material, its composition, its temperature, and the wavelength of the photons moving through it.

Energy transfer by radiation is most efficient in regions with low opacity. The radiative zone (see Figure 14.5) is the region in the inner part of the Sun where the opacity is relatively low, and radiation carries the energy produced in the core outward through the star. This radiative zone extends about 70 percent of the way out toward the surface of the Sun. Even though this region's opacity is low enough for radiation to dominate convection as an energy transport mechanism, photons still travel only a short distance within the region before being absorbed, emitted, or deflected by matter, much like a beach ball being batted about by a

**Astronomy in Action:** Random Walk

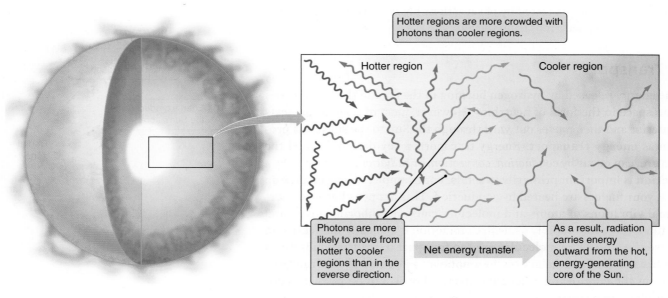

**Figure 14.6** Higher-temperature regions deep within the Sun produce more radiation than do lower-temperature regions farther out. Although radiation flows in both directions, more radiation flows from the hotter regions to the cooler regions than from the cooler regions to the hotter regions. Therefore, radiation carries energy outward from the inner parts of the Sun. For simplicity, we have only included in this illustration the most common photons (those at the peak of the blackbody curve). Photons of all colors are present in all regions, and there are more of all kinds in the hotter regions and fewer of all kinds in the cooler regions.

crowd of people as illustrated in **Figure 14.7**. Each interaction sends the photon in an unpredictable direction—not necessarily toward the surface of the star. The distances between interactions are so short that, on average, it takes the energy of a gamma-ray photon produced in the interior of the Sun about 100,000 years to find its way to the outer layers of the Sun. Opacity holds energy within the interior of the Sun and lets it seep away only slowly. As it travels, the gamma-ray photon gradually becomes converted to lower-energy photons, emerging mostly as optical and infrared radiation from the surface.

From a peak of 15 million K at the center of the Sun, the temperature falls to about 100,000 K at the outer margin of the radiative zone. At this cooler temperature, the opacity is higher, so radiation is less efficient at carrying energy from one place to another. The energy that is flowing outward through the Sun "piles up" against this edge of the radiative zone.

Nearer the surface of the Sun, transfer by radiation becomes inefficient and the temperature changes quickly. Instead, **convection** takes over. Convection carries energy from the interior of a planet to its surface or from the Sun-heated surface of Earth upward through Earth's atmosphere. Convection also plays an important role in the transport of energy outward from the interior of the Sun. It transports energy by moving packets of hot gas, like hot-air balloons, which become buoyant and rise up through the lower-temperature gas above them, carrying energy with them. The solar convective zone (see Figure 14.5) extends from the outer boundary of the radiative zone outward to just below the visible surface of the Sun where evidence of convection can be seen in the bubbling surface (**Figure 14.8**).

In the outermost layers of the Sun, radiation again takes over as the primary mode of energy transport, and it is radiation that transports energy from the Sun's outermost layers off into space.

## Observing Neutrinos from the Core of the Sun

The model of energy production and energy transport in the Sun discussed above correctly matches observed global properties of the Sun such as its size, temperature, and luminosity. The nuclear fusion model of the Sun predicts exactly which nuclear reactions should be occurring in the core of the Sun and at what rate. The nuclear reactions that make up the proton-proton chain produce a vast number of neutrinos. Because neutrinos barely interact with other ordinary matter, almost all of the neutrinos produced in the heart of the Sun travel freely through the outer parts of the Sun and on into space as if the outer layers of the Sun were not there. Solar neutrinos produced in the core of the Sun, traveling at nearly the speed of light, take only $8\frac{1}{3}$ minutes to reach Earth, much quicker than the 100,000-year journey of photons.

Neutrinos interact so weakly with matter that they are extremely difficult to observe. However, the extremely large number of nuclear reactions in the Sun means that the Sun produces an enormous number of neutrinos. As you read this sentence, about 400 trillion solar neutrinos pass through your body. This happens even at night, as neutrinos easily pass through Earth. With this many neutrinos about, a neutrino detector does not have to detect a very large percentage of them to be effective.

A neutrino telescope looks very different from other telescopes. The first apparatus designed to detect solar neutrinos was built 1,500 meters underground,

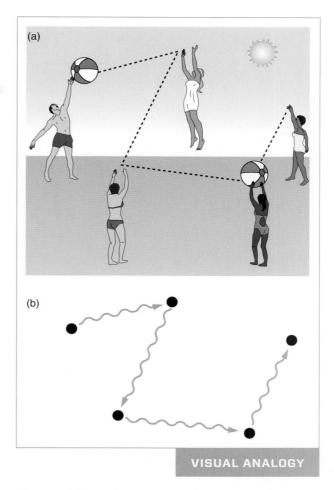

**VISUAL ANALOGY**

**Figure 14.7** (a) When a crowd of people plays with a beach ball, the ball never travels very far before someone hits it, turning it in another direction. It often takes a ball a long time to make its way from one edge of the crowd to the other. (b) Similarly, when a photon travels through the Sun, it takes a long time for a photon to make its way out of the Sun.

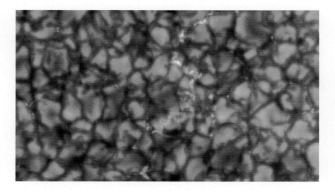

**Figure 14.8** The top of the Sun's convective zone shows the bubbling of the surface caused by rising and falling packets of gas.

within the Homestake Mine in Lead, South Dakota. Astronomers filled a tank with 100,000 gallons of dry-cleaning fluid—$C_2Cl_4$, or perchloroethylene. Over the course of 2 days, astronomers predicted that roughly $10^{22}$ solar neutrinos passed through the Homestake detector. Of these, on average only one neutrino interacted with a chlorine atom within the fluid to form a radioactive isotope of argon. Over time, a measurable amount of argon was produced.

The Homestake experiment operated from the late 1960s to the early 1990s and detected this argon isotope—evidence of neutrinos from the Sun, confirming that nuclear fusion powers the Sun. However, astronomers noticed that they were measuring only one-third to one-half as many solar neutrinos as predicted by solar models. The difference between the predicted and measured number of solar neutrinos was called the **solar neutrino problem**.

One possible explanation of the solar neutrino problem was that the working model of the structure of the Sun was somehow wrong. This possibility seemed unlikely, however, because of the many other successful predictions of the solar model. A second possibility was that an understanding of the neutrino itself was incomplete. The neutrino was long thought to have zero mass, like photons, and to travel at the speed of light. But if neutrinos actually do have a tiny amount of mass, then particle physics suggests that solar neutrinos should oscillate—alternate back and forth among three different kinds of neutrinos: the electron, muon, and tau neutrinos. According to this explanation, early neutrino experiments could detect only the electron neutrino and consequently observed only about a third of the expected number of neutrinos. Since then, many other neutrino detectors have been built, each using different reactions to detect neutrinos of different energies or different types. Experiments at high-energy physics labs, nuclear reactors, and neutrino telescopes around the world have shown that neutrinos do have a nonzero mass and do oscillate among neutrino types.

Solving the solar neutrino problem is a good example of how science works—how a better model of the neutrino showed that the solar neutrino problem was real and not merely an experimental mistake, and how a single set of anomalous observations was later confirmed by other, more sophisticated experiments. All of this effort has led to a better understanding of basic physics (**Process of Science Figure**).

## Probing the Sun's Interior

Models of Earth's interior predict how density and temperature change from place to place. These differences affect the seismic waves traveling through Earth, bending the paths that they travel. Geologists test models of Earth's interior by comparing measurements of seismic waves from earthquakes with model predictions of how seismic waves should travel through the planet.

Just as geologists use seismic waves from earthquakes to probe the interior of Earth, solar physicists use the surface oscillations of the Sun to probe the solar interior. The science that uses solar oscillations to study the Sun is called **helioseismology**. Detailed observations of motions of material from place to place across the surface of the Sun show that the Sun vibrates or rings, something like a bell that has been struck. Unlike a well-tuned bell—which vibrates primarily at one frequency—the vibrations of the Sun are very complex. In the Sun, many different frequencies of vibrations occur simultaneously, which cause some parts of

**The first detections of solar neutrinos raised more questions than they answered.**

The Hypothesis:
The Sun's energy comes
from nuclear fusion,
which produces neutrinos.

The Test:
A specific number of neutrinos
must be produced each day to account
for the brightness of the Sun.

The Experiment:
Homestake detects one-third
as many neutrinos as predicted.

The Conclusion:
One of these things is true...

Scientists don't understand
nuclear fusion.

But thousands of
experiments on Earth
support our understanding!

Scientists don't understand
neutrinos.

New Hypothesis:
What if neutrinos come
in three types and Homestake
can detect only one type?

Newer laboratory and solar measurements confirmed the new hypothesis. Part of the "scientific attitude" is to find failure exciting. When experiments do not turn out as expected, good scientists get excited—there is something new to understand!

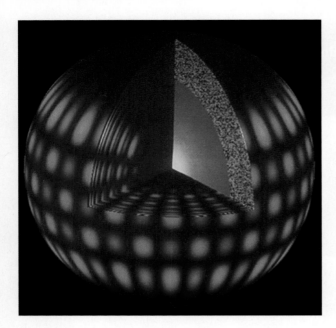

**Figure 14.9** The interior of the Sun rings like a bell as helioseismic waves move through it. This figure shows one particular mode of the Sun's vibration. Red indicates regions where gas is traveling inward; blue indicates regions where gas is traveling outward. Astronomers observe these motions via Doppler shifts.

the Sun to bulge outward and some to draw inward. These motions help us to probe what lies below. **Figure 14.9** illustrates the motions of the different parts of the Sun, with red and blue areas moving in opposite directions. Some waves are amplified and some are suppressed, depending on how they overlap as they travel through the Sun. Astronomers study these waves using the Doppler effect (see Chapter 5), which distinguishes between parts of the Sun that move toward the observer and parts that move away.

To detect the disturbances of helioseismic waves on the surface of the Sun, astronomers must measure Doppler shifts of less than 0.1 m/s while detecting changes in brightness of only a few parts per million at any given location on the Sun. Tens of millions of different wave motions are possible within the Sun. Some waves travel around the circumference of the Sun, providing information about the density of the upper convection zone. Other waves travel through the interior of the Sun, revealing the density structure of the Sun close to its core. Still others travel inward toward the center of the Sun, until they are bent by the changing solar density and return to the surface.

All of these wave motions are going on at the same time, so sorting them out requires computer analysis of long, unbroken strings of solar observations from several sources. The Global Oscillation Network Group (GONG) is a network of six solar observation stations spread around the world that enables astronomers to observe the surface of the Sun approximately 90 percent of the time.

To interpret helioseismology data, scientists compare the measurements of the strength, frequency, and wavelengths of the waves against predicted vibrations calculated from models of the solar interior. This technique provides a powerful test of models of the solar interior, and it has led both to some surprises and to improvements in the models. For example, some scientists proposed that the solar neutrino problem might be solved if the models had overestimated the amount of helium in the Sun. This explanation was ruled out by analysis of the waves that penetrate to the core of the Sun. Helioseismology showed that the value for opacity used in early solar models was too low. This realization led astronomers to recalculate the location of the bottom of the convective zone. Both theory and observation now put the base of the convective zone at 70 percent of the way out from the center of the Sun, with an uncertainty in this number of less than half a percent.

Working back and forth between observation and theory has enabled astronomers to probe the otherwise inaccessible interior of the Sun. We now know that the energy is produced by nuclear fusion deep in the core and that it moves outward by radiation to a point about 70 percent of the radius of the Sun. Then it travels outward by convection to the surface. We also know how the temperature, density, and pressure change with radius and how these factors change the opacity at different distances from the center. Even though it is usually not possible to sample directly or to set up controlled experiments, this kind of collaboration between theory and observation is essential to observational sciences like astronomy.

## CHECK YOUR UNDERSTANDING 14.3

How do neutrinos help us understand what is going on in the core of the Sun?
(a) Neutrinos from distant objects pass through the Sun, probing the interior.
(b) Neutrinos from the Sun pass easily through Earth. (c) Neutrinos from the interior of the Sun easily escape. (d) Neutrinos change form on their way to Earth.

# 14.3 The Atmosphere of the Sun

Beyond the convective zone lie the outer layers of the Sun, which are collectively known as the Sun's atmosphere. These layers, shown in **Figure 14.10a**, include the *photosphere*, the *chromosphere*, and the *corona*. We can observe these layers of the Sun directly using telescopes and satellites. Observations of the Sun's atmosphere are important because activity in the Sun's atmosphere has consequences for human infrastructure such as power grids and satellites in orbit around Earth.

The Sun is a large ball of gas, and so, unlike Earth, it has no solid surface. Its apparent surface is like a fog bank on Earth. Imagine watching some people walking into a fog bank. After they disappear from view, you would say they were inside the fog bank, even though they never passed through a definite boundary. The apparent surface of the Sun is similar. Light from the Sun's surface can escape directly into space, so we can see it. Light from below the Sun's surface cannot escape directly into space, so we cannot see it.

At the base of the atmosphere is the **photosphere**: the Sun's apparent surface. This is where features such as sunspots can be seen. Above this photosphere is the **chromosphere**, a region of strong emission lines. The top layer is the **corona**, which can be viewed during a solar eclipse as a halo around the Sun. In the Sun's atmosphere, the density of the gas drops very rapidly with increasing altitude. Figure 14.10b shows how density and temperature change across the atmosphere of the Sun. In this section, we will explore each of these layers, beginning at the bottom with the photosphere.

## The Photosphere

The **effective temperature** of the photosphere is calculated from the Sun's luminosity and radius using the Stefan-Boltzmann law (see Chapter 5). The photosphere has an effective temperature of 5780 K, ranging from 6600 K to 4500 K over a 500-km-thick zone. As you can see in the graphs in Figure 14.10b, the temperature

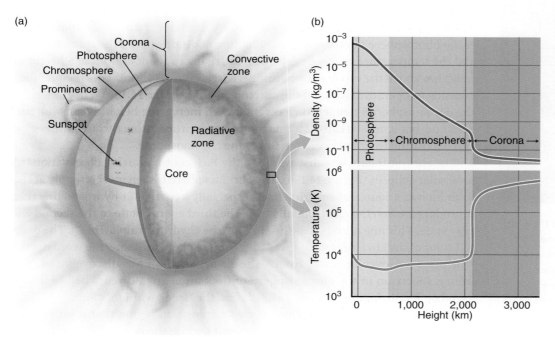

**Figure 14.10** (a) The components of the Sun's atmosphere are located above the convective zone. (b) The density and temperature of the Sun's atmosphere change abruptly at the boundary between the chromosphere and corona. Note that the *y*-axes are logarithmic.

(a)                 (b)

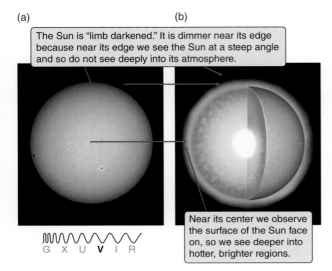

**Figure 14.11** (a) When viewed in visible light, the Sun appears to have a sharp outline, even though it has no true surface. The center of the Sun appears brighter, while the limb of the Sun is darker—an effect known as limb darkening. (b) Looking at the center of the Sun allows us to see deeper into the Sun's interior than we do when looking at the edge of the Sun. Because higher temperature means more luminous radiation, the center of the Sun appears brighter than its limb.

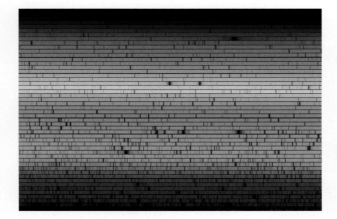

**Figure 14.12** This high-resolution spectrum of the Sun stretches from 400 nanometers (nm) in the lower left corner to 700 nm in the upper right corner and shows dark absorption lines. This spectrum was produced by passing the Sun's light through a prism-like device, and then cutting and folding the single long spectrum (from blue to red) into rows so that it will fit in a single image taken by a camera.

increases sharply across the boundary between the chromosphere and the corona, while the density falls sharply across the same boundary. The Sun appears to have a well-defined surface and a sharp outline when viewed from Earth because 500 km does not look very thick when viewed from a distance of 150 million km.

In **Figure 14.11a**, the Sun appears fainter near its edges than near its center, an effect called **limb darkening**. This effect is an artifact of the structure of the Sun's photosphere. When looking near the edge of the Sun, you are looking through the photosphere at a steep angle. As a result, you do not see as deeply into the interior of the Sun as when you are looking directly down through the photosphere near the center of the Sun's disk. The light from the limb of the Sun comes from a shallower layer that is cooler and fainter, as shown in Figure 14.11b.

In the Sun's atmosphere, the density of the gas drops very rapidly with increasing altitude. All visible solar phenomena take place in the Sun's atmosphere. Most of the radiation from below the Sun's photosphere is absorbed by matter and reemitted at the photosphere as a blackbody spectrum.

As we examine the structure of the Sun in more detail, however, we see that this simple description of the spectra of stars is incomplete. Light from the solar photosphere must escape through the upper layers of the Sun's atmosphere, which affects the spectrum we observe. In Chapter 13, we discussed the presence of absorption lines in the spectra of stars. Now we can take a closer look at how these absorption lines form. As photospheric light travels upward, atoms in the solar atmosphere absorb the light at discrete wavelengths, forming absorption lines. Because the Sun appears so much brighter than any other star, its spectrum can be studied in far more detail, so specially designed telescopes and high-resolution spectrometers have been built specifically to study the Sun's light. The solar spectrum is shown in **Figure 14.12**. Absorption lines from more than 70 elements have been identified. Analysis of these lines forms the basis for much of astronomers' knowledge of the solar atmosphere, including the composition of the Sun. This is also the starting point for an understanding of the atmospheres and spectra of other stars.

## The Chromosphere and Corona

Moving upward through the Sun's photosphere, the temperature falls from 6600 K at the photosphere's bottom to 4400 K at its top. At this point, the trend reverses and the temperature slowly begins to climb, rising to about 6000 K at a height of 1,500 km above the top of the photosphere (see Figure 14.10b). This region of increasing temperature is called the chromosphere (**Figure 14.13a**). The reason for the chromosphere's temperature reversal with increasing height is not well understood, but it may be caused by magnetic waves propagating through the region and depositing their energy at the top of the chromosphere.

The chromosphere was discovered in the 19th century during observations of total solar eclipses (Figure 14.13b). The chromosphere is seen most strongly at the solar limb as a source of emission lines, especially a particular hydrogen line that is produced when the electron falls from the third energy state to the second energy state. This line is known as the Hα line (the "hydrogen alpha line"). The deep red color of the Hα line is what gives the chromosphere its name; the word means "the place where color comes from." The element helium was discovered in 1868 from a spectrum of the chromosphere of the Sun nearly 30 years before it was found on Earth: helium is named after *helios*, the Greek word for "Sun."

At the top of the chromosphere, across a transition region that is only about 100 km thick, the temperature suddenly soars while the density abruptly drops

(a)

(b)

(c)

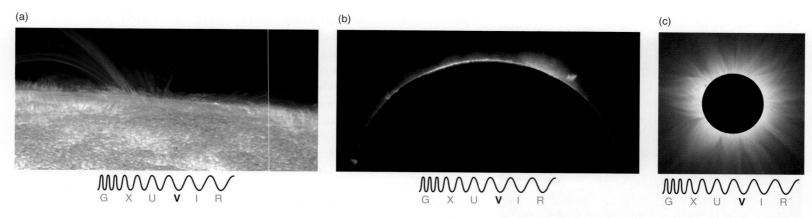

G X U V I R          G X U V I R          G X U V I R

**Figure 14.13** (a) This spacecraft image of the Sun shows fine structure in the chromosphere extending outward from the photosphere. (b) The chromosphere is visible during a total eclipse. (c) This eclipse image shows the Sun's corona, consisting of million-kelvin gas that extends for millions of kilometers beyond the surface of the Sun.

(see Figure 14.10b). Above this transition lies the outermost region of the Sun's atmosphere, the corona, where temperatures reach 1 million to 2 million K. The corona is thought to be heated by magnetic fields and micro solar flares.

The Sun's corona has been known since ancient times: it is visible during total solar eclipses as an eerie outer glow stretching a distance of several solar radii beyond the Sun's surface (Figure 14.13c). Because it is so hot, the solar corona is also a strong source of X-rays, and there is so much energy in these X-ray photons that many electrons are stripped away from nuclei, leaving atoms in the corona highly ionized.

**CHECK YOUR UNDERSTANDING 14.4**
The surface of the Sun appears sharp in visible light because: (a) the photosphere is cooler than the layers below it; (b) the photosphere is thin compared to the other layers in the Sun; (c) the photosphere is less dense than the convection zone; (d) the Sun has a distinct surface.

# 14.4 The Atmosphere of the Sun Is Very Active

The atmosphere of the Sun is a very turbulent place. The best-known features on the surface of the Sun are relatively dark blemishes in the solar photosphere, called **sunspots**. Sunspots come and go over time, though they remain long enough for us to determine the rotation rate of the Sun. These spots are associated with active regions: loops of material and explosions that fling particles far out into the Solar System. Long-term patterns have been observed in the variations of sunspots and active regions, revealing that the magnetic field of the Sun is constantly changing.

## Solar Activity Is Caused by Magnetic Effects

The magnetic field (see Chapter 5) of the Sun causes virtually all of the structure seen in the Sun's atmosphere. High-resolution images of the Sun show *coronal loops* that make up much of the Sun's lower corona (**Figure 14.14**). This texture is the result of magnetic structures called flux tubes. Magnetic fields are responsible for much of the structure of the corona as well. The corona is far too hot to be held in by the Sun's gravity, but over most of the surface of the Sun, coronal gas is

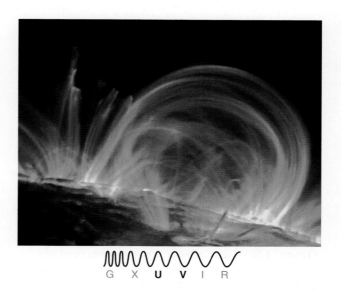

G X U V I R

**Figure 14.14** This close-up image of the Sun shows the tangled structure of coronal loops.

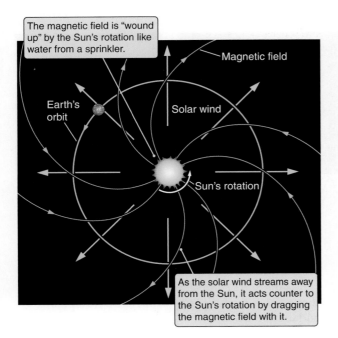

**Figure 14.15** The solar wind streams away from active areas and coronal holes on the Sun. As the Sun rotates, the solar wind takes on a spiral structure, much like the spiral of water that streams away from a rotating lawn sprinkler.

confined by magnetic loops with both ends firmly anchored deep within the Sun. The magnetic field in the corona acts almost like a network of rubber bands that coronal gas is free to slide along but cannot cross. In contrast, about 20 percent of the surface of the Sun is covered by an ever-shifting pattern of **coronal holes**, which are large regions where the magnetic field points outward, away from the Sun, and where coronal material is free to stream away into interplanetary space as the solar wind. In extreme ultraviolet images of the Sun we see coronal holes as dark regions, which indicates that they are cooler and lower in density than their surroundings (see the chapter-opening photograph).

The relatively steady part of the solar wind consists of lower-speed flows with velocities of about 350 km/s and higher-speed flows with velocities up to about 700 km/s. The higher-speed flows originate in coronal holes. Depending on their speed, particles in the solar wind take about 2–5 days to reach Earth. Frequently, 2–5 days after a coronal hole passes across the center of the face of the Sun, the speed and density of the solar wind reaching Earth increases. As you can see in **Figure 14.15**, the solar wind drags the Sun's magnetic field along with it. The magnetic field in the solar wind gets "wound up" by the Sun's rotation. Consequently, the solar wind has a spiral structure resembling the stream of water from a rotating lawn sprinkler.

The effects of the solar wind are felt throughout the Solar System. The solar wind blows the tails of comets away from the Sun, shapes the magnetospheres of the planets, and provides the energetic particles that power Earth's spectacular auroral displays. Using space probes, astronomers have been able to observe the solar wind extending out to 100 astronomical units (AU) from the Sun. But the solar wind does not go on forever. The farther it gets from the Sun, the more it has to spread out. Just like radiation, the density of the solar wind follows an inverse square law. At a distance of about 100 AU from the Sun, the solar wind stops abruptly. Here it piles up against the pressure of the **interstellar medium**, which is the gas and dust that lie between stars in a galaxy. **Figure 14.16** shows the

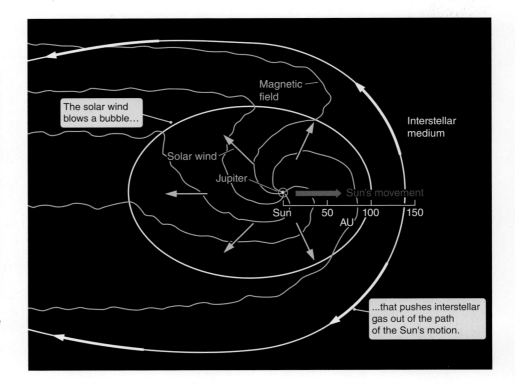

**Figure 14.16** The solar wind streams away from the Sun for about 100 AU, until it finally piles up against the pressure of the interstellar medium through which the Sun is traveling. The *Voyager 1* spacecraft has recently crossed this boundary.

region of space over which the solar wind is measured. The *Voyager 1* spacecraft has crossed the outer edge of this boundary and sent back the first direct measurements of true interstellar space. The *Interstellar Boundary Explorer* spacecraft, launched in 2008, is also exploring this region.

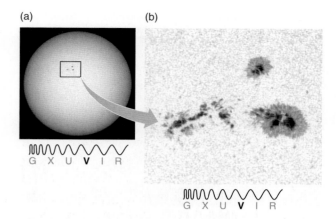

**Figure 14.17** (a) This image from the Solar Dynamics Observatory (SDO), taken in 2010, shows a large sunspot group. Sunspots are magnetically active regions that are cooler than the surrounding surface of the Sun. (b) This high-resolution view shows the sunspots in this group.

## Sunspots and Changes in the Sun

Sunspots have been noted since antiquity. Telescopic observations of sunspots date back almost 400 years, and there are records of naked-eye observations by Chinese, Greek, and medieval astronomers centuries before that. *But remember that you should never look directly at the Sun!* Direct viewing through a commercial solar filter is safe, as is projecting the image through a telescope or binoculars onto a surface such as paper and looking only at the projection. Many websites have live images of the Sun viewed through ground and space telescopes (see the "Using the Web" problems at the end of the chapter). Sunspots are places where material is trapped at the surface of the Sun by magnetic-field lines. When this material cools, convection cannot carry it downward, so it makes a cooler (therefore darker) spot on the surface of the Sun. **Figure 14.17** shows a large sunspot group. Sunspots appear dark, but only in contrast to the brighter surface of the Sun **(Working It Out 14.2)**.

Early telescopic observations of sunspots made during the 17th century led to the discovery of the Sun's rotation, which has an average period of about 27 days as seen from Earth and 25 days relative to the stars. Because Earth orbits the Sun in the same direction that the Sun rotates, observers on Earth see a slightly longer rotation period. Observations of sunspots also show that the Sun's rotation period

---

## 14.2 Working It Out Sunspots and Temperature

Sunspots are about 1500 K cooler than their surroundings. What does this lower temperature tell us about their luminosity? Think back to the Stefan-Boltzmann law in Chapter 5. The flux, $\mathcal{F}$, from a blackbody is proportional to the fourth power of the temperature, $T$. The constant of proportionality is the Stefan-Boltzmann constant, $\sigma$, which has a value of $5.67 \times 10^{-8} \text{ W/(m}^2 \text{ K}^4)$. We write this relationship as

$$\mathcal{F} = \sigma T^4$$

Remember that the flux is the amount of energy coming from a square meter of surface every second. How much less energy comes out of a sunspot than out of the rest of the Sun? Let's take round numbers for the temperature of a typical sunspot and the surrounding photosphere: 4500 K and 6000 K, respectively. We can set up two equations:

$$\mathcal{F}_{\text{spot}} = \sigma T_{\text{spot}}^4 \quad \text{and} \quad \mathcal{F}_{\text{surface}} = \sigma T_{\text{surface}}^4$$

We could solve each of these separately, and then divide the value of $\mathcal{F}_{\text{spot}}$ by $\mathcal{F}_{\text{surface}}$ to find out how much fainter the sunspot is, but it's much easier to solve for the ratio of the fluxes:

$$\frac{\mathcal{F}_{\text{spot}}}{\mathcal{F}_{\text{surface}}} = \frac{\sigma T_{\text{spot}}^4}{\sigma T_{\text{surface}}^4} = \frac{T_{\text{spot}}^4}{T_{\text{surface}}^4} = \left(\frac{T_{\text{spot}}}{T_{\text{surface}}}\right)^4$$

Plugging in our values for $T_{\text{spot}}$ and $T_{\text{surface}}$ gives

$$\frac{\mathcal{F}_{\text{spot}}}{\mathcal{F}_{\text{surface}}} = \left(\frac{4500 \text{ K}}{6000 \text{ K}}\right)^4 = 0.32$$

and multiplying both sides by $\mathcal{F}_{\text{surface}}$ gives

$$\mathcal{F}_{\text{spot}} = 0.32 \mathcal{F}_{\text{surface}}$$

So the amount of energy coming from a square meter of sunspot every second is about one-third as much as the amount of energy coming from a square meter of surrounding surface every second. In other words, the sunspot is about one-third as bright as the surrounding photosphere. If you could cut out the sunspot and place it elsewhere in the sky, it would be brighter than the full Moon.

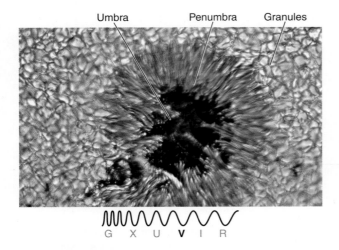

G X U V I R

**Figure 14.18** This very high-resolution view of a sunspot shows the dark umbra surrounded by the lighter penumbra. The solar surface around the sunspot bubbles with separate cells of hot gas called *granules*. The smallest features are about 100 km across.

is shorter at its equator than at higher latitudes, an effect called **differential rotation**. Differential rotation is possible only because the Sun is a large ball of gas rather than a solid object.

**Figure 14.18** shows the structure of a sunspot on the surface of the Sun. A sunspot consists of an inner dark core called the **umbra**, which is surrounded by a less dark region called the **penumbra**, which shows an intricate radial pattern, reminiscent of the petals of a flower. Sunspots are caused by magnetic fields thousands of times greater than the magnetic field at Earth's surface. They occur in pairs that are connected by loops in the magnetic field. Sunspots range in size from a few tens of kilometers across up to complex groups that may contain several dozen individual spots and span as much as 150,000 km. The largest sunspot groups are so large that they can be seen without a telescope.

Although sunspots occasionally last 100 days or longer, half of all sunspots come and go in about 2 days, and 90 percent are gone within 11 days. The number and distribution of sunspots change over time in a pattern averaging 11 years called the **sunspot cycle**. **Figure 14.19a** shows data for several recent cycles. At the beginning of a cycle, sunspots appear at solar latitudes of about 30° north and south of the solar equator. Over the following years, sunspots are found closer to the equator as their number increases to a maximum and then declines.

**Figure 14.19** (a) The number of sunspots varies with time, as shown in this graph of the past few solar cycles. (b) The "solar butterfly" diagram shows the fraction of the Sun covered by sunspots at each latitude. The data are color coded to show the percentage of the strip at that latitude that is covered in sunspots at that time: black, 0 to 0.1 percent; red, 0.1–1.0 percent; yellow, greater than 1.0 percent. (c) The Sun's magnetic poles flip every 11 years. Yellow indicates magnetic south; blue indicates magnetic north.

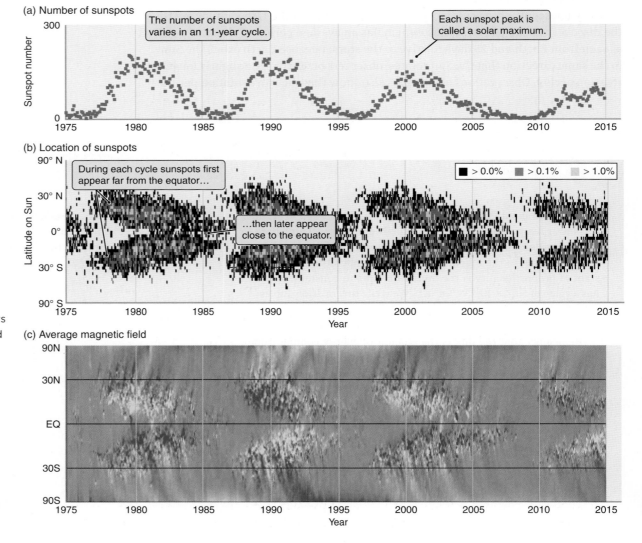

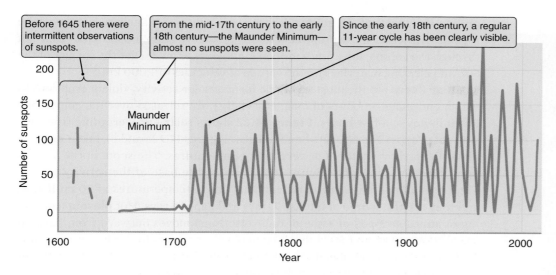

Before 1645 there were intermittent observations of sunspots.

From the mid-17th century to the early 18th century—the Maunder Minimum—almost no sunspots were seen.

Since the early 18th century, a regular 11-year cycle has been clearly visible.

Maunder Minimum

**Figure 14.20** Sunspots have been observed for hundreds of years. In this plot, the 11-year cycle in the number of sunspots (half of the 22-year solar magnetic cycle) is clearly visible. Sunspot activity varies greatly over time. The period from the middle of the 17th century to the early 18th century, when almost no sunspots were seen, is called the *Maunder Minimum*.

As the last few sunspots approach the equator, new sunspots again begin appearing at middle latitudes, and the next cycle begins. Figure 14.19b shows the number of sunspots at a given latitude plotted against time: this diagram of opposing diagonal bands is called the sunspot "butterfly diagram."

In the early 20th century, solar astronomer George Ellery Hale (1868–1938) was the first to show that the 11-year sunspot cycle is actually half of a 22-year magnetic cycle during which the direction of the Sun's magnetic field reverses after each 11-year sunspot cycle. Figure 14.19c shows how the average strength of the magnetic field at every latitude has changed over more than 35 years. The direction of the Sun's magnetic field flips at the maximum of each sunspot cycle. Sunspots tend to come in pairs, with one spot (the leading sunspot) in front of the other with respect to the Sun's rotation. In one 11-year sunspot cycle, the leading sunspot in each pair tends to be a north magnetic pole, whereas the trailing sunspot tends to be a south magnetic pole. In the next 11-year sunspot cycle, this polarity is reversed: the leading sunspot in each pair is a south magnetic pole, whereas the trailing sunspot tends to be a north magnetic pole. The transition between these two magnetic polarities occurs near the peak of each sunspot cycle. Magnetic activity on the Sun affects the photosphere, chromosphere, and corona.

Telescopic observations of sunspots date back almost 400 years. As you can see in **Figure 14.20**, the 11-year cycle is neither perfectly periodic nor especially reliable. The time between peaks in the number of sunspots actually varies between about 9.7 and 11.8 years. The number of spots seen during a given cycle fluctuates as well, and there have been periods when sunspot activity has disappeared almost entirely. An extended lull in solar activity, called the **Maunder Minimum**, lasted from 1645 to 1715. Typically, there are about six peaks of solar activity in 70 years, but virtually no sunspots were seen during the Maunder Minimum, and auroral displays were less frequent than usual.

Sunspots are only one of several phenomena that follow the Sun's 22-year cycle of magnetic activity. The peaks of the cycle, called **solar maxima**, are times of intense activity. Sunspots are often accompanied by a brightening of the solar chromosphere that is seen most clearly in emission lines such as H$\alpha$. These bright regions are known as solar active regions. The magnificent loops arching through the solar corona, shown in **Figure 14.21**, are solar **prominences**, magnetic flux tubes of relatively cool (5000–10,000 K) but dense gas extending through the

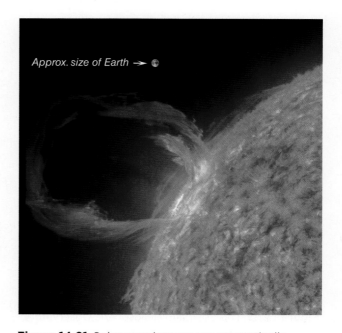

*Approx. size of Earth*

**Figure 14.21** Solar prominences are magnetically supported arches of hot gas that rise high above active regions on the Sun. Here, you can see a close-up view at the base of a large prominence. An image of Earth is included for scale (it is not actually that close to the Sun).

(a)       (b)       (c)

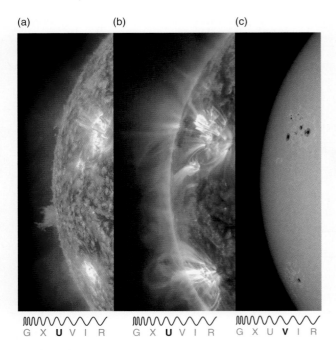

G X **U** V I R    G X **U** V I R    G X U **V** I R

**Figure 14.22** The Solar Dynamics Observatory (SDO) observed these active regions of the Sun that produced solar flares in August 2011. (a) Activity near the surface at 60,000 K is visible in extreme ultraviolet light (along with a prominence rising up from the Sun's edge). (b) Viewed at other ultraviolet wavelengths, many looping arcs and plasma heated to about 1 million K become visible. (c) The dark spots in this image are the magnetically intense sunspots that are the sources of all the activity.

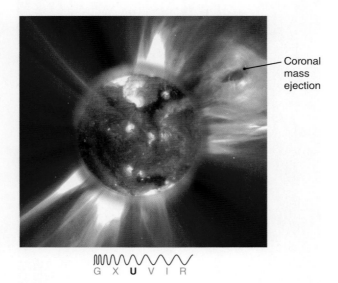

Coronal mass ejection

G X **U** V I R

**Figure 14.23** This Solar and Heliospheric Observatory (*SOHO*) image shows a coronal mass ejection (upper right): a simultaneously recorded ultraviolet image of the solar disk is superimposed.

million-kelvin gas of the corona. These prominences are anchored in the active regions. Although most prominences are relatively quiet, others can erupt out through the corona, towering a million kilometers or more over the surface of the Sun and ejecting material into the corona at velocities of 1,000 km/s.

**Solar flares** are the most energetic form of solar activity, violent eruptions in which enormous amounts of magnetic energy are released over the course of a few minutes to a few hours. **Figure 14.22** shows solar flares erupting from two sunspot groups. The two left-hand images (Figures 14.22a and b), taken in ultraviolet light, show material at very high temperatures. The spots in the visible-light image on the right (Figure 14.22c) are at the base of the activity seen in Figures 14.22a and b. Solar flares can heat gas to temperatures of 20 million K, and they are the source of intense X-rays and gamma rays. Hot **plasma** (consisting of atoms stripped of some of their electrons) moves outward from flares at speeds that can reach 1,500 km/s. Magnetic effects can then accelerate subatomic particles to almost the speed of light. Such events, called **coronal mass ejections (CMEs)** (**Figure 14.23**), send powerful bursts of energetic particles outward through the Solar System. Coronal mass ejections occur about once per week during the minimum of the sunspot cycle and as often as several times per day near the maximum of the cycle.

## Solar Activity Affects Earth

The amount of solar radiation received at the distance of Earth from the Sun has been measured to be, on average, 1,361 watts per square meter ($W/m^2$). As you can see in **Figure 14.24**, satellite measurements of the amount of radiation coming from the Sun show that this value varies by as much as 0.2 percent over periods of a few weeks, as dark sunspots in the photosphere and bright spots in the chromosphere move across the disk. Overall, however, the increased radiation from active regions on the Sun more than makes up for the reduction in radiation from sunspots. On average, the Sun seems to be about 0.1 percent brighter during the peak of a solar cycle than it is at its minimum.

Solar activity affects Earth in many ways. Solar active regions are the source of most of the Sun's extreme ultraviolet and X-ray emissions, energetic radiation that heats Earth's upper atmosphere and, during periods of increased solar activity, causes Earth's upper atmosphere to expand. When this happens, the swollen upper atmosphere can significantly increase the atmospheric drag on spacecraft orbiting at relatively low altitudes, such as that of the Hubble Space Telescope, causing their orbits to decay. Periodic boosts have been necessary to keep the Hubble Space Telescope in its orbit.

Earth's magnetosphere is the result of the interaction between Earth's magnetic field and the solar wind. Increases in the solar wind accompanying solar activity, especially coronal mass ejections directed at Earth, can disrupt Earth's magnetosphere. Spectacular auroras can accompany such events, as can magnetic storms that have been known to disrupt electric power grids and cause blackouts across large regions. Coronal mass ejections that are emitted in the direction of Earth also hinder radio communication and navigation, and they can damage sensitive satellite electronics, including communication satellites. In addition, energetic particles accelerated in solar flares pose one of the greatest dangers to human exploration of space.

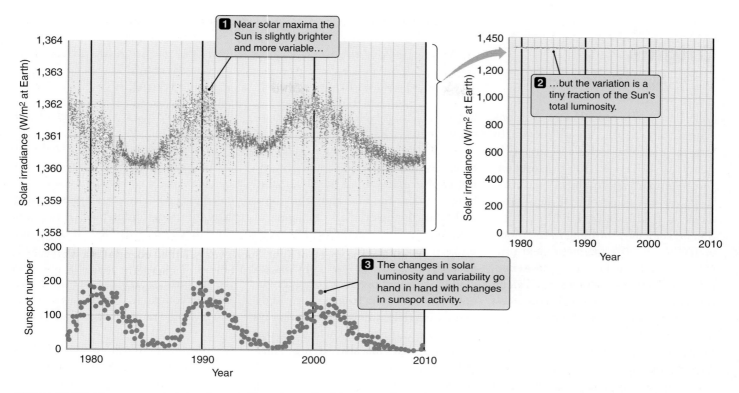

**Figure 14.24** Measurements taken by satellites above Earth's atmosphere show that the amount of light from the Sun changes slightly over time.

Detailed observations from the ground and from space help astronomers understand the complex nature of the solar atmosphere. The Solar and Heliospheric Observatory (*SOHO*) spacecraft is a joint mission between NASA and the European Space Agency (ESA). *SOHO* moves in lockstep with Earth at a location approximately 1,500,000 km from Earth that is almost directly in line between Earth and the Sun. *SOHO* carries 12 scientific instruments that monitor the Sun and measure the solar wind upstream of Earth. Additionally, NASA's Solar Dynamics Observatory (SDO) studies the solar magnetic field in order to predict when major solar events will occur, rather than simply responding after they happen.

## CHECK YOUR UNDERSTANDING 14.5

Sunspots appear dark because: (a) they have very low density; (b) magnetic fields absorb most of the light that falls on them; (c) they are regions of very high pressure; (d) they are cooler than their surroundings.

# Origins

## The Solar Wind and Life

Solar flares and coronal mass ejections can affect the space around Earth. In fact, energetic particles accelerated in solar flares pose one of the greatest dangers to human exploration of space, and they need to be considered when astronauts are orbiting Earth in a space station or, someday, traveling to the Moon or farther. Earth's magnetic field protects life on the surface from these energetic particles: the particles travel along the magnetic-field lines to Earth: poles, creating the auroras, without bombarding the surface, which could be harmful to life on Earth. But the Moon does not have this protection because its magnetic field is very weak. Astronauts on the lunar surface would be exposed to as much radiation as astronauts traveling in space. The strength of the solar wind varies with the solar cycle, as

noted in Section 14.4, so exposure danger varies as well.

As illustrated in **Figure 14.25**, the Solar System is surrounded by the **heliosphere**, in which the solar wind blows against the interstellar medium and clears out an area like the inside of a bubble. As the Sun and Solar System move through the Milky Way Galaxy, passing in and out of interstellar clouds, this heliosphere protects the entire Solar System from galactic high-energy particles known as cosmic rays that originate primarily in high-energy explosions of massive dying stars. When the Sun is in its lower-activity state, the heliosphere is weaker, so more galactic cosmic rays enter the Solar System. In addition, the intensity of these cosmic rays depends on where the Sun and Solar System are located

in their orbit about the center of the Milky Way Galaxy.

Some scientists have theorized that at times when the Sun was quiet and the heliosphere was weaker than average, and the Solar System was passing through a particular part of the galaxy, the cosmic-ray flux in the Solar System—and on Earth—increased. This increased flux possibly led to a disruption in Earth's ozone layer and possibly contributed to a mass extinction in which many species died out on Earth.

Thus, in addition to the obvious contribution of the Sun to heat and light on Earth, the extension of the Sun through the solar wind may have affected the evolution of life on Earth—and it may also affect the ability of humans to live and work in space.

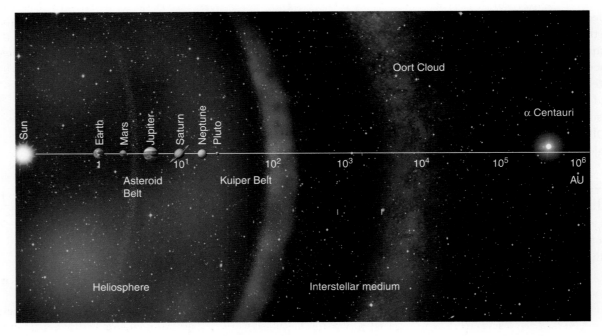

**Figure 14.25** The heliosphere of the Sun, a bubble of charged particles partially covering the Solar System that is formed by the solar wind blowing against the interstellar medium. The *Voyager* spacecraft are just past 100 AU. Notice that the scale is logarithmic.

# Carrington-Class CME Narrowly Misses Earth

By **DR. TONY PHILLIPS,** Science@NASA

Last month (April 8–11, 2014), scientists, government officials, emergency planners, and others converged on Boulder, Colorado, for NOAA's Space Weather Workshop—an annual gathering to discuss the perils and probabilities of solar storms.

The current solar cycle is weaker than usual, so you might expect a correspondingly low-key meeting. On the contrary, the halls and meeting rooms were abuzz with excitement about an intense solar storm that narrowly missed Earth.

"If it had hit, we would still be picking up the pieces," says Daniel Baker of the University of Colorado, who presented a talk entitled The Major Solar Eruptive Event in July 2012: Defining Extreme Space Weather Scenarios.

The close shave happened almost two years ago. On July 23, 2012, a plasma cloud or "CME" rocketed away from the Sun as fast as 3,000 km/s, more than four times faster than a typical eruption. The storm tore through the Earth's orbit, but fortunately Earth wasn't there. Instead it hit the *STEREO-A* spacecraft. Researchers have been analyzing the data ever since, and they have concluded that the storm was one of the strongest in recorded history.

"It might have been stronger than the Carrington Event itself," says Baker.

The Carrington Event of September 1859 was a series of powerful CMEs that hit Earth head-on, sparking Northern Lights as far south as Tahiti. Intense geomagnetic storms caused global telegraph lines to spark, setting fire to some telegraph offices and disabling the "Victorian Internet." A similar storm today could have a catastrophic effect on modern power grids and telecommunication networks. According to a study by the National Academy of Sciences, the total economic impact could exceed $2 trillion or 20 times greater than the costs of a Hurricane Katrina. Multi-ton transformers fried by such a storm could take years to repair and impact national security.

A recent paper in *Nature Communications* authored by UC Berkeley space physicist Janet G. Luhmann and former postdoc Ying D. Liu describes what gave the July 2012 storm Carrington-like potency. For one thing, the CME was actually *two* CMEs separated by only 10 to 15 minutes. This double storm cloud traveled through a region of space that had been cleared out by another CME only four days earlier. As a result, the CMEs were not decelerated as much as usual by their transit through the interplanetary medium.

Had the eruption occurred just one week earlier, the blast site would have been facing Earth, rather than off to the side, so it was a relatively narrow escape.

When the Carrington Event enveloped Earth in the 19th century, technologies of the day were hardly sensitive to electromagnetic disturbances. Modern society, on the other hand, is deeply dependent on Sun-sensitive technologies such as GPS, satellite communications, and the Internet.

"The effect of such a storm on our modern technologies would be tremendous," says Luhmann.

During informal discussions at the workshop, Nat Gopalswamy of the Goddard Space Flight Center noted that "without NASA's *STEREO* probes, we might never have known the severity of the 2012 superstorm. This shows the value of having 'space weather buoys' located all around the Sun."

It also highlights the potency of the Sun even during so-called "quiet times." Many observers have noted that the current solar cycle is weak, perhaps the weakest in 100 years. Clearly, even a weak solar cycle can produce a very strong storm. Says Baker, "We need to be prepared."

1. What is a CME?
2. Why would a CME cause disruptions on Earth?
3. Explain how this storm missed Earth by 1 week.
4. More sensationalistic headlines for this story claimed that Earth almost "was sent back to the Dark Ages." What did they mean by this exaggeration?
5. Go to the NASA press release for this story (http://science.nasa.gov/science-news/science-at-nasa/2014/23jul_superstorm/), and watch the 4-minute "ScienceCast" video. What happened during the Carrington CME in 1859? Is this video effective at communicating the science information to the nonspecialist?

# Summary

The forces due to pressure and gravity balance each other in hydrostatic equilibrium, maintaining the Sun's structure. Nuclear reactions converting hydrogen to helium are the source of the Sun's energy. Energy created in the Sun's core moves outward to the surface, first by radiation and then by convection. The solar wind may adversely affect astronauts located in space or on planets that lack a protective magnetic field, but it also has protected the Solar System from galactic, high-energy, cosmic-ray particles.

**LG 1** **Describe the balance between the forces that determine the structure of the Sun.** The outward pressure of the hot gas inside the Sun balances the inward pull of gravity at every point. This balance is dynamically maintained. An energy balance is also maintained, with the energy produced in the core of the Sun balancing the energy lost from the surface.

**LG 2** **Explain how mass is converted into energy in the Sun's core and how long it will take the Sun to use up its fuel.** In the core of the Sun, mass is converted to energy via the proton-proton cycle. When four hydrogen atoms fix to one helium atom, some mass is lost. This mass is released as energy, nearly all of which leaves the Sun either as photons or as neutrinos. Neutrinos are elusive, almost massless particles that interact only very weakly with other matter. Observations of neutrinos confirm that nuclear fusion is the Sun's primary energy source.

**LG 3** **Sketch a physical model of the Sun's interior, and list the different ways that energy moves outward from the Sun's core toward its surface.** The interior of the Sun is divided into zones that are defined by how energy is transported in that region. Energy moves outward through the Sun by radiation and by convection.

**LG 4** **Describe how observations of solar neutrinos and seismic vibrations on the surface of the Sun test astronomers' models of the Sun.** The Sun has multiple layers, each with a characteristic pressure, density, and temperature. Neutrinos directly probe the interior of the Sun. This model of the interior of the Sun has been tested by helioseismology, in much the same way that the model of Earth's interior has been tested by seismology.

**LG 5** **Describe the solar activity cycles of 11 and 22 years, and explain how these cycles are related to the Sun's changing magnetic field.** Activity on the Sun follows a cycle that peaks every 11 years but takes 22 full years for the magnetic field to reverse. Sunspots are photospheric regions that are cooler than their surroundings, and they reveal the cycles in solar activity. Material streaming away from the Sun's corona creates the solar wind, which moves outward through the Solar System until it meets the interstellar medium. Solar storms, including ejections of mass from the corona, produce auroras and can disrupt power grids and damage satellites.

## ? UNANSWERED QUESTIONS

- Will nuclear fusion become a major source of energy production on Earth? Scientists have been working on controlled nuclear fusion for more than 60 years, since the first hydrogen bombs were developed. But so far there are too many difficulties in replicating the conditions inside the Sun. Nuclear fusion requires that we have hydrogen isotopes at very high temperature, density, and pressure, just as is the case when a hydrogen bomb explodes. However, controlled nuclear fusion requires that we confine this material sufficiently long to get more energy out than we put in. Several major experiments have attempted to fuse isotopes of hydrogen. An alternative approach is to fuse an isotope of helium, $^3$He, which has only three particles in the nucleus (two protons and one neutron). On Earth, $^3$He is found in very limited supply. But $^3$He is in much greater abundance on the Moon, so some people propose setting up mining colonies on the Moon to extract $^3$He for use in fusion reactions on Earth or possibly even on the Moon (see question 49b at the end of the chapter).

- Are variations in Earth's climate related to solar activity? Solar activity affects Earth's upper atmosphere, and it may affect weather patterns as well. It has also been suggested that variations in the amount of radiation from the Sun might be responsible for past variations in Earth's climate. Current models indicate that observed variations in the Sun's luminosity could account for only about 0.1 K differences in Earth's average temperature—much less than the effects due to the ongoing buildup of carbon dioxide in Earth's atmosphere. Triggering the onset of an ice age may require a sustained drop in global temperatures of only about 0.2–0.5 K, so astronomers are continuing to investigate a possible link between solar variability and changes in Earth's climate.

# Questions and Problems

## Test Your Understanding

1. The physical model of the Sun's interior has been confirmed by observations of
   a. neutrinos and seismic vibrations.
   b. sunspots and solar flares.
   c. neutrinos and positrons.
   d. sample returns from spacecraft.
   e. sunspots and seismic vibrations.

2. Place in order the following steps in the fusion of hydrogen into helium. If two or more steps happen simultaneously, use an equals sign (=).
   a. A positron is emitted.
   b. One gamma ray is emitted.
   c. Two hydrogen nuclei are emitted.
   d. Two $^3$He collide and become $^4$He.
   e. Two hydrogen nuclei collide and become $^2$H.
   f. Two gamma rays are emitted.
   g. A neutrino is emitted.
   h. One deuterium nucleus and one hydrogen nucleus collide and become $^3$He.

3. Sunspots, flares, prominences, and coronal mass ejections are all caused by
   a. magnetic activity on the Sun.
   b. electrical activity on the Sun.
   c. the interaction of the Sun's magnetic field and the interstellar medium.
   d. the interaction of the solar wind and Earth's magnetic field.
   e. the interaction of the solar wind and the Sun's magnetic field.

4. The structure of the Sun is determined by both the balance between the forces due to _____ and gravity and the balance between energy generation and energy _____.
   a. pressure; production
   b. pressure; loss
   c. ions; loss
   d. solar wind; production

5. In the proton-proton chain, four hydrogen nuclei are converted to a helium nucleus. This does not happen spontaneously on Earth because the process requires
   a. vast amounts of hydrogen.
   b. very high temperatures and densities.
   c. hydrostatic equilibrium.
   d. very strong magnetic fields.

6. The solar neutrino problem pointed to a fundamental gap in our knowledge of
   a. nuclear fusion.
   b. neutrinos.
   c. hydrostatic equilibrium.
   d. magnetic fields.

7. Sunspots change in number and location during the solar cycle. This phenomenon is connected to
   a. the rotation rate of the Sun.
   b. the temperature of the Sun.
   c. the magnetic field of the Sun.
   d. the tilt of the axis of the Sun.

8. Suppose an abnormally large amount of hydrogen suddenly burned in the core of the Sun. Which of the following would be observed first?
   a. The Sun would become brighter.
   b. The Sun would swell and become larger.
   c. The Sun would become bluer.
   d. The Sun would emit more neutrinos.

9. The solar corona has a temperature of 1 million to 2 million K; the photosphere has a temperature of only about 6000 K. Why isn't the corona much, much brighter than the photosphere?
   a. The magnetic field traps the light.
   b. The corona emits only X-rays.
   c. The photosphere is closer to us.
   d. The corona has a much lower density.

10. The Sun rotates once every 25 days relative to the stars. The Sun rotates once every 27 days relative to Earth. Why are these two numbers different?
    a. The stars are farther away.
    b. Earth is smaller.
    c. Earth moves in its orbit during this time.
    d. The Sun moves relative to the stars.

11. Place the following regions of the Sun in order of increasing radius.
    a. corona
    b. core
    c. radiative zone
    d. convective zone
    e. chromosphere
    f. photosphere
    g. a sunspot

12. Coronal mass ejections
    a. carry away 1 percent of the mass of the Sun each year.
    b. are caused by breaking magnetic fields.
    c. are always emitted in the direction of Earth.
    d. are unimportant to life on Earth.

13. As energy moves out from the Sun's core toward its surface, it first travels by _____, then by _____, and then by _____.
    a. radiation; conduction; radiation
    b. conduction; radiation; convection
    c. radiation; convection; radiation
    d. radiation; convection; conduction

14. Energy is produced primarily in the center of the Sun because
    a. the strong nuclear force is too weak elsewhere.
    b. that's where neutrinos are created.
    c. that's where most of the helium is.
    d. the outer parts have lower temperatures and densities.

15. The solar wind pushes on the magnetosphere of Earth, changing its shape, because
    a. the solar wind is so dense.
    b. the magnetosphere is so weak.
    c. the solar wind contains charged particles.
    d. the solar wind is so fast.

## Thinking about the Concepts

16. Explain how hydrostatic equilibrium acts as a safety valve to keep the Sun at its constant size, temperature, and luminosity.

17. Two of the three atoms in a molecule of water ($H_2O$) are hydrogen. Why are Earth's oceans not fusing hydrogen into helium and setting Earth ablaze?

18. Why are neutrinos so difficult to detect?

19. Explain the proton-proton chain through which the Sun generates energy by converting hydrogen to helium.

20. On Earth, nuclear power plants use *fission* to generate electricity. In fission, a heavy element like uranium is broken into many atoms, where the total mass of the fragments is less than that of the original atom. Explain why fission could not be powering the Sun today.

21. If an abnormally large amount of hydrogen suddenly burned in the core of the Sun, what would happen to the rest of the Sun? Would the Sun change as seen from Earth?

22. Study the Process of Science Figure. If the follow-up experiments did not detect the other types of neutrinos, what would have been the next step for scientists at that point?

23. What is the solar neutrino problem, and how was it solved?

24. The Sun's visible "surface" is not a true surface, but a feature called the photosphere. Explain why the photosphere is not a true surface.

25. How are orbiting satellites and telescopes affected by the Sun?

26. Describe the solar corona. Under what circumstances can it be seen without special instruments?

27. In the proton-proton chain, the mass of four protons is slightly greater than the mass of a helium nucleus. Explain what happens to this "lost" mass.

28. What have sunspots revealed about the Sun's rotation?

29. Why are different parts of the Sun best studied at different wavelengths? Which parts are best studied from space?

30. Why is it important to study the interaction of the solar wind with the interstellar medium?

## Applying the Concepts

31. In Figure 14.10, density and temperature are both graphed versus height.
    a. Is the height axis linear or logarithmic? How do you know?
    b. Is the density axis linear or logarithmic? How do you know?
    c. Is the temperature axis linear or logarithmic? How do you know?

32. Using the data in Figures 14.19b and c, present an argument that sunspots occur in regions of strong magnetic field.

33. Study Figure 14.17a and the graph in Figure 14.20.
    a. Estimate the fraction of the Sun's surface that is covered by the large sunspot group in the image. (Remember that you are seeing only one hemisphere of the Sun.)
    b. From the graph, estimate the average number of sunspots that occurs at solar maximum.
    c. On average, what fraction of the Sun could be covered by sunspots at solar maximum? Is this a large fraction?
    d. Compare your conclusion to the graph of irradiance in Figure 14.24. Does this graph make sense to you?

34. The Sun has a radius equal to about 2.3 light-seconds. Explain why a gamma ray produced in the Sun's core does not emerge from the Sun's surface 2.3 seconds later.

35. Assume that the Sun's mass is about 300,000 Earth masses and that its radius is about 100 times that of Earth. The density of Earth is about 5,500 kg/m³.
    a. What is the average density of the Sun?
    b. How does this compare with the density of Earth? With the density of water?

36. The Sun shines by converting mass into energy according to $E = mc^2$. Show that if the Sun produces $3.85 \times 10^{26}$ J of energy per second, it must convert 4.3 million metric tons ($4.3 \times 10^9$ kg) of mass per second into energy.

37. Assume that the Sun has been producing energy at a constant rate over its lifetime of 4.6 billion years ($1.4 \times 10^{17}$ seconds).
    a. How much mass has it lost creating energy over its lifetime?
    b. The current mass of the Sun is $2 \times 10^{30}$ kg. What fraction of its current mass has been converted into energy over the lifetime of the Sun?

38. Suppose our Sun was an A5 main-sequence star, with twice the mass and 12 times the luminosity of the Sun, a G2 star. How long would this A5 star burn hydrogen to helium? What would this mean for Earth?

39. Imagine that the source of energy in the interior of the Sun changed abruptly.
    a. How long would it take before a neutrino telescope detected the event?
    b. When would a visible-light telescope see evidence of the change?

40. On average, how long does it take particles in the solar wind to reach Earth from the Sun if they are traveling at an average speed of 400 km/s?

41. A sunspot appears only 70 percent as bright as the surrounding photosphere. The photosphere has a temperature of approximately 5780 K. What is the temperature of the sunspot?

42. The hydrogen bomb represents an effort to create a similar process to what takes place in the core of the Sun. The energy released by a 5-megaton hydrogen bomb is $2 \times 10^{16}$ J.
    a. This textbook, *21st Century Astronomy*, has a mass of about 1.6 kg. If all of its mass were converted into energy, how many 5-megaton bombs would it take to equal that energy?
    b. How much mass did Earth lose each time a 5-megaton hydrogen bomb was exploded?

43. Verify the claim made at the start of this chapter that the Sun produces more energy per second than all the electric power plants on Earth could generate in a half-million years. Estimate or look up how many power plants there are on the planet and how much energy an average power plant produces. Be sure to account for different kinds of power; for example, coal, nuclear, wind.

44. Let's examine the reason that the Sun cannot power itself by chemical reactions. Using Working It Out 14.1 and the fact that an average chemical reaction between two atoms releases $1.6 \times 10^{-19}$ J, estimate how long the Sun could emit energy at its current luminosity. Compare that estimate to the known age of Earth.

45. The Sun could get energy from gravitational contraction for a time period of ($GM_{Sun}/R_{Sun}L_{Sun}$). How long would the Sun last at its current luminosity? (Be careful with units!)

## USING THE WEB

46. a. Go to *QUEST*'s "Journey into the Sun" Web page (http://science.kqed.org/quest/video/journey-into-the-sun) to watch a short video on the Solar Dynamics Observatory (SDO), launched in 2010. Why is studying the magnetic field of the Sun so important? What is new and different about this observatory? What is the "Music of the Sun"?
    b. Go to the SDO website (http://sdo.gsfc.nasa.gov). Under "Data," select "The Sun Now" and view the Sun at many wavelengths. What activity do you observe in the images at the location of any sunspots seen in the "HMI Intensity-gram" images? (You can download a free SDO app by Astra to get real-time images on your mobile device.) Look at a recent news story from the SDO website. What was observed, and why is it newsworthy?
    c. Go to the *STEREO* mission's website (http://stereo.gsfc.nasa.gov). What is *STEREO*? Where are the spacecraft located? How does this configuration enable observations of the entire Sun at once? (You can download the app "3-D Sun" to get the latest images on your mobile device.)

47. a. What are the science goals of NASA's *Interface Region Imaging Spectrograph* (*IRIS*) mission (www.nasa.gov/mission_pages/iris/)? What has it discovered?
    b. An older space mission, *SOHO* (Solar and Heliospheric Observatory; http://sohowww.nascom.nasa.gov), was launched in 1995 by NASA and ESA. Click on "The Sun Now" to see today's images. The Extreme Ultraviolet imaging Telescope (EIT) images are in the far ultraviolet and show violent activity. How do these images differ from the ones of SDO in question 46b?
    c. Go to the Daniel K. Inouye Solar Telescope (DKIST) website (http://atst.nso.edu). This adaptive-optics telescope under construction on Haleakala, Maui, will be the largest solar telescope. Why is it important to study the magnetic field of the Sun? What are some of the advantages of studying the Sun from a ground-based telescope instead of a space-based telescope? What wavelengths does the DKIST observe? Why is Maui a good location? When is the telescope scheduled to be completed?

48. a. Go to the Space Weather website (http://spaceweather.com). Are there any solar flares today? What is the sunspot number? Is it about what you would expect for this year? (Click on "What is the sunspot number?" to see a current graph.) Are there any coronal holes today?
    b. Citizen science: Go to the website for Sunspotter (http://www.sunspotter.org/), a Zooniverse project that evaluates the complexity of sunspots and how they change over time. Zooniverse projects offer an opportunity for people to

contribute to science by analyzing pieces of data. Create an account for Zooniverse if you don't already have one (you will use it again in this course). Log in and click on "Science" and skim through the sections. What are the goals of this project? Why is it useful to have multiple people looking at these data? Click on "Classify" and analyze some sunspots. Save a screen shot for your homework.

c.  Citizen science: Go to the Solar Stormwatch website (http://solarstormwatch.com), a Zooniverse project from the Royal Observatory in Greenwich, England. Create an account for Zooniverse if you don't already have one (you will use it again in this course). Log in and click on "Spot and Track Storms" and go through the Spot and Track training exercises. You are now ready to look at some real data. Click on an image to do the classification. Save a screen shot for your homework.

49. a.  Go to the National Ignition Facility (NIF) website (https://lasers.llnl.gov/about/). Under "Science," click on "How to Make a Star." How are lasers used in experiments to develop controlled nuclear fusion on Earth? How does the fusion reaction here differ from that in the Sun?

b.  An alternative approach is to fuse $^3$He + $^3$He instead of the hydrogen isotopes. But on Earth, $^3$He is in limited supply.

Helium-3 is in much greater abundance on the Moon, so some people propose setting up mining colonies on the Moon to extract $^3$He for fusion reactions on Earth. Do a search on "helium 3 Moon." Which countries are talking about going to the Moon for this purpose? What is the timeline for when this might happen? What are the difficulties?

50. a.  Go to http://voyager.jpl.nasa.gov/where/. Where are the *Voyager* spacecraft now? Has *Voyager 2* crossed into interstellar space?

b.  Go to the website for the Interstellar Boundary Explorer (*IBEX*) (http://www.nasa.gov/mission_pages/ibex/.) What has *IBEX* learned about the solar wind and the interstellar medium?

# smartw⊛rk5

If your instructor assigns homework in Smartwork5, access your assignments at digital.wwnorton.com/astro5.

digital.wwnorton.com/astro5

The proton-proton chain powers the Sun by fusing hydrogen into helium. This fusion process produces several different particles as by-products, as well as energy. In this Exploration, we will explore the steps of the proton-proton chain in detail, with the intent of helping you keep them straight.

Visit the Student Site at the Digital Landing Page, and open the "Proton-Proton Animation" for this chapter.

Watch the animation all the way through once.

Play the animation again, pausing after the first collision. Two hydrogen nuclei (both positively charged) have collided to produce a new nucleus with only one positive charge.

**1** Which particle carried away the other positive charge?

_____

_____

**2** What is a neutrino? Did the neutrino enter the reaction or was the neutrino produced in the reaction?

_____

_____

_____

Compare the interaction on the top with the interaction on the bottom.

**3** Did the same reaction occur in each instance?

_____

_____

Resume playing the animation, pausing it after the second collision.

**4** What two types of nuclei entered the collision? What type of nucleus resulted?

_____

_____

**5** Was charge conserved in this reaction or was it necessary for a particle to carry charge away?

_____

_____

_____

**6** What is a gamma ray? Did the gamma ray enter the reaction or was it produced by the reaction?

_____

_____

Resume the animation again, and allow it to run to the end.

**7** What nuclei enter the final collision? What nuclei are produced?

_____

_____

**8** In chemistry, a catalyst facilitates the reaction but is not used up in the process. Do any nuclei act like catalysts in the proton-proton chain?

_____

_____

_____

Make a table of inputs and outputs. Which of the particles in the final frame of the animation were inputs to the reaction? Which were outputs? Fill in your table with these inputs and outputs.

**9** Which outputs are converted into energy that leaves the Sun as light?

_____

_____

**10** Which outputs could become involved in another reaction immediately?

_____

_____

**11** Which output is likely to stay in that form for a very long time?

_____

_____

# 24 Life

Throughout history, many people have wondered if we are alone in the universe. But it is only in recent times that science has been able to address the underlying questions that must be answered, such as: How common are planets? What is the range of conditions under which life can thrive? How does life begin and evolve? Even answering the question "What is life?" is surprisingly complicated. Answering these questions is part of a science called **astrobiology**— the study of the origin, evolution, distribution, and future of life in the universe. Throughout this book, the "Origins" section in each chapter has discussed how astronomers think about these questions. In this chapter, we expand on some of these topics and provide a more systematic overview of how astronomers think about life in the universe and how they search for signs of it.

## LEARNING GOALS

By the conclusion of this chapter, you should be able to:

**LG 1** Explain our current understanding of how and when life began on Earth and how it has evolved.

**LG 2** Explain how life is a structure that has evolved through the action of the physical and chemical processes that shape the universe.

**LG 3** Describe the locations in our Solar System and around other stars where astronomers think life might be possible.

**LG 4** Describe some of the methods used to search for intelligent extraterrestrial life.

The Allen Telescope Array searches for signs of intelligent life. ▶ ▶ ▶

Are we
alone in the
universe?

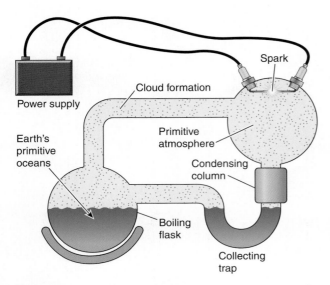

**Figure 24.1** The Urey-Miller experiment was designed to simulate conditions in an early-Earth atmosphere.

(a)

(b)

**Figure 24.2** (a) Life on Earth may have arisen near ocean hydrothermal vents like this one. Similar environments might exist elsewhere in the Solar System. (b) Living organisms around hydrothermal vents, such as the giant tube worms shown here, rely on hydrothermal rather than solar energy for their survival.

# 24.1 Life Evolves on Earth

What is **life**? Many scientists suggest that there is no single definition of life that would encompass all the different forms of life that may exist in the universe. To date, we have discovered only a single example of life: that found here on Earth. Life on Earth may be very different from life found in other places in the universe. Indeed, even comparing varied organisms on Earth leads to complications in the definition of life. Viruses, for example, meet some criteria for life, but not others. A complete definition of life may also have to take into account life-forms that have not yet been discovered. From studies of known life, we conclude that like planets, stars, and galaxies, life is a structure that has evolved in the universe. Life draws energy from the environment in order to survive and reproduce. On Earth, all life involves carbon-based chemistry and uses liquid water as its biochemical solvent, while specific biological molecules such as ribonucleic acid (RNA) and deoxyribonucleic acid (DNA) enable life to reproduce and evolve. In this section, we will briefly review what is known about the origin and evolution of life on Earth.

## The Origin of Life on Earth

How did life begin on Earth? Recall from Chapter 7 that Earth's secondary atmosphere was formed in part by carbon dioxide and water vapor emitted by volcanoes. Comets and asteroids likely added large quantities of water, methane, and ammonia to the mix. Liquid water is considered essential for any terrestrial-type life to get its start and evolve because it is an effective solvent that can move other atoms and molecules around, therefore making them more accessible to cells. Early Earth had abundant sources of energy, such as lightning and ultraviolet solar radiation, that fragmented these simple molecules. These fragments subsequently reassembled into molecules of greater mass and complexity. Some of these were organic molecules; that is, molecules that contained carbon. Rain carried the heavier molecules out of the atmosphere into Earth's oceans, forming a primordial soup.

In 1952, chemists Harold Urey (1893–1981) and Stanley Miller (1930–2007) attempted to create conditions similar to these early-Earth conditions. Using equipment illustrated in **Figure 24.1**, they placed water in a sterilized laboratory jar to represent the ocean, and then added methane, ammonia, and hydrogen as a primitive atmosphere; electric sparks simulated lightning as a source of energy. Within a week, the Urey-Miller experiment yielded molecules associated with life: amino acids and components of nucleic acids. Proteins, the structural molecules of life, are made of 20 amino acids. Eleven of these were synthesized in the Urey-Miller experiment. Nucleic acids are the precursors of RNA and DNA.

Additional sealed samples from this old experiment were examined 50 years later. These samples had added hydrogen sulfide to the "primitive atmosphere." When the samples were analyzed, 23 amino acids were found. This suggests that hydrogen sulfide, which would have come from volcanic plumes in the early Earth, was important. More recent experiments with carbon dioxide and nitrogen as the primitive atmosphere have produced results similar to those of Urey and Miller. A feasible atmospheric composition with an energy source can produce significant quantities of amino acids and other substances important to life.

From laboratory experiments such as these, scientists have developed various models to explain how life might have begun in an early-Earth environment.

However, the details of how these precursor molecules evolved into the molecules of life are not yet clear. Some biologists think life began in the ocean depths, where volcanic vents provided the hydrothermal energy needed to create the highly organized molecules responsible for biochemistry (**Figure 24.2**). Others think that life originated in tide pools, where lightning and ultraviolet radiation supplied the energy (**Figure 24.3**). In either case, short strands of molecules that could replicate themselves may have formed first, later evolving into RNA, and finally into DNA, the huge molecule that serves as the biological "blueprint" for self-replicating organisms.

A few scientists have suggested that life on Earth may have been "seeded" from space in the form of microorganisms brought here by meteoroids or comets. However, there is no scientific evidence at this time to support the "seeding" hypothesis. In addition, while this hypothesis might explain how life came to Earth, it does not explain how life itself began.

**Figure 24.3** Life may have begun in tide pools, where lightning and ultraviolet light provide energy for chemical processes.

## When Life Began

If life did indeed get its start in Earth's oceans, when did it happen? Recall from Chapter 7 that Earth was bombarded by Solar System debris for several hundred million years after it formed roughly 4.6 billion years ago. These conditions might have been too harsh for life to form and evolve on Earth. Once the bombardment abated and oceans formed, the opportunities for life to begin greatly improved. It seems that terrestrial life quickly took advantage of this more favorable environment. Scientists debate whether carbonized material in Greenland rocks dating back 3.65 billion to 3.85 billion years provides indirect evidence of early life. Stronger and more direct evidence for early life appears in the form of fossilized **stromatolites** (masses of simple microorganisms) that date back about 3.5 billion years. Fossilized stromatolites have been found in western Australia and southern Africa, and living examples still exist today (**Figure 24.4**). This evidence suggests that the earliest life formed less than a billion years after the formation of the Solar System, and within 500 million years of the end of the late heavy-bombardment period.

**Figure 24.4** These modern-day stromatolites are growing in colonies along an Australian shore.

All life on Earth shares a similar genetic code that originated from a common ancestor. Close comparison of DNA of different species enables biologists to trace backward to the time when different types of life first appeared on Earth and to identify the species from which these life-forms evolved. The earliest organisms were **extremophiles**—life-forms that not only survive, but thrive, under extreme environmental conditions. Extremophiles include organisms such as thermophiles, which flourish in water temperatures as high as 120°C and occur in the vicinity of deep-ocean hydrothermal vents, like the one shown in Figure 24.2a. Other extremophiles thrive in conditions of extraordinary cold, salinity, pressure, dryness, acidity, or alkalinity. Scientists today study extremophiles in boiling-hot sulfur springs in Yellowstone National Park, in salt crystals beneath the Atacama Desert in Chile, at the bottoms of glaciers, in ice fields in the Arctic, and in other extreme environments (**Figure 24.5**).

Among the early life-forms was an ancestral form of **cyanobacteria**, single-celled organisms otherwise known as *bluegreen algae*. These microorganisms form extensive sheets on the

**Figure 24.5** These thermophiles in the Grand Prismatic Spring in Yellowstone National Park live in temperatures of 70°C. The different colors result from different amounts of chlorophyll.

(a)

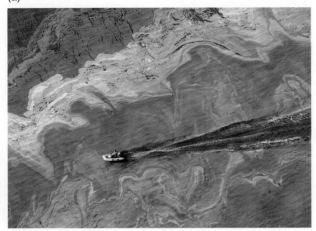

(b)

**Figure 24.6** (a) Cyanobacteria today form sheets on lakes and other bodies of water. (b) Under a microscope, the individual microorganisms are visible.

surface of bodies of water, as shown in **Figure 24.6a**. Under a microscope, it becomes clear that they are colonies of individual microorganisms, as seen in Figure 24.6b. Cyanobacteria **photosynthesize**, using sunlight and carbon dioxide as food and generating oxygen as a waste product. Initially, the highly reactive oxygen produced by cyanobacteria was quickly removed from Earth's atmosphere by oxidation, or rusting, of surface minerals. Once the exposed minerals could no longer absorb oxygen, atmospheric levels of oxygen began to rise. Oxygenation of Earth's atmosphere and oceans began about 2 billion years ago, and the current level was reached only about 250 million years ago, as shown in **Figure 24.7**. Without cyanobacteria and other photosynthesizing organisms, Earth's atmosphere would be as oxygen-free as the atmospheres of Venus and Mars.

Biologists comparing DNA sequences find that terrestrial life is divided into two types: prokaryotes and eukaryotes. Prokaryotes, which include Bacteria and Archaea, are simple organisms that consist of free-floating DNA inside a cell wall; as shown in **Figure 24.8a**, they lack both cell structure and a nucleus. Eukaryotes, which form the cells in animals, plants, and fungi, have a more complex form of DNA contained within the cell's membrane-enclosed nucleus, illustrated in Figure 24.8b. The first eukaryote fossils date from about 2 billion years ago, coincident with the rise of free oxygen in the oceans and atmosphere, although the first multicellular eukaryotes did not appear until a billion years later.

## Increasing Complexity

Scientists have used DNA sequencing to establish what is known as the "phylogenetic tree of life," shown in **Figure 24.9**. This complex tree describes the evolutionary interconnectivity of all species of Bacteria, Archaea, and Eukarya and has revealed some interesting relationships. For example, Archaea were initially thought to be the same as Bacteria, but genetic studies show they diverged long ago, and the Archaea have genes and metabolic pathways that are more similar to those of Eukarya than to those of Bacteria. On the macroscopic scale, the phylogenetic tree places animals closest to fungi, which branched off the evolutionary tree after slime molds and plants.

Living creatures in Earth's oceans remained much the same—a mixture of single-celled and relatively primitive multicellular organisms—for more than 3 billion years after the first appearance of terrestrial life. Between 540 million and 500 million years ago, the number and diversity of biological species increased spectacularly. Biologists call this event the **Cambrian explosion**. The trigger of this sudden surge in biodiversity remains unknown, but possibilities include rising oxygen levels, an increase in genetic complexity, major climate change, or a combination of these. The "Snowball Earth" hypothesis suggests that before the Cambrian explosion, Earth was in a period of extreme cold between about 750 million and 550 million years ago, and was covered almost entirely by ice. During this period of extreme cold, predatory animals died out, making it easier for new species to adapt and thrive. Another possibility is that the marked increase in atmospheric oxygen ($O_2$) would have been accompanied by a corresponding increase in stratospheric ozone ($O_3$), which shields Earth's surface from deadly solar ultraviolet radiation. With this protective ozone

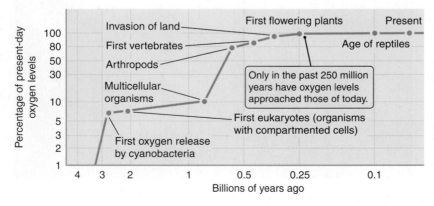

**Figure 24.7** The amount of oxygen in Earth's atmosphere has built up over time as a result of cyanobacteria and plant life on the planet.

layer in place, life was free to leave the oceans and move to land. *Tiktaalik*, a fish with limblike fins and ribs, was an animal in a midevolutionary step of leaving the water for dry land, as shown in the artist's illustration in **Figure 24.10**.

The first plants appeared on land about 475 million years ago. Large forests and insects go back 360 million years. The age of dinosaurs began 230 million years ago and ended abruptly 65 million years ago, when a small asteroid or comet collided with Earth. The collision threw so much dust into the atmosphere that the sunlight was dimmed for months, causing the extinction of more than 70 percent of all existing plant and animal species. Mammals were the big winners in the aftermath. Primates evolved from the last ancestor common with other mammals about 70 million years ago. The great apes (gorillas, chimpanzees, bonobos, and orangutans) split off from the lesser apes about 20 million years ago (Figure 24.9, inset). DNA tests show that humans and chimpanzees share about 98 percent of their DNA, indicating that they evolved from a common ancestor about 6 million years ago. By comparison, all humans share 99.9 percent of their DNA. The earliest human ancestors appeared a few million years ago, and the first civilizations occurred a mere 10,000 years ago. Present-day industrial society, barely more than two centuries old, is but a moment in the history of life on Earth.

Humans are here today because of a series of events that occurred throughout the history of the universe. Some of these events are common in the universe, such as the formation of heavy elements in earlier generations of stars and the formation of planets. Other events in Earth's history may have been less likely to

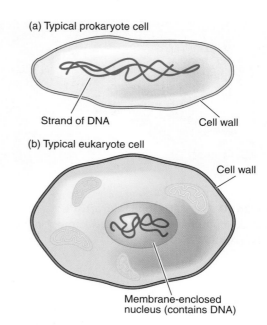

**Figure 24.8** (a) A simple prokaryote cell contains little more than the cell's genetic material. (b) A eukaryote cell contains several membrane-enclosed structures, including a nucleus that houses the cell's genetic material.

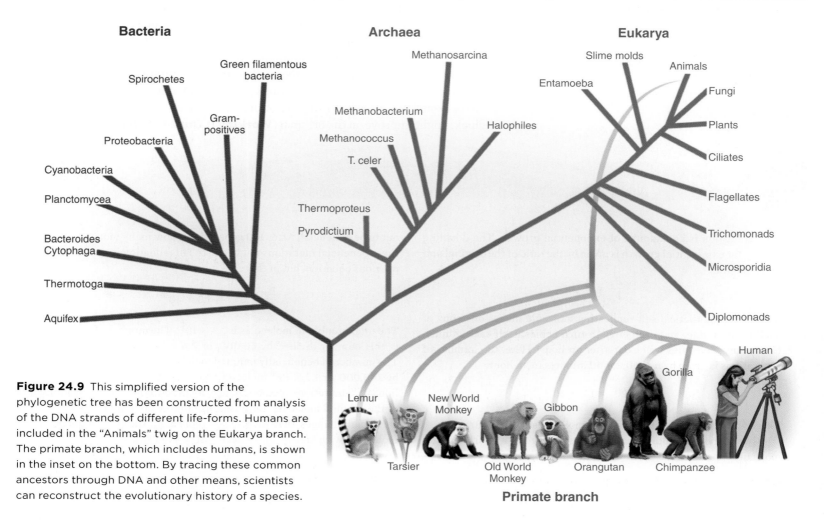

**Figure 24.9** This simplified version of the phylogenetic tree has been constructed from analysis of the DNA strands of different life-forms. Humans are included in the "Animals" twig on the Eukarya branch. The primate branch, which includes humans, is shown in the inset on the bottom. By tracing these common ancestors through DNA and other means, scientists can reconstruct the evolutionary history of a species.

**Figure 24.10** This illustration is an artist's reconstruction from a fossil of *Tiktaalik* found in the Canadian Arctic.

happen elsewhere, such as the formation of a planet with life-supporting conditions like Earth or the development of self-replicating molecules that led to Earth's earliest life. A few events stand out, such as major extinctions that allowed the evolution of mammalian life and, ultimately, human beings.

## Evolution as a Mechanism of Change

Imagine that just once during the first few hundred million years after the formation of Earth, a single molecule formed somewhere in Earth's oceans. That molecule had a very special property: chemical reactions between that molecule and other molecules in the surrounding water caused the molecule to make a copy of itself. The molecule became "self-replicating." Chemical reactions then produced copies of each of these two molecules, making four molecules. Four molecules became eight, eight became 16, 16 became 32, and so on. By the time the original molecule had copied itself just 100 times, more than a million trillion trillion ($10^{30}$) of these molecules existed. That is about 100 million times more of these molecules than there are stars in the observable universe. The molecules of DNA that make up the chromosomes in the nuclei of the cells of all advanced life today are direct descendants of those early self-replicating molecules that flourished in the oceans of the young Earth.

Over the course of time, not all replications are exact. The likelihood that a copying variation will occur while a molecule is replicating increases significantly with the number of copies being made. For DNA, which contains the genetic code for an entire organism, a change in the genetic code is called a **mutation**. In some cases, a mutation has no effect. In others, it can prevent an organism from flourishing. In still other cases, a mutation can make an organism better suited to its environment. Organisms with these advantageous mutations will survive to reproduce successfully. Even if mutations are rare, and only a small fraction turn out to be beneficial, after just 100 generations there are trillions of mutations that, by luck, might improve on the original (**Working It Out 24.1**). **Heredity**—the ability

## 24.1 Working It Out Exponential Growth

Self-replication is an example of exponential growth. The doubling time, $n$, for exponential growth is given by the ratio of the original and final amounts:

$$\frac{P_F}{P_O} = 2^n$$

Assume a hypothetical self-replicating molecule makes one copy of itself each minute, and each copy in turn copies itself each minute. How many molecules will exist after an hour? Here, the number of generations is given by $n = 60$, for 60 minutes in an hour:

$$\frac{P_F}{P_O} = 2^{60} = 1.2 \times 10^{18}$$

There will be a billion billion of these molecules after 1 hour.

Now suppose a mutation occurs once every 50,000 times that a molecule reproduces itself, and one out of 200,000 mutations turns

out to be beneficial. After 100 generations, how many molecules with these beneficial mutations might exist? This equation is similar to the previous equation, but in this case $n = 100$:

$$\frac{P_F}{P_O} = 2^{100} = 1.3 \times 10^{30}$$

The total number of molecules is $1.3 \times 10^{30}$. The number of mutations is this number divided by 50,000, or $2.6 \times 10^{25}$ mutated molecules. The number of beneficially mutated molecules is this number divided by 200,000, or $1.3 \times 10^{20}$ molecules. So there will be 100 million trillion ($10^{20}$) mutations that, by chance, might improve the survivability of the original molecule. Because this number does not count earlier beneficial changes that themselves replicated, the total number of molecules with beneficial changes will be even larger!

of one generation to pass on its genetic code to future generations—allow beneficial mutations to persist and be incorporated into a species' genetic code.

As the organisms of the early Earth continued to interact with their surroundings and make copies of themselves, mutations caused them to diversify into many different species. In some cases, the resources they needed to reproduce became scarce. In the face of this scarcity of resources, varieties that were more successful reproducers became more numerous. Competition for resources, predation by one species on another, and cooperation between organisms became important to the survival of different varieties. Some varieties were more successful and reproduced to become more numerous, while less successful varieties became less and less common. This process, in which better-adapted organisms reproduce and thrive, while less well-adapted organisms become extinct, is called **natural selection**.

Life has existed on Earth for about 4 billion years, which is a very long time—long enough for the combined effects of heredity and natural selection to shape the descendants of that early self-copying molecule into a huge variety of complex, competitive, successful structures. Geological processes on Earth have preserved a fossil record of the history of some of these structures (**Figure 24.11**). Among these descendants are human "structures" capable of thinking about their own existence and unraveling the mysteries of the stars.

**Figure 24.11** Fossils, such as this Parasaurolophys ("near crested lizard"), record the history of the evolution of life on Earth. This plant-eating dinosaur lived in North America about 75 million years ago.

## CHECK YOUR UNDERSTANDING 24.1

Extremophiles are organisms that: (a) are extremely reactive; (b) are extremely rare; (c) have an extreme quality, such as mass or size; (d) live in extreme conditions.

# 24.2 Life Involves Complex Chemical Processes

The evolution of life on Earth cannot be separated from the narrative of astronomy: it is one of many examples of the emergence of structure in an evolving universe (**Process of Science Figure**). This leads naturally to a profound question: Has life arisen elsewhere? Unlike the study of planets, stars, and galaxies, there is only one known case for the study of life—Earth—and scientists do not know how much can be generalized to other places. To explore this question, we need to take a closer look at the processes that have led to life on Earth. In this section, we explore the chemical and physical properties of life on Earth.

The infant universe was composed basically of hydrogen and helium and very little else. After 9 billion years of stellar nucleosynthesis, all the heavier chemical elements essential to life were present and available in the molecular cloud that gave birth to the Solar System. Those heavier elements were formed by nuclear fusion in the cores of earlier generations of stars and were then dispersed into space. At times, this dispersal was passive. For example, low-mass stars such as the Sun, when they become puffed up, dying red giants, may shed their extended atmospheres, sending some newly created carbon into space. Other dispersals were more violent. High-mass stars produce even heavier elements through nucleosynthesis in their cores—up to and including iron. But some of the trace elements essential to biology on Earth are even more massive than iron. They are

ALL OF SCIENCE IS INTERCONNECTED

More than many other subjects, astrobiology draws on all of science and makes clear that all the fields are interconnected.

Geology

Evolution

Timescales

Biology

Astronomy

Requirements for Life

**Astrobiology**

Habitability

Chemistry

Atmospheric Physics

Stability

Physics

Energy and Life

No science stands alone. All are connected. Interdisciplinary fields of study like astrobiology provide opportunities for new tests for theories from many fields.

produced within a matter of minutes during the violent supernova explosions that mark the death of high-mass stars and then are thrown into the chemical mix found in molecular clouds.

All known living organisms on Earth are composed of a more or less common suite of complex chemicals. Approximately two-thirds of the atoms in the human body are hydrogen (H), about one-fourth are oxygen (O), a tenth are carbon (C), and a few hundredths are nitrogen (N). Carbon, nitrogen, and oxygen are the three most abundant products of stellar nucleosynthesis after helium; see the "astronomer's periodic table" in Figure 5.17. The several dozen remaining atomic elements in the human body make up only 0.2 percent of the total. All known living creatures are assemblages of molecules composed almost entirely of these four elements, sometimes called CHON (carbon, hydrogen, oxygen, nitrogen), along with small amounts of phosphorus and sulfur. Some of these molecules are enormous. Consider DNA, which is responsible for genetic codes, illustrated in **Figure 24.12**. DNA is made up entirely of only five atomic elements: CHON and phosphorus. But the DNA in each cell of the human body is composed of combinations of *tens of billions* of atoms of these same five elements. Proteins, the huge molecules responsible for the structure and function of living organisms, are long chains of smaller molecules called amino acids. Terrestrial life uses 20 specific amino acids, which also consist of no more than five atomic elements—in this case CHON plus sulfur instead of phosphorus.

The chemistry of life is far too complex to have only a half-dozen atomic elements. Many of the other elements, which are present in smaller amounts, are essential to the chemical processes that living organisms carry out. These elements include sodium, chlorine, potassium, calcium, magnesium, iron, manganese, and iodine. Trace elements such as copper, zinc, selenium, and cobalt also play a crucial role in biochemistry but are needed in only tiny amounts.

Notice that carbon, which can bond to four other atoms or molecules, forms the backbone of the DNA molecule shown in Figure 24.12. This is why carbon is so important to life on Earth. There could be forms of extraterrestrial life that are also carbon based but have chemistries quite different from that of life on Earth. For example, there are countless varieties of amino acids in addition to the 20 used by terrestrial life. Most other atoms are more limited than carbon in the number of bonds they can make, but silicon, like carbon, can bind to four other atoms, so that a large number of combinations is possible. As a potential life-enabling atom, silicon has both advantages and disadvantages compared to carbon. Silicon-based molecules remain stable at much higher temperatures than carbon-based molecules, perhaps enabling possible silicon-based life to thrive in high-temperature environments, such as on planets that orbit close to their parent star. But silicon is also a larger and more massive atom than carbon: it cannot form molecules as complex as those based on carbon. Any silicon-based life probably would be simpler than life-forms here on Earth, but it might exist in high-temperature niches somewhere within the universe. Although carbon's unique properties make it readily adaptable to the chemistry of life on Earth, other types of life might be found elsewhere.

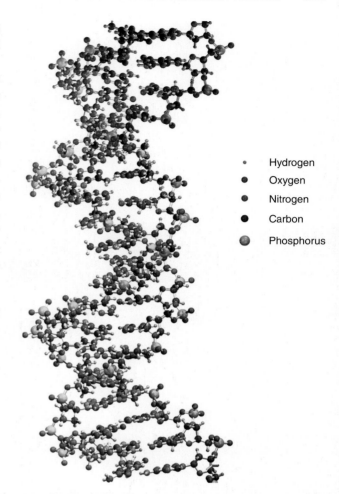

- • Hydrogen
- • Oxygen
- • Nitrogen
- • Carbon
- • Phosphorus

**Figure 24.12** DNA, the hereditable molecule that forms the basis for life on Earth, contains only five different atoms. Even so, these atoms are combined in billions of ways, giving rise to the diversity of life on Earth.

## CHECK YOUR UNDERSTANDING 24.2

Carbon is a favorable base for life because: (a) it can bond to many other atoms in long chains; (b) it is nonreactive; (c) it forms weak bonds that can be readily reorganized as needed; (d) it is organic.

## 24.3 Where Do Astronomers Look for Life?

One approach to the scientific search for extraterrestrial life is to use robotic spacecraft to explore the planets and moons of the Solar System (see Chapter 6). Spacecraft have visited all of the planets and some moons and sent back at least some information about the conditions on these worlds. Another approach is to use telescopes to detect planets outside of our Solar System (see Chapter 7). Telescopes on the ground and in space have detected a few thousand planets orbiting other stars. In this section, we will survey the locations where life might be found, both in our own Solar System and around other stars.

### Life within Our Solar System

Scientists start the search for evidence of extraterrestrial life here in our own Solar System. Early conjectures about life in our Solar System seem naïve, considering what we now know. Two centuries ago, the eminent astronomer Sir William Herschel, discoverer of Uranus, proclaimed, "We need not hesitate to admit that the Sun is richly stored with inhabitants." In 1877, astronomer Giovanni Schiaparelli (1835–1910) observed what appeared to be linear features on Mars and dubbed them *canali* ("channels" in Italian). The famous observer of Mars, Percival Lowell (1855–1916), misinterpreted Schiaparelli's *canali* as "canals," suggesting that they were constructed by intelligent beings.

Because Mercury and the Moon lacked atmospheres, astronomers determined they were not conducive to life. The giant planets and their moons were thought to be too remote and too cold to sustain life. The surface of Venus was far too hot, but Mars seemed more promising. During the mid-20th century, astronomers using ground-based telescopes discovered that Mars possesses an atmosphere, water ice, and carbon dioxide ice. During the 1960s, the United States and the Soviet Union sent reconnaissance spacecraft to the Moon, Venus, and Mars, but the instruments on these spacecraft probed the physical and geological properties of these astronomical bodies, rather than searching for life. Serious efforts to look for signs of life—past or present—require more advanced spacecraft with specialized instrumentation.

In the mid-1970s, two American *Viking* spacecraft were sent to Mars with detachable landers containing a suite of instruments designed to find evidence of a terrestrial type of life. When the *Viking* landers failed to find convincing evidence of life on Mars, hopes faded for finding life on any other body orbiting the Sun. Since that time, however, further exploration of the Solar System has generated renewed optimism. A better understanding of the history of Mars indicates the planet's climate has changed. Mars was once wetter and warmer than it is today.

In the 1990s, Mars missions began to map the planet's surface from the ground and from space. In 2008, NASA's *Phoenix* spacecraft landed at a far-northern latitude, inside the planet's arctic circle, where specialized instruments dug into and analyzed the martian water-ice permafrost. *Phoenix* found that the martian arctic soil has a chemistry similar to the Antarctic dry valleys on Earth, where life exists deep below the surface at the ice-soil boundary. Minerals that form in water, for example calcium carbonate, have been detected. This suggests that oceans existed in the past on Mars. However, *Phoenix* did not find evidence of life.

The *Curiosity* rover (originally known as the *Mars Science Laboratory*) landed in Gale Crater on Mars in 2012. This rover studies the rocks and soil of Mars to

**Figure 24.13** The Mars *Curiosity* rover detected evidence that Mars had a watery past. The rounded gravel surrounding the bedrock suggests there was an ancient, flowing stream.

provide data for a better understanding of the history of the planet's climate and geology. Shortly after landing, *Curiosity* found evidence that a stream of liquid water had once flowed in the crater. The rover observed rounded, gravelly pebbles stuck together, which have been interpreted as coming from a stream that varied at times from ankle-deep to hip-deep, and moved at about 1 meter per second (**Figure 24.13**). It found sedimentary rocks containing clay, which suggested that at one time there had been a freshwater lake bed. Later observations found that the surface soil contained up to 2 percent water by weight—or about 1 quart per cubic foot of martian dirt. This is too dry to support plant life, which permanently wilts in soil that is about 10% water by weight.

The *Mars Atmosphere and Volatile EvolutioN* (*MAVEN*) mission arrived at Mars in September 2014. It is studying the upper atmosphere in order to learn more about the escape of carbon dioxide, hydrogen, and nitrogen from the planet's atmosphere and how the loss of those gases affected surface pressure and the existence of liquid water.

In September 2015, NASA announced that spectroscopic observations from the orbiting Mars Reconnaissance Orbiter indicate that there is liquid water on Mars today. Darkish streaks on Mars that change seasonally contain hydrated salts and minerals, indicating that liquid water is important to their formation (**Figure 24.14**). This water is briny (salty), which has a much lower freezing point than non-briny water and thus could exist in a liquid state during the martian summer. Future experiments will look for liquid water—and fossil or living microorganisms—below the martian surface.

NASA's instrumented robotic spacecraft reached the outer Solar System starting in the 1980s, and many astrobiologists were surprised by the findings. Although the outer planets themselves did not appear to be habitats for life, some of their moons became objects of special interest. Jupiter's moon Europa is covered with a layer of water ice that appears to overlie a great ocean of briny liquid water (**Figure 24.15**). The water remains liquid because of high pressure and tidal heating by Jupiter. Impacts by comet nuclei may have added a mix of organic material, another essential ingredient for life. Once thought to be a frozen, inhospitable world, Europa is now a candidate for biological exploration. Recently, scientists using the Hubble Space Telescope observed water geysers taller than Mount Everest erupting from the icy surface of Europa. Ejected material from these geysers may make it possible to search for life on Europa without drilling down through the ice.

Saturn's moon Titan has an atmosphere that is rich in organic chemicals, many of which are thought to be precursor molecules of a type that existed on Earth before life appeared here. A probe from the *Cassini* spacecraft in orbit around Saturn descended through Titan's atmosphere and found additional evidence for a variety of molecules that might be necessary for life, as well as liquid lakes of methane on the surface and probably a liquid-water ocean under the surface. As noted in Chapter 1, the *Cassini* spacecraft also detected water-ice crystals spouting from cryovolcanoes (which erupt ice crystals instead of rocks) near the south pole of Saturn's tiny moon Enceladus. Liquid water must lie beneath its icy surface, and Enceladus therefore joins Europa as a possible habitat of extremophile life, perhaps life similar to that found near hydrothermal vents deep within Earth's oceans.

**Figure 24.14** This MRO image shows narrow, dark, downhill streaks, about 100 meters long, which are thought to indicate liquid water flowing on Mars today. Hydrated salts and minerals, including pyroxene (the blue color) were detected by spectroscopy.

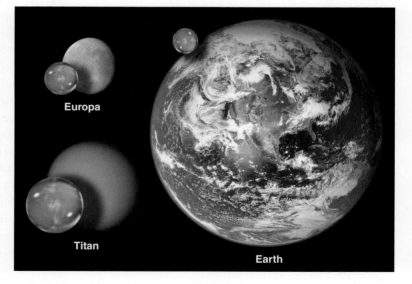

**Figure 24.15** The total amount of liquid water (blue ball) on Jupiter's moon Europa and Saturn's moon Titan compared to the amount on Earth. All figures are drawn to scale and assume average ocean depths of 4 km (Earth), 100 km (Europa), and 200 km (Titan).

The discovery of life on even one Solar System body beyond Earth would be exciting. If scientists discover that life arose independently *twice* in the same planetary system, this finding could suggest that the spontaneous appearance of life is not rare at all.

---

### CHECK YOUR UNDERSTANDING 24.3

Which of the following Solar System objects is *not* a good candidate for future searches for life? (a) Mars; (b) Jupiter's moon Europa; (c) Saturn's moon Titan; (d) Uranus

---

## Habitable Zones

Recall from Chapter 7 that there are more than 1,000 confirmed and thousands of candidate extrasolar planets within the Milky Way Galaxy. To decide which planets to focus on for further study, astronomers are narrowing the possibilities by searching for planets with environments conducive to the formation and evolution of life, as we understand it, while eliminating planets that are clearly unsuitable. Astronomers consider issues such as each planet's orbit, its inferred temperature, its distance from its star, and its location in the galaxy. One criterion astrobiologists look for is planetary systems that are stable. As noted in Chapters 2, 3, and 4, astronomers think about the effects of a planet's rotation and orbit. Planets in stable systems have nearly circular orbits that preserve relatively uniform climatological environments. Planets in very elliptical orbits or planets with a large axial tilt can experience more intense temperature swings that could be detrimental to the survival of life. A stable temperature that maintains the existence of water in a liquid state might be important. We know that liquid water was essential for the formation and evolution of life on Earth. Of course, we don't know if liquid water is an absolute requirement for life elsewhere, but it's a good starting point.

In Chapter 7, we discussed the idea of the habitable zone, the location of a planet relative to its parent star that provides a range of temperatures in which liquid water can exist. On planets that are too close to their parent stars, water would exist only as a vapor—if at all. On planets that are too far from their stars, water would be permanently frozen as ice. Planet size is another consideration: Large gas giants retain most of their light gases during formation and do not have a surface. Small planets may be rocky or a mix of water, rock, and ice: measurement of the mass and radius enables scientists to estimate the density. Planets that are very small may have insufficient surface gravity to retain their atmospheric gases and so end up like our Moon. Calculating whether any particular planet is in the habitable zone is complicated. Recall from Chapter 7 that even if a planet is located in the habitable zone, it only means that liquid water could exist on the surface: it does not mean that astronomers have confirmed the presence of liquid water or that the planet has inhabitants.

In our own Solar System, Venus, which orbits at 0.7 times Earth's distance from the Sun, has become an inferno because of its runaway greenhouse effect (see Chapter 9). Any liquid water that might once have existed on Venus has long since evaporated and been lost to space. Mars orbits about 1.5 times farther from the Sun than the orbit of Earth, and the water that we see on Mars today is nearly always frozen. But the orbit of Mars is more elliptical and variable than Earth's, giving the planet a greater variety of climate, including long-term cycles that might occasionally permit liquid water to exist. Most astrobiologists put the habitable zone of our Solar System at about 0.9–1.4 astronomical units (AU), which includes Earth but just misses Venus and Mars. However, this range excludes the possibility of liquid water under ice, as occurs on the moons of the outer Solar System.

Astronomers must also think about the type of star they are observing in their search for planets that could have liquid water. Stars that are less massive than the Sun and thus cooler will have narrower habitable zones. A planet in the habitable zone of a cool star is close in to its star. As a result, it is more likely to be tidally locked to the star so there is no day/night cycle. Stars that are more massive than the Sun are hotter and will have a larger and more distant habitable zone (**Figure 24.16**). However, massive stars have shorter main-sequence lifetimes and might not last long enough for evolution to take place. For example, a star of 3 $M_{Sun}$ has a lifetime of only a few hundred million years. On Earth, a billion years was long enough for bacterial life to form and cover the planet but insufficient for anything more advanced to evolve. On Earth, it took 3.5 billion years of evolution to reach the period known as the Cambrian explosion. Even though evolution might happen at a different pace elsewhere, stellar lifetime is still a sufficiently strong consideration, so astronomers focus their efforts on stars with longer lifetimes; specifically, stars of 0.6–1.4 $M_{Sun}$, which corresponds to spectral types F, G, K, and M.

Another factor to consider is a planet's atmosphere. The ability of a planet to keep an atmosphere depends on its mass and radius (and therefore its escape velocity) and its temperature. Planets that are very small may have insufficient surface gravity to retain their atmospheric gases. In the inner Solar System, the Moon and Mercury were too small to keep any atmosphere. Mars lost its atmosphere over time, but the larger Earth and Venus were able to keep a thick atmosphere. Another important consideration is the greenhouse effect, which traps heat underneath an atmosphere and raises the temperature on a planetary

**Nebraska Simulation:** Circumstellar Habitable Zone

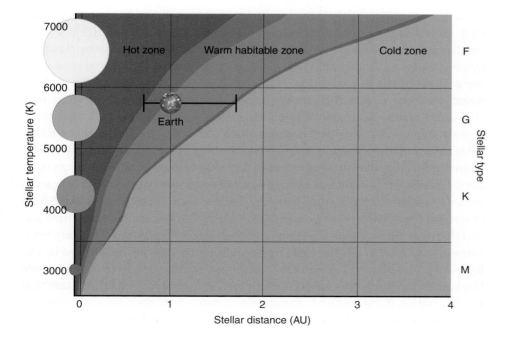

**Figure 24.16** The habitable zone changes with the mass and temperature of a star. Habitable zones around hot, high-mass stars are larger and more distant than the zones around cooler, lower-mass stars.

surface. This has happened on Venus, Earth, and Mars, each of which has a higher surface temperature because of its atmosphere. The thickness and chemical content of the atmosphere affect the strength of the greenhouse effect, so that, for example, Venus is much hotter than its distance from the Sun would suggest because of its thick atmosphere of carbon dioxide (see Figure 5.24 and the Chapter 5 "Origins"). The total amount of atmosphere affects the atmospheric pressure at the surface, which, along with the temperature, determines whether water (or other molecules) can be in a liquid state on the surface. The current thin atmosphere of Mars does not permit standing liquid water on its surface.

In the next few years, there will be many more observations from planet-finding projects that identify atmospheres on extrasolar planets. Water vapor, hydrogen, and carbon monoxide have been found on a few extrasolar planets already. In particular, the discovery of oxygen in the atmosphere of an extrasolar planetary atmosphere would be exciting, but not definitive. The oxygen in Earth's atmosphere makes it stand out from the rest of the planets and moons in the Solar System, and we know that most of the terrestrial oxygen was created by photosynthetic life. However, oxygen can also come from the breakup of water molecules, so the presence of oxygen alone doesn't necessarily mean life exists there.

Astrobiologists continue to develop ways to classify planets more clearly as they try to narrow down the possibilities about where life might exist. One such classification, the Earth Similarity Index, uses the currently available data on an extrasolar planet to estimate how much it is like Earth. Factors include the radius, density, escape velocity, and surface temperature. The ESI ranges from 0 to 1. A value of 0.8–1 is used for rocky planets that can retain an atmosphere at temperatures suitable for liquid water; that is, that are Earth-like. This is an Earth-centric approach based on the experience of life on Earth. The Planetary Habitability Index aims to be less Earth-centric and to broaden the options for habitability, but it depends on factors not yet measured or measurable for most extrasolar planets. This index depends on whether the planet has a surface on which organisms can grow, as well as the right kind of chemistry, a source of energy, and the ability to hold a liquid solvent. Saturn's moon Titan or Jupiter's moon Europa might satisfy these conditions, and Mars might have done so in the past. Over the next decade, improvements in observations will likely lead to enough information that at least some of the extrasolar planets can be classified by their Planetary Habitability Index. These indices are preliminary, representing various efforts to proceed with limited data. As observations of extrasolar planets become more complete, astrobiologists will undoubtedly develop new classification schemes that are more accurate and informative.

**Nebraska Simulation:** Milky Way Habitability Explorer

Astronomers also consider the *galactic habitable zone*—the idea that there may be some locations within the Milky Way Galaxy where planets might have a higher probability of hosting life. Stars that are situated too far from the galactic center may be without enough heavy elements—such as oxygen, silicon (silicates), iron, and nickel—in their protoplanetary disks to form rocky planets like Earth. Conversely, regions too close to the galactic center experience less star formation and therefore fewer opportunities to gather heavy elements into planetary environments. Stars that are too close to the galactic center may be affected by the high-energy radiation environment (X-rays and gamma rays from supermassive black holes or gamma-ray bursts), which can damage RNA and DNA. Stars that migrate within the galaxy and change their distance from the galactic center over time may move in and out of any galactic habitable zone.

The habitable zone around a star depends most on the star's: (a) mass and age; (b) radius and distance; (c) age and radius; (d) color and distance; (e) luminosity and velocity.

## 24.4 Scientists Are Searching for Signs of Intelligent Life

Are we alone? This question is approached by scientists from many directions. Biologists consider the origin and evolution of life and the definition of intelligence. Astronomers send messages and search for alien signals in the vast array of astronomical data. In this section, you will learn about the search for intelligent life and how scientists think about the probability of finding it.

### Sending Messages

During the 1970s, messages were sent from Earth to space. The *Pioneer 11* spacecraft, which will probably spend eternity drifting through interstellar space, carries the plaque shown in **Figure 24.17**. It pictures humans and the location of Earth for any future interstellar traveler who might happen to find it and understand its content. Another message to the cosmos accompanied the two *Voyager* spacecraft on identical phonograph records that contained greetings from planet Earth in 60 languages, samples of music, animal sounds, and a message from then-President Jimmy Carter. Some politicians were concerned that scientists were dangerously advertising our location in the galaxy, even though radio signals had already been broadcast into space for nearly 80 years. Some philosophers also worried that these messages contained anthropomorphic assumptions about aliens being sufficiently like us to decode the messages. However, sending messages on spacecraft is not an efficient way to make contact with extraterrestrial life. The probability that any of these messages will actually be found by an alien species is very, very small.

In 1974, astronomers used the 300-meter-wide dish of the Arecibo radio telescope to beam a message in binary code (**Figure 24.18**) toward the star cluster M13, located 25,000 light-years away. That is sufficiently far that by the time the message arrives, the core of M13 will have moved, and the radio signal will not actually arrive there. However, the intention of the experiment was to demonstrate that such a message could be sent, not to make contact, as it would take 50,000 years for a reply to come back. The experiment confirmed that such a message could be sent. In 2008, a radio telescope in Ukraine sent a message to the exoplanet Gliese 581c. The message was composed of 501 digitized images and text messages selected by users on a social networking site and will arrive at Gliese 581c in 2029.

### The Drake Equation

The first serious effort to quantify the probability of the existence of intelligent extraterrestrial life was made by astronomer Frank Drake in 1960. He developed the **Drake equation**, which estimates the likely number (*N*) of intelligent

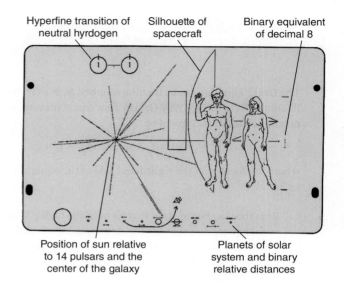

Hyperfine transition of neutral hydrogen | Silhouette of spacecraft | Binary equivalent of decimal 8

Position of sun relative to 14 pulsars and the center of the galaxy | Planets of solar system and binary relative distances

**Figure 24.17** This plaque is carried by the *Pioneer 11* probe, which was launched in 1973 and will eventually leave the Solar System to travel in interstellar space.

**Figure 24.18** This message was beamed toward the star cluster M13 in 1974. This binary-encoded message contains the numbers 1–10, hydrogen and carbon atoms, some interesting molecules, DNA, a human figure and its size, the basics of the Solar System, and a depiction of the Arecibo telescope.

## 24.2 Working It Out Putting Numbers into the Drake Equation

The Drake equation states that the number, $N$, of extraterrestrial civilizations in the Milky Way Galaxy that can communicate by electromagnetic radiation is given by

$$N = R^* \times f_p \times n_e \times f_l \times f_i \times f_c \times L$$

where the factors on the right-hand side of the equation are explained as follows:

1.  $R^*$ is the number of stars that form in the Milky Way Galaxy each year that are suitable for the development of intelligent life. Astronomers consider these to be F, G, K, or M spectral-type stars because their lifetimes are sufficiently long. This is about 5–7 stars per year.

2.  $f_p$ is the fraction of stars that form planetary systems. The discoveries of extrasolar planets over the past two decades have shown that planets form as a natural by-product of star formation and that many—perhaps most—stars have planets. For this calculation, astronomers assume that $f_p$ is between 0.5 and 1.

3.  $n_e$ is the number of planets and moons in each planetary system with an environment suitable for life. In the Solar System, this number is at least 1 (for Earth), but it could be more if Mars or an outer-planet moon or two has suitability for life. Only recently have stars with multiple planets been discovered, so astronomers are just starting to get data on this factor. Generally they estimate 0–3.

4.  $f_l$ is the fraction of suitable planets and moons on which life actually arises. Remember that just a single self-replicating molecule may be enough to get the ball rolling. Some biochemists think that if the right chemical and environmental conditions are present, then life *will* develop, but others disagree. Values of $f_l$ range from 100 percent (life always develops) to 1 percent (life is more rare). Astronomers use a range of 0.01–1.

5.  $f_i$ is the fraction of those planets harboring life that eventually develop intelligent life. Intelligence is certainly the kind of survival trait that might often be strongly favored by natural selection. Yet on Earth, it took about 4 billion years—roughly half the expected lifetime of the Sun—to evolve tool-building intelligence. The correct value for $f_i$ might be close to 0.01 or it might be closer to 1. The truth is, no one knows.

6.  $f_c$ is the fraction of intelligent life-forms that develop technologically advanced civilizations; that is, civilizations that send communications into space. With only one example of a technological civilization to work with, $f_c$ also is unknown. Astronomers estimate $f_c$ to be between 0.1 and 1.

7.  $L$ is the number of years that technologically advanced civilizations exist. This factor is certainly difficult to estimate because it depends on the long-term stability of these civilizations. On Earth, the longest-lived civilizations have existed for, at most, thousands of years. These civilizations, however, were not at the level of technology that allows interstellar communication—thus far, the first "technologically advanced civilization" on Earth is less than 100 years old. We do not know whether life that is intelligent enough to manipulate its environment technologically can maintain its planetary resources for any extended length of time. Astronomers usually put a value between 1,000 years and 1 million years in their estimates, but the average number could be much smaller or much larger.

civilizations currently existing in the Milky Way Galaxy. The Drake equation is different from the other equations in this book because the values for many of the variables are quite uncertain. However, it is a useful way to categorize some of the factors that relate to the conditions that must be met for a civilization to exist. The equation is discussed further in **Working It Out 24.2**.

As illustrated in **Figure 24.19**, the conclusions we draw using the Drake equation depend a great deal on the assumptions we make, therefore on the numbers used in the equation. For the most pessimistic estimates, the Drake equation sets the number of technological civilizations in our galaxy at about 1, in which case we are the *only* technological civilization in the Milky Way at this time. Such a universe could still be full of intelligent life. With 100 billion galaxies in the observable universe, even these pessimistic assumptions mean that there could be 100 billion technological civilizations out there somewhere. However, if the nearest neighbors are in another galaxy, they are *very* far away—millions of parsecs on average.

At the other extreme—with the most optimistic numbers, which assume that intelligent life arises and survives everywhere it gets the chance—there could be tens of millions of technological civilizations in the Milky Way alone! In this case, the nearest neighbors may be "only" 40 or 50 light-years away.

If humans did meet a technologically advanced civilization, what would it be like? The Drake equation suggests that it's highly unlikely there are neighbors nearby, unless civilizations typically live for many thousands or even millions of years (see Working It Out 24.2). If that were the case, any civilization we encountered would probably have been around for much longer than we have.

## Technologically Advanced Civilizations

During lunch with colleagues, the physicist Enrico Fermi (1901–1954), a firm believer in extraterrestrial life, is reported to have asked, "If the universe is teeming with aliens . . . where is everybody?" Fermi's question—first posed in 1950 and sometimes called the *Fermi paradox*—remains unanswered. If intelligent life-forms are common but interstellar travel is difficult or impossible, would the aliens send out messages—perhaps by electromagnetic waves instead? And if they did, why haven't astronomers detected their signals?

Drake used what was then astronomy's most powerful radio telescope to listen for signals from intelligent life around two nearby stars, but he found nothing unusual. His original project has grown over the years into a much more elaborate program called the Search for Extraterrestrial Intelligence, or **SETI**. Scientists from around the world have thought carefully about what strategies might be useful for finding life in the universe. Most of these use radio telescopes to listen for signals from space that bear an unambiguous signature of an intelligent source. Some have focused on significant parts of the spectrum, assuming that a civilization will broadcast on a channel that astronomers throughout the galaxy should find interesting; for example, the 21-centimeter (21-cm) line from hydrogen gas. More recent searches have made use of advances in technology to record as broad a range of radio signals from space as possible. Analysts search these databases for regular signals that might be intelligent in origin.

The SETI Institute's Allen Telescope Array (ATA) received much of its initial financing from Microsoft cofounder Paul Allen. The ATA consists of a "farm" of small, inexpensive radio dishes like those used to capture signals from orbiting television-broadcasting satellites (see the chapter-opening photograph). One of the key projects of the ATA is to observe the planets discovered by the Kepler Mission. Each dish has a diameter of 6.1 meters, but all of the telescopes working together have a total signal-receiving area greater than that of a 100-meter radio telescope. Just as your brain can sort out sounds coming from different directions, this array of radio telescopes is able to determine the direction a signal is coming from, allowing it to listen to many stars at the same time. Over several years' time, astronomers using the ATA are expected to survey as many as a million stars, hoping to find a civilization that has sent a signal toward Earth.

As we stated earlier in this chapter, finding even one nearby civilization in the Milky Way Galaxy—that is, a *second* technological civilization in Earth's small corner of the universe—will make scientists optimistic that the universe as a whole is teeming with intelligent life. The likelihood of SETI's success is difficult to predict, but its potential payoff is enormous. Few discoveries would have a more profound impact than the certain knowledge that we on Earth are not alone.

Science fiction is filled with tales of humans who leave Earth to "seek out new life and new civilizations." Unfortunately, these scenarios are not scientifically realistic. The distances to the stars and their planets are enormous: to explore a significant sample of stars would require extending the physical search over tens

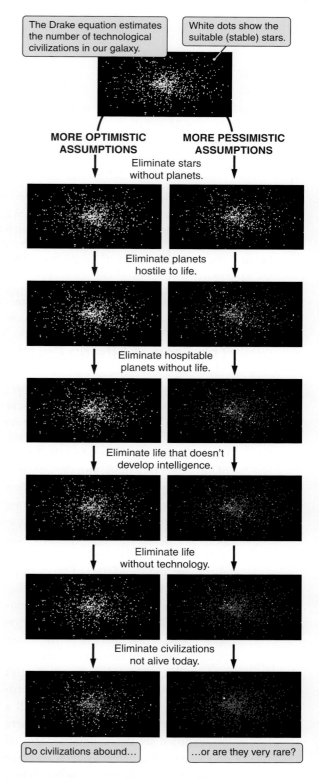

The Drake equation estimates the number of technological civilizations in our galaxy.

White dots show the suitable (stable) stars.

**MORE OPTIMISTIC ASSUMPTIONS**    **MORE PESSIMISTIC ASSUMPTIONS**

Eliminate stars without planets.

Eliminate planets hostile to life.

Eliminate hospitable planets without life.

Eliminate life that doesn't develop intelligence.

Eliminate life without technology.

Eliminate civilizations not alive today.

Do civilizations abound…    …or are they very rare?

**Figure 24.19** The two columns show estimates of the existence of intelligent, communicative civilizations in the Milky Way Galaxy using the Drake equation. White dots represent stars with possible civilizations. Notice how widely these estimates vary, given optimistic and pessimistic assumptions about the seven factors in the equation.

or hundreds of light-years. Special relativity limits how fast one can travel. The speed of light is the limit, and even at that rate it would take more than 4 years to reach the *nearest* star. The relativistic effect of time dilation would mean that time passes slower for astronauts traveling at very high speeds, and they would return to Earth younger than if they had stayed at home. For example, suppose astronauts visited a star 15 light-years distant. Even if they traveled at speeds close to the speed of light, by the time they returned to Earth, 30 years would have passed at home.

Some science fiction writers get around this problem by invoking "warp speed" or "hyperdrive," which enables travel faster than the speed of light, or by using wormholes as shortcuts across the galaxy—but there is absolutely no evidence that any of these options are possible. And most of these imaginative stories ignore the vast number of other complications that accompany human space travel. Humans are just beginning to learn to live in space even for short periods of time.

# Origins
## The Fate of Life on Earth

Astronomers have used their understanding of physics and cosmology to look back through time and watch as structure formed throughout the universe and to look forward to the future of our Sun, our galaxy, and the ultimate fate of the universe. About 5 billion years from now, the Sun will end its long period of relative stability. It will expand to become a red giant star, swelling to hundreds of times larger than it is at present and thousands of times more luminous. The giant planets, orbiting outside the extended red giant atmosphere, will probably survive. But at least some of the planets of the inner Solar System will not. Just as an artificial satellite is slowed by drag in Earth's tenuous outer atmosphere and eventually falls to the ground, so, too, will a planet caught in the Sun's atmosphere be engulfed by the expanding Sun. If this is what happens to Earth, no trace of this planet will remain other than a slight increase in the amount of massive elements in the Sun's atmosphere.

Another possibility is that the red giant Sun will lose mass in a powerful wind, its gravitational pull on the planets will weaken, and the orbits of both the inner and outer planets will spiral outward. If Earth moves out far enough, it may survive as a seared cinder, orbiting the small, hot, white dwarf star that the Sun will become. Barely larger than Earth and with its nuclear fuel exhausted, the white dwarf Sun will slowly cool, eventually becoming a cold sphere of densely packed carbon, orbited by what remains of its planets. The ultimate outcome for Earth—consumed in the heart of the Sun or left behind as a cold, burned rock orbiting a long-dead white dwarf—is not yet known.

In either case, however, Earth's status as a garden spot in the habitable zone will be at an end. If the Sun does not expand too far or Mars also migrates outward, Mars could become the habitable planet in the Solar System, at least for a while. As the dying Sun loses more and more of its atmosphere in a stellar wind, Earth's atoms might be expelled back into the reaches of interstellar space from which they came, perhaps to be recycled into new generations of stars, planets, and even life itself.

But even before the Sun's change into a red giant star, the Sun's luminosity will begin to rise. As solar luminosity increases, so will temperatures on all the planets. The inner edge of the Sun's habitable zone will slowly move out past the orbit of Earth. Eventually, Earth's temperatures will climb so high that all animal and plant life will perish. Even the extremophiles that inhabit the oceanic depths will die, as the oceans themselves boil away. Models of the Sun's evolution are still not precise enough for astronomers to predict with certainty when that fatal event will occur, but the end of all terrestrial life may be 1 billion to 4 billion years away. Additionally, the Milky Way Galaxy is headed for a collision with the Andromeda Galaxy in 4 billion to 5 billion years' time. Galaxies are mostly empty space, so the Sun is not likely to collide with another star, but one effect of the collision is that our Solar System may be gravitationally flung to a different part of our galaxy.

There is much work to be done before we can realistically contemplate even voyages within the Solar System.

Some people claim that aliens have already visited Earth: tabloid newspapers, books, and websites are filled with tales of UFO sightings, government conspiracies and cover-ups, alleged alien abductions, and UFO religious cults. However, none of these reports meet the basic standards of science. They are not falsifiable—they lack verifiable evidence and repeatability—and we must conclude that there is no scientific evidence for any alien visitations.

## CHECK YOUR UNDERSTANDING 24.5

The Drake equation enables astronomers to: (a) calculate precisely the number of alien civilizations; (b) organize their thoughts about probabilities; (c) locate the stars they should study to find life; (d) find new kinds of life.

It is far from certain, however, that the descendants of today's humanity will even be around a billion years from now. Some of the threats to life come from beyond Earth. For the remainder of the Sun's life, the terrestrial planets, including Earth, will continue to be bombarded by asteroids and comets. Perhaps a hundred or more of these impacts will involve kilometer-sized objects, capable of causing the kind of devastation that eradicated the dinosaurs. Although these events may create new surface scars and may be harmful to human life as we know it, they will have little effect on the integrity of Earth itself. Earth's geological record is filled with such events, and each time they happen, life manages to recover and reorganize.

Humans might protect themselves from the fate of the dinosaurs, but in the long run humanity will either leave this world or die out. Planetary systems are common to other stars, and many other Earth-like planets may well exist throughout the Milky Way Galaxy. Colonizing other planets is currently the stuff of science fiction, but if the descendants of modern-day humans are ultimately to survive the death of the home planet, off-Earth colonization must become science fact at some point in the future.

By studying astronomy, you have learned where you come from. You have learned about the self-correcting nature of science—how it continually adapts to new information to give us the ability to make better and better predictions about the behavior of the physical world. This predictive ability makes science an extremely powerful tool. No other species in the history of Earth has been able to understand its position, predict what will happen next, and therefore adapt its behavior to seek the best possible future for its members.

**Figure 24.20** shows Earth as seen by the *Cassini* spacecraft, looking near Saturn's rings. That tiny dot, which is Earth, is the only place in the entire universe where we know that life exists. Compare the size of that dot to the size of the universe. Compare the history of life on Earth to the history of the universe. Compare Earth's future to the fate of the universe. Astronomy is humbling. We occupy a tiny part of space and time. Yet we are unique, as far as we know. Think for a moment about what that means to you. This may be the most important lesson the universe has to offer.

**Figure 24.20** This image from the *Cassini* spacecraft shows Earth as seen from Saturn. The pale dot in the lower right (arrow) is Earth.

*In this article, scientists discuss how they will search for life beyond Earth.*

## Finding Life Beyond Earth Is within Reach

**NASA**

Many scientists believe we are not alone in the universe. It's probable, they say, that life could have arisen on at least some of the billions of planets thought to exist in our galaxy alone—just as it did here on planet Earth. This basic question about our place in the universe is one that may be answered by scientific investigations. What are the next steps to finding life elsewhere?

Experts from NASA and its partner institutions addressed this question on July 14, 2014, at a public talk held at NASA headquarters in Washington. They outlined NASA's road map to the search for life in the universe, an ongoing journey that involves a number of current and future telescopes.

"Sometime in the near future, people will be able to point to a star and say, 'that star has a planet like Earth,'" says Sara Seager, professor of planetary science and physics at the Massachusetts Institute of Technology in Cambridge, Massachusetts. "Astronomers think it is very likely that every single star in our Milky Way Galaxy has at least one planet."

NASA's quest to study planetary systems around other stars started with ground-based observatories, then moved to space-based assets like the Hubble Space Telescope, the Spitzer Space Telescope, and the Kepler Space Telescope. Today's telescopes can look at many stars and tell if they have one or more orbiting planets. Even more, they can determine if the planets are the right distance away from the star to have liquid water, the key ingredient to life as we know it.

The NASA road map will continue with the launch of the Transiting Exoplanet Surveying Satellite (TESS) in 2017, the James Webb Space Telescope (Webb telescope) in 2018, and perhaps the proposed Wide Field Infrared Survey Telescope–Astrophysics Focused Telescope Assets (WFIRST-AFTA) early in the next decade. These upcoming telescopes will find and characterize a host of new exoplanets—those planets that orbit other stars—expanding our knowledge of their atmospheres and diversity. The Webb telescope and WFIRST-AFTA will lay the groundwork, and future missions will extend the search for oceans in the form of atmospheric water vapor and for life as in carbon dioxide and other atmospheric chemicals on nearby planets that are similar to Earth in size and mass, a key step in the search for life.

"This technology we are using to explore exoplanets is real," said John Grunsfeld, astronaut and associate administrator for NASA's Science Mission Directorate in Washington. "The James Webb Space Telescope and the next advances are happening now. These are not dreams—this is what we do at NASA."

Since its launch in 2009, Kepler has dramatically changed what we know about exoplanets, finding most of the more than 5,000 potential exoplanets, of which more than 1,700 have been confirmed. The Kepler observations have led to estimates of billions of planets in our galaxy and shown that most planets within 1 astronomical unit are less than 3 times the diameter of Earth. Kepler also found the first Earth-size planet to orbit in the "habitable zone" of a star, the region where liquid water can pool on the surface.

"What we didn't know five years ago is that perhaps 10 to 20 percent of stars around us have Earth-size planets in the habitable zone," says Matt Mountain, director and Webb telescope scientist at the Space Telescope Science Institute in Baltimore. "It's within our grasp to pull off a discovery that will change the world forever. It is going to take a continuing partnership between NASA, science, technology, the U.S. and international space endeavors, as exemplified by the James Webb Space Telescope, to build the next bridge to humanity's future."

This decade has seen the discovery of more and more super-Earths, which are rocky planets that are larger and heftier than Earth. Finding smaller planets, the Earth twins, is a tougher challenge because they produce fainter signals. Technology to detect and image these Earth-like planets is being developed now for use with the future space telescopes. The ability to detect alien life may still be years or more away, but the quest is under way.

Said Mountain, "Just imagine the moment, when we find potential signatures of life. Imagine the moment when the world wakes up and the human race realizes that its long loneliness in time and space may be over—the possibility we're no longer alone in the universe."

1. Why is NASA so optimistic that they will find signatures of alien life?
2. What are some potential signatures they hope to find?
3. Why are telescopes in space needed to find these signatures?
4. Do a search to see what is the status of the TESS and James Webb space telescopes.
5. How do you think the world would react to a discovery of signatures of life on another planet?

# Summary

Astrobiology seeks to answer the question, Are we alone? All the sciences have a part to play in answering this question, from astronomy to zoology. Theories of how life began and evolved on Earth help to limit the number of targets astronomers study in their search for life elsewhere. Even before the Sun expands in size and evolves to a red giant star, its luminosity will increase enough to alter the location of its habitable zone. When that happens, Earth may no longer be the planet at the right location for maintaining liquid water.

LG 1 **Explain our current understanding of how and when life began on Earth and how it has evolved.** Life on Earth is a form of complex carbon-based chemistry, made possible by self-replicating molecules. Life likely formed in Earth's oceans, then evolved chemically from simple molecules into self-replicating organisms through a combination of mutation and heredity.

LG 2 **Explain how life is a structure that has evolved through the action of the physical and chemical processes that shape the universe.** All terrestrial life is composed primarily of only six elements: carbon, hydrogen, oxygen, nitrogen, sulfur, and phosphorus. Life-forms that are very different from those on Earth, including those based on silicon chemistry, cannot be ruled out.

LG 3 **Describe the locations in our Solar System and around other stars where astronomers think life might be possible.** Within the Solar System, Mars and some moons of Jupiter and Saturn are the most promising candidates for life. A habitable zone around a star is a location within which the temperature of a planet will support liquid water on its surface. Thus, astronomers look for extrasolar planets that orbit in habitable zones surrounding solar-type stars. This concept of habitable zone can be somewhat extended by special circumstances, such as tidal heating from a giant planet or atmospheres containing greenhouse gases.

LG 4 **Describe some of the methods used to search for intelligent extraterrestrial life.** The Drake equation includes different factors that astronomers consider when thinking about the possibility of life in the universe. Astronomers use radio telescopes to search for signals from extraterrestrial life, particularly in astronomically important regions of the spectrum. In recent times, this has been expanded to a search through as broad a region of the radio spectrum as possible. No intelligent extraterrestrial life has yet been detected.

## ? UNANSWERED QUESTIONS

- Will humans spread life into space? Some scientists have suggested that seeding from Earth may have already happened as Earth microorganisms scattered into space after giant impacts. It is also possible that humans have unintentionally sent microorganisms to space aboard our spacecraft. More intentional methods of seeding include sending microorganisms from Earth to other planets or moons to try to jump-start evolution, "terraforming" Mars or a moon to change conditions on it to make it more habitable for humanity, or sending humans in spaceships to colonize the galaxy.

- Will humans themselves spread into space? At some point, humans must leave planet Earth if the species is to survive. But space is a dangerous place, and many of the problems that humans encounter in space have not yet been solved. These range from purely physical issues such as the loss of bone density in low gravity to the societal problems that occur when a small number of people are confined together for long periods of time. We do not yet know if humanity can overcome these problems and journey to other planets.

# Questions and Problems

## Test Your Understanding

1. The study of life and the study of astronomy are connected because (select all that apply)
   a. life may be commonplace in the universe.
   b. studying other planets may help explain why there is life on Earth.
   c. explorations of extreme environments on Earth suggest where to look for life elsewhere.
   d. life is a structure that evolved through physical processes, and life on Earth may not be unique.
   e. life elsewhere is most likely to be found by astronomers.

2. Scientists look for water to indicate places where life might exits because
   a. water is a common molecule in interstellar space.
   b. life on Earth depends on it.
   c. no other molecules are solvents.
   d. the spectrum of water is very complex.

3. The Urey-Miller experiment produced _____ in a laboratory jar.
   a. life
   b. RNA and DNA
   c. amino acids
   d. proteins

4. A mutation is
   a. always a deadly change to DNA.
   b. always a beneficial change to DNA.
   c. a change to DNA that is sometimes beneficial and sometimes not.
   d. a change to DNA that is not inheritable.

5. Scientists think that terrestrial life probably originated in Earth's oceans, rather than on land, because (select all that apply)
   a. all the chemical pieces were in the ocean.
   b. energy was available in the ocean.
   c. the earliest evidence for life on Earth is from fossils of ocean-dwelling organisms.
   d. the deepest parts of the ocean have hydrothermal vents.

6. Any system with the processes of heredity, mutation, and natural selection will (choose all that apply)
   a. change over time.
   b. become larger over time.
   c. become more complex over time.
   d. develop intelligence.

7. The fact that no alien civilizations have yet been detected indicates that
   a. they are not there.
   b. they are rare.
   c. Earth is in a "blackout," and they are not talking to us.
   d. we don't know enough yet to draw any conclusions.

8. During the Cambrian explosion
   a. the dinosaurs were killed.
   b. all the carbon that is now here on Earth was produced.
   c. biodiversity increased significantly.
   d. a lot of carbon dioxide was released into the atmosphere.

9. The habitable zone is the place around a star where
   a. life has been found.
   b. atmospheres can contain oxygen.
   c. liquid water exists.
   d. liquid water can exist on the surface of a planet.

10. The difference between a prokaryote and a eukaryote is that prokaryotes
    a. have no DNA.
    b. have no cell wall.
    c. have no nucleus.
    d. do not exist today.

11. A thermophile is an organism that lives in extremely _____ environments.
    a. salty
    b. hot
    c. cold
    d. dry

12. The search for life elsewhere in the Solar System is carried out primarily by
    a. astronauts.
    b. robotic spacecraft.
    c. astronomers using optical telescopes.
    d. astronomers using radio telescopes.

13. Astronomers think that intelligent life is more likely to be found around stars of types F, G, K, and M because
    a. those stars are hot enough to have planets and moons with liquid water.
    b. those stars are cool enough to have planets and moons with liquid water.
    c. those stars live long enough for life to begin and evolve.
    d. those stars produce no ultraviolet radiation or X-rays.

14. Life first appeared on Earth
    a. billions of years ago.
    b. millions of years ago.
    c. hundreds of thousands of years ago.
    d. thousands of years ago.

15. In the phrase "theory of evolution," the word *theory* means that evolution
    a. is an idea that can't be tested scientifically.
    b. is an educated guess to explain natural phenomena.
    c. probably doesn't happen anymore.
    d. is a tested and corroborated scientific explanation of natural phenomena.

# Thinking about the Concepts

16. How does the evolution of life on Earth depend on RNA and DNA?

17. How do scientists think that amino acids first formed on Earth?

18. Today, most organisms on Earth enjoy relatively moderate climates and temperatures. Compare this environment to some of the conditions in which early life developed.

19. In the Process of Science Figure, many areas of science are represented that all inform the science of astrobiology. Choose any two of these, and explain how one of these areas of science depends on the other.

20. How was Earth's carbon dioxide atmosphere changed into today's oxygen-rich atmosphere? How long did that transformation take?

21. What was the Cambrian explosion, and what might have caused it?

22. Why did plants and forests appear in high numbers before large animals did?

23. What is a habitable zone? What defines its boundaries?

24. Which general conditions were needed on Earth for life to arise?

25. Is evolution under way on Earth today? If so, how might humans continue to evolve?

26. Where did all the atoms in your body come from?

27. The two *Viking* spacecraft did not find convincing evidence of life on Mars when they visited that planet in the late 1970s nor did the *Phoenix* lander when it examined the martian soil in 2009. Do these results imply that life never existed on the planet? Why or why not?

28. Some scientists believe that humans may be the only advanced life in the galaxy today. If this is indeed the case, which factors in the Drake equation must be extremely small?

29. The second law of thermodynamics says that the entropy (a measure of disorder) of the universe is always increasing. Yet complex living organisms exist. Why does this not violate the second law of thermodynamics?

30. Why is it likely that life on Earth as we know it will end long before the Sun runs out of nuclear fuel?

# Applying the Concepts

31. The Kepler Mission is currently searching for planets in the habitable zones of stars. Explain which factors in the Drake equation are affected by this search and how the final number $N$ will be affected if Kepler finds that most stars have planets in their habitable zones.

32. Study Figure 24.18. The white blocks at the top represent the numbers 1–10. The bottom row of the set is a placeholder, and the top three rows of the set represent the numbers, in order from left to right. Explain the "rule" for the kind of counting shown here. (For example, how do three white blocks represent the number 7?)

33. Suppose that an organism replicates itself each second. If you start with a single specimen, what will the final population be after 10 seconds?

34. Suppose that an organism replicates itself each second. After how many seconds will the population increase by a factor of 1,024?

35. As a general rule, you can find the doubling time for exponential growth by dividing 70 by the rate of increase. So, if the population increases by 7 percent per year, the doubling time is 10 years (70/7 = 10). Suppose Earth's human population continues to grow by 1 percent annually. What is the doubling time? How much time will pass before there are 4 times as many humans on Earth?

36. As discussed in Working It Out 24.1, the doubling time for exponential growth is given by $P_F/P_O = 2^n$. Assume a self-replicating molecule that makes one copy of itself each second. Make a graph of the number of molecules versus time for the first 60 seconds after the molecule begins replicating.

37. The doubling time for *Escherichia coli* is 20 minutes, and you start getting sick when just 10 bacteria enter your system. How many bacteria are in your body after 12 hours?

38. If the chance that a given molecule will mutate is one in 100,000, how many generations are needed before, on average, at least one mutation has occurred?

39. Study the Drake equation in Working It Out 24.2. Make your own most optimistic and most pessimistic assumptions for each of the variables in the equation. What values do you find for $N$?

40. Study the Drake equation in Working It Out 24.2. Keep the other variables as is, but put in different values for the lifetime of a technological civilization. How does this affect your value of $N$? How does this affect the possibility of making contact with extraterrestrials?

41. Look back at Figure 5.24 in Chapter 5. Trace (or photocopy) this graph, and then add horizontal lines for the temperatures at which water freezes and boils. Which planets in the Solar System have measured temperatures that fall within those lines? Which planets have predicted temperatures (based on the equilibrium model) that fall within those lines? What does this tell you about assumptions about the habitable zone?

42. Figure 5.24 shows that Mercury's measured range of temperatures overlaps the temperatures at which water is a liquid. Is Mercury in the habitable zone? Why or why not?

43. Suppose astronomers announce the discovery of a new planet around a star with a mass equal to the Sun's. This planet has an orbital period of 87 days. Is this planet in the habitable zone for a Sun-like star?

44. Why do you think astronomers sent a coded radio signal to the globular cluster M13 in 1974—rather than, say, to a nearby star?

45. As noted in Section 24.1, some scientists suspect the early Earth was "seeded" with primitive life stored in comets and meteoroids. Knowing when and how our Solar System, galaxy, and universe formed, what timeline is required for such seeding to be possible?

## USING THE WEB

46. a. Go to the online *Astrobiology Magazine* (http://astrobio.net), which covers many topics included in this chapter. Under "Origins," click on "Extreme Life." What are some new findings about extremophiles? Why is this important for astrobiologists? Under "Deep Space" click on "New Planets." What is a recent discovery?
    b. Go to the "Life, Unbounded" blog (http://blogs.scientificamerican.com/life-unbounded). What is a recent topic of interest? Is the discussion based on some new data? A new theory?

47. Solar System space missions:
    a. Go to the website for the *Cassini* mission to Saturn (http://saturn.jpl.nasa.gov). Is it still collecting data? What has been found recently on one of the moons that would be of interest to astrobiologists?
    b. Go to the website for the *Mars Science Laboratory* (http://mars.jpl.nasa.gov/msl/mission). The rover *Curiosity* landed on the surface in 2012. What are the science goals of this mission, and how do they relate to astrobiology? What are some recent results?

c. The *MAVEN* Mars mission (http://mars.nasa.gov/maven/) arrived at Mars in 2014. What are the science goals of this mission? Why are astrobiologists interested in the history of the climate and atmosphere of Mars? Are there any results?

48. Go to the "Habitable Exoplanets Catalog" website (http://phl.upr.edu/projects/habitable-exoplanets-catalog). Click on "Methods" to read about the different criteria for evaluating a planet's habitability. How many confirmed habitable exoplanets are there? How many candidates? How might the number of confirmed exoplanets affect the terms in the Drake equation?

49. a. Go to the website for the Kepler space telescope (http://kepler.nasa.gov): click on "News" and "Planet-finding News." What is a recent discovery of a planet in the habitable zone? What is a recent discovery of a planet with an interesting atmosphere?
    b. Go to the website for the European Space Agency (ESA) Gaia mission (http://sci.esa.int/gaia). What are the science objectives of this mission? Click on "Exoplanets" on the left. How will Gaia identify new planets? What properties of the planet will it be able to measure? What are some recent results?

50. Go to the website for "Super Planet Crash" (http://www.stefanom.org/spc/). Start with an "Earth" in the habitable zone and start adding in other planets. Does the system remain stable if you add in a second planet to the habitable zone? What happens if you add a third planet? What happens if you add planets inside or outside of the habitable zone?

## smartw⬡rk5

If your instructor assigns homework in Smartwork5, access your assignments at digital.wwnorton.com/astro5.

# EXPLORATION Fermi Problems and the Drake Equation

The Drake equation is a way of organizing ideas about other intelligent, communicating civilizations in the galaxy. This type of thinking is very useful for estimating a value, particularly when analyzing systems for which counting is not possible. The types of problems that can be solved in this way are often called Fermi problems after Enrico Fermi, who was mentioned in this chapter. For example, we might ask, What is the circumference of Earth?

You could Google this question, or you could already "know" the answer, or you might look it up in this textbook. Alternatively, you could very carefully measure the shadow of a stick in two locations at the same time on the same day. Or you could drive around the planet.

Or you could reason this way:

How many time zones are between New York and Los Angeles?
*3 time zones. You know this from traveling or from television.*

How many miles is it from New York to Los Angeles?
*3,000 miles. You know this from traveling.*

So, how many miles per time zone?
*3,000/3 = 1,000*

How many time zones in the world?
*24, because there are 24 hours in a day, and each time zone marks an hour.*

So, what is the circumference of Earth?
*24,000 miles, because there are 24 time zones, each 1,000 miles wide.*

The measured circumference is 24,900 miles, which agrees with our estimate to within 4 percent.

Here we list several Fermi problems. Time yourself for an hour, and work as many of them as possible. (You don't have to do them in order!)

**1** How much has the mass of the human population on Earth increased in the past year?

_____

**2** How much energy does a horse consume in its lifetime?

_____

**3** How many pounds of potatoes are consumed in the United States annually?

_____

**4** How many cells are there in the human body?

_____

**5** If your life earnings were given to you by the hour, how much is your time worth per hour?

_____

**6** What is the weight of solid garbage thrown out by American families each year?

_____

**7** How fast does human hair grow (in feet per hour)?

_____

**8** If all the people on Earth were crowded together, how much area would be covered?

_____

**9** How many people could fit on Earth if every person occupied 1 square meter of land?

_____

**10** How much carbon dioxide ($CO_2$) does an automobile emit each year?

_____

**11** What is the mass of Earth?

_____

**12** What is the average annual cost of an automobile, including overhead (maintenance, looking for parking, insurance, cleaning, and so on)?

_____

# APPENDIX 1: Mathematical Tools

Mathematics helps scientists to understand the patterns they see and to communicate that understanding to others. Appendix 1 presents some tools that will be useful in our study of astronomy.

## Powers of 10 and Scientific Notation

Astronomy is a science of both the very large and the very small. The mass of an electron, for example, is

0.000000000000000000000000000009109 kilograms (kg)

whereas the distance to a galaxy far, far away might be about

100,000,000,000,000,000,000,000,000 meters

All it takes is a quick glance at these two numbers to see why astronomers need a more convenient way to express numbers.

Our number system is based on **powers of 10**. Going to the left of the decimal place,

$$10 = 10 \times 1$$
$$100 = 10 \times 10 \times 1$$
$$1,000 = 10 \times 10 \times 10 \times 1$$

and so on. Going to the right of the decimal place,

$$0.1 = \frac{1}{10} \times 1$$

$$0.01 = \frac{1}{10} \times \frac{1}{10} \times 1$$

$$0.001 = \frac{1}{10} \times \frac{1}{10} \times \frac{1}{10} \times 1$$

and so on. In other words, each place to the right or left of the decimal place in a number represents a power of 10. For example, 1 million can be written as

$$1 \text{ million} = 1,000,000 = 1 \times 10 \times 10 \times 10 \times 10 \times 10 \times 10$$

That is, 1 million is "1 multiplied by six factors of 10." **Scientific notation** combines these factors of 10 in convenient shorthand.

Rather than all being written out, the six factors of 10 are expressed using an exponent:

$$1 \text{ million} = 1 \times 10^6$$

which also means "1 multiplied by six factors of 10."

Moving to the right of the decimal place, each step *removes* a power of 10 from the number. One-millionth can be written as

$$1 \text{ millionth} = 1 \times \frac{1}{10} \times \frac{1}{10} \times \frac{1}{10} \times \frac{1}{10} \times \frac{1}{10} \times \frac{1}{10}$$

This divides by powers of 10, so this number can be written by use of a negative exponent

$$\frac{1}{10} = 10^{-1}$$

as

$$1 \text{ millionth} = 1 \times 10^{-1} \times 10^{-1} \times 10^{-1} \times 10^{-1} \times 10^{-1} \times 10^{-1}$$
$$= 1 \times 10^{-6}$$

Returning to our earlier examples, the mass of an electron is $9.109 \times 10^{-31}$ kg, and the distant galaxy is located $1 \times 10^{26}$ meters away. These are much more convenient ways of writing these values. Notice that *the exponent in scientific notation gives you a feel for the size of a number at a glance*. The exponent of 10 in the electron mass is −31, which quickly indicates that it is a very small number. The exponent of 10 in the distance to a remote galaxy, +26, quickly indicates that it is a very large number. This exponent is often called the *order of magnitude* of a number. When you see a number written in scientific notation while reading *21st Century Astronomy* (or elsewhere), just remember to look at the exponent to understand the size of the number.

Scientific notation is also convenient because it makes multiplying and dividing numbers easier. For example, 2 billion multiplied by eight-thousandths can be written as

$$2,000,000,000 \times 0.008$$

but it is more convenient to write these two numbers using scientific notation:

$$(2 \times 10^9) \times (8 \times 10^{-3})$$

We can regroup these expressions in the following form:

$$(2 \times 8) \times (10^9 \times 10^{-3})$$

The first part of the problem is just $2 \times 8 = 16$. The more interesting part of the problem is the multiplication in the right-hand parentheses. The first number, $10^9$, is just shorthand for $10 \times 10 \times 10 \ldots$ nine times. That is, it represents nine factors of 10. The second number stands for three factors of 1/10—or removing three factors of 10 if you prefer to think of it that way. Altogether, that makes $9 - 3 = 6$ factors of 10. In other words,

$$10^9 \times 10^{-3} = 10^{9-3} = 10^6$$

Putting the problem together, we get

$$(2 \times 10^9) \times (8 \times 10^{-3}) = (2 \times 8) \times (10^9 \times 10^{-3})$$
$$= 16 \times 10^6$$

By convention, when a number is written in scientific notation, only one digit is placed to the left of the decimal point. In this case, there are two. However, 16 is $1.6 \times 10$, so we can add this additional factor of 10 to the exponent at right, making the final answer

$$1.6 \times 10^7$$

Dividing is just the inverse of multiplication. Dividing by $10^3$ means removing three factors of 10 from a number. Using the previous number,

$$(1.6 \times 10^7) \div (2 \times 10^3) = (1.6 \div 2) \times (10^7 \div 10^3)$$
$$= 0.8 \times 10^{7-3}$$
$$= 0.8 \times 10^4$$

This time we have only a zero to the left of the decimal point. To get the number into proper form, we can substitute $8 \times 10^{-1}$ for 0.8, giving

$$0.8 \times 10^4 = (8 \times 10^{-1}) \times 10^4 = 8 \times 10^3$$

Adding and subtracting numbers in scientific notation is somewhat more difficult because all numbers must be written as values multiplied by the *same* power of 10 before they can be added or subtracted. Therefore, you will need to use a calculator that has scientific notation. Most scientific calculators have a button that says EXP or EE. These mean "times 10 to the." So for $4 \times 10^{12}$, you would type [4][EXP][1][2] or [4][EE][1][2] into your calculator. Usually, this number shows up in the window on your calculator either just as you see it written in this book or as a 4 with a smaller 12 all the way over in the right side of the window.

## Significant Figures

In the previous example, we actually broke some rules in the interest of explaining how powers of 10 are treated in scientific notation. The rules we broke involve the *precision* of the numbers. When expressing quantities in science, it is extremely important to know not only the value of a number but also how precise that value is.

The most complete way to keep track of the precision of numbers is to actually write down the uncertainty in the number. For example, suppose you know that the distance to a store (call it $d$) is between 0.8 and 1.2 kilometers (km); you can then write

$$d = 1.0 \pm 0.2 \text{ km}$$

where the symbol "$\pm$" is pronounced "plus or minus." In this example, $d$ is between $1.0 - 0.2 = 0.8$ km and $1.0 + 0.2 = 1.2$ km. This is an unambiguous statement about the limitations on knowing the value of $d$, but carrying along the formal errors with every number written would be cumbersome at best. Instead, you keep track of the approximate precision of a number by using *significant figures*.

The convention for significant figures is this: Assume the written number has been rounded from a number that had one additional digit to the right of the decimal point. If a quantity $d$, which might represent the distance to the store, is "1.", then $d$ is close to 1. It is likely not as small as "0.", and it is likely not as large as "2.". If instead it is written as

$$d = 1.0$$

then $d$ is likely not 0.9 and is likely not 1.1. It is roughly 1.0 to the nearest tenth. The greater the number of significant figures, the more precisely the number is being specified. For example, 1.00000 is not the same number as 1.00. The first number, 1.00000, represents a value that is probably not as small as 0.99999 and probably not as large as 1.00001. The second number, 1.00, represents a value that is probably not as small as 0.99 nor as large as 1.01. The number 1.00000 is much more precise than the number 1.00.

In mathematical operations, significant figures are important. For example, $2.0 \times 1.6 = 3.2$. It does *not* equal 3.20000000000. *The product of two numbers cannot be known to any greater precision than the numbers themselves.* As a general rule, when you multiply and divide, the answer should have the same number of significant figures as the less precise of the numbers being multiplied or divided. In other words, $2.0 \times 1.602583475 = 3.2$. Because all you know is that the first factor is probably closer to 2.0 than to 1.9 or 2.1, all you know about the product is that it is between about 3.0 and 3.4. It is 3.2. It is not 3.205166950 (*even if that is the answer on your calculator*). The rest of the digits to the right of 3.2 just do not mean anything.

When two numbers are added or subtracted, if one number has a significant figure with a particular place value but another number does not, their sum or difference cannot have a significant figure in that place value. For example,

$$1,045.$$
$$\underline{+1.34567}$$
$$1,046.$$

The answer is "1,046." *not* "1,046.34567". Again, the extra digits to the right of the decimal place have no meaning because "1,045." is not known to that precision.

What is the precision of the number 1,000,000? As it is written, the answer is unclear. Are all those zeros really significant or are they placeholders? If the number is written in scientific notation, however, there is never a question. Instead of 1,000,000, you write $1.0 \times 10^6$ for a number that is known to the nearest hundred thousand or so; or you write $1.00000 \times 10^6$ for a number that is known to the nearest 10.

So the earlier example would have been more correct if written as

$$(2.0 \times 10^9) \times (8.0 \times 10^{-3}) = 1.6 \times 10^7$$

## Algebra

There are many branches of mathematics. The branch that focuses on the relationships between quantities is called **algebra**. Basically, algebra begins by using symbols to represent quantities.

For example, you could write the distance you travel in a day as *d*. As it stands, *d* has no value. It might be 10,000 miles. It might be 30 feet. It does, however, have **units**; in this case, the units of distance. The average speed at which you travel is equal to the distance you travel divided by the time you take. By using the symbol *v* to represent your average speed and the symbol *t* to represent the time you take, instead of writing out "Your average speed is equal to the distance you travel divided by the time you take" you can write

$$v = \frac{d}{t}$$

The meaning of this algebraic expression is exactly the same as the sentence quoted before it, but it is much more concise. As it stands, *v*, *d*, and *t* still have no specific values. There are no numbers assigned to them yet. However, this expression indicates what the relationship between those numbers will be when you look at a specific example. For example, if you go 500 km ($d = 500$ km) in 10 hours ($t = 10$ hours), this expression tells you that your average speed is

$$v = \frac{d}{t} = \frac{500 \,\text{km}}{10 \,\text{h}} = 50 \,\text{km/h}$$

Notice that the units in this expression act exactly like the numerical values. Dividing the two shows that the units of *v* are

kilometers divided by hours, or km/h (pronounced "kilometers per hour").

We introduced algebra as shorthand for expressing relations between quantities, but it is far more powerful than that. Algebra provides rules for manipulating the symbols used to represent quantities. We begin with a bit of notation for *powers* and *roots*. Similar to powers of ten, raising a quantity to a power means multiplying the quantity by itself some number of times. For example, if *S* is a symbol for something (anything), then $S^2$ (pronounced "S squared" or "S to the second power") means $S \times S$, and $S^3$ (pronounced "S cubed" or "S to the third power") means $S \times S \times S$. Suppose *S* represents the length of the side of a square. The area of the square is given by

$$\text{Area} = S \times S = S^2$$

If $S = 3$ meters (m), then the area of the square is

$$S^2 = 3 \,\text{m} \times 3 \,\text{m} = 9 \,\text{m}^2$$

(pronounced "9 square meters"). This is why raising a quantity to the second power is called *squaring* the quantity. We could have done the same thing for the sides of a cube and found that the volume of the cube is

$$\text{Volume} = S \times S \times S = S^3$$

If $S = 3$ meters, then the volume of the cube is

$$S^3 = 3 \,\text{m} \times 3 \,\text{m} \times 3 \,\text{m} = 27 \,\text{m}^3$$

(pronounced "27 cubic meters"). This is why raising a quantity to the third power is called *cubing* the quantity.

Roots are the reverse of this process. The square root of a number is the value that, when squared, gives the original quantity. The square root of 4 is 2, which means that $2 \times 2 = 4$. The square root of 9 is 3, which means that $3 \times 3 = 9$. Similarly, the cube root of a quantity is the value that, when cubed, gives the original quantity. The cube root of 8 is 2, which means that $2 \times 2 \times 2 = 8$. Roots are written with the symbol $\sqrt{\phantom{x}}$. For example, the square root of 9 is written as

$$\sqrt{9} = 3$$

and the cube root of 8 is written as

$$\sqrt[3]{8} = 2$$

If the volume of a cube is $V = S^3$, then

$$S = \sqrt[3]{V} = \sqrt[3]{S^3}$$

Roots can also be written as powers. Powers and roots behave exactly like the exponents of 10 in our discussion of scientific notation. (The exponents used in scientific notation are just powers of 10.) For example, if $a$, $n$, and $m$ are all algebraic quantities, then

$$a^n \times a^m = a^{n+m} \quad \text{and} \quad \frac{a^n}{a^m} = a^{n-m}$$

(The square root of $a$ can also be written $a^{1/2}$ and the cube root of $a$ can be written $a^{1/3}$.)

The rules of arithmetic can be applied to the symbolic quantities of algebra. As long as the rules of algebra are applied properly, then the relationships among symbols arrived at through algebraic manipulation remain true for the physical quantities that those symbols represent.

Here we summarize a few algebraic rules and relationships. In this summary, $a$, $b$, $c$, $m$, $n$, $r$, $x$, and $y$ are all algebraic quantities.

**Associative rule:**

$$a \times b \times c = (a \times b) \times c = a \times (b \times c)$$

**Commutative rule:**

$$a \times b = b \times a$$

**Distributive rule:**

$$a \times (b + c) = (a \times b) + (a \times c)$$

**Cross-multiplication:**

$$\text{If } \frac{a}{b} = \frac{c}{d}, \text{then } ad = bc$$

**Working with exponents:**

$$\frac{1}{a^n} = a^{-n} \qquad a^n a^m = a^{n+m}$$

$$\frac{a^n}{a^m} = a^{n-m} \qquad (a^n)^m = a^{n \times m} \qquad \left(\frac{a}{b}\right)^n = \frac{a^n}{b^n}$$

**Equation of a line with slope $m$ and $y$-intercept $b$:**

$$y = mx + b$$

**Equation of a circle with radius $r$ centered at $x = 0$, $y = 0$:**

$$x^2 + y^2 = r^2$$

## Angles and Distances

The farther away something is, the smaller it appears. This is common sense and everyday experience. Because astronomers cannot walk up to the object they are studying and measure it with a meterstick, their knowledge about the sizes of things usually depends on relating the size of an object, its distance, and the angle it covers in the sky.

One way to measure angles is to use a unit called the **radian**. As shown in **Figure A1.1a**, the size of an angle in radians is the length of the arc subtending the angle, divided by the radius of the circle. In the figure, the angle $x = S/r$ radians.

Because the circumference of a circle is $2\pi$ multiplied by the radius ($C = 2\pi r$), a complete circle has an angular measure of $(2\pi r)/r = 2\pi$ radians. In more conventional angular measure, a complete circle is 360°, so

$$360° = 2\pi \text{ radians}$$

or

$$1\,\text{radian} = \frac{360°}{2\pi} = 57.2958°$$

Often, seconds of arc (**arcseconds**) are used to measure angles for stars and galaxies. A degree is divided into 60 minutes of arc (**arcminutes**), each of which is divided into 60 seconds of arc—so there are 3,600 seconds of arc in a degree. Therefore,

$$3{,}600\frac{\text{arcseconds}}{\text{degree}} \times 57.2958\frac{\text{degree}}{\text{radian}} = 206{,}265\frac{\text{arcseconds}}{\text{radian}}$$

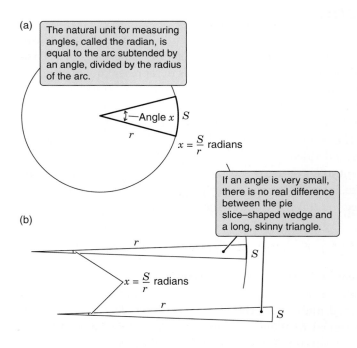

(a) The natural unit for measuring angles, called the radian, is equal to the arc subtended by an angle, divided by the radius of the arc.

—Angle $x$ | $S$

$x = \dfrac{S}{r}$ radians

If an angle is very small, there is no real difference between the pie slice–shaped wedge and a long, skinny triangle.

(b)

$x = \dfrac{S}{r}$ radians

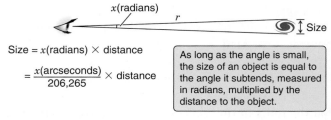

$x$(radians)

Size

Size $= x$(radians) $\times$ distance

$= \dfrac{x(\text{arcseconds})}{206{,}265} \times$ distance

As long as the angle is small, the size of an object is equal to the angle it subtends, measured in radians, multiplied by the distance to the object.

**Figure A1.1** Measuring angles.

If the angle is small enough (which it usually is in astronomy), there is very little difference between the pie slice just described and a long skinny triangle with a short side of length $S$, as Figure A1.1b illustrates. So, if you know the distance $d$ to an object and you can measure the angular size $x$ of the object, then the size of the object is given by

$$S = x(\text{in radians}) \times d = \frac{x(\text{in degrees})}{57.2958 \text{ degrees/radian}} \times d$$

$$S = \frac{x(\text{in arcseconds})}{206{,}265 \text{ arcseconds/radian}} \times d$$

which is how astronomers relate an object's angular size, distance, and physical size.

## Circles and Spheres

To round out these mathematical tools, here are a few useful formulas for circles and spheres. The circle or sphere in each case has a radius $r$.

$$\text{Circumference}_{\text{circle}} = 2\pi r$$

$$\text{Area}_{\text{circle}} = \pi r^2$$

$$\text{Surface area}_{\text{sphere}} = 4\pi r^2$$

$$\text{Volume}_{\text{sphere}} = \frac{4}{3}\pi r^3$$

## Working with Proportionalities

Most of the mathematics in *21st Century Astronomy* involves **proportionalities**—statements about how one physical quantity changes when another quantity changes. We began a discussion of proportionality in Working It Out 1.1; here, we offer a few examples of working with proportionalities.

To use a statement of proportionality to compare two objects, begin by turning the proportionality into a ratio. For example, the price of a bag of apples is **proportional** to the weight of the bag:

$$\text{Price} \propto \text{Weight}$$

Here, the symbol $\propto$ is pronounced "is proportional to." What this means is that the ratio of the prices of two bags of apples is equal to the ratio of the weights of the two bags:

$$\text{Price} \propto \text{Weight means}$$

$$\frac{\text{Price of A}}{\text{Price of B}} = \frac{\text{Weight of A}}{\text{Weight of B}}$$

Let's work a specific example. Suppose bag A weighs 2 pounds and bag B weighs 1 pound. That means bag A will cost twice as much as bag B. We can turn this proportionality into the following equation:

$$\frac{\text{Price of A}}{\text{Price of B}} = \frac{\text{Weight of A}}{\text{Weight of B}} = \frac{2 \text{ lb}}{1 \text{ lb}} = 2$$

In other words, the price of bag A is 2 times the price of bag B. The price per pound is an example of a **constant of proportionality**.

Now let's work another, more complicated example. In Chapter 13, we discuss how the luminosity, brightness, and distance of stars are related. The luminosity of a star—the total energy that the star radiates each second—is proportional to the star's brightness multiplied by the square of its distance:

$$\text{Luminosity} \propto \text{Brightness} \times \text{Distance}^2$$

We can turn this proportionality into a ratio for two stars, A and B:

$$\frac{\text{Luminosity of A}}{\text{Luminosity of B}} = \frac{\text{Brightness of A}}{\text{Brightness of B}} \times \left(\frac{\text{Distance of A}}{\text{Distance of B}}\right)^2$$

If we use the symbols $L$, $b$, and $d$ to represent luminosity, brightness, and distance, respectively, this equation becomes

$$\frac{L_A}{L_B} = \frac{b_A}{b_B} \times \left(\frac{d_A}{d_B}\right)^2$$

As an example, suppose that star A appears twice as bright in the sky as star B, but star A is located 10 times as far away as star B. How do the luminosities of the two stars compare? We know that

$$\text{Luminosity} \propto \text{Brightness} \times \text{Distance}^2$$

we write

$$\frac{\text{Luminosity of A}}{\text{Luminosity of B}} = \frac{\text{Brightness of A}}{\text{Brightness of B}} \times \left(\frac{\text{Distance of A}}{\text{Distance of B}}\right)^2$$

$$\frac{\text{Luminosity of A}}{\text{Luminosity of B}} = \frac{2}{1} \times \left(\frac{10}{1}\right)^2 = 200$$

In other words, star A is 200 times as luminous as star B.

Two quantities may also be inversely proportional, such that making one of them smaller makes the other larger. For example, when you are driving to another town, if you drive twice as fast, it takes half the time to get there. The travel time is inversely proportional to the travel speed. We write this relationship as

$$\text{Time} \propto \frac{1}{\text{Speed}}$$

Proportionalities are used to compare one object to another. Constants of proportionality are used to calculate actual values. In *21st Century Astronomy*, it is usually the proportionality that is important.

# APPENDIX 2: Physical Constants and Units

## Fundamental Physical Constants

| Constant | Symbol | Value |
|---|---|---|
| Speed of light in a vacuum | $c$ | $2.99792 \times 10^8$ m/s |
| Universal gravitational constant | $G$ | $6.6738 \times 10^{-11}$ m³/(kg s²) <br> $6.6738 \times 10^{-20}$ km³/(kg s²) |
| Planck constant | $h$ | $6.62607 \times 10^{-34}$ J-s |
| Boltzmann constant | $k$ | $1.38065 \times 10^{-23}$ J/K |
| Stefan-Boltzmann constant | $\sigma$ | $5.67037 \times 10^{-8}$ W/(m² K⁴) |
| Mass of electron | $m_e$ | $9.10938 \times 10^{-31}$ kg |
| Mass of proton | $m_p$ | $1.67262 \times 10^{-27}$ kg |
| Mass of neutron | $m_n$ | $1.67493 \times 10^{-27}$ kg |
| Electric charge of electron or proton | $e$ | $1.60218 \times 10^{-19}$ C |

SOURCE: Data from the Particle Data Group (http://pdg.lbl.gov).

## Unit Prefixes

| Prefix* | Name | Factor† |
|---|---|---|
| n | nano- | $10^{-9}$ |
| $\mu$ | micro- | $10^{-6}$ |
| m | milli- | $10^{-3}$ |
| k | kilo- | $10^{3}$ |
| M | mega- | $10^{6}$ |
| G | giga- | $10^{9}$ |
| T | tera- | $10^{12}$ |

These prefixes (*), when appended to a unit, change the size of the unit by the factor (†) given. For example, 1 km is $10^3$ meters.

## Units and Values

| Quantity | Fundamental Unit | Values |
|---|---|---|
| Length | meters (m) | radius of Sun ($R_{\text{Sun}}$) = $6.9551 \times 10^8$ m<br>astronomical unit (AU) = $1.49598 \times 10^{11}$ m<br>1 AU = 149,598,000 km<br>light-year (ly) = $9.4605 \times 10^{15}$ m<br>1 ly = $6.324 \times 10^4$ AU<br>1 parsec (pc) = 3.262 *ly*<br>= $3.0857 \times 10^{16}$ m<br>1 m = 3.281 feet |
| Volume | meters$^3$ (m$^3$) | 1 m$^3$ = 1,000 liters<br>= 264.2 gallons |
| Mass | kilograms (kg) | 1 kg = 1,000 grams<br>mass of Earth ($M_{\text{Earth}}$) = $5.9726 \times 10^{24}$ kg<br>mass of Sun ($M_{\text{Sun}}$) = $1.9885 \times 10^{30}$ kg |
| Time | seconds (s) | 1 hour (h) = 60 minutes (min) = 3,600 s<br>solar day (noon to noon) = 86,400 s<br>sidereal day (Earth rotation period) = 86,164.1 s<br>tropical year (equinox to equinox) = 365.24219 days<br>= $3.15569 \times 10^7$ s<br>sidereal year (Earth orbital period) = 365.25636 days<br>= $3.15581 \times 10^7$ s |
| Speed | meters/second (m/s) | 1 m/s = 2.237 miles/h<br>1 km/s = 1,000 m/s = 3,600 km/h<br>$c$ = $2.99792 \times 10^8$ m/s = 299,792 km/s |
| Acceleration | meters/second$^2$ (m/s$^2$) | $g$ = gravitational acceleration on Earth = 9.81 m/s$^2$ |
| Energy | joules (J) | 1 J = 1 kg m$^2$/s$^2$<br>1 megaton = $4.18 \times 10^{15}$ J |
| Power | watts (W) | 1 W = 1 J/s<br>solar luminosity ($L_{\text{Sun}}$) = $3.828 \times 10^{26}$ W |
| Force | newtons (N) | 1 N = 1 kg m/s$^2$<br>1 pound (lb) = 4.448 N<br>1 N = 0.22481 lb |
| Pressure | newtons/meter$^2$ (N/m$^2$) | atmospheric pressure at sea level = $1.013 \times 10^5$ N/m$^2$<br>= 1.013 bar |
| Temperature | kelvins (K) | absolute zero = 0 K = $-273.15°$C = $-459.67°$F |

SOURCE: Data from the Particle Data Group (http://pdg.lbl.gov).

# APPENDIX 3: Periodic Table of the Elements

**Legend:**
- 1 — Atomic number
- H — Symbol
- Hydrogen — Name
- 1.00794 — Average atomic mass
- Metals
- Metalloids
- Nonmetals

| 1 / 1A | 2 / 2A | 3 / 3B | 4 / 4B | 5 / 5B | 6 / 6B | 7 / 7B | 8 | 9 / 8B | 10 | 11 / 1B | 12 / 2B | 13 / 3A | 14 / 4A | 15 / 5A | 16 / 6A | 17 / 7A | 18 / 8A |
|---|---|---|---|---|---|---|---|---|---|---|---|---|---|---|---|---|---|
| 1 **H** Hydrogen 1.00794 | | | | | | | | | | | | | | | | | 2 **He** Helium 4.002602 |
| 3 **Li** Lithium 6.941 | 4 **Be** Beryllium 9.012182 | | | | | | | | | | | 5 **B** Boron 10.811 | 6 **C** Carbon 12.0107 | 7 **N** Nitrogen 14.0067 | 8 **O** Oxygen 15.9994 | 9 **F** Fluorine 18.9984032 | 10 **Ne** Neon 20.1797 |
| 11 **Na** Sodium 22.98976928 | 12 **Mg** Magnesium 24.3050 | | | | | | | | | | | 13 **Al** Aluminum 26.9815386 | 14 **Si** Silicon 28.0855 | 15 **P** Phosphorus 30.973762 | 16 **S** Sulfur 32.065 | 17 **Cl** Chlorine 35.453 | 18 **Ar** Argon 39.948 |
| 19 **K** Potassium 39.0983 | 20 **Ca** Calcium 40.078 | 21 **Sc** Scandium 44.955912 | 22 **Ti** Titanium 47.867 | 23 **V** Vanadium 50.9415 | 24 **Cr** Chromium 51.9961 | 25 **Mn** Manganese 54.938045 | 26 **Fe** Iron 55.845 | 27 **Co** Cobalt 58.933195 | 28 **Ni** Nickel 58.6934 | 29 **Cu** Copper 63.546 | 30 **Zn** Zinc 65.38 | 31 **Ga** Gallium 69.723 | 32 **Ge** Germanium 72.64 | 33 **As** Arsenic 74.92160 | 34 **Se** Selenium 78.96 | 35 **Br** Bromine 79.904 | 36 **Kr** Krypton 83.798 |
| 37 **Rb** Rubidium 85.4678 | 38 **Sr** Strontium 87.62 | 39 **Y** Yttrium 88.90585 | 40 **Zr** Zirconium 91.224 | 41 **Nb** Niobium 92.90638 | 42 **Mo** Molybdenum 95.96 | 43 **Tc** Technetium [98] | 44 **Ru** Ruthenium 101.07 | 45 **Rh** Rhodium 102.90550 | 46 **Pd** Palladium 106.42 | 47 **Ag** Silver 107.8682 | 48 **Cd** Cadmium 112.411 | 49 **In** Indium 114.818 | 50 **Sn** Tin 118.710 | 51 **Sb** Antimony 121.760 | 52 **Te** Tellurium 127.60 | 53 **I** Iodine 126.90447 | 54 **Xe** Xenon 131.293 |
| 55 **Cs** Cesium 132.9054519 | 56 **Ba** Barium 137.327 | 57 **La** Lanthanum 138.90547 | 72 **Hf** Hafnium 178.49 | 73 **Ta** Tantalum 180.94788 | 74 **W** Tungsten 183.84 | 75 **Re** Rhenium 186.207 | 76 **Os** Osmium 190.23 | 77 **Ir** Iridium 192.217 | 78 **Pt** Platinum 195.084 | 79 **Au** Gold 196.966569 | 80 **Hg** Mercury 200.59 | 81 **Tl** Thallium 204.3833 | 82 **Pb** Lead 207.2 | 83 **Bi** Bismuth 208.98040 | 84 **Po** Polonium [209] | 85 **At** Astatine [210] | 86 **Rn** Radon [222] |
| 87 **Fr** Francium [223] | 88 **Ra** Radium [226] | 89 **Ac** Actinium [227] | 104 **Rf** Rutherfordium [261] | 105 **Db** Dubnium [262] | 106 **Sg** Seaborgium [266] | 107 **Bh** Bohrium [264] | 108 **Hs** Hassium [277] | 109 **Mt** Meitnerium [268] | 110 **Ds** Darmstadtium [271] | 111 **Rg** Roentgenium [272] | 112 **Cn** Copernicium [285] | 113 **Uut** Ununtrium [284] | 114 **Fl** Flerovium [289] | 115 **Uup** Ununpentium [288] | 116 **Lv** Livermorium [292] | 117 **Uus** Ununseptium [294] | 118 **Uuo** Ununoctium [294] |

**6 Lanthanides**

| 58 **Ce** Cerium 140.116 | 59 **Pr** Praseodymium 140.90765 | 60 **Nd** Neodymium 144.242 | 61 **Pm** Promethium [145] | 62 **Sm** Samarium 150.36 | 63 **Eu** Europium 151.964 | 64 **Gd** Gadolinium 157.25 | 65 **Tb** Terbium 158.92535 | 66 **Dy** Dysprosium 162.500 | 67 **Ho** Holmium 164.93032 | 68 **Er** Erbium 167.259 | 69 **Tm** Thulium 168.93421 | 70 **Yb** Ytterbium 173.05 | 71 **Lu** Lutetium 174.967 |
|---|---|---|---|---|---|---|---|---|---|---|---|---|---|

**7 Actinides**

| 90 **Th** Thorium 232.03806 | 91 **Pa** Protactinium 231.03588 | 92 **U** Uranium 238.02891 | 93 **Np** Neptunium [237] | 94 **Pu** Plutonium [244] | 95 **Am** Americium [243] | 96 **Cm** Curium [247] | 97 **Bk** Berkelium [247] | 98 **Cf** Californium [251] | 99 **Es** Einsteinium [252] | 100 **Fm** Fermium [257] | 101 **Md** Mendelevium [258] | 102 **No** Nobelium [259] | 103 **Lr** Lawrencium [262] |
|---|---|---|---|---|---|---|---|---|---|---|---|---|---|

We have used the U.S. system as well as the system recommended by the International Union of Pure and Applied Chemistry (IUPAC) to label the groups in this periodic table. The system used in the United States includes a letter and a number (1A, 2A, 3B, 4B, etc.), which is close to the system developed by Mendeleev. The IUPAC system uses numbers 1–18 and has been recommended by the American Chemical Society (ACS). While we show both numbering systems here, we use the IUPAC system exclusively in the book. Elements with atomic numbers higher than 112 have been reported but not yet fully authenticated.

# APPENDIX 4: Properties of Planets, Dwarf Planets, and Moons

## Physical Data for Planets and Dwarf Planets

| Planet | EQUATORIAL RADIUS (km) | EQUATORIAL RADIUS ($R/R_{Earth}$) | MASS (kg) | MASS ($M/M_{Earth}$) | Average Density (relative to water*) | Rotation Period (days) | Tilt of Rotation Axis (degrees, relative to orbit) | Equatorial Surface Gravity (relative to Earth[†]) | Escape Velocity (km/s) | Average Surface Temperature (K) |
|---|---|---|---|---|---|---|---|---|---|---|
| Mercury | 2,440 | 0.383 | $3.30 \times 10^{23}$ | 0.055 | 5.427 | 58.65 | 0.01 | 0.378 | 4.3 | 340 (100, 700)[§] |
| Venus | 6,052 | 0.950 | $4.87 \times 10^{24}$ | 0.815 | 5.243 | 243.02[‡] | 177.3 | 0.907 | 10.36 | 735 |
| Earth | 6,378 | 1.000 | $5.97 \times 10^{24}$ | 1.000 | 5.513 | 1.000 | 23.44 | 1.000 | 11.19 | 288 (185, 331)[§] |
| Mars | 3,396 | 0.532 | $6.42 \times 10^{23}$ | 0.107 | 3.934 | 1.026 | 25.19 | 0.377 | 5.03 | 210 (120, 293)[§] |
| Ceres | 481.5 | 0.076 | $9.47 \times 10^{20}$ | 0.0002 | 2.09 | 0.378 | ~3 | 0.28 | 1.86 | 168 |
| Jupiter | 71,492 | 11.209 | $1.90 \times 10^{27}$ | 317.8 | 1.326 | 0.4135 | 3.13 | 2.528 | 6.02 | 125 |
| Saturn | 60,268 | 9.449 | $5.68 \times 10^{26}$ | 95.16 | 0.687 | 0.4440 | 26.73 | 1.065 | 36.1 | 134 |
| Uranus | 25,559 | 4.007 | $8.68 \times 10^{25}$ | 14.54 | 1.270 | 0.7183[‡] | 97.77 | 0.887 | 21.4 | 76 |
| Neptune | 24,764 | 3.883 | $1.02 \times 10^{26}$ | 17.148 | 1.638 | 0.6713 | 28.32 | 1.14 | 23.6 | 59 |
| Pluto | 1,195 | 0.187 | $1.30 \times 10^{22}$ | 0.0022 | 2.050 | 6.387[‡] | 122.53 | 0.070 | 1.23 | 44 |
| Haumea | ~650 | 0.11 | $4.0 \times 10^{21}$ | 0.0007 | ~3 | 0.163 | ? | 0.045 | 0.84 | <50 |
| Makemake | 715 | 0.11 | $4.0 \times 10^{21}$ | 0.0007 | ~1.7 | 0.937 | ? | ~0.5 | ~0.8 | ~30 |
| Eris | 1,163 | 0.182 | $1.67 \times 10^{22}$ | 0.0028 | 2.5 | 1.08 | ? | 0.084 | 1.38 | 43 |

*The density of water is 1,000 $kg/m^3$.
[†]The surface gravity of Earth is 9.81 $m/s^2$.
[‡]Venus, Uranus, and Pluto rotate opposite to the directions of their orbits. Their north poles are south of their orbital planes.
[§]Where provided, values in parentheses give extremes of recorded temperatures.

## Orbital Data for Planets and Dwarf Planets

| Planet | MEAN DISTANCE FROM SUN ($A$*) | | Orbital Period ($P$) (sidereal years) | Eccentricity | Inclination (degrees, relative to ecliptic) | Average Speed (km/s) |
| --- | --- | --- | --- | --- | --- | --- |
| | ($10^6$ km) | (AU) | | | | |
| Mercury | 57.9 | 0.387 | 0.241 | 0.2056 | 7.0 | 47.36 |
| Venus | 108.2 | 0.723 | 0.615 | 0.0068 | 3.39 | 35.02 |
| Earth | 149.6 | 1.000 | 1.000 | 0.0167 | 0.000 | 29.78 |
| Mars | 227.9 | 1.524 | 1.881 | 0.0934 | 1.85 | 24.08 |
| Ceres | 413.7 | 2.765 | 4.603 | 0.079 | 10.59 | 17.88 |
| Jupiter | 778.3 | 5.203 | 11.863 | 0.0484 | 1.30 | 13.06 |
| Saturn | 1,426.7 | 9.537 | 29.447 | 0.0539 | 2.49 | 9.64 |
| Uranus | 2,870.7 | 19.189 | 84.017 | 0.0473 | 0.77 | 6.80 |
| Neptune | 4,498.4 | 30.070 | 164.79 | 0.0086 | 1.769 | 5.43 |
| Pluto | 5,906.4 | 39.48 | 247.92 | 0.2488 | 17.14 | 4.67 |
| Haumea | 6,432.0 | 43.0 | 281.9 | 0.198 | 28.22 | 4.50 |
| Makemake | 6,783.3 | 45.5 | 305.3 | 0.164 | 29.00 | 4.39 |
| Eris | 10,180 | 68.0 | 561.4 | 0.434 | 43.17 | 3.43 |

*$A$ is the semimajor axis of the planet's elliptical orbit.

## Properties of Selected Moons*

| Planet | Moon | ORBITAL PROPERTIES | | PHYSICAL PROPERTIES | | Relative Density (g/cm³) (water = 1.00) |
|---|---|---|---|---|---|---|
| | | P (days) | A (10³ km) | R (km) | M (10²⁰ kg) | |
| Earth (1 moon) | Moon | 27.32 | 384.4 | 1,737.5 | 735 | 3.34 |
| Mars (2 moons) | Phobos | 0.32 | 9.38 | 13.4 × 11.2 × 9.2 | 0.0001 | 1.9 |
| | Deimos | 1.26 | 23.46 | 7.5 × 6.1 × 5.2 | 0.00002 | 1.5 |
| Jupiter (67 known moons) | Metis | 0.30 | 128 | 21.5 | 0.00012 | 3 |
| | Amalthea | 0.50 | 181.4 | 131 × 73 × 67 | 0.0207 | 0.8 |
| | Io | 1.77 | 421.8 | 1,822 | 893 | 3.53 |
| | Europa | 3.55 | 671.1 | 1,561 | 480 | 3.01 |
| | Ganymede | 7.15 | 1,070 | 2,631 | 1,482 | 1.94 |
| | Callisto | 16.69 | 1,883 | 2,410 | 1,076 | 1.83 |
| | Himalia | 250.56 | 11,461 | 85 | 0.067 | 2.6 |
| | Pasiphae | 744[†] | 23,624 | 30 | 0.0030 | 2.6 |
| | Callirrhoe | 759[†] | 24,102 | 4.3 | 0.00001 | 2.6 |
| Saturn (62 known moons) | Pan | 0.58 | 133.58 | 14 | 0.00005 | 0.42 |
| | Prometheus | 0.61 | 139.38 | 74 × 50 × 34 | 0.0016 | 0.48 |
| | Pandora | 0.63 | 141.70 | 55 × 44 × 31 | 0.0014 | 0.49 |
| | Mimas | 0.94 | 185.54 | 198 | 0.38 | 1.15 |
| | Enceladus | 1.37 | 238.04 | 252 | 1.08 | 1.6 |
| | Tethys | 1.89 | 294.67 | 533 | 6.18 | 0.97 |
| | Dione | 2.74 | 377.42 | 562 | 11.0 | 1.48 |
| | Rhea | 4.52 | 527.07 | 764 | 23.1 | 1.23 |
| | Titan | 15.95 | 1,222 | 2,575 | 1,346 | 1.88 |
| | Hyperion | 21.28 | 1,501 | 205 × 130 × 110 | 0.0559 | 0.54 |
| | Iapetus | 79.33 | 3,561 | 736 | 18.1 | 1.08 |
| | Phoebe | 550.3[†] | 12,948 | 107 | 0.08 | 1.6 |
| | Paaliaq | 687.5 | 15,024 | 11 | 0.0001 | 2.3 |
| Uranus (27 known moons) | Cordelia | 0.34 | 49.8 | 20 | 0.0004 | 1.3 |
| | Miranda | 1.41 | 129.9 | 236 | 0.66 | 1.21 |
| | Ariel | 2.52 | 190.9 | 579 | 12.9 | 1.59 |
| | Umbriel | 4.14 | 266.0 | 585 | 12.2 | 1.46 |

(continued)

# Properties of Selected Moons*

*(continued)*

| Planet | Moon | ORBITAL PROPERTIES | | PHYSICAL PROPERTIES | | Relative Density[†] (g/cm³) (water = 1.00) |
|---|---|---|---|---|---|---|
| | | P (days) | A (10³ km) | R (km) | M (10²⁰ kg) | |
| Uranus (27 known moons) | Titania | 8.71 | 436.3 | 789 | 34.2 | 1.66 |
| | Oberon | 13.46 | 583.5 | 761 | 28.8 | 1.56 |
| | Setebos | 2,225[†] | 17,420 | 24 | 0.0009 | 1.5 |
| Neptune (14 known moons) | Naiad | 0.29 | 48.2 | 48 × 30 × 26 | 0.002 | 1.3 |
| | Larissa | 0.56 | 73.5 | 108 × 102 × 84 | 0.05 | 1.3 |
| | Proteus | 1.12 | 117.6 | 218 × 208 × 201 | 0.5 | 1.3 |
| | Triton | 5.88[†] | 354.8 | 1,353 | 214 | 2.06 |
| | Nereid | 360.13 | 5,513.8 | 170 | 0.3 | 1.5 |
| Pluto (5 moons) | Charon | 6.39 | 17.54 | 604 | 15.5 | 1.68 |
| Haumea (2 moons) | Namaka | 18 | 25.66 | 85 | 0.018 | ~1 |
| | Hi'iaka | 49 | 49.88 | 170 | 0.179 | ~1 |
| Eris | Dysnomia | 15.8 | 37.4 | 100–500? | ? | ? |

*Innermost, outermost, largest, and/or a few other moons for each planet.
[†]Irregular moon (has retrograde orbit).

# APPENDIX 5: Space Missions

## Selected Recent and Current Solar System Missions

| Spacecraft | Sponsoring Nation(s)* | Destination | Launch Year | Type | Status (mid-2015) |
|---|---|---|---|---|---|
| *Voyager 1* and *2* | USA | Jupiter, Saturn, Uranus (2), Neptune (2) | 1977 | Flyby | Actively exploring outer edge of Solar System |
| *Galileo* | USA | Jupiter | 1989 | Orbiter/probe | Ended 2003 |
| *Ulysses* | USA, Europe | Sun | 1990 | Solar polar orbiter | Ended 2008 |
| *SOHO* | USA, Europe | Sun | 1995 | Orbiter | Active |
| *Mars Global Surveyor* | USA | Mars | 1996 | Orbiter | Ended 2006 |
| *Cassini-Huygens* | USA, Europe, Italy | Saturn, Titan | 1997 | Saturn orbiter, Titan probe/lander | Orbiter active |
| *Stardust* | USA | Comets | 1999 | Sample return/flyby | Ended 2011 |
| *Mars Odyssey* | USA | Mars | 2001 | Orbiter | Active |
| *Mars Exploration Rover* | USA | Mars | 2003 | Two landers | One rover active |
| *Hayabusa* | Japan | Asteroid | 2003 | Sample return | Ended 2010 |
| *Mars Express* | Europe | Mars | 2003 | Orbiter | Active |
| *Messenger* | USA | Mercury (2011) | 2004 | Orbiter | Ended 2015 |
| *Venus Express* | Europe | Venus | 2005 | Orbiter | Ended 2014 |
| *Mars Reconnaissance Orbiter (MRO)* | USA | Mars | 2005 | Orbiter | Active |
| *Deep Impact/EPOXI* | USA | Comet Hartley (2010) | 2005 | Impactor/flyby | Ended 2010 |
| *STEREO* | USA | Sun | 2006 | Two orbiters | Active |
| *New Horizons* | USA | Pluto (2015) | 2006 | Flyby | Active |
| *Chang'e 1* | China | Moon | 2007 | Orbiter | Ended 2009 |
| *Kayuga* | Japan | Moon | 2007 | Orbiter | Ended 2009 |

*(continued)*

## Selected Recent and Current Solar System Missions

*(continued)*

| Spacecraft | Sponsoring Nation(s)* | Destination | Launch Year | Type | Status (mid-2015) |
|---|---|---|---|---|---|
| *Artemis* | USA | Moon, solar wind | 2007 | | Active |
| *Dawn* | USA | Vesta (2011), Ceres (2015) | 2007 | Orbiter | Active |
| *Chandrayaan* | India | Moon | 2008 | Orbiter/impactor | Ended 2009 |
| *Lunar Reconnaissance Orbiter (LRO)* | USA | Moon | 2009 | Orbiter | Active |
| *Lunar Crater Observation and Sensing Satellite (LCROSS)* | USA | Moon | 2009 | Impactor | Ended 2009 |
| *Chang'e 2* | China | Moon | 2010 | Orbiter | Ended 2011 |
| *Juno* | USA | Jupiter (2016) | 2011 | Orbiter | En route |
| *Gravity Recovery and Interior Laboratory (GRAIL)* | USA | Moon | 2011 | Two orbiters | Ended 2012 |
| *Mars Science Laboratory (Curiosity* rover) | USA | Mars | 2011 | Lander | Active |
| *Mars Atmosphere and Volatile EvolutioN (MAVEN)* mission | USA | Mars | 2013 | Orbiter | Active |
| *Chang'e 3* | China | Moon | 2013 | Lander | On lunar surface |

*Countries are represented by the following agencies: China = CNSA (China National Space Administration); Europe = ESA (European Space Agency); India = ISRO (Indian Space Research Organisation); Italy = Italian Space Agency; Japan = JAXA (Japan Aerospace Exploration Agency); USA = NASA (National Aeronautics and Space Administration).

# APPENDIX 6: Nearest and Brightest Stars

## Stars within 12 Light-Years of Earth

| Name* | Distance (ly) | Spectral Type† | Relative Visual Luminosity‡ (Sun = 1.000) | Apparent Magnitude | Absolute Magnitude |
|---|---|---|---|---|---|
| Sun | $1.55 \times 10^{-5}$ | G2V | 1.000 | −26.74 | 4.83 |
| Alpha Centauri C (Proxima Centauri) | 4.24 | M5.0V | 0.000052 | 11.05 | 15.48 |
| Alpha Centauri A | 4.36 | G2V | 1.5 | 0.01 | 4.38 |
| Alpha Centauri B | 4.36 | K0V | 0.44 | 1.34 | 5.71 |
| Barnard's star | 5.96 | M4Ve | 0.00043 | 9.57 | 13.25 |
| CN Leonis | 7.78 | M5.5 | 0.000019 | 13.53 | 16.64 |
| BD +36-2147 | 8.29 | M2.0V | 0.0057 | 7.47 | 10.44 |
| Sirius A | 8.58 | A1V | 22.1 | −1.43 | 1.47 |
| Sirius B | 8.58 | DA2 | 0.0025 | 8.44 | 11.34 |
| BL Ceti | 8.73 | M5.5V | 0.000059 | 12.61 | 15.40 |
| UV Ceti | 8.73 | M6.0 | 0.000039 | 13.06 | 15.85 |
| V1216 Sagittarii | 9.68 | M3.5V | 0.00050 | 10.44 | 13.08 |
| HH Andromedae | 10.32 | M5.5V | 0.00010 | 12.29 | 14.79 |
| Epsilon Eridani | 10.52 | K2V | 0.28 | 3.73 | 6.20 |
| Lacaille 9352 | 10.74 | M1.0V | 0.011 | 7.34 | 9.76 |
| FI Virginis | 10.92 | M4.0V | 0.00033 | 11.16 | 13.53 |
| EZ Aquarii A | 11.26 | M5.0V | 0.000063 | 13.03 | 15.33 |
| EZ Aquarii B | 11.26 | M5e | 0.000050 | 13.27 | 15.58 |
| EZ Aquarii C | 11.26 | MV | 0.000010 | 15.07 | 17.37 |
| Procyon A | 11.40 | F5IV-V | 7.38 | 0.38 | 2.66 |
| Procyon B | 11.40 | DA | 0.00055 | 10.70 | 12.98 |

*(continued)*

## Stars within 12 Light-Years of Earth

*(continued)*

| Name* | Distance (ly) | Spectral Type† | Relative Visual Luminosity‡ (Sun = 1.000) | Apparent Magnitude | Absolute Magnitude |
|---|---|---|---|---|---|
| 61 Cygni A | 11.40 | K5.0V | 0.087 | 5.20 | 7.48 |
| 61 Cygni B | 11.40 | K7.0V | 0.041 | 6.03 | 8.31 |
| Gliese 725 A | 11.52 | M3.0V | 0.0029 | 8.90 | 11.17 |
| Gliese 725 B | 11.52 | M3.5V | 0.0014 | 9.69 | 11.96 |
| Andromedae GX | 11.62 | M1.5V | 0.0064 | 8.08 | 10.32 |
| Andromedae GQ | 11.62 | M3.5V | 0.00041 | 11.06 | 13.30 |
| Epsilon Indi A | 11.82 | K5Ve | 0.15 | 4.68 | 6.89 |
| Epsilon Indi B (brown dwarf) | 11.82 | T1.0 | — | — | — |
| Epsilon Indi C (brown dwarf) | 11.82 | T6.0 | — | — | — |
| DX Cancri | 11.82 | M6.0V | 0.000012 | 14.90 | 17.10 |
| Tau Ceti | 11.88 | G8.5V | 0.46 | 3.49 | 5.68 |
| Gliese 1061 | 11.99 | M5.5V | 0.000067 | 13.09 | 15.26 |

SOURCE: From the Research Consortium on Nearby Stars (http://www.recons.org).
*Stars may carry many names, including common names (such as Sirius), names based on their prominence within a constellation (such as Alpha Canis Majoris, another name for Sirius), or names based on their inclusion in a catalog (such as BD +36-2147). Addition of letters A, B, and so on, or superscripts indicates membership in a multiple-star system.
†Spectral types such as M3 are discussed in Chapter 13. Other letters or numbers provide additional information. For example, "V" after the spectral type indicates a main-sequence star, and "III" indicates a giant star. Stars of spectral type T are brown dwarfs.
‡Luminosity in this table refers only to radiation in "visual" light.

## The 25 Brightest Stars in the Sky

| Name | Common Name | Distance (ly) | Spectral Type | Relative Visual Luminosity* (Sun = 1.000) | Apparent Visual Magnitude | Absolute Visual Magnitude |
|---|---|---|---|---|---|---|
| Sun | Sun | $1.58 \times 10^{-5}$ | G2V | 1.000 | −26.74 | 4.83 |
| Alpha Canis Majoris | Sirius | 8.60 | A1V | 22.9 | −1.46 | 1.43 |
| Alpha Carinae | Canopus | 313 | F0II | 14,900 | −0.72 | −5.60 |
| Alpha¹ Centauri | Rigil Kentaurus A | 4.36 | G2V | 1.51 | −0.01 | 4.38 |
| Alpha² Centauri | Rigil Kentaurus B | 4.36 | K1V | 0.44 | 1.33 | 5.71 |
| Alpha Bootis | Arcturus | 36.7 | K1.5III | 113 | −0.04 | −0.30 |
| Alpha Lyrae | Vega | 25.3 | A0Va | 49.2 | 0.03 | 0.60 |
| Alpha Aurigae | Capella | 43 | G5IIIe+G0III | 137 | 0.08 | −0.51 |
| Beta Orionis | Rigel | 860 | B8Iab | 54,000 | 0.12 | −7.0 |
| Alpha Canis Minoris | Procyon | 11.5 | F5IV-V | 7.73 | 0.34 | 2.61 |
| Alpha Eridani | Achernar | 140 | B3Vpe | 1,030 | 0.46 | −2.70 |
| Beta Centauri | Hadar | 392 | B1III | 7,180 | 0.61 | −4.81 |
| Alpha Orionis | Betelgeuse | 570 | M1-2Iab | 13,600 | 0.7 | −5.5 |
| Alpha Aquilae | Altair | 16.7 | A7V | 11.1 | 0.77 | 2.22 |
| Alpha Crucis | Acrux | 325 | B0.5IV+B1V | 3,100 | 1.3 | −3.9 |
| Alpha Tauri | Aldebaran | 67 | K5III | 163 | 0.85 | −0.70 |
| Alpha Scorpii | Antares | 550 | M1.5Ib | 16,300 | 0.96 | −5.7 |
| Alpha Virginis | Spica | 250 | B1IV+B4V | 1,920 | 1.04 | −3.38 |
| Beta Geminorum | Pollux | 34 | K0III | 32.2 | 1.14 | 1.06 |
| Alpha Piscis | Fomalhaut | 25 | A3V | 17.4 | 1.16 | 1.73 |
| Beta Crucis | Mimosa | 280 | B0.5III | 1,980 | 1.25 | −3.41 |
| Alpha Cygni | Deneb | 1,425 | A2Ia | 58,600 | 1.25 | −7.09 |
| Alpha Leonis | Regulus | 79 | B7V | 146 | 1.35 | −0.58 |
| Epsilon Canis Majoris | Adhara | 405 | B2II | 3,400 | 1.50 | −4.0 |
| Alpha Gemini | Castor | 51 | A1V+A5Vm | 49 | 1.58 | 0.61 |
| Gamma Crucis | Gacrux | 88 | M3.5III | 138 | 1.63 | −0.52 |

SOURCES: Data from Jim Kaler's *STARS* page (http://stars.astro.illinois.edu/sow/bright.html); SIMBAD Astronomical Database (http://simbad.u-strasbg.fr/simbad).
*Luminosity in this table refers only to radiation in "visual" light.

# APPENDIX 7: Observing the Sky

The purpose of this appendix is to provide enough information so that you can make sense of a star chart or list of astronomical objects and find a few objects in the sky.

## Celestial Coordinates

In Chapter 2, we discuss the **celestial sphere**—the imaginary sphere with Earth at its center upon which celestial objects appear to lie. A number of different coordinate systems are used to specify the positions of objects on the celestial sphere. The simplest of these is the *altitude-azimuth coordinate system*. The altitude-azimuth coordinate system is based on the "map" direction to an object (the object's azimuth, with north = 0°, east = 90°, south = 180°, and west = 270°) combined with how high the object is above the horizon (the object's altitude, with the horizon at 0° and the zenith at 90°). For example, an object that is 10° above the eastern horizon has an altitude of 10° and an azimuth of 90°. An object that is 45° above the horizon in the southwest is at altitude 45°, azimuth 225°.

The altitude-azimuth coordinate system is the simplest way to tell someone where in the sky to look at the moment, but it is not a good coordinate system for cataloging the positions of objects. The altitude and azimuth of an object are different for each observer, depending on the observer's position on Earth, and they are constantly changing as Earth rotates on its axis. To specify the direction to an object in a way that is the same for everyone requires a coordinate system that is fixed relative to the celestial sphere. The most common such coordinates are called *celestial coordinates*.

Celestial coordinates are illustrated in **Figure A7.1**. Celestial coordinates are much like the traditional system of latitude and longitude used on the surface of Earth. On Earth, latitude specifies how far you are from Earth's equator, as discussed in Chapter 2. If you are on Earth's equator, your latitude is 0°. If you are at Earth's North Pole, your latitude is 90° north. If you are at Earth's South Pole, your latitude is 90° south.

The latitude-like coordinate on the celestial sphere is called **declination**, often signified with the lowercase Greek letter $\delta$ (delta). The celestial equator has $\delta = 0°$. The north celestial pole has $\delta = +90°$. The south celestial pole has $\delta = -90°$. (See Chapter 2 if you need to refresh your memory about the celestial equator or celestial poles.) Declination is usually expressed in degrees, minutes of arc, and seconds of arc. For example, Sirius, the brightest star in the sky, has $\delta = -16°42'58''$ meaning that it is located not quite 17° south of the celestial equator.

On Earth, east–west position is specified by longitude. Lines of constant longitude run north–south from one pole to the other. Unlike latitude, for which the equator provides a natural place to call "zero," there is no natural starting point for longitude, so one was invented. By arbitrary convention, the Royal Observatory in Greenwich, England, is defined to lie at a longitude of 0°. On the celestial sphere, the longitude-like coordinate is called **right ascension**, often signified with the lowercase Greek letter $\alpha$ (alpha). Unlike the case with longitude, there *is* a natural point on the celestial sphere to use as the starting point for right ascension: the vernal equinox, or the point at which the ecliptic crosses the celestial equator with the Sun moving from the southern sky into the northern sky. The (Northern Hemisphere) vernal equinox defines the line of right ascension at which $\alpha = 0°$. The (Northern Hemisphere) autumnal equinox, located on the opposite side of the sky, is at $\alpha = 180°$.

Normally, right ascension is measured in units of time rather than degrees. It takes Earth 24 hours (of sidereal time) to rotate on its axis, so the celestial sphere is divided into 24 hours of right ascension, with each hour of right ascension corresponding to 15°. Hours of right ascension are then subdivided into minutes and seconds of time. Right ascension increases going to the east. The right ascension of Sirius, for example, is $\alpha = 06^h45^m08.9^s$, meaning that Sirius is about 101° (that is, $06^h45^m$) east of the vernal equinox. Time is a natural unit for measuring right ascension because time naturally tracks the motion of objects due to Earth's rotation on its axis. If stars on the meridian at a certain time have $\alpha = 06^h$, then an hour later the stars on the meridian will have $\alpha = 07^h$, and an hour after that they will have $\alpha = 08^h$. The *local sidereal time*, or *star time*, at your location right now is equal to the right ascension of the stars that are on your meridian at the moment. Because of Earth's motion around the Sun, a sidereal day is about 4 minutes shorter than a solar day, so local sidereal time constantly gains on solar time. At midnight on September 22, the local sidereal time is $0^h$. By midnight on December 21, local sidereal time has advanced to $06^h$. On March 20, local sidereal time at midnight is $12^h$. And at midnight on June 20, local sidereal time is $18^h$.

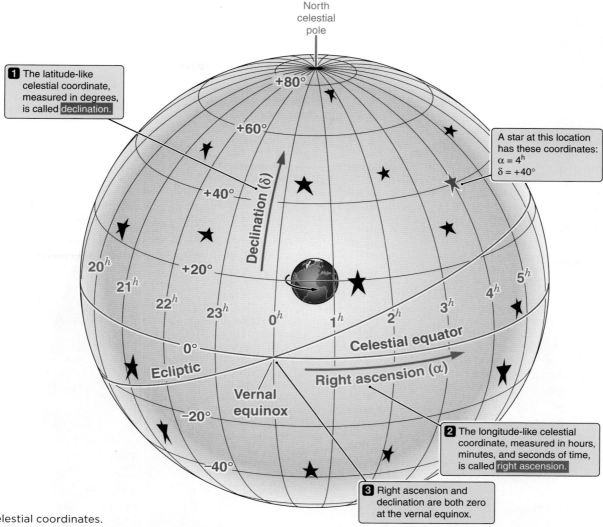

**Figure A7.1** Celestial coordinates.

Putting this all together, right ascension and declination provide a convenient way to specify the location of any object on the celestial sphere. Sirius is located at $\alpha = 06^h45^m08.9^s$, $\delta = -16°42'58''$, which means that at midnight on December 21 (local sidereal time = $06^h$), you will find Sirius about $45^m$ east of the meridian, not quite 17° south of the celestial equator.

There is just one final caveat. As we discussed in Chapter 2, the directions of the celestial equator, celestial poles, and vernal equinox are constantly changing as Earth's axis wobbles like the axis of a spinning top. In Chapter 2, we called this 26,000-year wobble the **precession of the equinoxes**, meaning that the location of the equinoxes is slowly advancing along the ecliptic. So when we specify the celestial coordinates of an object, we need to specify the date at which the positions of the vernal equinox and celestial poles were measured. By convention, coordinates are usually referred to with the position of the vernal equinox on January 1, 2000. A complete, formal specification of the coordinates of Sirius would then

be $\alpha(2000) = 06^h45^m08.9^s$, $\delta(2000) = -16°42'58''$, where the "2000" in parentheses refers to the equinox of the coordinates.

## Constellations and Names

Although it is certainly possible to specify any location on the surface of Earth exactly by giving its latitude and longitude, it is usually convenient to use a more descriptive address. We might say, for example, that one of the coauthors of this book works near latitude 37° north, longitude 122° west; but it would probably mean a lot more to you if we said that George Blumenthal works in Santa Cruz, California.

Just as the surface of Earth is divided into nations and states, the celestial sphere is divided into 88 **constellations**, the names of which are often used to refer to objects within their boundaries (see the star charts in **Figure A7.2**). The brightest stars within the boundaries of a constellation are named using a Greek letter

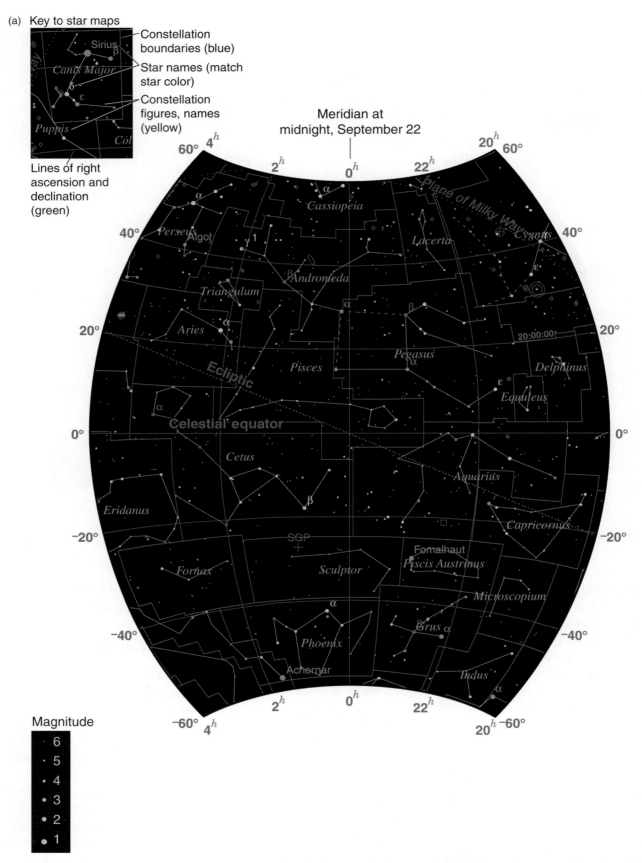

(a) Key to star maps

Constellation boundaries (blue)

Star names (match star color)

Constellation figures, names (yellow)

Lines of right ascension and declination (green)

Sirius
β
Canis Major
δ
ε
Puppis
Col

Meridian at midnight, September 22

Plane of Milky Way

Cassiopeia
α
Cygnus
Lacerta
Perseus
Algol
γ  1
β Andromeda
α
β
ε
Triangulum
θ
Aries
α
Pegasus
α
Delphinus
Ecliptic
Pisces
ε
Equuleus
Celestial equator
Cetus
β
Aquarius
Eridanus
Capricornus
SGP
Fomalhaut
Fornax
Sculptor
Piscis Austrinus
α
Microscopium
Grus  α
Phoenix
Achernar
Indus
α

60°  4ʰ
2ʰ
0ʰ
22ʰ
20ʰ  60°
40°
40°
20°
20:00:00
20°
0°
0°
−20°
−20°
−40°
−40°
2ʰ
0ʰ
22ʰ
−60°  4ʰ
20ʰ −60°

Magnitude
6
5
4
3
2
1

**Figure A7.2A** The sky from right ascension 20ʰ to 04ʰ and declination −60° to +60°.

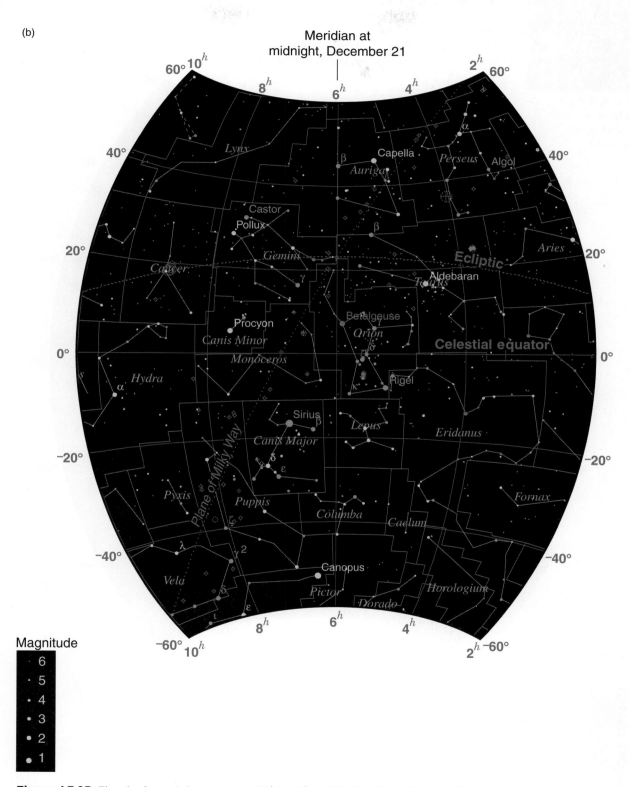

**Figure A7.2B** The sky from right ascension 02$^h$ to 10$^h$ and declination −60° to +60°.

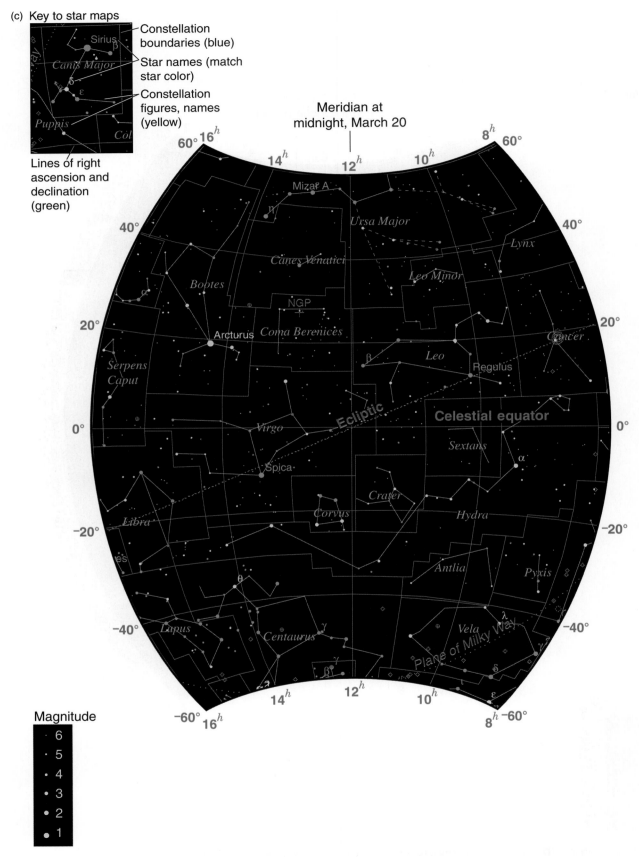

**Figure A7.2C** The sky from right ascension 08$^h$ to 16$^h$ and declination −60° to +60°.

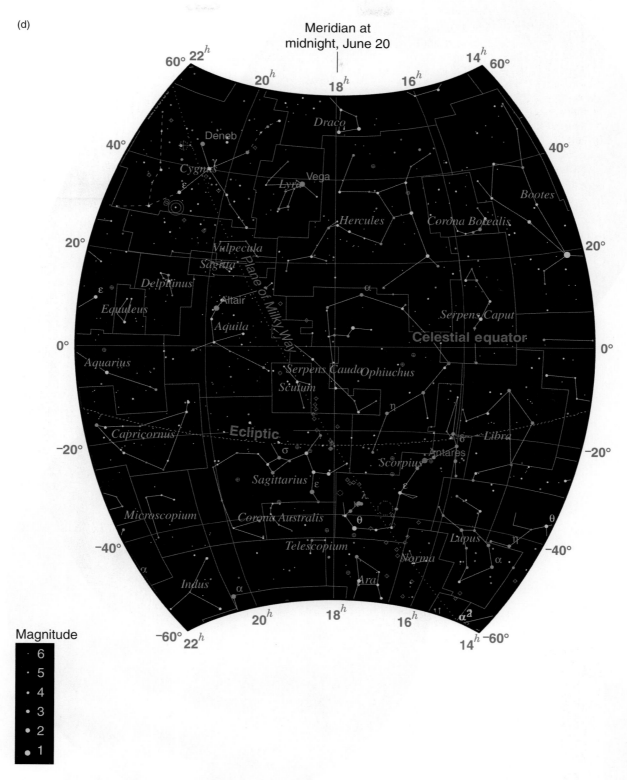

(d)

Meridian at
midnight, June 20

Magnitude
6
5
4
3
2
1

**Figure A7.2D** The sky from right ascension 14ʰ to 22ʰ and declination −60° to +60°.

(e)

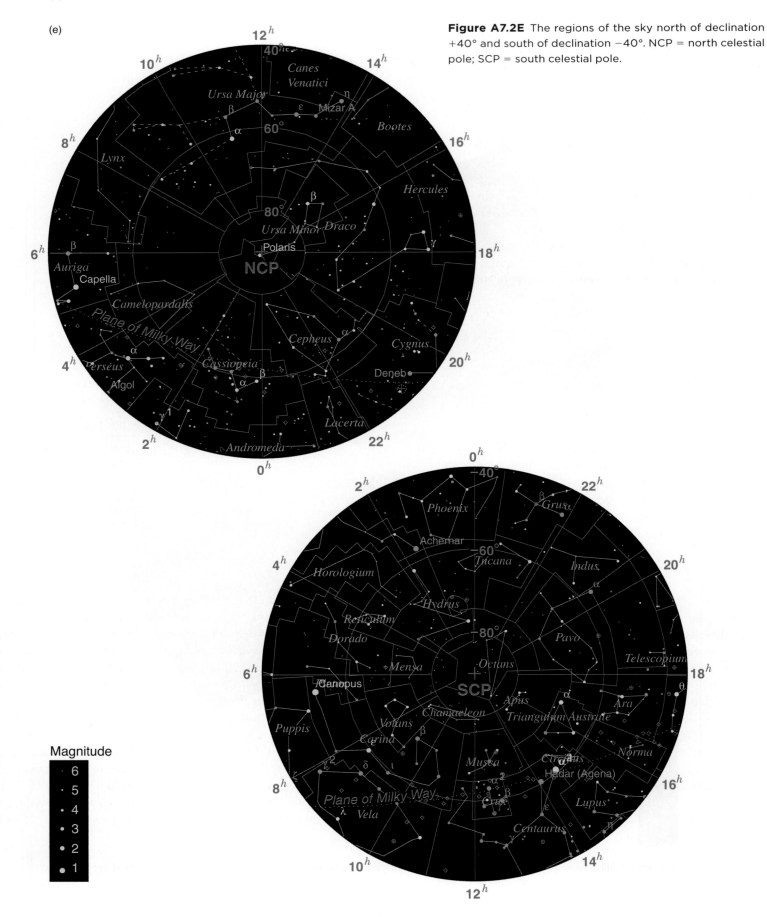

**Figure A7.2E** The regions of the sky north of declination +40° and south of declination −40°. NCP = north celestial pole; SCP = south celestial pole.

combined with the name of the constellation. For example, the star Sirius is the brightest star in the constellation Canis Major (literally, the "great dog"), so it is called "Alpha Canis Majoris." The bright red star in the northeastern corner of the constellation Orion is called "Alpha Orionis," also known as Betelgeuse. Rigel, the bright blue star in the southwest corner of Orion, is also called "Beta Orionis."

Astronomical objects can take on a bewildering range of names. For example, the bright southern star Canopus, also known as "Alpha Carinae" (the brightest star in the constellation Carina), has no fewer than 34 different names, most of which are about as memorable as "SAO 234480" (number 234,480 in the Smithsonian Astrophysical Observatory catalog of stars).

There is a slight difference in the way a constellation is spelled when it becomes part of a star's name. For example, Sirius is called "Alpha Canis Majoris," not "Alpha Canis Major"; Rigel is referred to as "Beta Orionis," not "Beta Orion"; and Canopus becomes "Alpha Carinae," not "Alpha Carina." This is because the Latin genitive, or possessive, case is used with star names; for example, *Orionis* means "of Orion."

## Astronomical Magnitudes

### Apparent Magnitudes

We first introduced magnitudes in Working It Out 13.2; here, we provide some additional information. You are most likely to see this system if you take a lab course in astronomy or if you use a star catalog. Astronomers use the logarithmic system of **apparent magnitudes** to compare the apparent brightness of objects in the sky. Other common systems of logarithmic measurements that you may have encountered include decibels for measuring sound levels and the Richter scale for measuring the strength of earthquakes. For example, an earthquake of magnitude 6 is not just a little stronger than an earthquake of magnitude 5; it is, in fact, 10 times stronger.

As discussed in Working It Out 13.2, a difference of five magnitudes between the apparent brightness of two stars (say, a star with $m = 6$ and a star with $m = 1$), corresponds to 100 times difference in brightness, and *the greater the magnitude, the fainter the object*. If five steps in magnitude correspond to a factor of 100 in brightness, then one step in magnitude must correspond to the fifth root of 100; that is, a factor of $100^{1/5}$ = approximately 2.512 in brightness ($100^{1/5} \times 100^{1/5} \times 100^{1/5} \times 100^{1/5} \times 100^{1/5} = 100$).

If star 1 has a brightness of $b_1$ and star 2 has a brightness of $b_2$, then the ratio of the brightness of the stars is given by

$$\frac{b_1}{b_2} = (2.512)^{m_2 - m_1} = 100^{\frac{(m_2 - m_1)}{5}}$$

We can put this into the more common base 10 by noting that $100 = 10^2$, so this becomes

$$\frac{b_1}{b_2} = 10^{2 \times \frac{(m_2 - m_1)}{5}} = 10^{0.4(m_2 - m_1)}$$

After taking the log of both sides and dividing by 0.4, the difference in magnitude ($m_2 - m_1$) between the two stars is given by

$$m_2 - m_1 = 2.5 \log_{10} \frac{b_1}{b_2}$$

The following table shows some examples using the preceding equations.

| Apparent Magnitude Difference ($m_2 - m_1$) | Ratio of Apparent Brightness ($b_1/b_2$) |
|---|---|
| 1 | 2.512 |
| 2 | $2.512^2 = 6.3$ |
| 3 | $2.512^3 = 15.8$ |
| 4 | $2.512^4 = 39.8$ |
| 5 | $2.512^5 = 100$ |
| 10 | $2.512^{10} = 100^2 = 10,000$ |
| 15 | $2.512^{15} = 100^3 = 1,000,000$ |
| 20 | $2.512^{20} = 100^4 = 10^8$ |
| 25 | $2.512^{25} = 100^5 = 10^{10}$ |

### Absolute Magnitudes

Recall that stars differ in their brightness for two reasons: the amount of light they are actually emitting, and their distance from Earth. The magnitude system is also used for **luminosity**, with the same scale as for brightness: a difference of five magnitudes corresponds to 100 times difference in luminosity. Astronomers call these **absolute magnitudes** ($M_{abs}$), and the idea is to imagine how bright the star would be if it were at a distance of 10 parsecs (pc). Absolute magnitudes enable comparison of how luminous two stars really are, without the factor of distance. The Sun is very bright because it is so close (apparent visual magnitude = −27), but if the Sun were at a distance of 10 pc, its magnitude would be only about 5.[1] Thus, the absolute magnitude of the Sun is $M_{abs} = 5$. Recall that the luminosity of a star is usually expressed by comparing it with the luminosity of the Sun. As with

---

[1]The apparent and absolute magnitudes of the Sun are −26.74 and +4.83, respectively. We use +5 for the Sun's absolute magnitude as an approximation.

apparent magnitudes, higher magnitude numbers corresponding to lower luminosity. Thus, a star that is 100 times less luminous than the Sun will be 5 absolute magnitudes fainter, or $M_{abs} = 10$. A star that is 10,000 times more luminous than the Sun will be 10 absolute magnitudes brighter, or $M_{abs} = -5$.

Absolute magnitudes and luminosities follow the same equations as those we provided already, using $L$ instead of $b$ and $M_{abs}$ instead of $m$:

$$\frac{L_1}{L_2} = 10^{2 \times \frac{(M_{abs(2)} - M_{abs(1)})}{5}} = 10^{0.4(M_{abs(2)} - M_{abs(1)})}$$

and

$$M_{abs(2)} - M_{abs(1)} = 2.5 \log_{10} \frac{L_1}{L_2}$$

Most often, astronomers think about the luminosity of a star compared to the luminosity of the Sun. In this case $L_1 = L_{star}$ and $L_2 = L_{Sun}$. The following table compares luminosity (where $L_{Sun} = 1$) with absolute magnitude of a star.

| $L_{Star}/L_{Sun}$ | $M_{abs}$ |
|---|---|
| 1,000,000 | −10 |
| 10,000 | −5 |
| 100 | 0 |
| 1 | 5 |
| 1/100 | 10 |
| 1/10,000 | 15 |

## Distance Modulus

The difference between the apparent magnitude and the absolute magnitude depends on the star's distance. By definition, a star at a distance of exactly 10 pc will have an apparent magnitude equal to its absolute magnitude. Astronomers can always measure the brightness of a star and thus its apparent magnitude and can estimate the luminosity of a star and thus its absolute magnitude using the Hertzsprung-Russell (H-R) diagram. This is the way the distances to most stars are found.

Using the preceding equations and the definition of absolute magnitude, we can get to the following relatively simple expression:

$$m - M_{abs} = 5 \log_{10} d - 5$$

where distance $d$ is in parsecs.

We can rewrite this equation to solve for distance as follows:

$$d = 10^{(\frac{m - M_{abs} + 5}{5})}$$

The following table shows how the difference between an object's apparent and absolute magnitudes leads to its distance in parsecs.

| $m - M_{abs}$ | Distance (pc) |
|---|---|
| −3 | 2.5 |
| −2 | 4.0 |
| −1 | 6.3 |
| 0 | 10 |
| 1 | 16 |
| 2 | 25 |
| 3 | 40 |
| 4 | 63 |
| 5 | 100 |
| 10 | 1,000 |
| 15 | 10,000 |
| 20 | 100,000 |

Although the system of astronomical magnitudes is convenient in many ways—which is why astronomers continue to use it—it can also be confusing to new students. Just remember three things and you will probably get by:

1. The greater the magnitude, the fainter the object.

2. One magnitude *smaller* means about two and a half times *brighter*.

3. The brightest stars in the sky have magnitudes of less than 1, and the faintest stars that can be seen with the naked eye on a dark night have magnitudes of about 6.

A final note: Astronomers sometimes use "colors" based on the ratio of the brightness of a star as seen in two different parts of the spectrum. The "$b_B/b_V$ color," for example, is the ratio of the brightness of a star seen through a blue filter, divided by the brightness of a star seen through a yellow-green (visual) filter. Normally, astronomers instead discuss the "$B - V$ color" of a star, which is equal to the difference between a star's blue magnitude and its visual magnitude. We can use the previous expression for a magnitude difference to write

$$B - V \text{ color} = m_B - m_V = -2.5 \log_{10} (b_B/b_V)$$

Thus, a star with a $b_B/b_V$ color of 1.0 has a $B - V$ color of 0.0, and a star with a $b_B/b_V$ color of 1.4 has a $B - V$ color of −0.37. Notice that, as with magnitudes, $B - V$ colors are "backward": the bluer a star, the greater its $b_B/b_V$ color but the less its $B - V$ color.

# APPENDIX 8: Uniform Circular Motion and Circular Orbits

## Uniform Circular Motion

In Chapter 4 (see Section 4.2 and Figure 4.7), we discuss the motion of an object moving in a circle at a constant speed. This motion, called **uniform circular motion**, is the result of the fact that centripetal force always acts toward the center of the circle. The key question when thinking about uniform circular motion is, How hard does something have to pull to keep the object moving in a circle? Part of the answer to this question is obvious: the more massive an object is, the harder it will be to keep it moving on its circular path. According to Newton's second law of motion, $F = ma$, or in this case, the centripetal force equals the mass multiplied by the centripetal acceleration. The larger the mass, the greater the force required to keep it moving in its circle.

The centripetal force needed to keep an object moving in constant circular motion also depends on two other quantities: the speed of the object and the size of the circle. The faster an object is moving, the more rapidly it has to change direction to stay on a circle of a given size. The second quantity that influences the needed acceleration is the radius of the circle. The smaller the circle, the greater the pull needed to keep it on track. You can understand this by looking at the motion. A small circle requires a continuous "hard" turn, whereas a larger circle requires a more gentle change in direction. It takes more force to keep an object moving faster in a smaller circle than it does to keep the same object moving more slowly in a larger circle. (To get a better feel for how this works, think about the difference between riding in a car that is taking a tight curve at high speed and a car that is moving slowly around a gentle curve.)

To arrive at the circular velocity and other results discussed in Chapter 4, these intuitive ideas about uniform circular motion are turned into a quantitative expression of exactly how much centripetal acceleration is needed to keep an object moving in a circle with radius $r$ at speed $v$. **Figure A8.1** shows a ball moving around a circle of radius $r$ at a constant speed $v$ at two different times. The centripetal acceleration that is keeping the ball on the circle is $a$. Remember that the acceleration is always directed toward the center of the circle, whereas the velocity of the ball is always perpendicular to the acceleration. The ball's velocity and its acceleration are always at right angles to each other. As the object moves around the circle, the direction of motion and the direction of the acceleration change together in lockstep.

Figure A8.1 contains two triangles. Triangle 1 shows the velocity (speed and direction) at each of the two times. The arrow labeled "$\Delta v$" connecting the heads of the two velocity arrows shows how much the velocity changed between time 1 ($t_1$) and time 2 ($t_2$). This change is the effect of the centripetal acceleration. If you imagine that points 1 and 2 are very close together—so close that the direction of the centripetal acceleration does not change

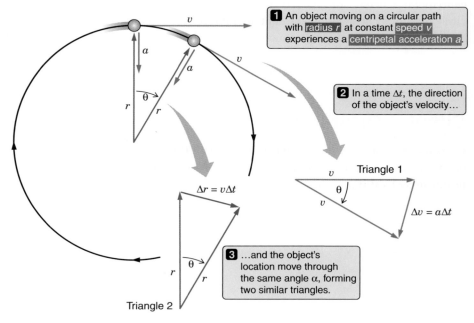

**1** An object moving on a circular path with radius $r$ at constant speed $v$ experiences a centripetal acceleration $a$.

**2** In a time $\Delta t$, the direction of the object's velocity…

Triangle 1

$\Delta v = a\Delta t$

$\Delta r = v\Delta t$

**3** …and the object's location move through the same angle $\alpha$, forming two similar triangles.

Triangle 2

**Figure A8.1** Similar triangles are used to find the centripetal force that is needed to keep an object moving at a constant speed on a circular path.

by much between the two—then the centripetal acceleration equals the change in the velocity divided by the time between the two, $\Delta t = t_2 - t_1$. So, $\Delta v = a\Delta t$.

Triangle 2 shows something similar. Here, the arrow labeled "$\Delta r$" indicates the change in the position of the ball between time 1 and time 2. Again, if you imagine that the time between the two points is very short, $\Delta r$ is equal to the velocity multiplied by the time, or $\Delta r = v\Delta t$.

The line between the center of the circle and the ball is always perpendicular to the velocity of the ball. So if the direction of the ball's velocity changes by an angle $\theta$, then the direction of the line between the ball and the center of the circle must also change by the same angle $\alpha$. In other words, triangles 1 and 2 are "similar triangles." They have the same *shape*. If the triangles are the same shape, the ratio of two sides of triangle 1 must equal the ratio of the two corresponding sides of triangle 2. Then,

$$\frac{a\Delta t}{v} = \frac{v\Delta t}{r}$$

If we divide by $\Delta t$ on both sides of the equation and then cross-multiply, we obtain

$$ar = v^2$$

which, after dividing both sides of the equation by $r$, becomes

$$a_{\text{centripetal}} = \frac{v^2}{r}$$

The subscript "centripetal" is added to $a$ to signify that this is the centripetal acceleration needed to keep the object moving in a circle of radius $r$ at speed $v$. The centripetal force required to keep an object of mass $m$ moving on such a circle is then

$$F_{\text{centripetal}} = ma_{\text{centripetal}} = \frac{mv^2}{r}$$

## Circular Orbits

In the case of an object moving in a circular orbit, there is no string to hold the ball on its circular path. Instead, this force is provided by **gravity**.

Think about an object with mass $m$ in orbit about a much larger object with mass $M$. The orbit is circular, and the distance between the two objects is given by $r$. The force needed to keep the smaller object moving at speed $v$ in a circle with radius $r$ is given by the previous expression for $F_{\text{centripetal}}$. The force actually provided by gravity (see Chapter 4) is

$$F_{\text{grav}} = G\frac{Mm}{r^2}$$

If gravity is responsible for holding the mass in its circular motion, then it should be true that $F_{\text{grav}} = F_{\text{centripetal}}$. That is, if mass $m$ is moving in a circle under the force of gravity, the force provided by gravity must equal the centripetal force needed to explain that circular motion. Setting the two expressions for $F_{\text{centripetal}}$ and $F_{\text{grav}}$ equal to each other gives

$$\frac{mv^2}{r} = G\frac{Mm}{r^2}$$

All that remains is a bit of algebra. Dividing by $m$ on both sides of the equation and multiplying both sides by $r$ gives

$$v^2 = G\frac{M}{r}$$

Taking the square root of both sides then yields the desired result:

$$v_{\text{circular}} = \sqrt{\frac{GM}{r}}$$

This is the **circular velocity** we presented in Chapter 4. It is the velocity at which an object in a circular orbit must be moving. If the object were not moving at this velocity, then gravity would not be providing the needed centripetal force, and the object would not move in a circle.

# APPENDIX 9*: IAU 2006 Resolutions: "Definition of a Planet in the Solar System" and "Pluto"

August 24, 2006, Prague

## Resolutions

Resolution 5 is the principal definition for the IAU usage of "planet" and related terms. Resolution 6 creates, for IAU usage, a new class of objects, for which Pluto is the prototype. The IAU will set up a process to name these objects.

## Resolution 5: Definition of a Planet in the Solar System

Contemporary observations are changing our understanding of planetary systems, and it is important that our nomenclature for objects reflect our current understanding. This applies, in particular, to the designation "planets." The word "planet" originally described "wanderers" that were known only as moving lights in the sky. Recent discoveries lead us to create a new definition, which we can make using currently available scientific information.

The IAU therefore resolves that planets and other bodies, except satellites, in our Solar System be defined into three distinct categories in the following way [**Figure A9.1**]:

(1) A "planet"[1] is a celestial body that

    (a) is in orbit around the Sun,

    (b) has sufficient mass for its self-gravity to overcome rigid body forces so that it assumes a hydrostatic equilibrium (nearly round) shape, and

    (c) has cleared the neighborhood around its orbit.

(2) A "dwarf planet" is a celestial body that

    (a) is in orbit around the Sun,

    (b) has sufficient mass for its self-gravity to overcome rigid body forces so that it assumes a hydrostatic equilibrium (nearly round) shape,[2]

    (c) has not cleared the neighborhood around its orbit, and

    (d) is not a satellite.

(3) All other objects,[3] except satellites, orbiting the Sun shall be referred to collectively as "Small Solar-System Bodies."

## Resolution 6: Pluto

The IAU further resolves:

Pluto is a "dwarf planet" by the above definition and is recognized as the prototype of a new category of Trans-Neptunian Objects.[4]

---

*SOURCE: International Astronomical Union (IAU).
[1]The eight planets are Mercury, Venus, Earth, Mars, Jupiter, Saturn, Uranus, and Neptune.
[2]An IAU process will be established to assign borderline objects to the dwarf planet or to another category.
[3]These currently include most of the Solar System asteroids, most Trans-Neptunian Objects (TNOs), comets, and other small bodies.
[4]An IAU process will be established to select a name for this category.

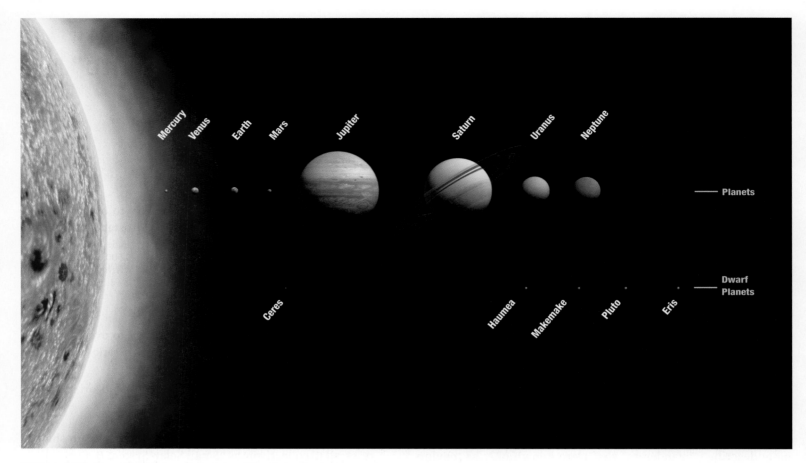

**Figure A9.1** Planets and dwarf planets of the Solar System.

# Glossary

## A

**aberration of starlight** The apparent displacement in the position of a star that is due to the finite speed of light and Earth's orbital motion around the Sun.

**absolute magnitude** A measure of the intrinsic brightness, or luminosity, of a celestial object, generally a star. Specifically, the apparent magnitude an object would have if it were located at a standard distance of 10 parsecs (pc). Compare *apparent magnitude*.

**absolute zero** The temperature at which thermal motions cease. The lowest possible temperature. Zero on the Kelvin temperature scale.

**absorption** The process by which an atom captures energy from a passing photon. Compare *emission*.

**absorption line** A minimum in the intensity of a spectrum that is due to the absorption of electromagnetic radiation at a specific wavelength determined by the energy levels of an atom or molecule. Compare *emission line*.

**acceleration** The rate at which the speed and/or direction of an object's motion is changing.

**accretion disk** A flat, rotating disk of gas and dust surrounding an object, such as a young stellar object, a forming planet, a collapsed star in a binary system, or a black hole.

**achondrite** A stony meteorite that does not contain chondrules. Compare *chondrite*.

**active comet** A comet nucleus that approaches close enough to the Sun to show signs of activity, such as the production of a coma and tail.

**active galactic nucleus (AGN)** A highly luminous, compact galactic nucleus whose luminosity may exceed that of the rest of the galaxy.

**active region** An area of the Sun's chromosphere anchoring bursts of intense magnetic activity.

**adaptive optics** Electro-optical systems that largely compensate for image distortion caused by Earth's atmosphere.

**AGB** See *asymptotic giant branch*.

**AGN** See *active galactic nucleus*.

**albedo** The fraction of electromagnetic radiation striking a surface that is reflected by that surface.

**algebra** A branch of mathematics in which numeric variables are represented by letters.

**alpha particle** A $^4$He nucleus, consisting of two protons and two neutrons. Alpha particles are given off in the type of radioactive decay referred to as *alpha decay*.

**altitude** The location of an object above the horizon, measured by the angle formed between an imaginary line from an observer to the object and a second line from the observer to the point on the horizon directly below the object.

**Amors** A group of asteroids whose orbits cross the orbit of Mars but not the orbit of Earth. Compare *Apollos* and *Atens*.

**amplitude** In a wave, the maximum deviation from its undisturbed or relaxed position. For example, in a water wave the amplitude is the vertical distance from the wave's crest to the undisturbed water level.

**angular momentum** A conserved property of a rotating or revolving system whose value depends on the velocity and distribution of the system's mass.

**angular resolution** The ability of an imaging device such as a telescope (or the eye) to separate two objects that appear close together.

**annular solar eclipse** The type of solar eclipse that occurs when the apparent diameter of the Moon is less than that of the Sun, leaving a visible ring of light ("annulus") surrounding the dark disk of the Moon. Compare *partial solar eclipse* and *total solar eclipse*.

**Antarctic Circle** The circle on Earth with latitude 66.5° south, marking the northern limit where at least one day per year is in 24-hour daylight. Compare *Arctic Circle*.

**anthropic principle** The idea that this universe (or this bubble in the universe) must have physical properties that allow for the development of intelligent life.

**anticyclonic motion** The rotation of a weather system resulting from the Coriolis effect as air moves outward from a region of high atmospheric pressure. Compare *cyclonic motion*.

**antimatter** Matter made up of antiparticles.

**antiparticle** An elementary particle of antimatter identical in mass but opposite in charge and all other properties to its corresponding ordinary matter particle.

**aperture** The clear diameter of a telescope's objective lens or primary mirror.

**aphelion** (pl. **aphelia**) The point in a solar orbit that is farthest from the Sun. Compare *perihelion*.

**Apollos** A group of asteroids whose orbits cross the orbits of both Earth and Mars. Compare *Amors* and *Atens*.

**apparent magnitude** A measure of the apparent brightness of a celestial object, generally a star. Compare *absolute magnitude*.

**arcminute (arcmin)** A minute of arc ('), a unit used for measuring angles. An arcminute is 1/60 of a degree of arc.

**arcsecond (arcsec)** A second of arc ("), a unit used for measuring very small angles. An arcsecond is 1/60 of an arcminute, or 1/3,600 of a degree of arc.

**Arctic Circle** The circle on Earth with latitude 66.5° north, marking the southern limit where at least one day per year is in 24-hour daylight. Compare *Antarctic Circle*.

**asteroid** Also called *minor planet*. A primitive rocky or metallic body (planetesimal) that has survived planetary accretion. Asteroids are parent bodies of meteoroids.

**asteroid belt** The region between the orbits of Mars and Jupiter that contains most of the asteroids in our Solar System.

**astrobiology** An interdisciplinary science combining astronomy, biology, chemistry, geology, and physics to study life in the cosmos.

**astrology** The belief that the positions and aspects of stars and planets influence human affairs and characteristics, as well as terrestrial events.

**astronomical seeing** A measurement of the degree to which Earth's atmosphere degrades the resolution of a telescope's view of astronomical objects.

**astronomical unit (AU)** The average distance from the Sun to Earth: approximately 150 million kilometers (km).

**astronomy** The scientific study of planets, stars, galaxies, and the universe as a whole.

**astrophysics** The application of physical laws to the understanding of planets, stars, galaxies, and the universe as a whole.

**asymptotic giant branch (AGB)** The path on the H-R diagram that goes from the horizontal branch toward higher luminosities and lower temperatures, asymptotically approaching and then rising above the red giant branch.

**Atens** A group of asteroids whose orbits cross the orbit of Earth but not the orbit of Mars. Compare *Amors* and *Apollos*.

**atmosphere** The gravitationally bound, outer gaseous envelope surrounding a planet, moon, or star.

**atmospheric greenhouse effect** A warming of planetary surfaces produced by atmospheric gases that transmit optical solar radiation but partially trap infrared radiation. Compare *greenhouse effect*.

**atmospheric probe** An instrumented package designed to provide on-site measurements of the chemical and/or physical properties of a planetary atmosphere.

**atmospheric window** A region of the electromagnetic spectrum in which radiation is able to penetrate a planet's atmosphere.

**atom** The smallest unit of a chemical element that retains the properties of that element. Each atom is composed of a nucleus (neutrons and protons) surrounded by a cloud of electrons.

**AU** See *astronomical unit*.

**aurora** Emission in the upper atmosphere of a planet from atoms that have been excited by collisions with energetic particles from the planet's magnetosphere.

**autumnal equinox** 1. One of two points where the Sun crosses the celestial equator. 2. The day on which the Sun appears at this location, marking the first day of autumn (about September 22 in the Northern Hemisphere and March 20 in the Southern Hemisphere). Compare *vernal equinox*. See also *summer solstice* and *winter solstice*.

**axion** A hypothetical elementary particle first proposed to explain certain properties of the neutron and now considered a candidate for cold dark matter.

# B

**backlighting** Illumination from behind a subject as seen by an observer. Fine material such as human hair and dust in planetary rings stands out best when viewed under backlighting conditions.

**bar** A unit of pressure. One bar is equivalent to $10^5$ newtons per square meter—approximately equal to Earth's atmospheric pressure at sea level.

**barred spiral galaxy** A spiral galaxy with a bulge having an elongated, barlike shape. Compare *elliptical galaxy*, *irregular galaxy*, *S0 galaxy*, and *spiral galaxy*.

**basalt** Gray to black volcanic rock, rich in iron and magnesium.

**beta decay** 1. The decay of a neutron into a proton by emission of an electron (beta ray) and an antineutrino. 2. The decay of a proton into a neutron by emission of a positron and a neutrino.

**Big Bang** The event that occurred 13.8 billion years ago that marks the beginning of time and the universe.

**Big Bang nucleosynthesis** The formation of low-mass nuclei (H, He, Li, Be) during the first few minutes after the Big Bang.

**Big Crunch** A hypothetical cosmic future in which the expansion of the universe reverses and the universe collapses onto itself.

**Big Rip** A hypothetical cosmic future in which all matter in the universe, from stars to subatomic particles, is progressively torn apart by expansion of the universe.

**binary star** A system in which two stars are in gravitationally bound orbits about their common center of mass.

**binding energy** The minimum energy required to separate an atomic nucleus into its component protons and neutrons.

**biosphere** The global sum of all living organisms on Earth (or any planet or moon). Compare *hydrosphere* and *lithosphere*.

**bipolar outflow** Material streaming away in opposite directions from either side of the accretion disk of a young star.

**black hole** An object so dense that its escape velocity exceeds the speed of light; a *singularity* in spacetime.

**blackbody** An object that absorbs and can reemit all electromagnetic energy it receives.

**blackbody spectrum** See *Planck spectrum*.

**blue-green algae** See *cyanobacteria*.

**blueshift** The Doppler shift toward shorter (bluer) wavelengths of light from an approaching object. Compare *redshift*.

**Bohr model** A model of the atom, proposed by Niels Bohr in 1913, in which a small positively charged nucleus is surrounded by orbiting electrons, similar to a miniature solar system.

**bolide** A very bright, exploding meteor.

**bound orbit** An orbit in which an object is gravitationally bound to the body it is orbiting. A bound orbit's velocity is less than the escape velocity. Compare *unbound orbit*.

**bow shock** The boundary at which the speed of the solar wind abruptly drops from supersonic to subsonic in its approach to a planet's magnetosphere; the boundary between the region dominated by the solar wind and the region dominated by a planet's magnetosphere.

**brightness** The apparent intensity of light from a luminous object. Brightness depends on both the *luminosity* of a source and its distance. Units at the detector: watts per square meter ($W/m^2$).

**brown dwarf** A "failed" star that is not massive enough to cause hydrogen fusion in its core. An object whose mass is intermediate between that of the least massive stars and that of supermassive planets.

**bulge** The central region of a spiral galaxy that is similar in appearance to a small elliptical galaxy.

# C

**C** See *Celsius*.

**C-type asteroid** An asteroid made of material that has remained mostly unmodified since the formation of the Solar System; the most primitive type of asteroid. Compare *M-type asteroid* and *S-type asteroid*.

**caldera** The summit crater of a volcano.

**Cambrian explosion** The spectacular rise in the number and diversity of biological species that occurred between 540 million and 500 million years ago.

**carbon-nitrogen-oxygen (CNO) cycle** One of the ways in which hydrogen is converted to helium (hydrogen burning) in the interiors of main-sequence stars. Compare *proton-proton chain*.

**carbon star** A cool red giant or asymptotic giant branch star that has an excess of carbon in its atmosphere.

**carbonaceous chondrite** A primitive stony meteorite that contains chondrules and is rich in carbon and volatile materials.

**Cassini Division** The largest gap in Saturn's rings, discovered by Jean-Dominique Cassini in 1675.

**catalyst** An atomic and molecular structure that permits or encourages chemical and nuclear reactions but does not change its own chemical or nuclear properties.

**CCD** See *charge-coupled device*.

**celestial equator** The imaginary great circle that is the projection of Earth's equator onto the celestial sphere.

**celestial sphere** An imaginary sphere with celestial objects on its inner surface and Earth at its center. The celestial sphere has no physical existence but is a convenient tool for picturing the directions in which celestial objects are seen from the surface of Earth.

**Celsius (C)** Also called *centigrade scale*. The arbitrary temperature scale, formulated by Anders Celsius (1701–1744), that defines 0°C as the freezing point of water and 100°C as the boiling point of water at sea level. Unit: degrees Celsius (°C). Compare *Fahrenheit* and *Kelvin scale*.

**center of mass** 1. The weighted average location of all the mass in a system of objects. The point in any isolated system that moves according to Newton's first law of motion. 2. In a binary star system, the point between the two stars that is the focus of both their elliptical orbits.

**centigrade scale** See *Celsius*.

**centripetal force** A force directed toward the center of curvature of an object's curved path.

**Cepheid variable** An evolved high-mass star with an atmosphere that is pulsating, leading to variability in the star's luminosity and color.

**Chandrasekhar limit** The upper limit on the mass of an object supported by electron degeneracy pressure; approximately 1.4 solar masses ($M_{Sun}$).

**chaotic** Behavior in complex, interrelated systems in which a small change in the initial state of a system can lead to a large change in the final state of the system.

**charge-coupled device (CCD)** A common type of solid-state detector of electromagnetic radiation that transforms the intensity of light directly into electric signals.

**chondrite** A stony meteorite that contains chondrules. Compare *achondrite*.

**chondrule** A small, crystallized, spherical inclusion of rapidly cooled molten droplets found inside some meteorites.

**chromatic aberration** A detrimental property of a lens in which rays of different wavelengths are brought to different focal distances from the lens.

**chromosphere** The region of the Sun's atmosphere located between the *photosphere* and the *corona*.

**circular velocity** The orbital velocity needed to keep an object moving in a circular orbit.

**circumpolar** Describing the part of the sky, near either celestial pole, that can always be seen above the horizon from a specific location on Earth.

**circumstellar disk** See *protoplanetary disk*.

**climate** The state of an atmosphere averaged over an extended time. Compare *weather*.

**closed universe** A finite universe with a curved spatial structure such that the sum of the angles of a triangle always exceeds 180 degrees. Compare *flat universe* and *open universe*.

**CMB** See *cosmic microwave background radiation*.

**CNO cycle** See *carbon-nitrogen-oxygen cycle*.

**cold dark matter** Particles of dark matter that move slowly enough to be gravitationally bound even in the smallest galaxies. Compare *hot dark matter*.

**color index** The color of a celestial object, generally a star, based on the ratio of its brightness in blue light to its brightness in "visual" (yellow-green) light. The difference between an object's blue ($B$) magnitude and visual ($V$) magnitude, $B - V$.

**coma** (pl. **comae**) The nearly spherical cloud of gas and dust surrounding the nucleus of an active comet.

**comet** A complex object consisting of a small, solid, icy nucleus; an atmospheric halo; and a tail of gas and dust.

**comet nucleus** A primitive planetesimal composed of ices and refractory materials that has survived planetary accretion. The "heart" of a comet, containing nearly the entire mass of the comet. A "dirty snowball."

**comparative planetology** The study of planets through comparison of their chemical and physical properties.

**composite volcano** A large, cone-shaped volcano formed by viscous, pasty lava flows alternating with pyroclastic (explosively generated) rock deposits. Compare *shield volcano*.

**compound lens** A lens made up of two or more elements of differing refractive index, the purpose of which is to minimize chromatic aberration.

**concave mirror** A telescope mirror with a surface that curves inward toward the incoming light.

**conduction** The transfer of energy in which the thermal energy of particles is transferred to adjacent particles by collisions or other interactions. Conduction is the most important way that thermal energy is transported in solid matter. Compare *convection*.

**conservation law** A physical law stating that the amount of a particular physical quantity (such as energy or angular momentum) of an isolated system does not change over time.

**conservation of angular momentum** The physical law stating that the amount of angular momentum of an isolated system does not change over time.

**conservation of energy** The physical law stating that the amount of energy of an isolated, closed system does not change over time.

**constant of proportionality** The number by which one quantity is multiplied to get another number.

**constellation** An imaginary image formed by patterns of stars; any of 88 defined areas on the celestial sphere used by astronomers to locate celestial objects.

**constructive interference** A state in which the amplitudes of two intersecting waves reinforce one another. Compare *destructive interference*.

**continental drift** The slow motion (centimeters per year) of Earth's continents relative to each other and to Earth's mantle. See also *plate tectonics*.

**continuous radiation** Electromagnetic radiation with intensity that varies smoothly over a wide range of wavelengths.

**continuous spectrum** A spectrum containing all wavelengths, without specific spectral lines.

**convection** The transport of thermal energy from the lower (hotter) to the higher (cooler) layers of a fluid by motions within the fluid driven by variations in buoyancy. Compare *conduction*.

**convective zone** A region within a star where energy is transported outward by convection. Compare *radiative zone*.

**co-orbital moons** Moons that occupy the same orbit.

**core** 1. The innermost region of a planetary interior. Compare *crust* and *mantle*. 2. The innermost part of a star.

**core accretion–gas capture** A process for forming giant planets, in which large amounts of surrounding hydrogen and helium gas are gravitationally captured onto a massive rocky core.

**Coriolis effect** The apparent displacement of objects in a direction perpendicular to their true motion as viewed from a rotating frame of reference. On a rotating planet, different latitudes rotating at different speeds cause this effect.

**corona** The hot, outermost part of the Sun's atmosphere. Compare *chromosphere* and *photosphere*.

**coronal hole** A low-density region in the solar corona containing "open" magnetic-field lines along which coronal material is free to stream into interplanetary space.

**coronal mass ejection (CME)** An eruption on the Sun that ejects hot gas and energetic particles at much higher speeds than are typical in the solar wind.

**cosmic microwave background radiation (CMB)** Also called simply *cosmic background radiation*. Isotropic microwave radiation from every direction in the sky having a 2.73-kelvin (K) blackbody spectrum. The CMB is residual radiation from the Big Bang.

**cosmic ray** A very fast-moving particle (usually an atomic nucleus) that originated in outer space; cosmic rays fill the disk of the Milky Way.

**cosmological constant** A constant, introduced into general relativity by Einstein, that characterizes an extra, repulsive force in the universe due to the vacuum of space itself.

**cosmological principle** The (testable) assumption that the same physical laws that apply here and now also apply everywhere and at all times, and that there are no special locations or directions in the universe.

**cosmological redshift** The redshift that results from the expansion of the universe rather than from the motions of galaxies or gravity. Compare *gravitational redshift*.

**cosmology** The study of the large-scale structure and evolution of the universe as a whole.

**crescent** Any phase of the Moon, Mercury, or Venus in which the object appears less than half illuminated by the Sun. Compare *gibbous*.

**Cretaceous-Tertiary (K-T) boundary** The boundary between the Cretaceous and Tertiary periods in Earth's history. This boundary corresponds to the time of the impact of an asteroid or comet and the extinction of the dinosaurs.

**critical density** The value of mass density of the universe that, ignoring any cosmological constant, is just barely capable of halting expansion of the universe.

**crust** The relatively thin, outermost, hard layer of a planet, which is chemically distinct from the interior. Compare *core* and *mantle*.

**cryovolcanism** Low-temperature volcanism in which the magmas are composed of molten ices rather than rocky material.

**cyanobacteria** Also called *blue-green algae*. Single-celled organisms that were responsible for creating oxygen in Earth's atmosphere by photosynthesizing carbon dioxide and releasing oxygen as a waste product.

**cyclonic motion** The rotation of a weather system resulting from the Coriolis effect as air moves toward a region of low atmospheric pressure. Compare *anticyclonic motion*.

# D

**Dark Ages** The epoch in the history of the universe during which there was no visible "light" from astronomical objects.

**dark energy** A form of energy that permeates all of space (including the vacuum), producing a repulsive force that accelerates the expansion of the universe.

**dark matter** Matter in galaxies that does not emit or absorb electromagnetic radiation. Dark matter is thought to constitute most of the mass in the universe. Compare *luminous matter*.

**dark matter halo** The centrally condensed, greatly extended dark matter component of a galaxy that accounts for up to 95 percent of the galaxy's mass.

**daughter product** An element resulting from radioactive decay of a more massive *parent element*.

**decay** 1. The process of a radioactive nucleus changing into its daughter product. 2. The process of an atom or molecule dropping from a higher energy state to a lower energy state. 3. The process of a satellite's orbit losing energy.

**declination** A measure, analogous to *latitude*, that tells you the angular distance of a celestial body north or south of the celestial equator (from 0° to ±90°). Compare *right ascension*.

**density** The measure of an object's mass per unit of volume. Possible units include kilograms per cubic meter ($kg/m^3$).

**destructive interference** A state in which the amplitudes of two intersecting waves cancel one another. Compare *constructive interference*.

**differential rotation** Rotation of different parts of a system at different rates.

**differentiation** The process by which materials of higher density sink toward the center of a molten or fluid planetary interior.

**diffraction** The spreading of a wave after it passes through an opening or beyond the edge of an object.

**diffraction grating** An optical component with many narrow parallel rules lines that separates the wavelengths of light to produce a spectrum.

**diffraction limit** The limit of a telescope's angular resolution caused by diffraction.

**diffuse ring** A sparsely populated planetary ring spread out both horizontally and vertically.

**dispersion** The separation of rays of light into their component wavelengths.

**distance ladder** A sequence of techniques for measuring cosmic distances: each method is calibrated using the results from other methods that have been applied to closer objects.

**Doppler effect** The change in wavelength of sound or light that is due to the relative motion of the source toward or away from the observer.

**Doppler shift** The amount by which the wavelength of light is shifted by the Doppler effect.

**Drake equation** A prescription for estimating the number of intelligent civilizations existing in the Milky Way Galaxy.

**dust devil** A small tornado-like column of air containing dust or sand.

**dust tail** A type of comet tail consisting of dust particles that are pushed away from the comet's head by radiation pressure from the Sun. Compare *ion tail*.

**dwarf galaxy** A small galaxy with a luminosity ranging from 1 million to 1 billion solar luminosities ($L_{Sun}$). Compare *giant galaxy*.

**dwarf planet** A body with characteristics similar to those of a planet except that it has not cleared smaller bodies from the neighboring regions around its orbit. Compare *planet* (definition 2).

**dynamic equilibrium** A state in which a system is constantly changing but its configuration remains the same because one source of change is exactly balanced by another source of change. Compare *static equilibrium*.

**dynamo theory** A theory postulating that Earth's magnetic field (and those of other planets) is generated from a rotating and electrically conducting liquid core.

# E

**Earth Similarity Index** (**ESI**) An index for quantifying the habitability of extrasolar planets in which currently available data on an extrasolar planet are used to estimate how much it is like Earth. Factors include the radius, density, escape velocity, and surface temperature. Compare *Planetary Habitability Index*.

**eccentricity (*e*)** The ratio of the distance between the two foci of an ellipse to the length of its major axis, which measures how noncircular the ellipse is.

**eclipse** 1. The total or partial obscuration of one celestial body by another. 2. The total or partial obscuration of light from one celestial body as it passes through the shadow of another celestial body.

**eclipse season** Any time during the year when the Moon's line of nodes is sufficiently close to the Sun for eclipses to occur.

**eclipsing binary** A binary system in which the orbital plane is oriented such that the two stars appear to pass in front of one another as seen from Earth. Compare *spectroscopic binary* and *visual binary*.

**ecliptic** 1. The apparent annual path of the Sun against the background of stars. 2. The projection of Earth's orbital plane onto the celestial sphere.

**ecliptic plane** The plane of Earth's orbit around the Sun. The ecliptic is the projection of this plane on the celestial sphere.

**effective temperature** The temperature at which a blackbody, such as a star, appears to radiate.

**Einstein ring** Light bent by gravitational lensing into a ring.

**ejecta** 1. Material thrown outward by the impact of an asteroid or comet on a planetary surface, leaving a crater behind. 2. Material thrown outward by a stellar explosion.

**electric field** A field that is able to exert a force on a charged object, whether at rest or moving. Compare *magnetic field*.

**electric force** The force exerted on electrically charged particles such as protons and electrons, arising from their electric charges. Compare *magnetic force*. See also *electromagnetic force*.

**electromagnetic force** The force, including both electric and magnetic forces, that acts on electrically charged particles. One of four fundamental forces of nature, along with the *strong nuclear force*, *weak nuclear force*, and *gravity* (definition 1). The force is mediated by the exchange of photons.

**electromagnetic radiation** A traveling disturbance in the electric and magnetic fields caused by accelerating electric charges. In quantum mechanics, a stream of photons. Light.

**electromagnetic spectrum** The spectrum made up of all possible frequencies or wavelengths of electromagnetic radiation, ranging from gamma rays through radio waves and including the portion our eyes can use.

**electromagnetic wave** A wave consisting of oscillations in the electric-field strength and the magnetic-field strength.

**electron (e⁻)** A subatomic particle having a negative electric charge of $1.6 \times 10^{-19}$ coulomb (C), a rest mass of $9.1 \times 10^{-31}$ kilogram (kg), and a rest energy of $8 \times 10^{-14}$ joule (J). The antiparticle of the *positron*. Compare *proton* and *neutron*.

**electron-degenerate** Describing matter, compressed to the point at which electron density reaches the limit imposed by the rules of quantum mechanics.

**electroweak theory** The quantum theory that combines descriptions of both the electromagnetic force and the weak nuclear force.

**element** One of 92 naturally occurring substances (such as hydrogen, oxygen, and uranium) and more than 20 human-made ones (such as plutonium). Each element is chemically defined by the specific number of protons in the nuclei of its atoms.

**elementary particle** One of the basic building blocks of nature that is not known to have substructure, such as the *electron* or the *quark*.

**ellipse** A conic section produced by the intersection of a plane with a cone when the plane is passed through the cone at an angle to the axis other than 0° or 90°. The shape that results when you attach the two ends of a piece of string to a piece of paper, stretch the string tight with the tip of a pencil, and then draw around those two points while keeping the string taut.

**elliptical galaxy** A galaxy of Hubble type "E" class, with a circular to elliptical outline on the sky, and containing almost no disk and a population of old stars. Compare *barred spiral galaxy*, *irregular galaxy*, *S0 galaxy*, and *spiral galaxy*.

**emission** The production of a photon when an atom decays to a lower energy state. Compare *absorption*.

**emission line** A peak in the intensity of a spectrum that is due to the emission of electromagnetic radiation at a specific wavelength determined by the energy levels of an atom or molecule. Compare *absorption line*.

**empirical science** Scientific investigation that is based primarily on observations and experimental data. It is descriptive rather than based on theoretical inference.

**energy** The conserved quantity that gives objects and systems the ability to do work. Possible units include: joules (J).

**energy transport** The transfer of energy from one location to another. In stars, energy transport is carried out mainly by radiation or convection.

**entropy** A measure of the disorder of a system related to the number of ways a system can be rearranged without its appearance being affected.

**equator** The imaginary great circle on the surface of a body midway between its poles that divides the body into northern and southern hemispheres. The equatorial plane passes through the center of the body and is perpendicular to its rotation axis. Compare *meridian*.

**equilibrium** The state of an object in which physical processes balance each other so that its properties or conditions remain constant.

**equinox** Literally, "equal night." 1. One of two positions on the ecliptic where it intersects the celestial equator. 2. Either of the two times of year (the *autumnal equinox* and *vernal equinox*) when the Sun is at one of these two positions. At this time, night and day are of the same length everywhere on Earth. Compare *solstice*.

**equivalence principle** The principle stating that there is no difference between a frame of reference that is freely floating through space and one that is freely falling within a gravitational field.

**erosion** The degradation of a planet's surface topography by the mechanical action of wind and/or water.

**escape velocity** The minimum velocity needed for an object to achieve a parabolic trajectory and thus permanently leave the gravitational grasp of another mass.

**ESI** See *Earth Similarity Index*.

**eternal inflation** The idea that a universe might inflate forever. In such a universe, quantum effects could randomly cause regions to slow their expansion, eventually stop inflating, and experience an explosion resembling the Big Bang.

**event** A particular location in spacetime.

**event horizon** The effective "surface" of a black hole. Nothing inside this surface—not even light—can escape from a black hole.

**evolutionary track** The path that a star follows across the H-R diagram as it evolves through its lifetime.

**excited state** Any energy level of a system or part of a system, such as an atom, molecule, or particle, that is higher than its ground state. Compare *ground state*.

**exoplanet** See *extrasolar planet*.

**exosphere** A very thin atmosphere or layer of atmosphere, where the molecules are bound by gravity to the moon or planet but their density is too low to behave like a gas of colliding particles

**extrasolar planet** Also called *exoplanet*. A planet orbiting a star other than the Sun.

**extremophile** A life-form that thrives under extreme environmental conditions.

**eyepiece** A lens that is closest to the eye in a telescope. Changing the eyepiece will change the magnification of the image in the telescope.

# F

**F** See *Fahrenheit*.

**Fahrenheit (F)** The arbitrary temperature scale, formulated by Daniel Gabriel Fahrenheit (1686–1736), that defines 32°F as the melting point of water and 212°F as the boiling point of water at sea level. Unit: degrees Fahrenheit (°F). Compare *Celsius* and *Kelvin scale*.

**falsified** A hypothesis that is shown to be false.

**fault** A fracture in the crust of a planet or moon along which blocks of material can slide.

**filter** An instrument element that transmits a limited wavelength range of electromagnetic radiation. For the optical range, such elements are typically made of different kinds of glass and take on the hue of the light they transmit.

**first quarter Moon** The phase of the Moon in which only the western half of the Moon, as viewed from Earth, is illuminated by the Sun. It occurs about a week after a new Moon. Compare *third quarter Moon*. See also *full Moon* and *new Moon*.

**fissure** A fracture in the planetary lithosphere from which magma emerges.

**flat rotation curve** A rotation curve of a spiral galaxy in which rotation rates do not decline in the outer part of the galaxy but remain relatively constant to the outermost points.

**flat universe** An infinite universe whose spatial structure obeys Euclidean geometry, such that the sum of the angles of a triangle always equals 180 degrees. Compare *closed universe* and *open universe*.

**flatness problem** The surprising result that the sum of $\Omega_m$ plus $\Omega_\Lambda$ is extremely close to 1 in the present-day universe; equivalent to saying that it is surprising the universe is so close to being exactly flat.

**flux** The total amount of energy passing through each square meter of a surface each second. Unit: watts per square meter ($W/m^2$).

**flux tube** A strong magnetic field contained within a tubelike structure. Flux tubes are found in the solar atmosphere and connecting the space between Jupiter and its moon Io.

**flyby** A spacecraft that first approaches and then continues flying past a planet or moon. Flybys can visit multiple objects, but they remain in the vicinity of their targets only briefly. Compare *orbiter*.

**focal length** The optical distance between a telescope's objective lens or primary mirror and the plane (called the focal plane) on which the light from a distant object is focused.

**focal plane** The plane, perpendicular to the optical axis of a lens or mirror, on which an image is formed.

**focus** (pl. **foci**) 1. One of two points that define an ellipse. 2. A point in the focal plane of a telescope.

**force** A push or a pull on an object.

**frame of reference** A coordinate system within which an observer measures positions and motions.

**free fall** The motion of an object when the only force acting on it is gravity.

**frequency** The number of times per second that a periodic process occurs. Unit: hertz (Hz), or cycles per second (1/s).

**full Moon** The phase of the Moon in which the near side of the Moon, as viewed from Earth, is fully illuminated by the Sun. It occurs about two weeks after a *new Moon*. See also *first quarter Moon* and *third quarter Moon*.

# G

**galaxy** A gravitationally bound system that consists of stars and star clusters, gas, dust, and dark matter; typically greater than 1,000 light-years across and recognizable as a discrete, single object.

**galaxy cluster** A large, gravitationally bound collection of galaxies containing hundreds to thousands of members; typically 3–5 megaparsecs (Mpc) across. Compare *galaxy group* and *supercluster*.

**galaxy group** A small, gravitationally bound collection of galaxies containing from several to a hundred members; typically 1–2 megaparsecs (Mpc) across. Compare *galaxy cluster* and *supercluster*.

**gamma ray** Also called *gamma radiation*. Electromagnetic radiation with higher frequency, higher photon energy, and shorter wavelength than all other types of electromagnetic radiation.

**gamma-ray burst (GRB)** A brief, intense burst of gamma rays from a distant energetic explosion.

**gas giant** A giant planet formed mostly of hydrogen and helium. In the Solar System, Jupiter and Saturn are the gas giants. Compare *ice giant*.

**gauss** A basic unit measuring the strength of a magnetic field.

**general relativistic time dilation** The verified prediction that time passes more slowly in a gravitational field than in the absence of a gravitational field. Compare *time dilation*.

**general relativity** See *general theory of relativity*.

**general theory of relativity** Sometimes referred to as simply *general relativity*. Einstein's theory explaining gravity as the distortion of spacetime by massive objects, such that particles travel on the shortest path between two events in spacetime. This theory deals with all types of motion. Compare *special theory of relativity*.

**geocentric model** A historical cosmological model with Earth at its center, and all of the other objects in the universe in orbit around Earth. Compare *heliocentric model*.

**geodesic** The path an object will follow through spacetime in the absence of external forces.

**geometry** A branch of mathematics that deals with points, lines, angles, and shapes.

**giant galaxy** A galaxy with luminosity greater than about 1 billion solar luminosities ($L_{Sun}$). Compare *dwarf galaxy*.

**giant molecular cloud** An interstellar cloud composed primarily of molecular gas and dust, having hundreds of thousands of solar masses.

**giant planet** Also called *Jovian planet*. One of the largest planets in the Solar System (Saturn, Jupiter, Uranus, or Neptune), typically 10 times the size and many times the mass of any *terrestrial planet* and lacking a solid surface.

**gibbous** Any phase of the Moon, Mercury, or Venus in which the object appears more than half illuminated by the Sun. Compare *crescent*.

**global circulation** The overall, planetwide circulation pattern of a planet's atmosphere.

**globular cluster** A spherically symmetric, highly condensed group of stars, containing tens of thousands to a million members. Compare *open cluster*.

**gluon** The particle that carries (or, equivalently, mediates) interactions due to the strong nuclear force.

**grand unified theory (GUT)** A unified quantum theory that combines the strong nuclear, weak nuclear, and electromagnetic forces but does not include gravity.

**granite** Rock that is cooled from magma and is relatively rich in silicon and oxygen.

**grating** An optical surface containing many narrow, closely and equally spaced parallel grooves or slits that spectrally disperse reflected or transmitted light.

**gravitational lens** A massive object that gravitationally focuses the light of a more distant object to produce multiple brighter, magnified, possibly distorted images.

**gravitational lensing** The bending of light by gravity.

**gravitational potential energy** The stored energy in an object that is due solely to its position within a gravitational field.

**gravitational redshift** The shifting to longer wavelengths of radiation from an object deep within a gravitational well. Compare *cosmological redshift*.

**gravitational wave** A wave in the fabric of spacetime emitted by accelerating masses.

**gravity** 1. The mutually attractive force between massive objects. One of four fundamental forces of nature, along with the *electromagnetic force*, the *strong nuclear force*, and the *weak nuclear force*. 2. An effect arising from the bending of spacetime by massive objects.

**GRB** See *gamma-ray burst*.

**great circle** Any circle on a sphere that has as its center the center of the sphere. The celestial equator, the meridian, and the ecliptic are all great circles on the sphere of the sky, as is any circle drawn through the zenith.

**Great Red Spot** The giant, oval, brick red anticyclone seen in Jupiter's southern hemisphere.

**greenhouse effect** The solar heating of air in an enclosed space, such as a closed building or car, resulting primarily from the inability of the hot air to escape. Compare *atmospheric greenhouse effect*.

**greenhouse gas** One of a group of atmospheric gases such as carbon dioxide that are transparent to visible radiation but absorb infrared radiation.

**greenhouse molecule** A molecule such as water vapor or carbon dioxide that transmits visible radiation but absorbs infrared radiation.

**Gregorian calendar** The modern calendar. A modification of the Julian calendar decreed by Pope Gregory XIII in 1582. By that time, the less accurate Julian calendar had developed an error of 10 days over the 13 centuries since its inception.

**ground state** The lowest possible energy state for a system or part of a system, such as an atom, molecule, or particle. Compare *excited state*.

**GUT** See *grand unified theory*.

# H

**H II region** A region of interstellar gas that has been ionized by ultraviolet radiation from nearby hot, massive stars.

**H-R diagram** The Hertzsprung-Russell diagram, a plot of the luminosities versus the surface temperatures of stars. The evolving properties of stars are plotted as tracks across the H-R diagram.

**habitable zone** The distance from its star at which a planet must be located in order to have a temperature suitable for water to exist in a liquid state.

**Hadley circulation** A simplified, and therefore uncommon, atmospheric global circulation that carries thermal energy directly from the equator to the polar regions of a planet.

**half-life** The time it takes for half a sample of a particular radioactive parent element to decay into a daughter product.

**halo** The spherically symmetric, low-density distribution of stars and dark matter that defines the outermost regions of a galaxy.

**harmonic law** See *Kepler's third law*.

**Hawking radiation** Radiation from a black hole.

**Hayashi track** The path that a protostar follows on the H-R diagram as it contracts toward the main sequence.

**head** The part of a comet that includes both the nucleus and the inner part of the coma.

**heat death** The possible eventual fate of an open universe, in which entropy has triumphed and all energy- and structure-producing processes have come to an end.

**heavy element** Also called *massive element*. Any element more massive than helium.

**Heisenberg uncertainty principle** The physical limitation that the product of the position and the momentum of a particle cannot be smaller than a well-defined value, Planck's constant ($h$).

**heliocentric model** A model of the Solar System, with the Sun at its center, and the planets, including Earth, in orbit around the Sun. Compare *geocentric model*.

**helioseismology** The use of solar oscillations to study the interior of the Sun.

**heliosphere** A region surrounding the Solar System in which the solar wind blows against the interstellar medium and clears out an area like the inside of a bubble. The heliosphere protects the Solar System from cosmic rays.

**helium flash** The runaway explosive burning of helium in the degenerate helium core of a red giant star.

**Herbig-Haro (HH) object** A glowing, rapidly moving knot of gas and dust that is excited by bipolar outflows in very young stars.

**heredity** The process by which one generation passes on its characteristics to future generations.

**hertz (Hz)** A unit of frequency equivalent to cycles per second.

**Hertzsprung-Russell diagram** See *H-R diagram*.

**HH object** See *Herbig-Haro object*.

**hierarchical clustering** The "bottom-up" process of forming large-scale structure. Small-scale structure first produces groups of galaxies, which in turn form clusters, which then form superclusters.

**high-mass star** A star with a main-sequence mass of greater than about 8 solar masses ($M_{Sun}$). Compare *low-mass star*.

**homogeneous** In cosmology, describing a universe in which observers in any location would observe the same properties. Compare *isotropic*.

**horizon** The boundary that separates the sky from the ground.

**horizon problem** The puzzling observation that the cosmic background radiation is so uniform in all directions, even though widely separated regions should have been "over the horizon" from each other in the early universe.

**horizontal branch** A region on the H-R diagram defined by stars burning helium to carbon in a stable core.

**hot dark matter** Particles of dark matter that move so fast that gravity cannot confine them to the volume occupied by a galaxy's normal luminous matter. Compare *cold dark matter*.

**hot Jupiter** A large, Jupiter-type extrasolar planet located very close to its parent star.

**hot spot** A place where hot plumes of mantle material rise near the surface of a planet.

**Hubble constant ($H_0$)** The constant of proportionality relating the recession velocities of galaxies to their distances. Compare *Hubble time*.

**Hubble time** An estimate of the age of the universe from the inverse of the *Hubble constant*, $1/H_0$.

**Hubble's law** The law stating that the speed at which a galaxy is moving away from Earth is proportional to the distance of that galaxy.

**hurricane** A large tropical cyclonic system circulating counterclockwise in the Northern Hemisphere and clockwise in the Southern Hemisphere. Hurricanes can extend outward from their center to more than 600 kilometers (km) and generate winds in excess of 300 kilometers per hour (km/h).

**hydrogen burning** The release of energy from the nuclear fusion of four hydrogen atoms into a single helium atom.

**hydrogen shell burning** The fusion of hydrogen in a shell surrounding a stellar core that may be either degenerate or fusing more massive elements.

**hydrosphere** The portion of Earth that is largely liquid water. Compare *biosphere* and *lithosphere*.

**hydrostatic equilibrium** The condition in which the weight bearing down at a particular point within an object is balanced by the pressure within the object.

**hypernova** (pl. **hypernovae**) A very energetic supernova from a very high-mass star.

**hypothesis** A well-considered idea, based on scientific principles and knowledge, that leads to testable predictions. Compare *theory*.

**Hz** See *hertz*.

# I

**ice** The solid form of a volatile material; sometimes the *volatile material* itself, regardless of its physical form.

**ice giant** A giant planet formed mostly of the liquid form of volatile substances (ices). In the Solar System, Uranus and Neptune are the ice giants. Compare *gas giant*.

**ideal gas law** The relationship of pressure ($P$) to density of particles ($n$) and temperature ($T$) expressed as $P = nkT$, where $k$ is Boltzmann's constant.

**igneous activity** The formation and action of molten rock (magma).

**impact crater** The scar of the impact left on a solid planetary or moon surface by collision with another object. Compare *secondary crater*.

**impact cratering** The process in which solid planetary objects collide with each other, leaving distinctive scars.

**index of refraction (*n*)** The ratio of the speed of light in a vacuum (*c*) to the speed of light in an optical medium (*v*).

**inert gas** A gaseous element that combines with other elements only under conditions of extreme temperature and pressure. Examples include helium, neon, and argon.

**inertia** The tendency for objects to retain their state of motion.

**inertial frame of reference** 1. A frame of reference that is moving in a straight line at constant speed; that is, not accelerating. 2. In general relativity, a frame of reference that is falling freely in a gravitational field.

**inertial reference frame** See *inertial frame of reference*.

**inferior planet** A Solar System planet that orbits the Sun at a closer distance than Earth's orbit. *See superior planet*.

**inflation** An extremely brief phase of ultra-rapid expansion of the very early universe. After inflation, the standard Big Bang models of expansion apply.

**infrared (IR) radiation** Electromagnetic radiation with frequencies, photon energies, and wavelengths between those of visible light and microwaves.

**instability strip** A region of the H-R diagram containing stars that pulsate with a periodic variation in luminosity.

**integration time** The time interval during which photons are collected and added up in a detecting device.

**intensity** Of light, the amount of radiant energy emitted per second per unit area. Units for electromagnetic radiation: watts per square meter ($W/m^2$).

**intercloud gas** A low-density region of the interstellar medium that fills the space between interstellar clouds.

**interference** The interaction of two sets of waves producing high and low intensity, depending on whether their amplitudes reinforce (*constructive interference*) or cancel (*destructive interference*).

**interferometer** Linked optical or radio telescopes whose overall separation determines the angular resolution of the system.

**interferometric array** An interferometer that is made up of several telescopes arranged in an array.

**interstellar cloud** A discrete, high-density region of the interstellar medium made up mostly of atomic or molecular hydrogen and dust.

**interstellar dust** Small particles or grains (0.01–10 microns [$\mu$m] in diameter) of matter, primarily carbon and silicates, distributed throughout interstellar space.

**interstellar extinction** The dimming of visible and ultraviolet light by interstellar dust.

**interstellar gas** The tenuous gas, far less dense than air, composing 99 percent of the matter in the interstellar medium.

**interstellar medium** The gas and dust that fill the space between the stars within a galaxy.

**inverse square law** The rule stating that a quantity or effect diminishes with the square of the distance from the source.

**ion** An atom or molecule that has lost or gained one or more electrons.

**ionize** see *ionization*

**ionization** The process by which electrons are stripped free from an atom or molecule, resulting in free electrons and a positively charged atom or molecule.

**ionosphere** A layer high in Earth's atmosphere in which most of the atoms are ionized by solar radiation.

**ion tail** A type of comet tail consisting of ionized gas. Particles in the ion tail are pushed directly away from the comet's head in the antisolar direction at high speeds by the solar wind. Compare *dust tail*.

**IR** Infrared. See *infrared radiation*.

**iron meteorite** A metallic meteorite composed mostly of iron-nickel alloys. Compare *stony-iron meteorite* and *stony meteorite*.

**irregular galaxy** A galaxy without regular or symmetric appearance. Compare *barred spiral galaxy*, *elliptical galaxy*, *S0 galaxy*, and *spiral galaxy*.

**irregular moon** A moon that has been captured by a planet rather than having formed along with that planet. Some irregular moons revolve in a direction opposite to the rotation of the planet, and many are in distant, unstable orbits. Compare *regular moon*.

**isotopes** Forms of the same chemical element that have the same number of protons but a different number of neutrons.

**isotropic** In cosmology, having the same appearance to an observer in all directions. Compare *homogeneous*.

## J

**J** See *joule*.

**jansky (Jy)** The basic unit of flux density. Unit: watts per square meter per hertz ($W/m^2/Hz$).

**jet** 1. A stream of gas and dust ejected from a comet nucleus by solar heating. 2. A stream of material that moves away from a protostar or active galactic nucleus at hundreds of kilometers per second.

**joule (J)** A unit of energy or work. 1 J = 1 newton meter.

**Jovian planet** See *giant planet*.

**Jy** See *jansky*.

## K

**K** See *kelvin*.

**K-T boundary** See *Cretaceous-Tertiary boundary*.

**KBO** See *Kuiper Belt object*.

**kelvin (K)** The basic unit of the Kelvin scale of temperature.

**Kelvin scale** The temperature scale, formulated by William Thomson, better known as Lord Kelvin (1824–1907), that uses Celsius-sized degrees, but defines 0 K as absolute zero instead of as the melting point of water. Unit: kelvins (K). Compare *Celsius* and *Fahrenheit*.

**Kepler's first law** A rule of planetary motion, inferred by Johannes Kepler, stating that planets move in elliptical orbits with the Sun at one focus.

**Kepler's laws** The three rules of planetary motion inferred by Johannes Kepler from the data collected by Tycho Brahe.

**Kepler's second law** Also called *law of equal areas*. A rule of planetary motion, inferred by Johannes Kepler, stating that a line drawn from the Sun to a planet sweeps out equal areas in equal times as the planet orbits the Sun.

**Kepler's third law** Also called *harmonic law*. A rule of planetary motion, inferred by Johannes Kepler, that describes the relationship between the period of a planet's orbit and its distance from the Sun. The law states that the square of the period of a planet's orbit, measured in years, is equal to the cube of the semimajor axis of the planet's orbit, measured in astronomical units: $(P_{years})^2 = (A_{AU})^3$.

**kiloparsec** A unit of distance equal to 1 thousand parsecs, or 3.26 thousand light-years

**kinetic energy ($E_K$)** The energy of an object due to its motions. $E_K = \frac{1}{2}mv^2$. Possible units include joules (J).

**Kirkwood gap** A gap in the main asteroid belt related to orbital resonances with Jupiter.

**Kuiper Belt** A disk-shaped population of comet nuclei extending from Neptune's orbit to perhaps several thousand astronomical units (AU) from the Sun. The highly populated innermost part of the Kuiper Belt has an outer edge approximately 50 AU from the Sun.

**Kuiper Belt object (KBO)** Also called *trans-Neptunian object*. An icy planetesimal (comet nucleus) that orbits within the Kuiper Belt beyond the orbit of Neptune.

## L

**Lagrangian equilibrium point** Also called simply *Lagrangian point*. One of five points of equilibrium in a system consisting of two massive objects in nearly circular orbit around a common center of mass. Only two Lagrangian points ($L_4$ and $L_5$) represent stable equilibrium. A third, smaller body located at one of the five points will move in lockstep with the center of mass of the larger bodies.

**Lambda-CDM** The standard model of the Big Bang universe in which most of the energy density of the universe is dark energy (similar to Einstein's cosmological constant), and most of the mass in the universe is cold dark matter.

**lander** An instrumented spacecraft designed to land on a planet or moon. Compare *rover*.

**large-scale structure** Observable aggregates on the largest scales in the universe, including galaxy groups, clusters, and superclusters.

**latitude** The angular distance north (+) or south (−) from the equatorial plane of a nearly spherical body. Compare *longitude*.

**lava** Molten rock flowing out of a volcano during an eruption; also the rock that solidifies and cools from this liquid.

**law of equal areas** See *Kepler's second law*.

**law of gravitation** See *universal law of gravitation*.

**leap year** A year that contains 366 days. Leap years occur every 4 years when the year is divisible by 4, correcting for the accumulated excess time in a normal year, which is approximately 365¼ days long.

**length contraction** The relativistic compression of moving objects in the direction of their motion.

**Leonids** A November meteor shower associated with the dust debris left by comet Tempel-Tuttle.

**life** A biochemical process in which living organisms can reproduce, evolve, and sustain themselves by drawing energy from their environment. All terrestrial life involves carbon-based chemistry, assisted by the self-replicating molecules ribonucleic acid (RNA) and deoxyribonucleic acid (DNA).

**light** All electromagnetic radiation, which composes the entire electromagnetic spectrum.

**light-year (ly)** The distance that light travels in 1 year—about 9.5 trillion kilometers (km).

**limb** The outer edge of the visible disk of a planet, moon, or the Sun.

**limb darkening** The darker appearance caused by increased atmospheric absorption near the limb of a planet or star.

**limestone** A common sedimentary rock composed of calcium carbonate.

**line of nodes** 1. A line defined by the intersection of two orbital planes. 2. The line defined by the intersection of Earth's equatorial plane and the plane of the ecliptic.

**lithosphere** The solid, brittle part of Earth (or any planet or moon), including the crust and the upper part of the mantle. Compare *biosphere* and *hydrosphere*.

**lithospheric plate** A separate piece of Earth's lithosphere capable of moving independently. See also *continental drift* and *plate tectonics*.

**Local Group** The group of galaxies that includes the Milky Way and Andromeda galaxies as members.

**long-period comet** A comet with an orbital period of greater than 200 years. Compare *short-period comet*.

**longitude** The angular distance east (+) or west (−) from the prime meridian at Greenwich, England. Compare *latitude*.

**longitudinal wave** A wave that oscillates parallel to the direction of the wave's propagation. Compare *transverse wave*.

**look-back time** The amount of time that the light from an astronomical object has taken to reach Earth.

**low-mass star** A star with a main-sequence mass of less than about 3 solar masses ($M_{Sun}$). Compare *high-mass star*.

**luminosity** The total amount of light emitted by an object. Unit: watts (W). Compare *brightness*.

**luminosity class** A spectral classification based on stellar size, ranging from supergiants at the large end to white dwarfs at the small end.

**luminosity-temperature-radius relationship** A relationship among these three properties of stars indicating that if any two are known, the third can be calculated.

**luminous matter** Also called *normal matter*. Matter in galaxies—including stars, gas, and dust—that emits electromagnetic radiation. Compare *dark matter*.

**lunar eclipse** An eclipse that occurs when the Moon is partially or entirely in Earth's shadow. Compare *solar eclipse*.

**lunar tide** A tide on Earth that is due to the differential gravitational pull of the Moon. Compare *solar tide*.

**ly** See *light-year*.

# M

**μm** See *micron*.

**M-type asteroid** An asteroid made of material that was once part of the metallic core of a larger, differentiated body that has since broken into pieces; made mostly of iron and nickel. Compare *C-type asteroid* and *S-type asteroid*.

**MACHO** Short for *massive compact halo object*. MACHOs include brown dwarfs, white dwarfs, and black holes and are candidates for dark matter. Compare *WIMP*.

**magma** Molten rock, often containing dissolved gases and solid minerals.

**magnetic field** A field that is able to exert a force on a moving electric charge. Compare *electric field*.

**magnetic force** The force exerted on electrically charged particles such as protons and electrons, arising from their motion. Compare *electric force*. See also *electromagnetic force*.

**magnetometer** A device that measures magnetic fields.

**magnetosphere** The region surrounding a planet that is filled with relatively intense magnetic fields and plasmas.

**magnitude** A system used by astronomers to describe the brightness or luminosity of stars. The brighter the star, the lower its magnitude.

**main asteroid belt** See *asteroid belt*.

**main sequence** The strip on the H-R diagram where most stars are found. Main-sequence stars are fusing hydrogen to helium in their cores.

**main-sequence lifetime** The amount of time a star spends on the main sequence, fusing hydrogen into helium in its core.

**main-sequence turnoff** The location on the main sequence of an H-R diagram made from a population of stars of the same age (such as a star cluster) where stars are just evolving off the main sequence. This location is determined by the age of the population of stars.

**mantle** The solid portion of a rocky planet that lies between the *crust* and the *core*.

**mare** (pl. **maria**) A dark region on the Moon composed of basaltic lava flows.

**mass** 1. Inertial mass: the property of matter that determines its resistance to changes in motion. Compare *weight*. 2. Gravitational mass: the property of matter defined by its attractive force on other objects. According to general relativity, the two are equivalent.

**mass-luminosity relationship** An empirical relationship between the luminosity ($L$) and mass ($M$) of main-sequence stars; for example, $L \propto M^{3.5}$.

**mass transfer** The transfer of mass from one member of a binary star system to its companion. Mass transfer occurs when one of the stars evolves to the point that it overfills its Roche lobe, so that its outer layers are pulled toward its binary companion.

**massive element** Also called *heavy element*. Any element more massive than helium.

**matter** 1. Objects made of particles that have mass, such as protons, neutrons, and electrons. 2. Anything that occupies space and has mass.

**Maunder Minimum** The period from 1645 to 1715, when very few sunspots were observed.

**medium** The substance that light travels through, such as air or glass. Compare *vacuum*.

**megabar** A unit of pressure equal to 1 million bars.

**megaparsec (Mpc)** A unit of distance equal to 1 million parsecs, or 3.26 million light-years.

**meridian** The imaginary arc in the sky running from the horizon at due north through the zenith to the horizon at due south. The meridian divides the observer's sky into eastern and western hemispheres. Compare *equator*.

**mesosphere** The layer of Earth's atmosphere immediately above the stratosphere, extending from an altitude of 50 kilometers (km) to about 90 km. Compare *troposphere*, *stratosphere*, and *thermosphere*.

**meteor** The incandescent trail produced by a small piece of interplanetary debris as it travels through the atmosphere at very high speeds. Compare *meteorite* and *meteoroid*.

**meteor shower** A larger-than-normal display of meteors, occurring when Earth passes through the orbit of a disintegrating comet, sweeping up its debris. Compare *sporadic meteor*.

**meteorite** A piece of rock or other fragment of material (a meteoroid) that survives to reach a planet's surface. Compare *meteor* and *meteoroid*.

**meteoroid** A small cometary or asteroidal fragment, ranging in size from 100 microns ($\mu m$) to 100 meters. When entering a planetary atmosphere, the meteoroid creates a *meteor*. Compare *meteor* and *meteorite*; also *planetesimal* and *zodiacal dust*.

**micrometer (μm)** See *micron*.

**micron (μm)** One-millionth ($10^{-6}$) of a meter; a unit of length used for the wavelength of infrared light.

**microwave radiation** Electromagnetic radiation with frequencies, photon energies, and wavelengths between those of infrared radiation and radio waves.

**Milky Way Galaxy** The galaxy in which the Sun and Solar System reside.

**minor planet** See *asteroid*.

**minute of arc** See *arcminute*.

**modern physics** Usually, the physical principles, including relativity and quantum mechanics, developed after 1900.

**molecular cloud** An interstellar cloud composed primarily of molecular hydrogen.

**molecular-cloud core** A dense clump within a molecular cloud that forms as the cloud collapses and fragments. Protostars form from molecular-cloud cores.

**molecule** Generally, the smallest particle of a substance that retains its chemical properties and is composed of two or more atoms.

**momentum** The product of the mass and velocity of a particle. Possible units include: kilograms times meters per second (kg m/s).

**moon** A less massive satellite orbiting a more massive object. Moons are found around planets, dwarf planets, asteroids, and Kuiper Belt objects. The term is usually capitalized when referring to Earth's Moon.

**Mpc** See *megaparsec*.

**multiverse** A collection of parallel universes that together comprise all that is.

**mutation** In biology, an imperfect reproduction of self-replicating material.

## N

**N** See *newton*.

**nadir** The point on the celestial sphere located directly below an observer, opposite the *zenith*.

**nanometer** (**nm**) One-billionth ($10^{-9}$) of a meter; a unit of length used for the wavelength of visible light.

**natural selection** The process by which forms of structure, ranging from molecules to whole organisms, that are best adapted to their environment become more common than less well-adapted forms.

**NCP** See *north celestial pole*.

**neap tide** An especially weak tide that occurs around the time of the first or third quarter Moon, when the gravitational forces of the Moon and the Sun on Earth are at right angles to each other. Compare *spring tide*.

**near-Earth asteroid** An asteroid whose orbit brings it close to the orbit of Earth. See also *near-Earth object*.

**near-Earth object** (**NEO**) An asteroid, comet, or large meteoroid whose orbit intersects Earth's orbit.

**nebula** (pl. **nebulae**) A cloud of interstellar gas and dust, either illuminated by stars (bright nebula) or seen in silhouette against a brighter background (dark nebula).

**nebular hypothesis** The first plausible theory of the formation of the Solar System, proposed by Immanuel Kant in 1734, which stated that the Solar System formed from the collapse of an interstellar cloud of rotating gas.

**NEO** See *near-Earth object*.

**neutrino** A very low-mass, electrically neutral particle emitted during beta decay. Neutrinos interact with matter only very feebly and so can penetrate through great quantities of matter.

**neutrino cooling** The process in which thermal energy is carried out of the center of a star by neutrinos rather than by electromagnetic radiation or convection.

**neutron** A subatomic particle having no net electric charge and a rest mass and rest energy nearly equal to that of the proton. Compare *electron* and *proton*.

**neutron star** The neutron-degenerate stellar core left behind by a Type II supernova.

**new Moon** The phase of the Moon in which the Moon is between Earth and the Sun, and from Earth we see only the side of the Moon not being illuminated by the Sun. Compare *full Moon*. See also *first quarter Moon* and *third quarter Moon*.

**newton** (**N**) The force required to accelerate a 1-kilogram (kg) mass at a rate of 1 meter per second per second (m/s²). Unit: kilograms multiplied by meters per second squared (kg m/s²).

**Newton's first law of motion** The law, formulated by Isaac Newton, stating that an object will remain at rest or will continue moving along a straight line at a constant speed until an unbalanced force acts on it.

**Newton's laws** The three physical laws of motion formulated by Isaac Newton.

**Newton's second law of motion** The law, formulated by Isaac Newton, stating that if an unbalanced force acts on a body, the body will have an acceleration proportional to the unbalanced force and inversely proportional to the object's mass: $a = F/m$. The acceleration will be in the direction of the unbalanced force.

**Newton's third law of motion** The law, formulated by Isaac Newton, stating that for every force there is an equal force in the opposite direction.

**nm** See *nanometer*.

**normal matter** See *luminous matter*.

**north celestial pole** (**NCP**) The northward projection of Earth's rotation axis onto the celestial sphere. Compare *south celestial pole*.

**North Pole** The location in the Northern Hemisphere where Earth's rotation axis intersects the surface of Earth. Compare *South Pole*.

**nova** (pl. **novae**) A stellar explosion that results from runaway nuclear fusion in a layer of material on the surface of a white dwarf in a binary system.

**nuclear burning** Release of energy by nuclear fusion of low-mass elements.

**nuclear fusion** The combination of two less massive atomic nuclei into a single more massive atomic nucleus.

**nucleosynthesis** The formation of more massive atomic nuclei from less massive nuclei, either in the Big Bang (Big Bang nucleosynthesis) or in the interiors of stars (stellar nucleosynthesis).

**nucleus** (pl. **nuclei**) 1. The dense, central part of an atom. 2. The central core of a galaxy, comet, or other diffuse object.

## O

**objective lens** The primary optical element in a telescope or camera that produces an image of an object.

**oblateness** The flattening of an otherwise spherical planet or star caused by its rapid rotation.

**obliquity** The inclination of a celestial body's equator to its orbital plane.

**observational uncertainty** The fact that real measurements are never perfect; all observations are uncertain by some amount.

**Occam's razor** The principle that the simplest hypothesis is the most likely, named after William of Occam (circa 1285–1349), the medieval English cleric to whom the idea is attributed.

**Oort Cloud** A spherical distribution of comet nuclei stretching from beyond the Kuiper Belt to more than 50,000 astronomical units (AU) from the Sun.

**opacity** A measure of how effectively a material blocks the radiation going through it.

**open cluster** A loosely bound group of a few dozen to a few thousand stars that formed together in the disk of a spiral galaxy. Compare *globular cluster*.

**open universe** An infinite universe with a negatively curved spatial structure (much like the surface of a saddle) such that the sum of the angles of a triangle is always less than 180 degrees. Compare *closed universe* and *flat universe*.

**orbit** The path taken by one object moving around another object under the influence of their mutual gravitational or electric attraction.

**orbital resonance** A situation in which the orbital periods of two objects are related by a ratio of small integers.

**orbiter** A spacecraft that is placed in orbit around a planet or moon. Compare *flyby*.

**organic** Containing the element carbon.

## P

**P wave** See *primary wave*.

**pair production** The creation of a particle-antiparticle pair from a source of electromagnetic energy.

**paleoclimatology** The study of changes in Earth's climate throughout its history.

**paleomagnetism** The record of Earth's magnetic field as preserved in rocks.

**palimpsest** The flat circular patch of bright terrain that remains after a crater has been deformed over time.

**parallax** Also called *parallactic angle*. The displacement in the apparent position of a nearby star caused by the changing location of Earth in its orbit.

**parent element** A radioactive element that decays to form more stable *daughter products*.

**parsec (pc)** Short for *parallax second*. The distance to a star with a parallax of 1 arcsecond (arcsec) using a base of 1 astronomical unit (AU). One parsec is approximately 3.26 light-years.

**partial lunar eclipse** An eclipse that occurs when the Moon is partially in Earth's shadow.

**partial solar eclipse** The type of eclipse that occurs when Earth passes through the penumbra of the Moon's shadow, so that the Moon blocks only a portion of the Sun's disk. Compare *annular solar eclipse* and *total solar eclipse*.

**pc** See *parsec*.

**peculiar velocity** The motion of a galaxy relative to the overall expansion of the universe.

**penumbra** (pl. **penumbrae**) 1. The outer part of a shadow, where the source of light is only partially blocked. Compare *umbra* (definition 1). 2. The region surrounding the umbra of a sunspot. The penumbra is cooler and darker than the surrounding surface of the Sun but not as cool or dark as the umbra. Compare *umbra* (definition 2).

**penumbral lunar eclipse** A lunar eclipse in which the Moon passes through the penumbra of Earth's shadow. Compare *total lunar eclipse*.

**perihelion** (pl. **perihelia**) The point in a solar orbit that is closest to the Sun. Compare *aphelion*.

**period** The time it takes for a regularly repetitive process to complete one cycle.

**period-luminosity relationship** The relationship between the period of variability of a pulsating variable star, such as a Cepheid or RR Lyrae variable, and the luminosity of the star. Longer-period pulsating variable stars are more luminous than shorter-period ones.

**Perseids** A prominent August meteor shower associated with the dust debris left by comet Swift-Tuttle.

**phase** One of the various appearances of the sunlit surface of the Moon or a planet caused by the change in viewing location of Earth relative to both the Sun and the object. Examples include crescent phase and gibbous phase.

**PHI** See *Planetary Habitability Index*.

**photino** An elementary particle related to the photon. One of the candidates for cold dark matter.

**photochemical** Resulting from the action of light on chemical systems.

**photodissociation** The breaking apart of molecules into smaller fragments or individual atoms by the action of photons. Compare *recombination* (definition 1).

**photoelectric effect** The emission of electrons from a substance that is illuminated by electromagnetic radiation greater than a certain critical frequency.

**photometry** The process of measuring the brightness of a source of light, generally over a specific range of wavelength.

**photino** A hypothetical subatomic particle.

**photon** Also called *quantum of light*. A discrete unit or particle of electromagnetic radiation. The energy of a photon is equal to Planck's constant ($h$) multiplied by the frequency ($f$) of its electromagnetic radiation: $E_{\text{photon}} = h \times f$. The photon is the carrier of the electromagnetic force.

**photosphere** The apparent surface of the Sun as seen in visible light. Compare *chromosphere* and *corona*.

**photosynthesize** The process by which plants and algae convert energy from sunlight into chemical energy.

**physical law** A broad statement that predicts a particular aspect of how the physical universe behaves and that is supported by many empirical tests. See also *theory*.

**pixel** The smallest picture element in a digital image array.

**Planck era** The early time, just after the Big Bang, when the universe as a whole must be described with quantum mechanics.

**Planck spectrum** Also called *blackbody spectrum*. The spectrum of electromagnetic energy emitted by a blackbody per unit area per second, which is determined only by the temperature of the object.

**Planck's constant (h)** The constant of proportionality between the energy and the frequency of a photon. This constant defines how much energy a single photon of a given frequency or wavelength has. Value: $6.63 \times 10^{-34}$ joule-second.

**planet** 1. A large body that orbits the Sun or other star that shines only by light reflected from the Sun or star. 2. In the Solar System, a body that orbits the Sun, has sufficient mass for self-gravity to overcome rigid body forces so that it assumes a spherical shape, and has cleared smaller bodies from the neighborhood around its orbit. Compare *dwarf planet*.

**planet migration** The theory that a planet can move to a location away from where it formed, through gravitational interactions with other bodies or loss of orbital energy from interaction with gas in the protoplanetary disk.

**Planetary Habitability Index (PHI)** An index for quantifying the habitability of extrasolar planets that aims to be less Earth-centric than the *Earth Similarity Index* and to broaden the options for habitability. It depends on factors not yet measured (or measurable) for most extrasolar planets, including the availability of energy, the presence of some kind of liquid, the type of surface, and the chemical makeup.

**planetary nebula** The expanding shell of material ejected by a dying asymptotic giant branch star. A planetary nebula glows from fluorescence caused by intense ultraviolet light coming from the hot, stellar remnant at its center.

**planetary system** A system of planets and other smaller objects in orbit around a star.

**planetesimal** A primitive body of rock and ice, 100 meters or more in diameter, that combines with others to form a planet. Compare *meteoroid* and *zodiacal dust*.

**plasma** A gas that is composed largely of charged particles but also may include some neutral atoms.

**plate tectonics** The geological theory concerning the motions of lithospheric plates, which in turn provides the theoretical basis for *continental drift*.

**positron** A positively charged subatomic particle; the antiparticle of the *electron*.

**power** The rate at which work is done or at which energy is delivered. Possible units include watts (W) and joules per second (J/s).

**precession of the equinoxes** The slow change in orientation between the ecliptic plane and the celestial equator caused by the wobbling of Earth's axis.

**pressure** Force per unit area. Possible units include newtons per square meter ($N/m^2$) and bars.

**primary atmosphere** An atmosphere, composed mostly of hydrogen and helium, that forms at the same time as its host planet. Compare *secondary atmosphere*.

**primary mirror** The principal optical mirror in a reflecting telescope. The primary mirror determines the telescope's light-gathering power and resolution. Compare *secondary mirror*.

**primary wave** Also called *P wave*. A longitudinal seismic wave, in which the oscillations involve compression and decompression parallel to the direction of travel. Compare *secondary wave*.

**principle** A general idea or sense about how the universe is that guides us in constructing new scientific theories. Principles can be testable theories.

**prograde motion** 1. Rotational or orbital motion of a moon that is in the same direction as the planet it orbits. 2. The counterclockwise orbital motion of Solar System objects as seen from above Earth's orbital plane. Compare *retrograde motion*.

**prominence** An archlike projection above the solar photosphere often associated with a sunspot.

**proportional** See *proportionality*

**proportionality** A relationship between two things whose ratio is a constant.

**proton ($p$ or $p^+$)** A subatomic particle having a positive electric charge of $1.6 \times 10^{-19}$ coulomb (C), a rest mass of $1.67 \times 10^{-27}$ kilogram (kg), and a rest energy of $1.5 \times 10^{-10}$ joule (J). Compare *electron* and *neutron*.

**proton-proton chain** One of the ways in which hydrogen burning can take place. This is the most important path for hydrogen burning in low-mass stars such as the Sun. Compare *carbon-nitrogen-oxygen cycle*.

**protoplanetary disk** The remains of the accretion disk around a young star from which a planetary system may form. Sometimes called *circumstellar disk*.

**protostar** A young stellar object that derives its luminosity from the conversion of gravitational energy to thermal energy, rather than from nuclear reactions in its core.

**pulsar** A rapidly rotating neutron star that beams radiation into space in two searchlight-like beams. To a distant observer, the star appears to flash on and off.

**pulsating variable star** A variable star that undergoes periodic radial pulsations.

**QCD** See *quantum chromodynamics*.

**QED** See *quantum electrodynamics*.

**quantized** Existing as discrete, irreducible units.

**quantum chromodynamics (QCD)** The quantum theory describing the strong nuclear force and its mediation by gluons. Compare *quantum electrodynamics*.

**quantum efficiency** The likelihood that a particular photon falling on a detector will actually produce a response in the detector.

**quantum electrodynamics (QED)** The quantum theory describing the electromagnetic force and its mediation by photons. Compare *quantum chromodynamics*.

**quantum mechanics** The branch of physics that deals with the quantized and probabilistic behavior of atoms and subatomic particles.

**quantum of light** See *photon*.

**quark** The building block of protons and neutrons.

**quasar** Short for *quasi-stellar radio source*. The most luminous of the active galactic nuclei, seen only at great distances from the Milky Way.

# R

**radial velocity** The component of velocity that is directed toward or away from the observer.

**radian** The angle at the center of a circle subtended by an arc equal to the length of the circle's radius; $2\pi$ radians equals 360°, and 1 radian equals approximately 57.3°.

**radiant** The direction in the sky from which the meteors in a meteor shower seem to come.

**radiation** Waves or particles of energy traveling through space or a medium.

**radiation belt** A toroidal ring of high-energy particles surrounding a planet.

**radiative transfer** The transport of energy from one location to another by electromagnetic radiation.

**radiative zone** A region within a star where energy is transported outward by radiation. Compare *convective zone*.

**radio galaxy** A type of elliptical galaxy that has an active galactic nucleus at its center and very strong emission ($10^{35}$ to $10^{38}$ watts [W]) in the radio part of the electromagnetic spectrum. Compare *Seyfert galaxy*.

**radio telescope** An instrument for detecting and measuring radio frequency emissions from celestial sources.

**radio wave** Electromagnetic radiation in the extreme long-wavelength region of the spectrum, beyond the region of microwaves.

**radioisotope** A radioactive element.

**radiometric dating** Use of the radioactive decay of elements to measure the ages of materials such as minerals.

**ratio** The relationship in quantity or size between two or more things.

**ray** 1. A beam of electromagnetic radiation. 2. A bright streak emanating from a young impact crater.

**recombination** 1. The combining of ions and electrons to form neutral atoms. Compare *photodissociation*. 2. An event early in the evolution of the universe in which hydrogen and helium nuclei combined with electrons to form neutral atoms. The removal of electrons caused the universe to become transparent to electromagnetic radiation.

**red giant** A low-mass star that has evolved beyond the main sequence and is now fusing hydrogen in a shell surrounding a degenerate helium core.

**red giant branch** A region on the H-R diagram defined by low-mass stars evolving from the main sequence toward the horizontal branch.

**reddening** The effect by which stars and other objects, when viewed through interstellar dust, appear redder than they actually are. Reddening is caused by the fact that blue light is more strongly absorbed and scattered than red light.

**redshift** The Doppler shift toward longer (redder) wavelengths of light from an approaching object. Compare *blueshift*.

**reflecting telescope** A telescope that uses mirrors for collecting and focusing incoming electromagnetic radiation to form an image in their focal planes. The size of a reflecting telescope is defined by the diameter of the primary mirror. Compare *refracting telescope*.

**reflection** The redirection of a beam of light that strikes, but does not cross, the surface between two media having different refractive indices. If the surface is flat and smooth, the angle of incidence equals the angle of reflection. Compare *refraction*.

**refracting telescope** A telescope that uses objective lenses for collecting and focusing incoming electromagnetic radiation to form an image. Compare *reflecting telescope*.

**refraction** The redirection or bending of a beam of light when it crosses the boundary between two media having different refractive indices. Compare *reflection*.

**refractory material** Material that remains solid at high temperatures. Compare *volatile material*.

**regular moon** A moon that formed together with the planet it orbits. Compare *irregular moon*.

**reionization** A period after the Dark Ages during which objects formed that radiated enough energy to ionize neutral hydrogen, at redshift $6 < z < 20$.

**relative humidity** The amount of water vapor held by a volume of air at a given temperature compared (stated as a percentage) to the total amount of water that could be held by the same volume of air at the same temperature.

**relative motion** The difference in motion between two individual frames of reference.

**relativistic** Describing systems that travel at nearly the speed of light or are located in the vicinity of very strong gravitational fields.

**relativistic beaming** The effect created when material moving at nearly the speed of light beams the radiation it emits in the direction of its motion.

**relativistic speed** A speed high enough that special relativity, rather than Newtonian physics, is needed to describe the motion. Speeds greater than about 10% the speed of light are relativistic.

**remote sensing** The use of images, spectra, radar, or other techniques to measure the properties of an object from a distance.

**resolution** The ability of a telescope to separate two point sources of light. Resolution is determined by the telescope's aperture and the wavelength of light it receives.

**rest wavelength** The wavelength of light that is seen coming from an object at rest with respect to the observer.

**retrograde motion** 1. Rotation or orbital motion of a moon that is in the opposite direction to the rotation of the planet it orbits. 2. The clockwise orbital motion of Solar System objects as seen from above Earth's orbital plane. Compare *prograde motion*. 3. Apparent retrograde motion is a motion of the planets with respect to the "fixed stars," in which the planets appear to move westward for a period of time before resuming their normal eastward motion.

**revolve** Motion of one object in orbit around another.

**rift zone** A region created by a geological fault, in which mantle material rises up, cools, and slowly spreads out, forming new crust.

**right ascension** A measure, analogous to *longitude*, that tells you the angular distance of a celestial body eastward along the celestial equator from the vernal equinox. Compare *declination*.

**ring** An aggregation of small particles orbiting a planet or star. The rings of the four giant planets of the Solar System are composed variously of silicates, organic materials, and ices.

**ring arc** A discontinuous, higher-density region within an otherwise continuous, narrow ring.

**ringlet** A narrowly confined concentration of ring particles.

**Roche limit** The distance at which a planet's tidal forces exceed the self-gravity of a smaller object—such as a moon, asteroid, or comet—causing the object to break apart.

**Roche lobes** The hourglass-shaped or figure eight–shaped volume of space surrounding two stars, which constrains material that is gravitationally bound by one or the other.

**rotation curve** A plot showing how the orbital velocity of stars and gas in a galaxy changes with radial distance from the galaxy's center.

**rover** A remotely controlled instrumented vehicle designed to move and explore the surface of a terrestrial planet or moon. Compare *lander*.

**RR Lyrae variable** A variable giant star whose regularly timed pulsations are good predictors of its luminosity. RR Lyrae variables are used for distance measurements to globular clusters.

# S

**S-type asteroid** An asteroid made of material that was once part of the outer layer of a larger, differentiated body that has since broken into pieces. Compare *C-type asteroid* and *M-type asteroid*.

**S wave** See *secondary wave*.

**S0 galaxy** A galaxy with a bulge and a disk-like spiral, but smooth in appearance like ellipticals. Compare *barred spiral galaxy*, *elliptical galaxy*, *irregular galaxy*, and *spiral galaxy*.

**satellite** An object in orbit around a more massive body; for example, a human-made satellite or a moon of any planet.

**scale factor ($R_U$)** A dimensionless number proportional to the distance between two points in space. The scale factor increases as the universe expands.

**scattering** The random change in the direction of travel of photons, caused by their interactions with molecules or dust particles.

**Schwarzschild radius** The distance from the center of a nonrotating, spherical black hole at which the escape velocity equals the speed of light.

**scientific method** The formal procedure—including hypothesis, prediction, and experiment or observation—used to test (attempt to falsify) the validity of scientific hypotheses and theories.

**scientific notation** The standard expression of numbers with one digit (which can be zero) to the left of the decimal point and multiplied by 10 to the exponent required to give the number its correct value. Example: $2.99 \times 10^8 = 299,000,000$.

**SCP** See *south celestial pole*.

**second law of thermodynamics** The law stating that the entropy or disorder of an isolated system always increases as the system evolves.

**second of arc** See *arcsecond*.

**secondary atmosphere** An atmosphere that forms—as a result of volcanism, comet impacts, or another process—sometime after its host planet has formed. Compare *primary atmosphere*.

**secondary crater** A crater formed from ejecta thrown from an *impact crater*.

**secondary mirror** A small mirror placed on the optical axis of a reflecting telescope that returns the beam back through a small hole in the *primary mirror*, thereby shortening the mechanical length of the telescope.

**secondary wave** Also called *S wave*. A transverse seismic wave, which involves the sideways motion of material. Compare *primary wave*.

**sedimentation** A process in which material carried by water or wind deposits layers of material and buries what lies below.

**seismic wave** A vibration due to an earthquake, a large explosion, or an impact on the surface that travels through a planet's interior.

**seismometer** An instrument that measures the amplitude and frequency of seismic waves.

**self-gravity** The gravitational attraction among all parts of the same object.

**semimajor axis** Half of the longer axis of an ellipse.

**SETI** The Search for Extraterrestrial Intelligence project, which uses advanced technology combined with radio telescopes to search for evidence of intelligent life elsewhere in the universe.

**Seyfert galaxy** A type of spiral galaxy with an active galactic nucleus at its center; first discovered in 1943 by Carl Seyfert. Compare *radio galaxy*.

**shepherd moon** A moon that orbits close to rings and gravitationally confines the orbits of the ring particles.

**shield volcano** A volcano formed by very fluid lava flowing from a single source and spreading out from that source. Compare *composite volcano*.

**short-period comet** A comet with an orbital period of less than 200 years. Compare *long-period comet*.

**sidereal day** Earth's period of rotation with respect to the stars—about 23 hours 56 minutes—which is the time it takes for Earth to make one rotation and face the exact same star on the meridian. It differs from the *solar day* because of Earth's motion around the Sun.

**sidereal period** An object's orbital or rotational period measured with respect to the stars. Compare *synodic period*.

**silicate** One of the family of minerals composed of silicon and oxygen in combination with other elements.

**singularity** The point where a mathematical expression or equation becomes meaningless, such as a fraction whose denominator approaches zero See also *black hole*.

**solar abundance** The relative amount of an element detected in the atmosphere of the Sun, expressed as the ratio of the number of atoms of that element to the number of hydrogen atoms.

**solar day** The day in common use—24 hours, which is Earth's period of rotation that brings the Sun back to the same local meridian where the rotation started. Compare *sidereal day*.

**solar eclipse** An eclipse that occurs when the Sun is partially or entirely blocked by the Moon. Compare *lunar eclipse*.

**solar flare** An explosion on the Sun's surface associated with a complex sunspot group and a strong magnetic field.

**solar maximum** (pl. **maxima**) The time, occurring about every 11 years, when the Sun is at its peak activity, meaning that sunspot activity and related phenomena (such as prominences, flares, and coronal mass ejections) are at their peak.

**solar neutrino problem** The historical observation that only about a third as many neutrinos as predicted by theory seemed to be coming from the Sun.

**Solar System** The gravitationally bound system made up of the Sun, planets, dwarf planets, moons, asteroids, comets, and Kuiper Belt objects, along with their associated gas and dust.

**solar tide** A tide on Earth that is due to the differential gravitational pull of the Sun. Compare *lunar tide*.

**solar wind** The stream of charged particles emitted by the Sun that flows at high speeds through interplanetary space.

**solstice** Literally, "Sun standing still." 1. One of the two most northerly and southerly points on the ecliptic. 2. Either of the two times of year (the *summer solstice* and *winter solstice*) when the Sun is at one of these two positions. Compare *equinox*.

**south celestial pole (SCP)** The southward projection of Earth's rotation axis onto the celestial sphere. Compare *north celestial pole*.

**South Pole** The location in the Southern Hemisphere where Earth's rotation axis intersects the surface of Earth. Compare *North Pole*.

**spacetime** A concept that combines space and time into a four-dimensional continuum with three spatial dimensions plus time.

**special relativity** See *special theory of relativity*.

**special theory of relativity** Sometimes referred to as simply *special relativity*. Einstein's theory explaining how the fact that the speed of light is a constant affects nonaccelerating frames of reference. Compare *general theory of relativity*.

**spectral type** A classification system for stars based on the presence and relative strength of absorption lines in their spectra. Spectral type is related to the surface temperature of a star.

**spectrograph** Also called *spectrometer*. A device that spreads out the light from an object into its component wavelengths.

**spectrometer** See *spectrograph*.

**spectroscopic binary** A binary star system whose existence and properties are revealed to astronomers only by the Doppler shift of its spectral lines. Most spectroscopic binaries are close pairs. Compare *eclipsing binary* and *visual binary*.

**spectroscopic parallax** Use of the spectroscopically determined luminosity and the observed brightness of a star to determine the star's distance.

**spectroscopy** The study of an object's electromagnetic radiation in terms of its component wavelengths.

**spectrum** (pl. **spectra**) Waves sorted by wavelength. See also *electromagnetic spectrum*.

**speed** The rate of change of an object's position with time, without regard to the direction of movement. Possible units include meters per second (m/s) and kilometers per hour (km/h). Compare *velocity*.

**spherically symmetric** Describing an object whose properties depend only on distance from the object's center, so that the object has the same form viewed from any direction.

**spin-orbit resonance** A relationship between the orbital and rotation periods of an object such that the ratio of their periods can be expressed by simple integers.

**spiral density wave** A stable, spiral-shaped change in the local gravity of a galactic disk that can be produced by periodic gravitational kicks from neighboring galaxies or from nonspherical bulges and bars in spiral galaxies.

**spiral galaxy** A galaxy of Hubble type "S" class, with a discernible disk in which large spiral patterns exist. Compare *barred spiral galaxy*, *elliptical galaxy*, *irregular galaxy*, and *S0 galaxy*.

**spoke** One of several narrow radial features seen occasionally in Saturn's B Ring. Spokes appear dark in backscattered light and bright in forward, scattering light, indicating that they are composed of tiny particles. Their origin is not well understood.

**sporadic meteor** A meteor that is not associated with a specific *meteor shower*.

**spreading center** A zone from which two tectonic plates diverge.

**spring tide** An especially strong tide that occurs around the time of a new or full Moon, when lunar tides and solar tides reinforce each other. Compare *neap tide*.

**stable equilibrium** An equilibrium state in which the system returns to its former condition after a small disturbance. Compare *unstable equilibrium*.

**standard candle** An object whose luminosity either is known or can be predicted in a distance-independent way, so its brightness can be used to determine its distance via the inverse square law of radiation.

**standard model** The theory of particle physics that combines electroweak theory with quantum chromodynamics to describe the structure of known forms of matter.

**star** A luminous ball of gas that is held together by gravity. A normal star is powered by nuclear reactions in its interior.

**star cluster** A group of stars that all formed at the same time and in the same general location.

**static equilibrium** A state in which the forces within a system are all in balance so that the system does not change. Compare *dynamic equilibrium*.

**Stefan-Boltzmann constant ($\sigma$)** The constant of proportionality that relates the flux emitted by an object to the fourth power of its absolute temperature. Value: $5.67 \times 10^{-8}$ W/(m² K⁴) (W = watts, m = meters, K = kelvins).

**Stefan-Boltzmann law** The law, formulated by Josef Stefan and Ludwig Boltzmann, stating that the amount of electromagnetic energy emitted from the surface of a body (flux), summed over the energies of all photons of all wavelengths emitted, is proportional to the fourth power of the temperature of the body: $\mathcal{F} = \sigma T^4$.

**stellar-mass loss** The loss of mass from the outermost parts of a star's atmosphere during the course of its evolution.

**stellar occultation** An event in which a planet or other Solar System body moves between the observer and a star, eclipsing the light emitted by that star.

**stellar population** A group of stars with similar ages, chemical compositions, and dynamic properties.

**stereoscopic vision** The way an animal's brain combines the different information from its two eyes to perceive the distances to objects around it.

**stony-iron meteorite** A meteorite consisting of a mixture of silicate minerals and iron-nickel alloys. Compare *iron meteorite* and *stony meteorite*.

**stony meteorite** A meteorite composed primarily of silicate minerals, similar to those found on Earth. Compare *iron meteorite* and *stony-iron meteorite*.

**stratosphere** The atmospheric layer immediately above the troposphere. On Earth, it extends upward to an altitude of 50 kilometers (km). Compare *troposphere*, *mesosphere*, and *thermosphere*.

**string theory** See *superstring theory*.

**stromatolite** A structure created by living or fossilized cyanobacteria..

**strong nuclear force** The attractive short-range force between protons and neutrons that holds atomic nuclei together. One of the four fundamental forces of nature, along with the *electromagnetic force*, the *weak nuclear force*, and *gravity* (definition 1). The force is mediated by the exchange of gluons.

**subduction zone** A region where two tectonic plates converge, with one plate sliding under the other and being drawn downward into the interior.

**subgiant** A giant star that is smaller and lower in luminosity than normal giant stars of the same spectral type. Subgiants evolve to become giants.

**subgiant branch** A region of the H-R diagram defined by stars that have left the main sequence but have not yet reached the red giant branch.

**sublimation** The process by which a solid becomes a gas without first becoming a liquid.

**subsonic** Moving within a medium at a speed slower than the speed of sound in that medium. Compare *supersonic*.

**summer solstice** 1. One of two points where the Sun is at its greatest distance from the celestial equator. 2. The day on which the Sun appears at this location, marking the first day of summer (about June 20 in the Northern Hemisphere and December 21 in the Southern Hemisphere). Compare *winter solstice*. See also *autumnal equinox* and *vernal equinox*.

**Sun** The star at the center of the Solar System.

**sungrazer** A comet whose perihelion is within a few solar diameters of the surface of the Sun.

**sunspot** A cooler, transitory region on the solar surface produced when loops of magnetic flux break through the surface of the Sun.

**sunspot cycle** The approximate 11-year cycle during which sunspot activity increases and then decreases. This is one-half of a full 22-year cycle, in which the magnetic polarity of the Sun first reverses and then returns to its original configuration.

**supercluster** A large conglomeration of galaxy clusters and galaxy groups; typically, more than 100–300 megaparsecs (Mpc) in size and containing tens of thousands to hundreds of thousands of galaxies. Compare *galaxy cluster* and *galaxy group*.

**super-Earth** An extrasolar planet with about 2–10 times the mass of Earth.

**superior planet** A Solar System planet that orbits the Sun at a greater distance than Earth's orbit.

**superluminal motion** The appearance (though not the reality) that a jet is moving faster than the speed of light.

**supermassive black hole** A black hole of 1,000 solar masses ($M_{Sun}$) or more that resides in the center of a galaxy, and whose gravity powers active galactic nuclei.

**supernova** (pl. **supernovae**) A stellar explosion resulting in the release of tremendous amounts of energy, including the high-speed ejection of matter into the interstellar medium. See also *Type Ia supernova* and *Type II supernova*.

**supernova remnant** The material ejected from the outer layers of a star following a supernova explosion.

**supersonic** Moving within a medium at a speed faster than the speed of sound in that medium. Compare *subsonic*.

**superstring theory** The theory that conceives of particles as strings in 10 dimensions of space and time; the current contender for a theory of everything.

**surface brightness** The amount of electromagnetic radiation emitted or reflected per unit area.

**surface wave** A seismic wave that travels on the surface of a planet or moon.

**symmetry** In theoretical physics, the properties of physical laws that remain constant when certain things change, such as the symmetry between matter and antimatter even though their charges may be different.

**synchronous rotation** The case that occurs when a body's rotation period equals its orbital period around another body. A special type of spin-orbit resonance.

**synchrotron radiation** Radiation from electrons moving at close to the speed of light as they spiral in a strong magnetic field; named because this kind of radiation was first identified on Earth in particle accelerators called synchrotrons.

**synodic period** An object's orbital or rotational period measured with respect to the Sun. Compare *sidereal period*.

# T

**T Tauri star** A young stellar object that has dispersed enough of the material surrounding it to be seen in visible light.

**tail** A stream of gas and dust swept away from the coma of a comet by the solar wind and by radiation pressure from the Sun.

**tectonism** Deformation of the lithosphere of a planet.

**telescope** The basic tool of astronomers. Working over the entire range from gamma rays to radio waves, astronomical telescopes collect and concentrate electromagnetic radiation from celestial objects.

**temperature** A measure of the average kinetic energy of the atoms or molecules in a gas, solid, or liquid.

**terrestrial planet** An Earth-like planet, made of rock and metal and having a solid surface. In the Solar System, the terrestrial planets are Mercury, Venus, Earth, and Mars. Compare *giant planet*.

**theoretical model** A detailed description of the properties of a particular object or system in terms of known physical laws or theories; often, a computer calculation of predicted properties based on such a description.

**theory** A well-developed idea or group of ideas that are tied solidly to known physical laws and make testable predictions about the world. A very well-tested theory may be called a *physical law*, or simply a fact. Compare *hypothesis*.

**theory of everything (TOE)** A theory that unifies all four fundamental forces of nature: strong nuclear, weak nuclear, electromagnetic, and gravitational forces.

**thermal conduction** See *conduction*.

**thermal energy** The energy that resides in the random motion of atoms, molecules, and particles, by which we measure their temperature.

**thermal equilibrium** The state in which the rate of thermal-energy emission by an object is equal to the rate of thermal-energy absorption.

**thermal motion** The random motion of atoms, molecules, and particles that gives rise to thermal radiation.

**thermal radiation** Electromagnetic radiation resulting from the random motion of the charged particles in every substance.

**thermosphere** The layer of Earth's atmosphere at altitudes greater than 90 kilometers (km), above the mesosphere. Near its top, at an altitude of 600 km, the temperature can reach 1000 K. Compare *troposphere*, *stratosphere*, and *mesosphere*.

**third quarter Moon** The phase of the Moon in which only the eastern half of the Moon, as viewed from Earth, is illuminated by the Sun. It occurs about one week after the full Moon. Compare *first quarter Moon*. See also *full Moon* and *new Moon*.

**tidal bulge** Distortion of a body resulting from tidal stresses.

**tidal force** A force caused by the change in the strength of gravity across an object.

**tidal locking** Synchronous rotation of an object caused by internal friction as the object rotates through its tidal bulge.

**tide** On Earth, the rise and fall of the oceans as Earth rotates through a tidal bulge caused by the Moon and the Sun. See also *lunar tide*, *neap tide*, *solar tide*, and *spring tide*.

**time dilation** The relativistic "stretching" of time. Compare *general relativistic time dilation*.

**TOE** See *theory of everything*.

**topographic relief** The differences in elevation from point to point on a planetary surface.

**tornado** A violent rotating column of air, typically 75 meters across with 200-kilometer-per-hour (km/h) winds. Some tornadoes can be more than 3 km across, and winds up to 500 km/h have been observed.

**torus** (pl. **tori**) A three-dimensional, doughnut-shaped ring.

**total lunar eclipse** A lunar eclipse in which the Moon passes through the umbra of Earth's shadow. Compare *penumbral lunar eclipse*.

**total solar eclipse** The type of eclipse that occurs when Earth passes through the umbra of the Moon's shadow, so that the Moon completely blocks the disk of the Sun. Compare *annular solar eclipse* and *partial solar eclipse*.

**transform fault** The actively slipping segment of a fracture zone between lithospheric plates.

**transit method** A method of detecting extrasolar planets by measuring the decrease in light from a star as its orbiting planet passes in front of the star as viewed from Earth.

**trans-Neptunian object** See *Kuiper Belt object*.

**transverse wave** A wave that oscillates perpendicular to the direction of the wave's propagation. Compare *longitudinal wave*.

**triple-alpha process** The nuclear fusion reaction that combines three helium nuclei (alpha particles) together into a single nucleus of carbon.

**Trojans** A group of asteroids orbiting in the $L_4$ and $L_5$ Lagrangian points of Jupiter's orbit.

**tropical year** The time between one crossing of the vernal equinox and the next. Because of the precession of the equinoxes, a tropical year is slightly shorter than the time that it takes for Earth to orbit once around the Sun. Compare *year*.

**Tropics** The region on Earth between latitudes 23.5° south and 23.5° north, where the Sun appears directly overhead twice during the year.

**tropopause** The top of a planet's troposphere.

**troposphere** The convection-dominated layer of a planet's atmosphere. On Earth, the atmospheric region closest to the ground within which most weather phenomena take place. Compare *stratosphere*, *mesosphere*, and *thermosphere*.

**tuning fork diagram** The two-pronged diagram showing Hubble's classification of galaxies into ellipticals, S0s, spirals, barred spirals, and irregular galaxies.

**turbulence** The random motion of blobs of gas within a larger cloud of gas.

**Type Ia supernova** A supernova explosion with a calibrated peak luminosity that occurs as a result of runaway carbon burning in a white dwarf star that accretes mass from a companion and approaches the Chandrasekhar mass limit of 1.4 $M_{\text{Sun}}$.

**Type II supernova** A supernova explosion in which the degenerate core of an evolved massive star suddenly collapses and rebounds.

# U

**ultrafaint dwarf galaxy** A dim dwarf galaxy with only 1,000–100,000 times the Sun's luminosity. Ultrafaint dwarf galaxies differ from globular clusters in that they are composed of large amounts of dark matter.

**ultraviolet (UV) radiation** Electromagnetic radiation having frequencies and photon energies greater than those of visible light but less than those of X-rays and having wavelengths shorter than those of visible light but longer than those of X-rays.

**umbra** (pl. **umbrae**) 1. The darkest part of a shadow, where the source of light is completely blocked. Compare *penumbra* (definition 1). 2. The darkest, innermost part of a sunspot. Compare *penumbra* (definition 2).

**unbalanced force** The nonzero net force acting on a body.

**unbound orbit** An orbit in which an object is no longer gravitationally bound to the body it was orbiting. An unbound orbit's velocity is greater than the escape velocity. Compare *bound orbit*.

**uncertainty principle** See *Heisenberg uncertainty principle*.

**unified model of AGN** A model in which many different types of activity in the nuclei of galaxies are all explained by accretion of matter around a supermassive black hole.

**uniform circular motion** Motion in a circular path at a constant speed.

**unit** A fundamental quantity of measurement. The meter is an example of a metric unit; the foot is an example of an English unit.

**universal gravitational constant (G)** The constant of proportionality in the universal law of gravitation. Value: $6.67 \times 10^{-11}$ meters cubed per kilogram second squared [$\text{m}^3/\text{kg s}^2 = \text{N m}^2/\text{kg}^2$].

**universal law of gravitation** The law, formulated by Isaac Newton, stating that the gravitational force between any two objects is proportional to the product of their masses and inversely proportional to the square of the distance between them:

$$F_{\text{grav}} = G \times \frac{m_1 \times m_2}{r^2}$$

**universe** 1. All of space and everything contained therein. 2. Our own universe in a collection of parallel universes that together comprise all that is.

**unstable equilibrium** An equilibrium state in which a small disturbance will cause a system to move away from equilibrium. Compare *stable equilibrium*.

**UV** Ultraviolet. See *ultraviolet radiation*.

# V

**vacuum** A region of space that contains very little matter. In quantum mechanics and general relativity, however, even a perfect vacuum has physical properties.

**variable star** A star with varying luminosity. Many periodic variables are found within the instability strip on the H-R diagram.

**velocity** The rate and direction of change of an object's position with time. Possible units include: meters per second (m/s) and kilometers per hour (km/h). Compare *speed*.

**vernal equinox** 1. One of two points where the Sun crosses the celestial equator. 2. The day on which the Sun appears at this location, marking the first day of spring (about March 20 in the Northern Hemisphere and September 22 in the Southern Hemisphere). Compare *autumnal equinox*. See also *summer solstice* and *winter solstice*.

**virtual particle** A particle that, according to quantum mechanics, comes into existence only momentarily. According to theory, fundamental forces are mediated by the exchange of virtual particles.

**visual binary** A binary system in which the two stars can be seen individually from Earth. Compare *eclipsing binary* and *spectroscopic binary*.

**void** A region in space containing little or no matter. Examples include regions in cosmological space that are largely empty of galaxies.

# W

**W** See *watt*.

**waning** Describing the changing phases of the Moon as it becomes less fully illuminated between full Moon and new Moon as seen from Earth. Compare *waxing*.

**waning gibbous Moon** The phases of the moon between gibbous and 3rd quarter

**water cycle** The flow of water on, above, and through Earth's surface.

**watt (W)** A measure of *power*. Unit: joules per second (J/s).

**wave** A disturbance moving along a surface or passing through a space or a medium.

**wavefront** The imaginary surface of an electromagnetic wave, either plane or spherical, oriented perpendicular to the direction of travel.

**wavelength** The distance on a wave between two adjacent points having identical characteristics. The distance a wave travels in one period. Possible units include meters (m).

**waxing** Describing the changing phases of the Moon as it becomes more fully illuminated between new Moon and full Moon as seen from Earth. Compare *waning*.

**waxing crescent Moon** The phases of the Moon between new and 1st quarter.

**waxing gibbous Moon** The phases of the Moon between 1st quarter and full.

**weak nuclear force** The force underlying some forms of radioactivity and certain interactions between subatomic particles. It is responsible for radioactive beta decay and for the initial proton-proton interactions that lead to nuclear fusion in the Sun and other stars. One of the four fundamental forces of nature, along with the *electromagnetic force*, the *strong nuclear force*, and *gravity* (definition 1). The force is mediated by the exchange of *W* and *Z* particles.

**weather** The state of an atmosphere at any given time and place. Compare *climate*.

**weight** The gravitational force acting on an object; that is, the force equal to the mass of an object multiplied by the local acceleration due to gravity. In general relativity, the force equal to the mass of an object multiplied by the acceleration of the frame of reference in which the object is observed. Compare *mass*.

**white dwarf** The stellar remnant left at the end of the evolution of a low-mass star. A typical white dwarf has a mass of 0.6 solar mass ($M_{Sun}$) and a size about equal to that of Earth; it is made of nonburning, electron-degenerate carbon.

**Wien's law** A law, named for Wilhelm Wien, stating that location of the peak wavelength in the electromagnetic spectrum of an object is inversely proportional to the temperature of the object.

**WIMP** Short for *weakly interacting massive particle*. A hypothetical massive particle that interacts through the weak nuclear force and gravity but not with electromagnetic radiation. WIMPs are candidates for dark matter. Compare *MACHO*.

**winter solstice** 1. One of two points where the Sun is at its greatest distance from the celestial equator. 2. The day on which the Sun appears at this location, marking the first day of winter (about December 21 in the Northern Hemisphere and June 20 in the Southern Hemisphere). Compare *summer solstice*. See also *autumnal equinox* and *vernal equinox*.

# X

**X-ray** Electromagnetic radiation having frequencies and photon energies greater than those of ultraviolet (UV) light but less than those of gamma rays and having wavelengths shorter than those of UV light but longer than those of gamma rays.

**X-ray binary** A binary system in which mass from an evolving star spills over onto a collapsed companion, such as a neutron star or black hole. The material falling in is heated to such high temperatures that it glows brightly in X-rays.

# Y

**year** The time it takes Earth to make one revolution around the Sun. A solar year is measured from equinox to equinox. A sidereal year, Earth's true orbital period, is measured relative to the stars. Compare *tropical year*.

# Z

**zenith** The point on the celestial sphere located directly overhead from an observer. Compare *nadir*.

**zero-age main sequence** The strip on the H-R diagram plotting where stars of all masses in a cluster begin their lives.

**zodiac** The 12 constellations lying along the plane of the ecliptic.

**zodiacal dust** Particles of cometary and asteroidal debris less than 100 microns ($\mu$m) in size that orbit the inner Solar System close to the plane of the ecliptic. Compare *meteoroid* and *planetesimal*.

**zodiacal light** A band of light in the night sky caused by sunlight reflected by zodiacal dust.

**zonal wind** The east–west component of a wind.

# V

**volatile material** Sometimes called *ice*. Material that remains gaseous at moderate temperature. Compare *refractory material*.

**volcanism** A form of geological activity on a planet or moon in which molten rock (magma) erupts at the surface.

# Selected Answers

These pages contain selected answers for all 24 chapters in the complete volume of *21st Century Astronomy,* Fifth Edition. The Solar System Edition does not include Chapters 15–23. The Stars and Galaxies Edition does not include Chapters 8–12.

## Chapter 1

### Check Your Understanding
1. Radius of Earth–light-minute–distance from Earth to Sun–light-hour–radius of Solar System–light-year.
2. b
3. c

### Thinking about the Concepts
19. 2.5 million years.
22. *Falsifiable* means that something can be tested and shown to be false/incorrect through an experiment or observation.
29. The use of mathematics is not the hallmark of good science. Rather, it is following the scientific method, which astrology does not employ.

### Applying the Concepts
35. 20 days
39. 1/16$^{th}$
44. The 18-inch pizza is more economical.

## Chapter 2

### Check Your Understanding
1. b
2. d
3. d
4. Full
5. If the lunar cycle was 30 days and the year was exactly 12 cycles, holidays like these based on a lunar calendar would stay on the same date.
6. b

### Thinking about the Concepts
17. Zenith.
24. Midnight, noon.
29. Longer and more severe seasonal differences.

### Applying the Concepts
35. (a) 23.5°, (b) 23.5°
37. 78.5°
44. 3.67 times larger

## Chapter 3

### Check Your Understanding
1. Planets wander through the sky.
2. c
3. b
4. c-b-d-a
5. d
6. b

### Thinking about the Concepts
17. *Empirical* means basing conclusions on what one has found in observations and data
21. A planet is always moving the fastest when it is closest to the Sun, and slowest when furthest.
24. Yes, the laws of physics are universal.

### Applying the Concepts
34. 30.1 AU, 367.5 days.
41. (a) 200 km/h. (b) 200 km/h.
44. 5 m/s$^2$

## Chapter 4

### Check Your Understanding
1. d
2. c, a
3. a
4. c-a-d-b
5. a

### Thinking about the Concepts
18. *Mass* is a measurement of an object's inertia. *Weight* is how strongly gravity acts on the object.
21. The Earth rotates the fastest at the equator, thus launching satellites from here gives them the highest initial speed. Launched toward the east because Earth rotates to the east.
25. Near sunrise or sunset during the new and full Moon phases.

### Applying the Concepts
34. 1.12 m/s
39. 7.7 km/s
44. (a) 0.71 yr. (b) 42.1 km/s

## Chapter 5

### Check Your Understanding
1. e-c-b-d-a
2. c
3. Spectral features correspond to elements present.
4. d
5. e
6. c

### Thinking about the Concepts
19. The red beam would have to be more intense.
24. The object is moving toward you.
26. The blue star is hotter than the yellow one.

### Applying the Concepts
33. 10,000 km/s, moving away.
38. 9 times more luminous.
42. 9.35 microns. 131 W

# Chapter 6

## Check Your Understanding
1. a, b, c, d
2. b
3. c
4. d
5. c
6. a, b, d

## Thinking about the Concepts
16. Simple lenses suffer from chromatic aberration, that is, they bend different colors by different amounts, so the resulting image is not nearly as crisp as in telescopes with multiple lenses.
23. An adaptive-optic system will use a star and sophisticated computer programs to control a deformable mirror in an attempt to keep the star's image steady and as small as possible, removing the effects of atmospheric blurring and providing higher-resolution images.
25. The nearest star is about 4 light-years away, while low-Earth orbit is about 400 kilometers above the ground. Putting a telescope in orbit gets the telescope closer to that closest star by such an infinitesimally small amount that there is no suitable analogy that immediately comes to mind.

## Applying the Concepts
33. 16 times.
40. (a) 35 milliarcsec. (b) 1 millarcsec.
45. 1.06 mm, microwave.

# Chapter 7

## Check Your Understanding
1. a, b, d
2. a
3. d
4. a
5. c

## Thinking about the Concepts
22. An accretion disk is the thin, rotating disk that forms as a gas cloud collapses on itself.
28. Stars are very bright and far away while planets appear very close to their stars. Thus it is difficult to mask out the light of the star and still see close enough to it to see the reflected light of a planet.
30. The Kepler satellite uses planetary transits to search for exoplanets. By "Earth-like," we first mean planets that are terrestrial and have roughly the same mass.

## Applying the Concepts
31. 80 mph, or 838 times smaller than Earth's orbital speed.
41. 1%
45. (a) $4.43 \times 10^{27}$ kg. (b) $1.02 \times 10^8$ m. (c) $4.45 \times 10^{24}$. (d) 996 kg/m$^3$, slightly less than that of water. The planet is gaseous.

# Chapter 8

## Check Your Understanding
1. b
2. d
3. d
4. c
5. c
6. b

## Thinking about the Concepts
18. We can select igneous rocks from various levels of the Grand Canyon and measure their ages using radiometric dating.
20. The Moon is smaller than the Earth, so it contained less thermal energy in its core to begin with, and a smaller core cools faster than a larger one.
29. Mars shows dry river beds, canyons, and teardrop-shaped erosions around craters,

## Applying the Concepts
34. Figure 8.23 shows a very smooth surface, with fewer small craters than Figure 8.18, suggesting this region is younger
38. The moon cools off 3.7 times faster.
41. (a) 6 half lives. (b) 34,200 y. (c) overestimate.

# Chapter 9

## Check Your Understanding
1. a, b, c, d
2. c
3. c-a-e-b-d
4. a
5. d, c, b, a
6. a-b-c

## Thinking about the Concepts
19. Impacts from icy comets, and outgassing from volcanoes.
24. (a) Artic ice floats, so it adds no extra mass to the ocean when it melts. (b) Glaciers reside on land, adding to the mass of water in the oceans.
25. Venus has perpetual, opaque cloud cover due to its thick atmosphere.

## Applying the Concepts
35. (a) 2000 N. (b) 200 kg. (c) 3. (d) Our bones and muscle hold us up.
42. (a) 0.85 times. (b) No.
45. 920 m

# Chapter 10

## Check Your Understanding
1. a
2. b
3. b
4. d
5. a-d-b-c

## Thinking about the Concepts
18. The size of a planet, and study the presence and properties of the planet's atmosphere and rings.
19. Coriolis forces
28. Synchrotron radiation.

## Applying the Concepts
35. 632 AU; 21 times further than Neptune.
36. (a) 51,100 km. (b) a lower limit to the size.
43. 60 K

# Chapter 11

## Check Your Understanding

1. a, b, c
2. a
3. a-d-b-c
4. a
5. b

## Thinking about the Concepts

16. Tidal stresses from Jupiter continuously flex Io by many tens of meters per orbit and keep the interior mantle molten.
24. Breakup of moons, asteroids, or comets; volcanic processes; impact events; atmospheric escape.
25. Gaps can be created by small moons; larger gaps are created by orbital resonances of shepherd moons.

## Applying the Concepts

31. 2.6 km/s
36. Increase; increase; increase
44. 0.69; separate particles.

# Chapter 12

## Check Your Understanding

1. Because they do not orbit another body.
2. c
3. c
4. a, b, c, e
5. Pre-collision orbit, structure, and chemical make-up.

## Thinking about the Concepts

18. Asteroids are loose agglomerations of refractory rock and metal held together only by self-gravity. Comet nuclei are "dirty snowballs" that are composed primarily of ices and organic materials held together by a loose rocky matrix
24. The meteorite probably came from Mars, or another object that was geologically active 1 billion years ago.
30. Comets and asteroids might have brought water and organic molecules to Earth, or their impacts might have created enough energy to cause chemical reactions that formed self-replicating proteins in our "primordial soup."

## Applying the Concepts

35. $3.68 \times 10^{20}$ J, or 87,600 H-bombs.
39. 2.5 AU.
43. $5 \times 10^{16}$ cm$^3$.

# Chapter 13

## Check Your Understanding

1. b
2. d
3. a, b, d
4. b, d

## Thinking about the Concepts

18. (a) The sapphire blue star has the higher temperature. (b) The golden star is larger.
19. A star with a 2,500 K blackbody emits mostly in the red and infrared but still gives off a measurable amount of blue light.
28. Eclipsing and spectroscopic binaries; visual binaries.

## Applying the Concepts

32. 1.5 times further
38. (a) 0.85 $M_{Sun}$. (b) 19 AU.
42. $4 \times 10^6$ AU. 6.32 ly.

# Chapter 14

## Check Your Understanding

1. c
2. b
3. c
4. b
5. d

## Thinking about the Concepts

18. Neutrinos are very hard to detect because they carry no charge, travel at nearly the speed of light, and interact very weakly with matter.
21. The core would heat and expand, cool down, fuse less, contract, and return to equilibrium. The Sun would brighten for a short time.
28. Sunspots tell us the Sun's overall rotation period, but more important, that the Sun has differential rotation, whereby the equator rotates faster than the poles

## Applying the Concepts

34. A photon scatters around in the Sun, traveling about 1 cm before hitting another atom. Thus, the distance required for a photon to escape the Sun is much longer than the straight path out.
37. (a) $6 \times 10^{26}$ kg. (b) 0.03%.
38. 1/6 as long as the Sun, i.e., 1.7 billion years.

# Chapter 15

## Check Your Understanding

1. d
2. c
3. a
4. a

## Thinking about the Concepts

16. Dust grains are comparable in size to visible-light wavelengths, so they act as a barrier to visible light. Gas molecules are much smaller than dust grains, so they do not obstruct light at visible wavelengths.
22. Tenuous gases have very few collisions with other atoms or dust; thus the atoms have a hard time getting rid of their energy, and as a result, they stay hot. On the other hand, dense gases have frequent collisions; thus they can cool easily.
27. As a protostar collapses, more mass implies higher gravity, but that higher gravity compresses the gas further. This compression raises its temperature and thus its pressure, so the star maintains hydrostatic equilibrium at all times.

## Applying the Concepts

33. $10^{-12}$ grains/m$^3$.
38. 43 times.
43. 39 times.

# Chapter 16

## Check Your Understanding

1. a
2. c
3. a
4. a
5. d

## Thinking about the Concepts

16. To burn helium, a star must have a massive amount of it in its core and the core must be hotter than about 100 million K. Neither are possible with a newly-born star.
19. Close binary stars share mass, shifting stars higher and lower on the main sequence.
27. A degenerate gas does not expand when you heat it, so there is no safety valve to slow down the nuclear reactions.

## Applying the Concepts

35. (a) 0.1 $L_{Sun}$, (b) 529 $L_{Sun}$, and (c) $1.7 \times 10^6$ $L_{Sun}$.
39. $3.16 \times 10^{13}$ km (about 1 pc)
40. 800 million years

# Chapter 17

### Check Your Understanding
1. b, d
2. b
3. d
4. b

### Thinking about the Concepts
24. The neutrinos from SN 1987A escaped directly from the core, while the shock wave from core collapse took a few hours to break out of the envelope.
26. The ejecta from a supernova loses energy as it blows into low-density interstellar gas and dust, and thus the ejecta slow down.
27. Under the assumption that all stars in a cluster formed at approximately the same time, the "turnoff point" where main sequence stars are beginning to evolve into supergiants gives the age, since these stars have just finished their main-sequence lifetime.

### Applying the Concepts
35. Approximately 3,000 $L_{Sun}$. Observed (or apparent) brightness.
37. $3.7 \times 10^9$ km/s$^2$ (380 trillion times stronger than on Earth)
40. $3.8 \times 10^{23}$ kg. (b) 5.2 times the mass of the moon.

# Chapter 18

### Check Your Understanding
1. d
2. a
3. d
4. c

### Thinking about the Concepts
17. $c, c$
24. It must be elongated (i.e., an oval or rugby-shaped ball) on the ship.
26. It will not affect us any differently than the stars of similar mass that are already at that distance.

### Applying the Concepts
38. You would be 87% of your twin's age.
40. $10^{-25}$ m.
44. at 99% C, 3.5 yrs, 25.3 yrs, 25.3 yrs

# Chapter 19

### Check Your Understanding
1. d
2. c
3. c
4. a
5. c

### Thinking about the Concepts
17. Cepheids found in the Andromeda Galaxy (M31) by Hubble showed that M31 was far too large and too far away to be part of our own Milky Way.
22. The bulge of a spiral has an older population of stars and the stars are on random-ized orbits, which mimics on a small scale the structure of an elliptical galaxy.
26. AGNs are about the size of our Solar System. We know because the time variability of an object constrains its size, that is, an object can't be much larger than the times-cales of variability, otherwise that signal would be washed out.

### Applying the Concepts
33. (a) 0.143. (b) $4.3 \times 10^4$ km/s. 613 Mpc.
37. 512 AU
41. 5,540 Mpc

# Chapter 20

### Check Your Understanding
1. The Milky Way contains gas and dust as well as ongoing star formation. Observa-tions of 21-cm radiation emitted from neu-tral hydrogen clouds show the existence of spiral arms. The galaxy has a well-defined rotation curve that flattens toward the vis-ible edge. Finally, most stars in our galac-tic neighborhood share the same relative motion around the galactic center.
2. The disk contains old and young stars; the halo contains old stars.
3. c
4. Andromeda is moving toward us because of our mutual gravitational attraction.

### Thinking about the Concepts
17. Globulars are distributed roughly sphe-roidally, showing that there is a large, spheroidal halo that encloses the disk of the galaxy. The center of that spheroid is roughly 8,000 pc from the Sun, showing that we are roughly that distance from the center of the galaxy.
21. Like most other spiral galaxies, ours has a flat rotation curve, which implies that up to 90 percent of the mass of our galaxy must be dark matter.
24. There is far too much dust along the path to the galactic center to see optical light from that region. Whereas dust is opaque to opti-cal and UV light, it is transparent to radio, X-ray, and infrared, hence we can (and do) observe this region in these wavelengths.

### Applying the Concepts
33. 1.8 billion years
36. (a) 92 kpc. (b) The halo is at least 3 times larger than the disk.
44. 4.5 million $M_{Sun}$

# Chapter 21

### Check Your Understanding
1. b, a
2. d
3. c
4. b
5. a

### Thinking about the Concepts
19. Quite the opposite, Hubble's law implies that there is no center of the universe, since galaxies are moving away from each other.
23. Hubble's redshift-distance relationship, the chemical abundances of stars and the inter-stellar medium, the presence of the cosmic-microwave background, inhomogeneities and fluctuations within the CMB.
27. The tiny variations in the CMB are the microscopic fluctuations that grew over time into the large-scale structure we observe and live in today.

### Applying the Concepts
35. $1.96 \times 10^{10}$ yr, $9.79 \times 10^9$ yr.
37. 9 times.
42. 0.24 atoms/m$^3$

# Chapter 22

*Check Your Understanding*

1. c
2. c
3. flatness problem and horizon problem
4. d
5. b

*Thinking about the Concepts*

19. Gravity holds them together.
20. The Hubble time does not take into account the changing dynamics that result from having different cosmological densities, a cosmological constant, or dark energy.
26. If a photon contains more energy than the combined mass energy of a particle and its antiparticle, then the photon can become a particle-antiparticle pair that flies off in different directions, conserving mass-energy and momentum.

*Applying the Concepts*

38. 5
40. $1.5 \times 10^{-10}$ J.
41. $2 \times 10^{-31}$ kg. We can ignore it.

# Chapter 23

*Check Your Understanding*

1. c-b-d-a
2. b
3. c
4. b

*Thinking about the Concepts*

18. In the absence of luminous matter, we could detect a dark matter galaxy only from the gravitational influence it exerts on its surroundings or by gravitational lensing.
20. Quantum-sized (e.g., ultramicroscopic) fluctuations in density after the Big Bang
28. Hot dark matter could not be gravitationally bound enough to form the halos needed to contain galaxies and to explain galactic rotation curves.

*Applying the Concepts*

31. $1.05 \times 10^7$ m/s.
37. $10^{18}$ times more time.
45. $6 \times 10^{14} M_{Sun}$.

# Chapter 24

*Check Your Understanding*

1. d
2. a
3. d
4. a
5. b

*Thinking about the Concepts*

17. Amino acids were probably built from simpler elements and molecules, using energy from the Sun or lightning as a catalyst.
22. Plants and forests use photosynthesis, which relies on $CO_2$, and which was abundant far earlier than oxygen. Large animals required oxygen.
27. Absence of evidence is not evidence of absence. Conditions on Mars have changed over time, and these spacecraft did not look deep under the surface of Mars.

*Applying the Concepts*

34. 10 sec
38. 17 generations.
43. This is at the location of Mercury, outside the habitable zone.

# Credits

These pages contain credits for all 24 chapters in the complete volume of *21st Century Astronomy*, Fifth Edition. The Solar System Edition does not include Chapters 15–23. The Stars and Galaxies Edition does not include Chapters 8–12.

## Photos

### Chapter 1
**2-3** NASA **5 top** NASA **6** NASA and STScI **6** NASA **7 bottom** Sebastian Kaulitzki / Alamy **7 center bottom** Subaru Telescope (NAOJ), Hubble Space Telescope; Processing & Copyright: Roberto Colombari & Robert Gendler **11** Courtesy of the Archives, California Institute of Technology **12** Matheisl / Getty Images **15** ESA/NASA **16 top** NASA/JPL-Caltech **16 bottom** NASA/ JPL-Caltech/Space Science Institute

### Chapter 2
**22-23** Raymond Patrick/National Geographic Creative/Corbis **24 bottom** Henry Westheim Photography / Alamy **27 left** Pekka Parviainen / Science Source **27 right** D. Nunuk / Science Source **38** Arnulf Husmo/Getty Images **44** akg-images / Rabatti - Domingie / The Image Works **46** Nick Quinn **48** GSFC/NASA **49 top** Johannes Schedler / Panther Observatory **49 bottom** Anthony Ayiomamitis (TWAN)

### Chapter 3
**58-59** © Damian Peach **60 top** Royal Astronomical Society / Science Source **60 bottom** Tunc Tezel **61 center** Nicolaus Copernicus Museum, Frombork, Poland / Bridgeman Images **61 bottom** Bettmann/ Corbis **64** The Granger Collection, NYC **65 top** De Mundi Aetherei Recentioribus Phaenomenis, 1588. The Tycho Brahe Museum, Sweden. www.tychobrahe.com **65 bottom** SSPL / The Image Works **69** Galleria Palatina, Palazzo Pitti, Florence, Italy / The Bridgeman Art Library **70 top** The Granger Collection, NYC **70 bottom** SSPL/ Jamie Cooper / The Image Works **71** Lebrecht Music and Arts Photo Library / Alamy

### Chapter 4
**82-83** NASA **97 both** © Christopher Mackay **101** NASA

### Chapter 5
**108-109** Calvin_Bradshaw/Wikimedia Commons **114** Yva Momatiuk & John Eastcott/ Minden

Pictures/National Geographic Stock **123** Nigel Sharp, NOAO/NSO/Kitt Peak FTS/AURA/NSF **135 (all)** NASA

### Chapter 6
**142-143** © Laurie Hatch **144** ASU Physics Instructional Resource Team **145** Vik Dhillon/ University of Sheffield **147 top** ASU Physics Instructional Resource Team **147 bottom** Jim Sugar/Corbis **149 top** © Laurie Hatch **149 center** The Space Telescope Science Institute **149 bottom** ASU Physics Instructional Resource Team **151 center** Courtesy TMT Observatory Corporation **151 bottom** NASA Earth Observatory/NOAA NGDC **151 top** NASA, ESA, HEIC, and The Hubble Heritage Team (STScI/AURA) **153 top** Science Photo Library **153 bottom** Jean-Charles Cuillandre (CFHT) **155** NASA **155a** © Laurie Hatch **155b** NASA **155c** NASA **155d** NASA **155e** East Asian Observatory **155f** Photo by Dave Finley, courtesy National Radio Astronomy Observatory and Associated Universities, Inc. **155g** NRAO/AUI, James J. Condon, John J. Broderick, and George A. Seielstad **155** David Parker/Science Source **156 top** Roger Ressmeyer/Corbis **156 center** David Parker/ Science Source **156 bottom** Photo by Dave Finley, courtesy National Radio Astronomy Observatory and Associated Universities, Inc. **157 top** ALMA (ESO/NAOJ/NRAO)/W. Garnier (ALMA) **157 center** NASA Photo / Carla Thomas, (bottom right): NASA/Tom Tschida **157 bottom** NASA Photo / Carla Thomas, (bottom right): NASA/Tom Tschida **159** NASA/JPL-Caltech/MSSS **160** NASA/ JPL-Caltech/MSSS **162 top** CERN **162 bottom** Jim Haugen/NSF **163 top** University of Maryland **163 center** UPPA/Photoshot **164** Patrik Jonsson, Greg Novak & Joel Primack, UC Santa Cruz, 2008

### Chapter 7
**172-173** Caltech/NASA/JPL **175 top** NASA **175 bottom** Photograph by Pelisson, SaharaMet **178** Reuters/Corbis **185** NASA/Johns Hopkins University Applied Physics Laboratory/Carnegie Institution of Washington **189** NASA **191 bottom** NASA **193** NASA Ames/SETI Institute/JPL-Caltech

### Chapter 8
**200-201** NASA/JPL-Caltech/Univ. of Arizona **202** NASA **203** Montes De Oca & Associates **204 top** Photograph by D.J. Roddy and K.A. Zeller, USGS, Flagstaff, AZ. **204 bottom** NASA/Goddard/ MIT/Brown **205** NASA/JPL/Caltech **209** NASA **213 top** Grant Heilman Photography **213 bottom** Art Directors & TRIP / Alamy **213 bottom** Art Directors & TRIP / Alamy **214** Donald Duckson/ Visuals Unlimited/Corbis **218 bottom right** NASA **218 top** NASA/JSC **218 bottom left** NSSDC/NASA **219** NASA/Magellan Image/JPL **221 top** NASA/ JSC **221 center** K.C. Pau **221 bottom** NASA/Johns Hopkins University Applied Physics Laboratory/ Arizona State University/Carnegie Institution of Washington. Image reproduced courtesy of Science/AAAS. **222** NASA/MOLA Science Team **223** NASA **224** NASA/JPL/University of Arizona **225 top** NASA **225 bottom** NASA/JPL-Caltech/ University of Arizona/Texas A&M University **225 top right** NASA/JPL **227** © Don Davis **228** NASA/GSFC/Arizona State University

### Chapter 9
**234-235** Stockli, Nelson, Hasler, Goddard Space Flight Center/NASA. **244** NOAA **249 top** NASA **249 center** Design Pics Inc / Alamy **249 bottom** NASA **252 top** Dmitri Titov/ESA **252 center** © Ted Stryk 2007 **252 bottom** Dmitri Titov/ESA **253** NASA/JPL/Caltech **254 top** NASA/JPL-Caltech/MSSS **254 bottom** NASA/STScI

### Chapter 10
**268-269** NASA **273** NASA/JPL/University of Arizona, **274** NASA **275** NASA, JPL; Digital processing: Bjorn Jonsson (IAAA) **276 top** NASA/ JPL/Caltech. **276 bottom left** Lawrence Sromocsky, University of Wisconsin-Madison W. M. Keck Obervatory **276 bottom right** NASA, L. Sromovsky, and P. Fry (University of Wisconsin-Madison) **276 bottom right** NASA **276 bottom right** NASA **280** NASA, ESA, L. Sromovsky and P. Fry (University of Wisconsin), H. Hammel (Space Science Institute), and K. Rages (SETI Institute) **285** http://www.freenaturepictures.com/ **286** NASA/STScI. **288** NASA's Goddard Space

## Chapter 21

**590-591** ESA and the Planck Collaboration **594** NASA, ESA, H. Teplitz and M. Rafelski (IPAC/ Caltech), et al. **605 top** Lucent Technologies, Bell Labs **605 center** Ann and Rob Simpson Nature Photography **605 bottom** Ann and Rob Simpson Nature Photography **607** ESA and the Planck Collaboration **615** Courtesy of Stacy Palen

## Chapter 22

**616-617** NASA/JPL/Space Science Institute **622** NASA, ESA, H. Teplitz and M. Rafelski (IPAC/ Caltech), A. Koekemoer (STScI), R. Windhorst (Arizona State University), and Z. Levay (STScI) **645** CERN PhotoLab

## Chapter 23

**646-647** NASA, ESA, H. Teplitz and M. Rafelski (IPAC/Caltech), A. Koekemoer (STScI), R. Windhorst (Arizona State University), and Z. Levay (STScI) **648 top** NASA/ESA/ESO/ NAOJ/G. Paglioli **648 bottom** R. Brent Tully (U. Hawaii) et al., SDvision, DP, CEA/Saclay **649** Max Tegmark/SDSS Collaboration **651 top** NASA/CXC/MIT/E.-H Peng et al; Optical: NASA/ STScI **651 bottom** NASA, A. Fruchter and the ERO team (STScI, ST-ECF). NASA, ESA **657** NASA/ WMAP Science Team **658** Matthew Turk, Tom Abel, Brian O'Shea Visualization: Matthew Turk, Samuel Skillman **659** NASA/JPL-Caltech/GSFC **660** NASA, ESA, P. Oesch and I. Momcheva (Yale University), and the 3D-HST and HUDF09/XDF Teams **662 bottom** NASA/Hubble Space Telescope Cosmic Assembly Near-infrared Deep Extragalactic Legacy Survey **662 top** Guedes, Javiera, et al. The Astrophysical Journal, Volume 742, Issue 2, article id. 76 (2011). **663** NASA/CXC/ SAO/P **664 top** NASA, ESA, and N. Pirzkal (STScI/ ESA) **664 bottom** X-ray (NASA/CXC/IfA/C. Ma et al.); Optical (NASA/STScI/IfA/C. Ma et al. **666 top** Courtesy Joel Primack and George Blumenthal **666 bottom** NASA, ESA, and A. Feild (STScI)

## Chapter 24

**674-675** SETI Institute **676 bottom** Woods Hole Oceanographic Institution, Deep Submergence Operations Group, Dan Fornari **676 top** Dr. Michael Perfit, University of Florida, Robert Embley/NOAA **677 top** Gaertner / Alamy **677 center** Chris Boydell/Australian Picture Library/Corbis **677 bottom** National Park Service Photo by Jim Peaco **678 bottom** Peter Essick/ Aurora Photos **678 top** Dr. Robert Calentine/ Visuals Unlimited/Corbis **680** Courtesy of National Science Foundation **681** With permission of the Royal Ontario Museum © ROM **685** NASA/ JPL-Caltech/MSSS **689** NASA **691** F. Drake (UCSC) et al., Arecibo Observatory (Cornell, NAIC) **693** NASA/JPL-Caltech/Space Science Institute

## Text

Appendix 9 text excerpt: Resolutions 5A and 6A from "Press Release - IAU 2006 General Assembly: Result of the IAU Resolution Votes." Reprinted by permission of IAU.

## Line Art

**188** (Figure 7.17) Graph: "Radial Velocity/ Year" from Exoplanets.org. Reprinted by permission of the Department of Terrestrial Magnetism, Carnegie Institution of Washington.
**659** (Figure 23.10) Figure from "The First Galaxies," by V. Bromm & N. Yoshida. Reproduced with permission of Annual Review of Astronomy & Astrophysics, Volume 49 (373-407) © by Annual Reviews, http://www.annualreviews.org
**687** (Figure 24.15) Figure 1 from "Summary of the Limits of the New Habitable Zone." PHL.UPR.edu, March 23, 2013. Reprinted by permission of PHL@ UPR Arecibo.

# Index

This index contains page references for all 24 chapters in the complete volume of *21st Century Astronomy,* Fifth Edition. The Solar System Edition does not include Chapters 15–23 (pp. 420–673). The Stars and Galaxies Edition does not include Chapters 8–12 (pp. 200–357).

Hawking radiation, 525
*Hayabusa,* 336
Hayashi, Chushiro, 437
Hayashi track, *437,* 438, 439, 440
HD 226868, 525–526
heavy elements, 15, 274, 369, 557, 658–659, 681
heliocentric theory, 61–62, 64, 65, *67,* 68, 70–71
helioseismology, 400, 402
*Helios II* spacecraft, 512
heliosphere, 412
helium
    from Big Bang nucleosynthesis, 608, *609*
    discovery in solar spectrum, 404
    emission spectrum, *123*
    on giant planets, 273, 274, 281, 282, 283
    helium cores in stars, 453–454, 455, 456–457,
        458, 460
    helium shell burning, 467, 482
    in interstellar medium, 422
    in primary atmospheres, 236
helium burning
    in binary stars, 467
    in cores of red giant stars, 456–460, 461
    helium flash, 458–459
    helium shell burning, 467, 482
    in high-mass stars, 481, 485, 490
    horizontal branch, 459, 460–461, 464, 467, 481
    medium-mass stars, 482
    triple-alpha process, 457–458, 461, 485, 490
    white dwarf formation from post-AGB star, 461
Helix Nebula, *463*
hematite, 224
Herbig-Haro objects (HH objects), 438–439
heredity, 680–681
Herman, Robert, 604
Herschel, Caroline, 332, 536
Herschel, William, 271, 307, 332, 536, 684
Herschel crater on Mimas, 307
Herschel Space Observatory, 158, 338, 344, 441
Hertz, Paul, 194
hertz (Hz), 113
Hertzsprung, Ejnar, 376, *378*
Hertzsprung-Russell diagram. *see* H-R diagram
Hewish, Anthony, *494*
HH objects. *see* Herbig-Haro objects
hierarchical clustering, 652, 659, 662, 663–664
Higgs, Peter, 631
Higgs boson, 631
Higgs field, 631
high-mass stars
    blue supergiant stars, 484
    convection in, 481, 486, 499
    defined, 450, 480
    first stars, 657–659, 667
    iron cores, *482*
    red supergiant stars, 481, 484
    spectral types, 450
    yellow supergiant stars, 482
high-mass stars, evolution, 478–505
    asymptotic giant branch (AGB), 482, 489, 499
    blue supergiant stars, 484
    carbon burning, 482
    CNO cycle, 480–481

core collapse, 486–488
final days of, *485–486*
helium burning, 481, 485, 490
helium shell burning, 482
hydrogen burning, 480–481
hydrogen shell burning, 481
layered structure, 482, 484
leaving the main sequence, 481–482
low-density winds and mass loss, 483–484
luminous blue variable (LBV) stars, 484, 489
red supergiant stars, 481, 484
triple-alpha process, 485, 490
yellow supergiant stars, 482
H II regions ("H two"), 426–427, 462, 573
Hipparchus, 364
Hipparcos satellite, 363, 377
Hobby-Eberly Telescope (HET), *148*
Homestake Mine experiment, 400, *401*
homogeneous universe, 592–593, 595, 602, 649
Hopkinsville, Kentucky, 52
horizon, defined, 25
horizon problem, 627–629
horizontal branch, 459, 460–461, 464, 467, 481, 483,
    569
Horsehead Nebula, *427*
hot dark matter, 654–655
    *see also* neutrinos
hot Jupiters, 191–192, 287–288
hot spots, 217, 219, 220, 221
H-R diagram (Hertzsprung-Russell diagram),
    376–381
    close binary systems, 467
    distance estimation, 379, 380, *381*
    evolutionary track of protostars, 436–438
    giants, *376,* 380
    Hayashi track, *437,* 438
    horizontal branch, 459, 460–461, 464, 467, 481,
        483, 569
    instability strip, 482–484, 569
    luminosity class, 380–381
    main sequence, 377, 379–380
    overview, 376–377
    red giant branch, 454–456, 457, *458,* 461, 481
    spectroscopic parallax, 379, 380, *381*
    star clusters, 495–498
    stars not on the main sequence, 380–381
    subgiant branch, 455
    supergiants, *376,* 380–381
    white dwarfs, *376,* 380, 463–464
    *see also* asymptotic giant branch
HST. *see* Hubble Space Telescope
Hubble, Edwin P.
    Andromeda Galaxy's blueshift, 582
    Cepheid variables, 537
    classification of galaxies, 537–538, *662*
    discovery of Hubble's law, 544–546, 603
    expanding universe, 544, 592, *596,* 620–621
    Local Group identified, 580
Hubble constant ($H_0$), 544–545, 595, 597, 598, 619,
    624
Hubble's law
    age of universe estimation, 597–599
    cosmological principle and, 592, 593–595

defined, 544, 592, 593–595
discovery of, 544–546
distance measurement and redshift, 595, 649
Hubble constant ($H_0$), 544–545, 595, 597, 598,
    619, 624
Hubble time, 597–598, *599,* 624
look-back time, 597, 603
mapping the universe, 649
*see also* Doppler shift; expanding universe
Hubble Space Telescope (HST)
    auroral rings on Jupiter and Saturn, *286*
    Cassini Division in rings of Saturn, *310*
    Cepheid variables as standard candles, 543
    Coma Cluster of galaxies, 563
    Comet Shoemaker-Levy 9 collision with Jupiter,
        348, 349
    diameter, 157
    disks around newly formed stars, *175*
    dust storm on Mars, *254*
    evidence of liquid water oceans on Europa, 302,
        685
    extrasolar planet imaging, 191
    filters and false colors, 154
    gravitational lensing, 521, *651*
    image, *155*
    Jupiter, *286, 290*
    low-magnitude star detection, 364
    M104 or Sombrero Galaxy, *540*
    merging galaxies, *664*
    Neptune, 276
    observing high-redshift objects, 660
    optical limitations, 364
    orbit above atmosphere, 150, 157
    parallax measurement, 383
    planetary nebulae, *463*
    quasars, *550, 556*
    resolution, 150
    Ring Nebula, *448–449*
    rings of Uranus, 309, 316, 318
    Saturn, *273, 274, 286*
    servicing missions, 158
    spatial scanning technique, 383
    Ultra Deep Field 2014, *646–647,* 666
Hubble time, 597–598, *599,* 624
Huygens, Christiaan, 308, 312
hydrogen
    atomic energy levels, 123
    comparison to oxygen, 236, 239, 243
    emission spectrum, *123*
    on giant planets, 273, 274, 281–282, 283
    H$\alpha$ emission line, 404, 409, 426, 547, *568*
    H⁻ ion and stellar temperature, 437
    in interstellar medium, 422
    isotopes, relative abundance of, 608
    metallic hydrogen, *281,* 282, 283
    in primary atmospheres, 236
    21-cm radiation of neutral hydrogen, 427–428,
        *429,* 566–567, 569, 577
hydrogen burning
    in binary stars, 467, 468
    CNO cycle, 480–481, 482, 658
    high-mass stars, 480–481
    hydrogen shell burning, 454, 455, 456, 481